This unique textbook presents a comprehensive overview of the important fundamental principles of geophysics. The book combines both the applied and theoretical aspects of the subject, in contrast to most other geophysics textbooks which tend to emphasize either one or the other. It is pitched at an intermediate level between introductory and advanced texts.

The need to explain many topics in geophysics using complex mathematics can be intimidating for a student. The author has overcome this problem by using abundant diagrams, a simplified mathematical treatment, and equations in which the student can follow each derivation step-by-step. The book begins by placing the Earth in the context of the solar system, and describes the major dynamical systems governing plate tectonics. It then proceeds to describe in detail each major branch of geophysics: the Earth's shape, rotation and gravitation, and analysis of gravity anomalies; seismology, earthquakes and the Earth's deep structure; the age, thermal and electrical properties of the Earth; geomagnetism and paleomagnetism; and lastly the geodynamics of isostasy, rheology and plate dynamics. Each chapter begins with a brief simplified summary of the basic physical principles, and a brief account of each topic's historical evolution.

The book will satisfy the needs of earth science students from a variety of backgrounds who may not pursue the topic of geophysics further than intermediate undergraduate level, while at the same time preparing geophysics majors for continued study to a higher level. The book is destined to become a core textbook for undergraduate geology and geophysics programs.

Fundamentals of Geophysics

Fundamentals of Geophysics

William Lowrie

Institute of Geophysics
Swiss Federal Institute of Technology
Zürich, Switzerland

PUBLISHED BY THE PRESS SYNDICATE OF THE UNIVERSITY OF CAMBRIDGE
The Pitt Building, Trumpington Street, Cambridge, United Kingdom

CAMBRIDGE UNIVERSITY PRESS
The Edinburgh Building, Cambridge CB2 2RU, UK
40 West 20th Street, New York, NY 10011–4211, USA
477 Williamstown Road, Port Melbourne, VIC 3207, Australia
Ruiz de Alarcón 13, 28014 Madrid, Spain
Dock House, The Waterfront, Cape Town 8001, South Africa

http://www.cambridge.org

First published 1997
Fifth printing 2004

Printed in the United Kingdom at the University Press, Cambridge

A catalogue record for this book is available from the British Library

Library of Congress Cataloguing in Publication data
Lowrie, William, 1939–
Fundamentals of geophysics / William Lowrie.
p. cm.
Includes bibliographical references and index.
ISBN 0 521 46164 2 (hc). – ISBN 1 521 46728 4 (pbk.)
1. Geophysics I. Title.
QC806.L67 1997
550–dc20 96-45966 CIP

ISBN 0 521 46164 2 hardback
ISBN 0 521 46728 4 paperback

This book is dedicated to

Mum, Dad and Marcia

Contents

Preface

During many years of teaching at universities I have often realized that, although there are several good textbooks to accompany geophysical lectures at introductory and advanced level, there are comparatively few at an intermediate standard (roughly equivalent to a junior year class at an American university). The aim of this book is to fill that gap. It is intended as a general geophysical textbook at a level between the simple and advanced tracts.

Fundamental knowledge in geophysics is presented in a way that should be suitable for earth science students with varied backgrounds and interests. Most of them do not ultimately become geophysicists but select other disciplines. They have different levels of education in physics and mathematics. I have tried to address these factors in writing the book. The need for equations to explain many topics in geophysics can be intimidating for a student, especially for a budding geologist whose talents and interests are often different but no less important than those of a fledgling geophysicist. To help overcome this problem I have developed fundamental equations in such a way that the student should be able to follow each derivation step-by-step. Where necessary, a chapter includes an explanation of the fundamental physical principles needed in the chapter.

Most texts emphasize either applied geophysics or general geophysics with rather little cross-fertilization between the two categories. Because many students do not know at this stage what direction their careers will take, I have tried to avoid this artificial segregation of topics. Moreover, the book is not intended as an introduction to current research, since several fine advanced textbooks and review articles already serve this purpose. As a result, the text is rather free of citations; most references are in the figure captions.

Each section of the book contains brief accounts, gleaned from several sources, of how knowledge of a particular topic evolved historically. The descriptions are not exhaustive or authoritative, but hopefully they may entice some readers to share the conviction of James Clerk Maxwell that *'Every student of science should, in fact, be an antiquary in his subject'* (from *A Treatise on Electricity and Magnetism*, 1873). Some of the figures are original, but many have been redrawn (and occasionally simplified) from books and articles. I am grateful to the original authors and their publishers for permission to use their illustrations.

Due to the pressures and obligations of office as a university professor this book has been written largely in my free time. This has meant sacrificing most weekends and evenings for the last few years. I have often wondered at my folly in undertaking the project. It began as a pleasant recreation in tandem with improving my class notes, but eventually I realized that so much time had been invested that there was no longer any way to gracefully abandon the venture. However, the process of writing has also involved much reading and studying, and this brought to me an awareness of some aspects of my subject that I had previously taken for granted. This rewarding aspect of preparing the book helped to rekindle my motivation whenever it flagged. I hope that the book will help its readers to understand and appreciate – and perhaps become enthusiastic about – the diverse ways in which geophysics has contributed to deciphering Earth's secrets, many of which are still only partially understood. The knowledge already gained has been put to use for mankind's benefit in finding important resources and solving environmental problems. The unanswered questions will provide research topics for many future generations of earth scientists.

Many friends and associates gave me helpful advice and suggestions that have sometimes been adopted, occasionally rejected, but always considered; all are sincerely appreciated. In particular, I thank Jörg Ansorge, Göran Ekström, Mike Fuller, Alan Green, Fritz Heinrich, Ann Hirt, Pierre Ihmlé, Hans Kahle, Dennis Kent, Emile Klingelé, Ken Kodama, Urs Kradolfer, Hans-Ruedi Maurer, Henry Pollack, Ladislaus Rybach, Peter Signer, Bob Smith, Rudi Steiger, and Tony Watts for sacrificing their time to help me improve the manuscript with constructive criticisms of individual chapters. Mike Fuller's interest inspired me to start writing and helped me to pass the point of no return.

Finally, I thank my wife Marcia for her understanding and tolerance of my resolve to accomplish this project, and above all for her loyal support, constant encouragement and welcome practical suggestions. Last but not least, I thank my parents, to whom the book is dedicated, for their many unselfish sacrifices, which helped me to get an education that has made all the difference.

William Lowrie,
Zürich,
January, 1997

Acknowledgements

The publishers and authors listed below are gratefully acknowledged for giving their permission to use redrawn figures based on illustrations in journals and books for which they hold the copyright. The original authors of the figures are cited in the figure captions. Every effort has been made to obtain permission to use copyrighted materials, and sincere apologies are rendered for any errors or omissions. The publishers would welcome these being brought to their attention.

Publisher	**Figure number(s)**
Academic Press Inc.	3.25a, 3.25b, 3.26, 3.73, 5.34, 5.52
American Geophysical Union	
Geodynamics Series	1.16
Geophysical Monographs	3.80
Geophysical Research Letters	4.26
Journal of Geophysical Research	2.23, 2.25, 2.26, 2.57, 2.59, 3.39, 3.41, 3.81, 3.85, 4.22, 4.25, 4.33, 5.39, 5.69, 5.77, 6.13, 6.14, 6.15, 6.17, 6.30, 6.31, 6.36
Maurice Ewing Series	3.46
Reviews of Geophysics	4.27, 4.28, 4.29, 5.67
American Association for the Advancement of Science	
Science	1.14, 1.15, 4.8, 5.76
Edward Arnold Publishers	2.41
Blackwell Scientific Publications Ltd.	1.21, 1.22, 6.31, 6.32
Geophysical Journal of the Royal Astronomical Society and *Geophysical Journal International*	2.56, 2.58, 4.33, 6.35
Brookfield Press	6.26
Cambridge Univ. Press	1.8, 2.38, 3.15, 4.46, 4.51a, 4.51b, 5.43, 5.55, 5.72, 5.85, 6.28a
Chapman & Hall	6.12
Earthquake Research Institute, Tokyo	5.35
Elsevier Science–NL	
Deep Sea Research	1.13
Earth and Planetary Science Letters	4.6, 4.11, 6.27, 6.29

Publisher (*cont.*)	**Figure number(s)**
Journal of Geodynamics	4.23
Physics of Earth and Planetary Interiors	4.41
Sedimentology	5.22
Tectonophysics	2.27, 5.82, 6.15, 6.16
W. H. Freeman & Co.	3.23, 3.44
Geological Society of America	1.23, 5.83
Gordon and Breach Scientific Publishers Inc.	6.23, 6.24, 6.25
Kluwer Academic Publishers	4.20
Macmillan Magazines Ltd.	
Nature	1.7, 1.18, 1.19, 1.20, 1.24a, 1.24b, 4.57
Maruzen, Tokyo	5.15a, 5.15b, 5.64
McGraw-Hill Inc.	2.46, 3,64, 6.10a, 6.10b
Naturforschende Gesellschaft in Zürich	3.82, 3.83
Oxford University Press	5.31a
Pergamon Press	4.5
Princeton University Press	6.19, 6.20, 6.21, 6.22
Schweizerische Geologische Gesellschaft	3.42
Schweizerische Geophysikalische Kommission	2.55, 6.5
Schweizerische Mineralogische und Petrologische Gesellschaft	3.42
Seismological Society of America	1.10, 3.40, 3.43
Society of Exploration Geophysicists	3.64, 5.44
Springer-Verlag	5.41
Stanford University Press	4.7
Terra Scientific Publishing Co.	5.17, 5.31b, 5.37, 5.38
The Royal Society, London	1.6, 2.14
University of Chicago Press	5.61
Van Nostrand Reinhold	2.15, 2.28, 2.29, 3.31, 3.32, 3.47, 3.84, 5.33
John Wiley & Sons Inc.	2.37, 2.43, 2.45, 2.54, 4.31, 4.40, 4.45

Authors	
C. Emiliani	4.25
A. N. Strahler	2.1, 2.2, 2.3, 2.16a, 3.22a, 3.22b, 5.30

1 The Earth as a planet

1.1 THE SOLAR SYSTEM

1.1.1 The discovery and description of the planets

To appreciate how impressive the night sky must have been to early man it is necessary today to go to a place remote from the distracting lights and pollution of urban centers. Viewed from the wilderness the firmaments appear to the naked eye as a canopy of shining points, fixed in space relative to each other. Early observers noted that the star pattern appeared to move regularly and used this as a basis for determining the timing of events. More than 3,000 years ago, in about the 13th century B.C., the year and month were combined in a working calendar by the Chinese, and about 350 B.C. the Chinese astronomer Shih Shen prepared a catalog of the positions of 800 stars. The ancient Greeks observed that several celestial bodies moved back and forth against this fixed background and called them the *planetes*, meaning 'wanderers'. In addition to the Sun and Moon, the naked eye could discern the planets Mercury, Venus, Mars, Jupiter and Saturn.

Geometrical ideas were introduced into astronomy by the Greek philosopher Thales in the 6th century B.C. This advance enabled the Greeks to develop astronomy to its highest point in the ancient world. Aristotle (384–322 B.C.) summarized the Greek work performed prior to his time and proposed a model of the universe with the Earth at its center. This *geocentric* model became imbedded in religious conviction and remained in authority until late into the Middle Ages. It did not go undisputed; Aristarchus of Samos (c.310–c.230 B.C.) determined the sizes and distances of the Sun and Moon relative to the Earth and proposed a *heliocentric* (sun-centered) cosmology. The methods of trigonometry developed by Hipparchus (190–120 B.C.) enabled the determination of astronomical distances by observing the angular positions of celestial bodies. Ptolemy, a Greco-Egyptian astronomer in the 2nd century A.D., applied these methods to the known planets and was able to predict their motions with remarkable accuracy considering the primitiveness of available instrumentation.

Until the invention of the telescope in the early 17th century the main instrument used by astronomers for determining the positions and distances of heavenly bodies was the *astrolabe*. This device consisted of a disk of wood or metal with the circumference marked off in degrees. At its center was pivoted a movable pointer called the *alidade*.

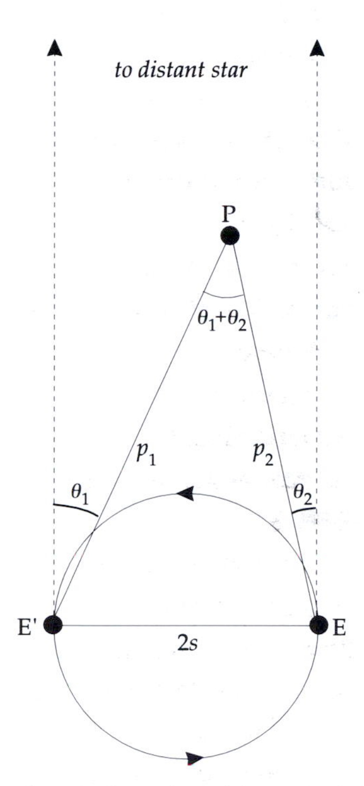

Fig. 1.1 Illustration of the method of parallax in which two measured angles (θ_1 and θ_2) are used to compute the distances (p_1 and p_2) of a planet from the Earth in terms of the Earth–Sun distance (s).

Angular distances could be determined by sighting on a body with the alidade and reading off its elevation from the graduated scale. The inventor of the astrolabe is not known, but it is often ascribed to Hipparchus (190–120 B.C.). It remained an important tool for navigators until the invention of the sextant in the 18th century.

The angular observations were converted into distances by applying the method of parallax. This is simply illustrated by the following example. Consider the planet P as viewed from the Earth at different positions in its orbit around the Sun (Fig. 1.1). The angle between a sighting on the planet and on a fixed star will appear to change because of the Earth's orbital motion around the Sun. Let the measured extreme angles be θ_1 and θ_2 and the distance of the Earth from the Sun be s; the distance between the extreme positions E and E′ of the orbit is then $2s$. The distances p_1 and p_2 of the planet from the Earth are computed in terms of the Earth–Sun distance by applying the trigonometric law of sines:

$$\frac{p_1}{2s} = \frac{\sin(90-\theta_2)}{\sin(\theta_1+\theta_2)} = \frac{\cos\,\theta_2}{\sin(\theta_1+\theta_2)}$$

$$\frac{p_2}{2s} = \frac{\cos\,\theta_1}{\sin(\theta_1+\theta_2)} \tag{1.1}$$

Further trigonometric calculations give the distances of the planets from the Sun. The principle of parallax was also used to determine relative distances in the Aristotlian geocentric system, which was propounded for centuries by Christian authorities who considered the fixed stars, Sun, Moon and planets to be in motion about the Earth.

In 1543, the year of his death, the Polish astronomer Nicolas Copernicus published a revolutionary work in which he asserted that the Earth was not the center of the universe. According to his model the Earth rotated about its own axis, and it and the other planets revolved about the Sun. Copernicus calculated the sidereal period of each planet about the Sun; this is the time required for a planet to make one revolution and return to the same angular position relative to a fixed star. He also determined the radii of their orbits about the Sun in terms of the Earth–Sun distance. The mean radius of the Earth's orbit about the Sun is called an *astronomical unit*; it equals 149,597,890 km. Accurate values of these parameters were calculated from observations compiled during an interval of 20 years by the Danish astronomer Tycho Brahe (1546–1601). On his death the records passed to his assistant, Johannes Kepler (1571–1630). Kepler succeeded in fitting the observations into a heliocentric model for the system of known planets. The three laws in which Kepler summarized his deductions were later to prove vital to Isaac Newton for verifying the law of Universal Gravitation. It is remarkable that the data base used by Kepler was founded on observations that were unaided by the telescope, which was not invented until early in the 17th century.

1.1.2 **Kepler's laws of planetary motion**

Kepler took many years to fit the observations of Tycho Brahe into three laws of planetary motion. The first and second laws (Fig. 1.2) were published in 1609 and the third law appeared in 1619. The laws may be formulated as follows:

(1) the orbit of each planet is an ellipse with the Sun at one focus;
(2) the orbital radius of a planet sweeps out equal areas in equal intervals of time;
(3) the ratio of the square of a planet's period (T^2) to the cube of the semi-major axis of its orbit (a^3) is a constant for all the planets, including the Earth.

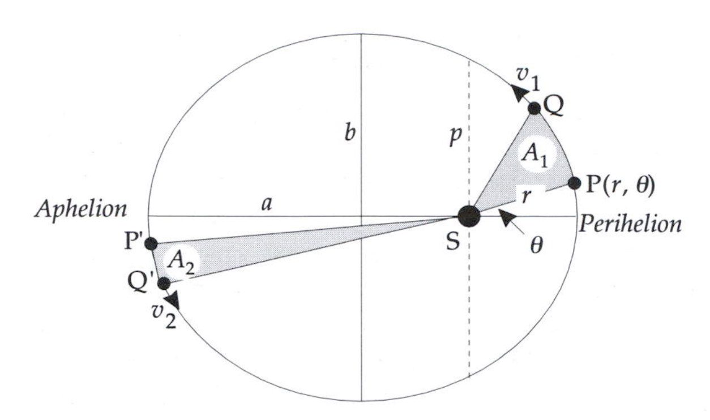

Fig. 1.2 Kepler's first two laws of planetary motion: (1) each planetary orbit is an ellipse with the Sun at one focus, and (2) the radius to a planet sweeps out equal areas in equal intervals of time.

Kepler's three laws are purely empirical, derived from accurate observations. In fact they are expressions of more fundamental physical laws. The elliptical shapes of planetary orbits described by the first law are a consequence of the *conservation of energy* of a planet orbiting the Sun under the effect of a central attraction that varies as the *inverse square* of distance. The second law describing the rate of motion of the planet around its orbit follows directly from the *conservation of angular momentum* of the planet. The third law results from the balance between the force of gravitation attracting the planet towards the Sun and the centrifugal force away from the Sun due to its orbital speed. The third law is easily proved for circular orbits (see § 2.3.2.3).

Kepler's laws were developed for the solar system but are applicable to any closed planetary system. They govern the motions of any natural and artificial satellite about a parent body. Kepler's third law relates the period (T) and the semi–major axis (a) of the orbit of the satellite to the mass (M) of the parent body through the equation

$$GM = \frac{4\pi^2}{T^2}a^3 \tag{1.2}$$

where G is the gravitational constant. This relationship was extremely important for determining the masses of those planets that have natural satellites. It can now be applied to determine the masses of planets using the orbits of artificial satellites.

Special terms are used in describing elliptical orbits. The nearest and furthest points of a planetary orbit around the Sun are called *perihelion* and *aphelion*, respectively. The terms *perigee* and *apogee* refer to the corresponding nearest and furthest points of the orbit of the Moon or a satellite about the Earth.

1.1.3 Characteristics of the planets

Galileo Galilei (1564–1642) is often regarded as a founder of modern science. He made fundamental discoveries in astronomy and physics, including the formulation of the laws of motion. He was one of the first scientists to use the telescope to acquire more detailed information about the planets. In 1610 Galileo discovered the four largest satellites of Jupiter (called Io, Europa, Ganymede and Callisto), and observed that (like the Moon) the planet Venus exhibited different phases of illumination, from full disk to partial crescent. This was persuasive evidence in favor of the Copernican view of the solar system.

In 1686 Newton applied his theory of Universal Gravitation to observations of the orbit of Callisto and calculated the mass of Jupiter (J) relative to that of the Earth (E). The value of the gravitational constant G was not yet known; it was first determined by Lord Cavendish in 1798. However, Newton calculated the value of GJ to be 124,400,000 km^3 s^{-2}. This was a very good determination; the modern value for GJ is 126,712,767 km^3 s^{-2}. Observations of the Moon's orbit about the Earth showed that the value GE was 398,600 km^3 s^{-2}. Hence Newton inferred the mass of Jupiter to be more than 300 times that of the Earth.

In 1781 William Herschel discovered Uranus, the first planet to be found by telescope. The orbital motion of Uranus was observed to have inconsistencies, and it was inferred that the anomalies were due to the perturbation of the orbit by a yet undiscovered planet. The predicted new planet, Neptune, was discovered in 1846. Although Neptune was able to account for most of the anomalies of the orbit of Uranus, it was subsequently realized that small residual anomalies remained. In 1914 Percival Lowell predicted the existence of an even more distant planet, the search for which culminated in the detection of Pluto in 1930.

The masses of the planets can be determined by applying Kepler's third law to the observed orbits of natural and artificial satellites and to the tracks of passing spacecraft. Estimation of the sizes and shapes of the planets depends on data from several sources. Early astronomers used occultations of the stars by the planets; an occultation is the eclipse of one celestial body by another, such as when a planet passes between the Earth and a star. The duration of an occultation depends on the diameter of the planet, its distance from the Earth and its orbital speed.

The dimensions of the planets have been determined with improved precision in modern times by the availability of data from spacecraft. Important innovations include radar-ranging and Doppler tracking from spacecraft. Radar-ranging involves measuring the distance between an orbiting (or passing) spacecraft and the planet's surface from the two-way travel-time of a pulse of electromagnetic waves in the radar-frequency range. The short wavelength of the radar waves enables precise determination of the altitude of the spacecraft. The Doppler effect is a shift in frequency of the reflected signal. If the signal is reflected from a part of the planet that is moving away from the spacecraft the frequency of the reflection is lower than that of the transmitted signal; the opposite effect is observed when the two approach each other. The Doppler effect yields the relative velocity of the spacecraft and planet. Together, these radar methods allow accurate determination of the path of the spacecraft, which is affected by the mass of the planet and the shape of its gravitational equipotential surfaces.

The rate of rotation of a planet about its own axis can be determined by observing the motion of features on its surface. Where this is not possible (e.g., the surface of Uranus is featureless) other techniques must be employed. In the case of Uranus the rotational period of 15.6 hr is estimated from the Doppler frequency shift of sunlight reflected from different parts of its surface. All planets revolve around the Sun in the same sense, which is anticlockwise when viewed from above the plane of the Earth's orbit (called the *ecliptic* plane). The planetary orbital planes are inclined at small angles to the ecliptic (Table 1.1). Most of the planets rotate about their rotation axis in the same sense, which is termed *prograde*. Venus rotates in the opposite, *retrograde* sense. The rotation axes of Uranus and Pluto lie close to the ecliptic; they are tilted away from the pole to the ecliptic at angles of *obliquity* greater than 90°, so that, strictly speaking, their rotations are also retrograde.

The relative sizes of the planets are shown in Fig. 1.3. They form three categories on the basis of their physical properties (Table 1.1). The *terrestrial planets* (Mercury, Venus, Earth, and Mars) resemble the Earth in size, composition and density; they rotate about their own axes at the same rate or slower than the Earth. The great, or *Jovian*, planets (Jupiter, Saturn, Uranus and Neptune) are much larger than the Earth, with different compositions and much lower densities; they rotate more rapidly than the Earth. Pluto is in a class by itself. Its large orbit is highly elliptical, but its physical properties resemble those of the terrestrial planets.

1.1.3.1 *The terrestrial planets and the Moon*

Mercury is the closest planet to the Sun. This nearness and its small size make it difficult to study telescopically. Until 1965 the rotational period was thought to be the same as the period of revolution (88 days). However, images from the close passage of the Mariner 10 spacecraft gave a period of

Table 1.1 *Dimensions of the planets and their orbits (source: Beatty and Chaikin, 1990)*

PLANET	Mean orbital radius (AU)	Eccen-tricity of orbit	Tilt of orbit to ecliptic (°)	Sidereal period of rev-olution	Equatorial radius (km)	Sidereal period of rotation	Polar flattening $f=(a-c)/a$	Obliquity of rotation axis (°)	Mass relative to Earth	Mean density (kg m^{-3})
Terrestrial planets and the Moon										
Mercury	0.3871	0.2056	7.004	87.969 d	2,439	58.65 d	0.0	~2	0.0553	5,426
Venus	0.7233	0.0068	3.394	224.701 d	6,051	243.01 d	0.0	177.3	0.8149	5,245
Earth	1.0000	0.0167	0.000	365.256 d	6,378	1 d	0.003353	23.45	1.000	5,515
Moon	0.00255 (about Earth)	0.055	5.15	27.3 d	1,738	27.32 d	0.002	5.1	0.0123	3,344
Mars	1.5237	0.0934	1.850	686.980 d	3,393	1.0288 d	0.0074	25.19	0.1074	3,934
Great planets and Pluto										
Jupiter	5.2028	0.0483	1.308	11.862 yr	71,492	9.841 hr	0.065	3.12	317.94	1,325
Saturn	9.5388	0.0560	2.488	29.458 yr	60,268	10.233 hr	0.098	26.73	95.18	685
Uranus	19.1914	0.0461	0.774	84.01 yr	25,559	17.9 hr	0.024	97.86	14.53	1,255
Neptune	30.0611	0.0097	1.774	164.79 yr	24,764	19.2 hr	0.022	29.6	17.14	1,627
Pluto	39.5294	0.2482	17.148	248.54 yr	1,150	6.387 d	—	122.46	0.0022	1,990

a = equatorial radius; c = polar radius.

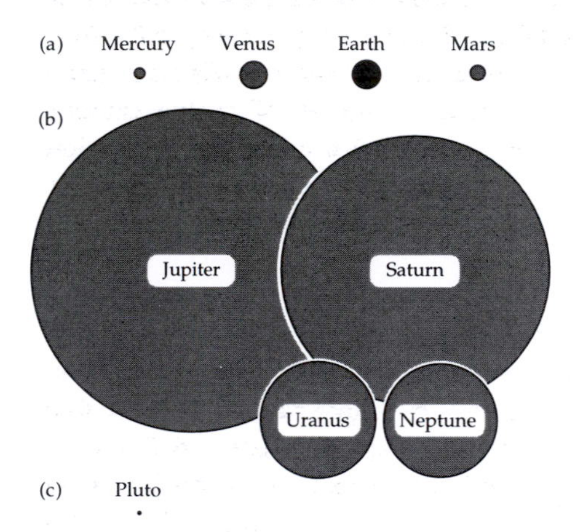

Fig. 1.3 The relative sizes of the planets: (a) the terrestrial planets, (b) the great (Jovian) planets and (c) Pluto, which is diminutive compared to the others.

rotation of about 59 days, and Doppler tracking gave a radius of 2440 km. Its high Earth-like density suggests that it is enriched in iron.

Venus is the brightest object in the sky after the Sun and Moon. Its orbit brings it closer to Earth than any other planet, which made it an early object of study by telescope. Its occultation with the Sun was observed telescopically as early as 1639. Estimates of its radius based on occultations indicated about 6120 km. Galileo observed that the apparent size of Venus changed with its position in orbit and, like the Moon, the appearance of Venus showed different phases from crescent-shaped to full. This was important evidence in favor of the Copernican heliocentric model of the solar system. Unfortunately, it brought Galileo into conflict with religious authority, which accused him of heresy for opposing the church-approved Aristotelian geocentric model. He was forced to recant his views and to retire into seclusion.

Venus is very similar in size and probable composition to the Earth. During a near-crescent phase the planet is ringed by a faint glow indicating the presence of an atmosphere. This has been confirmed by spacecraft that have visited the planet. Data from the Pioneer spacecraft in 1973 and 1974 established a mean radius of 6051 km and a rotational period of slightly over 243 days, longer than the period of revolution (225 days). The Magellan spacecraft orbiting the planet in 1990 obtained detailed geodetic data. Venus is unique among the planets in rotating in a *retrograde* sense about an axis that is almost normal to the ecliptic (Table 1.1). Like Mercury, it has a high Earth-like density.

The **Earth** moves around the Sun in a slightly elliptical orbit. The parameters of the orbital motion are important, because they define astronomical units of distance and time. The Earth's rotation about its own axis from one solar zenith to the next one defines the solar day (see §4.1.1.2). The length of time taken for it to complete one orbital revolution about the Sun defines the solar year, which is equal to 365.242 solar days. The eccentricity of the orbit is presently 0.0163 but it varies between a minimum of 0.001 and a maximum of 0.060 with a period of about 100,000 yr due to the influence of the other planets. The mean radius of the orbit (149,597,890 km) is called an *astronomical* *unit*(*AU*). Distances within the solar system are usually expressed as multiples of this unit. The distances to extra-galactic celestial bodies are expressed as multiples of a light-year (the distance travelled by light in one year). The Sun's light takes about 8 m 20 s to reach the Earth. Due to the difficulty of determining the gravitational constant the mass of the Earth (E) is not known with high precision, but is estimated to be 5.9737×10^{24} kg. In contrast, the product GE is known accurately; it is equal to $3.98600434 \times 10^{14}$ km^3 s^{-2}. The rotation axis of the Earth is presently inclined at 23.44° to the pole of the ecliptic. However, the effects of other planets also cause the angle of *obliquity* to vary between a minimum of 21.9° and a maximum of 24.3°, with a period of 40,000 yr.

The **Moon** is Earth's only natural satellite. The distance of the Moon from the Earth was first estimated with the method of parallax. Instead of observing the Moon from different positions of the Earth's orbit, as in Fig. 1.1, the Moon's position relative to a fixed star was observed at times 12 hours apart, close to moonrise and moonset, when the Earth had rotated through half a revolution. The baseline for the measurement is then the Earth's diameter. The distance of the Moon from the Earth was found to be about 60 times the Earth's radius. The orbit of the Moon about the Earth is slightly elliptical (eccentricity=0.0549).

The Moon rotates about its axis in the same sense as its orbital revolution about the Earth. Tidal friction resulting from the Earth's attraction has slowed down the Moon's rotation, so that it now has the same mean period as its revolution. As a result, the Moon always presents the same face to the Earth. In fact, slightly more than half (about 59%) of the lunar surface can be viewed from the Earth. Several factors contribute to this. First, the plane of the Moon's orbit around the Earth is inclined at 5°9′ to the ecliptic while the Moon's equator is inclined at 1°32′ to the ecliptic. The inclination of the Moon's equator varies by up to 6°41′ to the plane of its orbit. This is called the *libration of latitude*. It allows earth-based astronomers to see 6°41′ beyond each of Moon's poles. Secondly, the Moon moves with variable velocity around its elliptical orbit, while its own rotation is constant. Near perigee the Moon's orbital velocity is fastest (in accordance with Kepler's second law) and the rate of revolution exceeds slightly the constant rate of lunar rotation. Similarly, near apogee the Moon's orbital velocity is slowest and the rate of revolution is slightly less than the rate of rotation. The rotational differences are called the Moon's *libration of longitude*. Their effect is to expose zones of longitude beyond the average edges of the Moon. Finally, the Earth's diameter is quite large compared to the Moon's distance from Earth. During Earth's rotation the Moon is viewed from different angles that allow about one additional degree of longitude to be seen at the Moon's edge.

The distance to the Moon and its rotational rate are well known from laser-ranging using reflectors placed on the Moon by astronauts. The laser-ranging accuracy is about 2–3 cm. Curiously, the overall topography of the Moon is less well known than that of Venus or Mars; about 41% of the lunar surface, the far side, is essentially unknown. It cannot be seen from Earth and has only been viewed from lunar-orbiting spacecraft, so the topography is uncertain. Relative to its parent body the Moon is far larger than the natural satellites of the other planets. Its low density may be due to the absence of an iron core.

Mars, popularly called the red planet because of its hue when viewed from Earth, was also an object of early telescopic study. In 1666 Gian Domenico Cassini determined the rotational period at just over 24 hours; radio-tracking from Viking spacecrafts that landed on Mars more than three centuries later gave a period of 24.623 hr. Mars has two natural satellites, Phobos and Deimos. Observations of their orbits gave early estimates of the mass of the planet. Its size was established quite early telescopically from occultations. Its shape is known very precisely from spacecraft observations. The Earth-like rate of rotation causes a polar flattening about double that of the Earth. The low density of Mars may indicate that it also does not have an iron core.

1.1.3.2 *The great planets and Pluto*

Jupiter is the largest of all planets. It has 16 satellites. The outermost four are less than 35 km in radius, revolve in retrograde orbits and may be captured asteroids. Its four largest satellites – Io, Europa, Ganymede and Callisto – were discovered in 1610 by Galileo. Ganymede is the largest satellite in the solar system; with a radius of 2631 km it is slightly larger than the planet Mercury. In 1686 Newton calculated Jupiter's mass from observations of the orbit of Callisto. Jupiter has a powerful magnetic field that traps charged particles from the Sun. The motions of these electric charges cause radio emissions that are modulated by the rotation of the planet. This gives the best estimate of the period of rotation, which is about 9.9 hr. Despite its enormous size the planet has a very low density of only 1325 kg m^{-3}, which implies that its composition is dominated by hydrogen and helium.

Saturn is famous for the thin concentric rings in its equatorial plane. Each ring is composed of tiny particles in orbit around the planet. Saturn has 17 satellites, the largest of which, Titan, has a radius of 2575 km. Observations of the orbit of Titan allowed the first estimate of the mass of Saturn to be made in 1831. The period of rotation of Saturn has been deduced from modulated radio emissions associated with its magnetic field to be 10.2 hr. The shape of the planet is known from occultations of radio signals from the Voyager spacecrafts. As a result of its high rotational speed, Saturn has the greatest degree of polar flattening of any planet, almost 10%. Its mean density of 685 kg m^{-3} (less than that of water) is the lowest of all the planets, implying that Saturn, like Jupiter, is made up mainly of hydrogen and helium and contains few heavy elements. However, the strong gravitational field of Jupiter compresses hydrogen to a metallic state, which has a high density. This gives Jupiter a higher mean density than Saturn.

Uranus has 15 satellites, the largest of which, Titania, has a radius about half that of the Moon. Uranus is telescopically featureless, but its period of rotation of about 18 hr has been determined from the Doppler shift of sunlight reflected from its rims and from radio emissions associated with its magnetic field. The rapid rotation gives a polar flattening of about 5%, measured from stellar occultations. The prograde axis of rotation lies nearly in the plane of the ecliptic; its obliquity of 97.9° means that the rotation is effectively retrograde.

Neptune has 8 satellites. The planetary mass was determined from observations of the orbit of the larger satellite, Triton. Its dimensions come from stellar occultation observations. Its calculated mean density is higher than that of the other giant planets, which may indicate that it is the least centrally condensed. Its period of rotation near 19 hr was estimated by optical-tracking of cloud features.

Each of these great planets is encircled by a set of concentric rings, made up of numerous particles. The rings around Saturn, discovered by Galileo in 1610, are the most spectacular. For more than three centuries they appeared to be a feature unique to Saturn, but in 1977 a number of discrete rings was also detected around Uranus. In 1979 the Voyager 1 spacecraft detected faint rings around Jupiter, and in 1989 the Voyager 2 spacecraft confirmed that Neptune also has a ring system.

Pluto is too small and distant for its period of rotation to be determined directly by telescopic observation. Photometric detection of variations in its brightness gave a period of rotation of almost a week. In 1978 a large satellite, Charon, was discovered with an orbital period that nearly coincided with Pluto's rotation period. Like Uranus, Pluto's rotation axis is very oblique, making an angle of almost 120° to the ecliptic. Its orbit is tilted at the largest angle to the ecliptic and it has the highest eccentricity, which brings it at times inside the orbit of Neptune. Indeed, Pluto may once have been a satellite of Neptune that escaped into its own orbit.

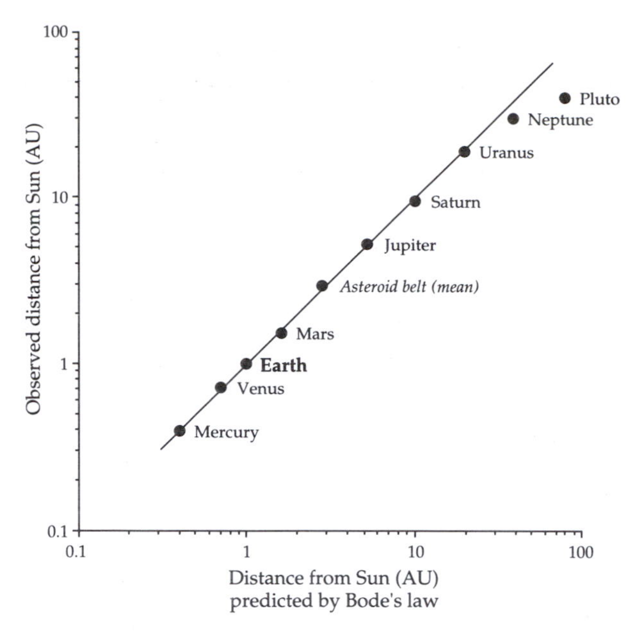

Fig. 1.4 Bode's empirical law for the distances of the planets from the Sun.

1.1.3.3 *Bode's law*

In 1772 the German astronomer Johann Bode devised an empirical formula to express the approximate distances of the planets from the Sun. A series of numbers is created in the following way: the first number is zero, the second is 0.3, and the rest are obtained by doubling the previous number. This gives the sequence 0, 0.3, 0.6, 1.2, 2.4, 4.8, 9.6, 19.2, 38.4, 76.8, etc. Each number is then augmented by 0.4 to give the sequence: 0.4, 0.7, 1.0, 1.6, 2.8, 5.2, 10.0, 19.6, 38.8, 77.2, etc. This series can be expressed mathematically as follows:

$$d_n = 0.4 \text{ for } n = 1$$
$$d_n = 0.4 + 0.3 \times 2^{n-2} \text{ for } n \geq 2 \qquad (1.3)$$

This expression gives the distance d_n in astronomical units (AU) of the nth planet from the Sun. It is usually known as Bode's law, but, as the same relationship had been suggested earlier by J. D. Titius of Wittenberg, it is sometimes called Titius–Bode's law. Examination of Fig. 1.4 and comparison with Table 1.1 show that this relationship holds remarkably well, except for Neptune and Pluto. A possible interpretation of the discrepancies is that the orbits of these planets are no longer their original orbits.

Bode's law predicts a fifth planet at 2.8 AU from the Sun, between the orbits of Mars and Jupiter. In the last years of the 18th century astronomers searched intensively for this missing planet. In 1801 a small planetoid, Ceres, only 750 km in diameter, was found at a distance of 2.77 AU from the

Table 1.2 *Distributions of mass and angular momentum in the solar system*

	Mass M (10^{24} kg)	Mean orbital radius r (10^{9} m)	Mean orbital rate ω (10^{-9} rad s^{-1})	Orbital angular momentum $M\omega r^2$ (10^{39} kg m^2 s^{-1})	Normalized moment of inertia C/MR^2	Polar moment of Inertia C (10^{40} kg m^2)	Rotation rate Ω (10^{-6} rad s^{-1})	Rotational angular momentum $C\Omega$ (10^{39} kg m^2 s^{-1})
Terrestrial planets								
Mercury	0.3303	57.9	826	0.91	—	—	1.24	—
Venus	4.870	108.2	323	18.4	—	—	0.299	—
Earth	5.976	149.6	199	26.6	0.3308	8.02×10^{-3}	72.7	5.83×10^{-6}
Mars	0.6421	227.9	106	3.54	0.376	2.74×10^{-4}	70.7	1.94×10^{-7}
Great planets and Pluto								
Jupiter	1,900	778.3	16.6	19,060	0.264	245	175.4	0.430
Saturn	568.8	1,427	6.75	7,820	0.207	39.9	163.4	0.0652
Uranus	86.84	2,871	2.37	1,690	0.26	1.46	101.3	0.00148
Neptune	102.4	4,497	1.21	2,520	0.26	1.62	99.6	0.00162
Pluto	0.0129	5,914	0.80	0.38	—	—	11.4	—
TOTAL	2,670	—	—	31,140	—	—	—	0.498
The Sun	1,989,000	—	—	—	—	5,700,000	2.87	164

a = moment of inertia about rotation axis; R = mean radius of Earth

Sun. Subsequently, it was found that numerous small planetoids occupied a broad band of solar orbits centered about 2.9 AU, now called the *asteroid belt*. Pallas was found in 1802, Juno in 1804, and Vesta, the only asteroid that can be seen with the naked eye, was found in 1807. By 1890 more than 300 asteroids had been identified. In 1891 astronomers began to record their paths on photographic plates. More than 2000 asteroids have since been tracked and cataloged. Scientific opinion is divided on what the asteroid belt represents. One idea is that it may represent fragments of an earlier planet that was broken up in some disaster. Alternatively, it may consist of material that was never able to consolidate into a planet.

Bode's law is not a true law in the scientific sense. It should be regarded as an intriguing empirical relationship. Some astronomers hold that the regularity of the planetary distances from the Sun cannot be mere chance but must be a manifestation of physical laws. However, this may be wishful thinking. No combination of physical laws has yet been assembled that accounts for Bode's law.

1.1.3.4 *Angular momentum*

An important characteristic that constrains models of the origin of the solar system is the difference between the distributions of mass and angular momentum. To determine the angular momentum of a rotating body it is necessary to know its moment of inertia. For a particle of mass m the moment of inertia (I) about an axis at distance r is defined as:

$$I=m\,r^2 \tag{1.4}$$

The angular momentum (h) is defined as the product of its moment of inertia (I) about an axis and its rate of rotation (Ω) about that axis:

$$h=I\,\Omega \tag{1.5}$$

Each planet revolves in a nearly circular orbit around the Sun and at the same time rotates about its own axis. Thus there are two contributions to its angular momentum. The angular momentum of a planet's revolution about the Sun is obtained quite simply. The solar system is so immense that the physical size of each planet is tiny compared to the size of its orbit. The moment of inertia of a planet about the Sun is computed by inserting the mass of the planet and its orbital radius (Table 1.2) in Eq. (1.4); the orbital angular momentum of the planet follows by combining the computed moment of inertia with the rate of orbital revolution in Eq. (1.5). To determine the moment of inertia of a solid body about an axis that passes through it (e.g., the rotational axis of a planet) is more complicated. Eq. (1.4) must be computed and summed for all particles in the planet. This requires knowledge of the density distribution within the planet. For some planets the variation of density with depth is not well known, but for most planets there is enough information to calculate the moment of inertia about the axis of rotation; combined with the rate of rotation in Eq. (1.5), this gives the rotational angular momentum.

The angular momentum of a planet's revolution about the Sun is much greater (on average about 60,000 times) than the angular momentum of its rotation about its own axis (Table 1.2). Whereas more than 99.9% of the total mass of the solar system is concentrated in the Sun, more than 99% of the angular momentum is carried by the orbital motion of the planets, especially the four great planets. Of these Jupiter is a special case: it accounts for over 70% of the mass and more than 60% of the angular momentum of the planets.

1.1.4 The origin of the solar system

There have been numerous theories for the origin of the solar system. Age determinations on meteorites indicate that the solar system originated about $4.5–4.6\times10^9$ years ago. A successful theory of how it originated must account satisfactorily for the observed characteristics of the planets. The most important of these properties are:

(1) Except for Pluto, the planetary orbits lie in or close to the same plane, which contains the Sun and the orbit of the Earth (the ecliptic plane).
(2) The planets revolve about the Sun in the same sense, which is anticlockwise when viewed from above the ecliptic plane. This sense of rotation is defined as prograde.
(3) The rotations of the planets about their own axes are also mostly prograde. The exceptions are Venus, which has a retrograde rotation; Uranus, whose axis of rotation lies nearly in the plane of its orbit; and Pluto, whose rotation axis and orbital plane are oblique to the ecliptic.
(4) Each planet is roughly twice as far from the Sun as its closest neighbor (Bode's law).
(5) The compositions of the planets make up two distinct groups: the terrestrial planets lying close to the Sun are small and have high densities, whereas the great planets far from the Sun are large and have low densities.
(6) The Sun has almost 99.9% of the mass of the solar system, but the planets account for more than 99% of the angular momentum.

The first theory based on scientific observation was the *nebular hypothesis* introduced by the German philosopher Immanuel Kant in 1755 and formulated by the French astronomer Pierre Simon de Laplace in 1796. According to this hypothesis the planets and their satellites were formed at the same time as the Sun. Space was filled by a rotating cloud (*nebula*) of hot primordial gas and dust that, as it cooled, began to contract. To conserve the angular momentum of the system, its rotation speeded up; a familiar analogy is the way a pirouetting skater spins more rapidly when he draws in his outstretched arms. Centrifugal force would have caused concentric rings of matter to be thrown off, which then condensed into planets. A serious objection to this hypothesis is that the mass of material in each ring would be too small to provide the gravitational attraction needed to cause the ring to condense into a planet. Moreover, as the nebula contracted, the largest part of the angular momentum would remain associated with the main mass that condensed to form the Sun, which disagrees with the observed distribution of angular momentum in the solar system.

Several alternative models were postulated subsequently, but have also fallen into disfavor. For example, the *collision hypothesis* assumed that the Sun was formed before the planets. The gravitational attraction of a closely passing star or the blast of a nearby exploding supernova drew out a filament of solar material that condensed to form the planets. However, a major objection to this scenario is that the solar material would have been so hot that it dissipated explosively into space rather than slowly condensing to form the planets.

Modern interpretations of the origin of the solar system are based on modifications of the nebular hypothesis. As the cloud of gas and dust contracted, its rate of rotation speeded up, flattening the cloud into a lens-shaped disk. When the core of the contracting cloud became dense enough, gravitation caused it to collapse upon itself to form a proto-Sun in which thermonuclear fusion was initiated. Hydrogen nuclei combined under the intense pressure to form helium nuclei, releasing huge amounts of energy. The material in the spinning disk was initially very hot and gaseous but, as it cooled, solid material condensed out of it as small grains. The grains coalesced as rocky or icy clumps called *planetesimals*. Asteroid-like planetesimals with a silicate, or rocky, composition formed near the Sun, while comet-like planetesimals with an icy composition formed far from the Sun's heat. In turn, helped by gravitational attraction, the planetesimals accreted to form the planets. Matter with a high boiling point (e.g., metals and silicates) could condense near to the Sun, forming the terrestrial planets. Volatile materials (e.g., water, methane) would vaporize and be driven into space by the stream of particles and radiation from the Sun. During the condensation of the large cold planets in the frigid distant realms of the solar system, the volatile materials were retained. The gravitational attractions of Jupiter and Saturn may have been strong enough to retain the composition of the original nebula.

It is important to keep in mind that this scenario is merely a hypothesis – a plausible but not unique explanation of how the solar system formed. It attributes the variable compositions of the planets to accretion at different distances from the Sun. The model can be embellished in many details to account for the characteristics of individual planets. However, the scenario is unsatisfactory because it is mostly qualitative. For example, it does not adequately explain the division of angular momentum. Physicists, astronomers, space scientists and mathematicians are constantly trying new methods of investigation and searching for additional clues that will improve the hypothesis of how the solar system formed.

1.2 THE DYNAMIC EARTH

1.2.1 Historical introduction

The Earth is a dynamic planet, perpetually changing both externally and internally. Its surface is constantly being altered by *endogenic* processes (i.e., of *internal* origin) resulting in volcanism and tectonism, as well as by *exogenic* processes (i.e., of *external* origin) such as erosion and deposition. These processes have been active throughout geological history. Volcanic explosions like the 1980 eruption of Mt St Helens in the northwestern United States can transform the surrounding landscape virtually instantaneously. Earthquakes also cause sudden changes in the landscape, sometimes producing faults with displacements of several meters in seconds. Weather-related erosion of surface features occasionally occurs at dramatic rates, especially if rivers overflow or landslides are triggered. The Earth's surface is also being changed constantly by less spectacular geological processes at rates that are extremely slow in human terms. Regions that have been depressed by the loads of past ice-sheets are still rebounding vertically at rates of up to several mm yr^{-1}. Tectonic forces cause mountains to rise at similar uplift rates, while the long-term average effects of erosion on a regional scale occur at rates of cm yr^{-1}. On a larger scale the continents move relative to each other at speeds of up to several cm yr^{-1} for time intervals lasting millions of years. The Earth's interior is also in motion. The mantle appears hard and solid to seismic waves, but is believed to exhibit a softer, plastic behavior over long geological time intervals, flowing (or 'creeping') at rates of several cm yr^{-1}. Deeper inside the Earth, the liquid core probably flows at a geologically rapid rate of a few tenths of a millimeter per second.

Geologists have long been aware of the Earth's dynamic condition. Several hypotheses have attempted to explain the underlying mechanisms. In the late 19th and early 20th centuries geological orthodoxy favored the hypothesis of a contracting Earth. Mountain ranges were thought to have formed on its shrinking surface like wrinkles on a desiccating apple. Horizontal tectonic displacements were known, but were considered to be a by-product of more important vertical motions. The realization that large overthrusts played an important role in the formation of nappe structures in the Alps implied amounts of horizontal shortening that were difficult to accommodate in the contraction hypothesis. A new school of thought emerged in which mountain-building was depicted as a consequence of horizontal displacements.

A key observation in this context was the congruity between the opposing coasts of the South Atlantic, especially the similar shapes of the coastlines of Brazil and Africa. As early as 1620, despite the inaccuracy and incompleteness of early 17th century maps, Francis Bacon drew attention to the parallelism of the Atlantic-bordering coastlines. In 1858 Antonio Snider constructed a map showing relative movements of the circum-Atlantic continents, although he did not maintain the shapes of the coastlines. In the late 19th century the Austrian geologist Eduard Suess coined the name *Gondwanaland* for a proposed great southern continent that existed during late Paleozoic times. It embodied Africa, Antarctica, Arabia, Australia, India and South America, and lay predominantly in the southern hemisphere. The Gondwana continents are now individual entities and some (e.g., India, Arabia) no longer lie in the southern hemisphere, but they are often still called the 'southern continents'. In the Paleozoic, the 'northern continents' of North America (including Greenland), Europe and most of Asia also formed a single continent, called *Laurasia*. Laurasia and Gondwanaland split apart in the Early Mesozoic. The Alpine–Himalayan mountain belt was thought to have developed from a system of geosynclines that formed in the intervening sea, which Suess called the *Tethys* ocean to distinguish it from the present Mediterranean Sea. Implicit in these reconstructions is the idea that the continents subsequently reached their present positions by slow horizontal displacements across the surface of the globe.

1.2.2 Continental drift

The 'displacement hypothesis' of continental movements matured early in the 20th century. In 1908 F. B. Taylor related the world's major fold-belts to convergence of the continents as they moved away from the poles, and in 1911

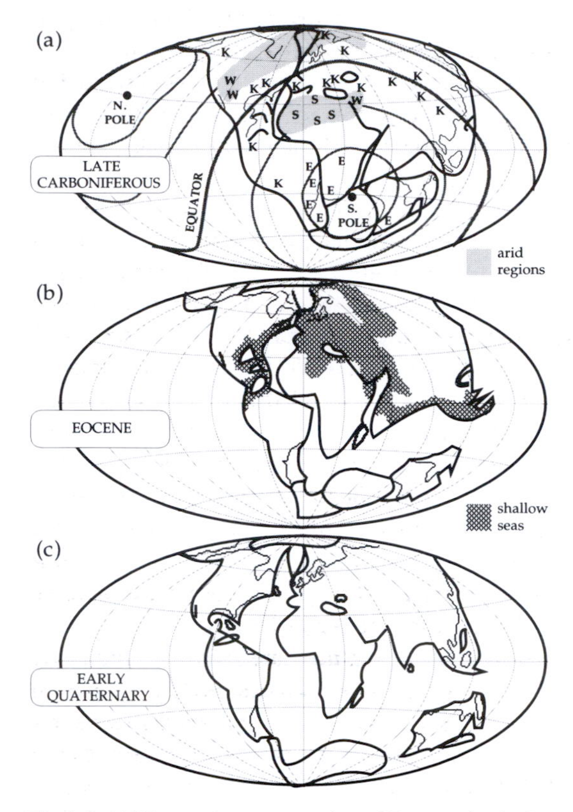

Fig. 1.5 (a) Wegener's reconstruction of Pangaea in the Late Carboniferous, showing estimated positions of the North and South poles and paleo-equator. *Shaded areas*: arid regions; K: coal deposits; S: salt deposits; W: desert regions; E: ice sheets (modified after Köppen and Wegener, 1924). Relative positions of the continents are shown in (b) the Eocene (*shaded areas*: shallow seas) and (c) the Early Quaternary (after Wegener, 1922). The latitudes and longitudes are arbitrary.

H. B. Baker reassembled the Atlantic-bordering continents together with Australia and Antarctica into a single continent; regrettably he omitted Asia and the Pacific. However, the most vigorous proponent of the displacement hypothesis was Alfred Wegener, a German meteorologist and geologist. In 1912 Wegener suggested that all of the continents were together in the Late Paleozoic, so that the land area of the Earth formed a single landmass (Fig. 1.5). He coined the name *Pangaea* (Greek for 'all Earth') for this supercontinent, which he envisioned was surrounded by a single ocean (*Panthalassa*). Wegener referred to the large-scale horizontal displacement of crustal blocks having continental dimensions as *Kontinentalverschiebung*. The anglicized form, *continental drift*, implies additionally that displacements of the blocks take place slowly over long time intervals.

1.2.2.1 *Pangaea*

As a meteorologist Wegener was especially interested in paleoclimatology. For the first half of the 20th century the best evidence for the continental drift hypothesis and the earlier existence of Pangaea consisted of geological indicators of earlier paleoclimates. In particular, Wegener observed a much better alignment of regions of Permo–Carboniferous glaciation in the southern hemisphere when the continents were in the reconstructed positions for Gondwanaland instead of their present positions. His reconstruction of Pangaea brought Carboniferous coal deposits into alignment and suggested that the positions of the continents relative to the Paleozoic equator were quite different from their modern ones. Together with W. Köppen, a fellow German meteorologist, he assembled paleoclimatic data that showed the distributions of coal deposits (evidence of moist temperate zones), salt, gypsum and desert sandstones (evidence of dry climate) for several geological eras (Carboniferous, Permian, Eocene, Quaternary). When plotted on Wegener's reconstruction maps, the paleoclimatic data for each era formed climatic belts just like today; namely, an equatorial tropical rain belt, two adjacent dry belts, two temperate rain belts, and two polar ice caps (Fig. 1.5a).

Wegener's continental drift hypothesis was bolstered in 1937 by the studies of a South African geologist, Alexander du Toit, who noted sedimentological, paleontological, paleoclimatic, and tectonic similarities between western Africa and eastern South America. These favored the Gondwanaland reconstruction rather than the present configuration of continents during the Late Paleozoic and Early Mesozoic.

Some of Wegener's theories were largely conjectural. On the one hand, he reasoned correctly that the ocean basins are not permanent. Yet he envisioned the sub-crustal material as capable of viscous yield over long periods of time, enabling the continents to drift through the ocean crust like ships through water. This model met with profound scepticism among geologists. He believed, in the face of strong opposition from physicists, that the Earth's geographic axis had moved with time, instead of the crust moving relative to the fixed poles. His timing of the opening of the Atlantic (Fig. 1.5b, c) was faulty, requiring a large part of the separation of South America from Africa to take place since the Early Pleistocene (i.e., in the last two million years or so). Moreover, he was unable to offer a satisfactory driving mechanism for continental drift. His detractors used the disprovable speculations to discredit his better-documented arguments in favor of continental drift.

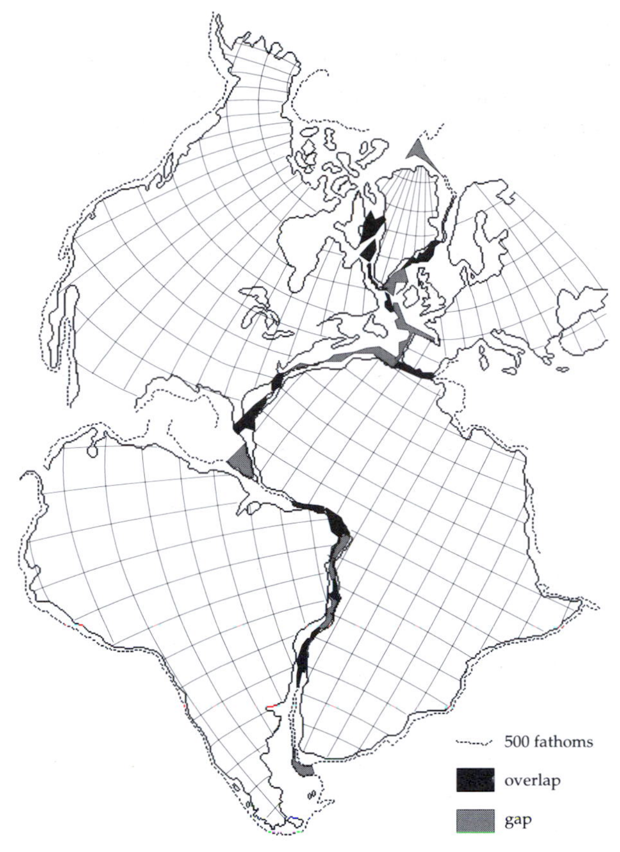

Fig. 1.6 Computer-assisted fit of the Atlantic-bordering continents at the 500 fathom (900 m) depth (after Bullard *et al.*, 1965).

1.2.2.2 *Computer-assisted reconstructions*

Wegener pointed out that it was not possible to fit the continents together using their present coastlines, which are influenced by recent sedimentary deposits at the mouths of major rivers as well as the effects of coastal erosion. The large areas of continental shelf must also be taken into account, so Wegener matched the continents at about the edges of the continental shelves, where the continental slopes plunge into the oceanic basins. The matching was visual and inexact by modern standards, but more precise methods only became available in the 1960s with the development of powerful computers.

In 1965 E. C. Bullard, J. E. Everett and A. G. Smith used a computer to match the relative positions of the continents bounding the Atlantic ocean (Fig. 1.6). They digitized the continental outlines at approximately 50 km intervals for different depth contours on the continental slopes, and selected the fit of the 500 fathom (900 m) depth contour as optimum. The traces of opposite continental margins were matched by an iterative procedure. One trace was rotated relative to the other (about a pole of relative rotation) until the differences between the traces were minimized; the procedure was then repeated with different rotation poles until the best fit was obtained. The optimum fit is not perfect, but has some overlaps and gaps. Nevertheless, the analysis gives an excellent geometric fit of the opposing coastlines of the Atlantic.

A few years later A. G. Smith and A. Hallam used the same computer-assisted technique to match the coastlines of the southern continents, also at the 500 fathom depth contour (Fig. 1.7). They obtained an optimum geometric reconstruction of Gondwanaland similar to the visual match suggested by du Toit in 1937; it probably represents the geometry of Gondwanaland that existed in the Late Paleozoic and Early Mesozoic. It is not the only possible good geometric fit, but it also satisfies other geological evidence. At various times in the Jurassic and Cretaceous, extensional plate margins formed within Gondwanaland, causing it to subdivide to form the present 'southern continents'. Their dispersal to their present positions took place largely in the Late Cretaceous and Tertiary.

Pangaea existed only in the Late Paleozoic and Early Mesozoic. Geological and geophysical evidence argues in favor of the existence of its northern and southern constituents – Laurasia and Gondwanaland – as separate entities in the Early Paleozoic and Precambrian. An important source of data bearing on continental reconstructions in ancient times and the drift of the continents is provided by *paleomagnetism*, which is the record of the Earth's ancient magnetic field. Paleomagnetism is described in §5.6 and summarized below.

1.2.2.3 *Paleomagnetism and continental drift*

In the late 19th century geologists discovered that rocks can carry a stable record of the geomagnetic field direction at the time of their formation. From the magnetization direction it is possible to calculate the position of the magnetic pole at that time; this is called the virtual geomagnetic pole (VGP) position. Averaged over a time interval longer than a few tens of thousands of years, the mean VGP position coincides with the geographic pole, as if the axis of the mean geomagnetic dipole field were aligned with the Earth's rotation axis. This correspondence can be proved for the present geomagnetic field, and a fundamental assumption of paleomagnetism – called the 'axial dipole hypothesis' – is that it has always been valid. The hypothesis can be verified for rocks and sediments up to a few million years old, but its validity has to be assumed for earlier geological epochs. However, the self-consistency of paleomagnetic data and their compatibility with continental reconstructions argue that the axial dipole hypothesis is also applicable to the Earth's ancient magnetic field.

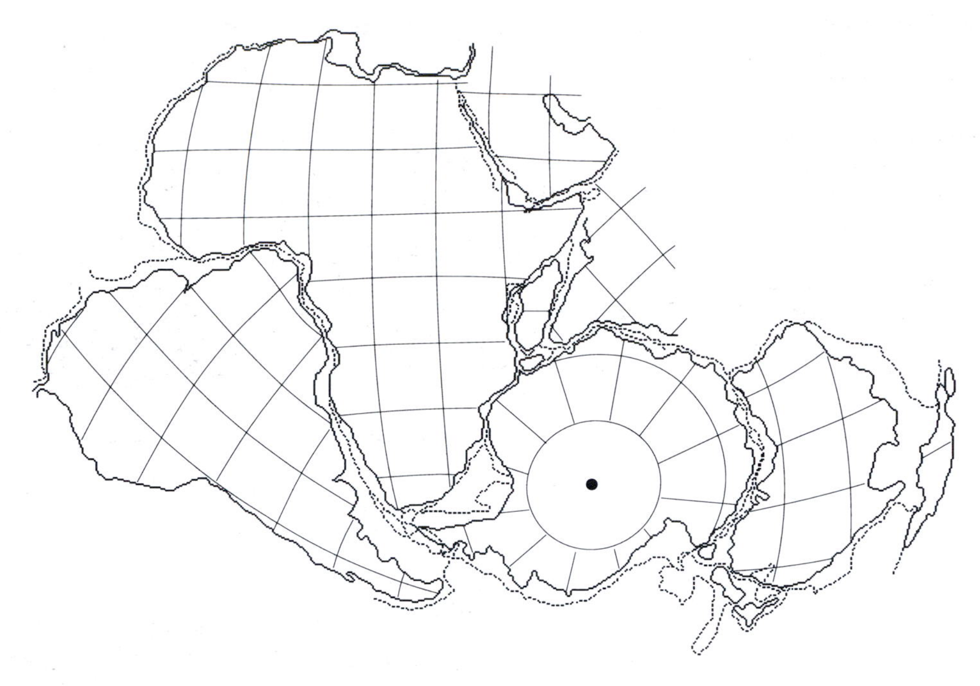

Fig. 1.7 Computer-assisted fit of the continents that form Gondwanaland (after Smith and Hallam, 1970).

For a particular continent, rocks of different ages give different mean VGP positions. The appearance that the pole has shifted with time is called apparent polar wander (APW). By connecting mean VGP positions of different ages for sites on the same continent a line is obtained, called the apparent polar wander path of the continent. Each continent yields a different APW path, which consequently cannot be the record of movement of the pole. Rather, each APW path represents the movement of the continent relative to the pole. By comparing APW paths the movements of the continents relative to each other can be reconstructed. The APW paths provide strong supporting evidence for continental drift.

Paleomagnetism developed as a geological discipline in the 1950s and 1960s. The first results indicating large-scale continental movement were greeted with some scepticism. In 1956 S. K. Runcorn demonstrated that the paleomagnetic data from Permian and Triassic rocks in North America and Great Britain agreed better if the Atlantic ocean were closed, i.e., as in the Laurasia configuration. In 1957 E. Irving showed that Mesozoic paleomagnetic data from the 'southern continents' were more concordant with du Toit's Gondwanaland reconstruction than with the present arrangement of the continents. Since these pioneering studies numerous paleomagnetic investigations have established APW paths for the different continents. The quality of the paleomagnetic record is good for most geological epochs since the Devonian.

The record for older geologic periods is less reliable for several reasons. In the Early Paleozoic the data become fewer and the APW paths become less well defined. In addition, the oldest parts of the paleomagnetic record are clouded by the increasing possibility of false directions due to undetected secondary magnetization. This happens when thermal or tectonic events alter the original magnetization, so that its direction no longer corresponds to that at the time of rock formation. Remagnetization can affect rocks of any age, but it is recognized more readily and constitutes a less serious problem in younger rocks.

Problems afflicting Precambrian paleomagnetism are even more serious than in the Early Paleozoic. APW paths have been derived for the Precambrian, especially for North America, but only in broad outline. In part this is because it is difficult to date Precambrian rocks precisely enough to determine the fine details of an APW path. It is often not possible to establish which is the north or south pole. In addition, the range of time encompassed by the Precambrian – more than 3.5 Ga – is about six times longer than the 570 Ma length of the Phanerozoic, and the probability of remagnetization events is correspondingly higher.

In spite of some uncertainties, Early Paleozoic paleomagnetism permits reassembly of the supercontinents Gondwanaland and Laurasia and traces their movements before they collided in the Carboniferous to form Pangaea. Geological and paleomagnetic evidence concur that, in the Cambrian period, Gondwanaland very likely existed as a supercontinent in essentially the du Toit configuration. It coexisted in the Early Paleozoic with three other cratonic centers: Laurentia (North America and Greenland), Baltica (northern Europe) and Siberia. Laurentia and

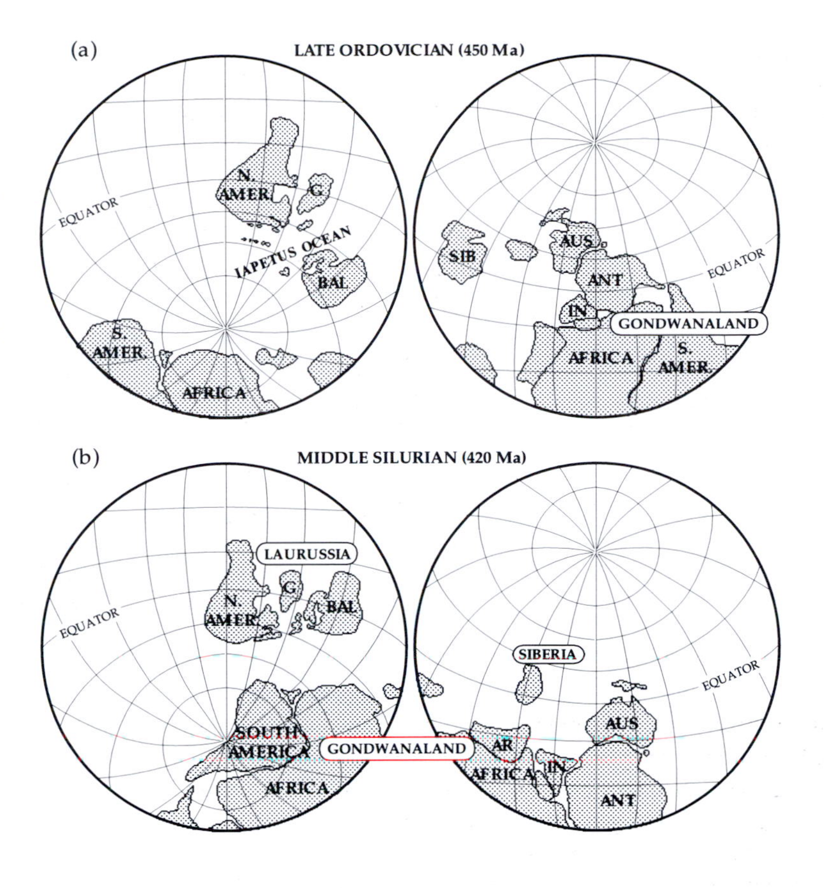

Fig. 1.8 Paleomagnetic reconstruction of the relative positions of (a) Laurentia (N. America and Greenland), Baltica and Gondwanaland (S. America, Africa, Arabia, Australia, India and Antarctica) in the Late Ordovician and (b) Laurussia (N. America and Baltica) and Gondwanaland in the Middle Silurian (after Van der Voo, 1993).

Baltica were separated by the Iapetus ocean (Fig. 1.8a), which began to close in the Ordovician (about 450 Ma ago). Paleomagnetic data indicate that Laurentia and Baltica fused together around Late Silurian time to form the supercontinent Laurussia; at that time the Siberian block remained a separate entity. The Laurentia–Baltica collision is expressed in the Taconic and Caledonian orogenies in North America and northern Europe. The gap between Gondwanaland and Laurussia in the Middle Silurian (Fig. 1.8b) closed about the time of the Silurian–Devonian boundary (about 410 Ma ago). Readjustments of the positions of the continental blocks in the Devonian produced the Acadian orogeny. Laurussia separated from Gondwanaland in the Late Devonian, but the two supercontinents began to collide again in the Early Carboniferous (about 350 Ma ago), causing the Hercynian orogeny. By the Late Carboniferous (300 Ma ago) Pangaea was almost complete, except for Siberia, which was probably appended in the Permian.

The general configuration of Pangaea from the Late Carboniferous to the Early Jurassic is supported by paleomagnetic results from the Atlantic-bordering continents. However, the paleomagnetic data suggest that the purely geometric 'Bullard-fit' is only appropriate for the later part of Pangaea's existence. The results for earlier times from the individual continents agree better for slightly different reconstructions (see §5.6.4.4). This suggests that some internal rearrangement of the component parts of Pangaea may have occurred. Also, the computer-assisted geometric assembly of Gondwanaland, similar to that proposed by du Toit, is not the only possible reconstruction, although paleomagnetic results confirm that it is probably the optimum one. Other models involve different relative placements of West Gondwanaland (i.e., South America and Africa) and East Gondwanaland (i.e., Antarctica, Australia and India), and imply that they may have moved relative to each other. The paleomagnetic data do not contradict the alternative models, but are not precise enough to discriminate definitively between them.

The consistency of paleomagnetic results leaves little room for doubt that the continents have changed position relative to each other throughout geological time. This lends justification to the concept of continental drift, but it does not account for the mechanism by which it has taken place. Another aspect of the paleomagnetic record – the history of magnetic field polarity rather than the APW paths – has played a key role in deducing the mechanism. The explanation requires an understanding of the Earth's internal struc-

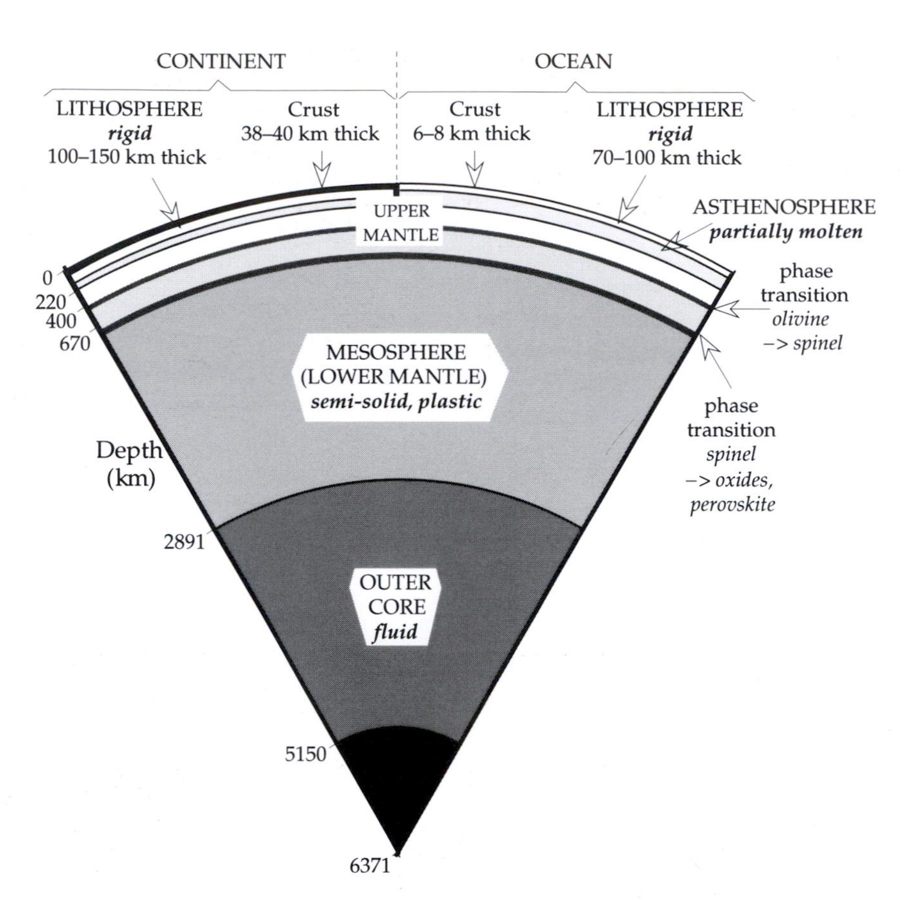

Fig. 1.9 Simplified layered structure of the Earth's interior showing the depths of the most important seismic discontinuities.

ture, the distribution of seismicity and the importance of the ocean basins.

1.2.3 Earth structure

Early in the 20th century it became evident from the study of seismic waves that the interior of the Earth has a radially layered structure, like that of an onion (Fig. 1.9). The boundaries between the layers are marked by abrupt changes in seismic velocity or velocity gradient. Each layer is characterized by a specific set of physical properties determined by the composition, pressure and temperature in the layer. The four main layers are the crust, mantle and the outer and inner cores. Their properties are described in detail in §3.7 and summarized briefly here.

At depths of a few tens of kilometers under continents and less than ten kilometers beneath the oceans seismic velocities increase sharply. This seismic discontinuity, discovered in 1909 by A. Mohorovičić, represents the boundary between the *crust* and *mantle*. R. D. Oldham noted in 1906 that the travel-times of seismic compressional waves that traversed the body of the Earth were greater than expected; the delay was attributed to a *fluid outer core*. Support for this idea came in 1914, when B. Gutenberg described a shadow zone for seismic waves at epicentral distances greater than about 105°. Just as light-waves cast a shadow of an opaque object, seismic waves from an earthquake cast a shadow of the core on the opposite side of the world. Compressional waves can in fact pass through the liquid core. They appear, delayed in time, at epicentral distances larger than 143°. In 1936 I. Lehmann observed the weak arrivals of compressional waves in the gap between 105° and 143°. They are interpreted as evidence for a *solid inner core*.

1.2.3.1 *Lithospheric plates*

The radially layered model of the Earth's interior assumes spherical symmetry. This is not valid for the crust and upper mantle. These outer layers of the Earth show important lateral variations. The crust and uppermost mantle down to a depth of about 70–100 km under deep ocean basins and 100–150 km under continents is rigid, forming a hard outer shell called the *lithosphere*. Beneath the lithosphere lies the *asthenosphere*, a layer in which seismic velocities often decrease, suggesting lower rigidity. It is about 150 km thick, although its upper and lower boundaries are not sharply defined. This weaker layer is thought to be partially molten; it may be able to flow over long periods of time like a viscous

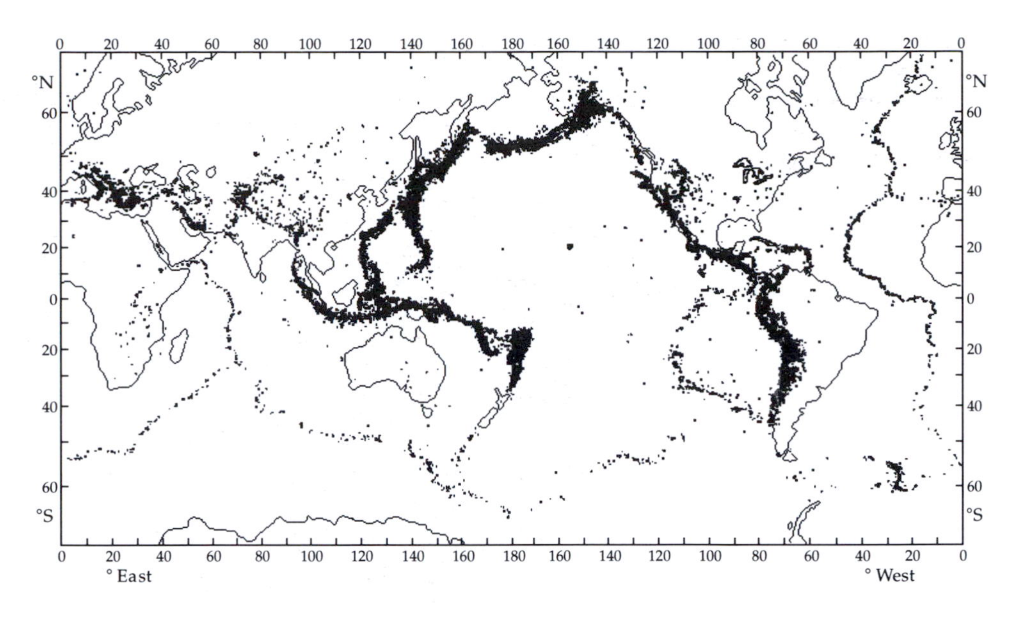

Fig. 1.10 The geographical distribution of epicenters for 30,000 earthquakes for the years 1961–1967 illustrates the tectonically active regions of the Earth (after Barazangi and Dorman, 1969).

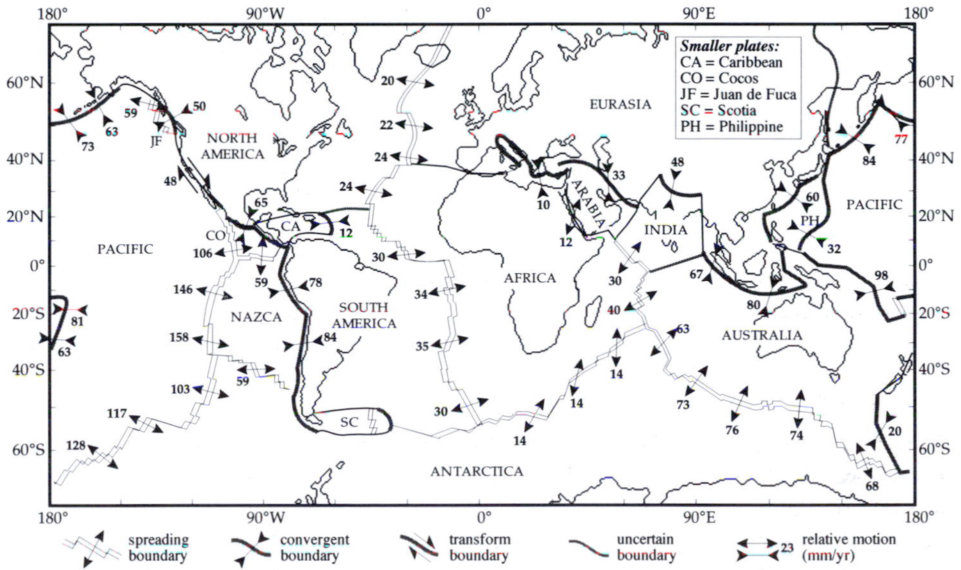

Fig. 1.11 The major and minor lithospheric plates. The arrows indicate relative velocities in mm yr^{-1} at active plate margins, as deduced from the model NUVEL-1 of current plate motions (data source: DeMets *et al.*, 1990).

liquid or plastic solid, in a way that depends on temperature and composition. The asthenosphere plays an important role in plate tectonics, because it makes possible the relative motions of the overlying lithospheric plates.

The brittle condition of the lithosphere causes it to fracture when strongly stressed. The rupture produces an earthquake, which is the violent release of elastic energy due to sudden displacement on a fault plane. Earthquakes are not distributed evenly over the surface of the globe, but occur predominantly in well-defined narrow seismic zones that are often associated with volcanic activity (Fig. 1.10). These are: (a) the circum-Pacific 'ring of fire'; (b) a sinuous belt running from the Azores through North Africa and the Alpine–Dinaride–Himalayan mountain chain as far as S.E. Asia; and (c) the world-circling system of oceanic ridges and rises. The seismic zones subdivide the lithosphere laterally into *tectonic plates* (Fig. 1.11). A plate may be as broad as 10,000 km (e.g., the Pacific plate) or as small as a few thousand km (e.g., the Philippines plate). There are twelve major plates (Antarctica, Africa, Eurasia, India, Australia, Arabia, Philippines, North America, South America, Pacific, Nazca, and Cocos) and several minor plates (e.g., Scotia, Caribbean, Juan de Fuca). The positions of the boundaries between the North American and South American plates and between the North American and Eurasian plates are uncertain. The boundary between the

Indian and Australian plates is not sharply defined, but may be a broad region of diffuse deformation.

A comprehensive model of current plate motions (called NUVEL-1), based on magnetic anomaly patterns and first-motion directions in earthquakes, shows rates of separation at plate boundaries that range from about 20 mm yr^{-1} in the North Atlantic to about 160 mm yr^{-1} on the East Pacific Rise (Fig. 1.11). The model also gives rates of closure ranging from about 10 mm yr^{-1} between Africa and Eurasia to about 80 mm yr^{-1} between the Nazca plate and South America.

1.2.4 Types of plate margin

An important factor in the evolution of modern plate tectonic theory was the development of oceanography in the years following World War II, when technology designed for warfare was turned to peaceful purposes. The bathymetry of the oceans was charted extensively by echo-sounding and within a few years several striking features became evident. Deep trenches, more than twice the depth of the ocean basins, were discovered close to island arcs and some continental margins; the Marianas Trench is more than 11 km deep. A prominent submarine mountain chain – called an oceanic ridge – was found in each ocean. The oceanic ridges rise to as much as 3000 m above the adjacent basins and form a continuous system, more than 60,000 km in length, that girdles the globe. Unlike continental mountain belts, which are usually less than several hundred kilometers across, the oceanic ridges are 2000–4000 km in width. The ridge system is offset at intervals by long horizontal faults forming fracture zones. These three features – trenches, ridges and fracture zones – originate from different plate tectonic processes.

The lithospheric plates are very thin in comparison to their breadth (compare Fig. 1.9 and Fig. 1.11). Most earthquakes occur at plate margins, and are associated with interactions between plates. Apart from rare intraplate earthquakes, which can be as large and disastrous as the earthquakes at plate boundaries, the plate interiors are aseismic. This suggests that the plates behave rigidly. Analysis of earthquakes allows the direction of displacement to be determined and permits interpretation of the relative motions between plates.

There are three types of plate margin, distinguished by different tectonic processes (Fig. 1.12). The world-wide pattern of earthquakes shows that the plates are presently moving apart at oceanic ridges. Magnetic evidence, discussed below, confirms that the separation has been going on for millions of years. New lithosphere is being formed at these *spreading centers*, so the ridges can be regarded as *constructive* plate margins. The seismic zones related to deep-sea trenches, island arcs and mountain belts mark places where lithospheric plates are converging. One plate is forced under another there in a so-called *subduction zone*. Because it is thin in relation to its breadth, the lower plate bends sharply before descending to depths of several hundred kilometers, where it is absorbed. The subduction zone marks a *destructive* plate margin.

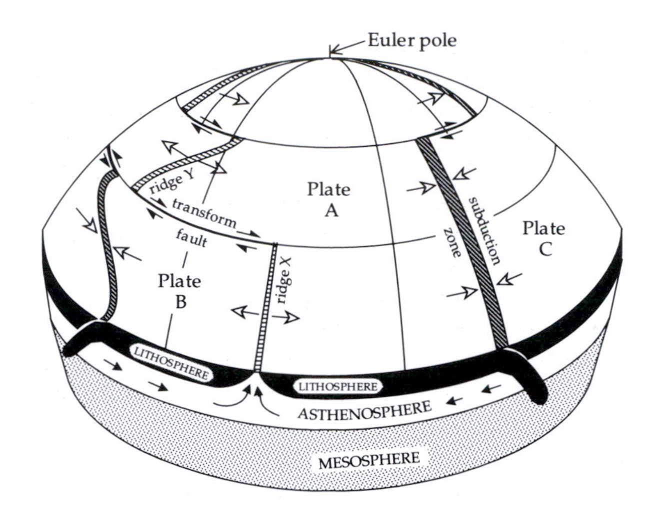

Fig. 1.12 Schematic model illustrating the three types of plate margin. Lightly hachured areas symbolize spreading ridges (constructive margins); darker shaded areas denote subduction zones (destructive margins); dark lines mark transform faults (conservative margins). The figure is drawn relative to the pole of relative motion between plates A and B. Small arrows denote relative motion on transform faults; large arrows show directions of plate motion, which can be oblique to the strike of ridge segments or subduction zones. Arrows in the asthenosphere suggest return flow from destructive to constructive margins.

Constructive and destructive plate margins may consist of many segments linked by horizontal faults. A crucial step in the development of plate tectonic theory was made in 1965 by a Canadian geologist, J. Tuzo Wilson, who recognized that these faults are not conventional transcurrent faults. They belong to a new class of faults, which Wilson called *transform faults*. The relative motion on a transform fault is opposite to what might be inferred from the offsets of bordering ridge segments. At the point where a transform fault meets an oceanic ridge it transforms the spreading on the ridge to horizontal shear on the fault. Likewise, where such a fault meets a destructive plate margin it transforms subduction to horizontal shear.

The transform faults form a *conservative* plate margin, where lithosphere is neither created nor destroyed; the boundary separates plates that move past each other horizontally. This interpretation was documented in 1967 by L. Sykes, an American seismologist. He showed that earthquake activity on an oceanic ridge system was confined almost entirely to the transform fault between ridge crests, where the neighboring plates rub past each other. Most importantly, Sykes found that the mechanisms of earthquakes on the transform faults agreed with the predicted sense of strike–slip motion.

Transform faults play a key role in determining plate motions. Spreading and subduction are often assumed to be perpendicular to the strike of a ridge or trench, as is the case for ridge X in Fig. 1.12. This is not necessarily the case. Oblique motion with a component along strike is possible at each of these margins, as on ridge Y. However, because lithosphere is neither created nor destroyed at a conservative margin, the relative motion between adjacent plates must be parallel to the strike of a shared transform fault. Pioneering independent studies by D. P. McKenzie and R. L. Parker (1967) and W. J. Morgan (1968) showed how transform faults could be used to locate the Euler pole of rotation for two plates (see §1.2.6.3). Using this method, X. Le Pichon in 1968 determined the present relative motions of the major tectonic plates. In addition, he derived the history of plate motions in the geological past by incorporating newly available magnetic results from the ocean basins.

1.2.5 Sea-floor spreading

One of the principal stumbling blocks of continental drift was the inability to explain the mechanism by which drift took place. Wegener had invoked forces related to gravity and the Earth's rotation, which were demonstrably much too weak to drive the continents through the resistant basaltic crust. A. Holmes proposed a model in 1944 that closely resembles the accepted plate tectonic model (Holmes, 1965). He noted that it would be necessary to remove basaltic rocks continuously out of the path of an advancing continent, and suggested that this took place at the ocean deeps where heavy eclogite 'roots' would sink into the mantle and melt. Convection currents in the upper mantle would return the basaltic magma to the continents as plateau basalts, and to the oceans through innumerable fissures. Holmes saw generation of new oceanic crust as a process that was dispersed throughout an ocean basin. At the time of his proposal the existence of the system of oceanic ridges and rises was not yet known.

The important role of oceanic ridges was first recognized by H. Hess in 1962. He suggested that new oceanic crust is generated from upwelling hot mantle material at the ridges. Convection currents in the upper mantle would rise to the surface at the ridges and then spread out laterally. The continents would ride on the spreading mantle material, carried along passively by the convection currents. In 1961 R. Dietz coined the expression 'sea-floor spreading' for the ridge process. It results in the generation of lineated marine magnetic anomalies at the ridges, which record the history of geomagnetic polarity reversals. Study of these magnetic effects led to the verification of sea-floor spreading.

1.2.5.1 *The Vine–Matthews–Morley hypothesis*

Paleomagnetic studies in the late 1950s and early 1960s of radiometrically dated continental lavas showed that the geomagnetic field has changed polarity at irregular time intervals. For tens of thousands to millions of years the polarity might be normal (as at present), then unaccountably the poles reverse within a few thousand years, so that the north magnetic pole is near the south geographic pole and the south magnetic pole is near the north geographic pole. This state may again persist for a long interval, before the polarity again switches. The ages of the reversals in the last 5 million years have been obtained radiometrically, giving an irregular but dated polarity sequence.

A magnetic anomaly is a departure from the theoretical magnetic field at a given location. If the field is stronger than expected, the anomaly is positive; if it is weaker than expected, the anomaly is negative. In the late 1950s magnetic surveys over the oceans revealed remarkable striped patterns of alternately positive and negative magnetic anomalies over large areas of oceanic crust (Fig. 1.13), for which conventional methods of interpretation gave no satisfactory account. In 1963 the English geophysicists F. J. Vine and D. H. Matthews, and, independently, the Canadian geologist L. W. Morley (Morley and Larochelle, 1964), formulated a landmark hypothesis that explains the origin of the oceanic magnetic anomaly patterns (see also §5.7.3).

Observations on dredged samples had shown that basalts in the uppermost oceanic crust carry a strong remanent magnetization (i.e., they are permanently magnetized, like a magnet). The Vine–Matthews–Morley hypothesis integrates this result with the newly acquired knowledge of geomagnetic polarity reversals and the Hess–Dietz concept of sea-floor spreading (Fig. 1.14). The basaltic lava is extruded in a molten state. When it solidifies and its temperature cools below the Curie temperature of its magnetic minerals, the basalt becomes strongly magnetized in the direction of the Earth's magnetic field at that time. Along an active spreading ridge, long thin strips of magnetized basaltic crust form symmetrically on opposite sides of the spreading center, each carrying the magnetic imprint of the field in which it formed. Sea-floor spreading can persist for many millions of years at an oceanic ridge. During this time the magnetic field changes polarity many times,

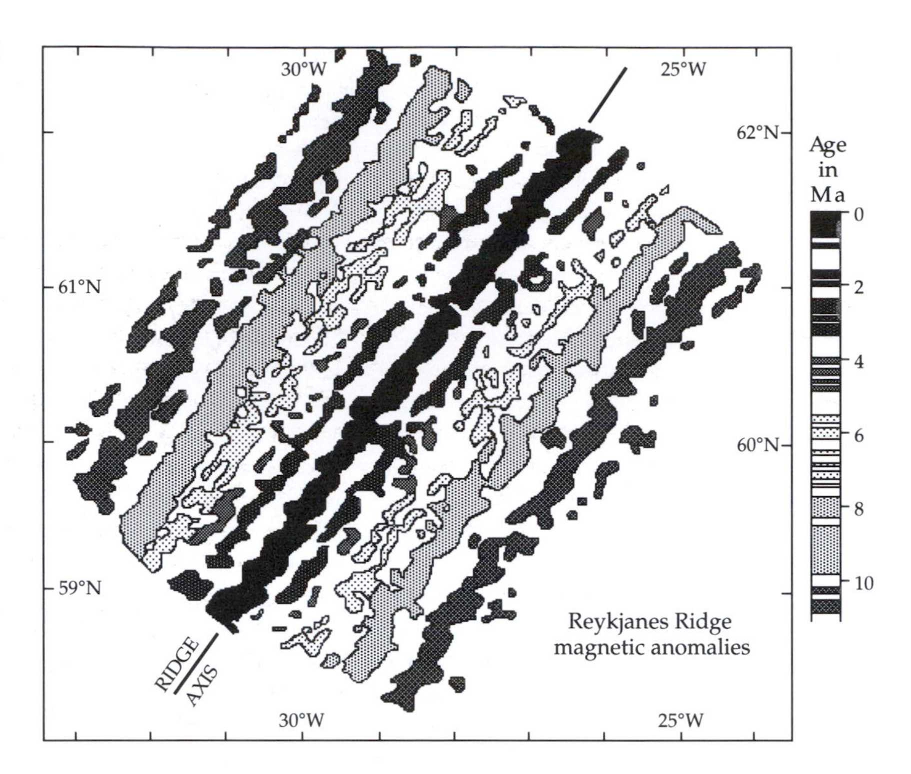

Fig. 1.13 Symmetric striped pattern of magnetic anomalies on the Reykjanes segment of the Mid-Atlantic Ridge southwest of Iceland. The positive anomalies are shaded according to their age, as indicated in the vertical column (after Heirtzler *et al.*, 1966).

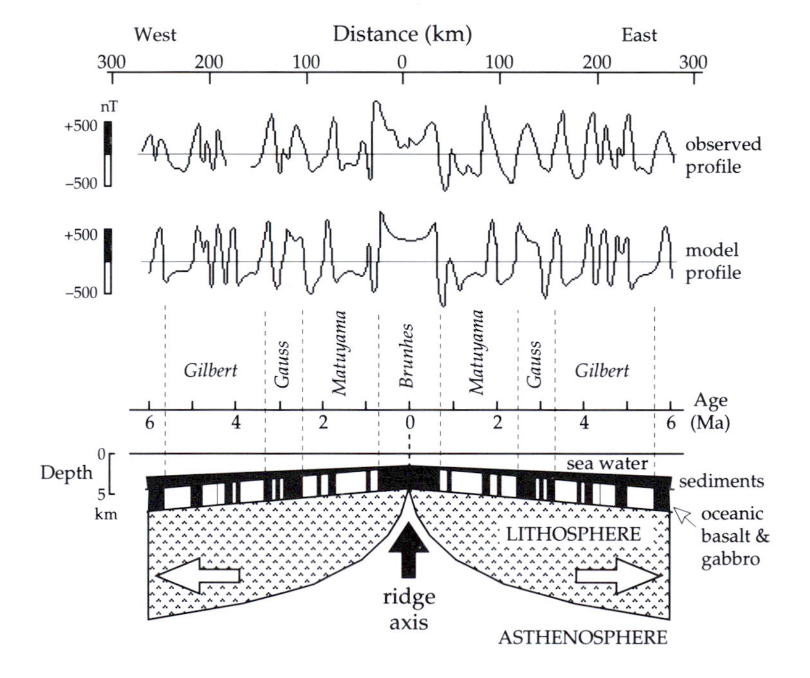

Fig. 1.14 *Upper*: observed and computed marine magnetic anomalies, in nanotesla (nT), across the Pacific–Antarctica ridge, and (*lower*) their interpreted origin in terms of the Vine–Matthews hypothesis (after Pitman and Heirtzler, 1966).

forming strips of oceanic crust that are magnetized alternately parallel and opposite to the present field, giving the observed patterns of positive and negative anomalies. Thus, the basaltic layer acts like a magnetic tape recorder, preserving a record of the changing geomagnetic field polarity.

1.2.5.2 *Rates of sea-floor spreading*

The width of a magnetic lineation (or stripe) depends on two factors: the speed with which the oceanic crust moves away from a spreading center, and the length of time that geomagnetic polarity is constantly normal or reversed. The distance between the edges of magnetized crustal stripes can be measured from magnetic surveys at the ocean surface, while the ages of the reversals can be obtained by correlating the oceanic magnetic record with the radiometrically dated reversal sequence determined in subaerial lavas for about the last 4 Ma. When the distance of a given polarity reversal from the spreading axis is plotted against the age of the reversal, a nearly linear relationship is obtained (Fig. 1.15). The slope of the best-fitting straight line gives the average *half-rate of spreading* at the ridge. These are of the order of 10 mm yr^{-1} in the North Atlantic ocean and 40–60 mm yr^{-1} in the Pacific ocean. The calculation applies to the rate of motion of crust on one side of the ridge only. In most cases spreading has been symmetric on each side of the ridge (i.e., the opposite sides are moving away from the ridge at equal speeds), so the full rate of separation at a ridge axis is double the calculated half-rate of spreading (Fig. 1.11).

The rates of current plate motion determined from axial anomaly patterns (Fig. 1.11) are average values over several million years. Modern geodetic methods allow these rates to be tested directly (see §2.4.6). Satellite laser-ranging (SLR) and very long baseline interferometry (VLBI) allow exceptionally accurate measurement of changes in the distance between two stations on Earth. Controlled over several years, the distances between pairs of stations on opposite sides of the Atlantic ocean are increasing slowly at a mean rate of 17 mm yr^{-1} (Fig. 1.16). This figure is close to the long-term value of about 20 mm yr^{-1} interpreted from model NUVEL-1 of current plate motions (Fig. 1.11).

Knowing the spreading rates at ocean ridges makes it possible to date the ocean floor. The direct correlation between polarity sequences measured in continental lavas and derived from oceanic anomalies is only possible for the last 4 Ma or so. Close to the axial zone, where linear spreading rates are observed (Fig. 1.15), simple extrapolation gives the ages of older anomalies, converting the striped pattern into an age map (Fig. 1.13). Detailed magnetic surveying of much of the world's oceans has revealed a continuous sequence of anomalies since the Late Cretaceous, preceded by an interval in which no reversals occurred; this Quiet Interval was itself preceded by a Mesozoic reversal sequence. Magnetostratigraphy in sedimentary rocks (§5.7.4) has enabled the identification, correlation and dating of key anomalies. The polarity sequence of the oceanic anomalies has been converted to a magnetic polarity timescale in which each polarity reversal is accorded an age (e.g., as in Fig. 5.78). In turn, this allows the pattern of magnetic anomalies in the ocean basins to be converted to a map of the age of the ocean basins (Fig. 5.82). The oldest areas of the oceans lie close to northwest Africa and eastern North America, as well as in the northwest Pacific. These

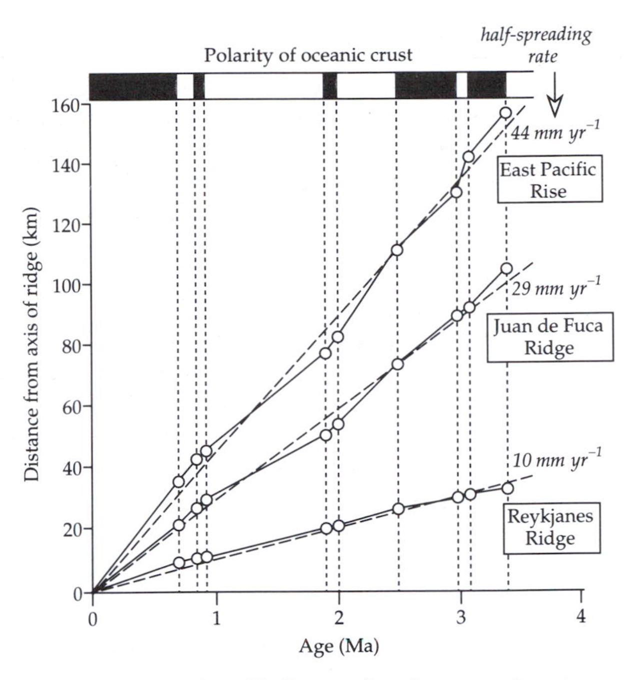

Fig. 1.15 Computation of half-rates of sea-floor spreading at different spreading centers by measuring the distances to anomalies with known radiometric ages (after Vine, 1966).

areas formed during the early stages of the breakup of Pangaea. They are of Early Jurassic age. The ages of the ocean basins have been confirmed by drilling through the sediment layers that cover the ocean floor and into the underlying basalt layer. Beginning in the late 1960s and extending until the present, this immensely expensive undertaking has been carried out in the Deep Sea Drilling Project (DSDP) and its successor the Ocean Drilling Project (ODP). These multinational projects, under the leadership of the United States, are prime examples of open scientific cooperation on an international scale.

1.2.6 Plate margins

It is important to keep in mind that the tectonic plates are not crustal units. They involve the entire thickness of the lithosphere, of which the crust is only the outer skin. Oceanic lithosphere is thin close to a ridge axis, but thickens with distance from the ridge, reaching a value of 80–100 km; the oceanic crust makes up only the top 5–10 km. Continental lithosphere may be up to 150 km thick, of which only the top 30–60 km is continental crust. Driven by mechanisms that are not completely understood, the lithospheric plates move relative to each other across the surface of the globe. This knowledge supplies the 'missing link' in Wegener's continental drift hypothesis, removing one of the

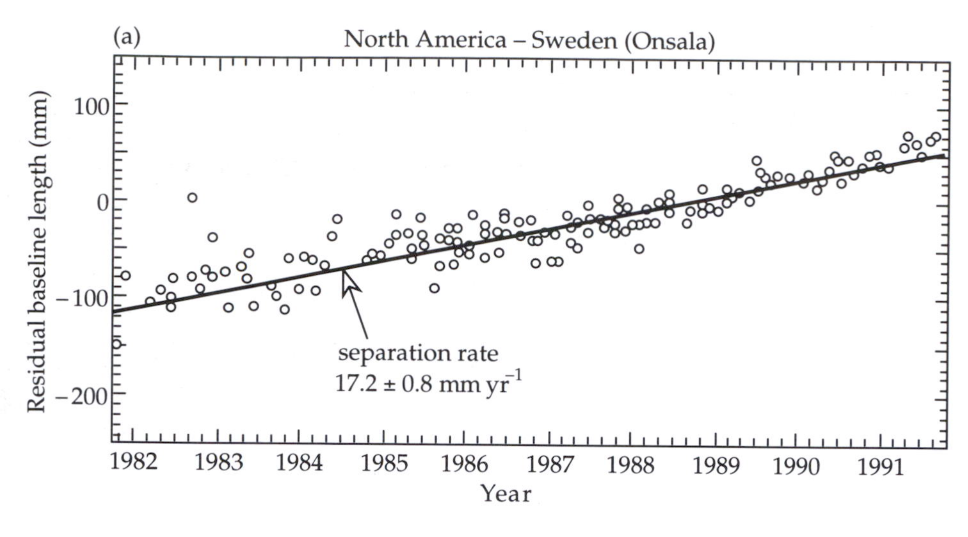

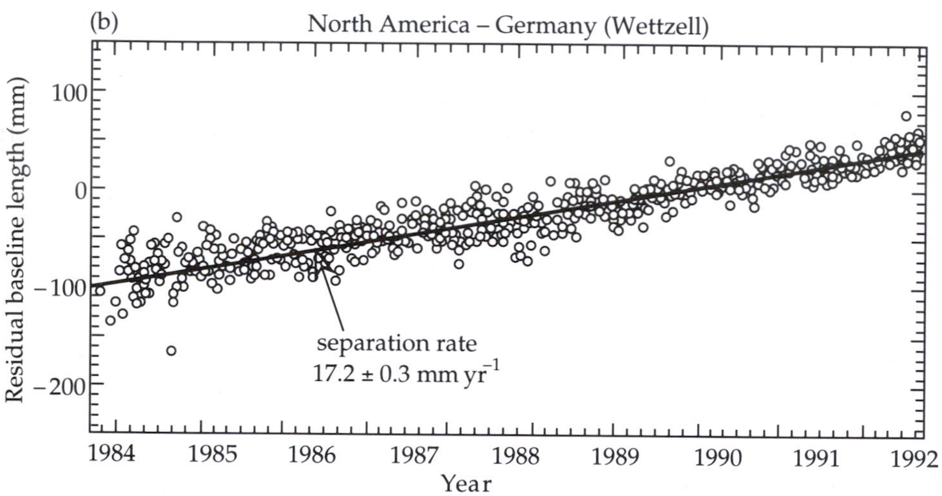

Fig. 1.16 Changes in separation between Westcott (Massachusetts, USA) and (a) Onsala (Sweden) and (b) Wettzell (Germany), as determined by very long baseline interferometry (after Ryan *et al.*, 1993).

most serious objections to it. It is not necessary for the continents to plow through the rigid ocean basins; they are transported passively on top of the moving plates, as logs float on a stream. Continental drift is thus a consequence of plate motions.

The plate tectonic model involves the formation of new lithosphere at a ridge and its destruction at a subduction zone (Fig. 1.17). Since the mean density of oceanic lithosphere exceeds that of continental lithosphere, oceanic lithosphere can be subducted under continental or oceanic lithosphere, whereas continental lithosphere cannot underride oceanic lithosphere. Just as logs pile up where a stream dives under a surface obstacle, a continent that is transported into a subduction zone collides with the deep-sea trench, island arc or adjacent continent. Such a collision results in an orogenic belt. In a continent–continent collision, neither plate can easily subduct, so relative plate motion may come to a halt. Alternatively, subduction may start at a new location behind one of the continents, leaving a mountain chain as evidence of the *suture zone* between the original colliding continents. The Alpine, Himalayan and Appalachian mountain chains are thought to have formed by this mechanism, the first two in Tertiary times, the last in several stages during the Paleozoic. Plate tectonic theory is supported convincingly by an abundance of geophysical, petrological and geological evidence from the three types of plate margin. A brief summary of the main geophysical observations at these plate margins is given in the following sections. Later chapters give more detailed treatments of the gravity (§2.6.4), seismicity (§3.5.3 and 3.5.4), geothermal (§4.2.5) and magnetic (§5.7.3) evidence.

1.2.6.1 *Constructive margins*

Although the ridges and rises are generally not centrally located in the ocean basins, they are often referred to as mid-ocean ridges. The type of oceanic basalt that is pro-

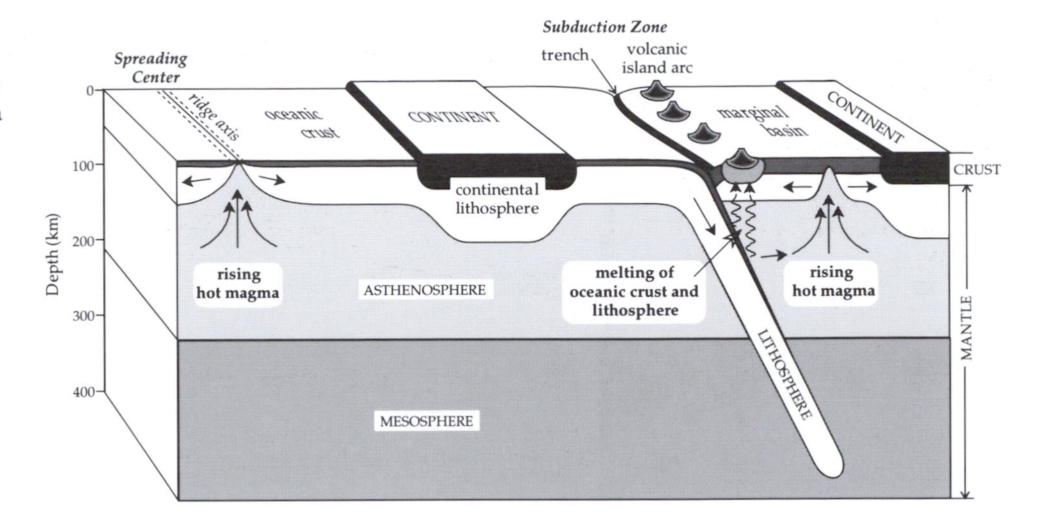

Fig. 1.17 Hypothetical vertical cross-section through a lithospheric plate from a spreading center to a subduction zone.

duced at an oceanic spreading center is even called a mid-ocean ridge basalt (MORB for short). Topographically, slow-spreading ridges have a distinct axial rift valley, which, for reasons that are not understood, is missing on faster-spreading ridges. Partially molten upper mantle rocks (generally assumed to be peridotites) from the asthenosphere rise under the ridges. The decrease of pressure due to the changing depth causes further melting and the formation of basaltic magma. Their chemical compositions and the concentrations of long-lived radioactive isotopes suggest that MORB lavas are derived by fractionation (i.e., separation of components, perhaps by precipitation or crystallization) from the upwelling peridotitic mush. Differentiation is thought to take place at about the depth of the lower crustal gabbroic layer beneath the ridge in a small, narrow magma chamber. Some of the fluid magma extrudes near the central rift or ridge axis and flows as lava across the ocean floor; part is intruded as dikes and sills into the thin oceanic crust. The Vine–Matthews–Morley hypothesis for the origin of oceanic magnetic anomalies requires fairly sharp boundaries between alternately magnetized blocks of oceanic crust. This implies that the zone of dike injection is narrow and close to the ridge axis.

The distribution of earthquakes defines a narrow band of seismic activity close to the crest of an oceanic ridge. These earthquakes occur at shallow depths of a few kilometers and are mostly small; magnitudes of 6 or greater are rare. The seismic energy released at ridges is an insignificant part of the world-wide annual release. Analyses show that the earthquakes are associated with normal faulting, implying extension away from the ridge axis (see §3.5.4).

Heat flow in the oceans is highest at the ocean ridges and decreases systematically with distance away from the ridge. The thermal data conform to the model of sea-floor spreading. High axial values are caused by the formation of new lithosphere from the hot uprising magma at the ridge axis. The associated volcanism on the floor of the axial rift zones has been observed directly from deep-diving submersibles. With time, the lithosphere spreads away from the ridge and gradually cools, so that the heat outflow diminishes with increasing age or distance from the ridge.

Oceanic crust is thin, so the high-density mantle rocks occur at shallower depths than under the continents. This causes a general increase of the Earth's gravity field over the oceans, giving positive gravity anomalies. However, over the ridge systems gravity decreases toward the axis so that a 'negative' anomaly is superposed on the normally positive oceanic gravity anomaly. The effect is due to the local density structure under the ridge. It has been interpreted in terms of anomalous mantle material with density slightly less than normal. The density is low because of the different mantle composition under the ridges and its high temperature.

The interpretation of magnetic anomalies formed by sea-floor spreading at constructive margins has already been discussed. The results provide direct estimates of the mean rates of plate motions over geological time intervals.

1.2.6.2 *Destructive margins*

Subduction zones are found where a plate plunges beneath its neighbor to great depths, until pressure and temperature cause its consumption. Density determines that the descending plate at a subduction zone is an oceanic one. The surface manifestation depends on the type of overriding plate. When this is another oceanic plate, the subduction zone is marked by a volcanic island arc and, parallel to it, a deep trench. The island arc lies near the edge of the overriding plate and is convex toward the underriding plate. The trench marks where the underriding plate turns down into the mantle (Fig. 1.17). It may be partly filled with carbonaceous and detrital sediments. Island arc and trench are a few hundred kilometers apart. Several examples are seen around the west and northwest margins of the Pacific plate (Fig. 1.11). Melting of the downgoing slab produces magma that rises to feed the volcanoes.

The intrusion of magma behind an island arc produces a back-arc basin on the inner, concave side of the arc. These basins are common in the Western Pacific. If the arc is close to a continent, the off-arc magmatism may create a marginal sea, such as the Sea of Japan. Back-arc basins and marginal seas are floored by oceanic crust.

A fine example of where the overriding plate is a continental one is seen along the west coast of South America. Compression between the Nazca and South American plates has generated the Andes, an arcuate-folded mountain belt near the edge of the continental plate. Active volcanoes along the mountain chain emit a type of lava, called andesite, which has a higher silica content than oceanic basalt. It does not originate from the asthenosphere-type of magma. A current theory is that it may form by melting of the subducting slab and overriding plate at great depths. If some siliceous sediments from the deep-sea trench are carried down with the descending slab, they might enhance the silica content of the melt, producing a magma with andesite-type composition.

The seismicity at a subduction zone provides the key to the processes active there. Where one plate is thrust over the other, the shear causes hazardous earthquakes at shallow depths. Below this region, earthquakes are systematically distributed within the subducting plate. They form an inclined *Wadati–Benioff seismic zone*, which may extend for several hundred kilometers into the mantle. The deepest earthquakes have been registered down to about 700 km.

Studies of the focal mechanisms (§3.5.4) show that at shallow depths the downgoing plate is in a state of down-dip extension (Fig. 1.18a). Subducting lithosphere is colder and denser than the underlying asthenosphere. This gives it negative buoyancy, which causes it to sink, pulling the plate downward. At greater depths the mantle is more rigid than the asthenosphere, and its strength resists penetration (Fig. 1.18b). While the upper part is sinking, the bottom part is being partly supported by the deeper layers; this results in down-dip compression in the lower part of the descending slab and down-dip extension in the upper part. A gap in the depth distribution of seismicity may arise where the deviatoric stress changes from extensional to compressional. In a very deep subduction zone the increase of resistance with depth causes down-dip compression throughout the descending slab (Fig. 1.18c). In some cases part of the slab may break off and sink to great depths,

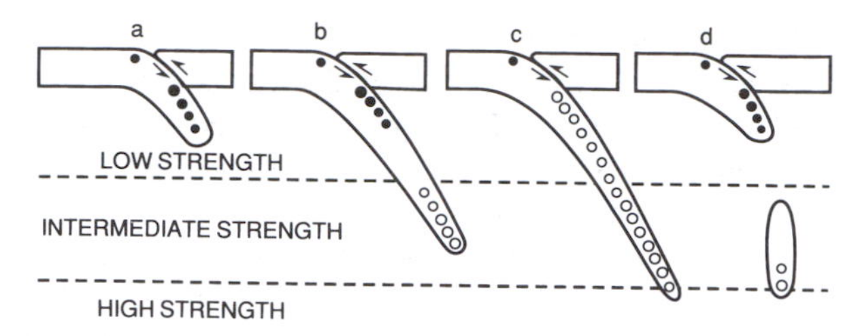

Fig. 1.18 Stresses acting on a subducting lithospheric plate. Arrows indicate shear where the underriding plate is bent downward. Solid and open circles within the descending slab denote extension and compression, respectively; the size of the circle represents qualitatively the seismic activity. In (a), (b) and (d) extensional stress in the upper part of the plate is due to the slab being pulled into low-strength asthenosphere. In (b) resistance of the more rigid layer under the asthenosphere causes compression within the lower part of the slab; if the plate sinks far enough, (c), the stress becomes compressional throughout; in some cases, (d), the deep part of the lower slab may break off (after Isacks and Molnar, 1969).

where the earthquakes have compressional-type mechanisms (Fig. 1.18d); a gap in seismicity exists between the two parts of the slab.

Heat flow at a destructive plate margin reflects to some extent the spreading history of the plate. The plate reaches its maximum age, and so has cooled the furthest, by the time it reaches a subduction zone. The heat flow values over deep ocean basins are uniformly low, but the values measured in deep-sea trenches are the lowest found in the oceans. In contrast, volcanic arcs and back-arc basins often have anomalously high heat flow due to the injection of fresh magma.

Gravity anomalies across subduction zones have several distinctive features. Seaward of the trench the lithosphere flexes upward slightly before it begins its descent, causing a weak positive anomaly; the presence of water or low-density sediments in a deep-sea trench gives rise to a strong negative gravity anomaly; and over the descending slab a positive anomaly is observed, due in part to the mineralogical conversion of subducted oceanic crust to higher-density eclogite.

Subduction zones have no particular magnetic signature. Close to an active or passive continental margin the contrast between the magnetic properties of oceanic and continental crust produces a magnetic anomaly, but this is not a direct result of the plate tectonic processes. Over marginal basins magnetic anomalies are not lineated except in some rare cases. This is because the oceanic crust in the basin does not originate by sea-floor spreading at a ridge, but by diffuse intrusion throughout the basin.

1.2.6.3 *Conservative margins*

Transform faults are strike–slip faults with steeply dipping fault planes. They may link segments of subduction zones, but they are mostly observed at constructive plate margins where they connect oceanic ridge segments. Transform faults are the most seismically active parts of a ridge system, because here the relative motion between neighboring plates is most pronounced. Seismic studies have confirmed that the displacements on transform faults agree with the relative motion between the adjacent plates.

The trace of a transform fault may extend away from a ridge on both sides as a *fracture zone*. Fracture zones are among the most dramatic features of ocean-floor topography. Although only some tens of kilometers wide, a fracture zone can be thousands of kilometers long. It traces the arc of a small circle on the surface of the globe. This important characteristic allows fracture zones to be used for the deduction of relative plate motions, which cannot be obtained from the strike of a ridge or trench segment, where oblique spreading or subduction is possible (note, for example, the direction of plate convergence relative to the strike of the Aleutian island arc in Fig. 1.11).

Any displacement on the surface of a sphere is equivalent to a small rotation about a pole. The motion of one plate relative to the other takes place as a rotation about the *Euler pole* of relative rotation between the plates (see §6.3.3). This pole can be located from the orientations of fracture zones, because the strike of a transform fault is parallel to the relative motion between two adjacent plates. Thus a great circle normal to a transform fault or fracture zone must pass through the Euler pole of relative rotation between the two plates. If several great circles are drawn at different places on the fracture zone (or normal to different transform faults offsetting a ridge axis) they intersect at the Euler pole. The current model of relative plate motions NUVEL-1 was obtained by determining the Euler poles of rotation between pairs of plates using magnetic anomalies, the directions of slip on earthquake fault planes at plate boundaries, and the topography that defines the strikes of transform faults. The rates of relative motion at different places on the plate boundaries (Fig. 1.11) were computed from the rates of rotation about the appropriate Euler poles.

There may be a large change in elevation across a fracture zone; this is related to the different thermal histories of the plates it separates. As a plate cools, it becomes more dense and less buoyant, so that it gradually sinks. Consequently, the depth to the top of the oceanic lithosphere increases with age, i.e., with distance from the spreading center. Places facing each other across a transform fault are at different distances from their respective spreading centers. They have different ages and so have subsided by different amounts relative to the ridge. This may result in a noticeable elevation difference across the fracture zone.

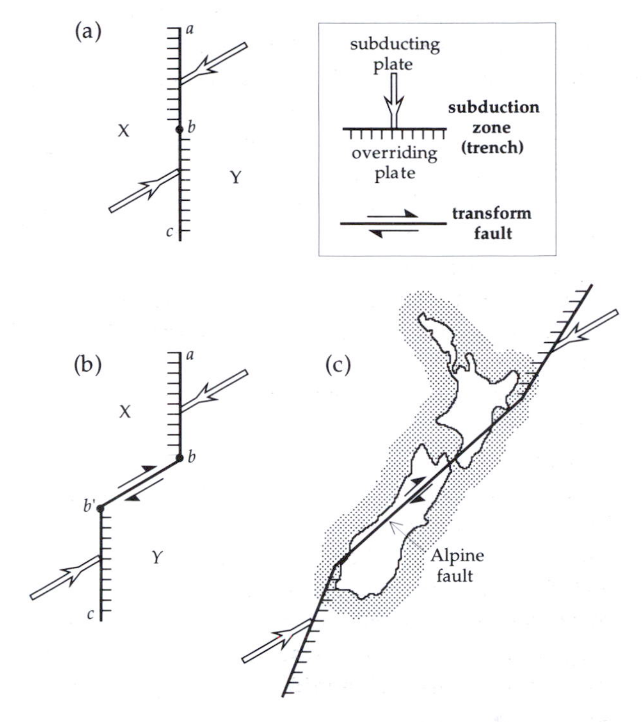

Fig. 1.19 (a) A consuming plate boundary consisting of two opposed subduction zones; along *ab* plate Y is consumed below plate X and along *bc* plate X is consumed beneath plate Y. (b) Development of a transform fault which displaces *bc* to the position *b′c*. (c) The Alpine fault in New Zealand is an example of such a transform boundary (after McKenzie and Morgan, 1969).

Ultrabasic rocks are found in fracture zones and there may be local magnetic anomalies. Otherwise, the magnetic effect of a transform fault is to interrupt the oceanic magnetic lineations parallel to a ridge axis, and to offset them in the same measure as ridge segments. This results in a very complex pattern of magnetic lineations in some ocean basins (e.g., in the northeast Pacific).

A transform fault can also connect subduction zones. Suppose a consuming plate boundary consisted originally of two opposed subduction zones (Fig. 1.19a). Plate Y is consumed below plate X along the segment *ab* of the boundary, whereas plate X is consumed beneath plate Y along segment *bc*. The configuration is unstable, because a trench cannot sustain subduction in opposite directions. Consequently, a dextral transform fault develops at the point *b*. After some time, motion on the fault displaces the lower segment to the position *b′c* (Fig. 1.19b). An example of such a transform boundary is the Alpine fault in New Zealand (Fig. 1.19c). To the northeast of North Island, the

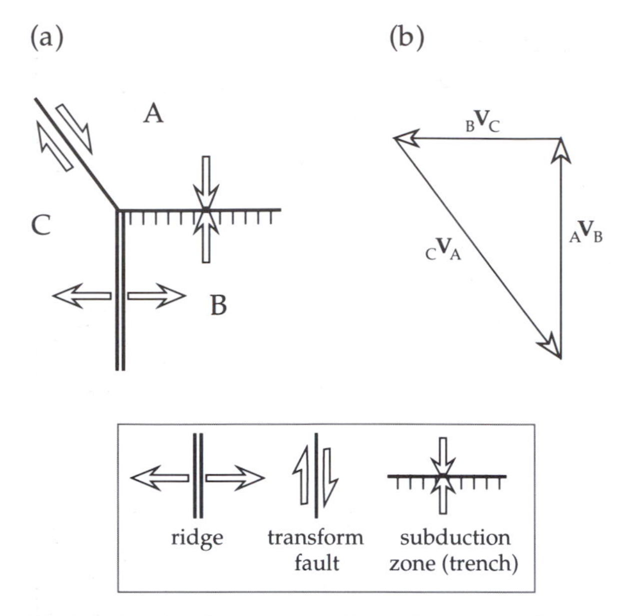

Fig. 1.20 (a) Triple junction formed by a ridge, trench and transform fault, and (b) vector diagram of the relative velocities at the three boundaries (after McKenzie and Parker, 1967).

Pacific plate is being subducted at the Tonga–Kermadec trench. To the southwest of South Island, the Pacific plate overrides the Tasman Sea at the anomalous Macquarie Ridge (earthquake analysis has shown that the plate margin at this ridge is compressive; the compression may be too slow to allow a trench to develop). The Alpine fault linking the two opposed subduction zones is therefore a dextral transform fault.

1.2.7 Triple junctions

It is common, although imprecise, to refer to a plate margin by its dominant topographic feature, rather than by the nature of the margin. A ridge (R) represents a constructive margin or spreading center, a trench (T) refers to a destructive margin or subduction zone, and a transform fault (F) stands for a conservative margin. Each margin is a location where two tectonic plates adjoin. Inspection of Fig. 1.11 shows that there are several places where three plates come together, but none where four or more plates meet. The meeting points of three plate boundaries are called *triple junctions*. They are important in plate tectonics because the relative motions between the plates that form a triple junction are not independent. This may be appreciated by considering the plate motions in a small plane surrounding the junction.

Consider the plate velocities at an RTF junction formed by all three types of boundary (Fig. 1.20a). If the plates are rigid, their relative motions take place entirely at their margins. Let $_A\mathbf{V}_B$ denote the velocity of plate B relative to plate A, $_B\mathbf{V}_C$ the velocity of plate C relative to plate B, and $_C\mathbf{V}_A$ the velocity of plate A relative to plate C. Note that these quantities are vectors; their directions are as important as their magnitudes. They can be represented on a vector diagram by straight lines with directions parallel to and lengths proportional to the velocities. In a circuit about the triple junction an observer must return to the starting point. Thus, a vector diagram of the interplate velocities is a closed triangle (Fig. 1.20b). The velocities are related by

$$_A\mathbf{V}_B + {_B\mathbf{V}_C} + {_C\mathbf{V}_A} = 0 \qquad (1.6)$$

This planar model is a 'flat Earth' representation. As discussed in §1.2.6.3, displacements on the surface of a sphere are rotations about Euler poles of relative motion. This can be taken into account by replacing each linear velocity $\mathbf{V}$ in Eq. (1.6) by the rotational velocity ω about the appropriate Euler pole.

1.2.7.1 *Stability of triple junctions*

The different combinations of three plate margins define ten possible types of triple junction. The combinations correspond to all three margins being of one type (RRR, TTT, FFF), two of the same type and one of the other (RRT, RRF, FFT, FFR, TTR, TTF), and all different (RTF). Different combinations of the sense of subduction at a trench increase the number of possible junctions to sixteen. Not all of these junctions are stable in time. For a junction to preserve its geometry, the orientations of the three plate boundaries must fulfil conditions which allow the relative velocities to satisfy Eq. (1.6). If they do so, the junction is stable and can maintain its shape. Otherwise, the junction is unstable and must evolve in time to a stable configuration.

The stability of a triple junction is assessed by considering how it can move along any of the plate boundaries that form it. The velocity of a plate can be represented by its coordinates in *velocity space*. Consider, for example, a trench or consuming plate margin (Fig. 1.21a). The point A in velocity space represents the consuming plate, which has a larger velocity than B for the overriding plate. A triple junction in which one plate margin is a trench can lie anywhere on this boundary, so the locus of its possible velocities is a line *ab* parallel to the trench. The trench is fixed relative to the overriding plate B, so the line *ab* must pass through B. Similar reasoning shows that a triple junction on a transform fault is represented in velocity space by a line *ab* parallel to the fault and passing through both A and B (Fig. 1.21b). A triple junction on a ridge gives a velocity line *ab* parallel to the ridge; in the case of symmetrical spreading

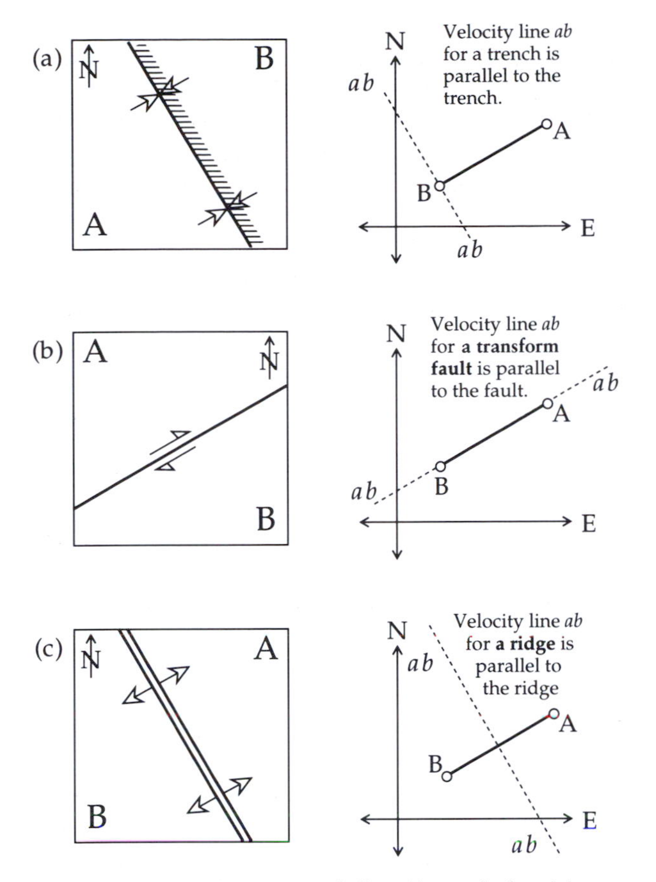

Fig. 1.21 Plate margin geometry (*left*) and locus *ab* of a triple junction in velocity space (*right*) for (a) a trench, (b) a transform fault, and (c) a ridge (after Cox and Hart, 1986).

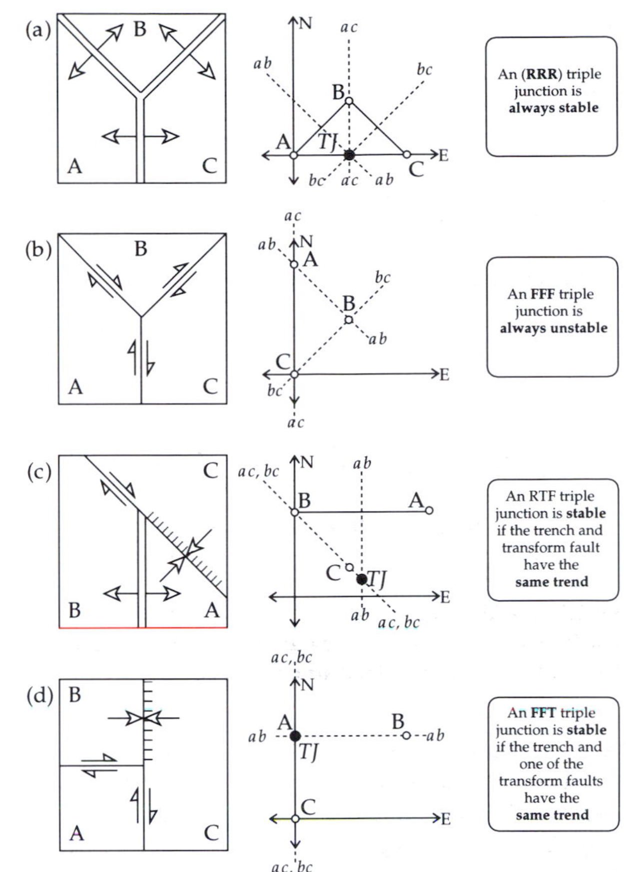

Fig. 1.22 Triple junction configuration (*left*), velocity lines of each margin in velocity space (*center*), and stability criteria (*right*) for selected triple junctions, TJ (after Cox and Hart, 1986).

normal to the trend of the ridge the line *ab* is the perpendicular bisector of AB (Fig. 1.21c).

Now consider the RRR-type of triple junction, formed by three ridges (Fig. 1.22a). The locus of the triple junction on the ridge between any pair of plates is the perpendicular bisector of the corresponding side of the velocity triangle ABC. The perpendicular bisectors of the sides of a triangle always meet at a point (the circumcenter). In velocity space this point satisfies the velocities on all three ridges simultaneously, so the RRR triple junction is always stable. Conversely, a triple junction formed by three intersecting transform faults (FFF) is always unstable, because the velocity lines form the sides of a triangle, which can never meet in a point (Fig. 1.22b). The other types of triple junction are conditionally stable, depending on the angles between the different margins. For example, in an RTF triple junction the velocity lines of the trench *ac* and transform fault *bc* must both pass through C, because this plate is common to both boundaries. The junction is stable if the velocity line *ab* of the ridge also passes through C, or if the trench and transform fault have the same trend (Fig. 1.22c). By similar reasoning, the FFT triple junction is only stable if the trench has the same trend as one of the transform faults (Fig. 1.22d).

In the present phase of plate tectonics only a few of the possible types of triple junction appear to be active. An RRR-type is formed where the Galapagos Ridge meets the East Pacific Rise at the junction of the Cocos, Nazca and Pacific plates. A TTT-type junction is formed by the Japan trench and the Bonin and Ryukyu arcs. The San Andreas fault in California terminates in an FFT-type junction at its northern end, where it joins the Mendocino Fracture Zone.

1.2.7.2 *Evolution of triple junctions in the northeast Pacific*

Oceanic magnetic anomalies in the northeast Pacific form a complex striped pattern. The anomalies can be identified by interpreting their shapes. Their ages can be found by comparison with a geomagnetic polarity timescale such as that shown in Fig. 5.78, which gives the age of each numbered chron since the Late Jurassic. In the northeast Pacific the anomalies become younger toward the North American continent in the east, and toward the Aleutian trench in the north. The anomaly pattern produced at a ridge is usually symmetric (as in Fig. 1.13), but in the northeast Pacific only the western half of an anomaly pattern is observed. The plate on which the eastern half of the anomaly pattern was formed is called the Farallon plate. It and the ridge itself are largely missing and have evidently been subducted under the American plate. Only two small remnants of the Farallon plate still exist: the Juan de Fuca plate off the coast of British Columbia, and the Rivera plate at the mouth of the Gulf of California. The magnetic anomalies also indicate that another plate, the Kula plate, existed in the Late Mesozoic but has now been entirely consumed under Alaska and the Aleutian trench. The anomaly pattern shows that in the Late Cretaceous the Pacific, Kula and Farallon plates were diverging from each other and thus met at an RRR-type triple junction. This type of junction is stable and preserved its shape during subsequent evolution of the plates. It is therefore possible to reconstruct the relative motions of the Pacific, Kula and Farallon plates in the Cenozoic (Fig. 1.23a–c).

The anomaly ages are known from the magnetic timescale so the anomaly spacing allows the half-rates of spreading to be determined. In conjunction with the trends of fracture zones, the anomaly data give the rates and directions of spreading at each ridge. The anomaly pattern at the mouth of the Gulf of California covers the last 4 Ma and gives a mean half-rate of spreading of 3 cm yr^{-1} parallel to the San Andreas fault. This indicates that the Pacific plate has moved northward past the American plate at this boundary with a mean relative velocity of about 6 cm yr^{-1} during the last 4 Ma. The half-rate of spreading on the remnant of the Farallon–Pacific ridge is 5 cm yr^{-1}, giving a relative velocity of 10 cm yr^{-1} between the plates. A vector diagram of relative velocities at the Farallon–Pacific–American triple junction (Fig. 1.23d) shows convergence of the Farallon plate on the American plate at a rate of 7 cm yr^{-1}. Similarly, the spacing of east–west trending magnetic anomalies in the Gulf of Alaska gives the half-rate of spreading on the Kula–Pacific ridge, from which it may be inferred that the relative velocity between the plates was 7 cm yr^{-1}. A vector diagram combining this value with the 6 cm yr^{-1} northward motion of the Pacific plate gives a velocity of 12 cm yr^{-1} for the Kula plate relative to the American plate.

Using these velocities the history of plate evolution in the Cenozoic can be deduced by extrapolation. The interpretation is tenuous, as it involves unverifiable assumptions. The most obvious is that the Kula–Pacific motion in the late Cretaceous (80 Ma ago) and the American–Pacific motion of the past 4 Ma have remained constant throughout the Cenozoic. With this proviso, it is evident that triple junctions formed and migrated along the American plate margin. The Kula–American–Farallon RTF junction was slightly north of the present location

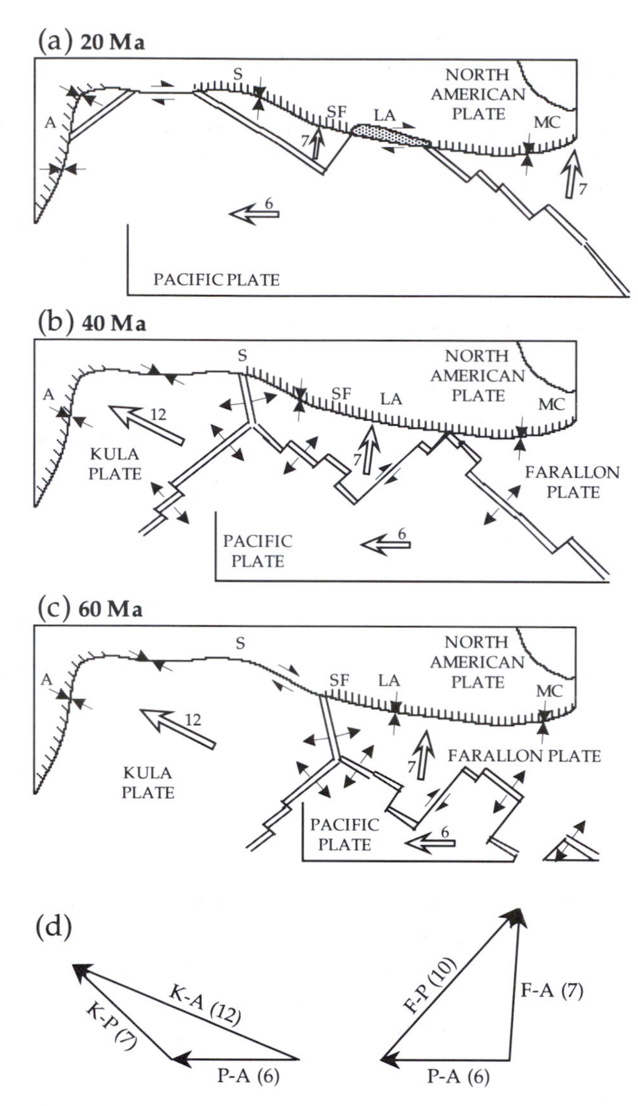

Fig. 1.23 (a)–(c) Extrapolated plate relationships in the northeast Pacific at different times in the Cenozoic (after Atwater, 1970). Letters on the American plate give approximate locations of some modern cities for reference: MC, Mexico City; LA, Los Angeles; SF, San Francisco; S, Seattle; A, Anchorage. The shaded area in (a) is an unacceptable overlap. (d) Vector diagrams of the relative plate velocities at the Kula–Pacific–American and Farallon–Pacific–American triple junctions (numbers are velocities in cm yr^{-1} relative to the American plate).

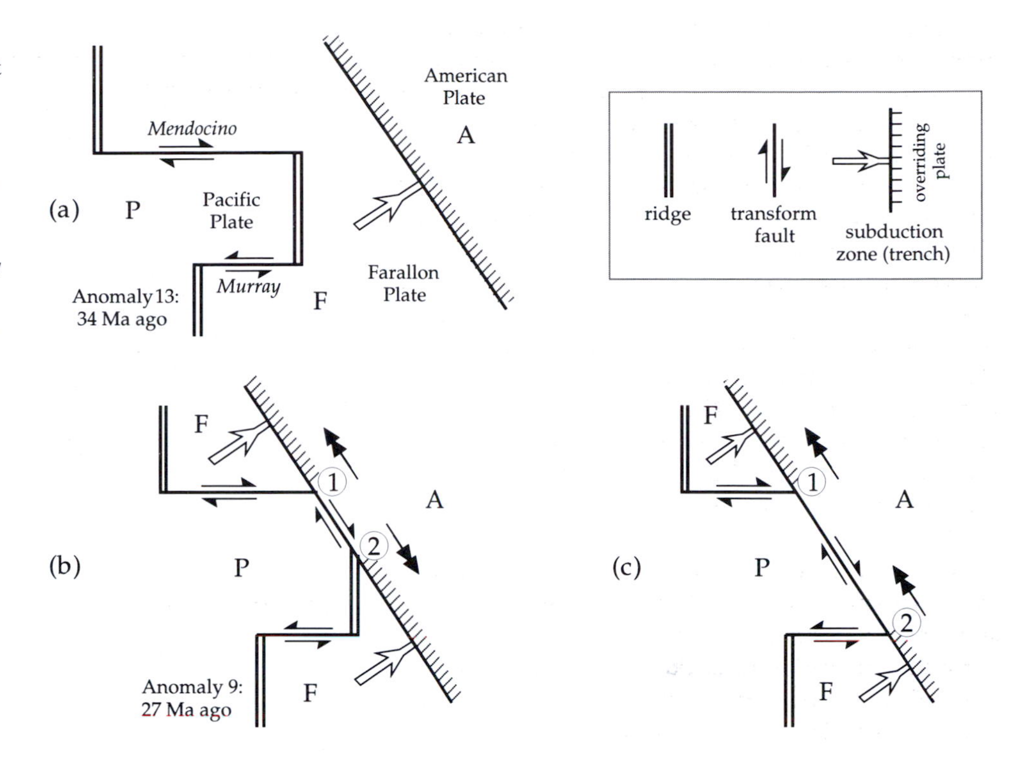

Fig. 1.24 Formation of the San Andreas fault as a result of the evolution of triple junctions in the northeast Pacific during the Oligocene: plate geometries at the times of (a) magnetic anomaly 13, about 34 Ma ago, (b) anomaly 9, about 27 Ma ago (after McKenzie and Morgan, 1969), and (c) further development when the Murray fracture zone collides with the trench. Double-headed arrows show directions of migration of triple junctions 1 and 2 along the consuming plate margin.

of San Francisco 60 Ma ago (Fig. 1.23c); it moved to a position north of Seattle 20 Ma ago (Fig. 1.23a). Around that time in the Oligocene an FFT junction formed between SF and LA, while the Farallon–Pacific–American RTF junction evolved to the south. The development of these two triple junctions is due to the collision and subduction of the Farallon–Pacific ridge at the Farallon–American trench.

At the time of magnetic anomaly 13, about 34 Ma ago, a north–south striking ridge joined the Mendocino and Murray transform faults as part of the Farallon–Pacific plate margin to the west of the American trench (Fig. 1.24a). By the time of anomaly 9, about 27 Ma ago, the ridge had collided with the trench and been partly consumed by it (Fig. 1.24b). The Farallon plate now consisted of two fragments: an FFT junction developed at point 1, formed by the San Andreas fault system, the Mendocino fault and the consuming trench to the north; and an RTF junction formed at point 2. Both junctions are stable when the trenches are parallel to the transform fault along the San Andreas system. Analysis of the velocity diagrams at each triple junction shows that point 1 migrated to the northwest and point 2 migrated to the southeast at this stage. Later, when the southern segment of the Farallon–Pacific ridge had been subducted under the American plate, the Murray transform fault changed the junction at point 2 to an FFT junction, which has subsequently also migrated to the northwest.

1.3 SUGGESTIONS FOR FURTHER READING

INTRODUCTORY LEVEL

Beatty, J. K. and Chaikin, A. 1990. *The New Solar System*, Cambridge Univ. Press, Cambridge, U.K. and Sky Publishing Corp., Cambridge, Mass., U.S.A.

Brown, G. C., Hawkesworth, C. J. and Wilson, R. C. L. (editors) 1992. *Understanding the Earth*, Cambridge Univ. Press, Cambridge.

Cox, A. and Hart, R. B. 1986. *Plate Tectonics*, Blackwell Scientific Publ., Oxford.

Kearey, P. and Vine, F. J. 1990. *Global Tectonics*, Blackwell Scientific Publ., Oxford.

Press, F. and Siever, R. 1985. *Earth*, W. H. Freeman, San Francisco.

Strahler, A. N. 1963. *The Earth Sciences*, Harper and Row, New York.

Uyeda, S. 1978. *The New View of the Earth*, W. H. Freeman, San Francisco.

INTERMEDIATE LEVEL

Bott, M. H. P. 1982. *The Interior of the Earth (2nd ed.)*, Edward Arnold, London.

Fowler, C. M. R. 1990. *The Solid Earth: an Introduction to Global Geophysics*, Cambridge Univ. Press, Cambridge.

ADVANCED LEVEL

Cox, A. (editor) 1973. *Plate Tectonics and Geomagnetic Reversals*, W. H. Freeman, San Francisco.

Le Pichon, X., Francheteau, J. and Bonnin, J. 1976. *Plate Tectonics*, Elsevier, Amsterdam.

2 Gravity and the figure of the Earth

2.1 THE EARTH'S SIZE AND SHAPE

2.1.1 Earth's size

The philosophers and savants in ancient civilizations could only speculate about the nature and shape of the world they lived in. The range of possible travel was limited and only simple instruments existed. Unrelated observations might have suggested that the Earth's surface was upwardly convex. For example, the Sun's rays continue to illuminate the sky and mountain peaks after its disk has already set, departing ships appear to sink slowly over the horizon, and the Earth's shadow can be seen to be curved during partial eclipse of the Moon. However, early ideas about the heavens and the Earth were intimately bound up with concepts of philosophy, religion and astrology. In Greek mythology the Earth was a disk-shaped region embracing the lands of the Mediterranean region and surrounded by a circular stream, *Oceanus*, the origin of all the rivers. In the 6th century B.C. the Greek philosopher Anaximander visualized the heavens as a celestial sphere that surrounded a flat Earth at its center. Pythagoras (582–507 B.C.) and his followers were apparently the first to speculate that the Earth was a sphere. This idea was further propounded by the influential philosopher Aristotle (384–322 B.C.). Although he taught the scientific principle that theory must follow fact, Aristotle is responsible for the logical device called syllogism, which can explain correct observations by apparently logical accounts that are based on false premises. His influence on scientific methodology was finally banished by the scientific revolution in the 17th century.

The first scientifically sound estimate of the size of the terrestrial sphere was made by Eratosthenes (275–195 B.C.), who was the head librarian at Alexandria, a Greek colony in Egypt during the 3rd century B.C. Eratosthenes had been told that in the city of Syene (modern Aswan) the Sun's noon rays on midsummer day shone vertically and were able to illuminate the bottoms of wells, whereas on the same day in Alexandria shadows were cast. Using a sun-dial Eratosthenes observed that at the summer solstice the Sun's rays made an angle of one-fiftieth of a circle (7.2°) with the vertical in Alexandria (Fig. 2.1). Eratosthenes believed that Syene and Alexandria were on the same meridian. In fact they are slightly displaced; their geographic coordinates are 24° 5′N 32° 56′E and 31° 13′N 29° 55′E, respectively. Syene is actually about half a degree north of the tropic of Cancer.

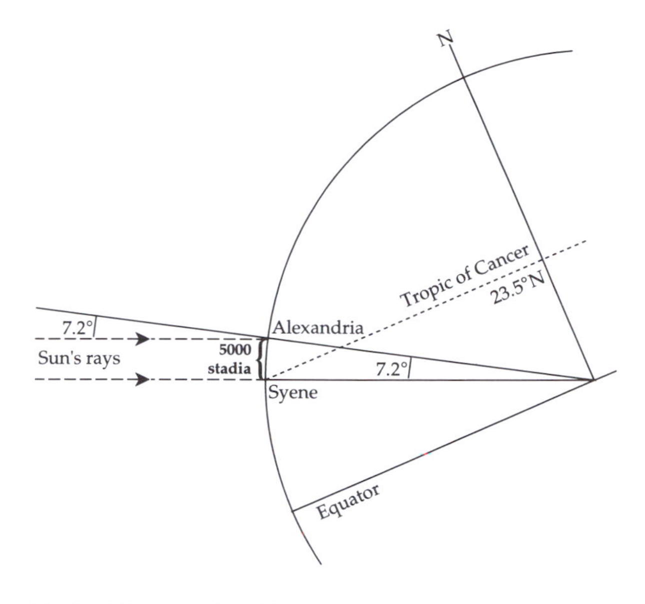

Fig. 2.1 The method used by Eratosthenes (275–195 B.C.) to estimate the Earth's circumference used the 7.2° difference in altitude of the Sun's rays at Alexandria and Syene, which are 5000 *stadia* apart (after Strahler, 1963).

Eratosthenes knew that the approximate distance from Alexandria to Syene was 5000 stadia, possibly estimated by travellers from the number of days ('10 camel days') taken to travel between the two cities. From these observations Eratosthenes estimated that the circumference of the global sphere was 250,000 stadia. The Greek *stadium* was the length (about 185 m) of the U-shaped racecourse on which footraces and other athletic events were carried out. Eratosthenes' estimate of the Earth's circumference is equivalent to 46,250 km, about 15% higher than the modern value of 40,030 km.

Estimates of the length of one meridian degree were made in the 8th century A.D. during the Tang dynasty in China, and in the 9th century A.D. by Arab astronomers in Mesopotamia. Little progress was made in Europe until the early 17th century. In 1662 the Royal Society was founded in London and in 1666 the Académie Royale des Sciences was founded in Paris. Both organizations provided support and impetus to the scientific revolution. The invention of the telescope enabled more precise geodetic surveying. In 1671 a French astronomer, Jean Picard (1620–1682), completed an accurate survey by triangulation of the length of a degree of

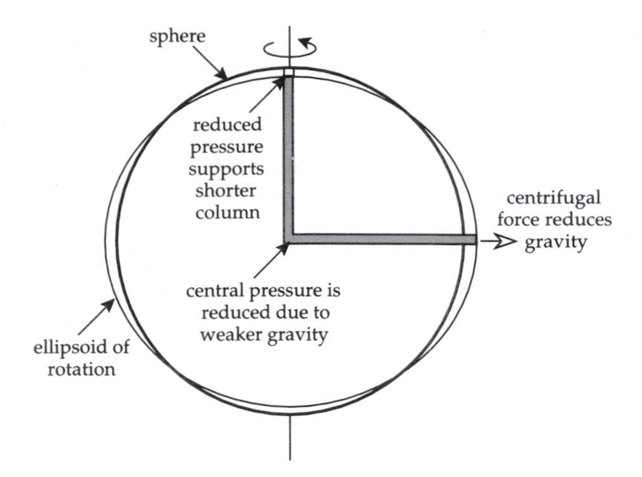

Fig. 2.2 Newton's argument that the shape of the rotating Earth should be flattened at the poles and bulge at the equator was based on hydrostatic equilibrium between polar and equatorial pressure columns (after Strahler, 1963).

meridian arc. From his results the Earth's radius was calculated to be 6372 km, remarkably close to the modern value of 6371 km.

2.1.2 Earth's shape

In 1672 another French astronomer, Jean Richer, was sent by Louis XIV to make astronomical observations on the equatorial island of Cayenne. He found that an accurate pendulum clock, which had been adjusted in Paris precisely to beat seconds, was losing about two and a half minutes per day, i.e. its period was now too long. The error was much too large to be explained by inaccuracy of the precise instrument. The observation aroused much interest and speculation, but was only explained some 15 years later by Sir Isaac Newton in terms of his laws of universal gravitation and motion.

Newton argued that the shape of the rotating Earth should be that of an *oblate ellipsoid*; compared to a sphere, it should be somewhat flattened at the poles and should bulge outward around the equator. This inference was made on logical grounds. Assume that the Earth does not rotate and that holes could be drilled to its center along the rotation axis and along an equatorial radius (Fig. 2.2). If these holes are filled with water, the hydrostatic pressure at the center of the Earth sustains equal water columns along each radius. However, the rotation of the Earth causes a centrifugal force at the equator but has no effect on the axis of rotation. At the equator the outward centrifugal force of the rotation opposes the inward gravitational attraction and pulls the water column upward. At the same time it reduces the hydrostatic pressure produced by the water column at the Earth's center. The reduced central pressure is unable to support the height of the water column along the polar radius, which subsides. If the Earth were a hydrostatic sphere, the form of the rotating Earth should be an oblate ellipsoid of revolution. Newton assumed the Earth's density to be constant and calculated that the flattening should be about 1:230 (roughly 0.5%). This is somewhat larger than the actual flattening of the Earth, which is about 1:298 (roughly 0.3%).

The increase in period of Richer's pendulum could now be explained. Cayenne was close to the equator, where the larger radius placed the observer further from the center of gravitational attraction, and the increased distance from the rotational axis resulted in a stronger opposing centrifugal force. These two effects resulted in a lower value of gravity in Cayenne than in Paris, where the clock had been calibrated.

There was no direct proof of Newton's interpretation. A corollary of his interpretation was that the degree of meridian arc should subtend a longer distance in polar regions than near the equator (Fig. 2.3). Early in the 18th century French geodesists extended the standard meridian from border to border of the country and found a puzzling result. In contrast to the prediction of Newton, the degree of meridian arc *decreased* northward. The French interpretation was that the Earth's shape was a *prolate* ellipsoid, elongated at the poles and narrowed at the equator, like the shape of a rugby football. A major scientific controversy arose between the 'flatteners' and the 'elongators'.

To determine whether the Earth's shape is oblate or prolate, the Académie Royale des Sciences sponsored two scientific expeditions. In 1736–1737 a team of scientists measured the length of a degree of meridian arc in Lapland, near the Arctic Circle. They found a length appreciably longer than the meridian degree measured by Picard near Paris. From 1735 to 1743 a second party of scientists measured the length of more than 3 degrees of meridian arc in Peru, near the equator. Their results showed that the equatorial degree of latitude was shorter than the meridian degree in Paris. Both parties confirmed convincingly the prediction of Newton that the Earth's shape is that of an oblate ellipsoid.

The ellipsoidal shape of the Earth resulting from its rotation has important consequences, not only for the variation with latitude of gravity on the Earth's surface, but also for the Earth's rate of rotation and the orientation of its rotational axis. These are modified by torques that arise from the gravitational attractions of the Sun, Moon and planets on the ellipsoidal shape.

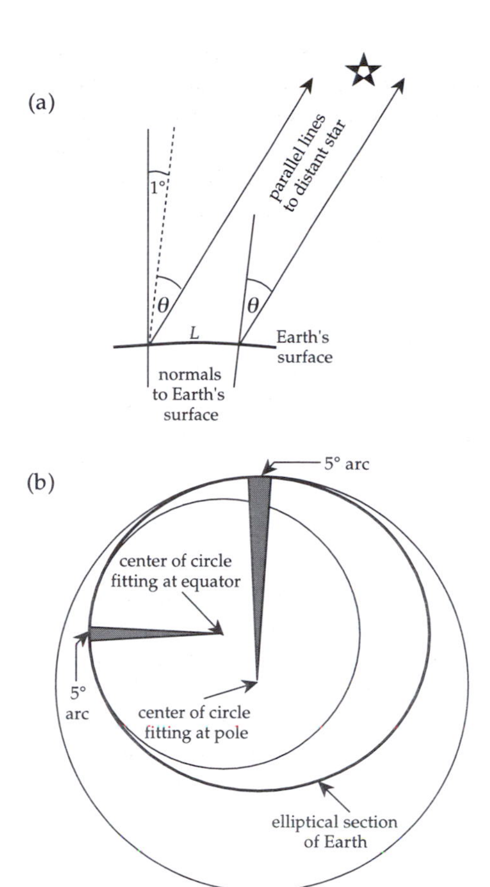

Fig. 2.3 (a) The length of a degree of meridian arc is found by measuring the distance between two points that lie one degree apart on the same meridian. (b) The larger radius of curvature at the flattened poles gives a longer arc distance than is found at the equator where the radius of curvature is smaller (after Strahler, 1963).

2.2 GRAVITATION

2.2.1 The law of universal gravitation

Sir Isaac Newton (1642–1727) was born in the same year in which Galileo died. Unlike Galileo, who relished debate, Newton was a retiring person and avoided confrontation. His modesty is apparent in a letter written in 1675 to his colleague Robert Hooke, famous for his experiments on elasticity. In this letter Newton made the famous disclaimer 'if I have seen further (than you and Descartes) it is by standing upon the shoulders of Giants'. In modern terms Newton would be regarded as a theoretical physicist. He had an outstanding ability to synthesize experimental results and incorporate them into his own theories. Faced with the need for a more powerful technique of mathematical analysis than existed at the time he invented differential and integral calculus, for which he is credited equally with Gottfried Wilhelm von Leibnitz (1646–1716) who discovered the same method independently. Newton was able to resolve many issues by formulating logical thought experiments; an example is his prediction that the shape of the Earth is an oblate ellipsoid. He was one of the most outstanding synthesizers of observations in scientific history, which is implicit in his letter to Hooke. His three-volume book *Philosophiae Naturalis Principia Mathematica*, published in 1687, ranks as the greatest of all scientific texts. The first volume of the *Principia* contains Newton's famous *Laws of Motion*, the third volume handles the *Law of Universal Gravitation*.

The first two laws of motion are generalizations from Galileo's results. As a corollary Newton applied his laws of motion to demonstrate that forces must be added as vectors and showed how to do this geometrically with a parallelogram. The second law of motion states that the rate of change of momentum of a mass is proportional to the force acting upon it and takes place in the direction of the force. For the case of constant mass, this law serves as the definition of force (**F**) in terms of the acceleration (**a**) given to a mass (m):

$$\mathbf{F} = m\mathbf{a} \tag{2.1}$$

The unit of force in the SI system of units is the Newton (N). It is defined as the force that gives a mass of one kilogram (1 kg) an acceleration of 1 m s^{-2}.

His celebrated observation of a falling apple may be a legend, but Newton's genius lay in recognizing that the type of gravitational field that caused the apple to fall was the same type that served to hold the Moon in its orbit around the Earth, the planets in their orbits around the Sun, and that acted between minute particles characterized only by their masses. Newton used Kepler's empirical third law (see §1.1.2 and Eq. (1.2)) to deduce that the force of attraction between a planet and the Sun varied with the 'quantities of solid matter that they contain' (i.e. their masses) and with the inverse square of the distance between them. Applying this law to two particles or point masses m and M separated by a distance r (Fig. 2.4a), we get for the gravitational attraction **F** exerted by M on m

$$\mathbf{F} = -G\frac{mM}{r^2}\hat{\mathbf{r}} \tag{2.2}$$

In this equation $\hat{\mathbf{r}}$ is a unit vector in the direction of increase of coordinate r, which is directed away from the centre of reference at the mass M. The negative sign in the equation indicates that the force **F** acts in the opposite direction, toward the attracting mass M. The constant G, which converts the physical law to an equation, is the constant of universal gravitation.

(a) point masses

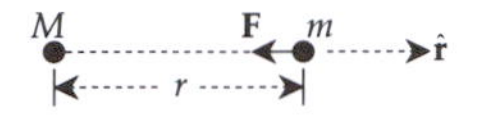

(b) point mass and sphere

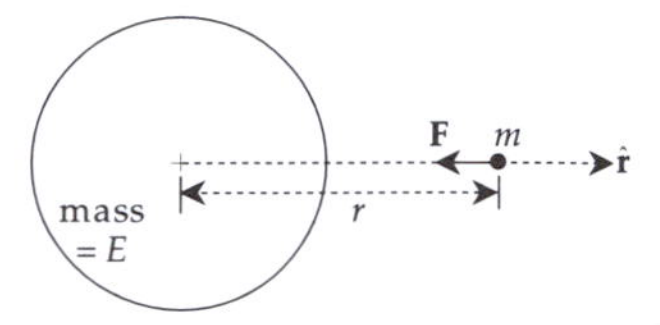

(c) point mass on Earth's surface

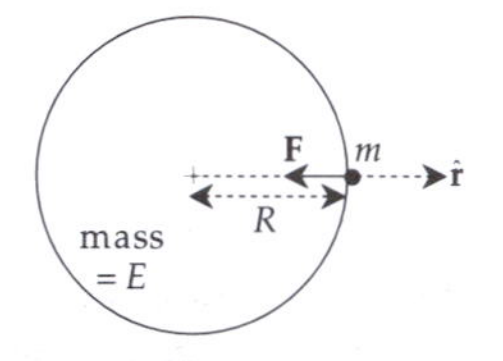

Fig. 2.4 Geometries for the gravitational attraction on (a) two point masses, (b) a point mass outside a sphere, and (c) a point mass on the surface of a sphere.

There was no way to determine the gravitational constant experimentally during Newton's lifetime. The method to be followed was evident, namely to determine the force between two masses in a laboratory experiment. However, 17th century technology was not yet up to this task. Experimental determination of G was extremely difficult, and was first achieved more than a century after the publication of *Principia* by Lord Charles Cavendish (1731–1810). From a set of painstaking measurements of the force of attraction between two spheres of lead Cavendish in 1798 determined the value of G to be 6.754×10^{-11} m^3 kg^{-1} s^{-2}. A modern value (Cohen and Taylor, 1986) is $6.672\,598\,5\times10^{-11}$ m^3 kg^{-1} s^{-2}. It has not yet been possible to determine G more precisely, due to experimental difficulty. Although other physical constants are now known to a precision of less than one part per million (p.p.m), the gravitational constant is known to only 128 p.p.m.

2.2.1.1 *Potential energy and work*

The law of conservation of energy means that the total energy of a closed system is constant. Two forms of energy need be considered here. The first is the potential energy, which an object has by virtue of its position relative to the origin of a force. The second is the work done against the action of the force during a change in position.

For example, when Newton's apple is on the tree it has a higher potential energy than when it lies on the ground. It falls because of the downward force of gravity and loses potential energy in doing so. To compute the change in potential energy we need to raise the apple to its original position. This requires that we apply a force equal and opposite to the gravitational attraction on the apple, and, because this force must be moved through the distance the apple fell, we have to expend energy in the form of work. If the original height of the apple above ground level was h and the value of the force exerted by gravity on the apple is F, the force we must apply to put it back is $(-F)$. Assuming that F is constant through the short distance of its fall, the work expended is $(-F)h$. This is the increase in potential energy of the apple when it is on the tree.

More generally, if the constant force F moves through a small distance dr in the same direction as the force, the work done is $\mathrm{d}W=F\,\mathrm{d}r$ and the change in potential energy $\mathrm{d}E_{\mathrm{p}}$ is given by

$$\mathrm{d}E_{\mathrm{p}}=-\mathrm{d}W=-F\,\mathrm{d}r \tag{2.3}$$

In the more general case we have to consider motions and forces that have components along three orthogonal axes. The displacement dr and the force F no longer need to be parallel to each other. We have to treat F and dr as vectors. In Cartesian coordinates the displacement vector dr has components (dx, dy, dz) and the force has components (F_x, F_y, F_z) along each of the respective axes. The work done by the x-component of the force when it is displaced along the x-axis is $F_x\,\mathrm{d}x$, and there are similar expressions for the displacements along the other axes. The change in potential energy $\mathrm{d}E_{\mathrm{p}}$ is now given by

$$\mathrm{d}E_{\mathrm{p}}=-\mathrm{d}W=-(F_x\,\mathrm{d}x+F_y\,\mathrm{d}y+F_z\,\mathrm{d}z) \tag{2.4}$$

The expression in brackets is called the *scalar product* of the vectors F and dr. It is equal to $F\,\mathrm{d}r\cos\theta$, where θ is the angle between the vectors.

2.2.2 **Gravitational acceleration**

In physics the *field* of a force is often more important than the absolute magnitude of the force. The field is defined as the force exerted on a material unit. For example, the electrical field of a charged body at a certain position is the force it exerts on a unit of electrical charge at that location. The *gravitational field* in the vicinity of an attracting mass is the force it exerts on a unit mass. Eq. (2.1) shows that this is equivalent to the acceleration vector.

In geophysical applications we are concerned with accelerations rather than forces. By comparing Eq. (2.1) and Eq. (2.2) we get the gravitational acceleration $\mathbf{a}_{\mathrm{G}}$ of the mass m due to the attraction of the mass M:

$$\mathbf{a}_{\mathrm{G}}=-G\frac{M}{r^2}\hat{\mathbf{r}} \tag{2.5}$$

The SI unit of acceleration is the m s^{-2}; this unit is unpractical for use in geophysics. In the now superseded c.g.s. system the unit of acceleration was the cm s^{-2}, which is called a *gal* in recognition of the contributions of Galileo. The small changes in the acceleration of gravity caused by geological structures are measured in thousandths of this unit, i.e., in milligal (*mgal*). Until recently, gravity anomalies due to geological structures were surveyed with field instruments accurate to about one-tenth of a milligal, which was called a *gravity unit*. Modern instruments are capable of measuring gravity differences to a millionth of a gal, or microgal (*μgal*), which is becoming the practical unit of gravity investigations. The value of gravity at the Earth's surface is about 9.8 m s^{-2}, and so the sensitivity of modern measurements of gravity is about 1 part in 10^9.

2.2.2.1 *Gravitational potential*

The gravitational potential is the potential energy of a unit mass in a field of gravitational attraction. Let the potential be denoted by the symbol U_G. The potential energy E_p of a mass m in a gravitational field is thus equal to (mU_G). Thus, a change in potential energy (dE_p) is equal to $(m\, dU_G)$. Eq. (2.3) becomes

$$m\, dU_G = -F\, dr$$

and by use of Eq. (2.1)

$$m\, dU_G = -m\, a_G\, dr \tag{2.6}$$

Rearranging this equation we get the gravitational acceleration

$$\mathbf{a}_G = -\frac{dU_G}{dr}\hat{\mathbf{r}} \tag{2.7}$$

In general, the acceleration is a three-dimensional vector. If we are using Cartesian coordinates (x, y, z), the acceleration will have components (a_x, a_y, a_z). These may be computed by separately calculating the derivatives of the potential with respect to x, y and z:

$$a_x = -\frac{\partial U_G}{\partial x};\ a_y = -\frac{\partial U_G}{\partial y};\ a_z = -\frac{\partial U_G}{\partial z} \tag{2.8}$$

Equating Eqs. (2.3) and (2.7) gives the gravitational potential of a point mass M:

$$\frac{dU_G}{dr} = G\frac{M}{r^2} \tag{2.9}$$

the solution of which is

$$U_G = -G\frac{M}{r} \tag{2.10}$$

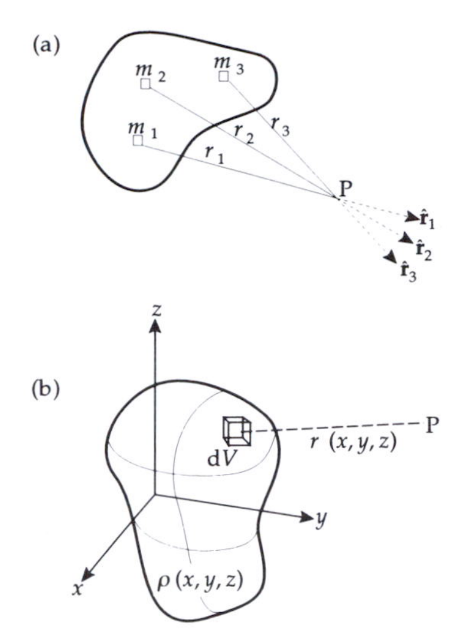

Fig. 2.5 (a) Each small particle of a solid body exerts a gravitational attraction in a different direction at an external point P. (b) Computation of the gravitational potential of a continuous mass distribution.

2.2.2.2 *Acceleration and potential of a distribution of mass*

Until now, we have considered only the gravitational acceleration and potential of point masses. A solid body may be considered to be composed of numerous small particles, each of which exerts a gravitational attraction at an external point P (Fig. 2.5a). To calculate the gravitational acceleration of the object at the point P we must form a vector sum of the contributions of the individual discrete particles. Each contribution has a different direction. Assuming m_i to be the mass of the particle at distance r_i from P, this gives an expression like

$$\mathbf{a}_G = -G\frac{m_1}{r_1^2}\hat{\mathbf{r}}_1 - G\frac{m_2}{r_2^2}\hat{\mathbf{r}}_2 - G\frac{m_3}{r_3^2}\hat{\mathbf{r}}_3 - \ldots \tag{2.11}$$

Depending on the shape of the solid, this vector sum can be quite complicated.

An alternative solution to the problem is by first calculating the gravitational potential, and then differentiating it as in Eq. (2.5) to get the acceleration. The expression for the potential at P is

$$U_G = -G\frac{m_1}{r_1} - G\frac{m_2}{r_2} - G\frac{m_3}{r_3} - \ldots \tag{2.12}$$

This is a scalar sum, which is usually more simple to calculate than a vector sum.

More commonly, the object is not represented as an assemblage of discrete particles but by a continuous mass distribution. However, we can subdivide the volume into discrete elements; if the density of the matter in each volume is known the mass of the small element can be calculated and its contribution to the potential at the external point P can be determined. By integrating over the volume of the body its gravitational potential at P can be calculated. At a point in the body with coordinates (x, y, z) let the density be $\rho(x, y, z)$ and let its distance from P be $r(x, y, z)$ as in Fig. 2.5b. The gravitational potential of the body at P is

$$U_G = -G \iiint_{x\,y\,z} \frac{\rho(x,y,z)}{r(x,y,z)} \mathrm{d}x\,\mathrm{d}y\,\mathrm{d}z \tag{2.13}$$

The integration readily gives the gravitational potential and acceleration at points inside and outside a hollow or homogeneous solid sphere. The values outside a sphere at distance r from its center are the same as if the entire mass E of the sphere were concentrated at its center (Fig. 2.4b):

$$U_G = -G\frac{E}{r}$$

$$\mathbf{a}_G = -G\frac{E}{r^2}\hat{\mathbf{r}} \tag{2.14}$$

2.2.2.3 *Mass and mean density of the Earth*

The relationships in Eq. (2.14) are valid everywhere outside a sphere, including on its surface where the distance from the center of mass is equal to the mean radius R (Fig. 2.4c). If we regard the Earth to a first approximation as a sphere with mass E and radius R, we can estimate the Earth's mass by rewriting Eq. (2.14) as a scalar equation in the form

$$E = \frac{R^2 a_G}{G} \tag{2.15}$$

The gravitational acceleration at the surface of the Earth is only slightly different from mean gravity, about 9.81 m s^{-2}, the Earth's radius is 6371 km, and the gravitational constant is 6.673×10^{-11} $m^3\,kg^{-1}\,s^{-2}$. The mass of the Earth is found to be 5.97×10^{24} kg. This large number is not so meaningful as the mean density of the Earth, which may be calculated by dividing the Earth's mass by its volume $(\frac{4}{3}\pi R^3)$. A mean density of 5515 kg m^{-3} is obtained, which is about double the density of crustal rocks. This indicates that the Earth's interior is not homogeneous, and implies that density must increase with depth in the Earth.

2.2.3 **The equipotential surface**

An equipotential surface is one on which the potential is constant. For a sphere of given mass the gravitational potential (Eq. (2.10)) varies only with the distance r from its center. A certain value of the potential, say U_1, is realized at a constant radial distance r_1. Thus, the equipotential surface on which the potential has the value U_1 is a sphere with radius r_1; a different equipotential surface U_2 is the sphere with radius r_2. The equipotential surfaces of the original spherical mass form a set of concentric spheres (Fig. 2.6a), one of which (e.g., U_0) coincides with the surface of the spherical mass. This particular equipotential surface describes the *figure* of the spherical mass.

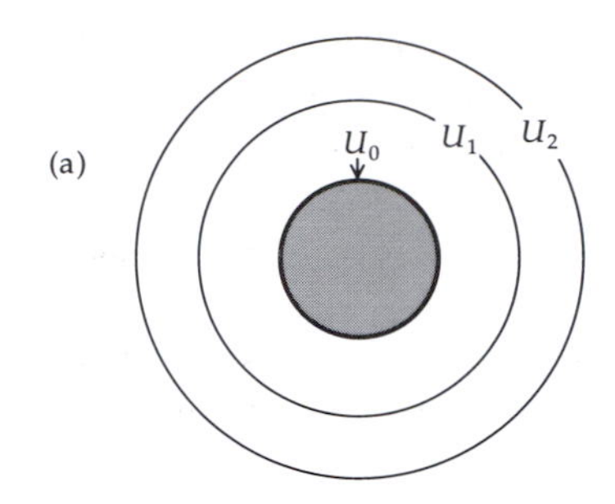

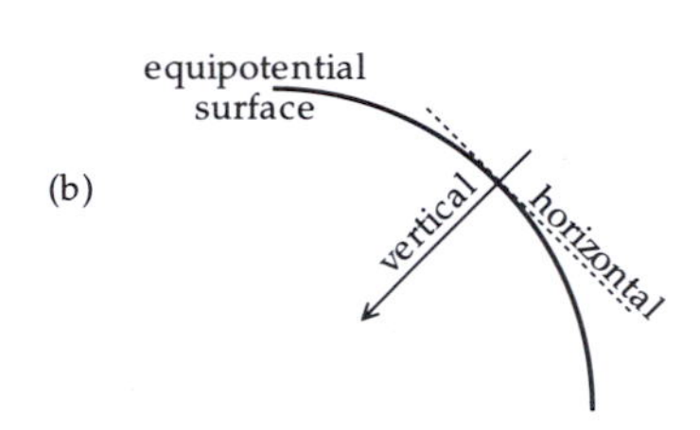

Fig. 2.6 (a) Equipotential surfaces of a spherical mass form a set of concentric spheres. (b) The normal to the equipotential surface defines the vertical direction; the tangential plane defines the horizontal.

By definition, no change in potential takes place (and no work is done) in moving from one point to another on an equipotential surface. The work done by a force F in a displacement dr is $F dr \cos\theta$, which is zero when $\cos\theta$ is zero, that is, when the angle θ between the displacement and the force is 90°. If no work is done in a motion along a gravitational equipotential surface, the force and acceleration of the gravitational field must act perpendicular to the surface. This normal to the equipotential surface defines the *vertical*, or plumb-line, direction (Fig. 2.6b). The plane tangential to the equipotential surface at a point defines the *horizontal* at that point.

2.3 EARTH'S ROTATION

2.3.1 **Introduction**

The rotation of the Earth is a vector, i.e., a quantity characterized by both magnitude and direction. The Earth behaves as an elastic body and deforms in response to the forces generated by its rotation, becoming slightly flattened at the poles with a compensating bulge at the equator. The gravitational attractions of the Sun, Moon and planets on the spinning, flattened Earth cause changes in its rate of rotation, in the orientation of the rotation axis, and in the shape of the Earth's orbit around the Sun. Even without extra-terrestrial influences the Earth reacts to tiny displacements of the rotation axis from its average position by acquiring a small, unsteady wobble. These perturbations reflect a balance between gravitation and the forces that originate in the Earth's rotational dynamics.

2.3.2 **Centripetal and centrifugal acceleration**

Newton's first law of motion states that every object continues in its state of rest or of uniform motion in a straight line unless compelled to change that state by forces acting on it. The continuation of a state of motion is by virtue of the *inertia* of the body. A framework in which this law is valid is called an *inertial system*. For example, when we are travelling in a car at constant speed, we feel no disturbing forces; reference axes fixed to the moving vehicle form an inertial frame. If traffic conditions compel the driver to apply the brakes, we experience decelerating forces; if the car goes around a corner, even at constant speed, we sense sideways forces toward the outside of the corner. In these situations the moving car is being forced to change its state of uniform rectilinear motion and reference axes fixed to the car form a *non-inertial system*.

Motion in a circle implies that a force is active that continually changes the state of rectilinear motion. Newton recognized that the force was directed inwards, towards the center of the circle, and named it the *centripetal* (meaning 'center-seeking') force. He cited the example of a stone being whirled about in a sling. The inward centripetal force exerted on the stone by the sling holds it in a circular path. If the sling is released, the restraint of the centripetal force is removed and the inertia of the stone causes it to continue its motion at the point of release. No longer under the influence of the restraining force, the stone flies off in a straight line. Arguing that the curved path of a projectile near the surface of the Earth was due to the effect of gravity, which caused it constantly to fall toward the Earth, Newton postulated that, if the speed of the projectile were exactly right, it might never quite reach the Earth's surface. If the projectile fell toward the center of the Earth at the same rate as the curved surface of the Earth fell away from it, the projectile would go into orbit around the Earth. Newton suggested that the Moon was held in orbit around the Earth by just such a centripetal force, which originated in the gravitational attraction of the Earth. Likewise, he visualized that a centripetal force due to gravitational attraction restrained the planets in their circular orbits about the Sun.

The passenger in a car going round a corner experiences a tendency to be flung outwards. He is restrained in position by the frame of the vehicle, which supplies the necessary centripetal acceleration to enable the passenger to go round the curve in the car. The inertia of the passenger's body causes it to continue in a straight line and pushes him outwards against the side of the vehicle. This outward force is called the *centrifugal* force. It arises because the car does not represent an inertial reference frame. An observer outside the car in a fixed (inertial) coordinate system would note that the car and passenger are constantly changing direction as they round the corner. The centrifugal force feels real enough to the passenger in the car, but it is called a pseudo-force, or *inertial force*. In contrast to the centripetal force, which arises from the gravitational attraction, the centrifugal force does not have a physical origin, but exists only because it is being observed in a non-inertial reference frame.

2.3.2.1 *Centripetal acceleration*

The mathematical form of the centripetal acceleration for circular motion with constant angular velocity ω about a point can be derived as follows. Define orthogonal Cartesian axes x and y relative to the center of the circle as in Fig. 2.7a. The linear velocity v at any point where the radius vector makes an angle $\theta=(\omega t)$ with the x-axis has components

$$\begin{aligned} & v_x=-v\cos(\omega t)=-\omega r\cos(\omega t) \\ \text{and} \quad & v_y=-v\sin(\omega t)=-\omega r\sin(\omega t) \end{aligned} \tag{2.16}$$

The x- and y- components of the acceleration are obtained by differentiating the velocity components with respect to time. This gives

$$\begin{aligned} & a_x=-\omega v\cos(\omega t)=-\omega^2 r\cos(\omega t) \\ \text{and} \quad & a_y=-\omega v\sin(\omega t)=-\omega^2 r\sin(\omega t) \end{aligned} \tag{2.17}$$

These are the components of the centripetal acceleration, which is directed radially inwards and has the magnitude $\omega^2 r$ (Fig. 2.7b).

2.3.2.2 *Centrifugal acceleration and potential*

In handling the variation of gravity on the Earth's surface we must operate in a non-inertial reference frame attached to the rotating Earth. Viewed from a fixed, external inertial frame, a stationary mass moves in a circle about the Earth's rotation axis with the same rotational speed as the Earth. However, within a rotating reference frame attached to the Earth, the mass is stationary. It experiences a centrifugal acceleration (a_c) that is exactly equal and opposite to the centripetal acceleration, and which can be written in the alternative forms

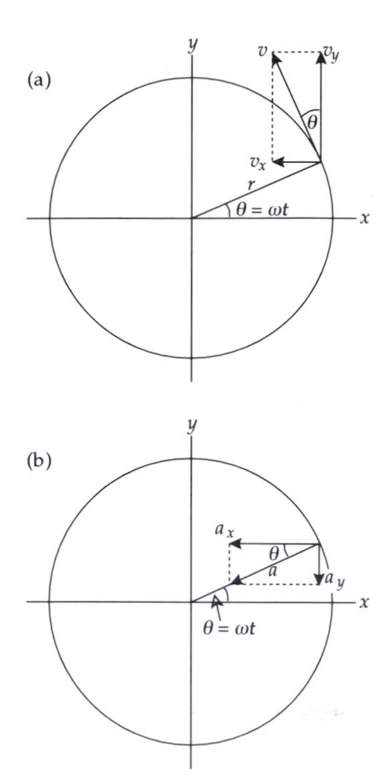

Fig. 2.7 (a) Components v_x and v_y of the linear velocity v where the radius makes an angle $\theta=(\omega t)$ with the x–axis, and (b) the components a_x and a_y of the centripetal acceleration, which is directed radially inward.

$$a_c=\omega^2 r$$

$$a_c=\frac{v^2}{r} \tag{2.18}$$

The centrifugal acceleration is not a centrally oriented acceleration like gravitation, but instead is defined relative to an axis of rotation. Nevertheless, potential energy is associated with the rotation and it is possible to define a *centrifugal potential.* Consider a point rotating with the Earth at a distance r from its center (Fig. 2.8). The angle θ between the radius to the point and the axis of rotation is called the colatitude; it is the angular complement of the latitude λ. The distance of the point from the rotational axis is x $(=r\sin\theta)$, and the centrifugal acceleration is $\omega^2 x$ outwards in the direction of increasing x. The centrifugal potential U_c is defined such that

$$\mathbf{a}_c=-\frac{\partial U_c}{\partial x}\hat{\mathbf{x}}=\omega^2 x\,\hat{\mathbf{x}} \tag{2.19}$$

where $\hat{\mathbf{x}}$ is the outward unit vector. On integrating, we obtain

$$U_c=-\tfrac{1}{2}\omega^2x^2=-\tfrac{1}{2}\omega^2r^2\sin^2\theta=-\tfrac{1}{2}\omega^2r^2\cos^2\lambda \tag{2.20}$$

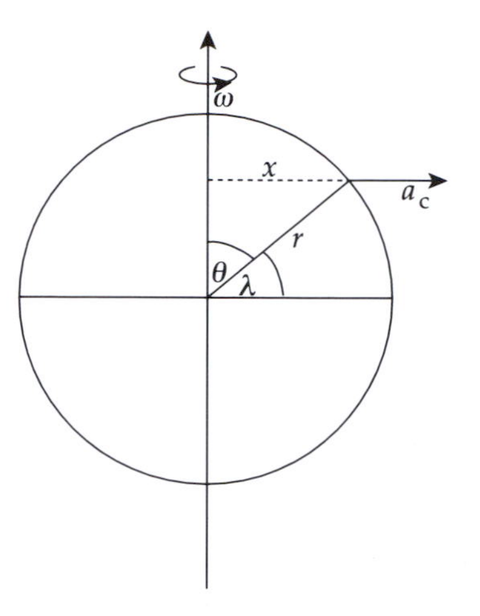

Fig. 2.8 The outwardly directed centrifugal acceleration a_c at latitude λ on a sphere rotating at angular velocity ω.

2.3.2.3 *Kepler's third law of planetary motion*

By comparing the centripetal acceleration of a planet about the Sun with the gravitational acceleration of the Sun, the third of Kepler's laws of planetary motion can be explained. Let S be the mass of the Sun, r_p the distance of a planet from the Sun, and T_p the period of orbital rotation of the planet around the Sun. Equating the gravitational and centripetal accelerations gives

$$G\frac{S}{r_p^2}=\omega_p^2 r_p=\left(\frac{2\pi}{T_p}\right)^2 r_p \tag{2.21}$$

Rearranging this equation we get Kepler's third law of planetary motion, which states that the square of the period of the planet is proportional to the cube of the radius of its orbit, or:

$$\frac{r_p^3}{T_p^2}=\frac{GS}{4\pi^2}=\text{constant} \tag{2.22}$$

2.3.2.4 *Verification of the inverse square law of gravitation*

Newton realized that the centripetal acceleration of the Moon in its orbit was supplied by the gravitational attraction of the Earth, and tried to use this knowledge to confirm the inverse square dependency on distance in his law of gravitation. The sidereal period (T_L) of the Moon about the Earth, a sidereal month, is equal to 27.3 days. Let the corresponding angular rate of rotation be ω_L. We can equate the gravitational acceleration of the Earth at the Moon with the centripetal acceleration due to ω_L:

$$G\frac{E}{r_L^2}=\omega_L^2 r_L \tag{2.23}$$

This equation can be rearranged as follows

$$\left(G\frac{E}{R^2}\right)\left(\frac{R}{r_L}\right)^2=\omega_L^2 R\left(\frac{r_L}{R}\right) \tag{2.24}$$

Comparison with Eq. (2.14) shows that the first quantity in parentheses is the mean gravitational acceleration on the Earth's surface, a_G. Therefore, we can write

$$a_G=G\frac{E}{R^2}=\omega_L^2 R\left(\frac{r_L}{R}\right)^3 \tag{2.25}$$

In Newton's time little was known about the physical dimensions of our planet. The distance of the Moon was known to be approximately 60 times the radius of the Earth (see §1.1.3.1) and its sidereal period was known to be 27.3 days. At first Newton used the accepted value 5500 km for the Earth's radius. This gave a value of only 8.4 m s^{-2} for gravity, well below the known value of 9.8 m s^{-2}. However, in 1671 Picard determined the Earth's radius to be 6372 km. With this value, the inverse square character of Newton's law of gravitation was confirmed.

2.3.3 The tides

The gravitational forces of Sun and Moon deform the Earth's shape, causing tides in the oceans, atmosphere and solid body of the Earth. The most visible tidal effects are the displacements of the ocean surface, which is a hydrostatic equipotential surface. The Earth does not react rigidly to the tidal forces. The solid body of the Earth deforms in a like manner to the free surface, giving rise to so-called *bodily Earth-tides*. These can be observed with specially designed instruments, which operate on a similar principle to the long-period seismometer.

The height of the marine equilibrium tide amounts to only half a meter or so over the free ocean. In coastal areas the tidal height is significantly increased by the shallowing of the continental shelf and the confining shapes of bays and harbors. Accordingly, the height and variation of the tide at any place is influenced strongly by complex local factors. Subsequent chapters deal with the tidal deformations of the Earth's hydrostatic figure.

2.3.3.1 *Lunar tidal periodicity*

The Earth and Moon are coupled together by gravitational attraction. Their common motion is like that of a pair of ballroom dancers. Each partner moves around the center of mass of the pair. For the Earth-Moon pair the location of the center of mass is easily found. Let E be the mass of the Earth, and m that of the Moon; let the separation of the centers of the Earth and Moon be r_L and let the distance of their common center of mass be d from the center of the Earth. The moment of the Earth about the center of mass is Ed and the moment of the Moon is $m(r_L-d)$. Setting these moments equal we get

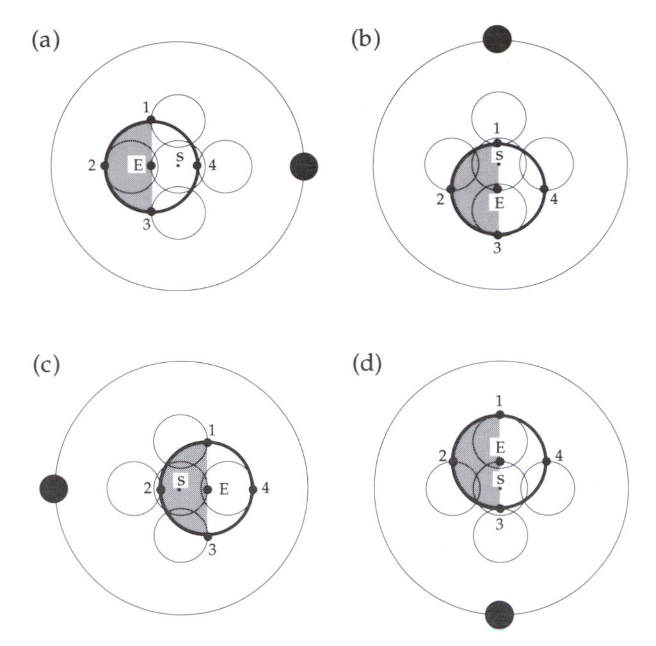

Fig. 2.9 Illustration of the 'revolution without rotation' of the Earth–Moon pair about their common center of mass at S.

$$d=\frac{m}{E+m}r_L \tag{2.26}$$

The mass of the Moon is 0.0123 that of the Earth and the distance between the centers is 382,000 km. These figures give d=4600 km, i.e., the center of revolution of the Earth–Moon pair lies within the Earth.

To understand the common revolution of the Earth–Moon pair we have to exclude the rotation of the Earth about its axis. The 'revolution without rotation' is illustrated in Fig. 2.9. The Earth–Moon pair revolves about S, the center of mass. Let the starting positions be as shown in Fig. 2.9a. Approximately one week later the Moon has advanced in its path by one-quarter of a revolution and the center of the Earth has moved so as to keep the center of mass fixed (Fig. 2.9b). The relationship is maintained in the following weeks (Fig. 2.9c, d) so that during one month the center of the Earth describes a circle about S. Now consider the motion of point number 2 on the left hand side of the Earth in Fig. 2.9. If the Earth revolves as a rigid body and the rotation about its own axis is omitted, after one week point 2 will have moved to a new position but will still be the furthest point on the left. Subsequently, during one month point 2 will describe a small circle with the same radius as the circle described by the Earth's center. Similarly points 1,

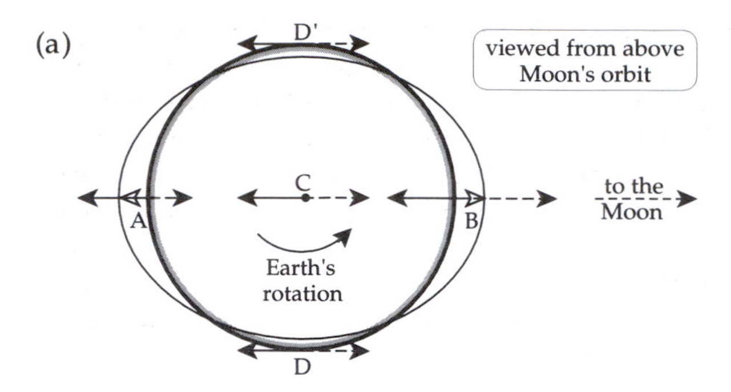

a_L ← constant centrifugal acceleration
a_G - - - -▶ variable lunar gravitation
a_T ⇾ residual tidal acceleration

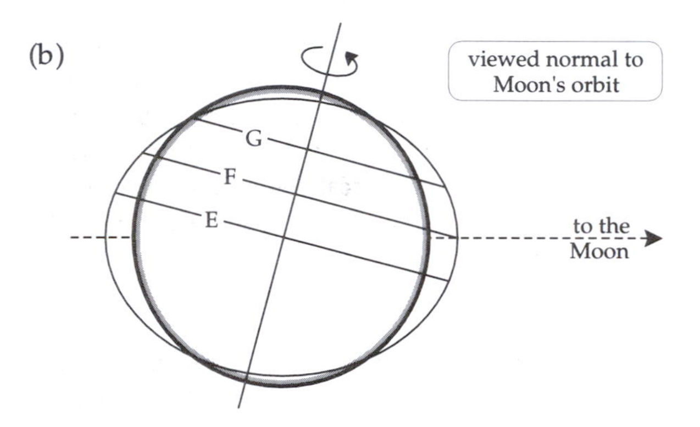

Fig. 2.10 (a) The relationships of the centrifugal, gravitational and residual tidal accelerations at selected points in the Earth. (b) Latitude effect that causes diurnal inequality of the tidal height.

3 and 4 will also describe circles of exactly the same size. A simple illustration of this point can be made by chalking the tip of each finger on one hand with a different color, then moving your hand in a circular motion while touching a blackboard; your fingers will draw a set of identical circles.

The 'revolution without rotation' causes each point in the body of the Earth to describe a circular path with identical radius. The centrifugal acceleration of this motion has therefore the same magnitude at all points in the Earth, and, as can be seen by inspection of Fig. 2.9 (a–d), it is directed away from the Moon parallel to the Earth–Moon line of centers. At C, the center of the Earth (Fig. 2.10a), this centrifugal acceleration exactly balances the gravitational attraction of the Moon. Its magnitude is given by

$$a_L = G\frac{m}{r_L^2} \tag{2.27}$$

At B, on the side of the Earth nearest to the Moon, the gravitational acceleration of the Moon is larger than at the center of the Earth and exceeds the centrifugal acceleration a_L. There is a residual acceleration *toward* the Moon, which raises a tide on this side of the Earth. The magnitude of the tidal acceleration at B is

$$a_T = Gm\left(\frac{1}{(r_L - R)^2} - \frac{1}{r_L^2}\right) = \frac{Gm}{r_L^2}\left[\left(1 - \frac{R}{r_L}\right)^{-2} - 1\right] \tag{2.28}$$

Expanding this equation with the binomial theorem and simplifying gives

$$a_T = \frac{Gm}{r_L^2}\left[2\frac{R}{r_L} + 3\left(\frac{R}{r_L}\right)^2 + \ldots\right] \tag{2.29}$$

At A, on the far side of the Earth, the gravitational acceleration of the Moon is less than the centrifugal acceleration a_L. The residual acceleration (Fig. 2.10a) is *away from* the Moon, and raises a tide on the far side of the Earth. The magnitude of the tidal acceleration at A is

$$a_T = Gm\left(\frac{1}{r_L^2} - \frac{1}{(r_L + R)^2}\right) \tag{2.30}$$

which reduces to

$$a_T = \frac{Gm}{r_L^2}\left[2\frac{R}{r_L} - 3\left(\frac{R}{r_L}\right)^2 + \ldots\right] \tag{2.31}$$

At points D and D′ the direction of the gravitational acceleration due to the Moon is not exactly parallel to the line of centers of the Earth–Moon pair. The residual tidal acceleration is almost along the direction toward the center of the Earth. Its effect is to lower the free surface in this direction.

The tidal deformation of the Earth produced by the Moon has a prolate ellipsoidal shape, like a rugby football, along the Earth–Moon line of centers. The daily tides are caused by superposing the Earth's rotation on this deformation. In the course of one day a point rotates past the points A, D, B and D′ and an observer experiences two full tidal cycles, called the *semi-diurnal* tides. The extreme tides are not equal at every latitude, because of the varying angle between the Earth's rotational axis and the Moon's orbit (Fig. 2.10b). At the equator E the semi-diurnal tides are equal; at an intermediate latitude F one tide is higher than the other; and at latitude G and higher there is only one (diurnal) tide per day. The difference in height between two successive high or low tides is called the *diurnal inequality*.

2.3.3.2 *Tidal effect of the Sun*

The Sun also has an influence on the tides. The theory of the solar tides can be followed in identical manner to the lunar tides by again applying the principle of 'revolution without rotation'. The Sun's mass is 333,340 times greater than that of the Earth, so the common center of mass is close to the center of the Sun at a radial distance of about 450 km from

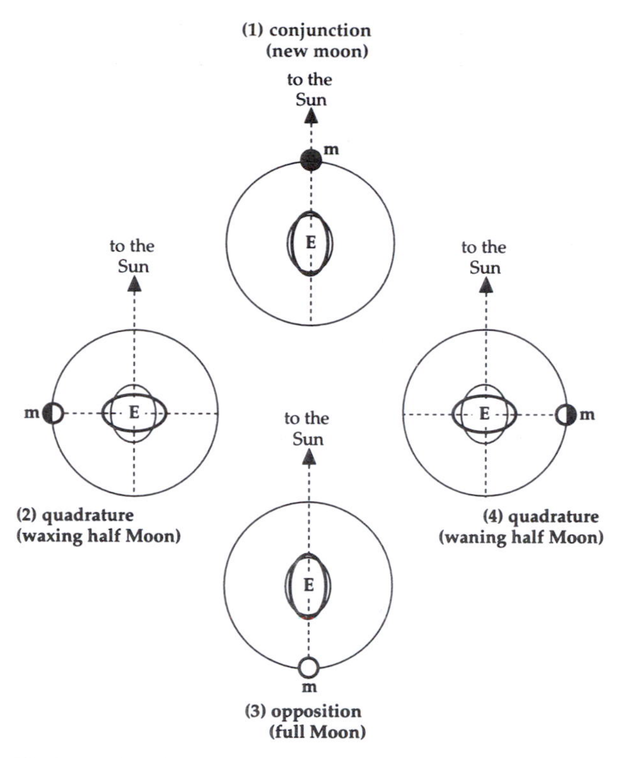

Fig. 2.11 The orientations of the solar and lunar tidal deformations of the Earth at different lunar phases.

its center. The period of the revolution is one year. As for the lunar tide, the imbalance between gravitational acceleration of the Sun and centrifugal acceleration due to the common revolution leads to a prolate ellipsoidal tidal deformation. The solar effect is smaller than that of the Moon. Although the mass of the Sun is vastly greater than that of the Moon, its distance from the Earth is also much greater, and, because gravitational acceleration varies inversely with the square of distance, the maximum tidal effect of the Sun is only about 45% that of the Moon.

2.3.3.3 *Spring and neap tides*

The superposition of the lunar and solar tides causes a modulation of the tidal amplitude. The ecliptic plane is defined by the Earth's orbit around the Sun. The Moon's orbit around the Earth is not exactly in the ecliptic but is inclined at a very small angle of about 5.2° to it. For discussion of the combination of lunar and solar tides we can assume the orbits to be coplanar. The Moon and Sun each produce a prolate tidal deformation of the Earth, but the relative orientations of these ellipsoids vary during one month (Fig. 2.11). At conjunction the (new) Moon is on the same side of the Earth as the Sun, and the ellipsoidal deformations augment each other. The same is the case half a month later at opposition, when the (full) Moon is on the opposite side of the Earth from the Sun. The unusually high tides at opposition and conjunction are called *spring tides*. In contrast, at the times of quadrature the waxing or waning half Moon causes a prolate ellipsoidal deformation out of phase with the solar deformation. The maximum lunar tide coincides with the minimum solar tide, and the effects partially cancel each other. The unusually low tides at quadrature are called *neap tides*. The superposition of the lunar and solar tides causes modulation of the tidal amplitude during a month (Fig. 2.12).

2.3.3.4 *Effect of the tides on gravity measurements*

The tides have an effect on gravity measurements made on the Earth. The combined effects of Sun and Moon cause an acceleration at the Earth's surface of up to about 0.3 mgal, of which about two-thirds are due to the Moon and one-third to the Sun. The sensitive modern instruments used for gravity exploration can readily detect gravity differences of 0.01 mgal. It is necessary to compensate gravity measurements for the tidal effects, which vary with location, date and time of day. Fortunately, tidal theory is so well established that the gravity effect can be calculated and tabulated for any place and time before beginning a survey.

2.3.3.5 *Bodily Earth-tides*

A simple way to measure the height of the marine tide might be to fix a stake to the sea-bottom at a suitably sheltered location and to record continuously the measured water level (assuming that confusion introduced by wave motion can be eliminated or taken into account). The observed amplitude of the marine tide, defined by the displacement of the free water surface, is found to be about 70% of the theoretical value. The difference is explained by the elasticity of the Earth. The tidal deformation corresponds to a redistribution of mass, which modifies the gravitational potential of the Earth and augments the elevation of the free surface. This is partially counteracted by a bodily tide in the solid Earth, which deforms elastically in response to the attraction of the Sun and Moon. The free water surface is raised by the tidal attraction, but the sea-bottom in which the measuring rod is implanted is also raised. The measured tide is the difference between the marine tide and the bodily Earth-tide.

In practice, the displacement of the equipotential surface is measured with a horizontal pendulum, which reacts to the tilt of the surface. The bodily Earth-tides also affect gravity measurements and can be observed with sensitive gravimeters. The effects of the bodily Earth-tides are incorporated into the predicted tidal corrections to gravity measurements.

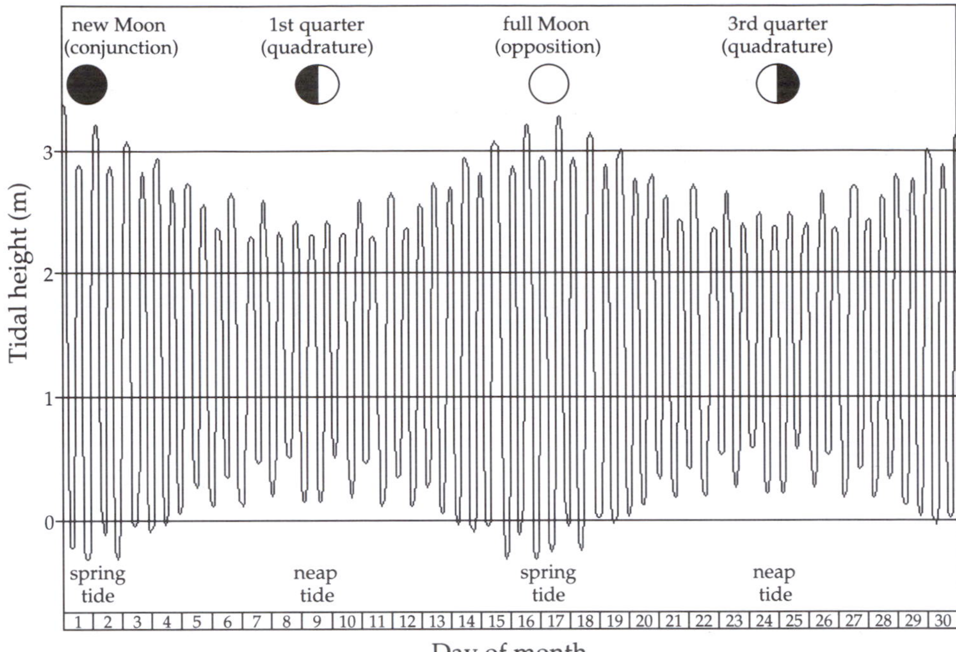

Fig. 2.12 Schematic representation of the modulation of the tidal amplitude as a result of superposition of the lunar and solar tides.

2.3.4 Changes in Earth's rotation

The Earth's rotational vector is affected by the gravitational attractions of the Sun, Moon and the planets. The rate of rotation and the orientation of the rotational axis change with time. The orbital motion around the Sun is also affected. The orbit rotates about the pole to the plane of the ecliptic and its ellipticity changes over long periods of time.

2.3.4.1 *Effect of lunar tidal friction on the length of the day*

If the Earth reacted perfectly elastically to the lunar tidal forces, the prolate tidal bulge would be aligned along the line of centers of the Earth–Moon pair (Fig. 2.13a). However, the motion of the seas is not instantaneous and the tidal response of the solid part of the Earth is partly anelastic. These features cause a slight delay in the time when high tide is reached, amounting to about 12 minutes. In this short interval the Earth's rotation carries the line of the maximum tides past the line of centers by a small angle of approximately 2.9° (Fig. 2.13b). A point on the rotating Earth passes under the line of maximum tides 12 minutes after it passes under the Moon. The small phase difference is called the tidal lag.

Because of the tidal lag the gravitational attraction of the Moon on the tidal bulges on the far side and near side of the Earth (F_1 and F_2, respectively) are not collinear (Fig. 2.13b). F_2 is stronger than F_1 so a torque is produced in the opposite sense to the Earth's rotation (Fig. 2.13c). The tidal torque acts as a brake on the Earth's rate of rotation, which is gradually slowing down.

The tidal deceleration of the Earth is manifested in a gradual increase in the length of the day. The effect is very small. Tidal theory predicts an increase in the length of the day of only 2.4 millisec per century. Observations of the phenomenon are based on ancient historical records of lunar and solar eclipses and on telescopically observed occultations of stars by the Moon. The current rate of rotation of the Earth can be measured with very accurate atomic clocks. Telescopic observations of the daily times of passage of stars past the local zenith are recorded with a camera controlled by an atomic clock. These observations give precise measures of the mean value and fluctuations of the length of the day.

The occurrence of a lunar or solar eclipse was a momentous event for ancient peoples, and was duly recorded in scientific and non-scientific chronicles. Untimed observations are found in non-astronomical works. They record, with variable reliability, the degree of totality and the time and place of observation. The unaided human eye is able to decide quite precisely just when an eclipse becomes total. Timed observations of both lunar and solar eclipses made by Arab astronomers around 800–1000 A.D. and Babylonian astronomers a thousand years earlier give two important groups of data (Fig. 2.14). By comparing the observed times of alignment of Sun, Moon and Earth with

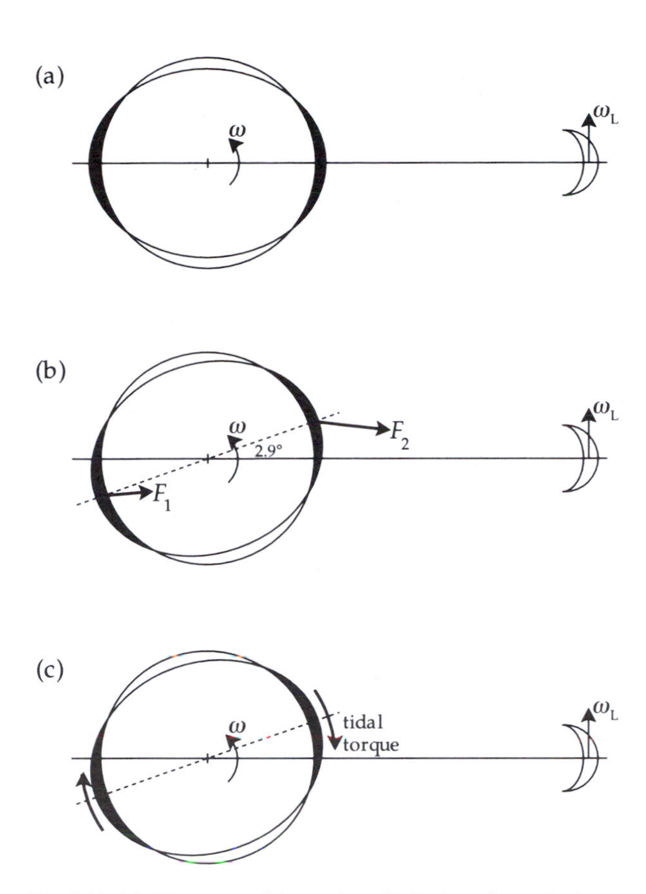

Fig. 2.13 (a) Alignment of the prolate tidal bulge of a perfectly elastic Earth along the line of centers of the Earth–Moon pair. (b) Tidal phase lag of 2.9° relative to the line of centers due to the Earth's partially anelastic response. (c) Tidal decelerating torque due to unequal gravitational attractions of the Moon on the far and near-sided tidal bulges.

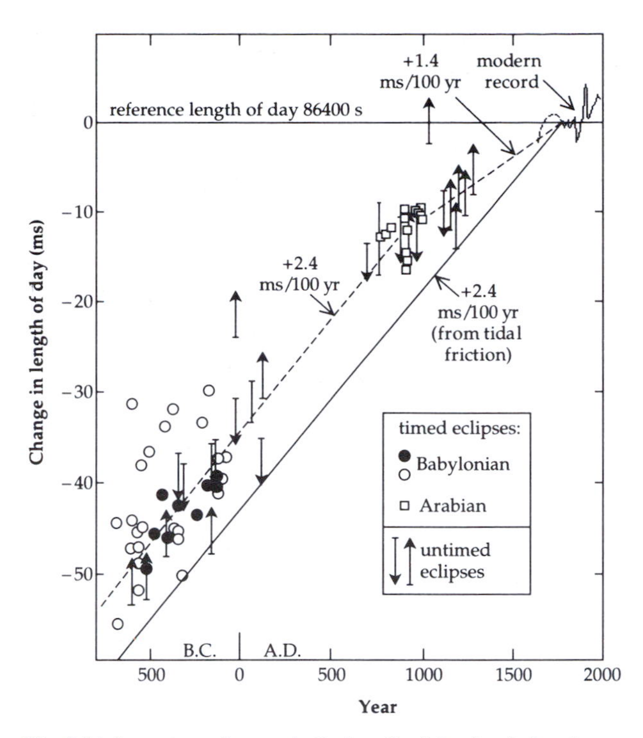

Fig. 2.14 Long-term changes in the length of the day deduced from observations of solar and lunar eclipses between 700 B.C. and 1980 A.D. (after Stephenson and Morrison, 1984).

times predicted from the theory of celestial mechanics, the differences due to change in length of the day may be computed. A straight line with slope equal to the rate of increase of the length of the day inferred from tidal theory, 2.4 millisec per century, connects the Babylonian and Arab data sets. Since the medieval observations of Arab astronomers the length of the day has increased on average by about 1.4 millisec per century. The data-set based on telescopic observations covers the time from A.D. 1620 to 1980. It gives a more detailed picture and shows that the length of the day fluctuates about the long-term trend of 1.4 millisec per century. A possible interpretation of the difference between the two slopes is that non-tidal causes have opposed the deceleration of the Earth's rotation since about 950 A.D. It would be wrong to infer that some sudden event at that epoch caused an abrupt change, because the data are equally compatible with a smoothly changing polynomial. The observations confirm the importance of tidal braking, but they also indicate that tidal friction is not the only mechanism affecting the Earth's rotation.

The short-term fluctuations in rotation rate are due to exchanges of angular momentum with the Earth's atmosphere and core. The atmosphere is tightly coupled to the solid Earth. An increase in average global wind speed corresponds to an increase in the angular momentum of the atmosphere and corresponding decrease in angular momentum of the solid Earth. Accurate observations by very long baseline interferometry (see §2.4.6.4) confirm that rapid fluctuations in the length of the day are directly related to changes in the angular momentum of the atmosphere. On a longer timescale of decades the changes in length of the day may be related to changes in the angular momentum of the core. The fluid in the outer core has a speed of the order of 0.1 mm s^{-1} relative to the overlying mantle. The mechanism for exchange of angular momentum between the fluid core and the rest of the Earth depends on the way the core and mantle are coupled. The coupling may be mechanical if topographic irregularities obstruct the flow of the core fluid along the core–mantle interface. The core fluid is a good electrical conductor, so, if the lower mantle also has an appreciable electrical conductivity, it is possible that the core and mantle are coupled electromagnetically.

2.3.4.2 *Increase of the Earth–Moon distance*

Further consequences of lunar tidal friction can be seen by applying the law of conservation of momentum to the Earth–Moon pair. Let the Earth's mass be E, its rate of rotation be ω and its moment of inertia about the rotation axis be C; let the corresponding parameters for the Moon be m, Ω_L, and C_L, and let the Earth–Moon distance be r_L. Further, let the distance of the common center of revolution be d from the center of the Earth, as given by Eq. (2.26). The angular momentum of the system is given by

$$C\omega+E\Omega_L d^2+m\Omega_L(r_L-d)^2+C_L\Omega_L=\text{constant} \qquad (2.32)$$

The fourth term is the angular momentum of the Moon about its own axis. Tidal deceleration due to the Earth's attraction has slowed down the Moon's rotation until it equals its rate of revolution about the Earth. Both Ω_L, and C_L are very small and the fourth term can be neglected. The second and third terms can be combined so that we get

$$C\omega+\left(\frac{E}{E+m}\right)m\Omega_L r_L^2=\text{constant} \qquad (2.33)$$

The gravitational attraction of the Earth on the Moon is equal to the centripetal acceleration of the Moon about the common center of revolution, thus

$$\frac{GE}{r_L^2}=\Omega_L^2(r_L-d)=\Omega_L^2 r_L\left(\frac{E}{E+m}\right) \qquad (2.34)$$

from which

$$\Omega_L r_L^2=\sqrt{G(E+m)r_L} \qquad (2.35)$$

Inserting this in Eq. (2.33) gives

$$C\omega+\frac{Em}{\sqrt{(E+m)}}\sqrt{Gr_L}=\text{constant} \qquad (2.36)$$

The first term in this equation decreases, because tidal friction reduces ω. To conserve angular momentum the second term must increase. Thus, lunar tidal braking of the Earth's rotation causes an increase in the Earth–Moon distance, r_L. At present this distance is increasing at about 3.7 cm yr^{-1}. As a further consequence Eq. (2.35) shows that the Moon's rate of revolution about the Earth (Ω_L) must decrease when r_L increases. Thus, tidal friction slows down the rates of Earth rotation and lunar revolution and increases the Earth–Moon distance.

2.3.4.3 *The Chandler wobble*

The Earth's rotation gives it the shape of a spheroid, or ellipsoid of revolution. This figure is symmetric with respect to the mean axis of rotation, about which the moment of inertia is greatest; this is also called the axis of figure (see §2.4). However, at any moment the instantaneous rotational axis is displaced by a few meters from the axis of figure. The orientation of the total angular momentum vector remains nearly constant but the axis of figure changes location with time and appears to meander around the rotation axis (Fig. 2.15).

The theory of this motion was described by Leonhard Euler (1707–1783), a Swiss mathematician. He showed that the displaced rotational axis of a rigid spheroid would execute a circular motion about its mean position, now called the *Euler nutation*. Because it occurs in the absence of an external driving torque, it is also called the *free nutation*. It is due to differences in the way mass is distributed about the axis of rotational symmetry and an axis at right angles to it in the equatorial plane. The mass distributions are represented by the moments of inertia about these axes. If C and A are the moments of inertia about the rotational axis and an axis in the equatorial plane, respectively, Euler's theory shows that the period of free nutation is $A/(C–A)$ days, or approximately 305 days.

Astronomers were unsuccessful in detecting a polar motion with this period. In 1891 an American geodesist and astronomer, S. C. Chandler, reported that the polar motion of the Earth's axis contained two important components. An annual component with amplitude about 0.10 seconds of arc is due to the transfer of mass between atmosphere and hydrosphere accompanying the changing of the seasons. A slightly larger component with amplitude 0.15 seconds of arc has a period of 435 days. This polar motion is now called the *Chandler wobble*. It corresponds to the Euler nutation in an elastic Earth. The increase in period from 305 days to 435 days is a consequence of the elastic yielding of the Earth. The superposition of the annual and Chandler frequencies results in a beat effect, in which the amplitude of the latitude variation is modulated with a period of 6–7 years (Fig. 2.15).

2.3.4.4 *Precession and nutation of the rotation axis*

During its orbital motion around the Sun the Earth's axis maintains an (almost) constant tilt, called the *obliquity*, to the pole to the ecliptic. The line of intersection of the plane of the ecliptic with the equatorial plane is called the line of equinoxes. Two times a year, when this line points directly at the Sun, day and night have equal duration over the entire globe.

In the theory of the tides the unequal lunar attractions on the near and far sided tidal bulges cause a torque about the rotation axis, which has a braking effect on the Earth's rotation. The attraction of the Moon (and Sun) on the equatorial bulge due to rotational flattening also produce torques on the spinning Earth. On the side of the Earth nearer to the Moon (or Sun) the gravitational attraction F_2 on the equatorial bulge is greater than the force F_1 on the

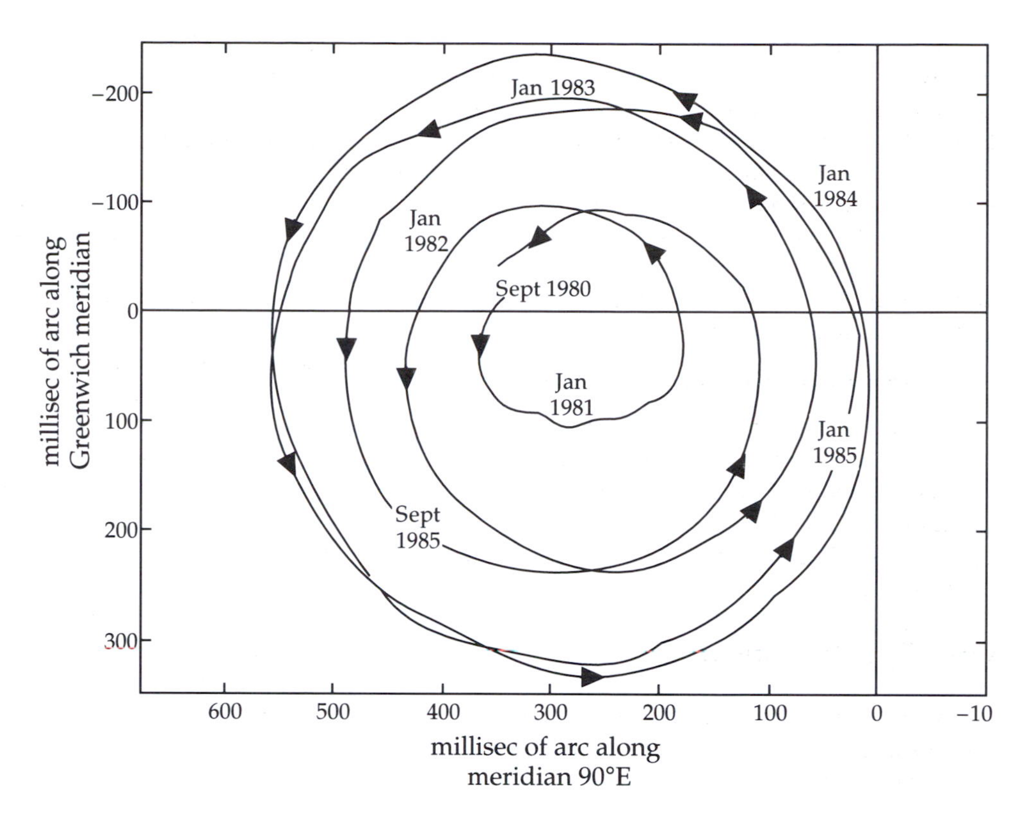

Fig. 2.15 Variation of latitude due to superposition of the 435 day Chandler wobble period and an annual seasonal component (after Carter, 1989).

distant side (Fig. 2.16a). Due to the tilt of the rotation axis to the ecliptic plane (23.5°), the forces are not collinear. A torque results which acts about a line in the equatorial plane, normal to the Earth–Sun line and normal to the spin axis. The magnitude of the torque changes as the Earth orbits around the Sun. It is minimum (and zero) at the spring and autumn equinoxes and maximum at the summer and winter solstices.

The response of a rotating system to an applied torque is to acquire an additional component of angular momentum parallel to the torque. In our example this will be perpendicular to the angular momentum (h) of the spinning Earth. The torque has a component (τ) parallel to the line of equinoxes (Fig. 2.16b) and a component normal to this line in the equatorial plane. The torque τ causes an increment Δh in angular momentum and shifts the angular momentum vector to a new position. If this exercise is repeated incrementally, the rotation axis moves around the surface of a cone whose axis is the pole to the ecliptic (Fig. 2.16a). The geographic pole P moves around a circle in the opposite sense from the Earth's spin. This motion is called *retrograde precession*. It is not a steady motion, but pulsates in sympathy with the driving torque. A change in orientation of the rotation axis affects the location of the line of equinoxes and causes the timing of the equinoxes to change slowly. The rate of change is only 50.4 arcsec per yr, but it has been recognized during centuries of observation. For example, the Earth's rotation axis now points at Polaris in the constellation Ursa Minor, but in the time of the Egyptians around 3000 B.C. the pole star was Alpha Draconis, the brightest star in the constellation Draco. Hipparchus is credited with discovering the precession of the equinoxes in 120 B.C. by comparing his own observations with those of earlier astronomers.

The theory of the phenomenon is well understood. The Moon also exerts a torque on the spinning Earth and contributes to the precession of the rotation axis (and equinoxes). As in the theory of the tides, the small size of the Moon compared to the Sun is more than compensated by its nearness, so that the precessional contribution of the Moon is about double the effect of the Sun. The theory of precession shows that the period of 25,700 yr is proportional to the Earth's *dynamical ellipticity*, $H=(C-A)/C$. This ratio (equal to 1/305.456) is an important indicator of the internal distribution of mass in the Earth.

The component of the torque in the equatorial plane adds an additional motion to the axis, called *nutation*, because it causes the axis to nod up and down (Fig. 2.16a). The solar torque causes a semi-annual nutation, the lunar torque a semi-monthly one. In fact the motion of the axis exhibits many *forced nutations*, so-called because they respond to external torques. All are tiny perturbations on

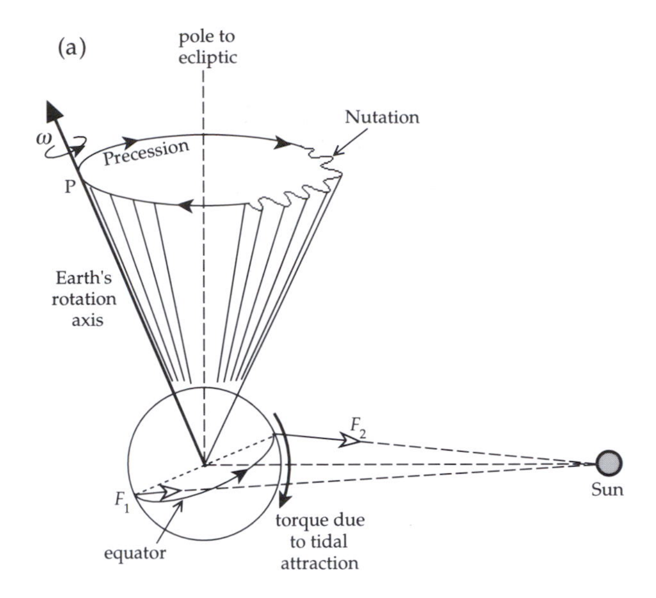

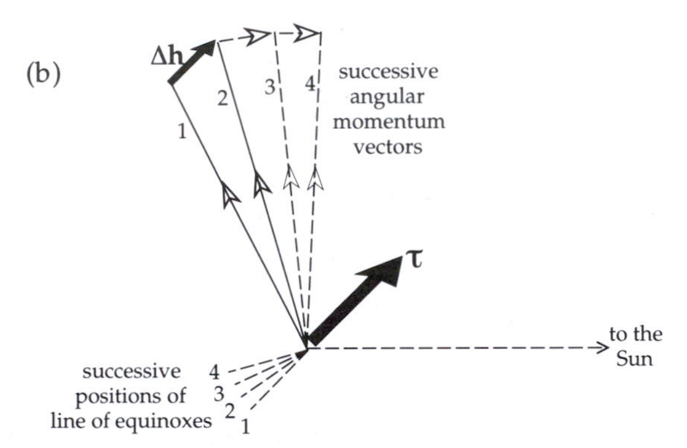

Fig. 2.16 (a) The precession and forced nutation (greatly exaggerated) of the rotation axis due to the lunar torque on the spinning Earth (after Strahler, 1963). (b) Torque and incremental angular momentum changes resulting in precession.

the precessional motion, the largest having an amplitude of only about 9 arcsec and a period of 18.6 yr. This nutation results from the fact that the plane of the lunar orbit is inclined at 5° 09′ to the plane of the ecliptic and (like the motion of artificial Earth satellites) precesses retrogradely. This causes the inclination of the lunar orbit to the equatorial plane to vary between about 18.4° and 28.6°, modulating the torque and forcing a nutation with a period of 18.6 yr.

It is important to note that the Euler nutation and Chandler wobble are polar motions about the rotation axis, but the precession and forced nutations are displacements of the rotation axis itself.

2.3.4.5 *Milankovitch climatic cycles*

The rotation of the Earth is also influenced by the gravitational attractions of the other planets, especially Jupiter. These cause effects with periods even longer than the precession. They include modulation of the angle of tilt of the rotation axis and variation in the shape and orientation of Earth's orbit.

The angle between the rotational axis and the pole to the ecliptic is called the obliquity. It is currently equal to 23° 26.5′ but varies slowly between a minimum of 21° 55′ and a maximum of 24° 18′. The obliquity determines the difference between summer and winter conditions. If the obliquity were zero, the only difference between summer and winter would be due to the elliptical shape of Earth's orbit. When the obliquity increases, the seasonal contrast in temperatures becomes more pronounced, and vice versa. The variation in obliquity causes a modulation in the seasonal contrast between summer and winter on a global scale. This effect is manifest as a cyclical change in climate with a period of about 41,000 yr.

A further effect of planetary attraction is that the eccentricity of the Earth's orbit is forced to change. At one extreme the orbit is almost circular, the closest distance from the Sun at perihelion being 99.8% of the furthest distance at aphelion. At the other extreme the orbit is more elongate, the perihelion distance being 88.6% of the aphelion distance. Solar energy can be imagined as flowing equally from the Sun in all directions; at distance r it floods a sphere with surface area $4\pi r^2$. Thus, solar insolation decreases as the inverse square of the distance. When the orbit is most circular, the insolation difference between summer and winter is negligible; when the orbit is most elliptical, the insolation in winter is only 78.3% of the summer insolation. The variation in eccentricity exhibits two cyclicities with periods around 109,000 yr and 413,000 yr, which engender matching fluctuations in paleoclimatic records.

Not only does planetary attraction force the shape of the orbit to change, it also causes the orbit to precess. Successive positions of the perihelion–aphelion axis differ slightly. The ellipse is not truly closed, and the path of the Earth describes a rosette with a period that is also 109,000 yr. Its effect is similar to the precession of the rotation axis, because it affects the times at which the seasons occur. Combining the 25,700 yr rotational precession with the 109,000 yr orbital precession gives a modulation of the seasons with a period of 20,500 yr.

Climatic effects related to cyclical changes in the Earth's rotational and orbital parameters were first studied between 1920 and 1938 by a Yugoslavian astronomer, Milutin Milankovitch. In recognition of his pioneering studies, periodicities of 20.5 kyr, 41 kyr, 109 kyr and 413 kyr detected in various sedimentary records are called the Milankovitch climatic cycles.

2.3.5 Coriolis and Eötvös accelerations

Every object on the Earth experiences the centrifugal acceleration due to the Earth's rotation. Moving objects on the rotating Earth experience additional accelerations related to the velocity at which they are moving. At latitude λ the distance d of a point on the Earth's surface from the rotational axis is equal to $R\cos\lambda$, and the rotational spin ω translates to an eastwards linear velocity v equal to $\omega R\cos\lambda$. Consider an object (e.g., a vehicle or projectile) that is moving at velocity v across the Earth's surface. In general v has a northward component v_N and an eastward component v_E. Consider first the effects related to the eastward velocity, which is added to the linear velocity of the rotation. The centrifugal acceleration increases by an amount Δa_c, which can be obtained by differentiating a_c in Eq. (2.18) with respect to ω

$$\Delta a_c = 2\omega (R\cos\lambda)\,\Delta\omega = 2\omega v_E \tag{2.37}$$

The extra centrifugal acceleration Δa_c can be resolved into a vertical component and a horizontal component (Fig. 2.17a). The vertical component, equal to $2\omega v_E \cos\lambda$, acts upward, opposite to gravity. It is called the *Eötvös acceleration*. Its effect is to *decrease* the measured gravity by a small amount. If the moving object has a westward component of velocity the Eötvös acceleration *increases* the measured gravity. If gravity measurements are made on a moving platform (for example, on a research ship or in an airplane), the measured gravity must be corrected to allow for the Eötvös effect. For a ship sailing eastward at 10 km h^{-1} at latitude 45° the Eötvös correction is 28.6 mgal; in an airplane flying eastward at 300 km h^{-1} the correction is 856 mgal. These corrections are far greater than the sizes of many important gravity anomalies. However, the Eötvös correction can be made satisfactorily in marine gravity surveys, and recent technical advances now make it feasible in aerogravimetry.

The horizontal component of the extra centrifugal acceleration due to v_E is equal to $2\omega v_E \sin\lambda$. In the northern hemisphere it acts to the south. If the object moves westward, the acceleration is northward. In each case it acts horizontally *to the right* of the direction of motion. In the southern hemisphere the sense of this acceleration is reversed; it acts to the left of the direction of motion. This acceleration is the northward component a_N of the *Coriolis acceleration*. Its eastward component a_E derives from the northward motion of the object.

Consider an object moving northward along a meridian of longitude (Fig. 2.17b, point 1). The linear velocity of a point on the Earth's surface decreases poleward, because the distance from the axis of rotation ($d = R\cos\lambda$) decreases. The

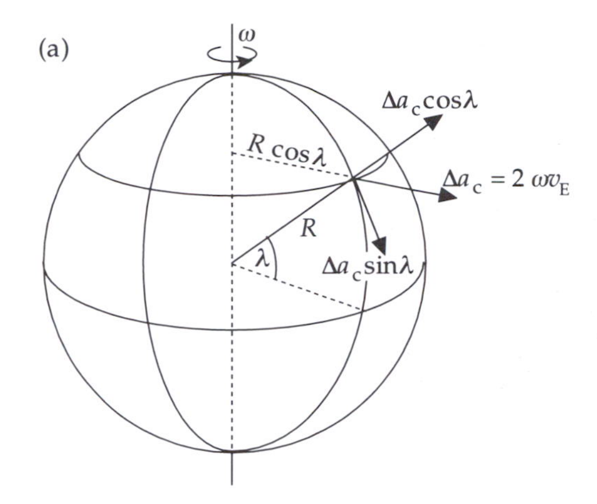

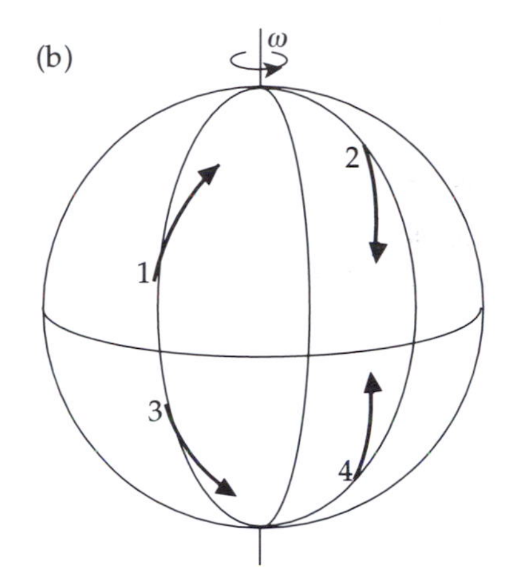

Fig. 2.17 (a) Resolution of the additional centrifugal acceleration Δa_c due to eastward velocity into vertical and horizontal components. (b) The horizontal deviations of the northward or southward trajectory of an object due to conservation of its angular momentum.

angular momentum of the moving object must be conserved, so the eastward velocity v_E must increase. As the object moves to the north its eastward velocity is faster than the circles of latitude it crosses and its trajectory deviates to the right. If the motion is to the south (Fig. 2.17b, point 2), the inverse argument applies. The body crosses circles of latitude with faster eastward velocity than its own and, in order to maintain angular momentum, its trajectory must deviate to the west. In each case the deviation is *to the right* of the direction of motion. A similar argument applied to the southern hemisphere gives a Coriolis effect *to the left* of the direction of motion (Fig. 2.17b, points 3 and 4).

The magnitude of the Coriolis acceleration is easily evaluated quantitatively. The angular momentum h of a mass m at latitude λ is equal to $m\omega R^2\cos^2\lambda$. Conservation of angular momentum gives

$$\frac{\partial h}{\partial t}=mR^2\cos^2\lambda\frac{\partial \omega}{\partial t}+m\omega R^2(-2\cos\lambda\sin\lambda)\frac{\partial \lambda}{\partial t}=0 \tag{2.38}$$

Rearranging and simplifying, we get

$$(R\cos\lambda)\frac{\partial \omega}{\partial t}=2\omega\sin\lambda\left(R\frac{\partial \lambda}{\partial t}\right) \tag{2.39}$$

The expression on the left of the equation is an acceleration, a_E, equal to the rate of change of the eastward velocity. The expression in brackets on the right is the northward velocity component v_N. We can write this component of the Coriolis acceleration as $2\omega v_N\sin\lambda$. The north and east components of the Coriolis acceleration are therefore:

$$a_N=-2\omega v_E\sin\lambda$$

$$a_E=2\omega v_N\sin\lambda \tag{2.40}$$

The Coriolis acceleration deflects the horizontal path of any object moving on the Earth's surface. It affects the directions of wind and ocean currents, eventually constraining them to form circulatory patterns about centers of high or low pressure, and thereby plays an important role in determining the weather.

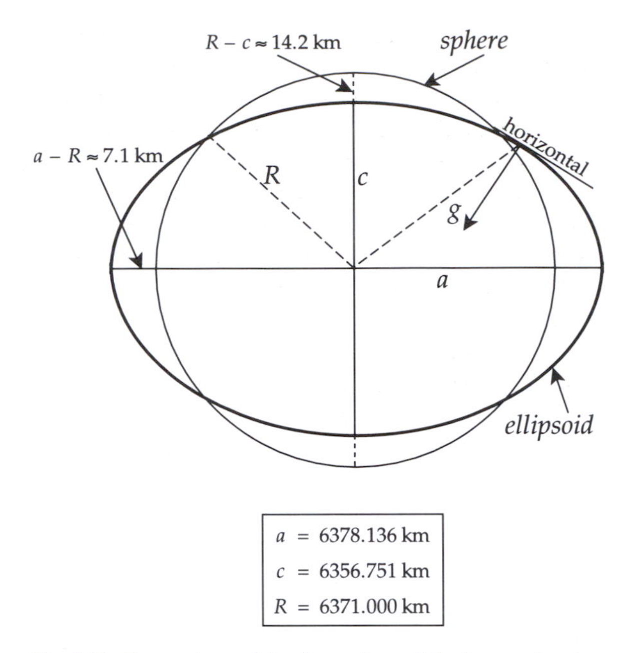

Fig. 2.18 Comparison of the dimensions of the *International Reference Ellipsoid* with a sphere of equal volume.

2.4 THE EARTH'S FIGURE AND GRAVITY

2.4.1 The figure of the Earth

The true surface of the Earth is uneven and irregular, partly land and partly water. For geophysical purposes the Earth's shape is represented by a smooth closed surface, which is called the figure of the Earth. Early concepts of the figure were governed by religion, superstition and non-scientific beliefs. The first circumnavigation of the Earth, completed in 1522 by Magellan's crew, established that the Earth was probably round. Before the era of scientific awakening the Earth's shape was believed to be a sphere. As confirmed by numerous photographs from spacecraft, this is in fact an excellent first approximation to Earth's shape that is adequate for solving many problems. The original suggestion that the Earth is a spheroid flattened at the poles is credited to Newton, who used a hydrostatic argument to account for the polar flattening. The slightly flattened shape permitted an explanation of why a clock that was precise in Paris lost time near to the equator (see §2.1).

Earth's shape and gravity are intimately associated. The figure of the Earth is the shape of an equipotential surface of gravity, in particular the one that coincides with mean sea level. The best mathematical approximation to the figure is an oblate ellipsoid, or spheroid (Fig. 2.18). The precise determination of the dimensions of the Earth (e.g., its polar and equatorial radii) is the main objective of the science of *geodesy*. It requires an exact knowledge of the Earth's gravity field, the description of which is the goal of *gravimetry*.

Modern analyses of the Earth's shape are based on precise observations of the orbits of artificial Earth satellites. These data are used to define a best-fitting oblate ellipsoid, called the *International Reference Ellipsoid*. In 1930 geodesists and geophysicists defined an optimum reference ellipsoid based on the best available data at the time. The dimensions of this figure have been subsequently refined as more exact data have become available. In 1980 geodesists and geophysicists agreed on a reference ellipsoid whose equatorial radius (a) is 6378.136 km and polar radius (c) is 6356.751 km. The radius of the equivalent sphere (R) is found from $R=(a^2c)^{1/3}$ to be 6371.000 km. Compared to the best-fitting sphere the spheroid is flattened by about 14.2 km at each pole and the equator bulges by about 7.1 km. The polar flattening f is defined as the ratio

$$f=\frac{a-c}{a} \tag{2.41}$$

The flattening of the optimum reference ellipsoid defined in 1930 was exactly 1/297. This ellipsoid, and the variation of gravity on its surface, served as the basis of gravimetric surveying for many years, until the era of satellite geodesy and highly sensitive gravimeters showed it to be too inexact. In 1980 the value of the flattening was revised to $f=3.352\,81\times10^{-3}$ (i.e., $f=1/298.257$).

If the Earth is assumed to be a rotating fluid in perfect hydrostatic equilibrium (as assumed by Newton's theory), the flattening should be 1/299.5, slightly smaller than the observed value. The hydrostatic condition assumes that the Earth has no internal strength. A possible explanation for the tiny discrepancy in f is that the Earth has sufficient strength to maintain a non-hydrostatic figure, and the present figure is inherited from a time of more rapid rotation. Alternatively, the slightly more flattened form of the Earth may be due to internal density contrasts, which could be the consequence of slow convection in the Earth's mantle. This would take place over long time intervals and could result in a non-hydrostatic mass distribution.

The cause of the polar flattening is the deforming effect of the centrifugal acceleration. This is maximum at the equator where the gravitational acceleration is smallest. The parameter m is defined as the ratio of the equatorial centrifugal acceleration to the equatorial gravity:

$$m=\frac{\omega^2 a}{g_e} \tag{2.42}$$

The currently accepted value is $m=3.467\ 75\times10^{-3}$ (i.e., $m=1/288.37$).

As a result of the flattening the distribution of mass within the Earth is not simply dependent on radius. The moments of inertia of the Earth about the rotation axis (C) and any axis in the equatorial plane (A) are unequal. As noted in the previous chapter the inequality affects the way the Earth responds to external gravitational torques and is a determining factor in perturbations of the Earth's rotation. The principal moments of inertia define the *dynamical ellipticity*:

$$H=\frac{C-A}{C} \tag{2.43}$$

The dynamical ellipticity is obtained from precise observations of the orbits of artificial satellites of the Earth (see §2.4.5.1). The currently accepted value is $3.273\ 79\times10^{-3}$ (i.e., $H=1/305.456$).

2.4.2 Gravitational potential of the spheroidal Earth

The ellipsoidal shape changes the gravitational potential of the Earth from that of an undeformed sphere. In 1849 J. MacCullagh developed the following formula for the gravitational potential of any body at large distances from its center of mass:

$$U_G=-G\frac{E}{r}-G\frac{(A+B+C-3I)}{2r^3}-\ldots \tag{2.44}$$

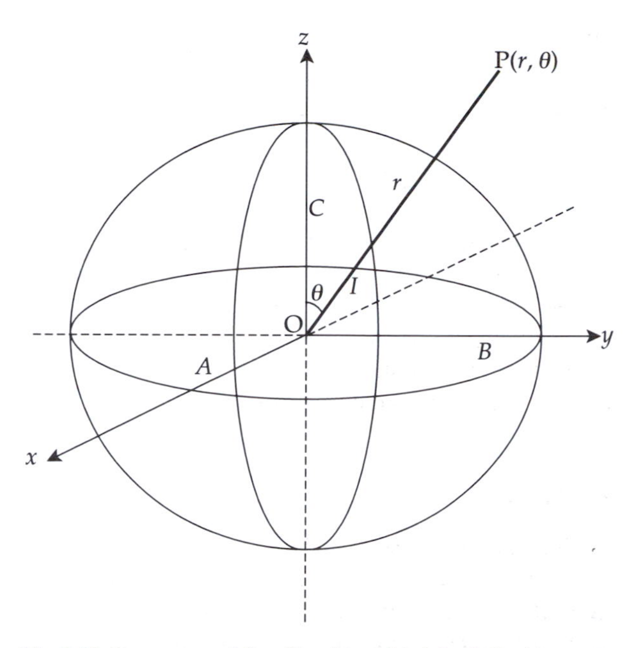

Fig. 2.19 Parameters of the ellipsoid used in MacCullagh's formula. A, B, and C are moments of inertia about the x-, y- and z-axes, respectively, and I is the moment of inertia about the line OP.

The first term, of order r^{-1}, is the gravitational potential of a point mass or sphere with mass E (Eqs. (2.10) and (2.14)); for the Earth it describes the potential of the undeformed globe. If the reference axes are centered on the body's center of mass, there is no term in r^{-2}. The second term, of order r^{-3}, is due to deviations from the spherical shape. For the flattened Earth it results from the mass displacements due to the rotational deformation. The parameters A, B, and C are the principal moments of inertia of the body and I is the moment of inertia about the line OP joining the center of mass to the point of observation (Fig. 2.19). In order to express the potential accurately an infinite number of terms of higher order in r are needed. In the case of the Earth these can be neglected, because the next term is about 1000 times smaller than the second term.

For a body with planes of symmetry, I is a simple combination of the principal moments of inertia. Setting A equal to B for rotational symmetry, and defining the angle between OP and the rotation axis to be θ, the expression for I is

$$I=A\sin^2\theta+C\cos^2\theta \tag{2.45}$$

MacCullagh's formula for the ellipsoidal Earth then becomes

$$U_G=-G\frac{E}{r}+G\frac{(C-A)}{r^3}\frac{(3\cos^2\theta-1)}{2} \tag{2.46}$$

The function $(3\cos^2\theta - 1)/2$ is a second-order polynomial in $\cos\theta$. It belongs to a family of functions called Legendre polynomials. The general member corresponds to a polynomial of degree n in $\cos\theta$, and is written in shorthand notation as $P_n(\cos\theta)$. The first few Legendre polynomials are

$$P_0(\cos\theta) = 1$$
$$P_1(\cos\theta) = \cos\theta$$
$$P_2(\cos\theta) = \tfrac{1}{2}(3\cos^2\theta - 1)$$
$$P_3(\cos\theta) = \tfrac{1}{2}(5\cos^3\theta - 3\cos\theta) \tag{2.47}$$

Using this notation MacCullagh's formula for the gravitational potential of the oblate ellipsoid becomes

$$U_G = -G\frac{E}{r} - G\frac{(C-A)}{r^3}P_2(\cos\theta) \tag{2.48}$$

This can be written in the alternative form

$$U_G = -G\frac{E}{r}\left[1 - \left(\frac{C-A}{ER^2}\right)\left(\frac{R}{r}\right)^2 P_2(\cos\theta)\right] \tag{2.49}$$

Gravitational acceleration is directed toward a center of mass. A French astronomer and mathematician Pierre Simon, marquis de Laplace (1749–1827), showed that, in order to fulfil this basic physical condition, the gravitational potential must satisfy a second-order differential equation. In Cartesian coordinates (x, y, z) the Laplace equation is

$$\frac{\partial^2 U_G}{\partial x^2} + \frac{\partial^2 U_G}{\partial y^2} + \frac{\partial^2 U_G}{\partial z^2} = 0 \tag{2.50}$$

In spherical polar coordinates (r, θ, ϕ) the equation becomes

$$\frac{1}{r^2}\frac{\partial}{\partial r}r^2\frac{\partial U_G}{\partial r} + \frac{1}{r^2\sin\theta}\frac{\partial}{\partial\theta}\sin\theta\frac{\partial U_G}{\partial\theta} + \frac{1}{r^2\sin^2\theta}\frac{\partial^2 U_G}{\partial\phi^2} = 0 \tag{2.51}$$

The variation with azimuth ϕ disappears for the axially symmetric Earth. The gravitational potential of the spheroidal Earth is the sum of an infinite number of terms of increasing order in r, each involving an appropriate Legendre polynomial:

$$U_G = -G\frac{E}{r}\left[1 - \sum_{n=2}^{n=\infty}\left(\frac{R}{r}\right)^n J_n P_n(\cos\theta)\right] \tag{2.52}$$

In this equation the coefficients J_n multiplying $P_n(\cos\theta)$ determine the relative importance of the term of nth order. The values of J_n are obtained from satellite geodesy: $J_2 = 1082.6\times10^{-6}$; $J_3 = -2.54\times10^{-6}$; $J_4 = -1.59\times10^{-6}$; higher orders are insignificant. The most important coefficient is the second-order one, which describes the effect of the polar flattening on the Earth's gravitational potential. Comparison of terms in Eqs. (2.48) and (2.51) gives the result

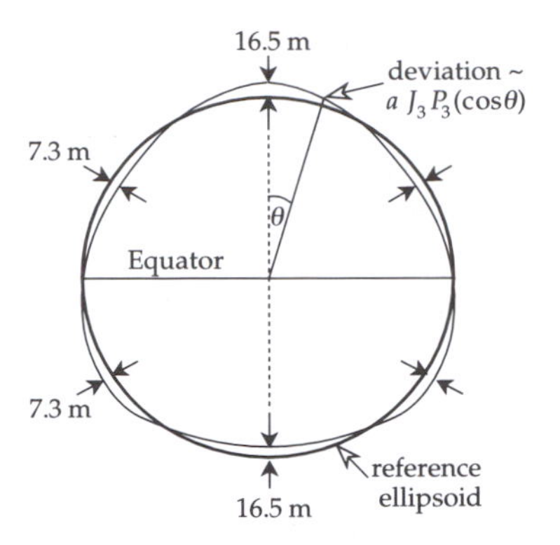

Fig. 2.20 The third-order term in the gravitational potential describes a pear-shaped Earth. The deviations from the reference ellipsoid are of the order of 10–20 m, much smaller than the deviations of the ellipsoid from a sphere, which are of the order of 10–20 km.

$$J_2 = \frac{C-A}{ER^2} \tag{2.53}$$

The term of next higher order ($n=3$) in Eq. (2.52) describes the deviations from the ellipsoid which correspond to a pear-shaped Earth (Fig. 2.20). The deviations from the reference ellipsoid are of the order of 7–17 m, much smaller than the deviations of the ellipsoid from a sphere, which are of the order of 7–14 km.

2.4.3 Gravity and its potential

The potential of gravity (U_g) is the sum of the gravitational and centrifugal potentials. It is often called the *geopotential*. At a point on the surface of the rotating spheroid it can be written

$$U_g = U_G - \tfrac{1}{2}\omega^2 r^2 \sin^2\theta \tag{2.54}$$

If the free surface is an equipotential surface of gravity, then U_g is everywhere constant on it. The shape of the equipotential surface is constrained to be that of the spheroid with flattening f. Under these conditions a simple relation is found between the constants f, m and J_2:

$$J_2 = \tfrac{1}{3}(2f - m) \tag{2.55}$$

By equating Eq. (2.53) and (2.55) and re-ordering terms slightly we obtain the following relationship

$$\frac{C-A}{C}\frac{C}{ER^2} = \tfrac{1}{3}(2f - m) \tag{2.56}$$

This yields useful information about the variation of density within the Earth. The quantities f, m and $(C-A)/C$ are each equal to approximately 1/300. Inserting their values in the equation gives $C \approx 0.33\ ER^2$. Compare this value with the principal moments of inertia of a hollow spherical shell ($0.66\ ER^2$) and a solid sphere with uniform density ($0.4\ ER^2$). The concentration of mass near the center causes a reduction in the multiplying factor from 0.66 to 0.4. The value of 0.33 for the Earth implies that, in comparison with a uniform solid sphere, the density must increase towards the center of the Earth.

2.4.4 Normal gravity

The direction of gravity at a point is defined as perpendicular to the equipotential surface through the point. This defines the *vertical* at the point, while the plane tangential to the equipotential surface defines the *horizontal* (Fig. 2.18). A consequence of the spheroidal shape of the Earth is that the vertical direction is generally not radial, except on the equator and at the poles.

On a spherical Earth there is no ambiguity in how we define latitude. It is the angle at the center of the Earth between the radius and the equator, the complement to the polar angle θ. This defines the geocentric latitude λ'. However, the geographic latitude λ in common use is not defined in this way. It is found by geodetic measurement of the angle of elevation of a fixed star above the horizon. But the horizontal plane is tangential to the ellipsoid, not to a sphere (Fig. 2.18), and the vertical direction (i.e., the local direction of gravity) intersects the equator at an angle λ that is slightly larger than the geocentric latitude λ' (Fig. 2.21). The difference $(\lambda - \lambda')$ is zero at the equator and poles and reaches a maximum at a latitude of 45°, where it amounts to only 0.19° (about 12⁻).

The *International Reference Ellipsoid* is the standardized reference figure of the Earth. The theoretical value of gravity on this rotating ellipsoid can be computed by differentiating the gravity potential (Eq. (2.48)). This yields the radial and transverse components of gravity, which are then combined to give the following formula for gravity normal to the ellipsoid:

$$g_n = g_e(1 + \beta_1 \sin^2\lambda + \beta_2 \sin^2 2\lambda) \qquad (2.57)$$

where

$$g_e = \frac{GE}{a^2}\left(1 + f - \frac{3}{2}m + f^2 - \frac{27}{14}fm\right) \quad = 9.780\,318\ \mathrm{ms}^{-2}$$

$$\beta_1 = \frac{5}{2}m - f + \frac{15}{4}m^2 - \frac{17}{14}fm \quad = 5.3024 \times 10^{-3}$$

$$\beta_2 = \frac{1}{8}f^2 - \frac{5}{8}fm \quad = -5.87 \times 10^{-6} \qquad (2.58)$$

Eq. (2.57) is known as the normal gravity formula. It is very important in the analysis of gravity measurements on the Earth, because it gives the theoretical variation of normal gravity (g_n) with latitude on the surface of the reference ellipsoid. The normal gravity is expressed in terms of g_e, the value of gravity on the equator. The second-order terms f^2, m^2 and fm are about 300 times smaller than the first-order terms f and m. The constant β_2 is about 1000 times smaller than β_1. If we drop second-order terms and use $\lambda = 90°$, the value of normal gravity at the pole is $g_p = g_e(1 + \beta_1)$, so by rearranging and retaining only first-order terms, we get

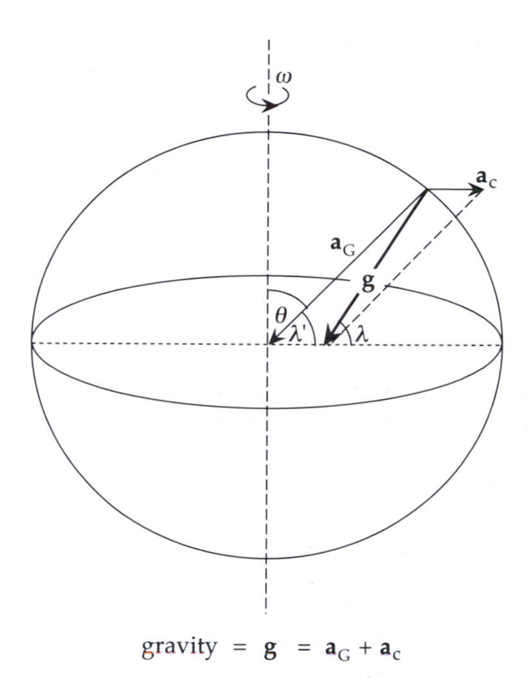

Fig. 2.21 Gravity on the ellipsoidal Earth is the vector sum of the gravitational and centrifugal accelerations and is not radial; consequently, geographic latitude (λ) is slightly larger than geocentric latitude (λ').

$$\frac{g_p - g_e}{g_e} = \frac{5}{2}m - f \qquad (2.59)$$

This expression is called Clairaut's theorem. It was developed in 1743 by a French mathematician, Alexis-Claude Clairaut, who was the first to relate the variation of gravity on the rotating Earth with the flattening of the spheroid. The normal gravity formula gives $g_p = 9.832\,177\ \mathrm{m\ s^{-2}}$. Numerically, this gives an increase in gravity from equator to pole of approximately $5.186 \times 10^{-2}\ \mathrm{m\ s^{-2}}$, or 5186 mgal.

There are two obvious reasons for the poleward increase of gravity. The distance to the center of mass of the Earth is shorter at the the poles than at the equator. This gives a stronger gravitational acceleration (a_G) at the poles. The difference is

$$\Delta a_G = \left(\frac{GE}{c^2} - \frac{GE}{a^2}\right) \tag{2.60}$$

This gives an excess gravity of approximately 6600 mgal at the poles. The effect of the centrifugal force in diminishing gravity is largest at the equator, where it equals (ma_G), and is zero at the poles. This also results in a poleward increase of gravity, amounting to about 3375 mgal. These figures indicate that gravity should increase by a total of 9975 mgal from equator to pole, instead of the observed difference of 5186 mgal. The discrepancy can be resolved by taking into account a third factor. The computation of the difference in gravitational attraction is not so simple as indicated by Eq. (2.61). The equatorial bulge places an excess of mass under the equator, increasing the equatorial gravitational attraction and thereby reducing the gravity decrease from equator to pole.

2.4.5 The geoid

The international reference ellipsoid is a close approximation to the equipotential surface of gravity, but it is really a mathematical convenience. The physical equipotential surface of gravity is called the *geoid*. It reflects the true distribution of mass inside the Earth and differs from the theoretical ellipsoid by small amounts. Far from land the geoid agrees with the free ocean surface, excluding the temporary perturbing effects of tides and winds. Over the continents the geoid is affected by the mass of land above mean sea level (Fig. 2.22a). The mass within the ellipsoid causes a downward gravitational attraction toward the center of the Earth, but a hill or mountain whose center of gravity is outside the ellipsoid causes an upward attraction. This causes a local elevation of the geoid above the ellipsoid. The displacement between the geoid and the ellipsoid is called a *geoid undulation*; the elevation caused by the mass above the ellipsoid is a positive undulation.

2.4.5.1 *Geoid undulations*

In computing the theoretical figure of the Earth the distribution of mass beneath the ellipsoid is assumed to be homogeneous. A local excess of mass under the ellipsoid will deflect and strengthen gravity locally. The potential of the ellipsoid is achieved further from the center of the Earth. The equipotential surface is forced to warp upward while remaining normal to gravity. This gives a positive geoid undulation over a mass excess under the ellipsoid (Fig. 2.22b). Conversely, a mass deficit beneath the ellipsoid will deflect the geoid below the ellipsoid, causing a negative geoid undulation. As a result of the uneven topography and heterogeneous internal mass distribution of the Earth, the geoid is a bumpy equipotential surface.

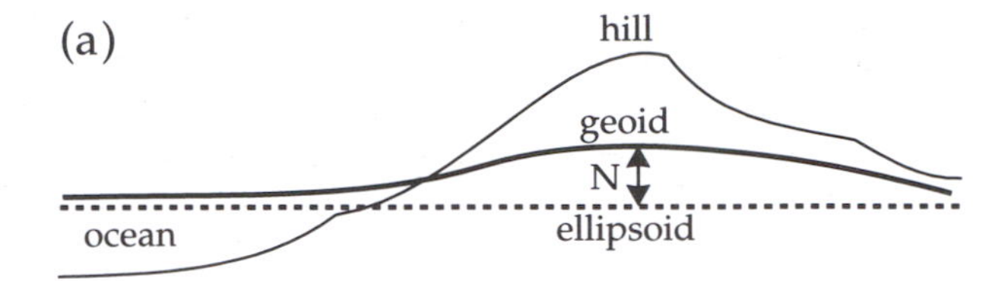

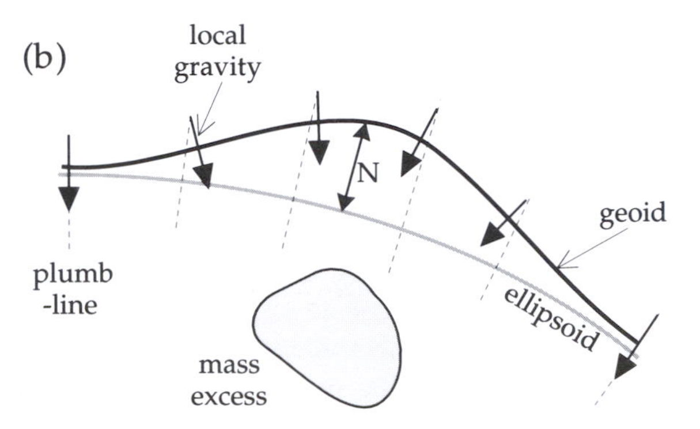

Fig. 2.22 (a) A mass outside the ellipsoid or (b) a mass excess below the ellipsoid elevates the geoid above the ellipsoid. N is the geoid undulation.

The potential on the geoid is represented mathematically by functions that are somewhat more complicated than the Legendre polynomials (Eq. (2.48)) used to describe the gravitational potential of the ellipsoid. In spherical coordinates the potential of the geoid is written

$$U = -G\frac{E}{r}\sum_{n=0}^{n=\infty}\left(\frac{R}{r}\right)^n \sum_{m=0}^{m=n}(C_{nm}\cos m\phi + S_{nm}\sin m\phi)P_{nm}(\cos\theta) \tag{2.61}$$

The combinations of functions in θ and ϕ are *spherical harmonic functions*, in which $P_{nm}(\cos\theta)$ are called *associated Legendre polynomials*. The spherical harmonics express the variation of potential with longitude as well as colatitude. An adequate treatment of these functions is beyond the scope of this text. The representation of the geopotential as a spherical harmonic expansion is analogous to the simpler representation of the gravitational potential of the rotationally symmetric Earth (Eq. (2.52)) by an infinite sum of terms of increasing order in $(1/r)$.

In modern analyses the coefficient of each term in the geopotential (like the coefficients J_n in Eq. (2.52)) can be calculated up to a high harmonic degree. The terms up to a selected degree are then used to compute a model of the

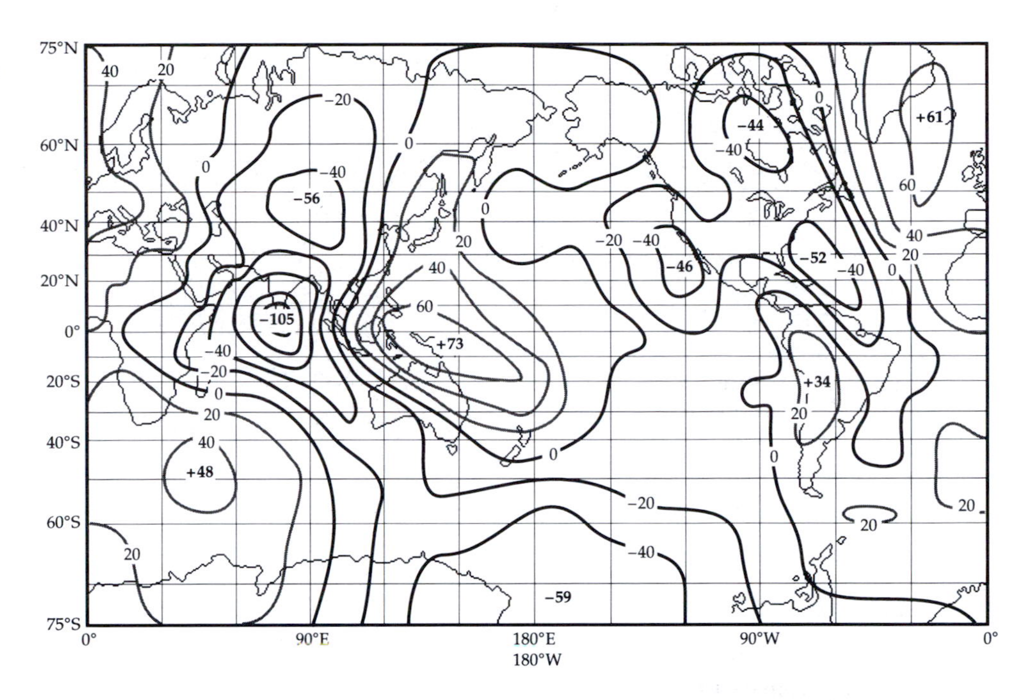

Fig. 2.23 World map of geoid undulations relative to a reference ellipsoid of flattening $f = 1/298.257$ (after Lerch *et al.*, 1979).

geoid and the Earth's gravity field. A combination of satellite data and surface gravity measurements was used to construct the Goddard Earth Model (GEM) 10. A global comparison between the reference ellipsoid with flattening 1/298.257 and the geoid surface computed from the GEM 10 model shows long-wavelength geoid undulations (Fig. 2.23). The largest negative undulation (−105 m) is in the Indian Ocean south of India, and the largest positive undulation (+73 m) is in the equatorial Pacific Ocean north of Australia. These large-scale features are too broad to be ascribed to shallow crustal or lithospheric mass anomalies. They are thought to be due to heterogeneities that extend deep into the lower mantle, but their origin is not yet understood.

2.4.6 Satellite geodesy

Since the early 1960s knowledge of the geoid has been dramatically enhanced by the science of satellite geodesy. The motions of artificial satellites in Earth orbits are influenced by the Earth's mass distribution. The most important interaction is the simple balance between the centrifugal force and the gravitational attraction of the Earth's mass, which determines the radius of the satellite's orbit. Analysis of the precession of the Earth's rotation axis (§2.3.4.4) shows that it is determined by the dynamical ellipticity H, which depends on the difference between the principal moments of inertia resulting from the rotational flattening. In principle, the gravitational attraction of an artificial satellite on the Earth's equatorial bulge also contributes to the precession, but the effect is too tiny to be measurable. However, the inverse attraction of the equatorial bulge on the satellite causes the orbit of the satellite to precess around the rotation axis. The plane of the orbit intersects the equatorial plane in the *line of nodes*. Let this be represented by the line CN_1 in Fig. 2.24. On the next passage of the satellite around the Earth the precession of the orbit has moved the nodal line to a new position CN_2. The orbital precession in this case is *retrograde*; the nodal line *regresses*. For a satellite orbiting in the same sense as the Earth's rotation the longitude of the nodal line shifts gradually westward; if the orbital sense is opposite to the Earth's rotation the longitude of the nodal line shifts gradually eastward. Because of the precession of its orbit the path of a satellite eventually covers the entire Earth between the north and south circles of latitude defined by the inclination of the orbit. The profusion of high-quality satellite data is the best source for calculating the dynamical ellipticity H or the related parameter J_2 in the gravity potential. Observations of satellite orbits are so precise that small perturbations of the orbit can be related to the gravitational field and to the geoid.

2.4.6.1 *Satellite laser-ranging*

The accurate tracking of a satellite orbit is achieved by satellite laser-ranging (SLR). The spherical surface of the target satellite is covered with numerous retro-reflectors. A retro-reflector consists of three orthogonal mirrors that form the corner of a cube; it reflects an incident beam of light back

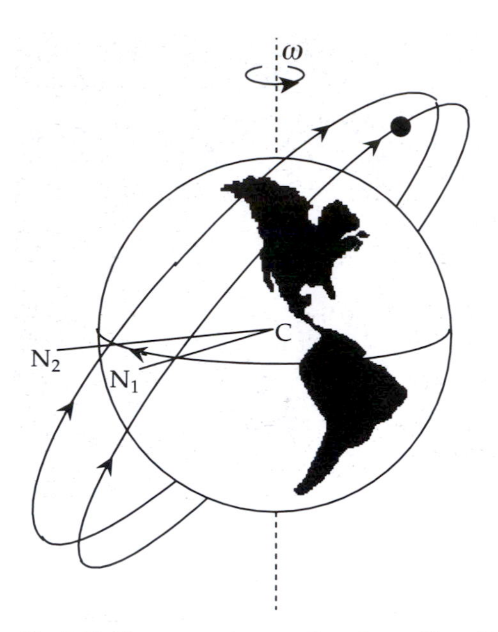

Fig. 2.24 The retrograde precession of a satellite orbit causes the line of nodes (CN_1, CN_2) to change position on successive equatorial crossings.

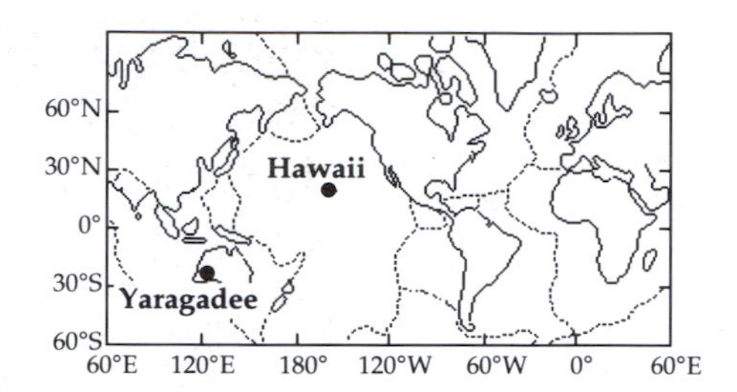

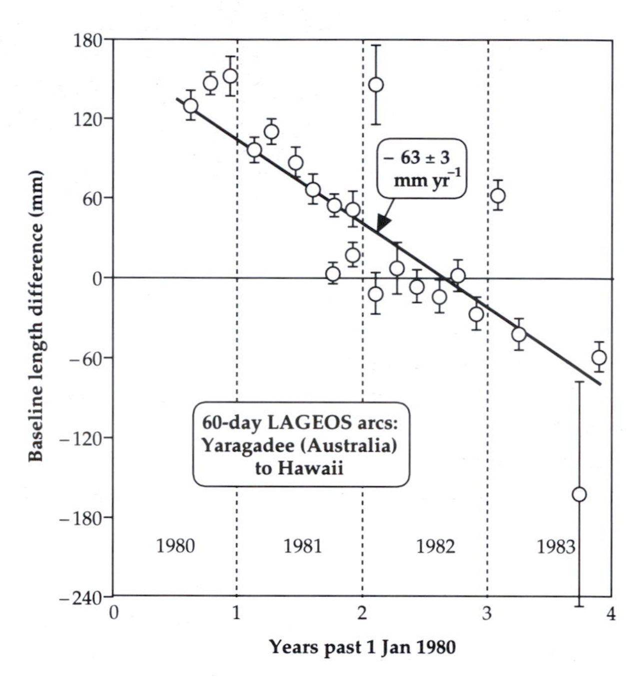

Fig. 2.25 Changes in the arc distance between satellite laser-ranging (SLR) stations in Australia and Hawaii determined from LAGEOS observations over a period of four years. The mean rate of convergence, 63±3 mm yr^{-1}, agrees well with the rate of 67 mm yr^{-1} deduced from plate tectonics (after Tapley *et al.*, 1985).

along its path. A brief pulse of laser light with a wavelength of 532 nm is sent from the tracking station on Earth to the satellite, and the two-way travel-time of the reflected pulse is measured. Knowing the speed of light, the distance of the satellite from the tracking station is obtained. The accuracy of a single range measurement is about 1 cm.

America's *Laser Geodynamics Satellite* (LAGEOS) and France's *Starlette* satellite have been tracked for many years. LAGEOS flies at 5858–5958 km altitude, the inclination of its orbit is 110° (i.e., its orbital sense is opposite to the Earth's rotation), and the nodal line of the orbit advances at 0.343° per day. *Starlette* flies at an altitude of 806–1108 km, its orbit is inclined at 50°, and its nodal line regresses at 3.95° per day.

The track of a satellite is perturbed by many factors, including the Earth's gravity field, solar and lunar tidal effects, and atmospheric drag. The perturbing influences of these factors can be computed and allowed for. For the very high accuracy that has now been achieved in SLR results variations in the coordinates of the tracking stations become detectable. The motion of the pole of rotation of the Earth can be deduced and the history of changes in position of the tracking station can be obtained. LAGEOS was launched in 1976 and has been tracked by more than twenty laser-tracking stations on five tectonic plates. The relative changes in position between pairs of stations can be compared with the rates of plate tectonic motion deduced from marine geophysical data. For example, a profile from the Yaragadee tracking station in Australia and the tracking station in Hawaii crosses the converging plate boundary between the Indo-Australian and Pacific plates (Fig. 2.25). The results of four years of measurement show a decrease of the arc distance between the two stations at a rate of 63 ± 3 mm yr^{-1}. This is in good agreement with the corresponding rate of 67 mm yr^{-1} inferred from the relative rotation of the tectonic plates.

2.4.6.2 *Satellite altimetry*

From satellite laser-ranging measurements the altitude of a spacecraft can be determined relative to the reference ellipsoid with a precision in the cm range. In satellite altimetry the tracked satellite carries a transmitter and receiver of microwave (radar) signals. A brief electromagnetic pulse is emitted from the spacecraft and reflected from the surface

Fig. 2.26 The mean sea surface as determined from SEASAT and GEOS-3 satellite altimetry, after removal of long-wavelength features of the GEM-10B geoid up to degree and order 12 (from Marsh *et al.*, 1992). The surface is portrayed as though illuminated from the northwest.

of the Earth. The two-way travel-time is converted using the speed of light to an estimate of the height of the satellite above the Earth's surface. The difference between the satellite's height above the ellipsoid and above the Earth's surface gives the height of the topography relative to the reference ellipsoid. The precision over land areas is poorer than over the oceans, but over smooth land features like deserts and inland water bodies an accuracy of better than a meter is achievable.

Satellite altimeters are best suited for marine surveys, where sub-meter accuracy is possible. The satellite GEOS-3 flew from 1975–1978, SEASAT was launched in 1978, and GEOSAT was launched in 1985. Specifically designed for marine geophysical studies, these satellite altimeters have revealed remarkable aspects of the marine geoid. The long-wavelength geoid undulations (Fig. 2.23) have large amplitudes up to several tens of meters and are maintained by mantle-wide convection. The short-wavelength features are accentuated by removing the computed geoid elevation up to a known degree and order. The data are presented in a way that emphasizes the elevated and depressed areas of the sea surface (Fig. 2.26).

There is a strong correlation between the short-wavelength anomalies in elevation of the mean sea surface and features of the sea-floor topography. Over the ocean ridge systems and seamount chains the mean sea surface (geoid) is raised. The locations of fracture zones, in which one side is elevated relative to the other, are clearly discernible. Very dark areas mark the locations of deep ocean trenches, because the mass deficiency in a trench depresses the geoid. Seaward of the deep ocean trenches the mean sea surface is raised as a result of the upward flexure of the lithosphere before it plunges downward in a subduction zone.

2.4.6.3 *Satellite determination of geodetic position*

Geodesy, the science of determining the three-dimensional coordinates of a position on the surface of the Earth, received an important boost with the advent of the satellite era. The Navy Navigation Satellite System TRANSIT consists of five or six satellites in polar orbits about 1100 km above the surface of the Earth. Signals transmitted from these satellites are combined in a receiver on Earth with a signal generated at the same frequency in the receiver. Because of the motion of the satellite, the frequency of its signal is modified by the Doppler effect and is slightly different from the receiver-generated signal, producing a beat frequency. Using the speed of light, the beat signal is converted to the oblique distance between the satellite and receiver. By integrating the beat signal over a chosen time interval the change in range to the satellite in the interval is obtained. This is repeated several times. The orbit of the satellite is known precisely, and so the position of the receiver may be calculated. This method is useful for navigation purposes, especially for fixing the position of a ship at sea.

The Navigation Satellite Timing and Ranging Global Positioning System (NAVSTAR GPS, or just GPS) utilizes satellites in much higher orbits, at around 20,000 km altitude, with an orbital period of half a sidereal day. The GPS system consists of 18 satellites. There are three satellites in each of six orbits, so that at least four are visible at any time and location on Earth. Each satellite broadcasts its own predetermined position and reference signal every six seconds. The time difference between emission and reception on Earth gives the 'pseudo-range' of the satellite, so-called because it must be corrected for errors in the clock of the receiver and for tropospheric refraction. Pseudo-range measurements to four satellites with known positions allows computation of the clock error and the exact position of the

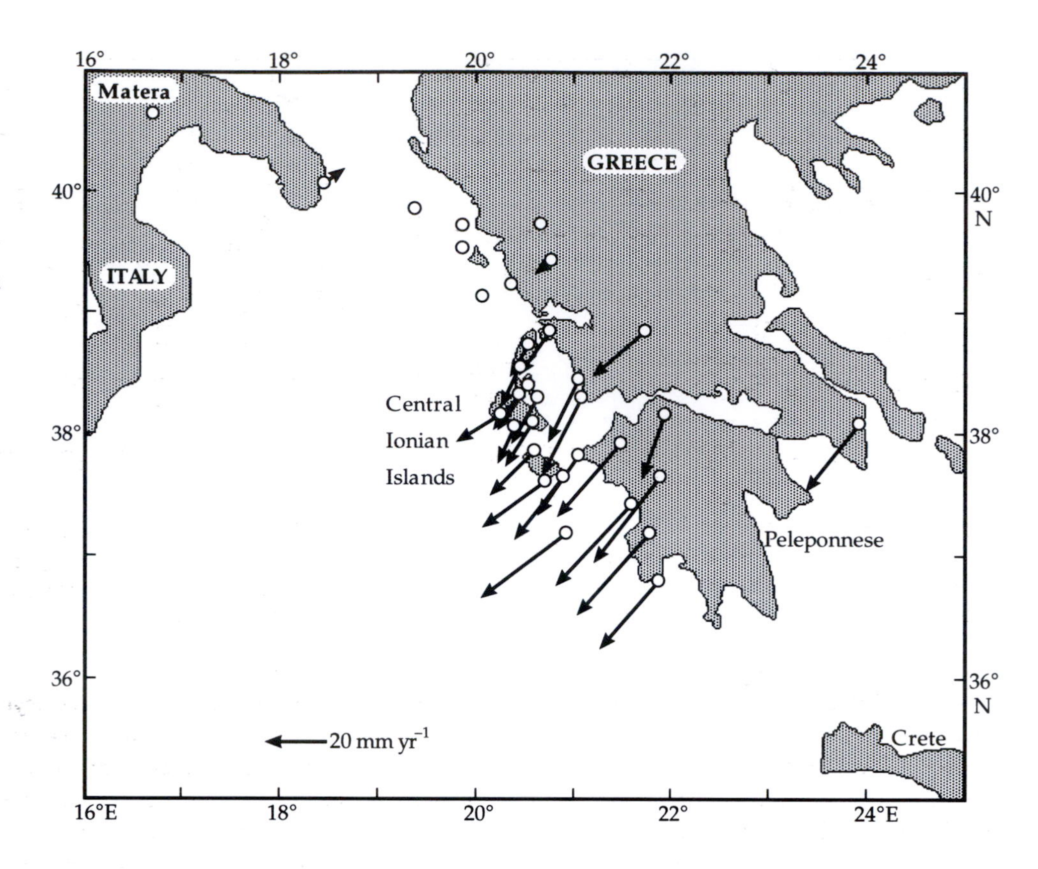

Fig. 2.27 Annual displacement rates in southeastern Italy, the Ionian Islands and western Greece relative to Matera (Italy), determined from GPS surveys in 1989 and 1993. The displacement arrows are much larger than the measurement errors, and indicate a distinct southwestward movement of western Greece relative to Italy (after Kahle *et al.*, 1995).

receiver. The precision is around 0.1–1 m, depending on the quality of the orbit and the length of the observation time.

The GPS system allows very precise determination of changes in the distance between observation points. For example, a dense network of GPS measurements was made in southeastern Italy, the Ionian Islands and western Greece in 1989 and 1993. The differences between the two measuring campaigns show that southwestern Greece moved systematically to the southwest relative to Matera in southeastern Italy at mean annual rates of 20–40 mm yr^{-1} (Fig. 2.27).

2.4.6.4 *Very Long Baseline Interferometry*

Extra-galactic radio sources (quasars) form the most stable inertial coordinate system yet known for geodetic measurements. The extra-galactic radio signals are detected almost simultaneously by radio-astronomy antennas at observatories on different continents. Knowing the direction of the incoming signal, the small differences in times of arrival of the signal wavefronts at the various stations are processed to give the lengths of the baselines between pairs of stations. This highly precise geodetic technique, called Very Long Baseline Interferometry (VLBI), allows determination of the separation of observatories several thousand kilometers apart with an accuracy of a few centimeters.

By combining VLBI observations from different stations the orientation of the Earth to the extra-galactic inertial coordinate system of the radio sources is obtained. Repeated determinations yield a record of the Earth's orientation and rotational rate with unprecedented accuracy. Motion of the rotation axis (e.g., the Chandler wobble, §2.3.4.3) can be described optically with a resolution of 0.5–1 m; the VLBI data have an accuracy of 3–5 cm. The period of angular rotation can be determined to better than 0.1 millisecond. This has enabled very accurate observation of irregularities in the rotational rate of the Earth, which are manifest as changes in the length of the day (LOD).

The most important, first-order changes in the LOD are due to the braking of the Earth's rotation by the lunar and solar marine tides (§2.3.4.1). The most significant non-tidal LOD variations are associated with changes in the angular momentum of the atmosphere due to shifts in the east–west component of the wind patterns. To conserve the total angular momentum of the Earth a change in the angular momentum of the atmosphere must be compensated by an equal and opposite change in the angular momentum of the crust and mantle. The largely seasonal transfers of angular momentum correlate well with high-frequency variations in the LOD obtained from VLBI results (Fig. 2.28).

If the effects of marine tidal braking and non-tidal

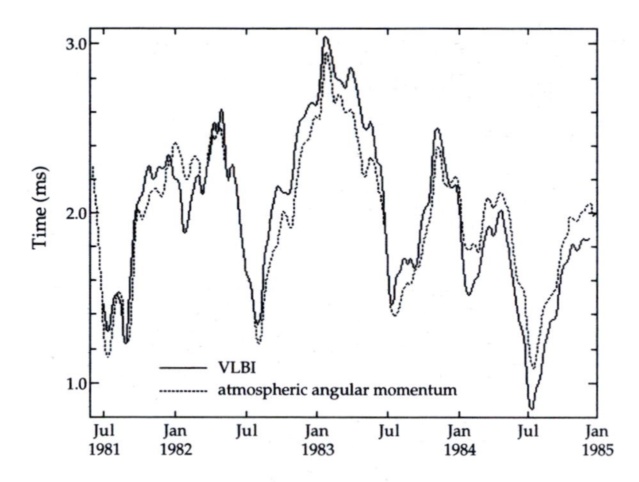

Fig. 2.28 Fine-scale fluctuations in the LOD observed by VLBI, and LOD variations expected from changes in the angular momentum of the atmosphere (after Carter, 1989).

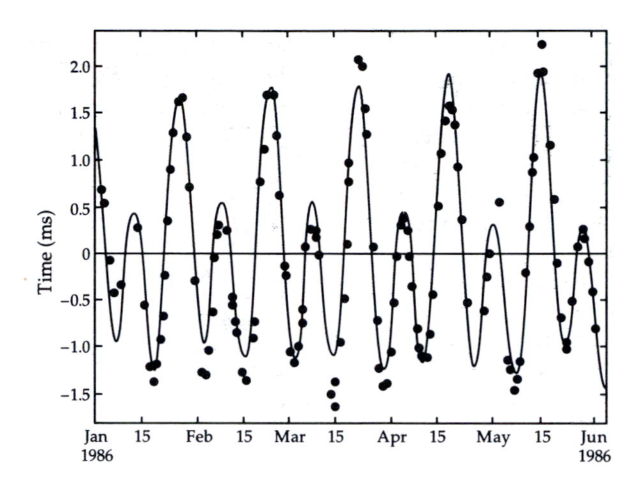

Fig. 2.29 High-frequency changes in the LOD after correction for the effects due to atmospheric angular momentum (points) and the theoretical variations expected from the solid bodily Earth-tides (after Carter, 1989).

transfers of atmospheric angular momentum variations are taken into account, small residual deviations in the LOD remain. These are related to the tides in the solid Earth (§2.3.3.4). The lunar and solar tidal forces deform the Earth elastically and change its ellipticity slightly. The readjustment of the mass distribution necessitates a corresponding change in the Earth's rate of rotation in order to conserve angular momentum. The expected changes in LOD due to the influence of tides in the solid Earth can be computed. The discrepancies in LOD values determined from VLBI results agree well with the fluctuations predicted by the theory of the bodily Earth-tides (Fig. 2.29).

2.5 GRAVITY ANOMALIES

2.5.1 Introduction

The mean value of gravity at the surface of the Earth is approximately 9.80 m s^{-2}, or 980,000 mgal. The Earth's rotation and flattening cause gravity to increase by roughly 5300 mgal from equator to pole, which is a variation of only about 0.5%. Accordingly, measurements of gravity are of two types. The first corresponds to determination of the absolute magnitude of gravity at any place; the second consists of measuring the change in gravity from one place to another. In geophysical studies, especially in gravity prospecting, it is necessary to measure accurately the small changes in gravity caused by underground structures. These require an instrumental sensitivity of the order of 0.01 mgal. It is very difficult to design an instrument to measure the absolute value of gravity that has this high precision and that is also portable enough to be used easily in different places. Gravity surveying is usually carried out with a portable instrument called a *gravimeter*, which determines the variation of gravity relative to one or more reference locations. In national gravity surveys the relative variations determined with a gravimeter may be converted to absolute values by calibration with absolute measurements made at selected stations.

2.5.2 Absolute measurement of gravity

The classical method of measuring gravity is with a pendulum. A simple pendulum consists of a heavy weight suspended at the end of a thin fiber. The compound (or reversible) pendulum, first described by Henry Kater in 1818, allows more exact measurements. It consists of a stiff metal or quartz rod, about 50 cm long, to which is attached a movable mass. Near each end of the rod is fixed a pivot, which consists of a quartz knife-edge resting on a flat quartz plane. The period of the pendulum is measured for oscillations about one of the pivots. The pendulum is then inverted and its period about the other pivot is determined. The position of the movable mass is adjusted until the periods about the two pivots are equal. The distance L between the pivots is then measured accurately. The period of the instrument is given by

$$T=2\pi\sqrt{\frac{I}{mgh}}=2\pi\sqrt{\frac{L}{g}} \tag{2.62}$$

where I is the moment of inertia of the pendulum about a pivot, h is the distance of the center of mass from the pivot, and m is the mass of the pendulum. Knowing the length L from Kater's method obviates knowledge of I, m and h.

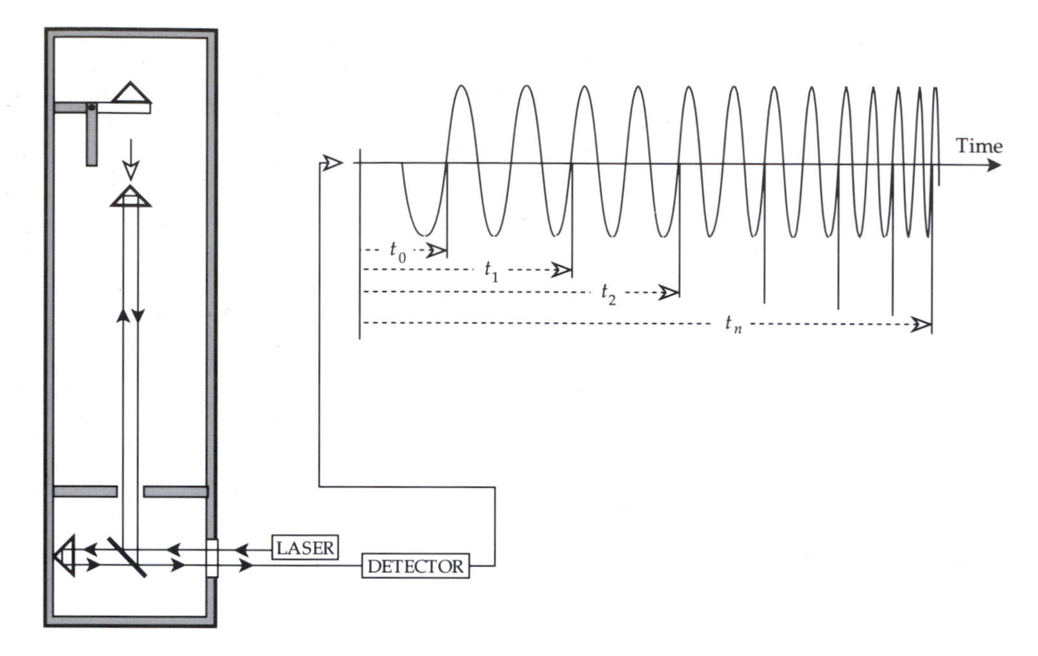

Fig. 2.30 The free-fall method of measuring absolute gravity.

The sensitivity of the compound pendulum is found by differentiating Eq. (2.62). This gives

$$\frac{\Delta g}{g} = -2\frac{\Delta T}{T} \tag{2.63}$$

To obtain a sensitivity of about 1 mgal it is necessary to determine the period with an accuracy of about 0.5 μs. This can be achieved easily today with precise atomic clocks. The compound pendulum was the main instrument for gravity prospecting in the 1930s, when timing the swings precisely was more difficult. It was necessary to time as accurately as possible a very large number of swings. As a result a single gravity measurement took about half an hour.

The performance of the instrument was handicapped by several factors. The inertial reaction of the housing to the swinging mass of the pendulum was compensated by mounting two pendulums on the same frame and swinging them in opposite phase. Air resistance was reduced by housing the pendulum assemblage in an evacuated thermostatically controlled chamber. Friction in the pivot was minimized by the quartz knife-edge and plane, but due to minor unevenness the contact edge was not exactly repeatable if the assemblage was set up in a different location, which affected the reliability of the measurements. The apparatus was bulky but was used until the 1950s as the main method of making absolute gravity measurements.

2.5.2.1 *Free-fall method*

Modern methods of determining the acceleration of gravity are based on observations of falling objects. For an object that falls from a starting position z_0 with initial velocity u the equation of motion gives the position z at time t as

$$z = z_0 + ut + \tfrac{1}{2}gt^2 \tag{2.64}$$

The absolute value of gravity is obtained by fitting a quadratic to the record of position versus time.

An important element in modern experiments is the accurate measurement of the change of position with a Michelson interferometer. In this device a beam of monochromatic light passes through a beam splitter, consisting of a semi-silvered mirror, which reflects half of the light incident upon it and transmits the other half. This divides the incident ray into two subrays, which subsequently travel along different paths and are then recombined to give an interference pattern. If the path lengths differ by a full wavelength (or a number of full wavelengths) of the monochromatic light, the interference is constructive. The recombined light has maximum intensity, giving a bright interference fringe. If the path lengths differ by half a wavelength (or by an odd number of half-wavelengths) the recombined beams interfere destructively, giving a dark fringe. In modern experiments the monochromatic light source is a laser beam of accurately known wavelength.

In an absolute measurement of gravity a laser beam is split along two paths that form a Michelson interferometer (Fig. 2.30). The horizontal path is of fixed length, while the vertical path is reflected off a corner-cube retro-reflector that is released at a known instant and falls freely. The path of free-fall is about 0.5 m long. The cube falls in an evacuated chamber to minimize air resistance. A photo-multiplier and counter permit the fringes within any time interval to be

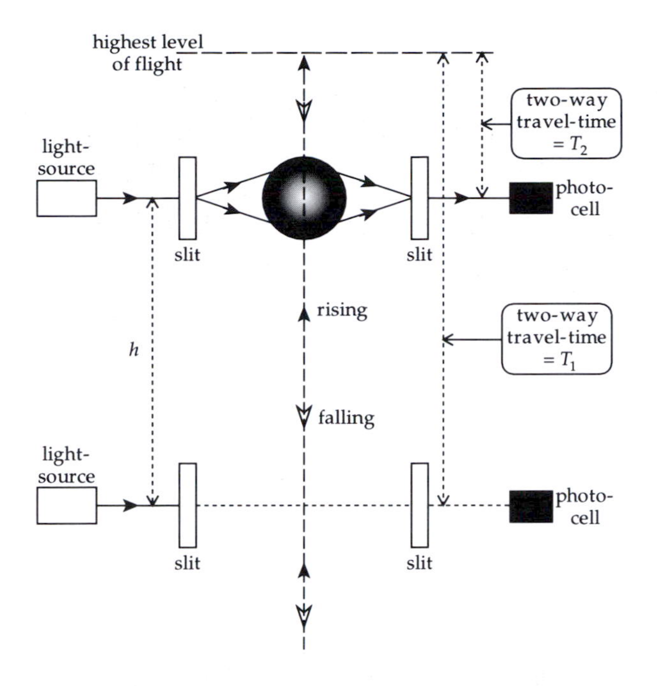

Fig. 2.31 The rise-and-fall method of measuring absolute gravity.

recorded and counted. The intensity of the recombined light fluctuates sinusoidally with increasing frequency the further and faster the cube falls. The distance between each zero crossing corresponds to half the wavelength of the laser light, and so the distance travelled by the falling cube in any time interval may be obtained. The times of the zero crossings must be measured with an accuracy of 0.1 ns (10^{-10} s) to give an accuracy of 1 μgal in the gravity measurement.

Although the apparatus is compact, it is not quite portable enough for gravity surveying. It gives measurements of the absolute value of gravity with an accuracy of about 0.005–0.010 mgal (5–10 μgal). A disadvantage of the free-fall method is the resistance of the residual air molecules left in the evacuated chamber. This effect is reduced by placing the retro-reflector in a chamber that falls simultaneously with the cube, so that in effect the cube falls in still air. Air resistance is further reduced in the rise-and-fall method.

2.5.2.2 *Rise-and-fall method*

In the original version of the rise-and-fall method a glass sphere was fired vertically upward and fell back along the same path (Fig. 2.31). Timing devices at two different levels registered the times of passage of the ball on the upward and downward paths. In each timer a light beam passed through a narrow slit. As the glass sphere passed the slit it acted as a lens and focussed one slit on the other. A photomultiplier and detector registered the exact passage of the ball past the timing level on the upward and downward paths. The distance h between the two timing levels (around 1 m) was measured accurately by optical interferometry.

Let the time spent by the sphere above the first timing level be T_1 and the time above the second level be T_2; further, let the distances from the zenith level to the timing levels be z_1 and z_2, respectively. The corresponding times of fall are $t_1 = T_1/2$ and $t_2 = T_2/2$. Then,

$$z_1 = \tfrac{1}{2} g \left(\frac{T_1}{2} \right)^2 \tag{2.65}$$

with a similar expression for the second timing level. Their separation is

$$h = z_1 - z_2 = \tfrac{1}{8} g (T_1^2 - T_2^2) \tag{2.66}$$

The following elegantly simple expression for the value of gravity is obtained:

$$g = \frac{8h}{(T_1^2 - T_2^2)} \tag{2.67}$$

Although the experiment is conducted in a high vacuum, the few remaining air molecules cause a drag that opposes the motion. On the upward path the air drag is downward, in the same direction as gravity; on the downward path the air drag is upward, opposite to the direction of gravity. This asymmetry helps to minimize the effects of air resistance.

In a modern variation Michelson interferometry is used as in the free-fall method. The projectile is a corner-cube retro-reflector, and interference fringes are observed and counted during its upward and downward paths. Sensitivity and accuracy are comparable to the free-fall method.

2.5.3 **Relative measurement of gravity: the gravimeter**

In principle, a gravity meter, or *gravimeter*, is a very sensitive balance. The first gravimeters were based on the straightforward application of Hooke's law (§3.2.1). A mass m suspended from a spring of length s_0 causes it to stretch to a new length s. The extension, or change in length, of the spring is proportional to the restoring force of the spring and so to the value of gravity, according to:

$$F = mg = -k(s - s_0) \tag{2.68}$$

where k is the elastic constant of the spring. The gravimeter is calibrated at a known location. If gravity is different at another location, the extension of the spring changes, and from this the change in gravity can be computed.

This type of gravimeter, based directly on Hooke's law, is called a stable type. It has been replaced by more sensitive

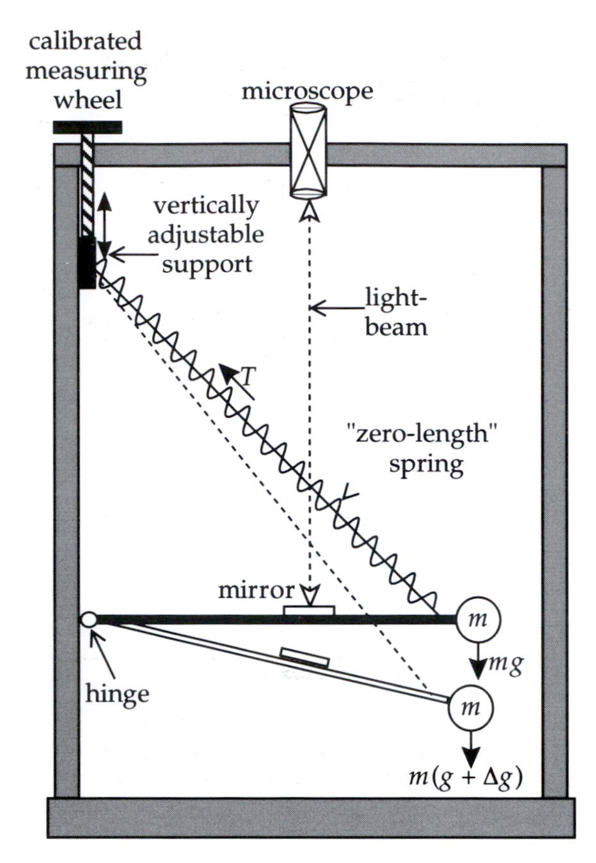

Fig. 2.32 The principle of operation of an unstable (astatic) type of gravimeter.

unstable or astatized types, which are constructed so that an additional force acts in the same direction as gravity and opposes the restoring force of the spring. The instrument is then in a state of unstable equilibrium. This condition is realized through the design of the spring. If the natural length s_0 can be made as small as possible, ideally zero, Eq. (2.68) shows that the restoring force is then proportional to the physical length of the spring instead of its extension. The *zero-length spring*, first introduced in the LaCoste–Romberg gravimeter, is now a common element in modern gravimeters. The spring is usually of the helical type. When a helical spring is stretched, the fiber of the spring is twisted; the total twist along the length of the fiber equals the extension of the spring as a whole. During manufacture of a zero-length spring the helical spring is given an extra twist, so that its tendency is to uncoil. An increase in gravity stretches the spring against its restoring force, and the extension is augmented by the built-in pre-tension.

The operation of a gravimeter is illustrated in Fig. 2.32. A mass is supported by a horizontal rod to which a mirror is attached. The position of the rod is observed with a light beam reflected into a microscope. If gravity changes, the zero-length spring is extended or shortened and the position of the rod is altered, which deflects the light beam. The null-deflection principle is utilized. An adjusting screw changes the position of the upper attachment of the spring, which alters its tension and restores the rod to its original horizontal position as detected by the light beam and microscope. The turns of the adjusting screw are calibrated in units of the change in gravity, usually in mgal.

The gravimeter is light, robust and portable. After initially levelling the instrument, an accurate measurement of a gravity difference can be made in a few minutes. The gravimeter has a sensitivity of about 0.01 mgal (10 μgal). This high sensitivity makes it susceptible to small changes in its own properties.

2.5.3.1 *Gravity surveying*

If a gravimeter is set up at a given place and monitored for an hour or so, the repeated readings are found to vary smoothly with time. The changes amount to several hundredths of a mgal. The *instrumental drift* is partly due to thermally induced changes in the elastic properties of the gravimeter spring, which are minimized by housing the critical elements in an evacuated chamber. In addition, the elastic properties of the spring are not perfect, but creep slowly with time. The effect is small in modern gravimeters and can be compensated by making a *drift correction*. This is obtained by repeated occupation of some measurement stations at intervals during the day (Fig. 2.33). Gravity readings at other stations are adjusted by comparison with the drift curve. In order to make this correction the time of each measurement must be noted.

During the day, while measurements are being made, the gravimeter is subject to tidal attraction, including vertical displacement due to the bodily Earth-tides. The theory of the tides is known well (see §2.3.3) and their time-dependent effect on gravity can be computed precisely for any place on Earth at any time. Again, the *tidal correction* requires that the time of each measurement be known.

The goal of gravity surveying is to locate and describe subsurface structures from the gravity effects caused by their anomalous densities. Most commonly, gravimeter measurements are made at a network of stations, spaced according to the purpose of the survey. In environmental studies a detailed high-resolution investigation of the gravity expression of a small area requires small distances of a few meters between measurement stations. In regional gravity surveys, as used for the definition of hidden structures of prospective commercial interest, the distance between stations may be several kilometers. If the area surveyed is not too large, a suitable site is selected as base station (or reference site), and the gravity differences

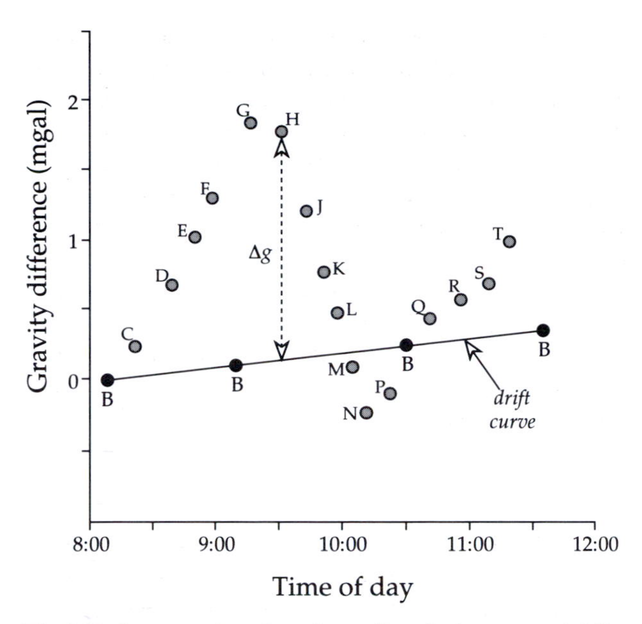

Fig. 2.33 Compensation of gravity readings for instrumental drift. The gravity stations B–T are occupied in sequence at known times. The repeated measurements at the base station B allow a drift correction to be made to the gravity readings at the other stations.

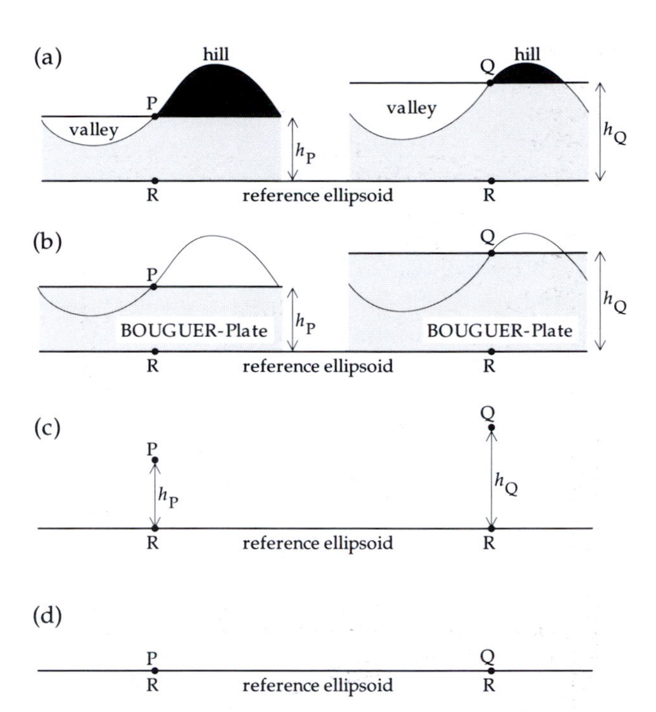

Fig. 2.34 After (a) terrain corrections, (b) the Bouguer plate correction, and (c) the free-air correction, the gravity measurements at stations P and Q can be compared to the theoretical gravity at R on the reference ellipsoid.

between the surveyed sites and this site are measured. In a gravity survey on a national scale, the gravity differences may be determined relative to a site where the absolute value of gravity is known.

2.5.4 Correction of gravity measurements

If the interior of the Earth were uniform, the value of gravity on the international reference ellipsoid would vary with latitude according to the normal gravity formula (Eq. (2.57)). This provides us with a reference value for gravity measurements. In practice, it is not possible to measure gravity on the ellipsoid at the place where the reference value is known. The elevation of a measurement station may be hundreds of meters above or below the ellipsoid. Moreover, the gravity station may be surrounded by mountains and valleys that perturb the measurement. For example, let P and Q represent gravity stations at different elevations in hilly terrain (Fig. 2.34a). The theoretical value of gravity is computed at the points R on the reference ellipsoid below P and Q. Thus, we must correct the measured gravity before it can be compared with the reference value.

The hill-top adjacent to stations P and Q has a center of mass that lies higher than the measurement elevation (Fig. 2.34a). The gravimeter measures gravity in the vertical direction, along the local plumb-line. The mass of the hill-top above P attracts the gravimeter and causes an acceleration with a vertically upward component at P. The measured gravity is reduced by the presence of the hill-top; to compensate for this a *terrain (or topographic) correction* is calculated and *added* to the measured gravity. A similar effect is observed at Q, but the hill-top above Q is smaller and the corresponding terrain correction is smaller. These corrections effectively level the topography to the same elevation as the gravity station. The presence of a valley next to each measurement station also requires a terrain correction. In this case, imagine that we could fill the valley up to the level of each station with rock of the same density ρ as under P and Q. The downward attraction on the gravimeter would be increased, so the terrain correction for a valley must also be *added* to the measured gravity, just as for a hill. Removing the effects of the topography around a gravity station requires making positive terrain corrections (Δg_T) for both hills and valleys.

After levelling the topography there is now a fictive uniform layer of rock with density ρ between the gravity station and the reference ellipsoid (Fig. 2.34b). The gravitational acceleration of this rock-mass is included in the measured gravity and must be removed before we can compare with the theoretical gravity. The layer is taken to be a flat disk or plate of thickness h_P or h_Q under each station; it is called the *Bouguer plate*. Its gravitational acceleration can be computed for known thickness and density r, and gives a *Bouguer plate correction* (Δg_{BP}) that must be subtracted from the measured gravity, if the gravity station is above sea-level. Note that, if the gravity station is below sea-level, we have to fill the space above it up to sea-level with rock of density ρ; this requires increasing the measured gravity correspondingly. The Bouguer plate correction (Δg_{BP}) is negative if the station is above sea-level but positive if it is below sea-level. Its size depends on the density of the local rocks, but typically amounts to about 0.1 mgal m^{-1}.

Finally, we must compensate the measured gravity for the elevation h_P or h_Q of the gravity station above the ellipsoid (Fig. 2.34c). The main part of gravity is due to gravitational attraction, which decreases proportionately to the inverse square of distance from the center of the Earth. The gravity measured at P or Q is smaller than it would be, if measured on the ellipsoid at R. A *free-air correction* (Δg_{FA}) for the elevation of the station must be added to the measured gravity. This correction ignores the effects of material between the measurement and reference levels, as this is taken care of in Δg_{BP}. Note that, if the gravity station were below sea-level, the gravitational part of the measured gravity would be too large by comparison with the reference ellipsoid; we would need to subtract Δg_{FA} in this case. The free-air correction is positive if the station is above sea-level but negative if it is below sea-level (as might be the case in Death Valley or beside the Dead Sea). It amounts to about 0.3 mgal m^{-1}.

The free-air correction is always of opposite sense to the Bouguer plate correction. For convenience, the two are often combined in a single *elevation correction*, which amounts to about 0.2 mgal m^{-1}. This must be added for gravity stations above sea-level and subtracted if gravity is measured below sea-level. In addition, a *tidal correction* (Δg_{tide}) must be made (§2.3.3), and, if gravity is measured in a moving vehicle, the *Eötvös correction* (§2.3.5) is also necessary.

After correction the measured gravity can be compared with the theoretical gravity on the ellipsoid (Fig. 2.34d). Note that the above procedure reduces the measured gravity to the surface of the ellipsoid. In principle it is equally valid to correct the theoretical gravity from the ellipsoid upward to the level where the measurement was made. This method is preferred in more advanced types of analysis of gravity anomalies where the possibility of an anomalous mass between the ellipsoid and ground surface must be taken into account.

2.5.4.1 *Latitude correction*

The theoretical gravity at a given latitude is given by the normal gravity formula (Eq. (2.57)). If the measured gravity is an absolute value, the correction for latitude is made by subtracting the value predicted by this formula. Often, however, the gravity survey is made with a gravimeter, and the quantity measured, g_m, is the gravity difference relative to a base station. The normal reference gravity g_n may then be replaced by a latitude correction, obtained by differentiating Eq. (2.57):

$$\frac{\partial g_n}{\partial \lambda}=g_e(\beta_1 \sin 2\lambda+2\beta_2 \sin 4\lambda) \tag{2.69}$$

After converting $\partial\lambda$ from radians to kilometers and neglecting the β_2 term, the latitude correction (Δg_{lat}) is 0.8140 sin 2λ mgal per kilometer of north–south displacement. Because gravity increases towards the poles, the correction for stations closer to the pole than the base station must be subtracted from the measured gravity.

2.5.4.2 *Terrain corrections*

The terrain correction (Δg_T) for a hill adjacent to a gravity station is computed by dividing the hill into a number of vertical prisms (Fig. 2.35a). The contribution of each vertical element to the vertical acceleration at the point of observation P is calculated by assuming cylindrical symmetry about P. The height of the prism is h, its inner and outer radii are r_1 and r_2, respectively, the angle subtended at P is ϕ_o, and the density of the hill is ρ (Fig. 2.35b). Let the sides of a small cylindrical element be dr, dz and (rdϕ); its mass is dm=r(rdϕ)drdz and its contribution to the upward acceleration caused by the prism at P is

$$\delta g=G\frac{dm}{r^2+z^2}\cos\theta=G\frac{\rho r dr\, dz\, d\phi}{r^2+z^2}\frac{z}{\sqrt{r^2+z^2}} \tag{2.70}$$

Combining and rearranging terms and the order of integration gives the upward acceleration at P due to the cylindrical prism:

$$\Delta g_T=G\rho\int_{\phi=0}^{\phi_o} d\phi\int_{r=r_1}^{r_2}\left[\int_{z=0}^{h}\frac{z\, dz}{(r^2+z^2)^{3/2}}\right]r\, dr \tag{2.71}$$

The integration over ϕ gives ϕ_o; after further integration over z we get:

$$\Delta g_T=G\rho\phi_o\int_{r=r_1}^{r_2}\left(\frac{r}{\sqrt{r^2+h^2}}-1\right)dr \tag{2.72}$$

Integration over r gives the upward acceleration produced at P by the cylinder:

$$\Delta g_T=G\rho\phi\left[\left(\sqrt{r_1^2+h^2}-r_1\right)-\left(\sqrt{r_2^2+h^2}-r_2\right)\right] \tag{2.73}$$

The direction of Δg_T in Fig. 2.35b is upward, opposite to

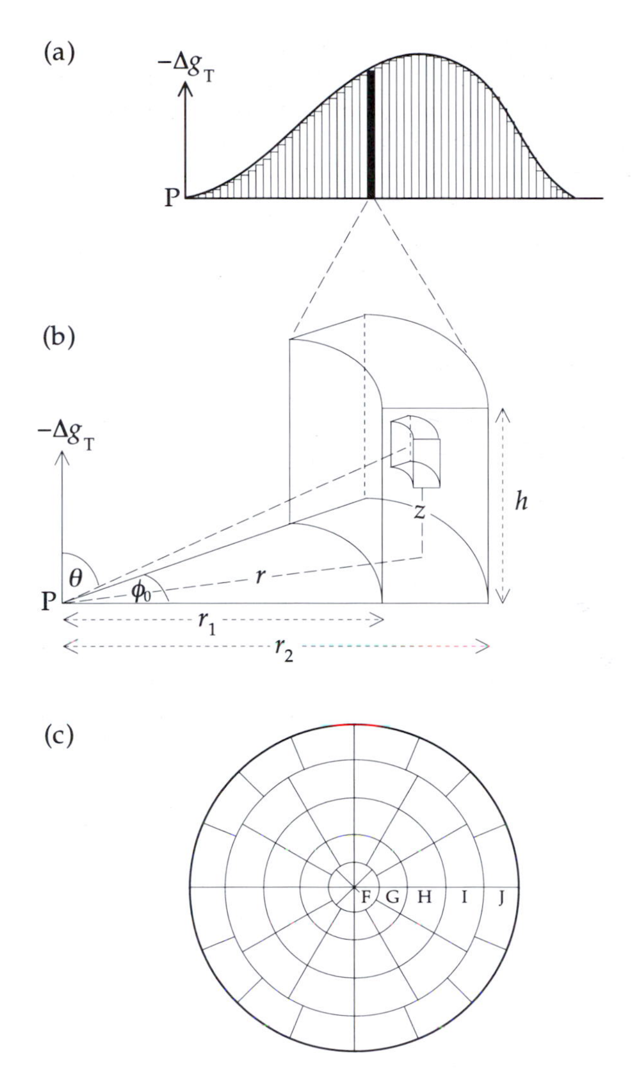

Fig. 2.35 Terrain corrections Δg_T are made by (a) dividing the topography into vertical elements, (b) computing the correction for each cylindrical element according to its height above or below the measurement station, and (c) adding up the contributions for all elements around the station with the aid of a transparent overlay on a topographic map.

gravity; the corresponding terrain correction must be *added* to the measured gravity.

In practice, terrain corrections can be made using a terrain chart (Fig. 2.35c) on which concentric circles and radial lines divide the area around the gravity station into sectors that have radial symmetry like the cross-section of the element of a vertical cylinder in Fig. 2.35b. The inner and outer radius of each sector correspond to r_1 and r_2, and the angle subtended by the sector is ϕ. The terrain correction for each sector within each zone is precalculated using Eq. (2.73) and tabulated. The chart is drawn on a transparent sheet that is overlaid on a topographic map at the same scale and centered on the gravity station. The mean elevation within each sector is estimated as accurately as possible, and the elevation difference (i.e., h in Eq. (2.73)) of the sector relative to the station is computed. This is multiplied by the correction factor for the sector to give its contribution to the terrain correction. Finally, the terrain correction at the gravity station is obtained by summing up the contributions of all sectors. The procedure must be repeated for each gravity station. When the terrain chart is centered on a new station, the mean topographic relief within each sector changes and must be computed anew. As a result, terrain corrections are time-consuming and tedious. The most important effects come from the topography nearest to the station. However, terrain corrections are generally necessary if a topographic difference within a sector is more than about 5% of its distance from the station.

2.5.4.3 *Bouguer plate correction*

The Bouguer plate correction (Δg_{BP}) compensates for the effect of a layer of rock whose thickness corresponds to the elevation difference between the measurement and reference levels. This is modelled by a solid disk of density r and infinite radius centered at the gravity station P. The correction is computed by extension of the calculation for the terrain correction. An elemental cylindrical prism is defined as in Fig. 2.35b. Let the angle ϕ subtended by the prism increase to 2π and the inner radius decrease to zero; the first term in brackets in Eq. (2.73) reduces to h. The gravitational acceleration at the center of a solid disk of radius r is then

$$\Delta g_{disk}=2\pi G\rho\left[h-\left(\sqrt{r^2+h^2}-r\right)\right] \tag{2.74}$$

Now let the radius r of the disk increase. The value of h gradually becomes insignificant compared to r; in the limit, when r is infinite, the second term in Eq. (2.74) tends to zero. Thus, the Bouguer plate correction (Δg_{BP}) is given by

$$\Delta g_{BP}=2\pi G\rho h \tag{2.75}$$

Inserting numerical values gives $0.0419\times10^{-3}\rho$ mgal m^{-1} for Δg_{BP}, where the density ρ is in kg m^{-3} (see §2.5.5). The correct choice of density is very important in computing Δg_{BP} and Δg_T. Some methods of determining the optimum choice are described in detail below.

An additional consideration is necessary in marine gravity surveys. Δg_{BP} requires uniform density below the surface of the reference ellipsoid. To compute Δg_{BP} over an oceanic region we must in effect replace the sea-water with rock of density ρ. However, the measured gravity contains a component due to the attraction of the sea-water (density 1030 kg m^{-3}) in the ocean basin. The Bouguer plate correction in marine gravity surveys is therefore made by replacing the density ρ in Eq. (2.75) by $(\rho - 1030)$ kg m^{-3}. When a shipboard gravity survey is made over a large deep lake, a similar allowance must be made for the depth of water in the lake using an assumed density of $(\rho - 1000)$ kg m^{-3}.

2.5.4.4 *Free-air correction*

The free-air correction (Δg_{FA}) has a rather colorful, but slightly misleading title, giving the impression that the measurement station is floating in air above the ellipsoid. The density of air at standard temperature and pressure is around 1.3 kg m^{-3} and a mass of air between the observation and reference levels would cause a detectable gravity effect of about 50 μgals at an elevation of 1000 m. In fact, the free-air correction pays no attention to the density of the material between the measurement elevation and the ellipsoid. It is a straightforward correction for the decrease of gravitational acceleration with distance from the center of the Earth:

$$\frac{\partial g}{\partial r}=\frac{\partial}{\partial r}\left(-G\frac{E}{r^2}\right)=+2G\frac{E}{r^3}=-\frac{2}{r}g \qquad (2.76)$$

On substituting the Earth's radius (6371 km) for r and the mean value of gravity (981,000 mgals) for g, we get a Δg_{FA} of 0.3086 mgal m^{-1}.

2.5.4.5 *Combined elevation correction*

The free-air and Bouguer plate corrections are often combined into a single elevation correction, which is $(0.3086 - (0.0419\rho\times10^{-3}))$ mgal m^{-1}. Substituting a typical density for crustal rocks, usually taken to be 2670 kg m^{-3}, gives a combined elevation correction of 0.197 mgal m^{-1}. This must be added to the measured gravity if the gravity station is above the ellipsoid and subtracted if it is below.

The high sensitivity of modern gravimeters allows an achievable accuracy of 0.01–0.02 mgal in modern gravity surveys. To achieve this accuracy the corrections for the variations of gravity with latitude and elevation must be made very exactly. This requires that the precise coordinates of a gravity station must be determined by accurate geodetic surveying. The necessary precision of horizontal positioning is indicated by the latitude correction. This is maximum at 45° latitude, where, in order to achieve a survey accuracy of ±0.01 mgal, the north–south positions of gravity stations must be known to about ±10 m. The requisite precision in vertical positioning is indicated by the combined elevation correction of 0.2 mgal m^{-1}. To achieve a survey accuracy of ±0.01 mgal the elevation of the gravimeter above the reference ellipsoid must be known to about ±5 cm.

The elevation of a site above the ellipsoid is often taken to be its altitude above mean sea-level. However, mean sea-level is equated with the geoid and not with the ellipsoid. Geoid undulations can amount to tens of meters (§2.4.5.1). They are long-wavelength features. Within a local survey the distance between geoid and ellipsoid is unlikely to vary much, and the gravity differences from the selected base station are unlikely to be strongly affected. In a national survey the discrepancies due to geoid undulations may be more serious. In the event that geoid undulations are large enough to affect a survey, the station altitudes must be corrected to true elevations above the ellipsoid.

2.5.5 **Density determination**

The density of rocks in the vicinity of a gravity profile is important for the calculation of the Bouguer plate and terrain corrections. Density is defined as the mass per unit of volume of a material. It has different units and different numerical values in the c.g.s. and SI systems. For example, the density of water is 1 g cm^{-3} in the c.g.s. system, but 1000 kg m^{-3} in the SI system. In gravity prospecting c.g.s. units are still in common use, but are slowly being replaced by SI units. The formulas given for Δg_T and Δg_{BP} in Eq. (2.73) and Eq. (2.75), respectively, require that density be given in kg m^{-3}.

A simple way of determining the appropriate density to use in a gravity study is to make a representative collection of rock samples with the aid of a geological map. The specific gravity of a sample may be found directly by weighing it first in air and then in water, and applying Archimedes' principle. This gives its density ρ_r relative to that of water:

$$\rho_r=\frac{W_a}{W_a-W_w} \qquad (2.77)$$

Typically, the densities found for different rock types by this method show a large amount of scatter about their means, and the ranges of values for different rock types overlap (Fig. 2.36). The densities of igneous and metamorphic rocks are generally higher than those of sedimentary rocks. This method is adequate for reconnaissance of an area. Unfortunately, it is often difficult to ensure that the surface collection of rocks is representative of the rock types in subsurface structures, so alternative methods of determining the appropriate density are usually employed. Density can be measured in vertical boreholes, drilled to explore the nature of a presumed structure. The density determined in the borehole is used to refine the interpretation of the structure.

2.5.5.1 *Density from seismic velocities*

Measurements on samples of water-saturated sediments and sedimentary rocks, and on igneous and metamorphic

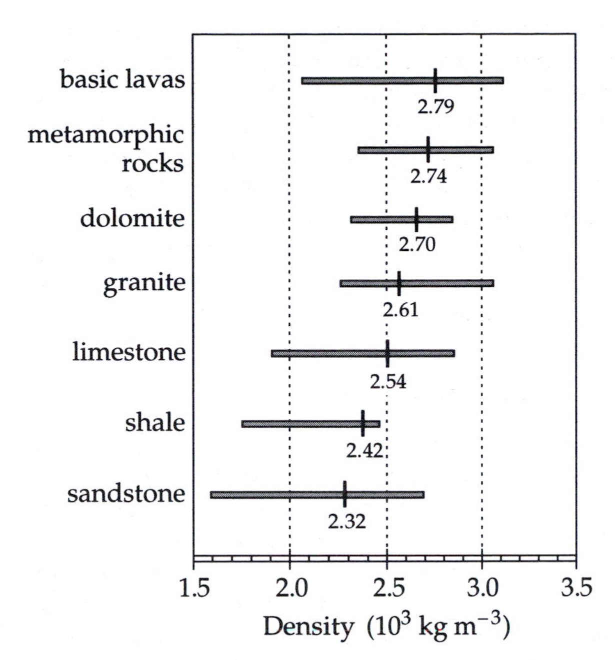

Fig. 2.36 Typical mean values and ranges of density for some common rock types (after Dobrin, 1976).

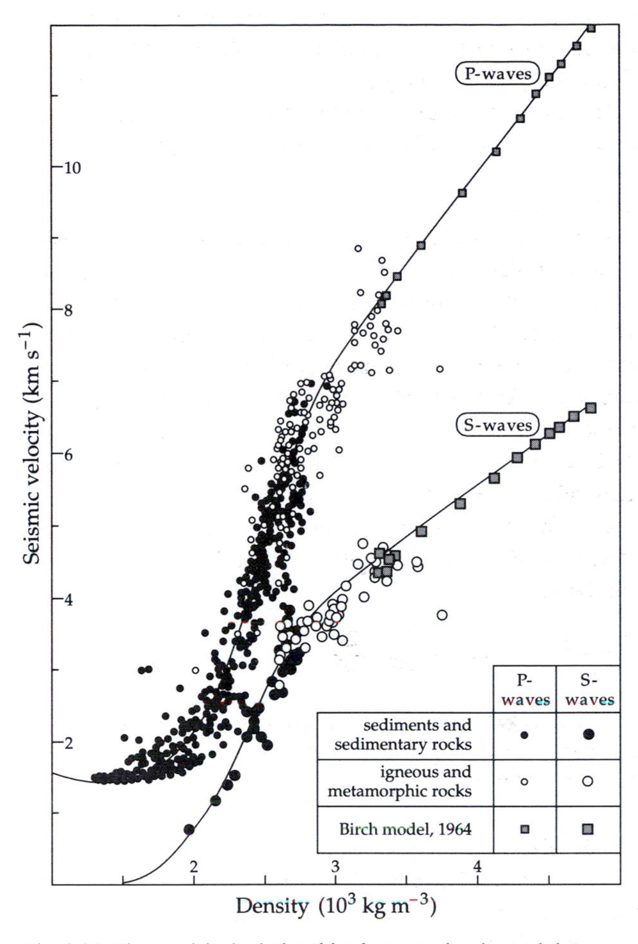

Fig. 2.37 The empirical relationships between density and the seismic P-wave and S-wave velocities in water-saturated sediments and sedimentary rocks, igneous and metamorphic rocks (after Ludwig *et al.*, 1970).

rocks show that density and the seismic P-wave and S-wave velocities are related. The optimum fit to each data-set is a smooth curve (Fig. 2.37). Each curve is rather idealized, as the real data contain considerable scatter. For this reason the curves are best suited for computing the mean density of a large crustal body from its mean seismic velocity. Adjustments must be made for the higher temperatures and pressures at depth in the Earth, which affect both the density and the elastic parameters of rocks. However, the effects of high pressure and temperature can only be examined in laboratory experiments on small specimens. It is not known to what extent the results are representative of the *in-situ* velocity–density relationship in large crustal blocks.

The velocity–density curves are empirical relationships that do not have a theoretical basis. The P-wave data are used most commonly. In conjunction with seismic refraction studies, they have been used for modelling the density distributions in the Earth's crust and upper mantle responsible for large-scale, regional gravity anomalies (see §2.6.4).

2.5.5.2 *Gamma–gamma logging*

The density of rock formations adjacent to a borehole can be determined from an instrument in the borehole. The principle makes use of the Compton scattering of γ-rays by loosely bound electrons in the rock adjacent to a borehole. An American physicist, Arthur H. Compton, discovered in 1923 that radiation scattered by loosely bound electrons experienced an increase in wavelength. This simple observation can not be explained at all if the radiation is treated as a wave; the scattered radiation would have the same wavelength as the incident radiation. The Compton effect is easily explained by regarding the radiation as particles or photons, i.e., particles of quantized energy, rather than as waves. The energy of a photon is inversely proportional to its wavelength. The collision of a γ-ray photon with an electron is like that between billiard balls; part of the photon's energy is transferred to the electron. The scattered photon has lower energy and hence a longer wavelength than the incident photon. The Compton effect was an important verification of quantum theory.

The density logger, or *gamma–gamma* logger (Fig. 2.38), is a cylindrical device that contains a radioactive source of γ-rays, such as ^{137}Cs, which emits radiation through a narrow slit. The γ-ray photons collide with the loosely bound electrons of atoms near the hole, and are scattered. A scintilla-

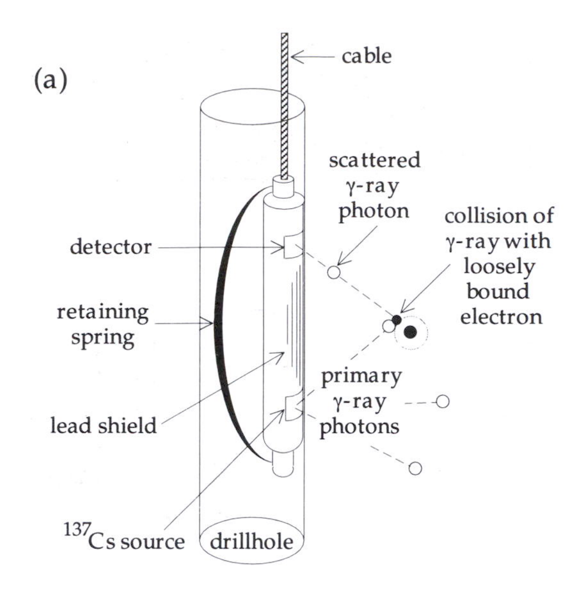

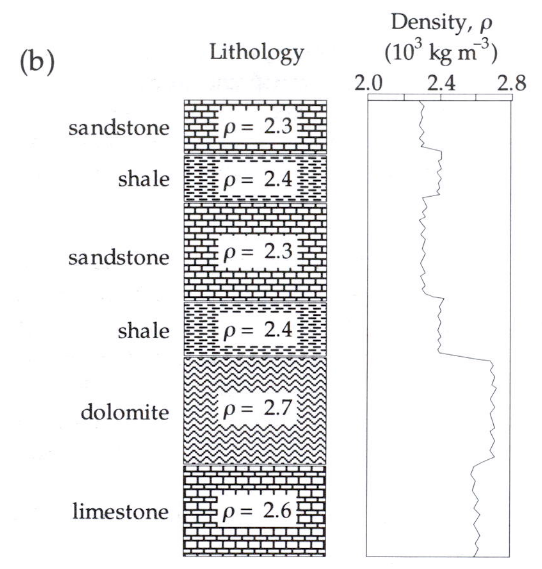

Fig. 2.38 (a) The design of a gamma–gamma logging device for determining density in a borehole (after Telford *et al.*, 1990), and (b) a schematic gamma–gamma log calibrated in terms of the rock density.

tion counter to detect and measure the intensity of γ-rays is located in the tool about 45–60 cm above the emitter; the radiation reaching it also passes through a slit. Emitter and detector are shielded with lead, and the tool is pressed against the wall of the borehole by a strong spring, so that the only radiation registered is that resulting from the Compton scattering in the surrounding formation. The intensity of detected radiation is determined by the density of electrons, and so by the density of rock near to the logging tool. The γ-rays penetrate only about 15 cm into the rock.

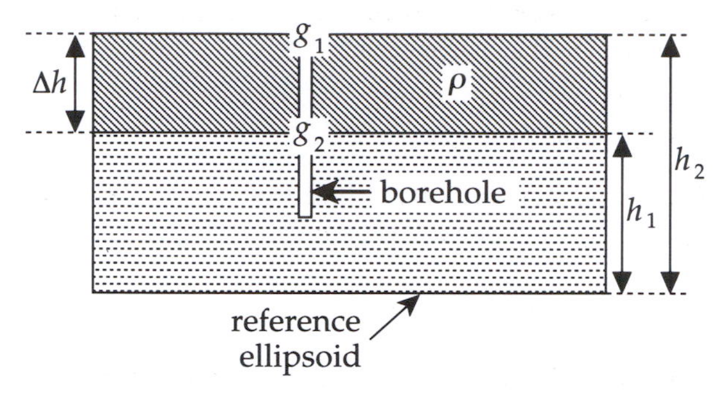

Fig. 2.39 Geometry for computation of the density of a rock layer from gravity measurements made in a vertical borehole.

Calibrated gamma–gamma logs give the bulk density of the rock surrounding a borehole. This information is also needed for calculating porosity, which is defined as the fractional volume of the rock represented by pore spaces. Most sedimentary rocks are porous, the amount depending on the amount of compaction experienced. Igneous and metamorphic rocks generally have low porosity, unless they have been fractured. Usually the pores are filled with air, gas or a fluid, such as water or oil. If the densities of the matrix rock and pore fluid are known, the bulk density obtained from gamma–gamma logging allows the porosity of the rock to be determined.

2.5.5.3 *Borehole gravimetry*

Modern instrumentation allows gravity to be measured accurately in boreholes. One type of borehole gravimeter is a modification of the LaCoste–Romberg instrument, adapted for use in the narrow borehole and under conditions of elevated temperature and pressure. Alternative instruments have been designed on different principles; they have a comparable sensitivity of about 0.01 mgal. Their usage for down-hole density determination is based on application of the free-air and Bouguer plate corrections.

Let g_1 and g_2 be the values of gravity measured in a vertical borehole at heights h_1 and h_2, respectively, above the reference ellipsoid (Fig. 2.39). The difference between g_1 and g_2 is due to the different heights and to the material between the two measurement levels in the borehole. The value g_2 will be larger than g_1 for two reasons. First, because the lower measurement level is closer to the Earth's center, g_2 will be greater than g_1 by the amount of the combined elevation correction, namely $(0.3086 - (0.0419\rho\times10^{-3}))\Delta h$ mgal, where $\Delta h = h_1 - h_2$. Second, at the lower level h_2 the gravimeter experiences an upward Bouguer attraction due to the material between the two measurement levels. This reduces the measured gravity at h_2 and requires a compensating increase to g_2 of amount $(0.0419\rho\times10^{-3})\Delta h$ mgal. The difference Δg between the *corrected* values of g_1 and g_2 after reduction to the level h_2 is then

$$\Delta g = (0.3086 - 0.0419\rho \times 10^{-3})\Delta h - 0.0419\rho \times 10^{-3}\Delta h = (0.3086 - 0.0838\rho \times 10^{-3})\Delta h \tag{2.78}$$

Rearranging this equation gives the density ρ of the material between the measurement levels in the borehole.

$$\rho = \left(3.683 - 11.93\frac{\Delta g}{\Delta h}\right) \times 10^3 \text{ kg m}^{-3} \tag{2.79}$$

If borehole gravity measurements are made with an accuracy of ± 0.01 mgal at a separation of about 10 m, the density of the material near the borehole can be determined with an accuracy of about ± 10 kg m^{-3}. More than 90% of the variation in gravity in the borehole is due to material within a radius of about $5\Delta h$ from the borehole (about 50 m for a distance $\Delta h \approx 10$ m between measurement levels). This is much larger than the lateral range penetrated by gamma–gamma logging. As a result, effects related to the borehole itself are unimportant.

2.5.5.4 *Nettleton's method for near-surface density*

The near-surface density of the material under a hill can be determined by a method devised by L. Nettleton that compares the shape of a Bouguer gravity anomaly (see §2.5.6) with the shape of the topography along a profile. The method makes use of the combined elevation correction ($\Delta g_{FA} + \Delta g_{BP}$) and the terrain correction (Δg_T), which are density dependent. The terrain correction is less important than the Bouguer plate correction and can usually be neglected.

A profile of closely spaced gravity stations is measured across a small hill (Fig. 2.40). The combined elevation correction is applied to each measurement. Suppose that the true average density of the hill is 2600 kg m^{-3}. If the value assumed for ρ is too small (say, 2400 kg m^{-3}), Δg_{BP} at each station will be too small. The discrepancy is proportional to the elevation, so the Bouguer gravity anomaly is a *positive* image of the topography. If the value assumed for ρ is too large (say, 2800 kg m^{-3}), the opposite situation occurs. Too much is subtracted at each point, giving a computed anomaly that is a *negative* image of the topography. The optimum value for the density is found when the gravity anomaly has minimum correlation with the topography.

2.5.6 Free-air and Bouguer gravity anomalies

Suppose that we can measure gravity on the reference ellipsoid. If the distribution of density inside the Earth is homogeneous, the measured gravity should agree with the theoretical gravity given by the normal gravity formula. The

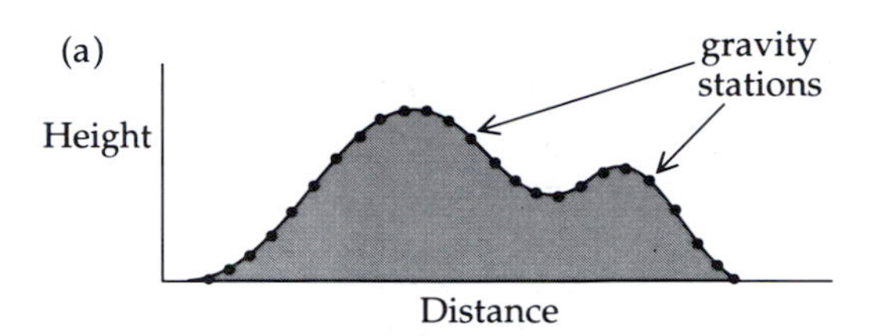

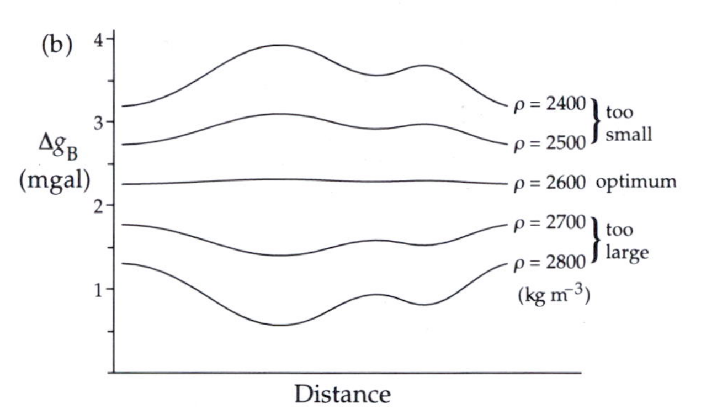

Fig. 2.40 Determination of the density of near-surface rocks by Nettleton's method. (a) Gravity measurements are made on a profile across a small hill. (b) The data are corrected for elevation with various test values of the density. The optimum density gives minimum correlation between the gravity anomaly (Δg_B) and the topography.

gravity corrections described in §2.5.4 compensate for the usual situation that the point of measurement is not on the ellipsoid. A discrepancy between the corrected, measured gravity and the theoretical gravity is called a *gravity anomaly*. It arises because the density of the Earth's interior is not homogeneous as assumed. The most common types of gravity anomaly are the *Bouguer anomaly* and the *free-air anomaly.*

The Bouguer gravity anomaly (Δg_B) is defined by applying all the corrections described individually in §2.5.4:

$$\Delta g_B = g_m + (\Delta g_{FA} - \Delta g_{BP} + \Delta g_T + \Delta g_{tide}) - g_n \tag{2.80}$$

In this formula g_m and g_n are the measured and normal gravity values; the corrections in parentheses are the free air correction (Δg_{FA}), Bouguer plate correction (Δg_{BP}), terrain correction (Δg_T) and tidal correction (Δg_{tide}).

The free-air anomaly Δg_F is defined by applying only the free-air, terrain and tidal corrections to the measured gravity:

$$\Delta g_F = g_m + (\Delta g_{FA} + \Delta g_T + \Delta g_{tide}) - g_n \tag{2.81}$$

The Bouguer and free-air anomalies across the same structure can look quite different. Consider first the topographic block (representing a mountain range) shown in Fig. 2.41a. For this simple structure we neglect the terrain and tidal corrections. The difference between the Bouguer

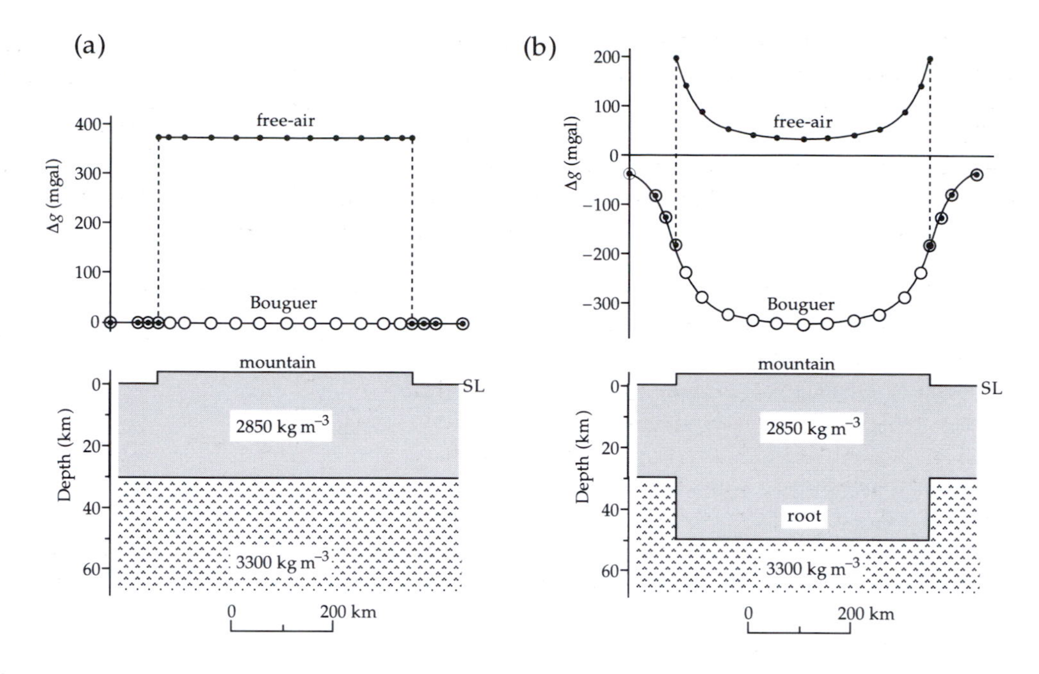

Fig. 2.41 Free-air and Bouguer anomalies across a mountain range. In (a) the mountain is modelled by a fully supported block, and in (b) the mass of the mountain above sea-level (SL) is compensated by a less-dense crustal root, which projects down into the denser mantle (based on Bott, 1982).

anomaly and the free-air anomaly arises from the Bouguer plate correction. In computing the Bouguer anomaly the simple elevation of the measurement station is taken into account together with the free-air correction. The measured gravity contains the attraction of the landmass above the ellipsoid, which is compensated with the Bouguer plate correction. The underground structure does not vary laterally, so the corrected measurement agrees with the theoretical value and the Bouguer anomaly is everywhere zero across the mountain range. In computing the free-air anomaly only the free-air correction is applied; the part of the measured gravity due to the attraction of the landmass above the ellipsoid is not taken into account. Away from the mountain-block the Bouguer and free-air anomalies are both equal to zero. Over the mountain the mass of the mountain-block increases the measured gravity compared to the reference value and results in a positive free-air anomaly across the mountain range.

In fact, seismic data show that the Earth's crust is usually much thicker than normal under a mountain range. This means that a block of less-dense crustal rock projects down into the denser mantle (Fig. 2.41b). After making the free-air and Bouguer plate corrections there remains a Bouguer anomaly due to a block that represents the 'root-zone' of the mountain range. As this is less dense than the adjacent and underlying mantle it constitutes a mass deficit. The attraction on a gravimeter at stations on a profile across the mountain range will be less than in Fig. 2.41a, so the corrected measurement will be less than the reference value. A strongly negative Bouguer anomaly is observed along the profile. At some distance from the mountain-block the Bouguer and free-air anomalies are equal but they are no longer zero, because the Bouguer anomaly now contains the effect of the root-zone. Over the mountain-block the free-air anomaly has a constant positive offset from the Bouguer anomaly, as in the previous example. Note that, although the free-air anomaly is positive, it falls to a very low value over the center of the block. At this point the attraction of the mountain is partly cancelled by the missing attraction of the less-dense root-zone.

2.6 INTERPRETATION OF GRAVITY ANOMALIES

2.6.1 Regional and residual anomalies

A gravity anomaly results from the inhomogeneous distribution of density in the Earth. Suppose that the density of rocks in a subsurface body is ρ and the density of the rocks surrounding the body is ρ_0. The difference $\Delta\rho=\rho-\rho_0$ is called the *density contrast* of the body with respect to the surrounding rocks. If the body has a higher density than the host rock, it has a positive density contrast; a body with lower density than the host rock has a negative density contrast. Over a high-density body the measured gravity is augmented; after reduction to the reference ellipsoid and subtraction of the normal gravity a positive gravity anomaly is obtained. Likewise a negative anomaly results over a region of low density. The presence of a gravity anomaly indicates a body or structure with anomalous density; the sign of the anomaly is the same as that of the density contrast and shows whether the density of the body is higher or lower than normal.

The appearance of a gravity anomaly is affected by the dimensions, density contrast and depth of the anomalous body. The horizontal extent of an anomaly is often called its apparent 'wavelength'. The wavelength of an anomaly is a measure of the depth of the anomalous mass. Large, deep bodies give rise to broad (long-wavelength), low-amplitude anomalies, while small, shallow bodies cause narrow (short-wavelength), sharp anomalies.

Usually a map of Bouguer gravity anomalies contains superposed anomalies from several sources. The long-wavelength anomalies due to deep density contrasts are called *regional* anomalies. They are important for understanding the large-scale structure of the Earth's crust under major geographic features, such as mountain ranges, oceanic ridges and subduction zones. Short-wavelength *residual* anomalies are due to shallow anomalous masses that may be of interest for commercial exploitation. Geological knowledge is essential for interpreting the residual anomalies. In eroded shield areas, like Canada or Scandinavia, anomalies with very short wavelengths may be due to near-surface mineralized bodies. In sedimentary basins, short- or intermediate-wavelength anomalies may arise from structures related to reservoirs for petroleum or natural gas.

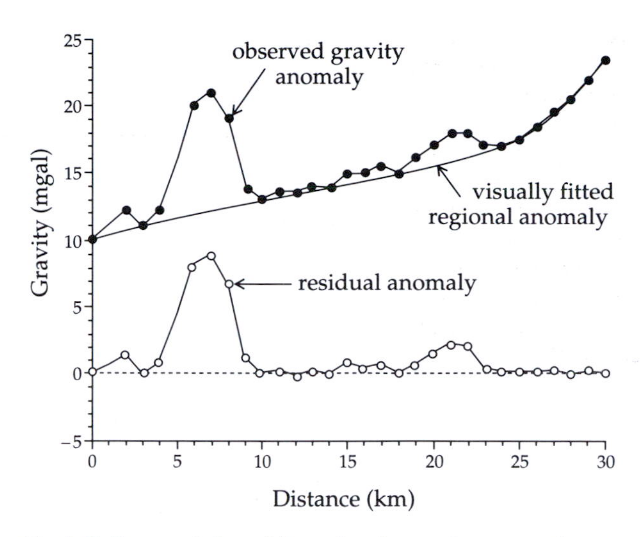

Fig. 2.42 Representation of the regional anomaly on a gravity profile by visually fitting the large-scale trend with a smooth curve.

2.6.2 **Separation of regional and residual anomalies**

The separation of anomalies of regional and local origin is an important step in the interpretation of a gravity map. The analysis may be based on selected profiles across some structure, or it may involve the two-dimensional distribution of anomalies in a gravity map. Numerous techniques have been applied to the decomposition of a gravity anomaly into its constituent parts. They range in sophistication from simple visual inspection of the anomaly pattern to advanced mathematical analysis. A few examples of these methods are described below.

2.6.2.1 *Visual analysis*

The simplest way of representing the regional anomaly on a gravity profile is by visually fitting the large-scale trend with a smooth curve (Fig. 2.42). The value of the regional gravity given by this trend is subtracted point by point from the Bouguer gravity anomaly. This method allows the interpreter to fit curves that leave residual anomalies with a sign appropriate to his interpretation of the density distribution.

This approach may be adapted to the analysis of a gravity map by visually smoothing the contour lines. In Fig. 2.43a the contour lines of equal Bouguer gravity curve sharply around a local abnormality. The more gently curved contours have been continued smoothly as dotted lines. They indicate how the interpreter thinks the regional gravity field (Fig. 2.43b) would continue in the absence of the local abnormality. The values of the regional and original Bouguer gravity are interpolated from the corresponding maps at points spaced on a regular grid. The regional value is subtracted from the Bouguer anomaly at each point and the computed residuals are contoured to give a map of the local gravity anomaly (Fig. 2.43c). The experience and skill of the interpreter are important factors in the success of visual methods.

2.6.2.2 *Polynomial representation*

In an alternative method the regional trend is represented by a straight line, or, more generally, by a smooth polynomial curve. If x denotes the horizontal position on a gravity profile, the regional gravity Δg_R may be written

$$\Delta g_R = \Delta g_0 + \Delta g_1 x + \Delta g_2 x^2 + \Delta g_3 x^3 + \ldots + \Delta g_n x^n \qquad (2.82)$$

The polynomial is fitted by the method of least squares to the observed gravity profile. This gives optimum values for the coefficients Δg_n. The method also has drawbacks. The higher the order of the polynomial, the better it fits the observations (Fig. 2.44). The ludicrous extreme is when the order of the polynomial is one less than the number of observations; the curve then passes perfectly through all the data points, but the regional gravity anomaly has no meaning geologically. The interpreter's judgement is important in selecting the order of the polynomial, which is usually chosen to be the lowest possible order that repre-

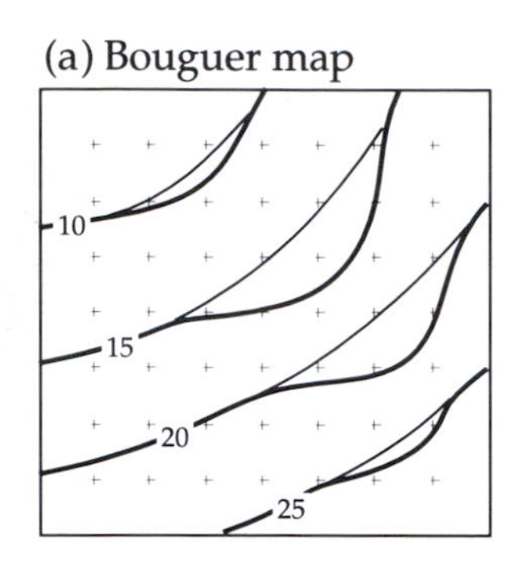

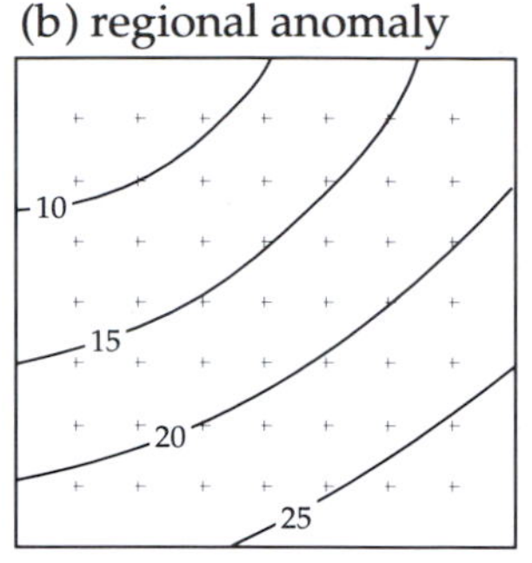

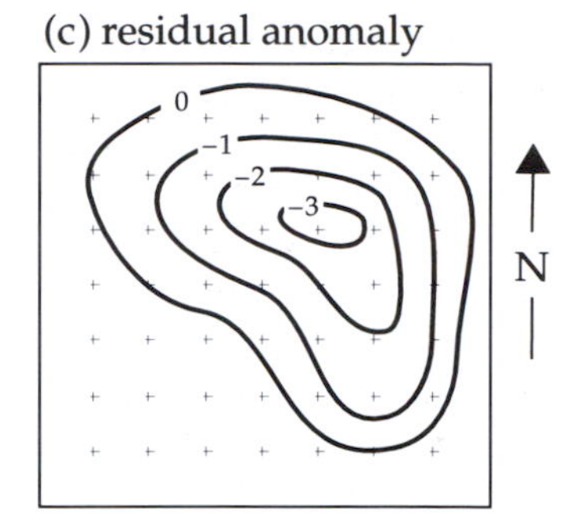

Fig. 2.43 Removal of regional trend from a gravity map by contour smoothing: (a) hand-drawn smoothing of contour lines on original Bouguer gravity map, (b) map of regional gravity variation, (c) residual gravity anomaly after subtracting the regional variation from the Bouguer gravity map (after Robinson and Çoruh, 1988). Values are in mgal.

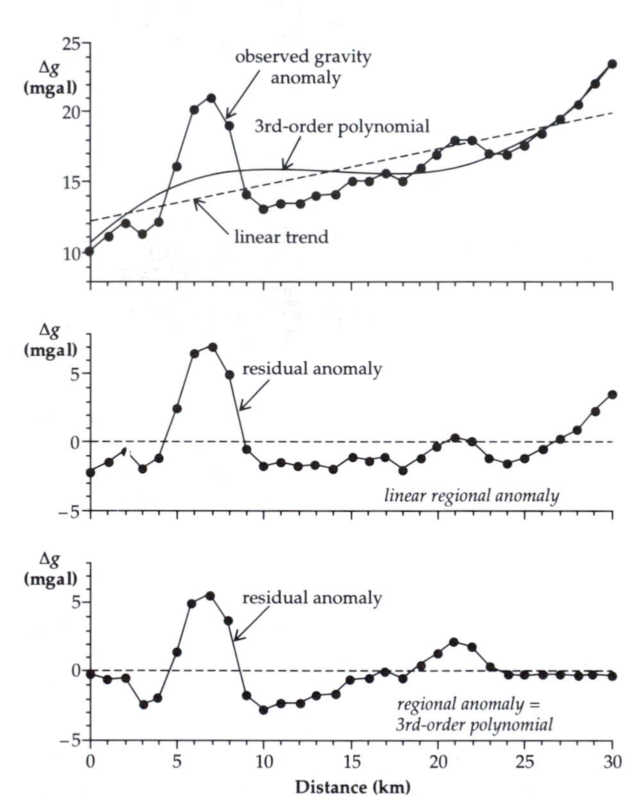

Fig. 2.44 Representation of the regional trend by a smooth polynomial curve fitted to the observed gravity profile by the method of least squares.

sents most of the regional trend. Moreover, a curve fitted by least squares must pass through the mean of the gravity values, so that the residual anomalies are divided equally between positive and negative values. Each residual anomaly is flanked by anomalies of opposite sign (Fig. 2.44), which are due to the same anomalous mass that caused the central anomaly and so have no significance of their own.

Polynomial fitting can also be applied to gravity maps. It is assumed that the regional anomaly can be represented by a smooth surface, $\Delta g(x, y)$, which is a low-order polynomial of the horizontal position coordinates x and y. In the simplest case the regional anomaly is expressed as a first-order polynomial, or plane. To express changes in the gradient of gravity a higher-order polynomial is needed. For example, the regional gravity given by a second-order polynomial is

$$\Delta g(x,y)=\Delta g_0+\Delta g_{x1}x+\Delta g_{y1}y+\Delta g_{x2}x^2+\Delta g_{y2}y^2+\Delta g_{xy}xy \tag{2.83}$$

As in the analysis of a profile, the optimum values of the coefficients Δg_{x1}, Δg_{y1}, etc., are determined by least-squares fitting. The residual anomaly is again computed point by point by subtracting the regional from the original data.

2.6.2.3 *Fourier analysis*

The gravity anomaly along a profile can be analyzed with techniques developed for investigating time series. Instead of varying with time, as the seismic signal does in a seismometer, the gravity anomaly $\Delta g(x)$ varies with position x along the profile. For a spatial distribution the wave number, $k=2\pi/\lambda$, is the counterpart of the frequency of a time series. If it can be assumed that its variation is periodic, the function $\Delta g(x)$ can be expressed as the superposition of discrete harmonics. Each harmonic is a sine or cosine function whose argument is a multiple of the fundamental wave number, which is defined by a wavelength λ that is twice the signal length of the anomaly.

$$\Delta g(x)=\sum_{n=1}^{N}\left(a_n\cos\frac{2n\pi x}{\lambda}+b_n\sin\frac{2n\pi x}{\lambda}\right) \tag{2.84}$$

The expression for $\Delta g(x)$ is called a *Fourier series*. In this example it is truncated after N sine and N cosine terms. In principle, the value of N can be as large as necessary to describe the regional anomaly adequately, a decision that must be made on practical grounds by the interpreter. The importance of any individual term of order n is described by the relative value of the coefficients a_n and b_n, which play the role of weighting functions.

The properties of sine and cosine functions allow the coefficients a_n and b_n to be calculated. If integrated over a full cycle, the resulting value of a sine or cosine function is zero; only the squared values of sines and cosines do not integrate to zero. If Eq. (2.84) is multiplied by $\cos(2m\pi x/\lambda)$, we get $\Delta g(x)\cos(2m\pi x/\lambda)$ on the left side and product terms $\cos(2n\pi x/\lambda)\cos(2m\pi x/\lambda)$ and $\sin(2n\pi x/\lambda)\cos(2m\pi x/\lambda)$ on the right side. Each product can be written as the sum or difference of two sines or cosines. Thus, if $\Delta g(x)\cos(2m\pi x/\lambda)$ is integrated over a full cycle, all terms on the right side vanish except for the case where n equals m. This yields the values of the coefficients a_n and b_n, namely

$$a_n=\frac{2}{\lambda}\int_0^{\lambda}\Delta g(x)\cos\frac{2n\pi x}{\lambda}\,dx$$

$$b_n=\frac{2}{\lambda}\int_0^{\lambda}\Delta g(x)\sin\frac{2n\pi x}{\lambda}\,dx \tag{2.85}$$

The two-dimensional variation of a mapped gravity anomaly can be expressed in a similar way with the aid of *double Fourier series*. In this case the gravity anomaly is a function of both the x- and y-coordinates and can be written

$$\Delta g(x,y)=\sum_{n=1}^{N}\sum_{m=1}^{M}(a_{nm}C_nC^*_m+b_{nm}C_nS^*_m+c_{nm}S_nC^*_m+d_{nm}S_nS^*_m) \tag{2.86}$$

where

$$C_n=\cos\frac{2n\pi x}{\lambda};\ S_n=\sin\frac{2n\pi x}{\lambda}$$

$$C^*_m=\cos\frac{2m\pi y}{\lambda};\ S^*_m=\sin\frac{2m\pi y}{\lambda} \tag{2.87}$$

The derivation of the coefficients a_{nm}, b_{nm}, c_{nm} and d_{nm} is similar to the one-dimensional case, but is somewhat more complicated.

As in the simpler one-dimensional case of a gravity anomaly on a profile the expression of two-dimensional gravity anomalies by double Fourier series is analogous to summing weighted sinusoidal functions. These can be visualized as corrugations of the x–y plane (Fig. 2.45), with each corrugation weighted according to its importance to $\Delta g(x, y)$.

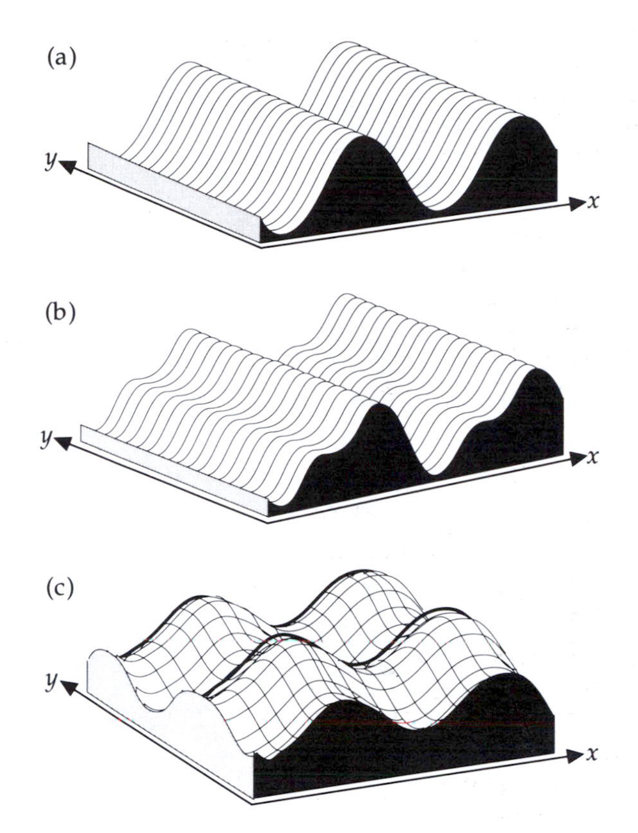

Fig. 2.45 Expression of the two-dimensional variation of a gravity anomaly using double Fourier series: (a) a single harmonic in the x-direction, (b) two harmonics in the x-direction, (c) superposed single harmonics in the x- and y-directions, respectively (after Davis, 1973).

2.6.2.4 *Anomaly enhancement and filtering*

The above discussion shows how a function that is periodic can be expressed as a Fourier sum of harmonics of a fundamental wavelength. The requirements of periodic behavior and discreteness of harmonic content are often not met. For example, the variation of gravity from one point to another is usually not periodic. Moreover, if the harmonic content of a function is made up of distinct multiples of a fundamental frequency or wave number, the wavelength spectrum consists of a number of distinct values. Yet many functions of geophysical interest are best represented by a continuous spectrum of wavelengths. To handle this kind of problem the Fourier sum of individual terms in Eq. (2.84) is replaced by a *Fourier integral*, which consists of a continuous set of frequencies or wave numbers instead of a discrete set. The Fourier integral can be used to represent non-periodic functions. It uses complex numbers (i.e., numbers that involve i, the square-root of −1). Instead of Eq. (2.84) we have

$$\Delta g(x)=\int_{-\infty}^{\infty}G(u)e^{iux}du$$

where

$$G(u)=\frac{1}{2\pi}\int_{-\infty}^{\infty}\Delta g(x)e^{-iux}dx \tag{2.88}$$

The complex function $G(u)$ defined by Eq. (2.88) is called the *Fourier transform* of the real-valued function $\Delta g(x)$. An adequate treatment of Fourier transforms is beyond the scope of this book. However, the use of this powerful mathematical technique can be illustrated without delving deeply into the theory.

A map of gravity anomalies can be represented by a function $\Delta g(x, y)$ of the Cartesian map coordinates. The Fourier transform of $\Delta g(x, y)$ is a two-dimensional complex function that involves wave numbers k_x and k_y defined by wavelengths of the gravity field with respect to the x- and y-axes ($k_x=2\pi/\lambda_x$, $k_y=2\pi/\lambda_y$). It is

$$G(x, y)=\int_{-\infty}^{\infty}\int_{-\infty}^{\infty} \Delta g(x, y)\{\cos(k_x x+k_y y) - \mathrm{i}\sin(k_x x+k_y y)\}\,dx\,dy \tag{2.89}$$

This equation assumes that the observations $\Delta g(x, y)$ can be represented by a continuous function defined over an infinite x–y plane, whereas in fact the data are of finite extent and are known at discrete points of a measurement grid. In practice, these inconsistencies are usually not important. Efficient computer algorithms permit the rapid computation of the Fourier transform $G(x, y)$ of the gravity anomaly $\Delta g(x, y)$.

The two-dimensional Fourier transform simplifies the operation of digitally filtering the gravity anomalies. A filter is a spatial function of the coordinates x and y. When the function $\Delta g(x, y)$ representing the gravity data is multiplied by the filter function, a new function is produced. The process is called *convolution* and the output is a map of the filtered gravity data. The computation in the spatial domain defined by the x- and y-coordinates can be time-consuming. It is often faster to compute the Fourier transforms of the gravity and filter functions, multiply these together in the Fourier domain and inverse Fourier transform the product back to the spatial domain.

The nature of the filter applied in the Fourier domain can be chosen to eliminate certain wavelengths. For example, it can be designed to cut out all wavelengths shorter than a selected wavelength and to pass longer wavelengths. This is called a *low-pass filter*; it passes long wavelengths that have low wave numbers. The irregularities in a Bouguer gravity anomaly map (Fig. 2.46a) are removed by low-pass filtering, leaving a filtered map (Fig. 2.46b) that is much smoother than the original. Alternatively, the filter in the Fourier domain can be designed to eliminate wavelengths longer than a selected wavelength and to pass shorter wavelengths. The application of such a *high-pass filter* enhances the short-wavelength (high wave number) component of the gravity map (Fig. 2.46c).

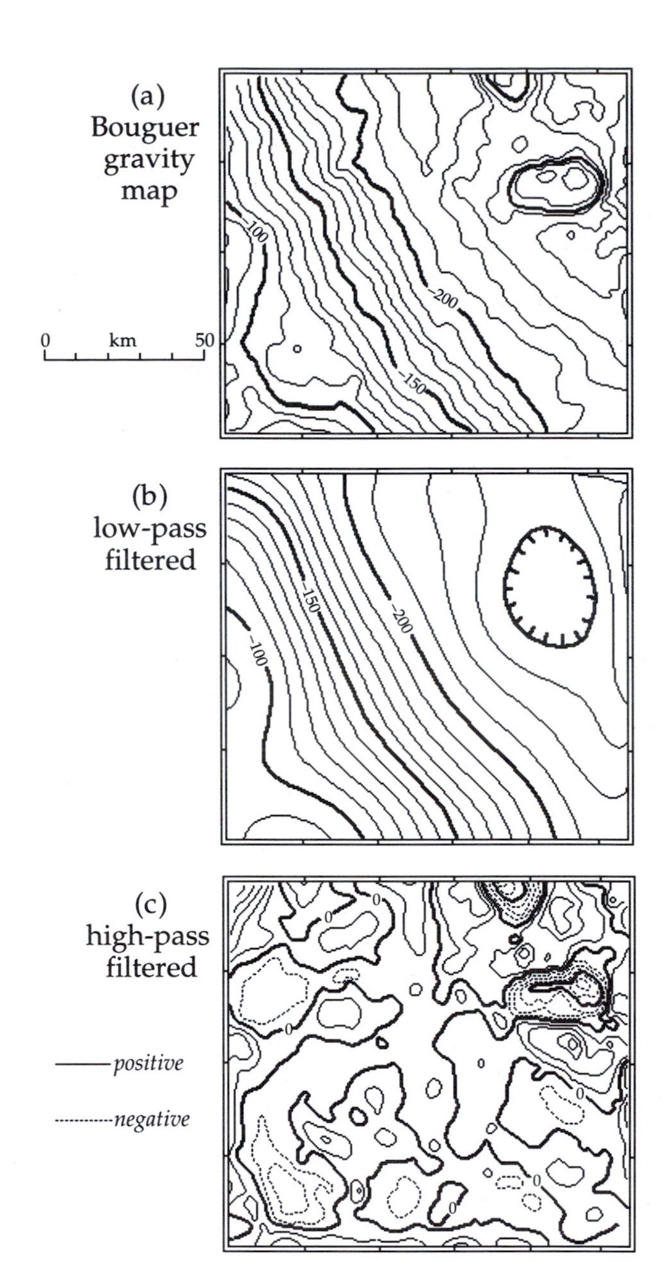

Fig. 2.46 The use of wavelength filtering to emphasize selected anomalies in the Sierra Nevada, California: (a) unfiltered Bouguer gravity map, (b) low-pass filtered gravity map with long-wavelength regional anomalies, and (c) high-pass filtered gravity map enhancing short-wavelength local anomalies. Contour interval: (a) and (b) 10 mgal, (c) 5 mgal (after Dobrin and Savit, 1988).

Wavelength filtering can be used to emphasize selected anomalies. For example, in studying large-scale crustal structure the gravity anomalies due to local small bodies are of less interest than the regional anomalies, which can be enhanced by applying a low-pass filter. Conversely, in the investigation of anomalies due to shallow crustal sources the regional effect can be suppressed by high-pass filtering.

2.6.3 Modelling gravity anomalies

After removal of regional effects the residual gravity anomaly must be interpreted in terms of an anomalous density distribution. Modern analyses are based on iterative modelling using high-speed computers. Earlier methods of interpretation utilized comparison of the observed gravity anomalies with the computed anomalies of geometric shapes. The success of this simple approach is due to the insensitivity of the shape of a gravity anomaly to minor variations in the anomalous density distribution. Some fundamental problems of interpreting gravity anomalies can be learned from the computed effects of geometric models. In particular, it is important to realize that the interpretation of gravity anomalies is not unique; different density distributions can give the same anomaly.

2.6.3.1 *Uniform sphere: model for a diapir*

Diapiric structures introduce material of different density into the host rock. A low-density salt dome (ρ=2150 kg m^{-3}) intruding higher-density carbonate rocks (ρ_0=2500 kg m^{-3}) has a density contrast $\Delta\rho$=– 350 kg m^{-3} and causes a negative gravity anomaly. A volcanic plug (ρ=2800 kg m^{-3}) intruding a granite body (ρ_0=2600 kg m^{-3}) has a density contrast $\Delta\rho$=+ 200 kg m^{-3}, which causes a positive gravity anomaly. The contour lines on a map of the anomaly are centered on the diapir, so all profiles across the center of the structure are equivalent. The anomalous body can be modelled equally by a vertical cylinder or by a sphere, which we will evaluate here because of the simplicity of the model.

Assume a sphere of radius R and density contrast Δr with its center at depth z below the surface (Fig. 2.47). The attraction Δg of the sphere is as though the anomalous mass M of the sphere were concentrated at its center. If we measure horizontal position from a point above its center, at distance x the vertical component Δg_z is given by

$$\Delta g_z = \Delta g \sin\theta = G\frac{M}{r^2}\frac{z}{r} \tag{2.90}$$

where

$$M = \tfrac{4}{3}\pi R^3 \Delta\rho \text{ and } r^2 = z^2 + x^2$$

Substituting these expressions into Eq. (2.90) and rearranging terms gives

$$\Delta g_z = \tfrac{4}{3}\pi G\left(\frac{\Delta\rho R^3}{z^2}\right)\left[\frac{z^3}{(z^2+x^2)^{3/2}}\right]$$

$$= \tfrac{4}{3}\pi G\left(\frac{\Delta\rho R^3}{z^2}\right)\left[\frac{1}{1+(x/z)^2}\right]^{3/2} \tag{2.91}$$

The terms in the first pair of parentheses depend on the size, depth and density contrast of the anomalous sphere.

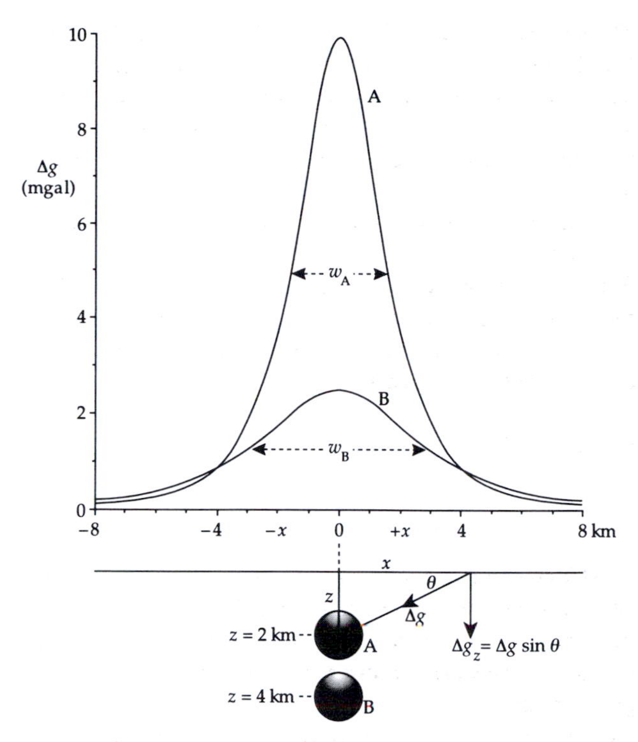

Fig. 2.47 Gravity anomalies for buried spheres with the same radius R and density contrast $\Delta\rho$ but with their centers at different depths z below the surface. The anomaly of the deeper sphere B is flatter and broader than the anomaly of the shallower sphere A.

They determine the maximum amplitude of the anomaly, Δg_0, which is reached over the center of the sphere at x=0. The peak value is given by

$$\Delta g_0 = \tfrac{4}{3}\pi G\frac{\Delta\rho R^3}{z^2} \tag{2.92}$$

This equation shows how the depth to the center of the sphere affects the peak amplitude of the anomaly; the greater the depth to the center, the smaller the amplitude (Fig. 2.47). For a given depth the same peak anomaly can be produced by numerous combinations of $\Delta\rho$ and R; a large sphere with a low density contrast can give an identical anomaly to a small sphere with a high density contrast. The gravity data alone do not allow us to resolve this ambiguity.

The terms in the second pair of parentheses in Eq. (2.91) describe how the amplitude of the anomaly varies with distance along the profile. The anomaly is symmetrical with respect to x, reaches a maximum value Δg_0 over the center of the sphere (x=0) and decreases to zero at great distances (x=∞). Note that the larger the depth z, the more slowly the amplitude decreases laterally with increasing x. A deep source produces a smaller but broader anomaly than the same source at shallower depth. The width w of the anomaly where the amplitude has one-half its maximum value is called the 'half-height width'. The depth z to the center of the sphere is deduced from this anomaly width from the relationship $z=0.652w$.

2.6.3.2 *Horizontal line element*

Many geologically interesting structures extend to great distances in one direction but have the same cross-sectional shape along the strike of the structure. If the length along strike were infinite, the two-dimensional variation of density in the area of cross-section would suffice to model the structure. However, this is not really valid as the lateral extent is never infinite. As a general rule, if the length of the structure normal to the profile is more than twenty times its width or depth, it can be treated as two-dimensional (2-D). Otherwise, the end effects due to the limited lateral extent of the structure must be taken into account in computing its anomaly. An elongate body that requires end corrections is sometimes referred to as a 2.5-D structure. For example, the mass distribution under elongated bodies like anticlines, synclines and faults should be modelled as 2.5-D structures. Here, we will handle the simpler two-dimensional models of these structures.

Let an infinitely long linear mass distribution with mass m per unit length extend horizontally along the y-axis at depth z (Fig. 2.48). The contribution $d(\Delta g_z)$ to the vertical gravity anomaly Δg_z at a point on the x-axis due to a small element of length dy is

$$d(\Delta g_z)=G\frac{m\,dy}{r^2}\sin\theta=G\frac{m\,dy}{r^2}\frac{z}{r} \tag{2.93}$$

The line element extends to infinity along the positive and negative y-axis, so its vertical gravity anomaly is found by integration:

$$\Delta g_z=Gmz\int_{-\infty}^{\infty}\frac{dy}{r^3}=Gmz\int_{-\infty}^{\infty}\frac{dy}{(u^2+y^2)^{3/2}} \tag{2.94}$$

where $u^2=x^2+z^2$. The integration is simplified by changing variables, so that $y=u\tan\varphi$; then $dy=u\sec^2\varphi\,d\varphi$ and $(u^2+y^2)^{3/2}=u^3\sec^3\varphi$. This gives

$$\Delta g_z=\frac{Gmz}{u^2}\int_{-\frac{\pi}{2}}^{\frac{\pi}{2}}\cos\varphi\,d\varphi \tag{2.95}$$

which, after evaluation of the integral, gives

$$\Delta g_z=\frac{2Gmz}{z^2+x^2} \tag{2.96}$$

This expression can be written as the derivative of a potential function Ψ:

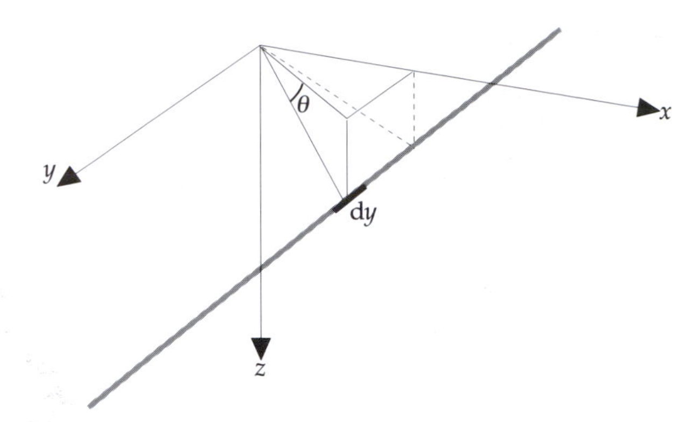

Fig. 2.48 Geometry for calculation of the gravity anomaly of an infinitely long linear mass distribution with mass m per unit length extending horizontally along the y-axis at depth z.

$$\Delta g_z=Gm\frac{2z}{u^2}=-\frac{\partial\Psi}{\partial z} \tag{2.97}$$

$$\Psi=Gm\log_e\left(\frac{1}{u}\right)=Gm\log_e\left(\frac{1}{\sqrt{x^2+z^2}}\right) \tag{2.98}$$

Ψ is called the *logarithmic potential*. Eqs. (2.96) and (2.98) are useful results for deriving formulas for the gravity anomaly of linear structures like an anticline (or syncline) or a fault.

2.6.3.3 *Horizontal cylinder: model for anticline or syncline*

The gravity anomaly of an anticline can be modelled by assuming that the upward folding of strata brings rocks with higher density nearer to the surface (Fig. 2.49a), thereby causing a positive density contrast. A syncline is modelled by assuming that its core is filled with strata of lower density that cause a negative density contrast. In each case the geometric model of the structure is an infinite horizontal cylinder (Fig. 2.49b).

A horizontal cylinder may be regarded as composed of numerous line elements parallel to its axis. The cross-sectional area of an element (Fig. 2.50) gives a mass anomaly per unit length, $m=\Delta\rho r\,d\theta\,dr$. The contribution $d\Psi$ of a line element to the potential at the surface is

$$d\Psi=2G\Delta\rho\log_e\frac{1}{u}r\,dr\,d\theta \tag{2.99}$$

Integrating over the cross-section of the cylinder gives its potential Ψ; the vertical gravity anomaly of the cylinder is then found by differentiating Ψ with respect to z. Noting that $du/dz=z/u$ we get

$$\Delta g_z=-\frac{\partial\Psi}{\partial z}=-\frac{z}{u}\frac{\partial\Psi}{\partial u}$$

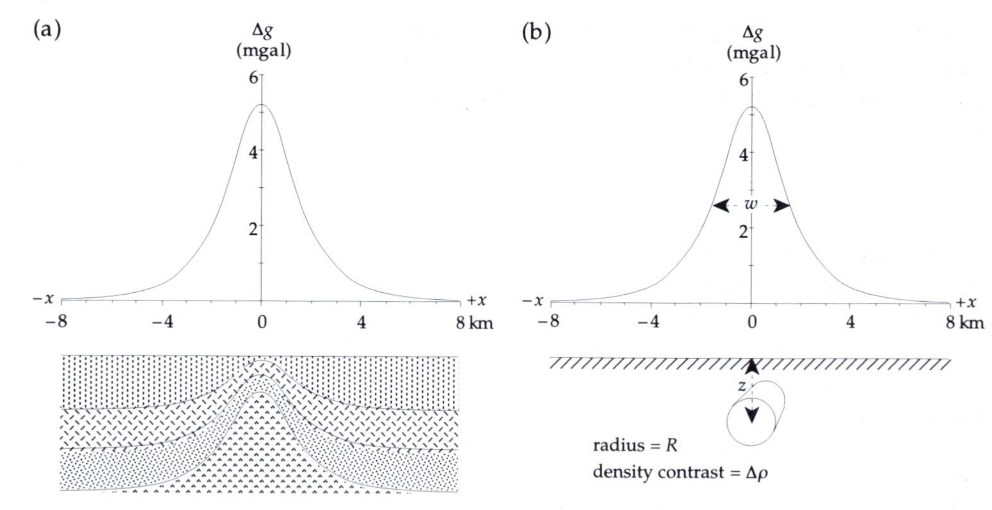

Fig. 2.49 Calculation of the gravity anomaly of an anticline: (a) structural cross-section, and (b) geometric model by an infinite horizontal cylinder.

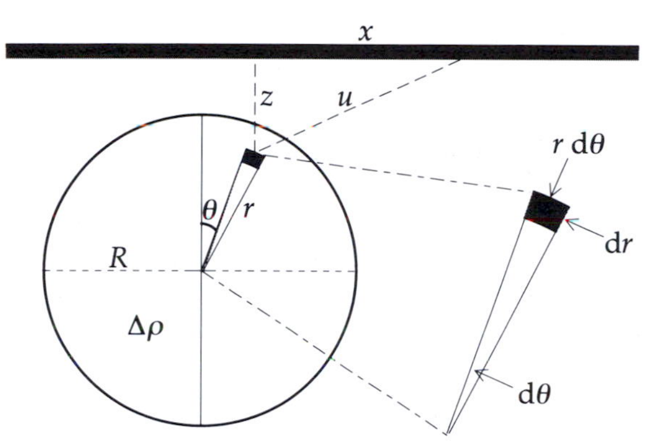

Fig. 2.50 Cross-sectional geometry for calculating the gravity anomaly of a buried horizontal cylinder made up of line elements parallel to its axis.

$$\Delta g_z = -\frac{2G\Delta\rho z}{u}\int_0^{2\pi}\int_0^R \frac{\partial}{\partial u}\log_e\frac{1}{u}\, r dr\, d\theta \tag{2.100}$$

After first carrying out the differentiation within the integral, this simplifies to

$$\Delta g_z = \frac{2G\Delta\rho z}{u}\int_0^{2\pi}\int_0^R r dr\, d\theta = \frac{2G\pi R^2\Delta\rho z}{x^2+z^2} \tag{2.101}$$

Comparing Eq. (2.101) and Eq. (2.96) it is evident that the gravity anomaly of the cylinder is the same as that of a linear mass element concentrated along its axis, with mass $m=\pi R^2\Delta\rho$ per unit length along the strike of the structure. The anomaly can be written

$$\Delta g_z = 2\pi G\left(\frac{\Delta\rho R^2}{z}\right)\left[\frac{1}{1+(x/z)^2}\right] \tag{2.102}$$

The shape of the anomaly on a profile normal to the structure (Fig. 2.49b) resembles that of a sphere (Fig. 2.47). The central peak value Δg_0 is given by

$$\Delta g_0 = 2\pi G\frac{\Delta\rho R^2}{z} \tag{2.103}$$

The anomaly of a horizontal cylinder decreases laterally less rapidly than that of a sphere, due to the long extent of the cylinder normal to the profile. The 'half-height width' w of the anomaly is again dependent on the depth z to the axis of the cylinder; in this case the depth is given by $z=0.5w$.

2.6.3.4 *Horizontal thin sheet*

We next compute the anomaly of a thin horizontal ribbon of infinite length normal to the plane of the profile. This is done by fictively replacing the ribbon with numerous infinitely long line elements laid side by side. Let the depth of the sheet be z, its thickness t and its density contrast $\Delta\rho$ (Fig. 2.51a); the mass per unit length in the y-direction of a line element of width dx is $(\Delta\rho t\, dx)$. Substituting in Eq. (2.96) gives the gravity anomaly of the line element; the anomaly of the thin ribbon is then computed by integrating between the limits x_1 and x_2 (Fig. 2.51b)

$$\Delta g_z = 2G\Delta\rho t z\int_{x_1}^{x_2}\frac{dx}{x^2+z^2}$$

$$= 2G\Delta\rho t\left[\tan^{-1}\left(\frac{x_2}{z}\right) - \tan^{-1}\left(\frac{x_1}{z}\right)\right] \tag{2.104}$$

Writing $\tan^{-1}(x_1/z)=\phi_1$ and $\tan^{-1}(x_2/z)=\phi_2$ as in Fig. 2.51b the equation becomes

$$\Delta g_z = 2G\Delta\rho t[\phi_2 - \phi_1] \tag{2.105}$$

i.e., the gravity anomaly of the horizontal ribbon is proportional to the angle it subtends at the point of measurement.

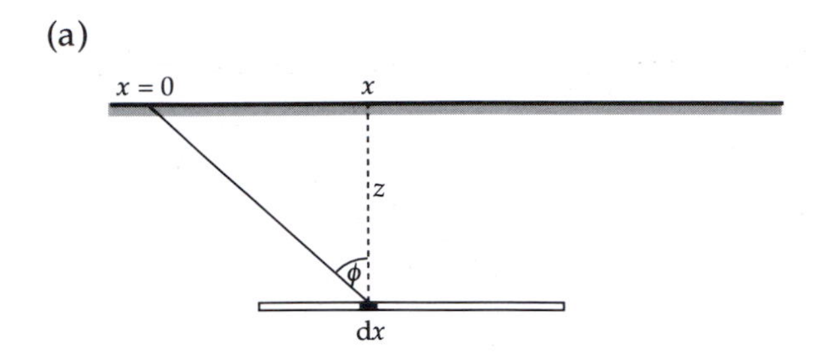

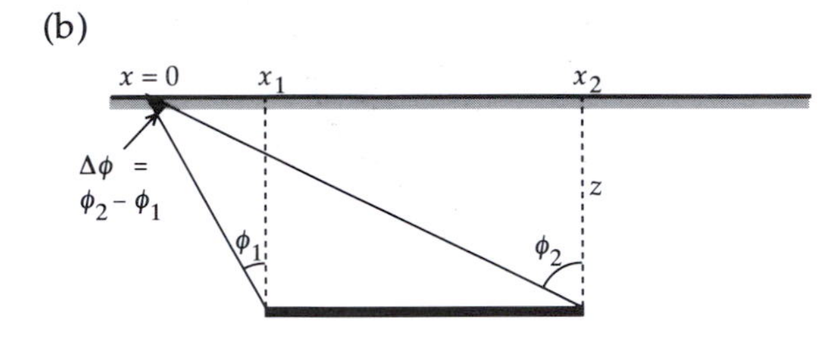

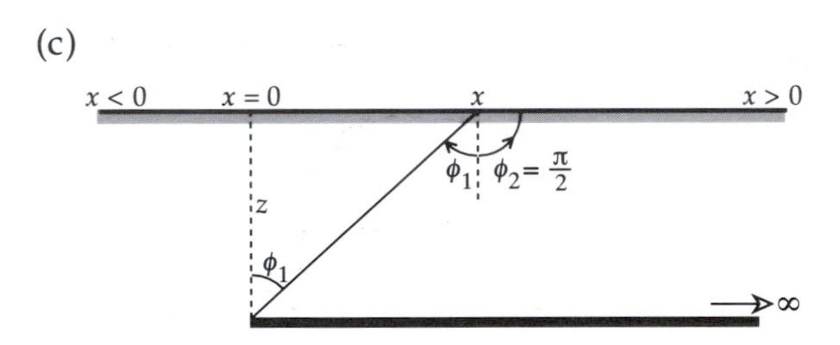

Fig. 2.51 Geometry for computation of the gravity anomaly across a horizontal thin sheet: (a) subdivision of the ribbon into line elements of width dx, (b) thin ribbon between the horizontal limits x_1 and x_2, and (c) semi-infinite horizontal thin sheet.

The anomaly of a semi-infinite horizontal sheet is a limiting case of this result. For easier reference, the origin of x is moved to the edge of the sheet, so that distances to the left are negative and those to the right are positive (Fig. 2.51c). This makes $\phi_1 = -\tan^{-1}(x/z)$. The remote end of the sheet is at infinity, and $\phi_2 = \pi/2$. The gravity anomaly is then

$$\Delta g_z = 2G\Delta\rho t\left[\frac{\pi}{2} + \tan^{-1}\left(\frac{x}{z}\right)\right] \tag{2.106}$$

A further example is the infinite horizontal sheet, which extends to infinity in the positive and negative x- and y-directions. With $\phi_2 = \pi/2$ and $\phi_1 = -\pi/2$ the anomaly is

$$\Delta g_z = 2\pi G\Delta\rho t \tag{2.107}$$

which is the same as the expression for the Bouguer plate correction (Eq. (2.75)).

2.6.3.5 *Horizontal slab: model for a vertical fault*

The gravity anomaly across a vertical fault increases progressively to a maximum value over the uplifted side

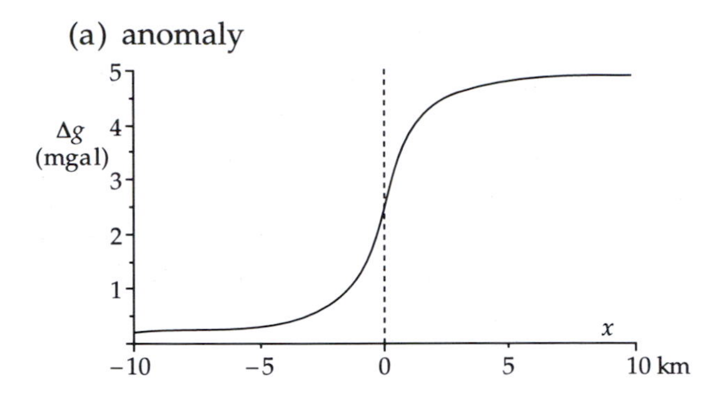

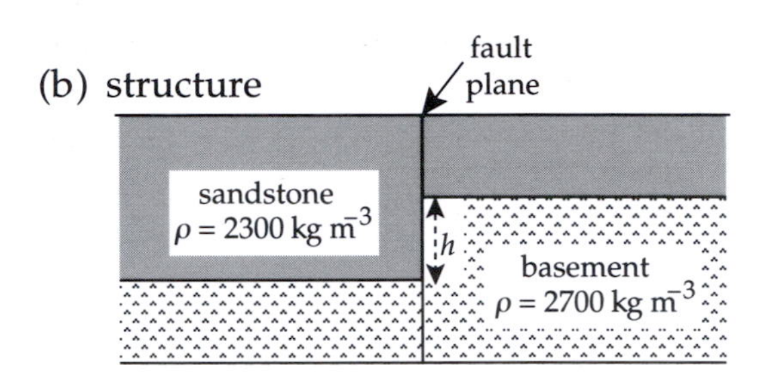

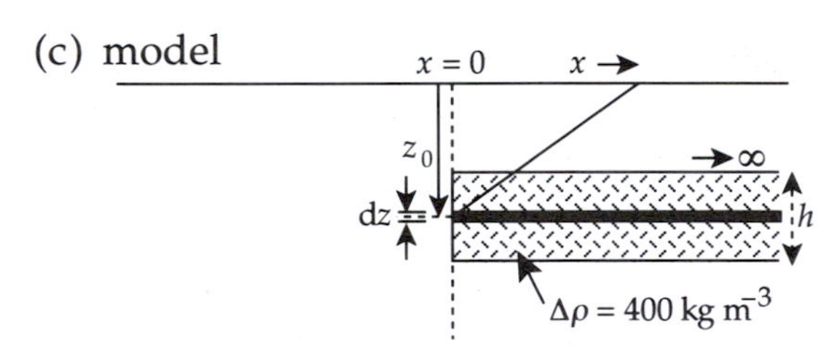

Fig. 2.52 (a) The gravity anomaly across a vertical fault; (b) structure of a fault with vertical displacement h, and (c) model of the anomalous body as a semi-infinite horizontal slab of height h.

(Fig. 2.52a). This is interpreted as due to the upward displacement of denser material, which causes a horizontal density contrast across a vertical step of height h (Fig. 2.52b). The faulted block can be modelled as a semi-infinite horizontal slab of height h and density contrast $\Delta\rho$ with its mid-point at depth z_0 (Fig. 2.52c).

Let the slab be divided into thin, semi-infinite horizontal sheets of thickness dz at depth z. The gravity anomaly of a given sheet is given by Eq. (2.106) with dz for the thickness t. The anomaly of the semi-infinite slab is found by integrating with respect to z over the thickness of the slab; the limits of integration are $z-(h/2)$ and $z+(h/2)$. After slightly rearranging terms this gives

$$\Delta g_z = 2G\Delta\rho h\left[\frac{\pi}{2}+\frac{1}{h}\int_{z_0-\frac{h}{2}}^{z_0+\frac{h}{2}} \tan^{-1}\left(\frac{x}{z}\right)dz\right] \quad (2.108)$$

The second expression in the brackets is the mean value of the angle $\tan^{-1}(x/z)$ averaged over the height of the fault step. This can be replaced to a good approximation by the value at the mid-point of the step, at depth z_0. This gives

$$\Delta g_z = 2G\Delta\rho h\left[\frac{\pi}{2}+\tan^{-1}\left(\frac{x}{z_0}\right)\right] \quad (2.109)$$

Comparison of this expression with Eq. (2.106) shows that the anomaly of the vertical fault (or a semi-infinite thick horizontal slab) is the same as if the anomalous slab were replaced by a thin sheet of thickness h at the mid-point of the vertical step. Eq. 2.109 is called the 'thin-sheet approximation'. It is accurate to about 2% provided that $z_0>2h$.

2.6.3.6 *Iterative modelling*

The simple geometric models used to compute the gravity anomalies in the previous sections are crude representations of the real anomalous bodies. Modern computer algorithms have radically changed modelling methods by facilitating the use of an *iterative* procedure. A starting model with an assumed geometry and density contrast is postulated for the anomalous body. The gravity anomaly of the body is then computed and compared with the residual anomaly. The parameters of the model are changed slightly and the computation is repeated until the discrepancies between the model anomaly and the residual anomaly are smaller than a determined value. However, as in the case of the simple models, this does not give a unique solution for the density distribution.

Two- and three-dimensional iterative techniques are in widespread use. The two-dimensional (2-D) method assumes that the anomalous body is infinitely long parallel to the strike of the structure, but end corrections for the possibly limited horizontal extent of the body can be made. We can imagine that the cross-sectional shape of the body is replaced by countless thin rods or line elements aligned parallel to the strike. Each rod makes a contribution to the vertical component of gravity at the origin (Fig. 2.53a). The gravity anomaly of the structure is calculated by adding up the contributions of all the line elements; mathematically, this is an integration over the end surface of the body. Although the theory is beyond the scope of this chapter, the gravity anomaly has the following simple form:

$$\Delta g_z = 2G\Delta\rho \oint z\,d\theta \quad (2.110)$$

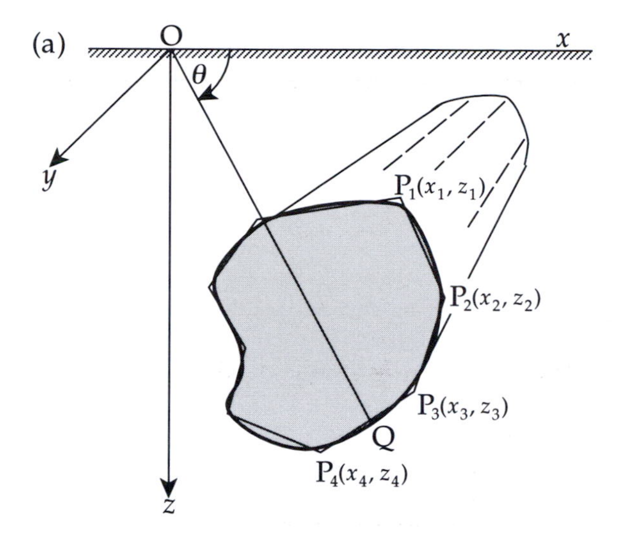

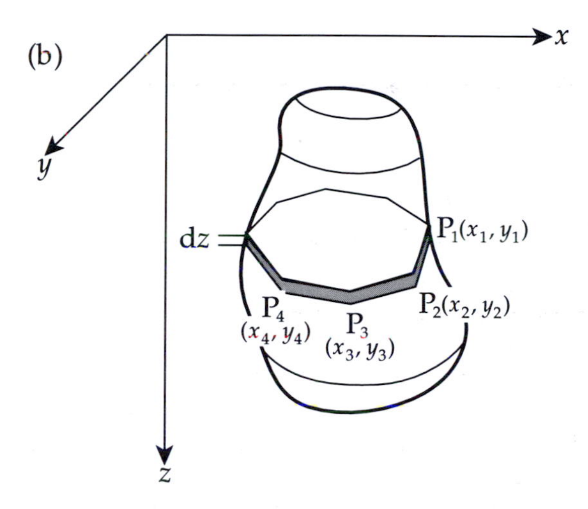

Fig. 2.53 Methods of computing gravity anomalies of irregular bodies: (a) the cross-section of a two-dimensional structure is replaced with a multi-sided polygon, and (b) a three-dimensional body is replaced with thin horizontal laminae.

The angle θ is defined to lie between the positive x-axis and the radius from the origin to a line element (Fig. 2.53a), and the integration over the end-surface has been changed to an integration around its boundary. The computer algorithm for the calculation of this integral is greatly speeded up by replacing the true cross-sectional shape with an N-sided polygon (Fig. 2.53a). Apart from the assumed density contrast, the only important parameters for the computation are the (x, z) coordinates of the corners of the polygon. The origin is now moved to the next point on a profile across the structure. This move changes only the x-coordinates of the corners of the polygon. The calculations are repeated for each successive point on the profile. Finally, the calculated gravity anomaly profile across the structure is compared to the observed anomaly and the residual differences are evaluated. The coordinates of the corners of the polygon are adjusted and the calculation is reiterated until the residuals are less than a selected tolerance level.

The gravity anomaly of a three-dimensional (3-D) body is modelled in a similar way. Suppose that we have a contour map of the body; the contour lines show the smooth outline of the body at different depths. We could construct a close replica of the body by replacing the material between successive contour lines with thin laminae. Each lamina has the same outline as the contour line and has a thickness equal to the contour separation. As a further approximation the smooth outline of each lamina is replaced by a multi-sided polygon (Fig. 2.53b). The gravity anomaly of the polygon at the origin is computed as in the 2-D case, using the (x, y) coordinates of the corners, the thickness of the lamina and an assumed density contrast. The gravity anomaly of the 3-D body at the origin is found by adding up the contributions of all the laminae. As in the simpler 2-D example, the origin is now displaced to a new point and the computation is repeated. The calculated and observed anomalies are now compared and the coordinates of the corners of the laminae are adjusted accordingly; the assumed density distribution can also be adjusted. The iterative procedure is repeated until the desired match between computed and observed anomalies is obtained.

2.6.4 Some important regional gravity anomalies

Without auxiliary information the interpretation of gravity anomalies is ambiguous, because the same anomaly can be produced by different bodies. An independent data source is needed to restrict the choices of density contrast, size, shape and depths in the many possible gravity models. The additional information may be in the form of surface geological observations, from which the continuation of structures at depth is interpreted. Seismic refraction or reflection data provide better constraints.

The combination of seismic refraction experiments with precise gravity measurements has a long, successful history in the development of models of crustal structure. Refraction seismic profiles parallel to the trend of elongate geological structures give reliable information about the vertical velocity distribution. However, refraction profiles normal to the structural trend give uncertain information about the tilt of layers or lateral velocity changes. Lateral changes of crustal structure can be interpreted from several refraction profiles more or less parallel to the structural trend or from seismic reflection data. The refraction results give layer velocities and the depths to refracting interfaces. To compute the gravity effect of a structure the velocity distribution must first be converted to a density model using a P-wave velocity–density relationship like the curve shown in Fig. 2.37. The theoretical gravity anomaly over the structure is computed using a 2-D or 3-D method. Comparison with the observed gravity anomaly (e.g., by calculating the residual differences point by point) indicates the plausibility of the model. It is always important to keep in mind that, because of the non-uniqueness of gravity modelling, a plausible structure is not necessarily the true structure. Despite the ambiguities some characteristic features of gravity anomalies have been established for important regions of the Earth.

2.6.4.1 *Continental and oceanic gravity anomalies*

In examining the shape of the Earth we saw that the ideal reference figure is a spheroid, or ellipsoid of rotation. It is assumed that the reference Earth is in hydrostatic equilibrium. This is supported by observations of free-air and isostatic anomalies which suggest that, except in some unusual locations such as deep-sea trenches and island arcs in subduction zones, the continents and oceans are in approximate isostatic equilibrium with each other. By applying the concepts of isostasy (§6.1) we can understand the large-scale differences between Bouguer gravity anomalies over the continents and those over the oceans. In general, Bouguer anomalies over the continents are negative, especially over mountain ranges where the crust is unusually thick; in contrast, strongly positive Bouguer anomalies are found over oceanic regions where the crust is very thin.

The inverse relationship between Bouguer anomaly amplitude and crustal thickness can be explained with the aid of a hypothetical example (Fig. 2.54). Continental crust that has not been thickened or thinned by tectonic processes is considered to be 'normal' crust. It is typically 30–35 km thick. Under location A on an undeformed continental coastal region a thickness of 34 km is assumed. The theoretical gravity used in computing a gravity anomaly is defined on the reference ellipsoid, the surface of which corresponds to mean sea-level. Thus, at coastal location A on normally thick continental crust the Bouguer anomaly is close to zero. Isostatic compensation of the mountain range gives it a root-zone that increases the crustal thickness at location B. Seismic evidence shows that continental crustal density increases with depth from about 2700 kg m^{-3} in the upper granitic crust to about 2900 kg m^{-3} in the lower gabbroic crust. Thus, the density in the root-zone is much lower than the typical mantle density of 3300–3400 kg m^{-3} at the same depth under A. The low-density root beneath B causes a negative Bouguer anomaly, which typically reaches -150 to -200 mgal.

At oceanic location C the vertical crustal structure is very different. Two effects contribute to the Bouguer anomaly. A 5 km thick layer of sea-water (density 1030 kg m^{-3}) overlies the thin basic oceanic crust (density 2900 kg m^{-3}) which has an average thickness of only about 6 km. To compute the Bouguer anomaly the sea-water must be replaced by oceanic crustal rock. The attraction of the water layer is inherent in the measured gravity so the density used in correcting for the Bouguer plate and the topography of the ocean bottom is the reduced density of the oceanic crust (i.e., 2900−1030=1870 kg m^{-3}). However, a more important effect is that the top of the mantle is at a depth of only 11 km. In a vertical section below this depth the mantle has a density of 3300–3400 kg m^{-3}, much higher than the density of the continental crust at equivalent depths below coastal site A. The lower 23 km of the section beneath C represents a large excess of mass. This gives rise to a strong positive Bouguer anomaly, which can amount to 300–400 mgal.

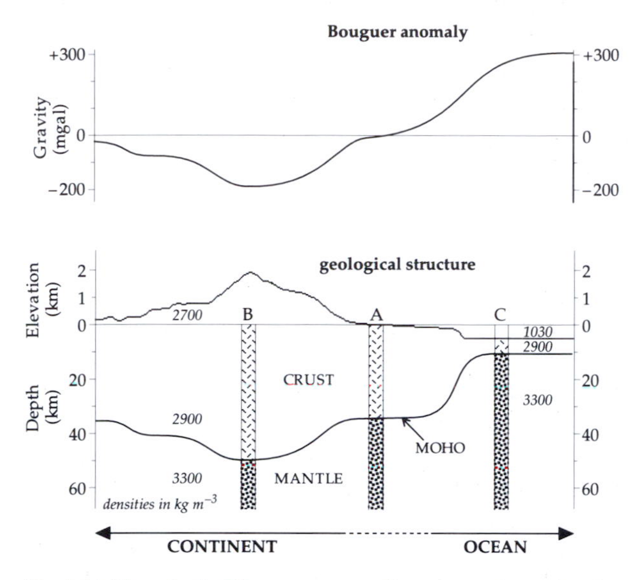

Fig. 2.54 Hypothetical Bouguer anomalies over continental and oceanic areas. The regional Bouguer anomaly varies roughly inversely with crustal thickness and topographic elevation (after Robinson and Çoruh, 1988).

2.6.4.2 *Gravity anomalies across mountain chains*

The typical gravity anomaly across a mountain chain is strongly negative due to the large low-density root-zone. The Swiss Alps provide a good example of the interpretation of such a gravity anomaly with the aid of seismic refraction and reflection results. A precise gravity survey of Switzerland carried out in the 1970s yielded an accurate Bouguer gravity map (Fig. 2.55). The map contains effects specific to the Alps. Most obviously, the contour lines are parallel to the trend of the mountain range. In the south a strong positive anomaly overrides the negative anomaly. It is the northern extension of the positive anomaly of the so-called Ivrea body, which is a high-density wedge of mantle material that was forced into an uplifted position within the western Alpine crust during an earlier continental collision. In addition, the Swiss gravity map contains the effects of

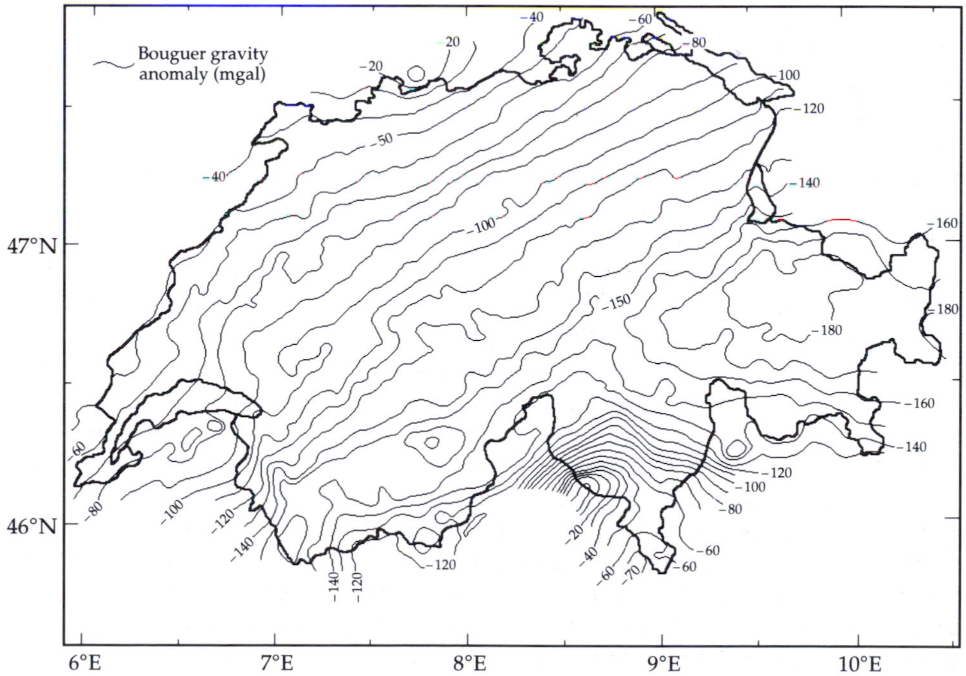

Fig. 2.55 Bouguer gravity map of Switzerland (after Klingelé and Olivier, 1980).

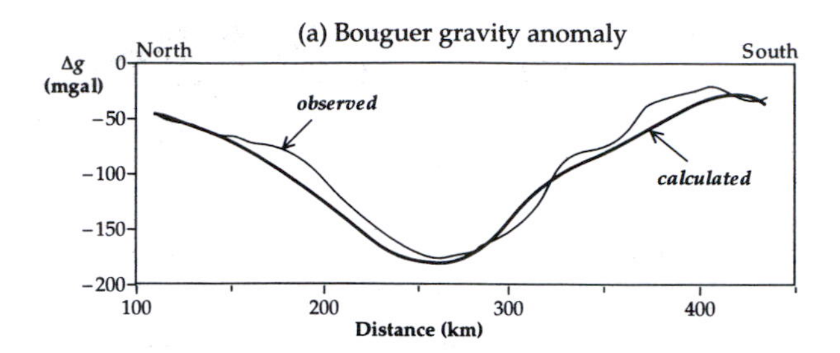

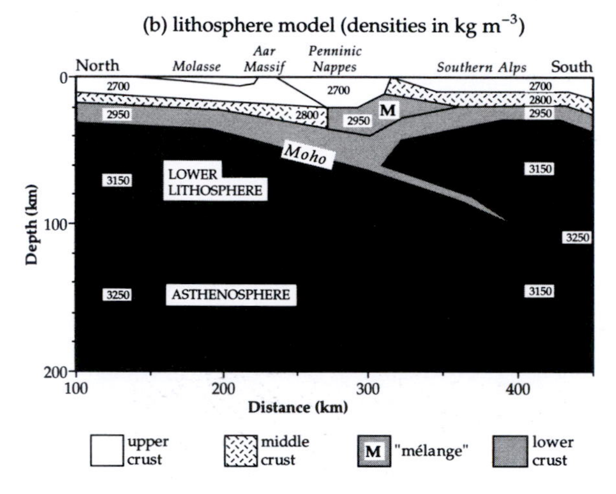

Fig. 2.56 Lithosphere density model for the Central Swiss Alps along the European Geotraverse transect, compiled from seismic refraction and reflection profiles. The 2.5-D gravity anomaly calculated for this lithospheric structure is compared to the observed Bouguer anomaly after removal of the effects of the high-density Ivrea Body and the low-density sediments in the Molasse Basin, Po Plain and larger Alpine valleys (after Holliger and Kissling, 1992)

low-density sediments that fill the Molasse basin north of the Alps, the Po Plain to the south and the major Alpine valleys.

In the late 1980s a coordinated geological and geophysical study – the European Geotraverse (EGT) – was made along and adjacent to a narrow path stretching from northern Scandinavia to the northern Africa. Detailed reflection seismic profiles along its transect through the Central Swiss Alps complemented a large amount of new and extant refraction data. The seismic results gave the depths to important interfaces. Assuming a velocity–density relationship, a model of the density distribution in the lithosphere under the traverse was obtained. Making appropriate corrections for end effects due to limited extent along strike, a 2.5-D gravity anomaly was calculated for this lithospheric structure (Fig. 2.56). After removal of the effects of the high-density Ivrea Body and the low-density sediments, the corrected Bouguer gravity profile is reproduced well by the anomaly of the lithospheric model. Using the geometric constraints provided by the seismic data, the lithospheric gravity model favors a subduction zone, dipping gently to the south, with a high-density wedge of deformed lower crustal rock ('mélange') in the middle crust. As already noted, the fact that a density model delivers a gravity anomaly that agrees well with observation does not establish the reality of the interpreted structure, which can only be confirmed by further seismic imaging. However, the gravity model provides an important check on the reasonableness of suggested models. A model of crustal or lithospheric structure that does not give an appropriate gravity anomaly can reasonably be excluded.

2.6.4.3 *Gravity anomalies across an oceanic ridge*

An oceanic ridge system is a gigantic submarine mountain range. The difference in depth between the ridge crest and adjacent ocean basin is about 3 km. The ridge system extends laterally for several hundred kilometers on either side of the axis. Gravity and seismic surveys have been carried out across several oceanic ridge systems. Some common characteristics of continuous gravity profiles across a ridge are evident on a WNW–ESE transect crossing the Mid-Atlantic Ridge at 32°N (Fig. 2.57). The free-air gravity anomalies are small, around 50 mgal or less, and correlate closely with the variations in ocean-bottom topography. This indicates that the ridge and its flanks are nearly compensated isostatically. As expected for an oceanic profile, the Bouguer anomaly is strongly positive. It is greater than 350 mgal at distances beyond 1000 km from the ridge, but decreases to less than 200 mgal over the axis of the ridge.

The depths to important refracting interfaces and the P-wave layer velocities are known from seismic refraction profiles parallel to the ridge. The seismic structure away from the ridge is layered, with P-wave velocities of 4–5 km s^{-1} for the basalts and gabbros in oceanic Layer 2 and 6.5–6.8 km s^{-1} for the meta-basalts and meta-gabbros in Layer 3; the Moho is at about 11 km depth, below which typical upper mantle velocities of 8–8.4 km s^{-1} are found. However, the layered structure breaks down under the ridge at distances less than 400 km from its axial zone. Unusual velocities around 7.3 km s^{-1} occurred at several places, suggesting the presence of anomalous low-density mantle material at comparatively shallow depth beneath the ridge. The seismic structure was converted to a density model using a velocity–density relationship. Assuming a 2-D structure, several density models were found that closely reproduced the Bouguer anomaly. However, to satisfy the Bouguer anomaly on the ridge flanks each model requires a flat body in the upper mantle beneath the ridge; it extends

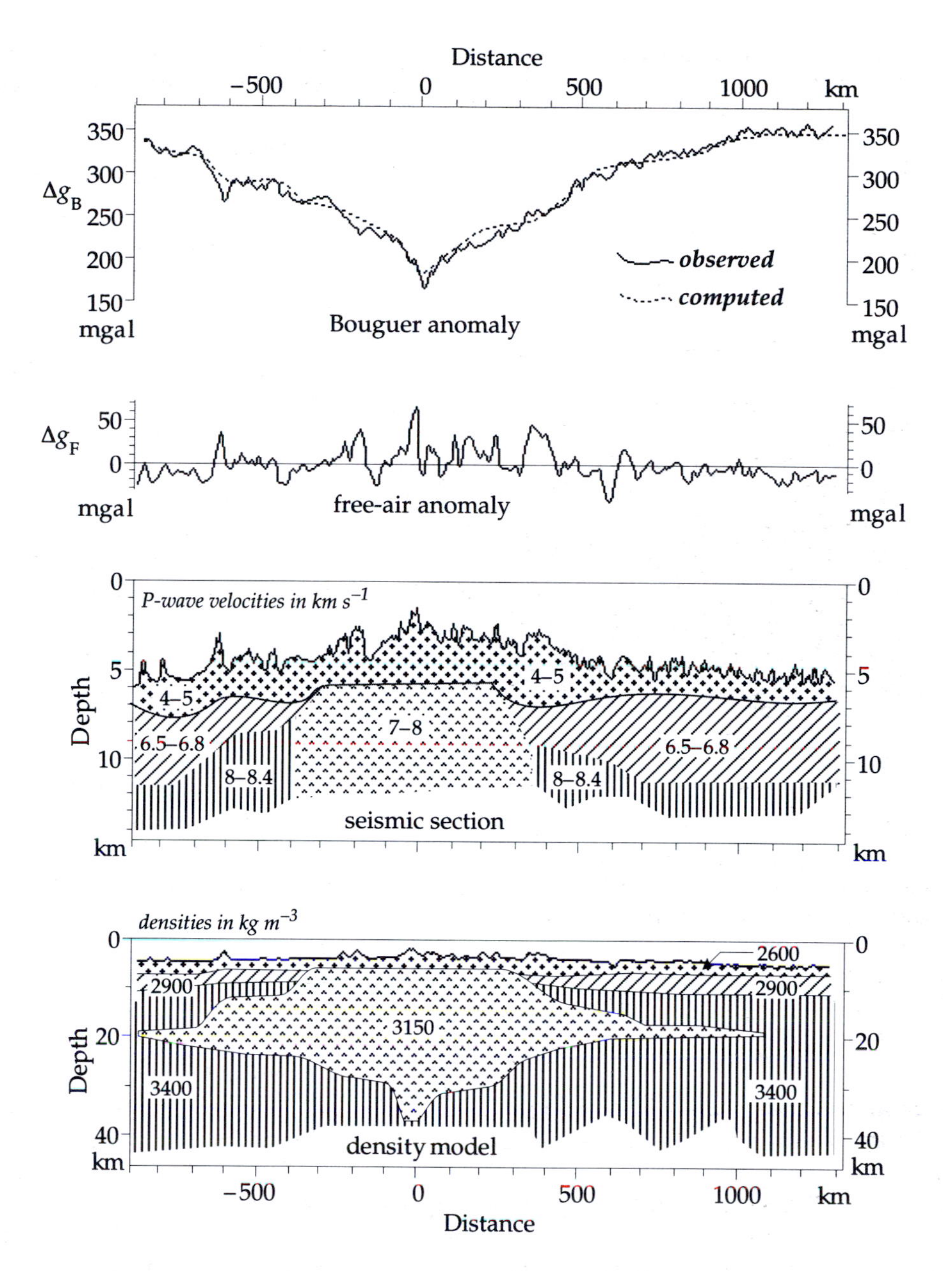

Fig. 2.57 Bouguer and free-air gravity anomalies over the Mid-Atlantic Ridge near 32°N. The seismic section is projected onto the gravity profile. The gravity anomaly computed from the density model fits the observed anomaly well, but is non-unique (after Talwani *et al.*, 1965).

down to about 30 km depth and for nearly 1000 km on each side of the axis (Fig. 2.57). The density of the anomalous structure is only 3150 kg m^{-3} instead of the usual 3400 kg m^{-3}. The model was proposed before the theory of plate tectonics was accepted. The anomalous upper mantle structure satisfies the gravity anomaly but has no relation to the known physical structure of a constructive plate margin. A broad zone of low upper-mantle seismic velocities was not found in later experiments, but narrow low-velocity zones are sometimes present close to the ridge axis.

A further combined seismic and gravity study of the Mid-Atlantic Ridge near 46°N gave a contradictory density model. Seismic refraction results yielded P-wave velocities of 4.6 km s^{-1} and 6.6 km s^{-1} for Layer 2 and Layer 3, respectively, but did not show anomalous mantle velocities beneath the ridge except under the median valley. A simpler density model was postulated to account for the gravity anomaly (Fig. 2.58). The small-scale free-air anomalies are accounted for by the variations in ridge topography seen in the bathymetric plot. The large-scale free-air gravity anomaly is reproduced well by a wedge-shaped structure that extends to 200 km depth. Its base extends to hundreds of kilometers on each side of the axis. A very small density contrast of only – 40 kg m^{-3} suffices to explain the broad

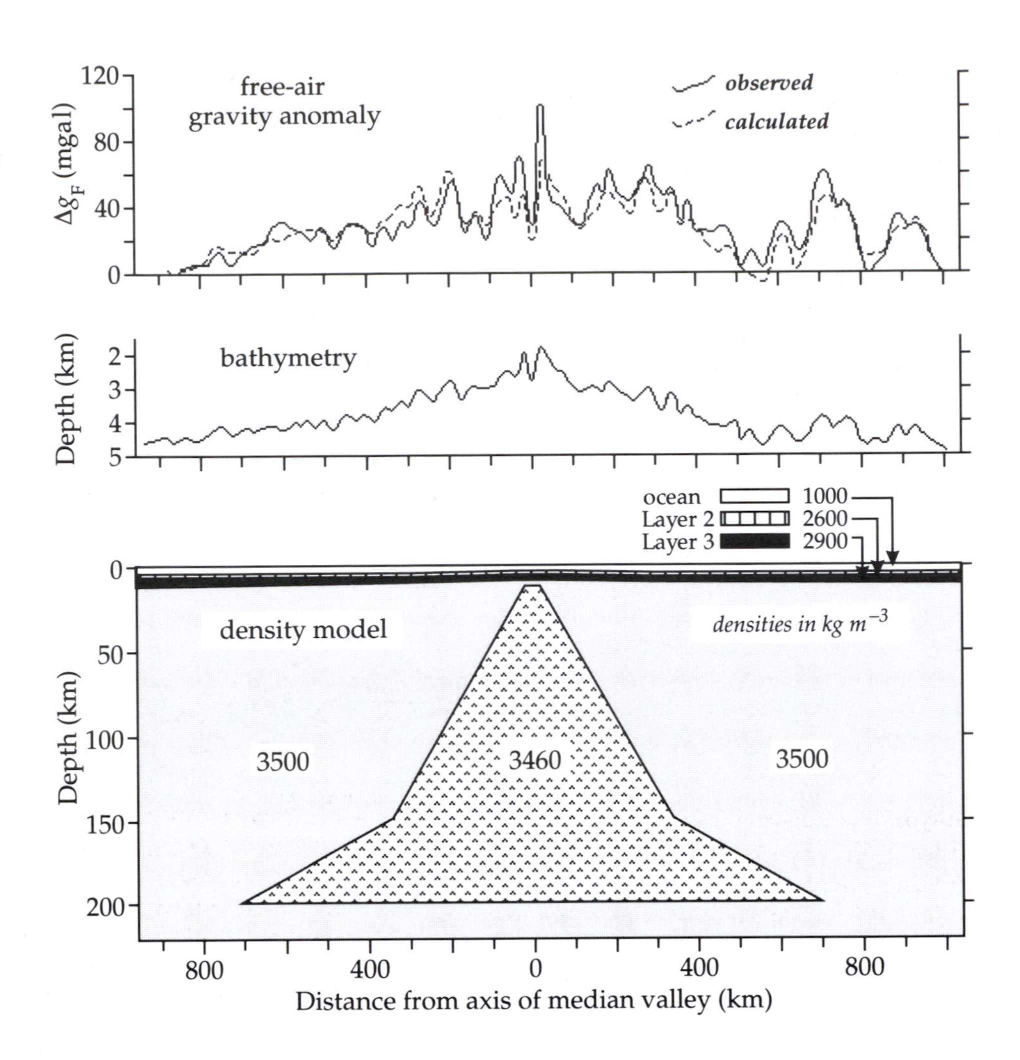

Fig. 2.58 Free-air gravity anomaly across the Mid-Atlantic Ridge near 46°N, and the lithospheric density model for the computed anomaly (after Keen and Tramontini, 1970)

free-air anomaly. The model is compatible with the thermal structure of an accreting plate margin. The low-density zone may be associated with hot material from the asthenosphere, which rises beneath the ridge, melts and accumulates in a shallow magma chamber within the oceanic crust.

2.6.4.4 *Gravity anomalies at subduction zones*

Subduction zones are found primarily at continental margins and island arcs. Elongate, narrow and intense isostatic and free-air gravity anomalies have long been associated with island arcs. The relationship of gravity to the structure of a subduction zone is illustrated by the free-air anomaly across the Chile trench at 23°S (Fig. 2.59). Seismic refraction data define the thicknesses of the oceanic and continental crust. Thermal and petrological data are integrated to give a density model for the structure of the mantle and the subducting lithosphere.

The continental crust is about 65 km thick beneath the Andes mountains, and gives large negative Bouguer anomalies. The free-air gravity anomaly over the Andes is positive, averaging about+50 mgal over the 4 km high plateau. Even stronger anomalies up to+100 mgal are seen over the east and west boundaries of the Andes. This is largely due to the edge effect of the low-density Andean crustal block (see Fig. 2.41b and §2.5.6).

A strong positive free-air anomaly of about +70 mgal lies between the Andes and the shore-line of the Pacific ocean. This anomaly is due to the subduction of the Nazca plate beneath South America. The descending slab is old and cool. Subduction exposes it to higher temperatures and pressures, but the slab descends faster than it can be heated up. The increase in density accompanying greater depth and pressure outweighs the decrease in density due to hotter temperatures. There is a positive density contrast between the subducting lithosphere and the surrounding mantle. Also, petrological changes accompanying the subduction result in mass excesses. Peridotite in the upper lithosphere changes phase from plagioclase-type to the higher-density

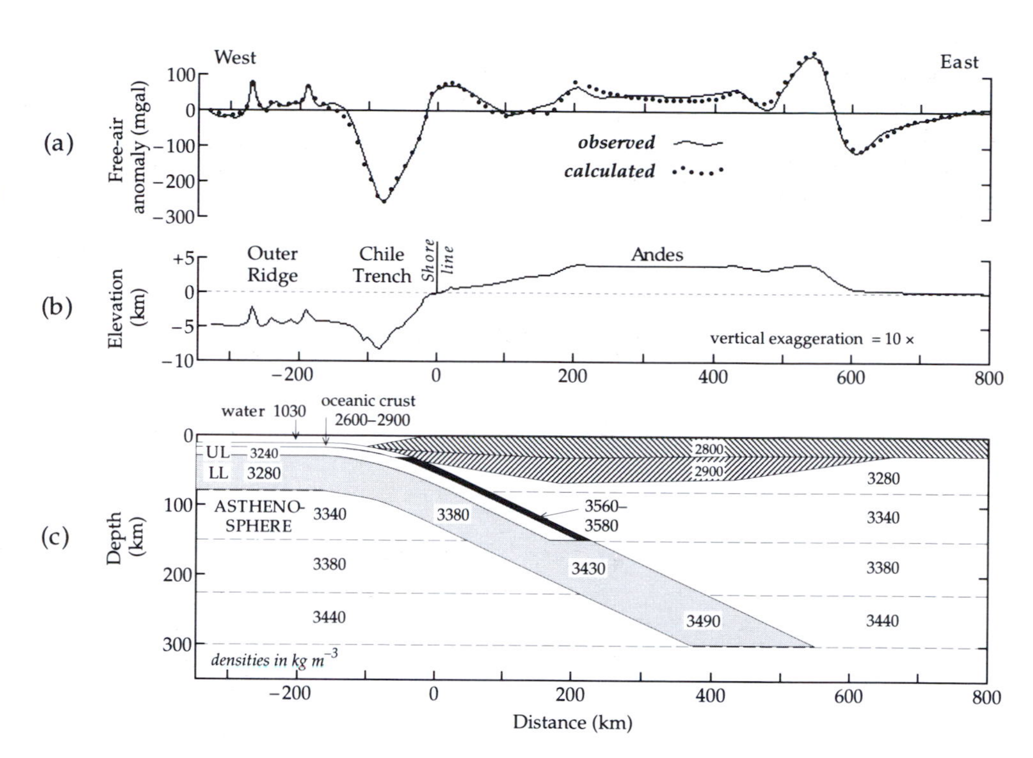

Fig. 2.59 Observed and computed free-air gravity anomalies across a subduction zone. The density model for the computed anomaly is based on seismic, thermal and petrological data. The profile crosses the Chile trench and Andes mountains at 23°S (after Grow and Bowin, 1975).

garnet-type. When oceanic crust is subducted to depths of 30–80 km, basalt changes phase to eclogite, which has a higher density (3560–3580 kg m^{-3}) than upper mantle rocks. These effects combine to produce the positive free-air anomaly.

The Chile trench is more than 2.5 km deeper than the ocean basin to the west. The sediments flooring the trench have low density. The mass deficiency of the water and sediments in the trench cause a strong negative free-air anomaly, which parallels the trench and has an amplitude greater than -250 mgal. A small positive anomaly of about $+20$ mgal is present about 100 km seaward of the trench axis. This anomaly is evident also in the mean level of the ocean surface as mapped by SEASAT (Fig. 2.26), which shows that the mean sea surface is raised in front of deep ocean trenches. This is due to upward flexure of the lithosphere before its downward plunge into the subduction zone. The flexure elevates higher-density mantle rocks and thereby causes the small positive free-air anomaly.

2.7 SUGGESTIONS FOR FURTHER READING

INTRODUCTORY LEVEL

Kearey, P. and Brooks, M. 1991. *An Introduction to Geophysical Exploration (2nd ed.)*, Blackwell Scientific Publ., Oxford.

Robinson, E. S. and Çoruh, C. 1988. *Basic Exploration Geophysics*, John Wiley & Sons, New York.

INTERMEDIATE LEVEL

Bott, M. H. P. 1982. *The Interior of the Earth (2nd ed.)*, Edward Arnold, London.

Dobrin, M. B. and Savit, C. H. 1988. *Introduction to Geophysical Prospecting (4th ed.)*, McGraw-Hill, New York.

Fowler, C. M. R. 1990. *The Solid Earth: an Introduction to Global Geophysics*, Cambridge Univ. Press, Cambridge.

Heiskanen, W. A. and Vening Meinesz, F. A. 1958. *The Earth and its Gravity Field*, McGraw-Hill, New York.

ADVANCED LEVEL

Bullen, K. E. 1975. *The Earth's Density*, Chapman and Hall, London.

Garland, G. D. 1979. *Introduction to Geophysics*, W. B. Saunders., Philadelphia.

Grant, F. S. and West, G. F. 1965. *Interpretation Theory in Applied Geophysics*, McGraw-Hill, New York.

Officer, C. B. 1974. *Introduction to Theoretical Geophysics*, Springer-Verlag, New York.

Stacey, F. D. 1992. *Physics of the Earth (3rd ed.)*, Brookfield Press, Brisbane, Australia

3 Seismology and the internal structure of the Earth

3.1 INTRODUCTION

Seismology is a venerable science with a long history. The Chinese scientist Chang Heng is credited with the invention in 132 A.D., nearly two thousand years ago, of the first functional seismoscope, a primitive but ingenious device of elegant construction and beautiful design that registered the arrival of seismic waves and enabled the observer to infer the direction from where they came. The origins of earthquakes were not at all understood. For centuries these fearsome events were attributed to supernatural powers. The accompanying destruction and loss of life were often understood in superstitious terms and interpreted as punishment inflicted by the gods on a sinful society. Biblical mentions of earthquakes – e.g., in the destruction of Sodom and Gomorrah – emphasize this vengeful theme. Although early astronomers and philosophers sought to explain earthquakes as natural phenomena unrelated to spiritual factors, the belief that earthquakes were an expression of divine anger prevailed until the advent of the Age of Reason in the 18th century. The path to a logical understanding of natural phenomena was laid in the 17th century by the systematic observations of scientists like Galileo, the discovery and statement of physical laws by Newton and the development of rational thought by contemporary philosophers.

In addition to the development of the techniques of scientific observation, an understanding of the laws of elasticity and the limited strength of materials was necessary before seismology could progress as a science. In a pioneering study, Galileo in 1638 described the response of a beam to loading, and in 1660 Hooke established the law of the spring. However, another 150 years were to pass before the generalized equations of elasticity were set down by Navier. During the early decades of the 19th century Cauchy and Poisson completed the foundations of modern elasticity theory.

Early descriptions of earthquake characteristics were necessarily restricted to observations and measurements in the 'near-field' region of the earthquake, i.e. in comparatively close proximity to the place where it occurred. A conspicuous advance in the science of seismology was accomplished with the invention of a sensitive and reliable seismograph by John Milne in 1892. Although massive and primitive by comparison with modern instruments, the precision and sensitivity of this revolutionary new device permitted accurate, quantitative descriptions of earthquakes at large distances from their source, in their 'far-field' region. The accumulation of reliable records of distant earthquakes (designated as 'teleseismic' events) made possible the systematic study of the Earth's seismicity and its internal structure.

The great San Francisco earthquake of 1906 was intensively studied and provided an impetus to efforts at understanding the origin of these natural phenomena, which were clarified in the same year by the elastic rebound model of H. F. Reid. Also, in 1906, R. D. Oldham proposed that the best explanation for the travel-times of teleseismic waves through the body of the Earth required a large, dense and probably fluid core; the depth to its outer boundary was calculated in 1913 by B. Gutenberg. From the analysis of the travel-times of seismic body waves from near earthquakes in Yugoslavia, A. Mohorovičić in 1909 inferred the existence of the crust–mantle boundary, and in 1936 the existence of the solid inner core was deduced by I. Lehmann. The definitions of these and other discontinuities associated with the deep internal structure of the Earth have since been greatly refined.

The needs of the world powers to detect incontrovertibly the testing of nuclear bombs by their adversaries provided considerable stimulus to the science of seismology in the 1950s and 1960s. The amount of energy released in a nuclear explosion is comparable to that of an earthquake, but the phenomena can be discriminated by analyzing the directions of first motion recorded by seismographs. The accurate location of the event required improved knowledge of seismic body-wave velocities throughout the Earth's interior. These political necessities of the cold war led to major improvements in seismological instrumentation, and to the establishment of a new world-wide network of seismic stations with the same physical characteristics. These developments had an important feedback to the earth sciences, because they resulted in more accurate locations of earthquake epicenters and a better understanding of the Earth's structure. The pattern of global seismicity, with its predominant concentration in narrow active zones, was an important factor in the development of the theory of plate tectonics, as it allowed the identification of plate margins and the sense of relative plate motions.

The techniques of refraction and reflection seismology using artificial, controlled explosions as sources, were developed in the search for petroleum. Since the 1960s these methods have been applied with notable success to the resolution of detailed crustal structure under continents and oceans. The development of powerful computer technology enabled refinements in earthquake location and in the determination of travel-times of seismic body waves. These advances led to the modern field of seismic tomography, a powerful and spectacular technique for revealing regions of the Earth's interior that have anomalous seismic velocities. In the field of earthquake seismology, the need to protect populations and man-made structures has resulted in the investment of considerable effort in the study of earthquake prediction and the development of construction codes to reduce earthquake damage.

To appreciate how seismologists have unravelled the structure of the Earth's interior it is necessary to understand what types of seismic waves can be generated by an earthquake or man-made source (such as a controlled explosion). The propagation of a seismic disturbance through the Earth is governed by physical properties such as density, and by the way in which the material of the Earth's interior reacts to the disturbance. Material within the seismic source suffers permanent deformation, but outside the source the passage of a seismic disturbance takes place predominantly by elastic displacement of the medium; that is, it suffers no permanent deformation. Before taking up the analysis of the different kinds of seismic waves, it is important to have a good grasp of elementary elasticity theory. This requires understanding the concepts of stress and strain, and the various elastic constants that relate them.

3.2 ELASTICITY THEORY

3.2.1 Elastic, anelastic and plastic behavior of materials

When a force is applied to a material, it deforms. This means that the particles of the material are displaced from their original positions. Provided the force does not exceed a critical value, the displacements are reversible; the particles of the material return to their original positions when the force is removed, and no permanent deformation results. This is called *elastic* behavior.

The laws of elastic deformation are illustrated by the following example. Consider a right cylindrical block of height h and cross-sectional area A, subjected to a force F which acts to extend the block by the amount Δh (Fig. 3.1). Experiments show that for elastic deformation Δh is directly proportional to the applied force and to the unstretched dimension of the block, but is inversely proportional to the cross-section of the block. That is, $\Delta h \propto Fh/A$, or

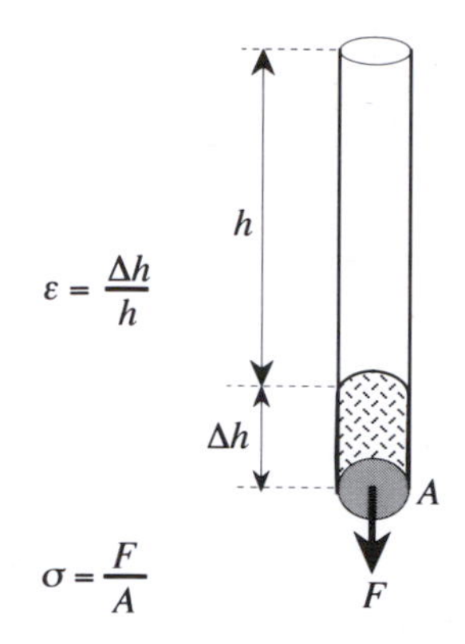

Fig. 3.1 A force F acting on a bar with cross-sectional area A extends the original length h by the amount Δh. Hooke's law of elastic deformation states that $\Delta h/h$ is proportional to F/A.

$$\frac{F}{A} \propto \frac{\Delta h}{h} \tag{3.1}$$

When the area A becomes infinitesimally small, the limiting value of the force per unit area (F/A) is called the *stress* σ. The units of stress are the same as the units of pressure. The SI unit is the pascal, equivalent to a force of 1 newton per square meter (1 Pa$=$1 N m^{-2}); the c.g.s. unit is the bar, equal to 10^6 dyne cm^{-2}.

When h is infinitesimally small, the fractional change in dimension ($\Delta h/h$) is called the *strain* ϵ, which is a dimensionless quantity. Eq. (3.1) states that, for elastic behavior, the strain in a body is proportional to the stress applied to it. This linear relationship is called *Hooke's law*. It forms the basis of elasticity theory.

Beyond a certain value of the stress, called the *proportionality limit*, Hooke's law no longer holds (Fig. 3.2a). Although the material is still elastic (it returns to its original shape when stress is removed), the stress–strain relationship is non-linear. If the solid is deformed beyond a certain point, known as the *elastic limit*, it will not recover its original shape when stress is removed. In this range a small increase in applied stress causes a disproportionately large increase in strain. The deformation is said to be *plastic*. If the applied stress is removed in the plastic range, the strain does not return to zero; a permanent strain has been produced. Eventually the applied stress exceeds the strength of the material and *failure* occurs. In some rocks failure can occur abruptly within the elastic range; this is called *brittle* behavior.

The non-brittle, or *ductile*, behavior of materials under stress depends on the time-scale of the deformation (Fig. 3.2b). An *elastic* material deforms immediately upon application of a stress and maintains a constant strain until the stress is removed, upon which the strain returns to its

(a)

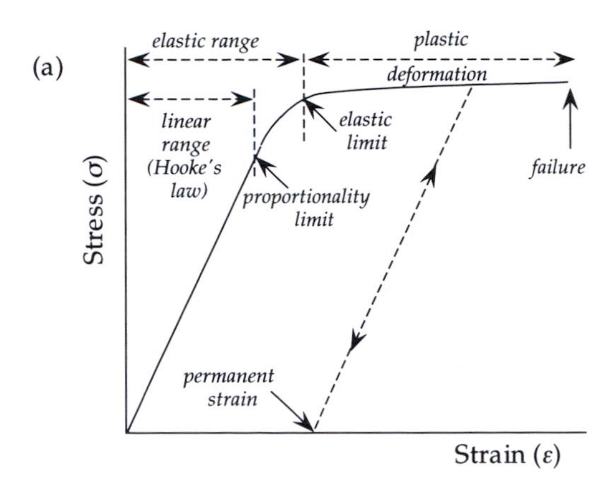

(b)

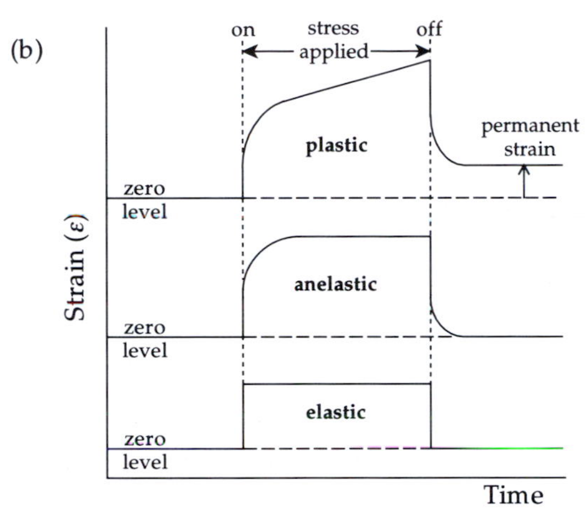

Fig. 3.2 (a) The stress–strain relation for a hypothetical solid is linear (Hooke's law) until the proportionality limit, and the material deforms elastically until it reaches the elastic limit; plastic deformation produces further strain until failure occurs. (b) Variations of elastic, anelastic and plastic strains with time, during and after application of a stress.

original state. A strain–time plot has a box-like shape. However, in some materials the strain does not reach a stable value immediately after application of a stress, but rises gradually to a stable value. This type of strain response is characteristic of *anelastic* materials. After removal of the stress, the time-dependent strain returns reversibly to the original level. In *plastic* deformation the strain keeps increasing as long as the stress is applied. When the stress is removed, the strain does not return to the original level; a permanent strain is left in the material.

Our knowledge of the structure and nature of the Earth's interior has been derived in large part from studies of seismic waves released by earthquakes. An earthquake occurs in the crust or upper mantle when the tectonic stress exceeds the local strength of the rocks and failure occurs. Away from the region of failure seismic waves spread out

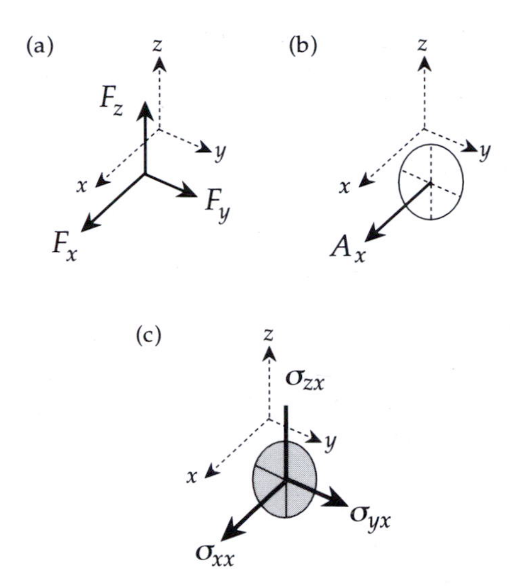

Fig. 3.3 (a) Components F_x, F_y and F_z of the force F acting in a reference frame defined by orthogonal Cartesian coordinate axes x, y and z. (b) The orientation of a small surface element with area A_x is described by the direction normal to the surface. (c) The components of force parallel to the x-axis result in the normal stress σ_{xx}; the components parallel to the y- and z-axes cause shear stresses σ_{xy} and σ_{xz}.

from an earthquake by elastic deformation of the rocks through which they travel. Their propagation depends on elastic properties that are described by the relationships between stress and strain.

3.2.2 The stress matrix

Consider a force **F** acting on a rectangular prism P in a reference frame defined by orthogonal Cartesian coordinate axes x, y and z (Fig. 3.3a). The component of **F** which acts in the direction of the x-axis is designated F_x; the force **F** is fully defined by its components F_x, F_y and F_z. The size of a small surface element is characterized by its area A, while its orientation is described by the direction normal to the surface (Fig. 3.3b). The small surface with area normal to the x-axis is designated A_x. The component of force F_x acting normal to the surface A_x produces a *normal stress*, denoted by σ_{xx}. The components of force along the y- and z-axes result in *shear stresses* σ_{yx} and σ_{zx} (Fig. 3.3c), given by

$$\sigma_{xx}=\lim_{A_x\to 0}\left(\frac{F_x}{A_x}\right),\quad \sigma_{yx}=\lim_{A_x\to 0}\left(\frac{F_y}{A_x}\right),\quad \sigma_{zx}=\lim_{A_x\to 0}\left(\frac{F_z}{A_x}\right) \qquad (3.2)$$

Similarly, the components of the force **F** acting on an element of surface A_y normal to the y-axis define a normal stress σ_{yy} and shear stresses σ_{xy} and σ_{zy}, while the compo-

σ_{ab} | a = direction of stress on given SA
b = surface Area (x, or y, or z)
acted on

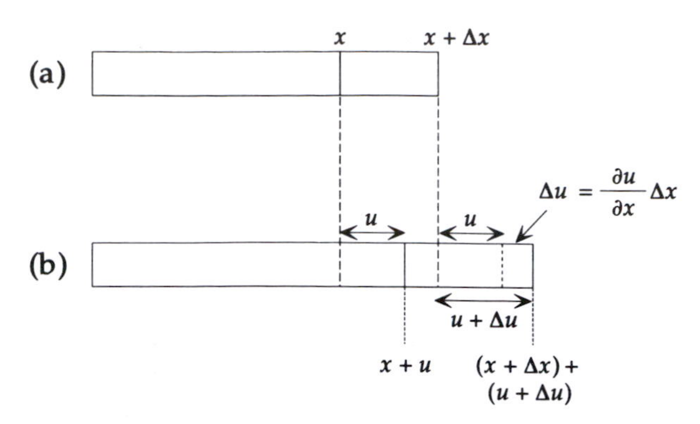

Fig. 3.4 Infinitesimal displacements u and $(u+\Delta u)$ of two points in a body that are located close together at the positions x and $(x+\Delta x)$, respectively.

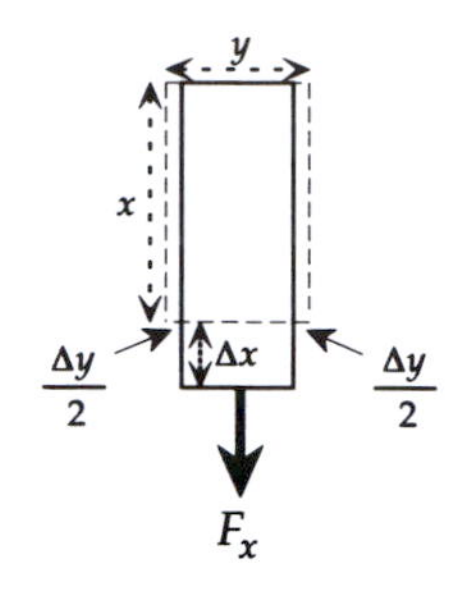

Poisson's ratio:

$$\nu = -\frac{\Delta y/y}{\Delta x/x} = -\frac{\varepsilon_{yy}}{\varepsilon_{xx}}$$

Fig. 3.5 Change of shape of a rectangular bar under extension. When stretched parallel to the x-axis, it becomes thinner parallel to the y-axis and z-axis.

nents of **F** acting on an element of surface A_z normal to the z-axis define a normal stress σ_{zz} and shear stresses σ_{xz} and σ_{yz}. The nine stress components completely define the state of stress of a body. They are described conveniently by the *stress matrix*

$$\begin{bmatrix} \sigma_{xx} & \sigma_{xy} & \sigma_{xz} \\ \sigma_{yx} & \sigma_{yy} & \sigma_{yz} \\ \sigma_{zx} & \sigma_{zy} & \sigma_{zz} \end{bmatrix} \quad (3.3)$$

If the forces on a body are balanced so as to give no rotation, this 3×3 matrix is symmetric (i.e. $\sigma_{xy}=\sigma_{yx}$; $\sigma_{yz}=\sigma_{zy}$; $\sigma_{zx}=\sigma_{xz}$) and contains only six independent elements.

3.2.3 The strain matrix

3.2.3.1 *Longitudinal strain*

The strains produced in a body can also be expressed by a 3×3 matrix. Consider first the one-dimensional case shown in Fig. 3.4 of two points in a body located close together at the positions x and $(x+\Delta x)$. If the point x is displaced by an infinitesimally small amount u in the direction of the x-axis, the point $(x+\Delta x)$ will be displaced by $(u+\Delta u)$, where Δu is equal to $(\partial u/\partial x)\Delta x$ to first order. The *longitudinal strain* or *extension* in the x-direction is the fractional change in length of an element along the x-axis. The original separation of the two points was Δx; one point was displaced by u, the other by $(u+\Delta u)$, so the new separation of the points is $(\Delta x+\Delta u)$. The component of strain parallel to the x-axis resulting from a small displacement parallel to the x-axis is denoted ϵ_{xx}, and is given by

$$\epsilon_{xx}=\frac{\text{change in separation}}{\text{original separation}}=\frac{\left(\Delta x+\frac{\partial u}{\partial x}\Delta x\right)-\Delta x}{\Delta x}$$

$$\epsilon_{xx}=\frac{\partial u}{\partial x} \quad (3.4)$$

The description of longitudinal strain can be expanded to three dimensions. If a point (x, y, z) is displaced by an infinitesimal amount to $(x+u, y+v, z+w)$, two further longitudinal strains ϵ_{yy} and ϵ_{zz} are defined by

$$\epsilon_{yy}=\frac{\partial v}{\partial y} \text{ and } \epsilon_{zz}=\frac{\partial w}{\partial z} \quad (3.5)$$

In an elastic body the transverse strains ϵ_{yy} and ϵ_{zz} are not independent of the strain ϵ_{xx}. Consider the change of shape of the bar in Fig. 3.5. When it is stretched parallel to the x-axis, it becomes thinner parallel to the y-axis and parallel to the z-axis. The transverse longitudinal strains ϵ_{yy} and ϵ_{zz} are of opposite sign but proportional to the extension ϵ_{xx} and can be expressed as

$$\epsilon_{yy}=-\nu\epsilon_{xx} \text{ and } \epsilon_{zz}=-\nu\epsilon_{xx} \quad (3.6)$$

The constant of proportionality ν is called *Poisson's ratio*. The values of the elastic constants of a material constrain ν to lie between 0 (no lateral contraction) and a maximum value of 0.5 (no volume change) for an incompressible fluid. In very hard, rigid rocks like granite ν is about 0.45, while in soft, poorly consolidated sediments it is about 0.05. In the interior of the Earth, ν commonly has a value around 0.24–0.27. A body for which the value of ν equals 0.25 is sometimes called an ideal Poisson body.

3.2.3.2 *Dilatation*

The *dilatation* is defined as the fractional change in volume of an element in the limit when its surface area decreases to zero. Consider an undeformed volume element (as in the description of longitudinal strain) which has sides Δx, Δy and Δz and undistorted volume $V=\Delta x\,\Delta y\,\Delta z$. As a result of the infinitesimal displacements Δu, Δv and Δw the edges

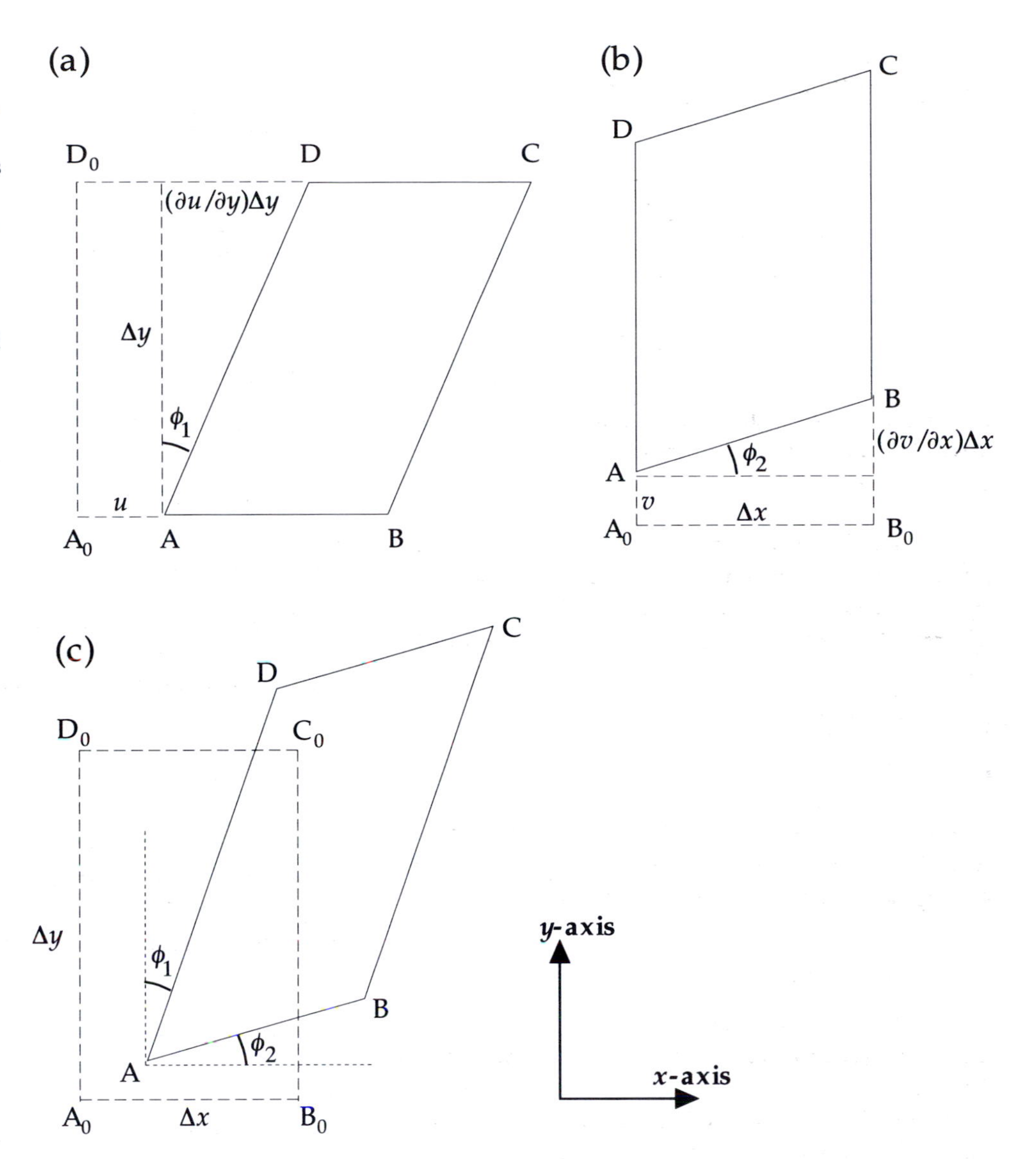

Fig. 3.6 (a) When a square is sheared parallel to the x-axis, side AD parallel to the y-axis rotates through a small angle ϕ_1; (b) when it is sheared parallel to the y-axis, side AB parallel to the x-axis rotates through a small angle ϕ_2. (c) In general, shear causes both sides to rotate, giving a total angular deformation $(\phi_1+\phi_2)$. In each case the diagonal AC is extended.

increase to $\Delta x+\Delta u$, $\Delta y+\Delta v$, and $\Delta z+\Delta w$, respectively. The fractional change in volume is

$$\frac{\Delta V}{V}=\frac{(\Delta x+\Delta u)(\Delta y+\Delta v)(\Delta z+\Delta w)-\Delta x\Delta y\Delta z}{\Delta x\Delta y\Delta z}$$

$$=\frac{\Delta x\Delta y\Delta z+\Delta u\Delta y\Delta z+\Delta v\Delta z\Delta x+\Delta w\Delta x\Delta y-\Delta x\Delta y\Delta z}{\Delta x\Delta y\Delta z}$$

$$=\frac{\Delta u}{\Delta x}+\frac{\Delta v}{\Delta y}+\frac{\Delta w}{\Delta z} \tag{3.7}$$

where very small quantities like $\Delta u\,\Delta v$, $\Delta v\,\Delta w$, $\Delta w\,\Delta u$ and $\Delta u\,\Delta v\,\Delta w$ have been ignored. In the limit, as Δx, Δy and Δz all approach zero we get the dilatation

$$\theta=\frac{\partial u}{\partial x}+\frac{\partial v}{\partial y}+\frac{\partial w}{\partial z}$$

$$\theta=\epsilon_{xx}+\epsilon_{yy}+\epsilon_{zz} \tag{3.8}$$

3.2.3.3 *Shear strain*

During deformation a body generally experiences not only longitudinal strains as described above. The shear components of stress (σ_{xy}, σ_{yz}, σ_{zx}) produce *shear strains*, which are manifest as changes in the angular relationships between parts of a body. This is most easily illustrated in two dimensions. Consider a rectangle ABCD with sides Δx and Δy and its distortion due to shear stresses acting in the x–y plane (Fig. 3.6). As in the earlier example of longitudinal strain, the point A is displaced parallel to the x-axis by an amount u. Because of the shear deformation, points between A and D experience larger x-displacements the further they are from A. The point D which is at a vertical distance Δy above A is displaced by the amount $(\partial u/\partial y)\,\Delta y$ in the direction of the x-axis. This causes a clockwise rotation of side AD through a small angle ϕ_1 given by

$$\tan\phi_1 = \frac{(\partial u/\partial y)\Delta y}{\Delta y} = \frac{\partial u}{\partial y} \tag{3.9}$$

Similarly, the point A is displaced parallel to the y-axis by an amount v, while the point B which is at a horizontal distance Δx from A is displaced by the amount $(\partial v/\partial x)\,\Delta x$ in the direction of the y-axis. As a result side AB rotates anticlockwise through a small angle ϕ_2 given by

$$\tan\phi_2 = \frac{(\partial v/\partial x)\Delta x}{\Delta x} = \frac{\partial v}{\partial x} \tag{3.10}$$

Elastic deformation involves infinitesimally small displacements and distortions, and for small angles we can write $\tan\phi_1 = \phi_1$ and $\tan\phi_2 = \phi_2$. The shear strain in the x–y plane (ϵ_{xy}) is defined as half the total angular distortion:

$$\epsilon_{xy} = \frac{1}{2}\left(\frac{\partial v}{\partial x} + \frac{\partial u}{\partial y}\right) \tag{3.11a}$$

By transposing x and y, and the corresponding displacements u and v, the shear component ϵ_{yx} is obtained:

$$\epsilon_{yx} = \frac{1}{2}\left(\frac{\partial u}{\partial y} + \frac{\partial v}{\partial x}\right) \tag{3.11b}$$

This is identical to ϵ_{xy}. The total angular distortion in the x–y plane is $(\epsilon_{xy} + \epsilon_{yx})$. Similarly, strain components ϵ_{yz} (= ϵ_{zy}) and ϵ_{xz} (= ϵ_{zx}) are defined for angular distortions in the y–z and z–x planes, respectively.

$$\epsilon_{yz} = \epsilon_{zy} = \frac{1}{2}\left(\frac{\partial w}{\partial y} + \frac{\partial v}{\partial z}\right)$$

$$\epsilon_{zx} = \epsilon_{xz} = \frac{1}{2}\left(\frac{\partial u}{\partial z} + \frac{\partial w}{\partial x}\right) \tag{3.12}$$

The longitudinal and shear strains define the symmetric 3×3 *strain matrix*

$$\begin{bmatrix} \epsilon_{xx} & \epsilon_{xy} & \epsilon_{xz} \\ \epsilon_{yx} & \epsilon_{yy} & \epsilon_{yz} \\ \epsilon_{zx} & \epsilon_{zy} & \epsilon_{zz} \end{bmatrix} \tag{3.13}$$

3.2.4 **The elastic constants**

According to Hooke's law, when a body deforms elastically, there is a linear relationship between stress and strain. The ratio of stress to strain defines an elastic constant (or elastic modulus) of the body. Strain is itself a ratio of lengths and therefore dimensionless. Thus the elastic moduli must have the units of stress (N m^{-2}). The elastic moduli, defined for different types of deformation, are Young's modulus, the rigidity modulus and the bulk modulus.

Young's modulus is defined from the extensional deformations. Each longitudinal strain is proportional to the corresponding stress component, that is,

$$\sigma_{xx} = E\epsilon_{xx},\ \sigma_{yy} = E\epsilon_{yy},\ \sigma_{zz} = E\epsilon_{zz} \tag{3.14}$$

where the constant of proportionality, E, is the Young's modulus.

The rigidity modulus (or *shear modulus*) is defined from the shear deformation. Like the longitudinal strains, each shear strain is proportional to the corresponding shear stress component, that is,

$$\sigma_{xy} = \mu\epsilon_{xy},\ \sigma_{yz} = \mu\epsilon_{yz},\ \sigma_{zx} = \mu\epsilon_{zx} \tag{3.15}$$

where the proportionality constant, μ, is the rigidity modulus.

The *bulk modulus* (or *incompressibility*) is defined from the dilatation experienced by a body under hydrostatic pressure. Shear components of stress are zero for hydrostatic conditions ($\sigma_{xy} = \sigma_{yz} = \sigma_{zx} = 0$), and the inwards pressure (negative normal stress) is equal in all directions ($\sigma_{xx} = \sigma_{yy} = \sigma_{zz} = -p$). The bulk modulus, K, is the ratio of the hydrostatic pressure to the dilatation; that is,

$$p = -K\theta \tag{3.16}$$

The inverse of the bulk modulus (K^{-1}) is called the *compressibility*.

3.2.4.1 *Bulk modulus in terms of Young's modulus and Poisson's ratio*

Consider a rectangular volume element subjected to normal stresses σ_{xx}, σ_{yy} and σ_{zz} on its end surfaces. Each longitudinal strain ϵ_{xx}, ϵ_{yy} and ϵ_{zz} results from the combined effects of σ_{xx}, σ_{yy} and σ_{zz} . For example, applying Hooke's law, the stress σ_{xx} produces an extension equal to σ_{xx}/E in the x-direction. The stress σ_{yy} causes an extension σ_{yy}/E in the y-direction, which results in an accompanying transverse strain $-\nu\,(\sigma_{yy}/E)$ in the x-direction, where ν is Poisson's ratio. Similarly, the stress component σ_{zz} makes a contribution $-\nu\,(\sigma_{zz}/E)$ to the total longitudinal strain ϵ_{xx} in the x-direction. Therefore,

$$\epsilon_{xx} = \frac{\sigma_{xx}}{E} - \nu\frac{\sigma_{yy}}{E} - \nu\frac{\sigma_{zz}}{E} \tag{3.17}$$

Similar equations describe the total longitudinal strains ϵ_{yy} and ϵ_{zz}. They can be rearranged as

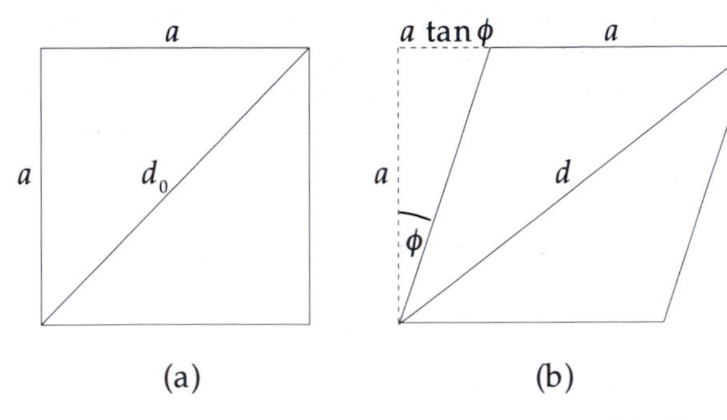

Fig. 3.7 (a) In the undeformed state d_0 is the length of the diagonal of a square with side length a. (b) When the square is deformed by shear through an angle ϕ, the diagonal is extended to the new length d.

$$E\epsilon_{xx}=\sigma_{xx}-\nu\sigma_{yy}-\nu\sigma_{zz}$$
$$E\epsilon_{yy}=\sigma_{yy}-\nu\sigma_{zz}-\nu\sigma_{xx}$$
$$E\epsilon_{zz}=\sigma_{zz}-\nu\sigma_{xx}-\nu\sigma_{yy} \tag{3.18}$$

Adding these three equations together we get

$$E(\epsilon_{xx}+\epsilon_{yy}+\epsilon_{zz})=(1-2\nu)(\sigma_{xx}+\sigma_{yy}+\sigma_{zz}) \tag{3.19}$$

Consider now the effect of a constraining hydrostatic pressure, p, where $\sigma_{xx}=\sigma_{yy}=\sigma_{zz}=-p$. Using the definition of dilatation (θ) in Eq. (3.8) we get

$$E\theta=(1-2\nu)(-3p)$$

$$E=(1-2\nu)\left(-3\frac{p}{\theta}\right) \tag{3.20}$$

from which, using the definition of bulk modulus (K) in Eq. (3.16),

$$K=\frac{E}{3(1-2\nu)} \tag{3.21}$$

3.2.4.2 *Shear modulus in terms of Young's modulus and Poisson's ratio*

The relationship between μ and E can be appreciated by considering the shear deformation of a rectangular prism that is infinitely long in one dimension and has a square cross-section in the plane of deformation. The shear causes shortening of one diagonal and extension of the other. Let the length of the side of the square be a (Fig. 3.7a) and that of its diagonal be d_0 ($= a\sqrt{2}$). The small shear through the angle ϕ displaces one corner by the amount ($a \tan \phi$) and stretches the diagonal to the new length d (Fig. 3.7b), which is given by Pythagoras' law:

$$\begin{aligned} d^2&=a^2+(a+a\tan\phi)^2\\ &=a^2+a^2+a^2\tan^2\phi+2a^2\tan\phi\\ &=2a^2(1+\tan\phi+\tfrac{1}{2}\tan^2\phi)\\ &\cong d_0^2(1+\phi)\end{aligned}$$
$$d\cong d_0(1+\tfrac{1}{2}\phi) \tag{3.22}$$

where for an infinitesimally small strain $\tan\phi=\phi$, and powers of ϕ higher than first order are negligibly small. The extension of the diagonal is

$$\frac{\Delta d}{d_0}=\frac{d-d_0}{d_0}=\frac{\phi}{2} \tag{3.23}$$

This extension is related to the *normal* stresses σ_{xx} and σ_{yy} in the x–y plane of the cross-section (Fig. 3.8a), which are in general unequal. Let p represent their average value: $p=(\sigma_{xx}+\sigma_{yy})/2$. The change of shape of the square cross-section results from the differences Δp between p and σ_{xx} and σ_{yy}, respectively (Fig. 3.8b). The outwards stress difference Δp along the x-axis produces an x-extension equal to $\Delta p/E$, while the inwards stress difference along the y-axis causes contraction along the y-axis and a corresponding contribution to the x-extension equal to $\nu(\Delta p/E)$, where ν is Poisson's ratio as before. The total x-extension $\Delta x/x$ is therefore given by

$$\frac{\Delta x}{x}=\frac{\Delta p}{E}(1+\nu) \tag{3.24}$$

Let each edge of the square represent an arbitrary area A normal to the plane of the figure. The stress differences Δp produce forces $f=\Delta pA$ on the edges of the square, which resolve to shear forces $f/\sqrt{2}$ parallel to the sides of the inner square defined by joining the mid-points of the sides of the original square (Fig. 3.8c). Normal to the plane of the figure the surface area represented by each inner side is $A/\sqrt{2}$, and therefore the tangential (shear) stress acting on these sides simply equals Δp (Fig. 3.8d). The inner square shears through an angle ϕ, and so we can write

$$\Delta p=\mu\phi \tag{3.25}$$

One diagonal becomes stretched in the x-direction while the other diagonal is shortened in the y-direction. The extension of the diagonal of a sheared square was shown above to be $\phi/2$. Thus,

$$\frac{\Delta p}{E}(1+\nu)=\frac{\phi}{2}=\frac{\Delta p}{2\mu} \tag{3.26}$$

Rearranging terms we get the relationship between μ, E and ν:

$$\mu=\frac{E}{2(1+\nu)} \tag{3.27}$$

3.2.4.3 *The Lamé constants*

Elasticity theory can be handled elegantly with the aid of tensor notation. The components of stress and strain are

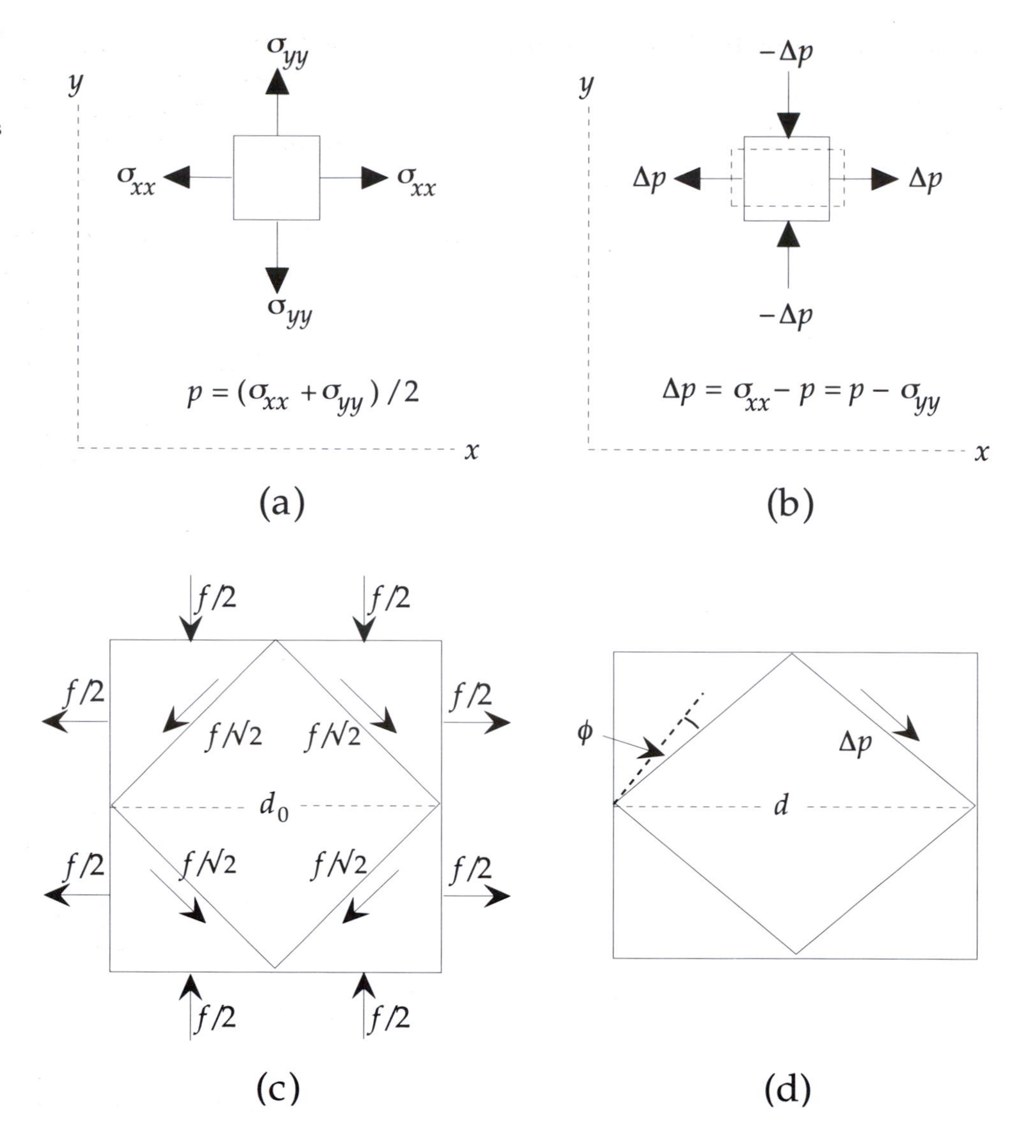

Fig. 3.8 (a) Unequal *normal* stresses σ_{xx} and σ_{yy} in the x–y plane, and their average value p. (b) Stress differences Δp between p and σ_{xx} and σ_{yy}, respectively, cause elongation parallel to x and shortening parallel to y. (c) Forces $f=\Delta pA$ along the sides of the original square give shear forces $f/\sqrt{2}$ along the edges of the inner square, each of which has area $A\sqrt{2}$. (d) The shear stress on each side of the inner square has value Δp and causes extension of the diagonal of the inner square and shear deformation through an angle ϕ.

expressed respectively as σ_{ij} and ϵ_{ij}, where each of the subscripts i and j can take the values x, y or z. Using this notation Hooke's law can be written for an isotropic elastic solid as

$$\sigma_{ij}=\lambda\theta\delta_{ij}+2\mu\epsilon_{ij} \qquad (3.28)$$

In this expression θ is the dilatation and δ_{ij} is called the Krönecker delta, for which the values are $\delta_{ij}=0$ when $i\neq j$ and $\delta_{ij}=1$ when $i=j$. The constants λ and μ are known as the Lamé constants. They are related to the elastic constants defined physically above; μ is equivalent to the rigidity modulus, and K and E can be expressed in terms of both λ and μ.

3.2.4.4 *Bulk modulus in terms of the Lamé constants*

The bulk modulus describes volumetric shape changes of a material under the effects of the *normal* stresses σ_{xx}, σ_{yy} and σ_{zz}. Expanding the above tensor equation for Hooke's law gives

$$\begin{aligned}\sigma_{xx}&=\lambda\theta+2\mu\epsilon_{xx}\\ \sigma_{yy}&=\lambda\theta+2\mu\epsilon_{yy}\\ \sigma_{zz}&=\lambda\theta+2\mu\epsilon_{zz}\end{aligned} \qquad (3.29)$$

Adding these equations together, and assuming hydrostatic conditions ($\sigma_{xx}=\sigma_{yy}=\sigma_{zz}=-p$) gives

$$\begin{aligned}\sigma_{xx}+\sigma_{yy}+\sigma_{zz}&=3\lambda\theta+2\mu(\epsilon_{xx}+\epsilon_{yy}+\epsilon_{zz})\\ -3p&=3\lambda\theta+2\mu\theta\end{aligned} \qquad (3.30)$$

from which, using the definition of $K=-p/\theta$, we get

$$K=\lambda+\tfrac{2}{3}\mu \tag{3.31}$$

3.2.4.5 *Young's modulus, Poisson's ratio and the Lamé constants*

Young's modulus describes the longitudinal strains when a uniaxial normal stress is applied to a material. When only the longitudinal stress σ_{xx} is applied (i.e., $\sigma_{yy}=\sigma_{zz}=0$), Hooke's law becomes

$$\begin{aligned}\sigma_{xx}&=\lambda\theta+2\mu\epsilon_{xx}\\0&=\lambda\theta+2\mu\epsilon_{yy}\\0&=\lambda\theta+2\mu\epsilon_{zz}\end{aligned} \tag{3.32}$$

Expanding these equations gives

$$\begin{aligned}(\lambda+2\mu)\epsilon_{xx}+\lambda\epsilon_{yy}+\lambda\epsilon_{zz}&=\sigma_{xx}\\\lambda\epsilon_{xx}+(\lambda+2\mu)\epsilon_{yy}+\lambda\epsilon_{zz}&=0\\\lambda\epsilon_{xx}+\lambda\epsilon_{yy}+(\lambda+2\mu)\epsilon_{zz}&=0\end{aligned} \tag{3.33}$$

These three simultaneous equations can be solved for ϵ_{xx}, ϵ_{yy} and ϵ_{zz} in terms of σ_{xx}. Keeping in mind the definitions of Young's modulus, $E=\sigma_{xx}/\epsilon_{xx}$, and Poisson's ratio, $\nu=-\epsilon_{yy}/\epsilon_{xx}=-\epsilon_{zz}/\epsilon_{xx}$, we get for Young's modulus and Poisson's ratio, respectively

$$E=\frac{\mu(3\lambda+2\mu)}{(\lambda+\mu)}$$

$$\nu=\frac{\lambda}{2(\lambda+\mu)} \tag{3.34}$$

The values of λ and μ are almost equal in some materials, and it is possible to assume $\lambda=\mu$, from which it follows that $\nu=0.25$. The approximation is called Poisson's relation; it applies to most rocks of the Earth.

3.2.4.6 *Anisotropy*

The foregoing discussion presents the elastic parameters as constants. In fact they are dependent on pressure and temperature and so can only be considered constant for specified conditions. The variations of temperature and pressure in the Earth ensure that the elastic parameters vary with depth. Moreover, it has been assumed that the relationships between stress and strain hold equally for all directions, a property called *isotropy*. This condition is not fulfilled in many minerals. For example, if a mineral has uniaxial symmetry in the arrangement of the atoms in its unit cell, the physical properties of the mineral parallel and perpendicular to the axis of symmetry are different. The mineral is *anisotropic*. The relations between components of stress and strain in an anisotropic substance are more complex than in the perfectly elastic, isotropic case examined in this chapter. The elastic parameters of an isotropic body are fully specified by the two parameters λ and μ, but as many as 21 parameters may be needed to describe anisotropic elastic behavior. Seismic velocities, which depend on the elastic parameters, vary with direction in an anisotropic medium.

Normally, a rock contains so many minerals that it can be assumed that they are oriented at random and the rock can be treated as isotropic. This assumption can also be made, at least to first order, for large regions of the Earth's interior. However, if anisotropic minerals are subjected to stress they develop a preferred alignment with the stress field. For example, platy minerals tend to align with their tabular shapes normal to the compression axis, or parallel to the direction of flow of a fluid. Preferential grain alignment results in seismic anisotropy. This has been observed in seismic studies of the upper mantle, especially at oceanic ridges, where anisotropic velocities have been attributed to the alignment of crystals by convection currents.

3.2.5 **Imperfect elasticity in the Earth**

A seismic wave passes through the Earth as an elastic disturbance of very short duration lasting only some seconds or minutes. Elasticity theory is used to explain seismic wave propagation. However, materials may react differently to brief, sudden stress than they do to long-lasting steady stress. The stress response of rocks and minerals in the Earth is affected by various factors, including temperature, hydrostatic confining pressure, and time. As a result, elastic, anelastic and plastic behavior occur with various degrees of importance at different depths.

Anelastic behavior in the Earth is related to the petrophysical properties of rocks and minerals. If a material is not perfectly elastic, a seismic wave passing through it loses energy to the material (e.g., as frictional heating) and the amplitude of the wave gradually diminishes. The amplitude decrease is called *attenuation*, and it is due to anelastic *damping* of the vibration of particles of the material (see §3.3.2.7). For example, the passage of seismic waves through the asthenosphere is damped due to anelastic behavior at the grain level of the minerals. This may consist of time-dependent slippage between grains; alternatively, fluid phases may be present at the grain boundaries.

A material that reacts elastically to a sudden stress may deform and flow plastically under a stress that acts over a long time interval. Plastic behavior in the asthenosphere and in the deeper mantle may allow material to flow, perhaps due to the motion of dislocations within crystal grains. The flow takes place over times on the order of hun-

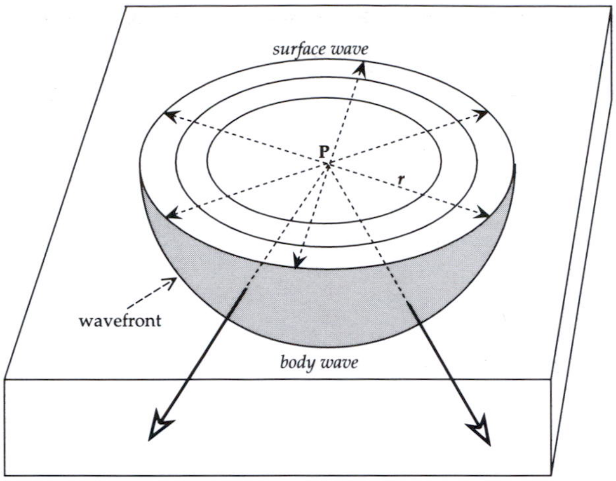

Fig. 3.9 Propagation of a seismic disturbance from a point source P near the surface of a homogeneous medium; the disturbance travels as a body wave through the medium and as a surface wave along the free surface.

dreds of millions of years, but it provides an efficient means of transporting heat out of the deep interior.

3.3 SEISMIC WAVES

3.3.1 Introduction

The propagation of a seismic disturbance through a heterogeneous medium is extremely complex. In order to derive equations that describe the propagation adequately, it is necessary to make simplifying assumptions. The heterogeneity of the medium is often modelled by dividing it into parallel layers, in each of which homogeneous conditions are assumed. By suitable choice of the thickness, density and elastic properties of each layer, the real conditions can be approximated. The most important assumption about the propagation of a seismic disturbance is that it travels by elastic displacements in the medium. This condition certainly does not apply close to the seismic source. In or near an earthquake focus or the shot point of a controlled explosion the medium is destroyed. Particles of the medium are displaced permanently from their neighbors; the deformation is anelastic. However, when a seismic disturbance has travelled some distance away from its source, its amplitude decreases and the medium deforms elastically to permit its passage. The particles of the medium carry out simple harmonic motions, and the seismic energy is transmitted as a complex set of wave motions.

When seismic energy is released suddenly at a point P near the surface of a homogeneous medium (Fig. 3.9), part of the energy propagates through the body of the medium as seismic *body waves*. The remaining part of the seismic energy spreads out over the surface as a seismic *surface wave*, analogous to the ripples on the surface of a pool of water into which a stone has been thrown.

3.3.2 Seismic body waves

When a body wave reaches a distance r from its source in a homogeneous medium, the *wavefront* (defined as the surface in which all particles vibrate with the same phase) has a spherical shape, and the wave is called a *spherical wave*. As the distance from the source increases, the curvature of the spherical wavefront decreases. At great distances from the source the wavefront is so flat that it can be considered to be a plane and the seismic wave is called a *plane wave*. The direction perpendicular to the wavefront is called the seismic *ray path*. The description of the harmonic motion in plane waves is simpler than for spherical waves, because for plane waves we can use orthogonal Cartesian coordinates. Even for plane waves the mathematical description of the three-dimensional displacements of the medium is fairly complex. However, we can learn quite a lot about body-wave propagation from a simpler, less rigorous description.

3.3.2.1 *Compressional waves*

Let Cartesian reference axes be defined such that the x-axis is parallel to the direction of propagation of the plane wave; the y- and z-axes then lie in the plane of the wavefront (Fig. 3.10). A generalized vibration of the medium can be reduced to components parallel to each of the reference axes. In the x-direction the particle motion is back and forward parallel to the direction of propagation. This results in the medium being alternately stretched and condensed in this direction (Fig. 3.11a). This harmonic motion produces a body wave that is transmitted as a sequence of rarefactions and condensations parallel to the x-axis.

Consider the disturbance of the medium shown in Fig. 3.11b. The area of the wavefront normal to the x-direction is A_x, and the wave propagation is treated as one-dimensional. At an arbitrary position x (Fig. 3.11c), the passage of the wave produces a displacement u and a force F_x in the x-direction. At the position $x+dx$ the displacement is $u+du$ and the force is F_x+dF_x. Here dx is the infinitesimal length of a small volume element which has mass $\rho\, dx\, A_x$. The net force acting on this element in the x-direction is given by

$$(F_x+dF_x)-F_x=dF_x=\frac{\partial F_x}{\partial x}dx \qquad (3.35)$$

The force F_x is caused by the stress element σ_{xx} acting on the area A_x, and is equal to $\sigma_{xx}A_x$. This allows us to write the one-dimensional equation of motion

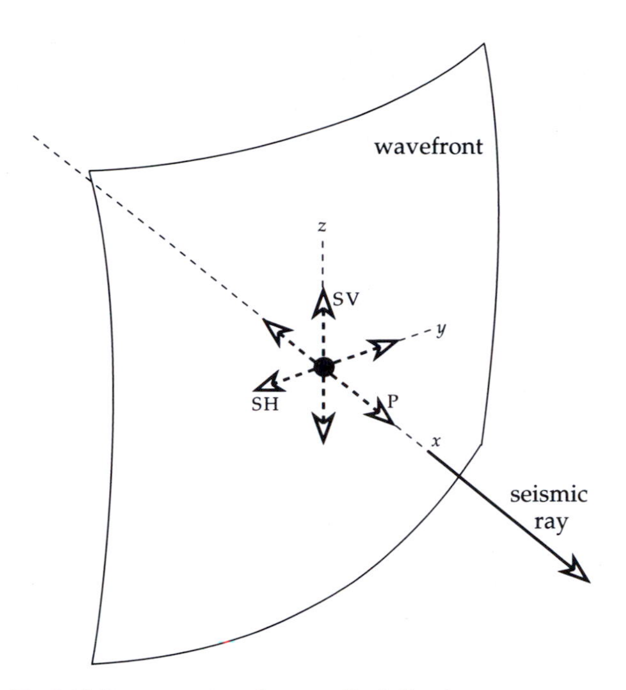

Fig. 3.10 Representation of a generalized vibration as components parallel to three orthogonal reference axes. Particle motion in the x-direction is back and forth parallel to the direction of propagation, corresponding to the P-wave. Vibrations along the y- and z-axes are in the plane of the wavefront and normal to the direction of propagation. The z-vibration in a vertical plane corresponds to the SV-wave; the y-vibration is horizontal and corresponds to the SH-wave.

$$(\rho\,\mathrm{d}xA_x)\frac{\partial^2 u}{\partial t^2}=\mathrm{d}xA_x\frac{\partial \sigma_{xx}}{\partial x} \tag{3.36}$$

The definitions of Young's modulus, E, (Eq. (3.14)) and the normal strain ϵ_{xx} (Eq. (3.4)) give, for a one-dimensional deformation

$$\sigma_{xx}=E\epsilon_{xx}=E\frac{\partial u}{\partial x} \tag{3.37}$$

Substitution of Eq. (3.37) into Eq. (3.36) gives the one-dimensional wave equation

$$\frac{\partial^2 u}{\partial t^2}=V^2\frac{\partial^2 u}{\partial x^2} \tag{3.38}$$

where V is the velocity of the wave, given by

$$V=\sqrt{\frac{E}{\rho}} \tag{3.39}$$

A one-dimensional wave is rather restrictive. It represents the stretching and compressing in the x-direction as effects that are independent of what happens in the y- and z-directions. In an elastic solid the elastic strains in any direc-

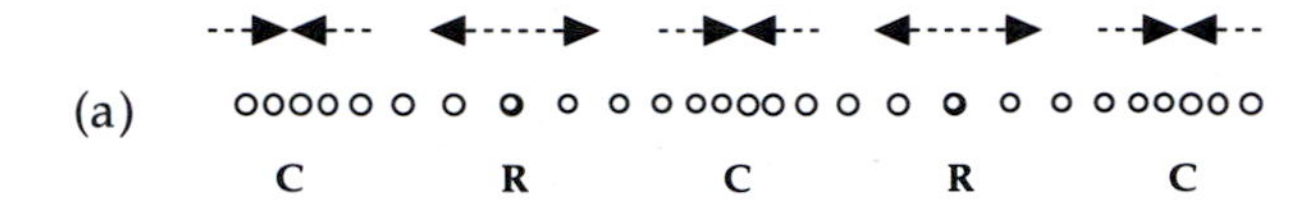

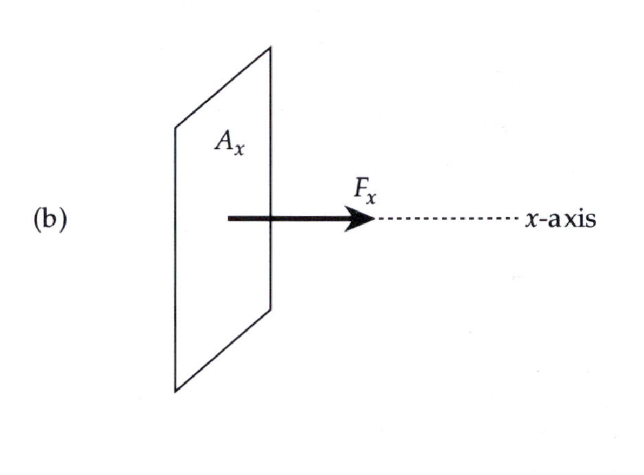

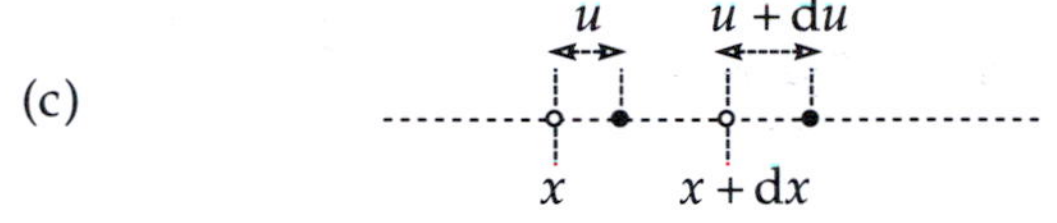

Fig 3.11 (a) The particle motion in a one-dimensional P-wave transmits energy as a sequence of rarefactions (R) and condensations (C) parallel to the x-axis. (b) Within the wavefront the component of force F_x in the x-direction of propagation is distributed over an element of area A_x normal to the x-axis. (c) A particle at position x experiences a longitudinal displacement u in the x-direction, while at the nearby position $x+\mathrm{d}x$ the corresponding displacement is $u+\mathrm{d}u$.

tion are coupled to the strains in transverse directions by Poisson's ratio for the medium. A more sophisticated three-dimensional analysis takes into account the simultaneous changes perpendicular to the direction of propagation. In this case the area A_x can no longer be considered constant. Instead of looking at the displacements in one direction only, all three axes must be taken into account. This is achieved by analyzing the changes in volume. The longitudinal (or compressional) body wave passes through a medium as a series of *dilatations* and *compressions*. The equation of the compressional wave in the x-direction is

$$\frac{\partial^2 \theta}{\partial t^2}=\alpha^2\frac{\partial^2 \theta}{\partial x^2} \tag{3.40}$$

where α is the wave velocity and (using Eq. (3.31)) is given by

$$\alpha=\sqrt{\frac{\lambda+2\mu}{\rho}}=\sqrt{\frac{K+\frac{4}{3}\mu}{\rho}} \tag{3.41}$$

The longitudinal wave is the fastest of all seismic waves. When an earthquake occurs, this wave is the first to arrive at a recording station. As a result it is called the *primary wave*, or *P-wave*. Equation 3.41 shows that P-waves can travel through solids, liquids and gases, all of which are compressible ($K \neq 0$). Liquids and gases do not allow shear. Consequently, $\mu = 0$, and the compressional wave velocity in a fluid is given by

$$\alpha = \sqrt{\frac{K}{\rho}} \tag{3.42}$$

3.3.2.2 *Transverse waves*

The vibrations along the y- and z-axes (Fig. 3.10) are parallel to the wavefront and transverse to the direction of propagation. If we wish, we can combine the y- and z-components into a single transverse motion. It is more convenient, however, to analyze the motions in the vertical and horizontal planes separately. Here we discuss the disturbance in the vertical plane defined by the x- and z-axes; an analogous description applies to the horizontal plane.

The transverse wave motion is akin to that seen when a rope is shaken. Vertical planes move up and down and adjacent elements of the medium experience shape distortions (Fig. 3.12a), changing repeatedly from a rectangle to a parallelogram and back. Adjacent elements of the medium suffer vertical shear.

Consider the distortion of an element bounded by vertical planes separated by a small horizontal distance dx (Fig. 3.12b) at an arbitrary horizontal position x. The passage of a wave in the x-direction produces a displacement w and a force F_z in the z-direction. At the position $x+dx$ the displacement is $w+dw$ and the force is F_z+dF_z. The mass of the small volume element bounded by the vertical planes is $\rho\, dx\, A_x$, where A_x is the area of the bounding plane. The net force acting on this element in the z-direction is given by

$$(F_z+dF_z)-F_z=dF_z=\frac{\partial F_z}{\partial x}dx \tag{3.43}$$

The force F_z arises from the shear stress σ_{xz} on the area A_x, and is equal to $\sigma_{xz}A_x$. The equation of motion of the vertically sheared element is

$$(\rho\, dx A_x)\frac{\partial^2 w}{\partial t^2}=dx A_x\frac{\partial \sigma_{xz}}{\partial x} \tag{3.44}$$

We now have to modify the Lamé expression for Hooke's law and the definition of shear strain so that they apply to the passage of a one-dimensional shear wave in the x-direction. In this case, because the areas of the parallelograms between adjacent vertical planes are equal, there is no volume change. The dilatation θ is zero, and Hooke's law (Eq. (3.28)) becomes

$$\sigma_{xz}=2\mu\epsilon_{xz} \tag{3.45}$$

Following the definition of shear-strain components in Eq. (3.12) we have

$$\epsilon_{xz}=\frac{1}{2}\left(\frac{\partial w}{\partial x}+\frac{\partial u}{\partial z}\right)$$

For a one-dimensional shear wave there is no change in the distance dx between the vertical planes; du and $\partial u/\partial z$ are zero and ϵ_{xz} is equal to $(\partial w/\partial x)/2$. On substitution into Eq. (3.45) this gives

$$\sigma_{xz}=\mu\frac{\partial w}{\partial x} \tag{3.46}$$

and on further substitution into Eq. (3.44) and rearrangement of terms we get

$$\frac{\partial^2 w}{\partial t^2}=\beta^2\frac{\partial^2 w}{\partial x^2} \tag{3.47}$$

where β is the velocity of the shear wave, given by

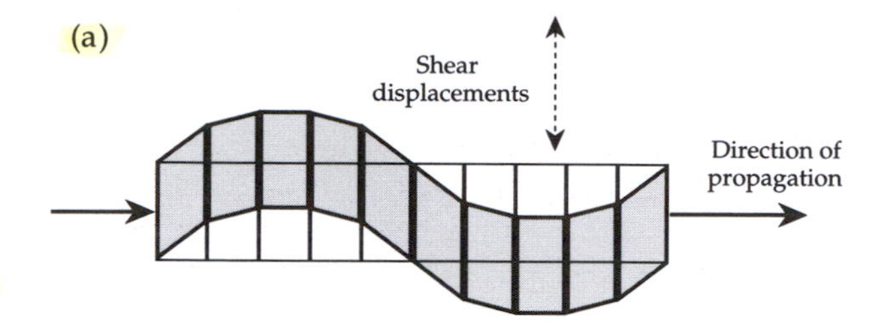

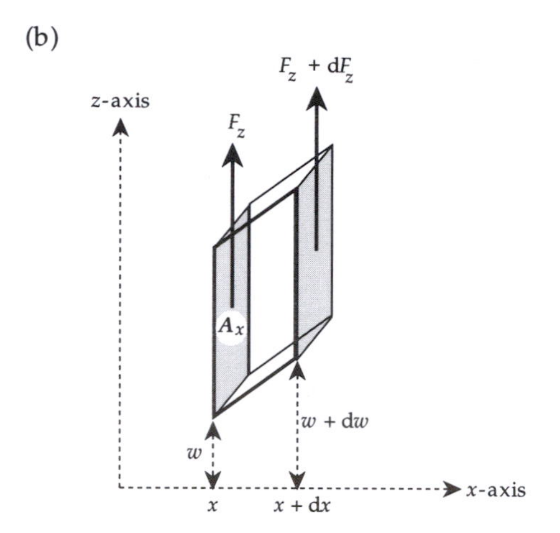

Fig. 3.12 (a) Shear distortion caused by the passage of a one-dimensional S-wave. (b) Displacements and forces in the z-direction at the positions x and $x+dx$ bounding a small sheared element.

$$\beta=\sqrt{\frac{\mu}{\rho}} \tag{3.48}$$

The only elastic property that determines the velocity of the shear wave is the rigidity or shear modulus, μ. In liquids and gases μ is zero and shear waves cannot propagate. In solids, a quick comparison of Eqs. (3.41) and (3.48) gives

$$\alpha^2-\tfrac{4}{3}\beta^2=\frac{K}{\rho} \tag{3.49}$$

By definition, the bulk modulus K is positive (if it were negative, an increase in confining pressure would cause an increase in volume), and therefore α is always greater than β. Shear waves from an earthquake travel more slowly than P-waves and are recorded at an observation station as later arrivals. Shear waves are often referred to as *secondary waves* or *S-waves.*

The general shear-wave motion within the plane of the wavefront can be resolved into two orthogonal components, one being horizontal and the other lying in the vertical plane containing the ray path (Fig. 3.10). Eq. (3.47) describes a one-dimensional shear wave which travels in the x-direction, but which has particle displacements (w) in the z-direction. This wave can be considered to be polarized in the vertical plane. It is called the *SV-wave*. A similar equation describes the shear wave in the x-direction with particle displacements (v) in the y-direction. A shear wave that is polarized in the horizontal plane is called an *SH-wave*.

As for the description of longitudinal waves, this treatment of shear wave transmission is over-simplified. The passage of a shear wave involves rotations of volume elements within the plane normal to the ray path, without changing their volume. For this reason, shear waves are also sometimes called rotational (or equivoluminal) waves. The rotation is a vector, $\boldsymbol{\psi}$, with x-, y- and z-components given by

$$\psi_x=\frac{\partial w}{\partial y}-\frac{\partial v}{\partial z};\ \psi_y=\frac{\partial u}{\partial z}-\frac{\partial w}{\partial x};\ \psi_z=\frac{\partial v}{\partial x}-\frac{\partial u}{\partial y} \tag{3.50}$$

The more complete equation for the shear wave in the x-direction is

$$\frac{\partial^2\psi}{\partial t^2}=\beta^2\frac{\partial^2\psi}{\partial x^2} \tag{3.51}$$

where β is again the shear-wave velocity as given by Eq. (3.48).

Until now we have chosen the direction of propagation along one of the reference axes so as to simplify the mathematics. If we remove this restriction, additional second-order differentiations with respect to the y- and z-coordinates must be introduced. The P-wave and S-wave equations become, respectively,

$$\frac{\partial^2\theta}{\partial t^2}=\alpha^2\left\{\frac{\partial^2\theta}{\partial x^2}+\frac{\partial^2\theta}{\partial y^2}+\frac{\partial^2\theta}{\partial z^2}\right\} \tag{3.52}$$

$$\frac{\partial^2\psi}{\partial t^2}=\beta^2\left\{\frac{\partial^2\psi}{\partial x^2}+\frac{\partial^2\psi}{\partial y^2}+\frac{\partial^2\psi}{\partial z^2}\right\} \tag{3.53}$$

By introducing the Laplace operator ∇ these equations can be written in the following convenient short form:

$$\frac{\partial^2\theta}{\partial t^2}=\alpha^2\nabla^2\theta \tag{3.54}$$

$$\frac{\partial^2\psi}{\partial t^2}=\beta^2\nabla^2\psi \tag{3.55}$$

$$\text{where } \nabla^2=\frac{\partial^2}{\partial x^2}+\frac{\partial^2}{\partial y^2}+\frac{\partial^2}{\partial z^2} \tag{3.56}$$

3.3.2.3 *The solution of the seismic wave equation*

Two important characteristics of a wave motion are: (1) it transmits energy by means of elastic displacements of the particles of the medium i.e., there is no net transfer of mass, and (2) the wave pattern repeats itself in both time and space. The harmonic repetition allows us to express the amplitude variation by a *sine* or *cosine* function. As the wave passes any point, the amplitude of the disturbance is repeated at regular time intervals, T, the *period* of the wave. The number of times the amplitude is repeated per second is the *frequency*, f. which is equal to the inverse of the period ($f=1/T$). At any instant in time, the disturbance in the medium is repeated along the direction of travel at regular distances, λ, the *wavelength* of the wave. During the passage of a P-wave in the x-direction, the harmonic displacement (u) of a particle from its mean position can be written

$$u=A\sin 2\pi\left(\frac{x}{\lambda}-\frac{t}{T}\right) \tag{3.57}$$

where A is the amplitude.

The quantity in brackets is called the *phase* of the wave. Any value of the phase corresponds to a particular amplitude and direction of motion of the particles of the medium. The *wave number* (k), *angular frequency* (ω) and velocity (c) are defined and related by

$$k=\frac{2\pi}{\lambda};\ \omega=2\pi f=\frac{2\pi}{T};\ c=\lambda f=\omega/k \tag{3.58}$$

Equation (3.57) for the displacement (u) can then be written

$$u=A\sin(kx-\omega t)=A\sin k(x-ct) \tag{3.59}$$

The velocity c introduced here is called the *phase velocity*. It is the velocity with which a constant phase (e.g., the 'peak' or 'trough', or one of the zero displacements) is transmitted. This can be seen by equating the phase to a constant and then differentiating the expression with respect to time, as follows:

$$kx-\omega t=\text{constant}$$

$$k\frac{dx}{dt}-\omega=0$$

$$\frac{dx}{dt}=\frac{\omega}{k}=c \tag{3.60}$$

To demonstrate that the displacement given by Eq. (3.59) is a solution of the one-dimensional wave equation (Eq. (3.38)) we must partially differentiate u in Eq. (3.59) twice with respect to time (t) and twice with respect to position (x):

$$\frac{\partial u}{\partial x}=Ak\cos(kx-\omega t);\ \frac{\partial^2 u}{\partial x^2}=-Ak^2\sin(kx-\omega t)=-k^2u$$

$$\frac{\partial u}{\partial t}=-A\omega\cos(kx-\omega t);\ \frac{\partial^2 u}{\partial t^2}=-A\omega^2\sin(kx-\omega t)=-\omega^2u \tag{3.61}$$

$$\frac{\partial^2 u}{\partial t^2}=\frac{\omega^2}{k^2}\frac{\partial^2 u}{\partial x^2}=c^2\frac{\partial^2 u}{\partial x^2} \tag{3.62}$$

For a P-wave travelling along the x-axis the dilatation θ is given by an equation similar to Eq. (3.59), with substitution of the P-wave velocity (α) for the velocity c. Similarly, for an S-wave along the x-axis the rotation ψ is given by an equation like Eq. (3.59) with appropriate substitutions of ψ for u and the S-wave velocity (β) for the velocity c. More generally, the solutions of the three-dimensional compressional and shear wave equations (Eqs. (3.52) and (3.53), respectively) are considerably more complicated than those given by Eq. (3.59).

3.3.2.4 *D'Alembert's principle*

Eq. (3.59) describes the particle displacement during the passage of a wave that is travelling in the direction of the positive x-axis with velocity c. Because the velocity enters the wave equation as c^2, the one-dimensional Eq. (3.38) is also satisfied by the displacement

$$u=B\sin k(x+ct) \tag{3.63}$$

which corresponds to a wave travelling with velocity c in the direction of the negative x-axis.

In fact, any function of $(x\pm ct)$ that is itself continuous and that has continuous first and second derivatives is a solution of the one-dimensional wave equation. This is known as *D'Alembert's principle*. It can be simply demonstrated for the function $\mathrm{F}=\mathrm{f}(x-ct)=\mathrm{f}(\phi)$ as follows.

$$\frac{\partial \mathrm{F}}{\partial x}=\frac{\partial \mathrm{F}}{\partial \phi}\frac{\partial \phi}{\partial x}=\frac{\partial \mathrm{F}}{\partial \phi};\quad \frac{\partial \mathrm{F}}{\partial t}=\frac{\partial \mathrm{F}}{\partial \phi}\frac{\partial \phi}{\partial t}=-c\frac{\partial \mathrm{F}}{\partial \phi}$$

$$\frac{\partial^2 \mathrm{F}}{\partial x^2}=\frac{\partial}{\partial x}\frac{\partial \mathrm{F}}{\partial x}=\frac{\partial \phi}{\partial x}\frac{\partial}{\partial \phi}\frac{\partial \mathrm{F}}{\partial x}=\frac{\partial^2 \mathrm{F}}{\partial \phi^2}$$

$$\frac{\partial^2 \mathrm{F}}{\partial t^2}=\frac{\partial}{\partial t}\frac{\partial \mathrm{F}}{\partial t}=\frac{\partial \phi}{\partial t}\frac{\partial}{\partial \phi}\frac{\partial \mathrm{F}}{\partial t}=-c\frac{\partial}{\partial \phi}\left(-c\frac{\partial \mathrm{F}}{\partial \phi}\right)=c^2\frac{\partial^2 \mathrm{F}}{\partial \phi^2}$$

$$\frac{\partial^2 \mathrm{F}}{\partial t^2}=c^2\frac{\partial^2 \mathrm{F}}{\partial x^2} \tag{3.64}$$

Because Eq. (3.64) is valid for positive and negative values of c, its general solution F represents the superposition of waves travelling in opposite directions along the x-axis, and is given by

$$\mathrm{F}=\mathrm{f}(x-ct)+\mathrm{g}(x+ct) \tag{3.65}$$

3.3.2.5 *The eikonal equation*

Consider a wave travelling with constant velocity c along the axis x' which has direction cosines (l, m, n). If x' is measured from the center of the coordinate axes (x, y, z) we can substitute $x'=lx+my+nz$ for x in Eq. (3.65). If we consider for convenience only the wave travelling in the direction of $+x'$, we get as the general solution to the wave equation

$$\mathrm{F}=\mathrm{f}(lx+my+nz-ct) \tag{3.66}$$

The wave equation is a second-order differential equation. However, the function F is also a solution of a first-order differential equation. This is seen by differentiating F with respect to x, y, z, and t, respectively, which gives

$$\frac{\partial \mathrm{F}}{\partial x}=\frac{\partial \mathrm{F}}{\partial \phi}\frac{\partial \phi}{\partial x}=l\frac{\partial \mathrm{F}}{\partial \phi};\quad \frac{\partial \mathrm{F}}{\partial y}=\frac{\partial \mathrm{F}}{\partial \phi}\frac{\partial \phi}{\partial y}=m\frac{\partial \mathrm{F}}{\partial \phi}$$

$$\frac{\partial \mathrm{F}}{\partial z}=\frac{\partial \mathrm{F}}{\partial \phi}\frac{\partial \phi}{\partial z}=n\frac{\partial \mathrm{F}}{\partial \phi};\quad \frac{\partial \mathrm{F}}{\partial t}=\frac{\partial \mathrm{F}}{\partial \phi}\frac{\partial \phi}{\partial t}=-c\frac{\partial \mathrm{F}}{\partial \phi} \tag{3.67}$$

The direction cosines (l, m, n) are related by $l^2+m^2+n^2=1$, and so, as can be verified by substitution, the expressions in Eq. (3.67) satisfy the equation

$$\left(\frac{\partial \mathrm{F}}{\partial x}\right)^2+\left(\frac{\partial \mathrm{F}}{\partial y}\right)^2+\left(\frac{\partial \mathrm{F}}{\partial z}\right)^2=\left(\frac{1}{c}\right)^2\left(\frac{\partial \mathrm{F}}{\partial t}\right)^2 \tag{3.68}$$

In seismic wave theory, the progress of a wave is described by successive positions of its wavefront, defined as the surface in which all particles at a given instant in time are moving with the same phase. For a particular value of t a constant phase of the wave equation solution given by Eq. (3.66) requires that

$$lx+my+nz=\text{constant} \tag{3.69}$$

From analytical geometry we know that Eq. (3.69) represents a family of planes perpendicular to a line with direction cosines (l, m, n). We began this discussion by describing a wave moving with velocity c along the direction x', and now we see that this direction is normal to the plane wavefronts. This is the direction that we earlier defined as the ray path of the wave.

In a medium like the Earth the elastic properties and density – and therefore also the velocity – vary with position. The ray path is no longer a straight line and the wavefronts are not planar. Instead of Eq. (3.66) we write

$$F=f\,[S(x,y,z)-c_0 t] \tag{3.70}$$

where $S(x, y, z)$ is a function of position only and c_0 is a constant reference velocity. Substitution of Eq. (3.70) into Eq. (3.68) gives

$$\left(\frac{\partial S}{\partial x}\right)^2+\left(\frac{\partial S}{\partial y}\right)^2+\left(\frac{\partial S}{\partial z}\right)^2=\left(\frac{c_0}{c}\right)^2=\zeta^2 \tag{3.71}$$

where ζ is known as the refractive index of the medium. Eq. (3.71) is called the *eikonal equation*. It establishes the equivalence of treating seismic wave propagation by describing the wavefronts or the ray paths. The surfaces $S(x, y, z)$=constant represent the wavefronts (no longer planar). The direction cosines of the ray path (normal to the wavefront) are in this case given by

$$\lambda=\zeta\frac{\partial S}{\partial x};\ \mu=\zeta\frac{\partial S}{\partial y};\ \nu=\zeta\frac{\partial S}{\partial z} \tag{3.72}$$

3.3.2.6 *The energy in a seismic disturbance*

It is important to distinguish between the velocity with which a seismic disturbance travels through a material and the speed with which the particles of the material vibrate during the passage of the wave. The vibrational speed (v_p) is obtained by differentiating Eq. (3.59) with respect to time, which yields

$$v_p=\frac{\partial u}{\partial t}=-\omega A\cos(kx-\omega t) \tag{3.73}$$

The *intensity* or *energy density* of a wave is the energy per unit volume in the wavefront and consists of kinetic and potential energy. The kinetic part is given by

$$I=\tfrac{1}{2}\rho v_p^2=\tfrac{1}{2}\rho\omega^2A^2\cos^2(kx-wt) \tag{3.74}$$

The energy density averaged over a complete harmonic cycle consists of equal parts of kinetic and potential energy; it is given by

$$I_{av}=\tfrac{1}{2}\rho\omega^2A^2 \tag{3.75}$$

i.e., the mean intensity of the wave is proportional to the square of its amplitude.

3.3.2.7 *Attenuation of seismic waves*

The further a seismic signal travels from its source the weaker it becomes. The decrease of amplitude with increasing distance from the source is referred to as *attenuation*. It is partly due to the geometry of propagation of seismic waves, and partly due to anelastic properties of the material through which they travel.

The most important reduction is due to *geometric attenuation*. Consider the seismic body waves generated by a seismic source at a point P on the surface of a uniform half-space (see Fig. 3.9). If there is no energy loss due to friction, the energy (E_b) in the wavefront at distance r from its source is distributed over the surface of a hemisphere with area $2\pi r^2$. The *intensity* (or energy density, I_b) of the body waves is the energy per unit area of the wavefront, and at distance r is:

$$I_b(r)=\frac{E_b}{2\pi r^2} \tag{3.76}$$

The surface wave is constricted to spread out laterally. The disturbance affects not only the free surface but extends downwards into the medium to a depth d, which we can consider to be constant for a given wave (Fig. 3.9). When the wavefront of a surface wave reaches a distance r from the source, the initial energy (E_s) is distributed over a circular cylindrical surface with area $2\pi rd$. At a distance r from its source the intensity of the surface wave is given by:

$$I_s(r)=\frac{E_s}{2\pi rd} \tag{3.77}$$

These equations show that the decrease in intensity of body waves is proportional to $1/r^2$ while the decrease of surface wave intensity is proportional to $1/r$. As found in Eq. (3.75), the intensity of a wave-form, or harmonic vibration, is proportional to the square of its amplitude. The corresponding amplitude attenuations of body waves and surface waves are proportional to $1/r$ and $1/\sqrt{r}$, respectively. Thus, seismic body waves are attenuated more rapidly than surface waves with increasing distance from the source. This explains why, except for the records of very deep earthquakes that do not generate strong surface waves, the surface-wave train on a seismogram is more prominent than that of the body waves.

Another reason for attenuation is the *absorption* of energy due to imperfect elastic properties. If the particles of a medium do not react perfectly elastically with their neighbors, part of the energy in the wave is lost (converting, for example, to frictional heat) instead of being transferred through the medium. This type of attenuation of the seismic wave is referred to as anelastic *damping*.

The damping of seismic waves is described by a parameter called the *quality factor* (Q), a concept borrowed from electric circuit theory where it describes the performance of an oscillatory circuit. It is defined as the fractional loss of energy per cycle

$$\frac{2\pi}{Q} = -\frac{\Delta E}{E} \tag{3.78}$$

In this expression ΔE is the energy lost in one cycle and E is the total elastic energy stored in the wave. If we consider the damping of a seismic wave as a function of the distance that it travels, a cycle is represented by the wavelength (λ) of the wave. Eq. (3.78) can be rewritten for this case as

$$\frac{2\pi}{Q} = -\frac{1}{E}\lambda\frac{dE}{dr}$$

$$\frac{dE}{E} = -\frac{2\pi}{Q}\frac{dr}{\lambda} \tag{3.79}$$

It is conventional to measure damping by its effect on the amplitude of a seismic signal, because that is what is observed on a seismic record. We have seen that the energy in a wave is proportional to the square of its amplitude A (Eq. (3.75)). Thus we can write $dE/E=2dA/A$ in Eq. (3.79), and on solving we get the damped amplitude of a seismic wave at distance (r) from its source:

$$A = A_0\exp\left(-\frac{\pi}{Q}\frac{r}{\lambda}\right) = A_0\exp\left(-\frac{r}{D}\right) \tag{3.80}$$

In this equation D is the distance within which the amplitude falls to 1/e (36.8%, or roughly a third) of its original value. The inverse of this distance (D^{-1}) is called the *absorption coefficient*. For a given wavelength, D is proportional to the Q-factor of the region through which the wave travels. A rock with a high Q-factor transmits a seismic wave with relatively little energy loss by absorption, and the distance D is large. For body waves D is generally of the order of 10,000 km and damping of the waves by absorption is not a very strong effect. It is slightly stronger for seismic surface waves, for which D is around 5000 km.

Equation (3.80) shows that the damping of a seismic wave is dependent on the Q-factor of the region of the Earth that the wave has travelled through. In general the Q-factor for P-waves is higher than the Q-factor for S-waves. This may indicate that anelastic damping is determined primarily by the shear component of strain. In solids with low rigidity, the shear strain can reach high levels and the damping is greater than in materials with high rigidity. In fluids the Q-factor is high and damping is low, because shear strains are zero and the seismic wave is purely compressional. The values of Q are quite variable in the Earth: values of around 10^2 are found for the mantle, and around 10^3 for P-waves in the liquid core. Because Q is a measure of the deviation from perfect elasticity, it is also encountered in the theory of natural oscillations of the Earth, and has an effect on fluctuations of the Earth's free rotation, as in the damping of the Chandler wobble.

The attenuation of a seismic wave by absorption is dependent upon the frequency of the signal. High frequencies are attenuated more rapidly than are lower frequencies. As a result, the frequency spectrum of a seismic signal changes as it travels through the ground. Although the original signal may be a sharp pulse (resulting from a shock or explosion), the preferential loss of high frequencies as it travels away from the source causes the signal to assume a smoother shape. This selective loss of high frequencies by absorption is analogous to removing high frequencies from a sound source by the use of a filter. Because the low frequencies are not affected so markedly, they pass through the ground with less attenuation. The ground acts as a low-pass filter to seismic signals.

3.3.3 Seismic surface waves

A disturbance at the free surface of a medium propagates away from its source partly as seismic surface waves. Just as seismic body waves can be classified as P- or S-waves, there are two categories of seismic surface waves, sometimes known collectively as L-waves (§3.4.4.3), and subdivided into *Rayleigh waves* (L_R) and *Love waves* (L_Q), which are distinguished from each other by the types of particle motion in their wavefronts. In the description of body waves, the motion of particles in the wavefront was resolved into three orthogonal components – a longitudinal vibration parallel to the ray path (the P-wave motion), a transverse vibration in the vertical plane containing the ray path (the vertical shear or SV-wave) and a horizontal transverse vibration (the horizontal shear or SH-wave). These components of motion, restricted to surface layers, also determine the particle motion and character of the two types of surface waves.

3.3.3.1 *Rayleigh waves* (L_R)

In 1885 Lord Rayleigh described the propagation of a surface wave along the free surface of a semi-infinite elastic half-space. The particles in the wavefront of the Rayleigh wave are polarized to vibrate in the vertical plane. The resulting particle motion can be regarded as a combination of the P- and SV-vibrations. If the direction of propagation of the Rayleigh wave is to the right of the viewer (as in Fig. 3.13), the particle motion describes a *retrograde ellipse* in the vertical plane with its major axis vertical and minor axis in the direction of wave propagation. If Poisson's relation holds for a solid (i.e., Poisson's ratio $\nu=0.25$) the theory of Rayleigh waves gives a speed (V_{LR}) equal to $\sqrt{(2-2/\sqrt{3})}=0.9194$ of the speed (β) of S-waves (i.e., $V_{LR}=0.9194\,\beta$). This is approximately the case in the Earth.

The particle displacement is not confined entirely to the

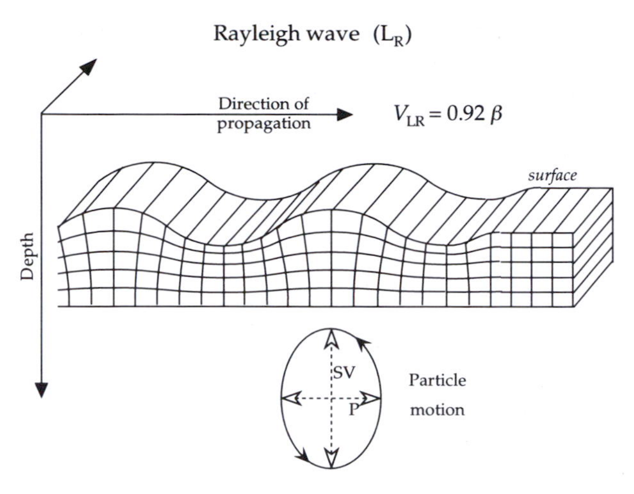

Fig. 3.13 The particle motion in the wavefront of a Rayleigh wave consists of a combination of P- and SV-vibrations in the vertical plane. The particles move in retrograde sense around an ellipse that has its major axis vertical and minor axis in the direction of wave propagation.

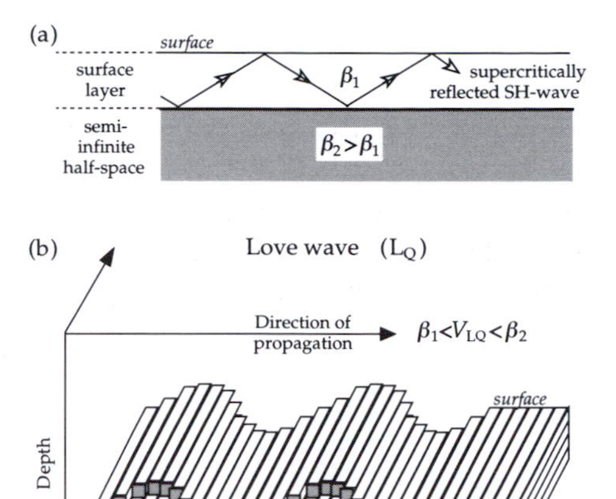

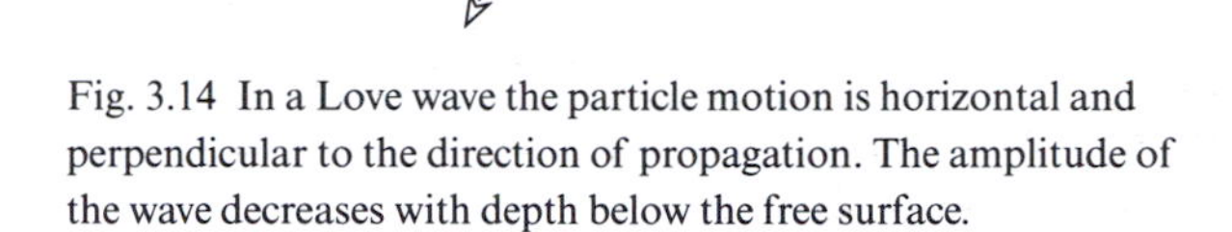

Fig. 3.14 In a Love wave the particle motion is horizontal and perpendicular to the direction of propagation. The amplitude of the wave decreases with depth below the free surface.

surface of the medium. Particles below the free surface are also affected by the passage of the Rayleigh wave; in a uniform half-space the amplitude of the particle displacement decreases exponentially with increasing depth. The penetration depth of the surface wave is typically taken to be the depth at which the amplitude is attenuated to (e^{-1}) of its value at the surface. For Rayleigh waves with wavelength λ the characteristic penetration depth is about 0.4 λ.

3.3.3.2 *Love waves* (L_Q)

The boundary conditions which govern the components of stress at the free surface of a semi-infinite elastic half-space prohibit the propagation of SH-waves along the surface. However, A .E. H. Love showed in 1911 that if a horizontal layer lies between the free surface and the semi-infinite half-space (Fig. 3.14a), SH-waves within the layer that are reflected at supercritical angles (see §3.6) from the top and bottom of the layer can interfere constructively to give a surface wave with horizontal particle motions (Fig. 3.14b). The velocity (β_1) of S-waves in the near-surface layer must be lower than in the underlying half-space (β_2). The velocity of the Love waves (V_{LQ}) lies between the two extreme values: $\beta_1 < V_{LQ} < \beta_2$.

Theory shows that the speed of Love waves with very short wavelengths is close to the slower velocity β_1 of the upper layer, while long wavelengths travel at a speed close to the faster velocity β_2 of the lower medium. This dependence of velocity on wavelength is termed *dispersion*. Love waves are always dispersive, because they can only propagate in a velocity-layered medium.

3.3.3.3 *The dispersion of surface waves*

The dispersion of surface waves provides an important tool for determining the vertical velocity structure of the lower crust and upper mantle. Love waves are intrinsically dispersive even when the surface layer and underlying half-space are uniform. Rayleigh waves over a uniform half-space are non-dispersive. However, horizontal layers with different velocities are usually present or there is a vertical velocity gradient. Rayleigh waves with long wavelengths penetrate more deeply into the Earth than those with short wavelengths. The speed of Rayleigh waves is proportional to the shear-wave velocity ($V_{LR} \approx 0.92\,\beta$), and in the crust and uppermost mantle β generally increases with depth. Thus, the deeper penetrating long wavelengths travel with faster seismic velocities than the short wavelengths. As a result, the Rayleigh waves are dispersive.

The packet of energy that propagates as a surface wave contains a spectrum of wavelengths. The energy in the wave propagates as the envelope of the wave packet (Fig. 3.15a), at a speed that is called the *group velocity* (U). The individual waves that make up the wave packet travel with *phase velocity* (c), as defined in Eq. (3.58). If the phase velocity is dependent on the wavelength, the group velocity is related to it by

$$U = \frac{\partial \omega}{\partial k} = \frac{\partial}{\partial k}(ck) = c + k\frac{\partial c}{\partial k} = c - \lambda\frac{\partial c}{\partial \lambda} \tag{3.81}$$

The situation in which phase velocity increases with increasing wavelength (i.e. the longer wavelengths propagate

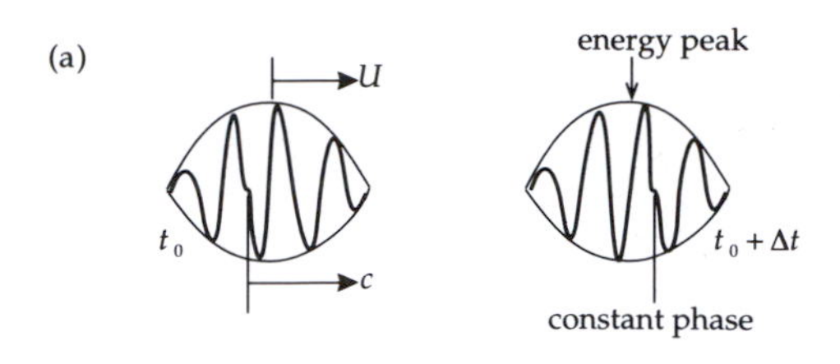

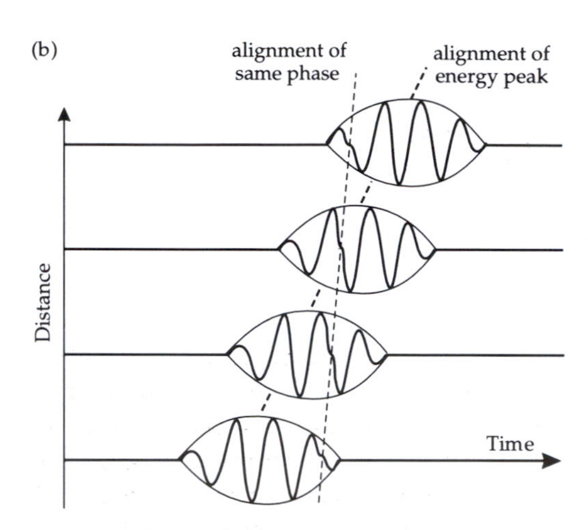

Fig. 3.15 (a) The surface-wave energy propagates as the envelope of the wave packet with the group velocity U, while the individual wavelengths travel with the phase velocity c. (b) Change of shape of a wave packet due to normal dispersion as the faster-moving long wavelengths pass through the packet. At large distances from the source, the long wavelengths arrive as the first part of the surface-wave train (modified from Telford *et al.*, 1990).

faster than the short wavelengths) is called *normal dispersion*. In this case, because $\partial c/\partial\lambda$ is positive, the group velocity U is slower than the phase velocity c. The shape of the wave packet changes systematically as the faster-moving long wavelengths pass through the packet (Fig. 3.15b). As time elapses, an initially concentrated pulse becomes progressively stretched out into a long train of waves. Consequently, over a medium in which velocity increases with depth the long wavelengths arrive as the first part of the surface-wave record at large distances from the seismic source.

3.3.4 Free oscillations of the Earth

When a bell is struck with a hammer, it vibrates freely at a number of natural frequencies. The combination of natural oscillations that are excited gives each bell its particular sonority. In an analogous way, the sudden release of energy in a very large earthquake can set the entire Earth into vibration, with natural frequencies of oscillation that are

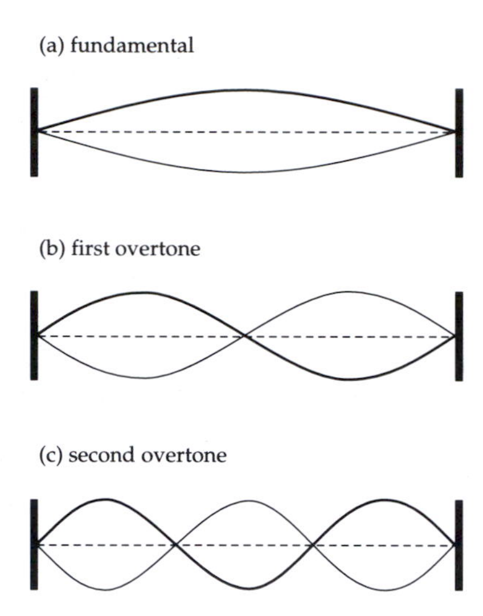

Fig. 3.16 Normal modes of vibration for a standing wave on a string fixed at both ends.

determined by the elastic properties and structure of the Earth's interior. The free oscillations involve three-dimensional deformation of the Earth's spherical shape and can be quite complex. Before discussing the Earth's free oscillations it is worthwhile to review some concepts of vibrating systems that can be learned from the one-dimensional excitation of a vibrating string that is fixed at both ends.

Any complicated vibration of the string can be represented by the superposition of a number of simpler vibrations, called the *normal modes* of vibration. These arise when travelling waves reflected from the boundaries at the ends of the string interfere with each other to give a *standing wave*. Each normal mode corresponds to a standing wave with frequency and wavelength determined by the condition that the length of the string must always equal an integral number of half-wavelengths (Fig. 3.16). As well as the fixed ends, there are other points on the string that have zero displacement; these are called the *nodes* of the vibration. The first normal (or *fundamental*) *mode* of vibration has no nodes. The second normal mode (sometimes called the *first overtone*) has one node; its wavelength and period are half those of the fundamental mode. The third normal mode (second overtone) has three times the frequency of the first mode, and so on. Modes with one or more nodes are called *higher-order* modes.

The concepts of modes and nodes are also applicable to a vibrating sphere. The complex general vibration of a sphere can be resolved into the superposition of a number of normal modes. The nodes of zero displacement become *nodal surfaces* on which the amplitude of the vibration is zero. The free oscillations of the Earth can be divided into three categories. In *radial* oscillations the displacements are

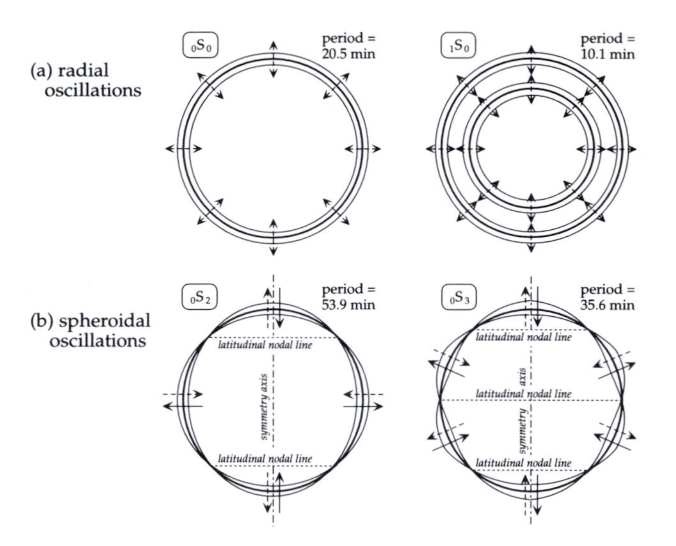

Fig. 3.17 (a) Modes of radial oscillation with their periods: in the fundamental mode $_0S_0$ (also called the balloon mode) the entire Earth expands and contracts in unison; higher modes, such as $_1S_0$, have internal spherical nodal surfaces concentric with the outer surface. (b) Modes of spheroidal oscillation $_0S_2$ ('football mode') and $_0S_3$ with their periods; the nodal lines are small circles perpendicular to the symmetry axis.

purely radial, in *spheroidal* oscillations they are partly radial and partly tangential, and in *toroidal* oscillations they are purely tangential.

3.3.4.1 *Radial oscillations*

The simplest kind of free oscillations are the *radial* oscillations, in which the shape of the Earth remains 'spherical' and all particles vibrate purely radially (Fig. 3.17a). In the fundamental mode of this type of oscillation the entire Earth expands and contracts in unison with a period of about 20.5 minutes. The second normal mode (first overtone) of radial oscillations has a single internal spherical nodal surface. While the inner sphere is contracting, the part outside the nodal surface is expanding, and vice versa. The nodal surfaces of higher modes are also spheres internal to the Earth and concentric with the outer surface.

3.3.4.2 *Spheroidal oscillations*

A general spheroidal oscillation involves both radial and tangential displacements that can be described by *spherical harmonic functions*. These functions are referred to an axis through the Earth at the point of interest (e.g., an earthquake epicenter), and to a great circle which contains the axis. With respect to this reference frame they describe the latitudinal and longitudinal variations of the displacement of a surface from a sphere. They allow complete mathematical description and concise identification of each mode of oscillation with the aid of three indices. The *longitudinal order m* is the number of nodal lines on the sphere that are great circles, the *order n* is determined from the $(n-m)$ latitudinal nodal lines, and the *overtone number l* describes the number of internal nodal surfaces. The notation $_lS_n^m$ denotes a spheroidal oscillation of order n, longitudinal order m, and overtone number l. In practice, only oscillations with longitudinal order $m=0$ (rotationally symmetric about the reference axis) are observed, and this index is usually dropped. Also, the oscillation of order $n=1$ does not exist; it would have only a single equatorial nodal plane and the vibration would involve displacement of the center of gravity. The spheroidal oscillations $_0S_2$ and $_0S_3$ are shown in Fig. 3.17b. Spheroidal oscillations displace the Earth's surface and alter the internal density distribution. After large earthquakes they produce records on highly sensitive gravity meters used for bodily Earth-tide observations, and also on strain gauges and tilt meters.

The *radial oscillations* can be regarded as a special type of spheroidal oscillation with $n=0$. The fundamental radial oscillation is also the fundamental spheroidal oscillation, denoted $_0S_0$; the next higher mode is called $_1S_0$ (Fig. 3.17a).

3.3.4.3 *Toroidal oscillations*

The third category of oscillation is characterized by displacements that are purely tangential. The spherical shape and volume of the Earth are unaffected by a toroidal oscillation, which involves only longitudinal displacements about an axis (Fig. 3.18a). The amplitude of the longitudinal displacement varies with latitude. (Note that, as for spheroidal oscillations, 'latitude' and 'longitude' refer to the symmetry axis and are different from their geographic definitions). The toroidal modes have nodal planes that intersect the surface on circles of 'latitude', on which the toroidal displacement is zero. Analogously to the spheroidal oscillations, the notation $_lT_n$ is used to describe the spatial geometry of toroidal modes. The mode $_0T_0$ has zero displacement and $_0T_1$ does not exist, because it describes a constant azimuthal twist of the entire Earth, which would change its angular momentum. The simplest toroidal mode is $_0T_2$, in which two hemispheres oscillate in opposite senses

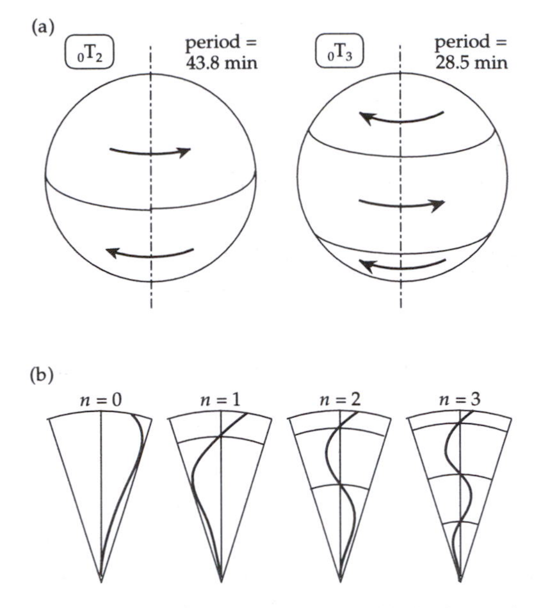

Fig. 3.18 (a) The modes of toroidal oscillation $_0T_2$ and $_0T_3$ with their periods; these modes involve oscillation in opposite senses across nodal planes normal to the symmetry axis. (b) In toroidal oscillations, displacements of internal spherical surfaces from their equilibrium positions vary with depth and are zero at internal nodal surfaces.

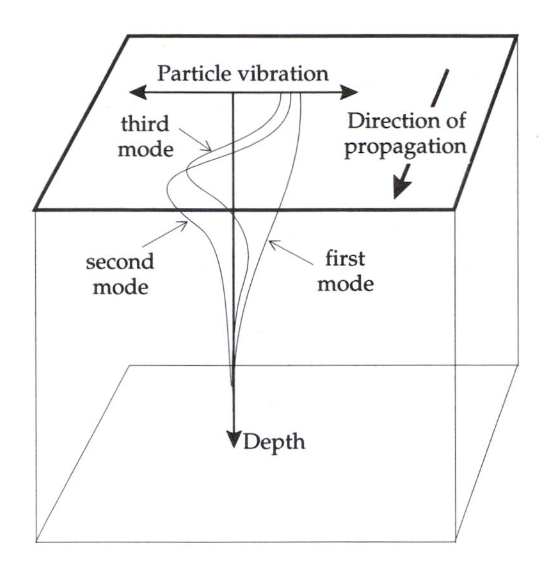

Fig. 3.19 The attenuation with depth of some low-order modes of Love waves.

across a single nodal plane (Fig. 3.18a). Higher toroidal modes of order n oscillate across (n–1) nodal planes perpendicular to the symmetry axis.

The amplitudes of toroidal oscillations inside the Earth change with depth. The displacements of internal spherical surfaces from their equilibrium positions are zero at internal nodal surfaces (Fig. 3.18b). The nomenclature for toroidal oscillations also represents these internal nodal surfaces. Thus, $_1T_l$ denotes a general toroidal mode with one internal nodal surface; the internal sphere twists in an opposite sense to the outer spherical shell; $_2T_l$ has two internal nodal surfaces, etc. (Fig. 3.18b).

The twisting motion in toroidal oscillations does not alter the radial density distribution in the Earth, and so they do not show in gravity meter records. They cause changes in strain and displacement parallel to the Earth's surface and can be recorded by strain meters. Toroidal oscillations are dependent on the shear strength of the Earth's interior. The Earth's fluid core cannot take part in these oscillations, and they are therefore restricted to the Earth's rigid mantle and crust.

3.3.4.4 *Comparison with surface waves*

The higher-order free oscillations of the Earth are related directly to the two types of surface wave. (i) In a Rayleigh wave the particle vibration is polarized in the vertical plane and has radial and tangential components (see Fig. 3.13). The higher-order *spheroidal* oscillations are equivalent to the standing wave patterns that arise from the interference of trains of long-period Rayleigh waves travelling in opposite directions around the Earth. (ii) In a Love wave the particle vibration is polarized horizontally. The *toroidal* oscillations may be regarded as the standing wave patterns due to the interference of oppositely travelling Love waves.

The similarity between surface waves and higher-order natural oscillations of the Earth is evident in the variations of displacement with depth. Like any vibration, a train of surface waves is made up of different modes. Theoretical analysis of surface waves shows that the amplitudes of different modes decay with depth in the Earth (Fig. 3.19) in an equivalent manner to the depth attenuation of natural oscillations (Fig. 3.18b).

The periods of the normal modes of free oscillations were calculated before they were observed. They have long periods – the period of $_0S_0$ is 20 minutes, that of $_0T_2$ is 44 minutes and that of $_0S_2$ is 54 minutes – and pendulum seismographs are not suitable for recording them. Their recognition had to await the development of long-period seismographs. The spheroidal oscillations have a radial component of displacement and can be recorded with long-period, vertical-motion seismographs. Continually recording gravimeters used for the observation of bodily Earth-tides also record the spheroidal oscillations but not the toroidal oscillations, which have no vertical component. These usually must be recorded with an instrument that is sensitive to horizontal displacements, such as the strain meter designed by H. Benioff in 1935. Long-period, horizontal-motion seismographs are capable of recording the toroidal oscillations induced by great earthquakes.

Strain meter records of the November 4, 1952, magnitude 8 earthquake in Kamchatka exhibited a long-period surface wave with a period of 57 minutes (Benioff, 1958). This is much longer than the known periods of travelling surface waves, and was interpreted as a free oscillation of the Earth. Several independent investigators, using bodily Earth-tide gravity meters and different kinds of seismograph, recorded long-period waves excited by the massive 1960 earthquake in Chile (which had a surface-wave magnitude M_s=8.5). These were conclusively identified with spheroidal and toroidal oscillations.

The study of the natural oscillations of the Earth set up by large earthquakes is an important branch of seismology, because the normal modes are strongly dependent on the Earth's internal structure. The low-order modes are affected by the entire interior of the Earth, while the higher-order modes react primarily to movements of the upper mantle. The periods of free oscillation are determined by the radial distributions of elastic properties and densities in the Earth. Comparison of the observed periods of different modes with values computed for different models of the Earth's velocity and density structure provides an important check on the validity of the Earth model. The oscillations are damped by the anelasticity of the Earth. The low-order mode oscillations with periods up to 40–50 minutes persist much longer than the higher-order modes with short periods of only a few minutes. By studying the decay times of different modes, a profile of the anelastic quality factor (Q) within the Earth can be calculated that is compatible with results obtained for body waves.

3.4 THE SEISMOGRAPH

3.4.1 **Introduction**

The earliest known instrument for indicating the arrival of a seismic tremor from a distant source is reputed to have been invented by a Chinese astronomer called Chang Heng in 132 A.D. The device consisted of eight inverted dragons placed at equal intervals around the rim of a vase. Under each dragon sat an open-mouthed metal toad. Each dragon held a bronze ball in its mouth. When a slight tremor shook the device, an internal mechanism opened the mouth of one dragon, releasing its bronze ball, which fell into the open mouth of the metal toad beneath, thereby marking the direction of arrival of the tremor. The principle of this instrument was used in 18th century European devices that consisted of brimful bowls of water or mercury with grooved rims under which tiny collector bowls were placed

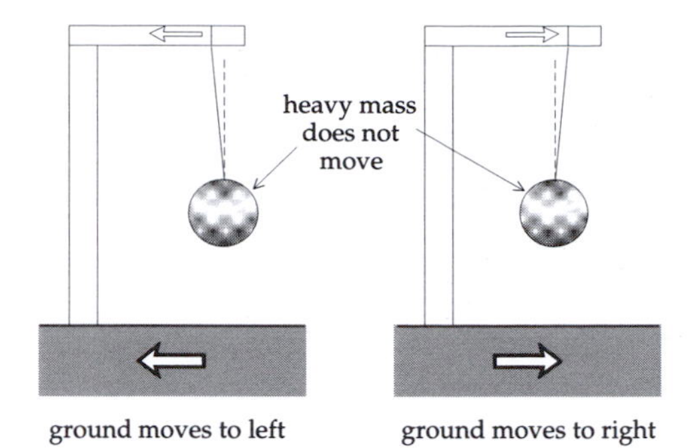

Fig. 3.20 The principle of the seismometer. Because of its inertia, a suspended heavy mass remains almost stationary when the ground and suspension move to the left or to the right.

to collect the overflow occasioned by a seismic tremor. These instruments gave visible evidence of a seismic event but were unable to trace a permanent record of the seismic wave itself. They are classified as *seismoscopes.*

The science of seismology dates from the invention of the *seismograph* by the English scientist John Milne in 1892. Its name derives from its ability to convert an unfelt ground vibration into a visible record. The seismograph consists of a receiver and a recorder. The ground vibration is detected and amplified by a sensor, called the *seismometer* or, in exploration seismology, the *geophone*. In modern instruments the vibration is amplified and filtered electronically. The amplified ground motion is converted to a visible record, called the *seismogram*.

The seismometer makes use of the principle of inertia. If a heavy mass is only loosely coupled to the ground (for example, by suspending it from a wire like a pendulum as in Fig. 3.20), the motion of the Earth caused by a seismic wave is only partly transferred to the mass. While the ground vibrates, the inertia of the heavy mass assures that it does not move as much, if at all. The seismometer amplifies and records the relative motion between the mass and the ground.

Early seismographs were undamped and reacted only to a limited band of seismic frequencies. Seismic waves with inappropriate frequencies were barely recorded at all, but strong waves could set the instrument into resonant vibration. In 1903, the German seismologist Emil Wiechert substantially increased the accuracy of the seismograph by improving the amplification method and by damping the instrument. These early instruments relied on mechanical levers for amplification and recording signals on smoked paper. This made them both bulky and heavy, which severely restricted their application.

A major technological improvement was achieved in 1906, when Prince Boris Galitzin of Russia introduced the electromagnetic seismometer, which allowed galvanometric recording on photographic paper. This electrical method had the great advantage that the recorder could now be separated from the seismometer. The seismometer has evolved constantly, with improvements in seismometer design and recording method, culminating in modern broad-band instruments with digital recording on magnetic tape.

3.4.2 Principle of the seismometer

Seismometers are designed to react to motion of the Earth in a given direction. Mechanical instruments record the amplified displacement of the ground; electromagnetic instruments respond to the velocity of ground motion. Depending on the design, either type may respond to vertical or horizontal motion. Some modern electromagnetic instruments are constructed so as to record simultaneously three orthogonal components of motion. Most designs employ variations on the pendulum principle.

3.4.2.1 *Vertical-motion seismometer*

In the mechanical type of vertical-motion seismometer (Fig. 3.21a), a large mass is mounted on a horizontal bar hinged at a pivot so that it can move only in the vertical plane. A pen attached to the bar writes on a horizontal rotating drum that is fixed to the housing of the instrument. The bar is held in a horizontal position by a weak spring. This assures a loose coupling between the mass and the housing, which is connected rigidly to the ground. Vertical ground motion, as sensed during the passage of a seismic wave, is transmitted to the housing but not to the inertial mass and the pen, which remain stationary. The pen inscribes a trace of the vertical vibration of the housing on a paper fixed to the rotating drum. This trace is the vertical-motion seismogram of the seismic wave.

The electromagnetic seismometer responds to the relative motion between a magnet and a coil of wire. One of these members is fixed to the housing of the instrument and thereby to the Earth. The other is suspended by a spring and forms the inertial member. Two basic designs are possible. In the moving-magnet type, the coil is fixed to the housing and the magnet is inertial. In the moving-coil type the roles are reversed (Fig. 3.21b). A coil of wire fixed to the inertial mass is suspended between the poles of a strong magnet, which in turn is fixed to the ground by the rigid housing. Any motion of the coil within the magnetic field induces a voltage in the coil proportional to the rate of change of magnetic flux. During a seismic arrival the vibration of the ground relative to the mass is converted to an electrical voltage by induction in the coil. The voltage is amplified and transmitted through an electrical circuit to the recorder.

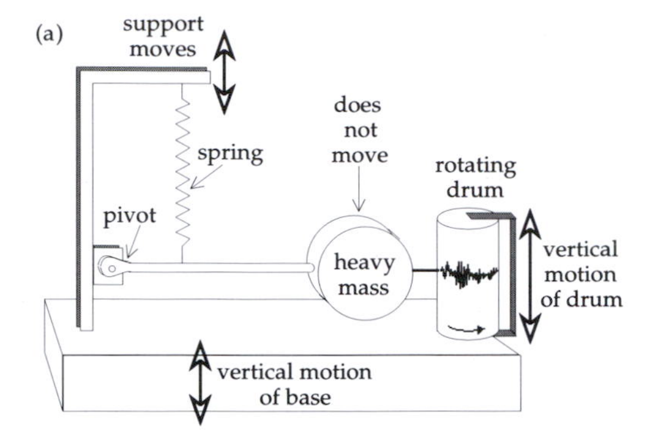

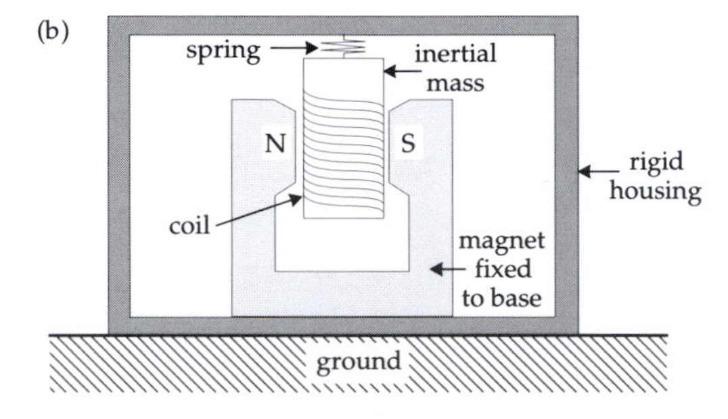

Fig. 3.21 Schematic diagrams illustrating the principle of operation of the vertical-motion seismometer: (a) mechanical pendulum type (after Strahler, 1963), (b) electromagnetic, moving-coil type.

3.4.2.2 *Horizontal-motion seismometer*

The principle of the mechanical type of horizontal-motion seismometer is similar to that of the vertical-motion instrument. As before the inertial mass is mounted on a horizontal bar, but the fulcrum is now hinged almost vertically so that the mass is confined to swing sideways in a nearly horizontal plane (Fig. 3.22). The behavior of the system is similar to that of a gate when its hinges are slightly out of vertical alignment. If the hinge axis is tilted slightly forward, the stable position of the gate is where its center of mass is at its lowest point. In any displacement of the gate, the restoring gravitational forces try to return it to this stable position. Similarly, the horizontal-motion seismometer swings about its equilibrium position like a horizontal pendulum (in fact it is the housing of the instrument that moves and not the inertial mass). As in the vertical-motion seismometer, a pen or light-beam attached to the stationary inertial mass writes on a rotating drum (which in this case has a horizontal axis) and records the relative motion between the mass and the instrument housing. The trace of the ground motion detected with this instrument is the horizontal-motion seismogram of the seismic wave.

The design of an electromagnetic horizontal-motion seismometer is similar to that of the vertical-motion type,

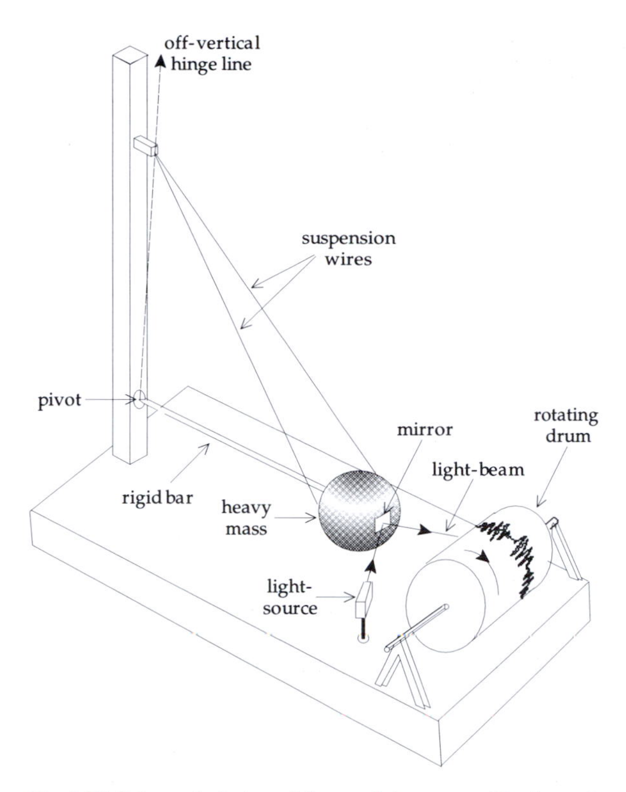

Fig. 3.22 Schematic design of the pendulum type of horizontal-motion seismometer (after Strahler, 1963).

with the exception that the axis of the moving member (coil or magnet) is horizontal.

3.4.2.3 *Strain seismometer*

The pendulum seismometers described above are inertial devices, which depend on the resistance of a loosely coupled mass to a change in its momentum. At about the same time that he developed the inertial seismograph, Milne also conducted experiments with a primitive strain seismograph that measured the change in distance between two posts during the passage of a seismic wave. The gain of early strain seismographs was low. However, in 1935 H. Benioff invented a sensitive strain seismograph from which modern versions are descended.

The principle of the instrument is shown in Fig. 3.23. It can record only horizontal displacements. Two collinear horizontal rods made of fused quartz so as to be insensitive to temperature change are attached to posts about 20 m apart, fixed to the ground; their near ends are separated by a small gap. The changes in separation of the two fixed posts result in changes in the gap width, which are detected with a capacitance or variable-reluctance transducer. In modern instruments the variation in gap width may be observed optically, using the interference between laser light-beams

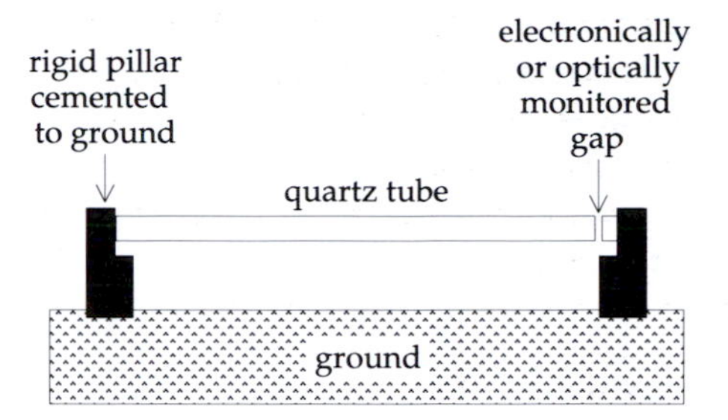

Fig. 3.23 Schematic design of a strain seismometer (after Press and Siever, 1985).

reflected from mirrors attached to the opposite sides of the gap. The strain instrument is capable of resolving strains of the order of 10^{-8} to 10^{-10}.

3.4.3 **The equation of the seismometer**

Inertial seismometers for recording horizontal and vertical ground motion function on the pendulum principle. When the instrument frame is displaced from its equilibrium position relative to the inertial mass, a restoring force arises that is, to first order, proportional to the displacement. Let the vertical or horizontal displacement, dependent on the type of seismometer, be u and the restoring force $-ku$, and let the corresponding displacement of the ground be q. The total displacement of the inertial mass M is then $u+q$, and the equation of motion is

$$M\frac{\partial^2}{\partial t^2}(u+q)=-ku \qquad (3.82)$$

We now divide throughout by M, write $k/M=\omega_0^2$, and after rearranging the equation, we get the familiar equation of forced simple harmonic motion

$$\frac{\partial^2 u}{\partial t^2}+\omega_0^2 u=-\frac{\partial^2 q}{\partial t^2} \qquad (3.83)$$

In this equation ω_0 is the *natural frequency* (or *resonant frequency*) of the instrument. For a ground motion with this frequency, the seismometer would execute large uncontrolled vibrations and the seismic signal could not be accurately recorded. To get around this problem, the seismometer motion is damped by providing a velocity-dependent force that opposes the motion. A damping term enters into the equation of motion, which becomes

$$\frac{\partial^2 u}{\partial t^2}+2\lambda\omega_0\frac{\partial u}{\partial t}+\omega_0^2 u=-\frac{\partial^2 q}{\partial t^2} \qquad (3.84)$$

The constant λ in this equation is called the *damping factor* of the instrument. It plays an important role in determining how the seismometer responds to a seismic wave.

A seismic signal is generally composed of numerous superposed harmonic vibrations with different frequencies. We can determine how a seismometer with natural frequency ω_0 responds to a seismic signal with any frequency ω by solving Eq. (3.84) with $q=A\cos\omega t$. Here, A is the magnified amplitude of the ground motion, equal to the true ground motion multiplied by a magnification factor that depends on the sensitivity of the instrument. Let the displacement recorded by the seismometer be $u=U\cos(\omega t-\Delta)$, where U is the maximum amplitude of the record and Δ is the phase difference between the record and the ground motion. Substituting in Eq. (3.84) we get the following sequence of equations:

$$-\omega^2 U\cos(\omega t-\Delta)-2\lambda\omega\omega_0 U\sin(\omega t-\Delta)+\omega_0^2 U\cos(\omega t-\Delta)=A\omega^2\cos\omega t$$

i.e., $U[(\omega_0^2-\omega^2)\cos(\omega t-\Delta)-\lambda\omega\omega_0\sin(\omega t-\Delta)]=A\omega^2\cos\omega t$

If we now write

$$(\omega_0^2-\omega^2)=R\cos\varphi \text{ and } 2\lambda\omega\omega_0=R\sin\varphi$$

so that $R=[(\omega_0^2-\omega^2)^2+4\lambda^2\omega^2\omega_0^2]^{1/2}$

the equations reduce to

$$U[R\cos\varphi\cos(\omega t-\Delta)-R\sin\varphi\sin(\omega t-\Delta)]=A\omega^2\cos\omega t$$

$$UR\cos(\omega t-\Delta+\varphi)=A\omega^2\cos\omega t \tag{3.85}$$

The simplest solution for the maximum amplitude U in Eq. 3.85 is when

$$\Delta=\varphi=\tan^{-1}\left(\frac{2\lambda\omega\omega_0}{\omega_0^2-\omega^2}\right) \tag{3.86}$$

from which we get for the amplitude u of the seismic record

$$u=\frac{A\omega^2}{[(\omega_0^2-\omega^2)^2+4\lambda^2\omega^2\omega_0^2]^{1/2}}\cos(\omega t-\Delta) \tag{3.87}$$

3.4.3.1 *Effect of instrumental damping*

The ground motion caused by a seismic wave contains a broad spectrum of frequencies. Equation (3.87) shows that the response of the seismometer to different signal frequencies is strongly dependent on the value of the damping factor λ (Fig. 3.24). A completely *undamped* seismometer has $\lambda=0$, and for small values of λ the response of the seismometer is said to be *underdamped.* An undamped or greatly underdamped seismometer preferentially amplifies signals near the natural frequency, and therefore cannot make an accurate record of the ground motion; the undamped instrument will resonate at its natural frequency ω_0. For all damping factors $\lambda<1/\sqrt{2}$ the instrument response function has a peak, indicating preferential amplification of a particular frequency.

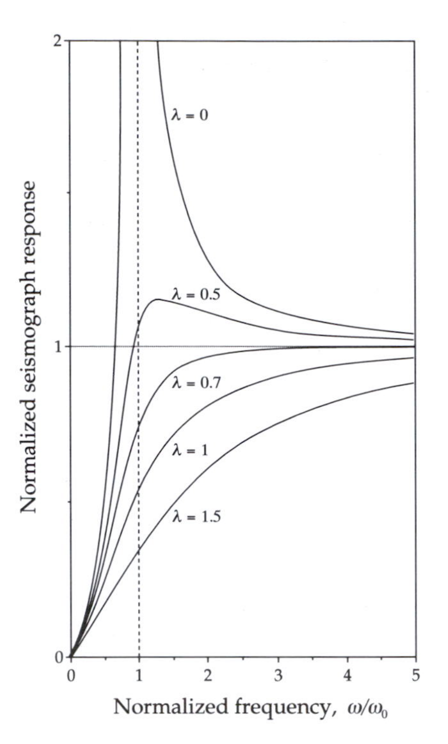

Fig. 3.24 Effect of the damping factor λ on the response of a seismometer to different signal frequencies. Critical damping corresponds to $\lambda=1$. Satisfactory operation corresponds to a damping factor between 0.7 and 1 (i.e., 70–100% of critical damping).

The value $\lambda=1$ corresponds to *critical damping*, so-called because it delineates two different types of seismometer response in the absence of a forcing vibration. If $\lambda<1$, the damped, free seismometer responds to a disturbance by swinging *periodically* with decreasing amplitude about its rest position. If $\lambda\geq1$, the disturbed seismometer behaves *aperiodically*, moving smoothly back to its rest position. However, if the damping is too severe ($\lambda \gg 1$), the instrument is *overdamped* and all frequencies in the ground motion are suppressed.

The optimum behavior of a seismometer requires that the instrument should respond to a wide range of frequencies in the ground motion, without preferential amplification or excessive suppression of frequencies. This requires that the damping factor should be close to the critical value. It is usually chosen to be in the range 70% to 100% of critical damping (i.e., $1/\sqrt{2}\leq\lambda<1$). At critical damping the response of the seismometer to a periodic disturbing signal with frequency ω is given by

$$u=\frac{A\omega^2}{\omega_0^2+\omega^2}\cos(\omega t-\Delta) \tag{3.88}$$

3.4.3.2 *Long-period and short-period seismometers*

The natural period ($2\pi/\omega_0$) of a seismometer is an important factor in determining what it actually records. Two examples of special interest correspond to instruments with very long and very short natural periods, respectively.

The long-period seismometer is an instrument in which the resonant frequency ω_0 is very low. For all but the lowest frequencies we can write $\omega_0 \ll \omega$. The phase lag Δ between the seismometer and the ground motion (see (Eq. 3.86)) becomes zero, and the amplitude of the seismometer displacement becomes equal to the amplified ground displacement q:

$$u = A\cos\omega t = q \tag{3.89}$$

The long-period seismometer is sometimes called a *displacement meter*. It is usually designed to record seismic signals with frequencies of 0.01–0.1 Hz (i.e., periods in the range 10–100 s).

The short-period seismometer is constructed so as to have a very short natural period and a correspondingly high resonant frequency ω_0, which is higher than most frequencies in the seismic wave. Under these conditions we have $\omega_0 \gg \omega$, the phase difference Δ is again small and Eq. (3.88) becomes

$$u = \frac{\omega^2}{\omega_0^2} A\cos\omega t = -\frac{1}{\omega_0^2}\ddot{q} \tag{3.90}$$

This equation shows that the displacement of the short-period seismometer is proportional to the acceleration of the ground, and the instrument is accordingly called an *accelerometer*. It is usually designed to respond to seismic frequencies of 1–10 Hz (periods in the range 0.1–1 s). An accelerometer is particularly suitable for recording strong motion earthquakes, when the amplitude of the ground motion would send a normal type of displacement seismometer off-scale.

3.4.3.3 *Broadband seismometers*

Short-period seismometers operate with periods of 0.1–1 s and long-period instruments are designed for periods greater than 10 s. The resolution of seismic signals with intermediate frequencies of 0.1–1 Hz (periods of 1–10 s) is hindered by the presence in this range of a natural form of seismic background noise. The noise derives from a nearly continuous succession of small ground movements that are referred to as *microseisms*. Some microseismic noise is of local origin, related to such effects as vehicular traffic, rainfall, wind action on trees, etc. However, an important source is the action of storm waves at sea, which is detectable on seismic records far inland. The drumming of rough surf on a shoreline and the interference of sea waves over deep water are thought to be the principal causes of microseismic noise.

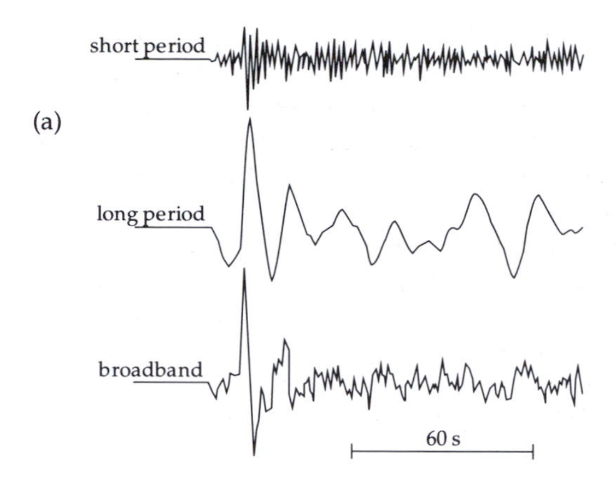

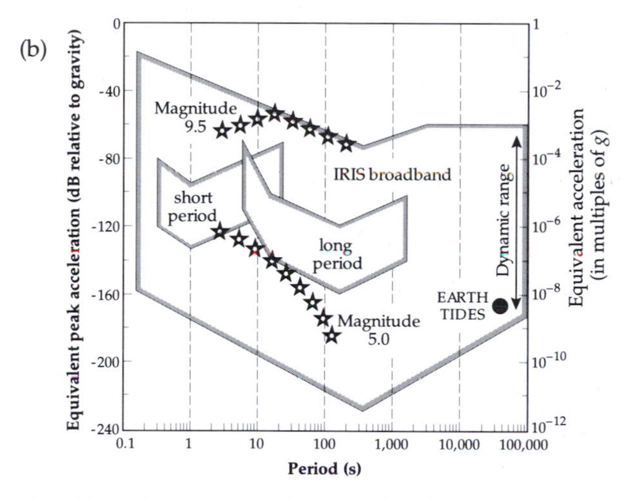

Fig. 3.25 (a) Comparison of short-period and long-period records of a teleseismic P-wave with a broadband seismometer recording of the same event, which contains more information than the other two records separately or combined. (b) Ranges of the ground acceleration (in dB) and periods of ground motion encompassed by the very broadband seismic system of the IRIS Global Seismic Network, compared with the responses of short-period and long-period seismometers and expected ground accelerations from magnitude 5.0 and 9.5 earthquakes (stars) and from bodily Earth-tides (redrawn from Lay and Wallace, 1995).

The microseismic noise has a low amplitude on a seismogram, but it may be as strong as a weak signal from a distant earthquake, which cannot be selectively amplified without also magnifying the noise. The problem is exacerbated by the limited dynamic range of short- or long-period seismometers. Short-period instruments yield records dominated by high frequencies while long-period devices smooth these out, giving a record with only a low-frequency content (Fig. 3.25a).

The range between the strongest and weakest signals that can be recorded without distortion by a given instrument is called its dynamic range. Dynamic range is measured by the power (or energy density) of a signal, and is expressed in units of decibels (dB). A decibel is defined as $10\log_{10}$(power). Because power is proportional to the square of amplitude (§3.3.2.6), a decibel is equivalent to $20\log_{10}$(amplitude). So, for example, a range of 20 dB in power corresponds to a factor 10 variation in acceleration, and a dynamic range of 100 dB corresponds to a 10^5 variation in amplitude. Short- and long-period seismometers have narrow dynamic ranges because they are designed to give an optimum performance in limited frequency ranges, below or above the band of ground noise. This handicap was overcome by the design of broadband seismometers that have high sensitivity over a very wide dynamic range.

The broadband seismometer has basically an inertial pendulum-type design, with enhanced capability due to a force-feedback system. This works by applying a force proportional to the displacement of the inertial mass to prevent it from moving significantly. The amount of feedback force applied is determined by using an electrical transducer to convert motion of the mass into an electrical signal. The force needed to hold the mass stationary corresponds to the ground acceleration. The signal is digitized with 16-to-24 bit resolution, synchronized with accurate time signals, and recorded on magnetic tape. The feedback electronics are critical to the success of this instrument, the most widely used presently being the Wielandt–Streckeisen model developed at the ETH–Zürich.

Broadband design results in a seismometer with great bandwidth and linear response. It is no longer necessary to avoid recording in the 1–10 s bandwidth of ground noise avoided by short-period and long-period seismometers. The recording of an earthquake by a broadband seismometer contains more useable information than can be obtained from the short-period or long-period recordings individually or in combination (Fig. 3.25a).

Broadband seismometers can be utilized for registering a wide range of signals (Fig. 3.25b). The dynamic range extends from ground noise up to the strong acceleration that would result from an earthquake with magnitude 9.5, and the periods that can be recorded range from high-frequency body waves to the very long period oscillations of the ground associated with bodily Earth-tides (§2.3.3.5). The instrument is employed world-wide in modern standardized seismic networks, replacing short-period and long-period seismometers. Combinations of seismometers, operating alone or in networks, were first organized systematically in the 1960s to monitor nuclear disarmament treaties. For this purpose it is essential to discriminate between surface waves generated by a nuclear explosion with low yield and those arising from weak distant earthquakes (§3.5.11).

3.4.4 The seismogram

A seismogram represents the conversion of the signal from a seismometer into a time record of a seismic event. The commonest method of obtaining a directly visible record, in use since the earliest days of modern seismology, uses a drum that rotates at a constant speed to provide the time axis of the record, as shown schematically in Fig. 3.21a and Fig. 3.22. In early instruments a mechanical linkage provided the coupling between sensor and record. The invention of the electromagnetic seismometer by Galitzin allowed transmission of the seismic signal to the recorder as an electrical signal. For many years, a galvanometer was used to convert the electrical signal back to a mechanical form for visual display.

In a galvanometer a small coil is suspended on a fine wire between the poles of a magnet. The current in the coil creates a magnetic field that interacts with the field of the permanent magnet and causes a deflection of the coil. The electrical circuitry of the galvanometer is designed with appropriate damping so that the galvanometer deflection is a faithful record of the seismic signal. The deflection was transferred to a visible record in a variety of ways.

Mechanical and electromagnetic seismometers delivered continuous analog recordings of seismic events. These types of seismometer are now of mainly historic interest, having been largely replaced by broadband seismometers. Galvanometer-based analog recording has been superseded by digital recording.

3.4.4.1 *Analog recording*

In an early method of recording, a smoked paper sheet was attached to the rotating drum. A fine stylus was connected to the pendulum by a system of levers. The point of the stylus scratched a fine trace on the smoked paper. Later instruments employed a pen instead of the stylus and plain paper instead of the smoked paper. These methods make a 'wiggly line' trace of the vibration.

In a further development of galvanometric recording a light-beam was reflected from a small mirror attached to the coil or its suspension to trace the record on photographic paper attached to the rotating drum. Photographic methods of recording are free of the slight friction of mechanical contact and allow inventive modifications of the form of the record. In the *variable density* method, the galvanometer current modulated the intensity of a light-bulb, so that the fluctuating signal showed on the photographic record as successive light and dark bands. A *variable area* trace was obtained by using the galvanometer current to vary the aperture of a narrow slit, through which a light-beam passed on to the photographic paper.

Every seismogram carries an accurate record of the elapsed time. This may be provided by a tuned electrical circuit whose frequency of oscillation is controlled by the natural frequency of vibration of a quartz crystal. At regular intervals the timing system delivers a short impulse to the instrument, causing a time signal to be imprinted on the seismogram. Modern usage is to employ a time signal transmitted by radio. The timing lines appear as regularly spaced blips on a wiggly line trace, or as bright lines on photographic records.

An important development after the early 1950s, especially in commercial seismology, was the replacement of photographic recording by *magnetic recording*. The electrical current from the seismometer was sent directly to the recording head of a tape recorder. The varying current in the recording head imprinted a corresponding magnetization on a magnetic tape. Magnetic recorders might have 24 to 50 parallel channels, each able to record from a different source.

Many years of development resulted in sophisticated methods of analog recording, but they have now been superseded universally by digital methods.

3.4.4.2 *Digital recording*

In digital recording, the analog signal from a seismometer is passed through an electronic device called an analog-to-digital converter, which samples the continuous input signal at discrete, closely spaced time-intervals and represents it as a sequence of *binary numbers*. Conventional representation on the familiar decimal scale expresses a number as a sum of powers of 10 multiplied by the digits 0–9. In contrast, a number is represented on the binary scale as a sum of powers of 2 multiplied by one of only two digits, 0 or 1. For example, in decimal notation the number 153 represents $1\times10^2+5\times10^1+3\times10^0$. In binary notation, the same number is represented as

$$\begin{aligned}153 &= 128+16+8+1\\ &= 1\times2^7+0\times2^6+0\times2^5+1\times2^4+1\times2^3+0\times2^2+0\times2^1+1\times2^0\\ &= 10011001.\end{aligned}$$

Each of the digits in a binary number is called a *bit* and the combination that expresses the digitized quantity is called a *word*; the binary number 10011001 is an *eight-bit word*. A binary number is much longer and thus more cumbersome for everyday use than the decimal form, but it is suitable for computer applications. Because it involves only two digits, a binary number can be represented by a simple condition, such as whether a switch is off or on, or the presence or absence of an electrical voltage or current.

Digital recording of seismic signals was developed in the 1960s and since the early 1970s it has virtually replaced analog recording. The analog method records the signal, usually employing the galvanometer principle, on photographic film or on magnetic tape as a continuous time-varying voltage whose amplitude is proportional to a characteristic of the ground disturbance (displacement, velocity or acceleration). The dynamic range of the analog method is limited and the system has to be adapted specifically to the characteristics of the signal to be recorded. For example, an analog device for recording strong signals lacks the sensitivity needed to record weak signals, whereas an analog recorder of weak signals will be overloaded by a strong signal. The digital recording technique samples the amplified output of the seismometer at time increments of a millisecond or so, and writes the digitized voltage directly to magnetic tape or to a computer hard-disk. This avoids possible distortion of the signal that can result in mechanical or optical recording. Digital recording has greater fidelity than the analog method, its dynamic range is wide, and the data are recorded in a form suitable for numerical processing by high-speed computers.

After processing, the digital record is usually converted back to an analog form for display and interpretation. The processed digital signal is passed through a digital-to-analog converter and displayed as a wiggle trace or variable density record. The familiar continuous paper record is still a common form of displaying earthquake records. However, instead of employing galvanometers to displace the pen in response to a signal, modern devices utilize a motor-driven pen; in this case the electrical signal from the earthquake record powers a small servo-motor which controls the pen displacement on the paper record.

3.4.4.3 *Phases on a seismogram*

The seismogram of a distant earthquake contains the arrivals of numerous seismic waves that have travelled along different paths through the Earth from the source to the receiver. The appearance of the seismogram can therefore be very complicated, and its interpretation demands considerable experience. The analysis of seismic waves that have been multiply reflected and refracted in the Earth's interior will be treated in §3.7. Each event that is recorded in the seismogram is referred to as a *phase*.

As described in §3.3.2.1, the fastest seismic waves are the longitudinal waves. The first phase on the seismogram corresponds to the arrival of a longitudinal body wave, identified as the primary wave, or *P-wave* (Fig. 3.26). The next phase is the bodily shear wave, referred to as the secondary wave or *S-wave*, which usually has a stronger amplitude than the P-wave. It is followed by the large-amplitude disturbance associated with surface waves, which are sometimes designated *L-waves* because of their much longer wavelengths. Dispersion (§ 3.3.3.3) causes the wavelengths at

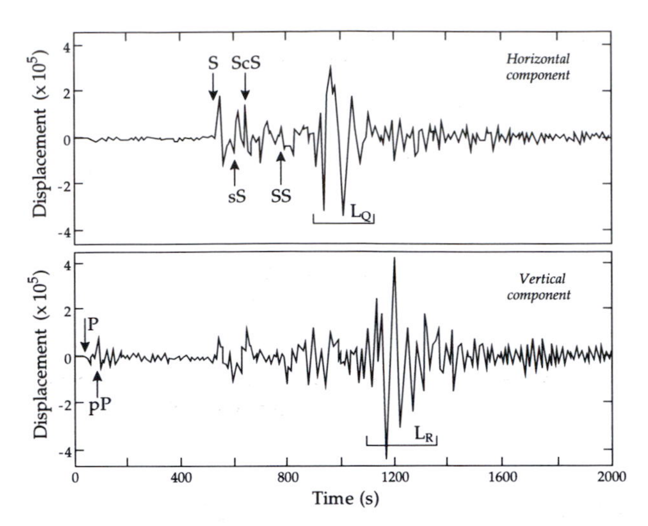

Fig. 3.26 Broadband seismograms of an earthquake in Peru recorded at Harvard, Massachusetts. *Top*: the SH-body wave and Love (L_Q) surface wave are prominent on the horizontal component record; *bottom*: the P- and SV- body waves and the Rayleigh (L_R) surface waves are clear on the vertical component record. Both seismograms also show several other phases (redrawn from Lay and Wallace, 1995).

the head of the surface-wave train to be longer than those at the tail. Conventionally, Rayleigh waves are referred to as L_R waves, while Love waves are called L_Q waves.

The arrivals recorded on any seismogram depend on the type of sensor used. For example, a vertical-component seismometer responds to P-, SV- and Rayleigh waves but does not register SH- or Love waves; a horizontal-component seismometer can register P-, SH-, Rayleigh and Love waves. The amplitudes of the different phases on a seismogram are influenced by several factors: the orientation of the instrument axis to the wave path, the epicentral distance (see §3.5.2), the focal mechanism and the structure traversed by the waves.

Representative seismograms for a distant earthquake are shown in Fig. 3.26. They were recorded at Harvard, Massachusetts, for an earthquake that occurred deep beneath Peru on May 24, 1991. The seismic body waves from this earthquake have travelled through deep regions of the mantle and several seismic phases are recorded. The upper seismogram is the trace of a horizontal-component seismometer oriented almost transverse to the seismic path, so the P-wave arrival is barely discernible. The first strong signal is the S- body wave (in this case, an SH-wave), closely followed by several other phases (defined in §3.7.2) and the Love (L_Q) surface-wave train. The lower seismogram, recorded by a vertical component seismometer, shows the arrivals of P- and SV- body waves and the Rayleigh (L_R) surface-wave train. Love waves travel along the surface at close to the near-surface S-wave velocity ($V_{LQ}\approx\beta$); Rayleigh waves are slower, having $V_{LR}\approx 0.92\,\beta$, so they reach the recording station later than the Love waves.

3.5 EARTHQUAKE SEISMOLOGY

3.5.1 Introduction

Most of the earthquakes which shake the Earth each year are so weak that they are only registered by sensitive seismographs, but some are strong enough to have serious, even catastrophic, consequences for mankind and the environment. About 90% of all earthquakes result from tectonic events, primarily movements on faults. The remaining 10% are related to volcanism, collapse of subterranean cavities, or man-made effects.

Our understanding of the processes that lead to earthquakes derives to a large extent from observations of seismic events on the San Andreas fault in California. The average relative motion of the plates adjacent to the San Andreas fault is about 5 cm/yr, with the block to the west of the fault moving northward. On the fault-plane itself, this motion is not continuous but takes place spasmodically. According to modern plate tectonic theory this extensively studied fault system is a transform fault. This is a rather special type, so it cannot be assumed that the observations related to the San Andreas fault are applicable without reservation to all other faults. However, the *elastic rebound model* proposed by H. F. Reid after the 1906 San Francisco quake, is a useful guide to how an earthquake may occur.

The model is illustrated in Fig. 3.27 by the changes to five parallel lines, drawn normal to the trace of the fault in the unstrained state and intersecting it at the points A–E. Strain due to relative motion of the blocks adjacent to the fault accumulates over several years. Far from the trace of the fault the five lines remain straight and parallel, but close to it they are bent. When the breaking point of the crustal rocks at C is exceeded, rupture occurs and there is a violent displacement on the fault-plane. The relative displacement that has been taking place progressively between the adjacent plates during years or decades is achieved on the fault-plane in a few seconds. The strained rocks adjacent to the fault 'rebound' suddenly. The accumulated strain energy is released with the seismic speed of the ruptured rocks, which is several km/s. The segments BC and C'D undergo compression, while CD and BC' experience dilatation. The points A and E do not move; the stored strain energy at these points is not released. The entire length of the fault-plane is not displaced, only the region in which the breaking

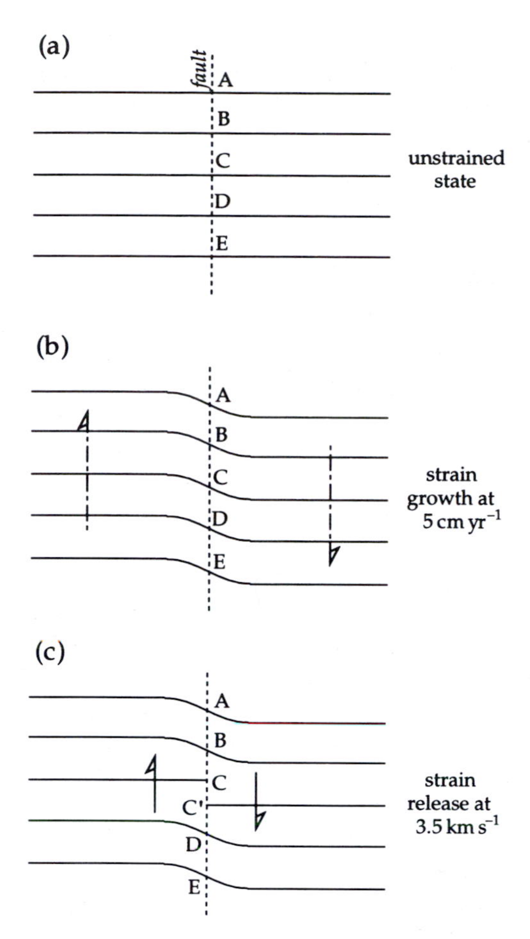

Fig. 3.27 Elastic rebound model of the origin of earthquakes: (a) unstrained state of a fault segment, (b) accumulation of strain close to the fault due to relative motion of adjacent crustal blocks, and (c) 'rebound' of strained segment as an earthquake with accompanying release of seismic energy.

point has been exceeded. The greater the length of the fault-plane that is activated, the larger is the ensuing earthquake.

The occurrence of a large earthquake is not necessarily as abrupt as described in the preceding paragraph, although it can be very sudden. In 1976 a major earthquake with magnitude 7.8 struck a heavily populated area of northern China near the city of Tangshan. Although there were known faults in the area, they had long been seismically inactive, and the large earthquake struck without warning. It completely devastated the industrial region and caused an estimated 243,000 fatalities. However, in many instances the accumulating strain is partially released locally as small earthquakes, or *foreshocks*. This is an indicator that strain energy is building up to the rupture level and is sometimes a premonition that a larger earthquake is imminent.

When an earthquake occurs, most of the stored energy is

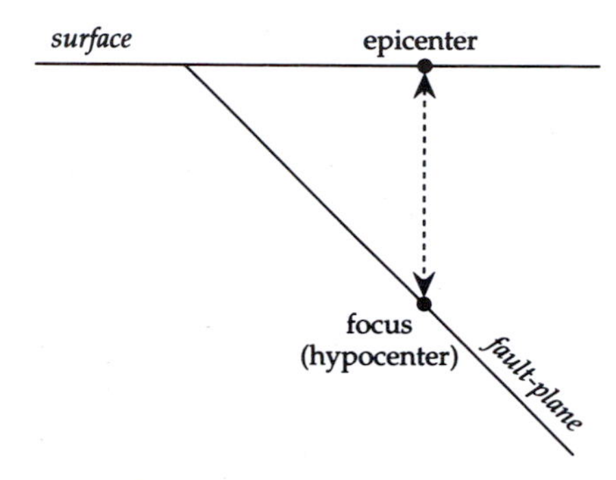

Fig. 3.28 Vertical section perpendicular to the plane of a normal fault, defining the epicenter and hypocenter (focus) of an earthquake.

released in the main shock. However, for weeks or months after a large earthquake there may be numerous lesser shocks, known as *aftershocks*, some of which can be comparable in size to the main earthquake. Structures weakened by the main event often collapse in large aftershocks, which can cause physical damage as severe as the main shock. The death toll from aftershocks is likely to be less, because people have evacuated damaged structures.

Although in fact the earthquake involves a part of the fault-plane measuring many square kilometers in area, from the point of view of an observer at a distance of hundreds or even thousands of kilometers the earthquake appears to happen at a point. This point is called the *focus* or *hypocenter* of the earthquake (Fig. 3.28). It generally occurs at a *focal depth* many kilometers below the Earth's surface. The point on the Earth's surface vertically above the focus is called the *epicenter* of the earthquake.

3.5.2 Location of the epicenter of an earthquake

The distance of a seismic station from the epicenter of an earthquake (the *epicentral distance*) may be expressed in kilometers Δ_{km} along the surface, or by the angle $\Delta° = (180/\pi)(\Delta_{km}/R)$ subtended at the Earth's center. The travel-times of P- and S-waves from an earthquake through the body of the Earth to an observer are dependent on the epicentral distance (Fig. 3.29a). The travel-time vs distance plots are not linear, because the ray paths of waves travelling to distant seismographs are curved. However, the standard seismic velocity profile of the Earth's interior is well enough known so that the travel-times for each kind of wave can be tabulated or graphed as a function of epicentral distance. In computing epicentral distance from earthquake data the total travel-time is not at first known, because an observer is rarely at the epicenter to record the exact time of occurrence t_0 of the earthquake. However, the difference in travel-times

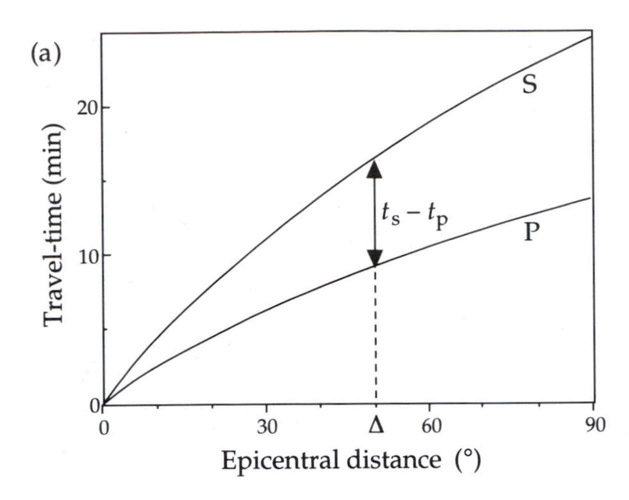

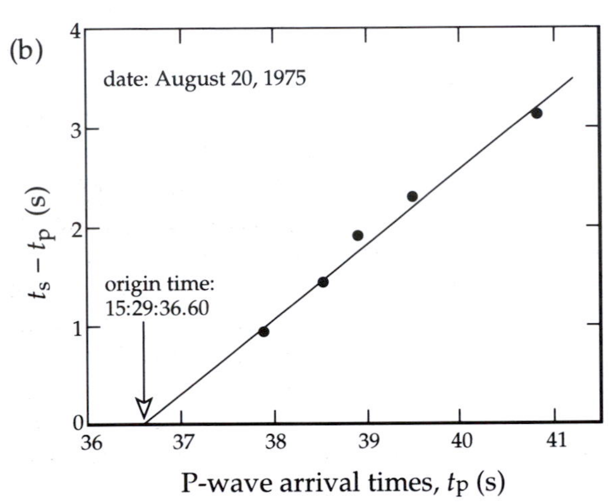

Fig. 3.29 (a) Travel-times of P- and S-waves from an earthquake through the body of the Earth to an observer at epicentral distances up to 90°. The epicentral distance (Δ) of the earthquake is found from the difference in travel times (t_s–t_p). (b) Wadati-diagram for determining the time of occurrence of an earthquake.

for P-and S-waves ($t_s - t_p$) can be obtained directly from the seismogram; it increases with increasing epicentral distance (Fig. 3.29a).

For local earthquakes we can assume that the seismic velocities a and b are fairly constant in the near-surface layers. The time when the earthquake occurred, t_0, can then be obtained by plotting the differences ($t_s - t_p$) against the arrival times t_p of the P-wave at different stations. The plot, called a *Wadati diagram*, is a straight line (Fig. 3.29b). If D is the distance travelled by the seismic wave, the travel-times of P- and S-waves are respectively $t_p = D/\alpha$ and $t_s = D/\beta$, so

$$t_s - t_p = D\left(\frac{1}{\beta} - \frac{1}{\alpha}\right) = t_p\left(\frac{\alpha}{\beta} - 1\right) \tag{3.91}$$

The intercept t_0 of the straight line with the arrival-time axis is the time of occurrence of the earthquake and the slope of the line is $[(\alpha/\beta) - 1]$. Knowing the P-wave velocity α, the distance to the earthquake is obtained from $D = \alpha(t_p - t_0)$.

In order to determine the location of an earthquake, epicenter travel-times of P-and S-waves to at least three seismic stations are necessary (Fig. 3.30). The data from one station give only the distance of the epicenter from that station. It could lie anywhere on a circle centered at the station. The data from an additional station define a second circle which intersects the first circle at two points, each of which could be the true epicenter. Data from a third station remove the ambiguity: the common point of intersection of the three circles is the epicenter.

Generally the circles do not intersect at a point, but form a small spherical triangle. The optimum location of the epicenter is at the center of the triangle. If data are available from more than three seismic stations, the epicentral location is improved; the triangle is replaced by a small polygon. This situation arises in part from observational errors, and because the theoretical travel-times are imperfectly known. The interior of the Earth is neither homogeneous nor isotropic, as must be assumed. The exact location of earthquake epicenters requires detailed knowledge of the seismic velocities along the entire path, but especially under the source area and the receiving station. The main reason for the intersection triangle or polygon is, however, that the seismic rays travel to the seismograph from the focus, and not from the epicenter. The focal depth of the earthquake, d, which may be up to several hundred kilometers, must be taken into account. It can be estimated from simple geometry. If Δ_{km} is the epicentral distance and D the distance travelled by the wave, to a first approximation $d = (D^2 - \Delta_{km}^2)^{1/2}$. Combining several values of d from different recording stations gives a reasonable estimate of the focal depth.

3.5.3 Global seismicity

The epicenters of around 30,000 earthquakes are now reported annually by the International Seismological Center. The geographical distribution of world seismicity (see Fig. 1.10) dramatically illustrates the tectonically active regions of the Earth. The seismicity map is important evidence in support of plate tectonic theory, and delineates the presently active plate margins.

Earthquake epicenters are not uniformly distributed over the Earth's surface, but occur predominantly along rather narrow zones of *interplate* seismic activity. The *circum-Pacific zone*, in which about 75–80% of the annual release of seismic energy takes place, forms a girdle that

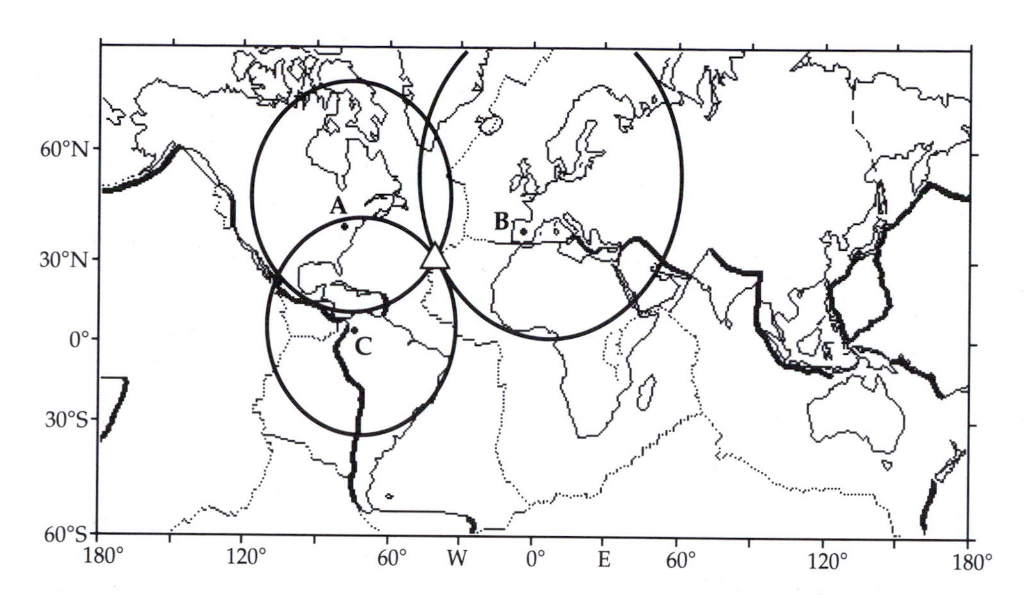

Fig. 3.30 Location of an earthquake epicenter using epicentral distances of three seismic stations (at A, B and C). The epicentral distance of each station defines the radius of a circle centered on the station. The epicenter (triangle) is located at the common intersection of the circles; their oval appearance is due to the map projection.

encompasses the mountain ranges on the west coast of the Americas and the island arcs along the east coast of Asia and Australasia. The *Mediterranean-transasiatic zone*, responsible for about 15–20% of the annual seismic energy release, begins at the Azores triple junction in the Atlantic Ocean and extends along the Azores–Gibraltar ridge; after passing through North Africa it makes a loop through the Italian peninsula, the Alps and the Dinarides; it then runs through Turkey, Iran, the Himalayan mountain chain and the island arcs of southeast Asia, where it terminates at the circum-Pacific zone. The system of *oceanic ridges and rises* form the third most active zone of seismicity, with about 3–7% of the annually released seismic energy. In addition to their seismicity, each of these zones is also characterized by active volcanism.

The remainder of the Earth is considered to be *aseismic*. However, no region of the Earth can be regarded as completely earthquake-free. About 1% of the global seismicity is due to *intraplate* earthquakes, which occur remote from the major seismic zones. These are not necessarily insignificant: some very large and damaging earthquakes (e.g. the New Madrid, Missouri, earthquakes of 1811 and 1812 in the Mississippi river valley) have been of the intraplate variety.

Earthquakes can also be classified according to their focal depths. Earthquakes with *shallow* focal depths less than 70 km occur in all the seismically active zones; only shallow earthquakes occur on the oceanic ridge systems. The largest proportion (about 85%) of the annual release of seismic energy is liberated in shallow-focus earthquakes. The remainder is set free by earthquakes with *intermediate* focal depths of 70–300 km (about 12%) and by earthquakes with *deep* focal depths greater than 300 km (about 3%). These occur only in the circum-Pacific and Mediterranean-transasiatic seismic zones, and accompany the process of plate subduction.

The distributions of epicentral locations and focal depths of intermediate and deep earthquakes give important evidence for the processes at a subduction zone. When the earthquake foci along a subduction zone are projected onto a cross-section normal to the strike of the plate margin, they are seen to define a zone of seismicity about 30–40 km thick in the upper part of the 80–100 km thick subducting oceanic plate, which plunges at roughly 30–60°beneath the overriding plate (Fig. 3.31). For many years the inclined seismic zone was referred to in Western literature as a *Benioff zone* in recognition of the Californian scientist, Hugo Benioff. In the years following World War II Benioff carried out important pioneering studies that described the distribution of deep earthquakes on steeply dipping surfaces of seismicity. Many characteristics of the occurrence of deep earthquakes had been described in the late 1920s by a Japanese seismologist, Kiyoo Wadati. He discovered that the closer the epicenters of earthquakes lay to the Asian continent, the greater were their focal depths; the deep seismicity appeared to lie on an inclined plane. It was Benioff, however, who in 1954 proposed as an explanation of the phenomenon that the ocean floor was being 'subducted' underneath the adjacent land. This was a bold proposal well in advance of the advent of plate tectonic theory. Today the zone of active seismicity is called a *Wadati–Benioff zone* in recognition of both discoverers.

In three dimensions the Wadati–Benioff zone constitutes an inclined slab dipping underneath the overriding plate. It marks the location and orientation of the upper surface of the subducting plate. The dip-angle of the zone varies between about 30° and 60°, becoming steeper with increas-

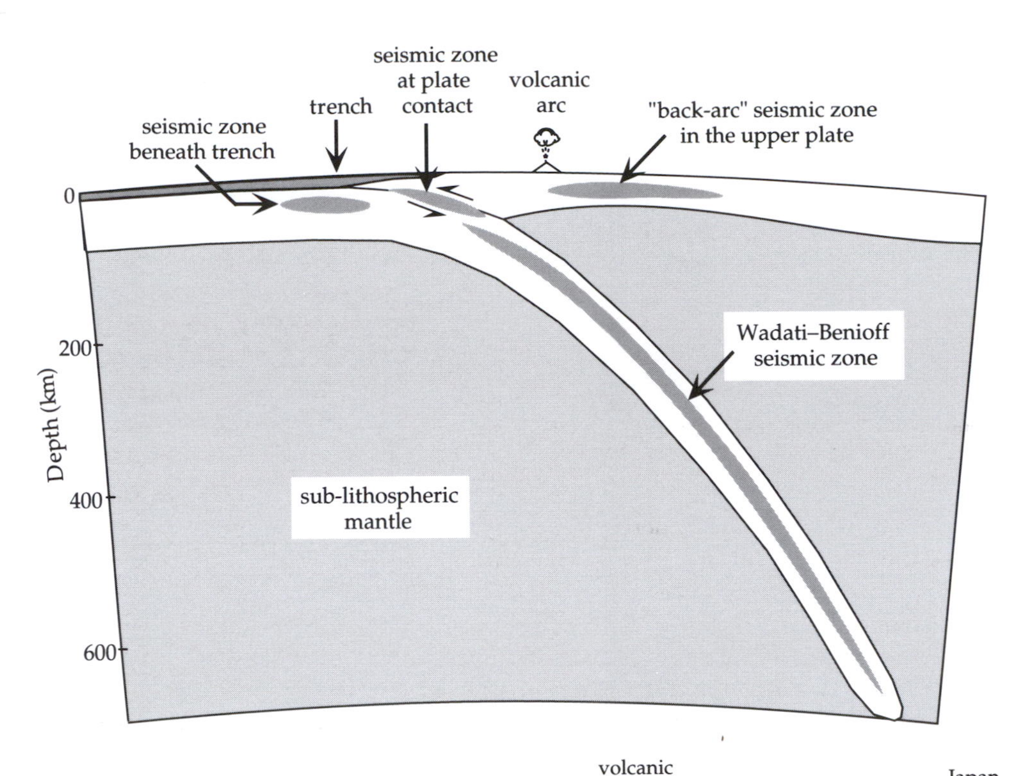

Fig. 3.31 Schematic cross-section through a subduction zone. The most active region is the zone of contact between the converging plates at depths of 10–60 km. There may be a 'back-arc' seismic zone in the overriding plate. Below about 70 km depth a Wadati–Benioff seismic zone is described within the subducting plate (after Isacks, 1989).

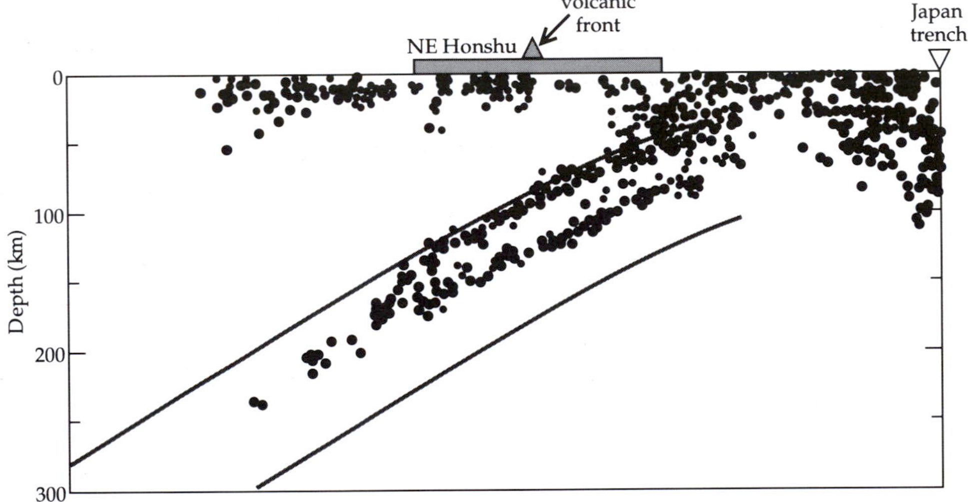

Fig. 3.32 Distribution of microearthquakes in a vertical section across a double subduction zone under the island of Honshu, Japan. Below 100 km the seismicity defines two parallel planes, one at the top of the subducting plate, the other in the middle. The upper plane is in a state of extension, the lower is under compression (after Hasegawa, 1989).

ing depth, and it can extend to depths of several hundred kilometers into the Earth. The deepest reliably located focal depths extend down to about 670 km. Important changes in the crystalline structure of mantle minerals occur below this depth.

The structure of a subducting plate is not always as simple as described. A detailed study of the subducting Pacific plate revealed a double Wadati–Benioff zone under northeast Honshu, Japan (Fig. 3.32). The seismicity at depths below 100 km defines two parallel planes about 30–40 km apart. The upper plane, identified with the top of the subducting plate, experiences a drag from the overriding plate and is in a state of extensional stress; the lower plane, in the middle of the slab, is in a state of compression. This information is inferred from analysis of the mechanisms by which the earthquakes occur.

3.5.4 Analysis of earthquake focal mechanisms

During an earthquake the accumulated elastic energy is released suddenly by physical displacement of the ground, as heat and as seismic waves that travel outwards from the

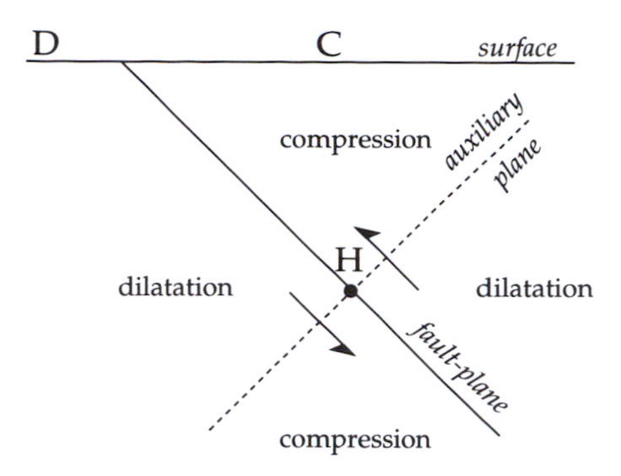

Fig. 3.33 Regions of compression and dilatation around an earthquake focus, separated by the fault-plane and the auxiliary plane.

focus. By studying the first motions recorded by seismographs at distant seismic stations, the focal mechanism of the earthquake can be inferred and the motion on the fault-plane interpreted.

Consider a vertical section perpendicular to the plane of a normal fault on which the hypocenter of an earthquake is located at the point H (Fig. 3.33). When the region above the fault moves up-slope, it produces a region of compression ahead of it and a region of dilatation (or expansion) behind it. In conjunction with the compensatory down-slope motion of the lower block, the earthquake produces two regions of compression and two regions of dilatation surrounding the hypocenter. These are separated by the fault-plane itself, and by an auxiliary plane through the focus and normal to the fault-plane. When a seismic P-wave travelling out from a region of compression reaches an observer at C, its first effect is to *push* the Earth's surface upwards; the initial effect of a P-wave that travels out from a region of dilatation to an observer at D is to *tug* the surface downwards. The P-wave is the earliest seismic wave to reach a seismograph at C or D and therefore the initial motion recorded by the instrument allows us to distinguish whether the first arrival was compressional or dilatational.

3.5.4.1 *Single-couple and double-couple radiation patterns*

The amplitudes of P-waves and S-waves vary with distance from their source because of the effects of physical damping and geometric dispersion. The amplitudes also depend geometrically on the angle at which the seismic ray leaves the source. This geometric factor can be calculated mathematically, assuming a model for the source mechanism. The simplest is to represent the source by a single pair of anti-parallel motions. Analysis of the amplitude of the P-wave as a function of the angle θ between a ray and the plane of the fault (Fig. 3.34a) gives an equation of the form

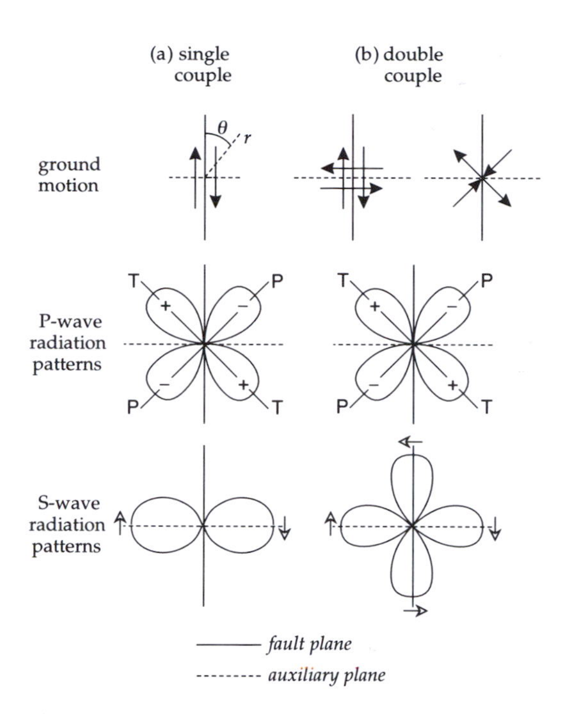

Fig. 3.34 Azimuthal patterns of amplitude variation for (a) single-couple and (b) double-couple earthquake source models are the same for P-waves but differ for S-waves. The radius of each pattern at azimuth θ from the fault-plane is proportional to the amplitude of the seismic wave in this direction. For P-waves the fault plane and auxiliary plane are nodal planes of zero displacement. The maximum compressional amplitude is along the T-axis at an angle of 45° to the fault plane. The dilatational amplitude is maximum along the P-axis, also at 45° to the fault plane.

$$A(r,t,\alpha,\theta)=A_0(r,t,\alpha)\sin^2 2\theta \qquad (3.92)$$

in which $A_0(r, t, \alpha)$ describes the decrease of amplitude with distance r, time t, and seismic P-wave velocity α. A plot of the amplitude variation with θ is called the radiation pattern of the P-wave amplitude, which for the single-couple model has a quadrupolar character (Fig. 3.34a). It consists of four lobes, two corresponding to the angular variation of amplitude where the first motion is compressional, and two where the first motion is dilatational. The lobes are separated by the fault-plane and the auxiliary plane.

The radiation pattern for S-waves from a single-couple source is described by an equation of the form

$$B(r,t,\beta,\theta)=B_0(r,t,\beta)\sin^2\theta \qquad (3.93)$$

where the amplitude B is now dependent on the S-wave velocity β. The radiation pattern has a dipolar character consisting of two lobes in which the first motions are of opposite sense.

An alternative model of the earthquake source is to represent it by a *pair* of orthogonal couples (Fig. 3.34b).

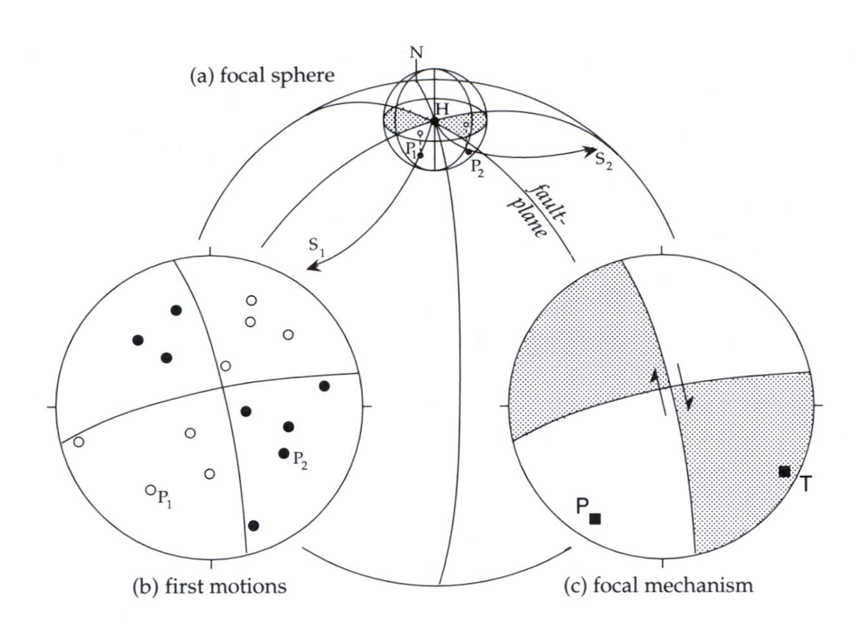

Fig. 3.35 Method of determining the focal mechanism of an earthquake. (a) The focal sphere surrounding the earthquake focus, with two rays S_1 and S_2 that cut the sphere at P_1 and P_2, respectively. (b) The points P_1 and P_2 are plotted on a lower-hemisphere stereogram as first-motion pushes (solid points) or tugs (open points). (c) The best-fitting great circles define regions of compression (shaded) and tension (unshaded). The P- and T-axes are located on the bisectors of the angles between the fault-plane and auxiliary plane.

The double-couple source gives the same form of radiation pattern for P-waves as the single-couple source, but the radiation pattern for S-waves is quadrupolar instead of dipolar. This difference in the S-wave characteristics enables the seismologist to determine which of the two earthquake source models is applicable. S-waves arrive later than P-waves, so their first motions must be resolved from the background noise of earlier arrivals. They can be observed and are consistent with the double-couple model.

Note that the maximum P-wave amplitudes occur at 45° to the fault plane. The directions of maximum amplitude in the compressional and dilatational fields define the T-axis and P-axis, respectively. Here T and P imply 'tension' and 'compression', respectively, the stress conditions before faulting. Geometrically the P- and T-axes are the bisectors of the angles between the fault-plane and auxiliary plane. The orientations of these axes and of the fault-plane and auxiliary plane can be obtained even for distant earthquakes by analyzing the directions of first motions recorded in seismograms of the events. The analysis is called a fault-plane solution, or focal mechanism solution.

3.5.4.2 *Fault-plane solutions*

The ray path along which a P-wave travels from an earthquake to the seismogram is curved because of the variation of seismic velocity with depth. The first step in the fault-plane solution is to trace the ray back to its source. A fictitious small sphere is imagined to surround the focus (Fig. 3.35a) and the point at which the ray intersects its surface is computed with the aid of standardized tables of seismic P-wave velocity within the Earth. The azimuth and dip of the angle of departure of the ray from the earthquake focus are calculated and plotted as a point on the lower hemisphere of the small sphere. This direction is then projected on to the horizontal plane through the epicenter. The projection of the entire lower hemisphere is called a stereogram. The direction of the ray is marked with a solid point if the first motion was a push away from the focus (i.e., the station lies in the field of compression). An open point indicates that the first motion was a tug towards the focus (i.e., the station lies in the field of dilatation).

First-motion data of any event are usually available from several seismic stations that lie in different directions from the focus. The solid and open points on the stereogram usually fall in distinct fields of compression and dilatation (Fig. 3.35b). Two mutually orthogonal planes are now drawn so as to delineate these fields as well as possible. The fit is best made mathematically by a least-squares technique, but often a visual fit is obvious and sufficient. The two mutually orthogonal planes correspond to the fault-plane and the auxiliary plane, although it is not possible to decide which is the active fault-plane on the seismic data alone. The regions of the stereogram corresponding to compressional first motions are usually shaded to distinguish them from the regions of dilatational first motions (Fig. 3.35c). The P- and T-axes are the lines that bisect the angles between the fault-plane and auxiliary plane in the fields of dilatation and compression, respectively. To attach a physical meaning to the P- and T-axes we will have to take a closer look at the mechanics of faulting.

3.5.4.3 *Mechanics of faulting*

As discussed in elasticity theory, the state of stress can be represented by the magnitudes and directions of the three principal stresses $\sigma_1>\sigma_2>\sigma_3$. The directions of these principal stresses are by definition parallel to the coordinate axes and are therefore positive for *tensional stress*. The theory of faulting of homogeneous materials has been developed by studying the failure of materials under *compressional stress*, which is directed inwards toward the origin of the coordinate axes. The minimum tensional stress corresponds to the maximum compressional stress, and vice versa. The reason for taking this view is that geologists are interested in the behavior of materials within the Earth, where pressure builds up with increasing depth and faulting occurs under high confining pressures.

We can combine both points of view if we consider stress to consist of a part that causes change of volume and a part that causes distortion. The first of these is called the *hydrostatic stress*, and is defined as the mean (σ_m) of the three principal stresses: $\sigma_m=(\sigma_1+\sigma_2+\sigma_3)/3$. If we now subtract this value from each of the principal stresses we get the *deviatoric stresses*: $\sigma'_1>\sigma'_2>\sigma'_3$. Note that σ'_1 is positive; it is a tensional stress, directed outward. However, σ'_3 is negative; it is a compressional stress, directed inward.

Failure of a material occurs on the plane of maximum *shear stress*. For a perfectly homogeneous material this is the plane that contains the intermediate principal stress axis σ_2 (or σ'_2). The fault-plane is oriented at 45° to the axes of maximum and minimum compressional stress, i.e., it bisects the angle between the σ'_1 and σ'_3 axes. In real materials inhomogeneity and the effects of internal friction result in failure on a fault-plane inclined at 20°–30° to the axis of maximum compression. Seismologists often ignore this complication when they make fault-plane analyses. The axis of maximum tensional stress σ'_1 is often equated with the T-axis, in the field of compressional first motions on the bisector of the angle between the two principal planes. The axis of maximum compressional stress σ'_3 is equated with the P-axis, in the field of dilatational first motions. In reality the direction of σ'_3 will lie between the P-axis and the fault plane.

The locations of P and T may at first seem strange, for the axes appear to lie in the wrong quadrants. However, one must keep in mind that the orientations of the principal stress axes correspond to the stress pattern *before* the earthquake, while the fault-plane solution shows the ground motions that occurred *after* the earthquake. The focal mechanism analysis makes it possible to interpret the directions of the principal axes of stress in the Earth's crust that led to the earthquake.

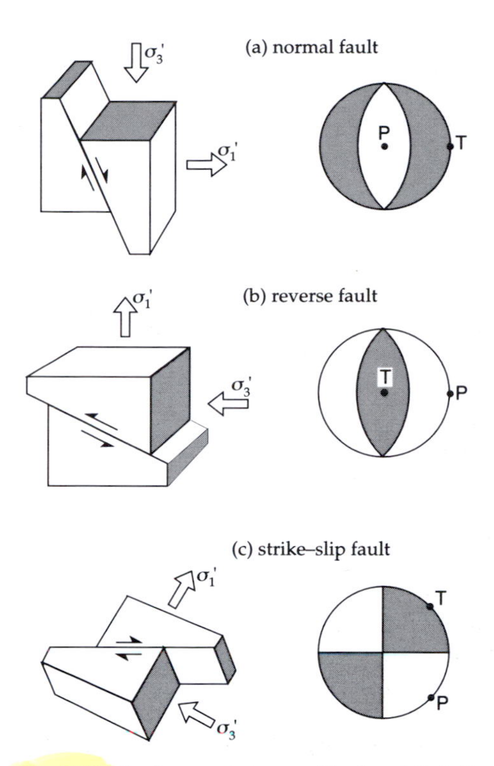

Fig. 3.36 The three main types of fault and their focal mechanisms. *Left*: the orientations of each fault-plane and the principal deviatoric stresses, σ'_1 and σ'_2. *Right*: focal mechanisms and orientations of P- and T- axes.

There are only three types of tectonic fault. These can be distinguished by the orientations of the principal axes of stress to the horizontal plane (Fig. 3.36). The focal solutions of earthquakes associated with each type of fault have characteristic geometries. When motion on the fault occurs up or down the fault plane it is called a *dip–slip* fault, and when the motion is horizontal, parallel to the strike of the fault, it is called a *strike–slip* fault.

Two classes of dip–slip fault are distinguished depending on the sense of the vertical component of motion. In a *normal fault*, the block on the upper side of the fault drops down an inclined plane of constant steepness relative to the underlying block (Fig. 3.36a). The corresponding fault-plane solution has regions of compression at the margins of the stereogram. The T-axis is horizontal and the P-axis is vertical. A special case of this type is the *listric fault,* in which the steepness of the fault surface is not constant but decreases with increasing depth.

In the second type of dip–slip fault, known as a *reverse fault* or *thrust fault*, the block on the upper side of the fault moves up the fault-plane, overriding the underlying block

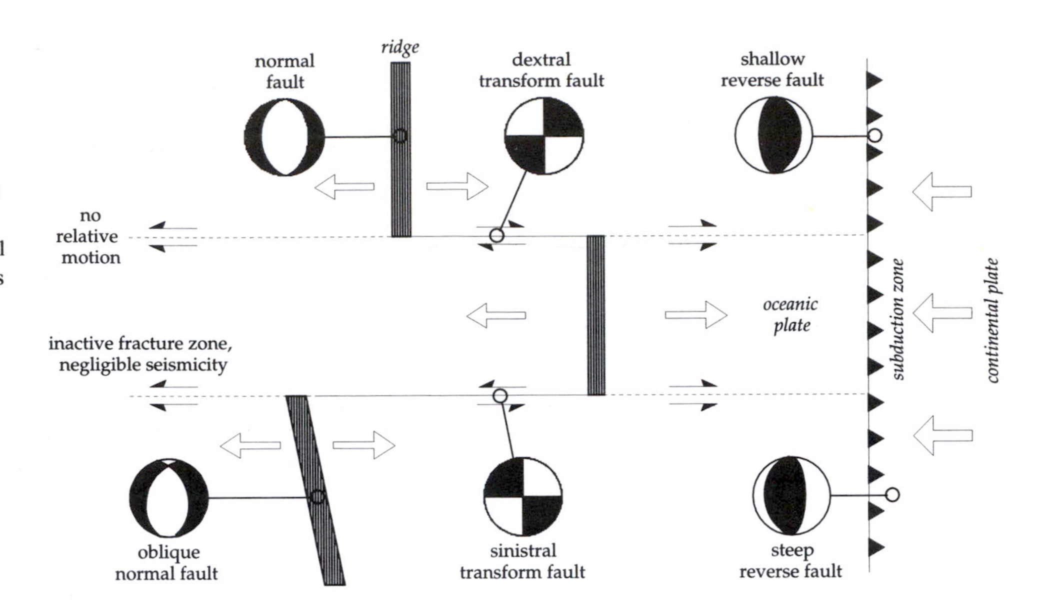

Fig. 3.37 Fault-plane solutions for hypothetical earthquakes at an ocean ridge and transform fault system. Note that the sense of movement on the fault is not given by the apparent offset of the ridge. The focal mechanisms of earthquakes on the transform fault reflect the relative motion between the plates. Note that in this and similar figures the sector with *compressional* first motions is shaded.

(Fig. 3.36b). The fault-plane solution is typified by a central compressional sector. The orientations of the axes of maximum tension and compression are the inverse of the case for a normal fault. The T-axis is now vertical and the P-axis is horizontal. When the fault-plane is inclined at a very flat angle, the upper block can be transported over large horizontal distances. This special type of *overthrust fault* is common in regions of continental collision, as for example in the Alpine–Himalayan mountain belts.

The simplest type of strike–slip fault is the *transcurrent* fault, in which the fault-plane is steep or vertical (Fig. 3.36c). To cause this type of fault the T- and P- axes must both lie in the horizontal plane. The fault-plane solution shows two compressional and two dilatational quadrants. Each side of the fault moves in opposite horizontal directions. If the opposite side of the fault to an observer is perceived to move to the left, the fault is said to be *sinistral*, or left-handed; if the opposite side moves to the right, the fault is *dextral*, or right-handed.

A variant of the strike–slip fault plays a very important role in the special context of plate tectonics. A *transform fault* allows horizontal motion of one plate relative to its neighbor. It joins segments of a spreading ocean ridge, or segments of a subducting plate margin. It is important in plate tectonics because it constitutes a conservative plate margin at which the tectonic plates are neither formed nor destroyed. The relative motion on the transform fault therefore reveals the direction of motion on adjacent segments of an active plate margin. The sense of motion is revealed by the pattern of compressional and dilatational sectors on the fault-plane solution.

3.5.4.4 *Focal mechanisms at active plate margins*

Some of the most impressive examples of focal mechanism solutions have been obtained from active plate margins. The results fully confirm the expectations of plate tectonic theory and give important evidence for the directions of plate motions. We can first ask what types of focal mechanism should be observed at each of the three types of active plate margin. In the theory of plate tectonics these are the constructive (or 'accreting'), conservative (or 'transform') and destructive (or 'consuming') margins.

Oceanic spreading systems consist of both constructive and conservative margins. The seismicity at these plate margins forms narrow belts on the surface of the globe. The focal depths are predominantly shallow, generally less than 10 km below the ocean bottom. Active ridge segments are separated by transform faults (Fig. 3.37). New oceanic lithosphere is generated at the spreading oceanic ridges; the separation of the plates at the spreading center is accompanied by extension. The plates appear to be pulled apart by the plate tectonic forces. The extensional nature of the ridge tectonics is documented by fault-plane solutions indicative of normal faulting, as is seen for some selected earthquakes along the Mid-Atlantic Ridge (Fig. 3.38). In each case the fault-plane is oriented parallel to the strike of the ridge. On a ridge segment that is nearly normal to the nearest transform faults the focal mechanism solution is symmetric, with shaded compressional quadrants at the margins of the stereogram. Note that where the ridge is inclined to the strike of the transform fault the focal mechanism solution is not symmetric. This means that the plates are not being pulled apart perpendicular to the ridge. The fault-plane is still parallel to the strike of the ridge, but the slip-vector is oblique; the plate motion has a component perpendicular to the ridge and a component parallel to the ridge. We can understand why the direction of plate motion is not determined by the strike of the ridge axis by examining the motion on the adjacent transform faults.

The special class of strike–slip fault that joins active segments of a ridge or subduction zone is called a transform fault because it marks a conservative plate margin, where the lithospheric plates are being neither newly generated nor destroyed. The adjacent plates move past each other on the active fault. Relative plate motion is present only between the ridge segments. Almost the entire seismicity on the transform fault is concentrated in this region. On the parts of the fracture zone outside the segment of active faulting the plates move parallel to each other and there is little or no seismicity.

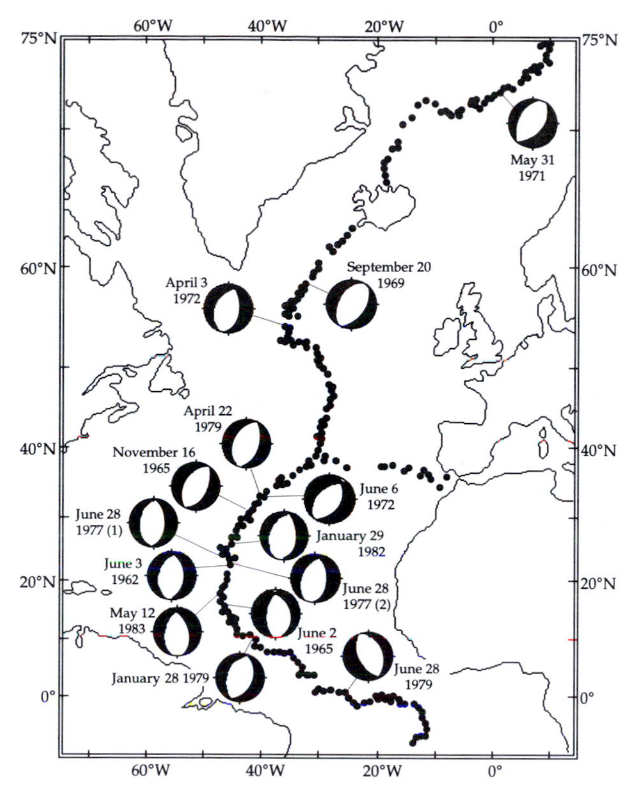

Fig. 3.38 Fault-plane solutions for earthquakes along the Mid-Atlantic oceanic Ridge, showing the prevalence of extensional tectonics with normal faulting in the axial zone of the spreading center (based on data from Huang *et al.*, 1986).

Because the relative motion is horizontal, the fault-plane solution is typical of a strike–slip fault. However, if the visible offset of the ridge segments were used to interpret the sense of motion on this fault, the wrong conclusion would be drawn. The conventional interpretation of this class of faults as a transcurrent fault was an early stumbling block to the development of plate tectonic theory. As indicated by arrows on the focal mechanism diagrams (Fig. 3.37), the relative motion on a transform fault is opposite to what one would expect for a transcurrent fault. It is determined by the opposite motions of the adjacent plates and not by the offset of the ridge segments. Hence, the focal mechanisms for a number of earthquakes on transform faults at the Mid-Atlantic Ridge in the central Atlantic reflect the eastwards motion of the African plate and westwards motion of the American plate (Fig. 3.39).

If we ignore changes in plate motion (which can, however, occur sporadically on some ridge systems), the offset of the neighboring ridge segments is a permanent feature that reflects how the plates first split apart. The orientations of the transform faults are very important, because the plates must move parallel to these faults. Thus the transform faults provide the key to determining the directions of plate motion. Where a ridge axis is not perpendicular to the transform fault the plate motion will have a component parallel to the ridge segment, which gives an oblique focal mechanism on the ridge.

A destructive (or consuming) plate margin is marked by a subduction zone, where a plate of oceanic lithosphere is

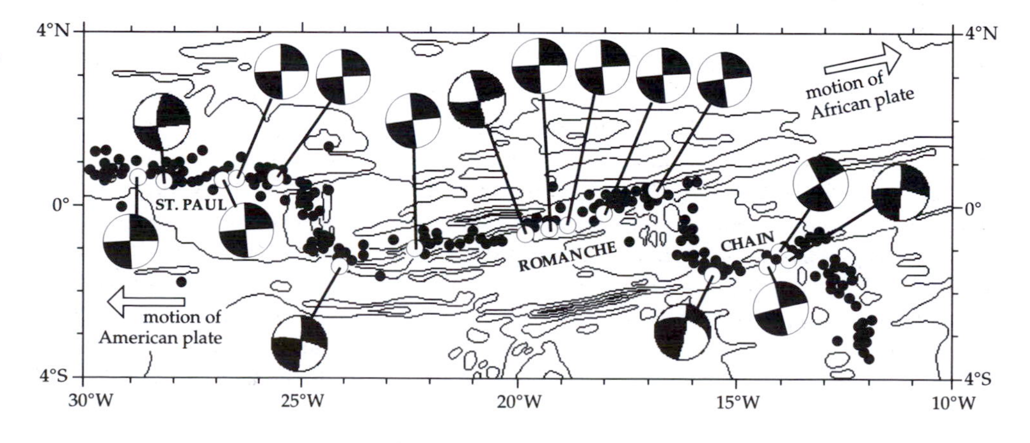

Fig. 3.39 Fault-plane solutions for earthquakes on the St. Paul, Romanche and Chain transform faults in the Central Atlantic ocean (after Engeln *et al.*, 1986). Most focal mechanisms show right lateral (dextral) motions on these faults, corresponding to the relative motion between the African and American plates.

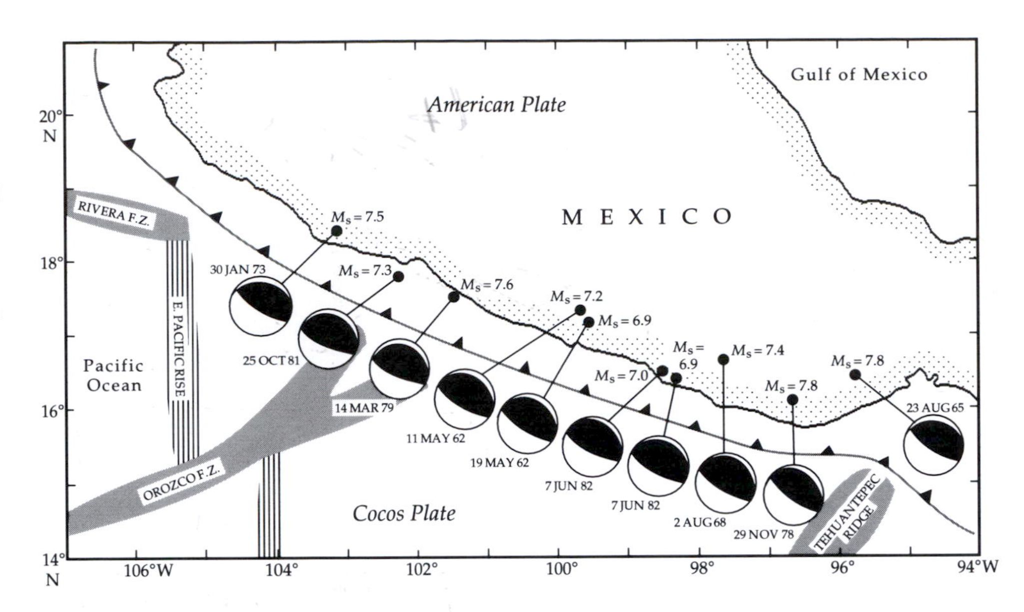

Fig. 3.40 Fault-plane solutions for selected large shallow earthquakes in the subduction zone along the west coast of Mexico (after Singh *et al.*, 1984). The focal mechanisms indicate low-angle overthrusting, as the Cocos plate is subducted to the northeast under Mexico.

destroyed by plunging under another plate of oceanic or continental lithosphere. Because this is a margin of convergence of the adjacent plates, the earthquake fault-plane solutions are typical of a compressional regime (Fig. 3.37). The regions of compressional first motion are in the center of the stereogram, indicating reverse faulting; the P-axes of maximum compressive stress are perpendicular to the strike of the subduction zone.

The type of focal mechanism observed at a subduction zone is dependent on the focal depth of the earthquake. This is because the state of stress varies within the subducting plate. At first the overriding plate is thrust at a shallow angle over the subducting plate. In the seismic zone at the contact between the two plates earthquake focal mechanisms are typical of low-angle reverse faulting. The focal mechanisms of earthquakes along the west coast of Mexico illustrate the first of these characteristics (Fig. 3.40). The selected earthquakes have large magnitudes ($6.9 \leq M_s \leq 7.8$) and shallow focal depths. The strikes of the fault-planes follow the trend of the oceanic trench along the Mexican coastline. The focal mechanisms have a central sector of compressional first motions and the fault-plane is at a low angle to the northeast, typical of overthrusting tectonics. The seismicity pattern documents the subduction of the Cocos plate under the North American plate.

As a result of the interplate collision the subducting oceanic plate is bent downwards and its state of stress changes. Deeper than 60–70 km the seismicity does not arise from the contact between the converging plates. It is caused by the stress pattern within the subducted plate itself. The focal mechanisms of some intermediate-depth earthquakes (70–300 km) show down-dip extension (i.e., the T-axes are parallel to the surface of the dipping slab) but some show down-dip compression (i.e., the P-axes are parallel to the dip of the slab). At great depths in most Wadati–Benioff zones the focal mechanisms indicate down-dip compression.

3.5.4.5 *Focal mechanisms in continental collisional zones*

When the continental portions of converging plates collide, they resist subduction into the denser mantle. The deformational forces are predominantly horizontal and lead to the formation of folded mountain ranges. The associated seismicity tends to be diffuse, spread out over a large geographic area. The focal mechanisms of earthquakes in the folded mountain belt reflect the ongoing collision. The collision of the northward-moving Indian plate with the Eurasian plate in the Late Tertiary led to the formation of the Himalayas. Focal mechanisms of earthquakes along the arcuate mountain belt show that the present style of deformation consists of two types (Fig. 3.41). Two fault-plane solutions south of the main mountain belt correspond to an extensional regime with normal faulting. To the north, in southern Tibet, the fault-plane solutions also show normal faulting on north–south oriented fault planes. Beneath the Lesser Himalaya seismic events distributed along the entire 1800 km length of the mountain chain indicate a deformational regime with some strike–slip faulting but predominantly low-angle thrust faulting. The Indian continental crust appears to be thrusting at a shallow angle under the Tibetan continental crust. This causes crustal thickening toward the north. In the main chain of the Lesser Himalaya the *minimum* compressive stress is vertical. The increased vertical load due to crustal thickening causes the

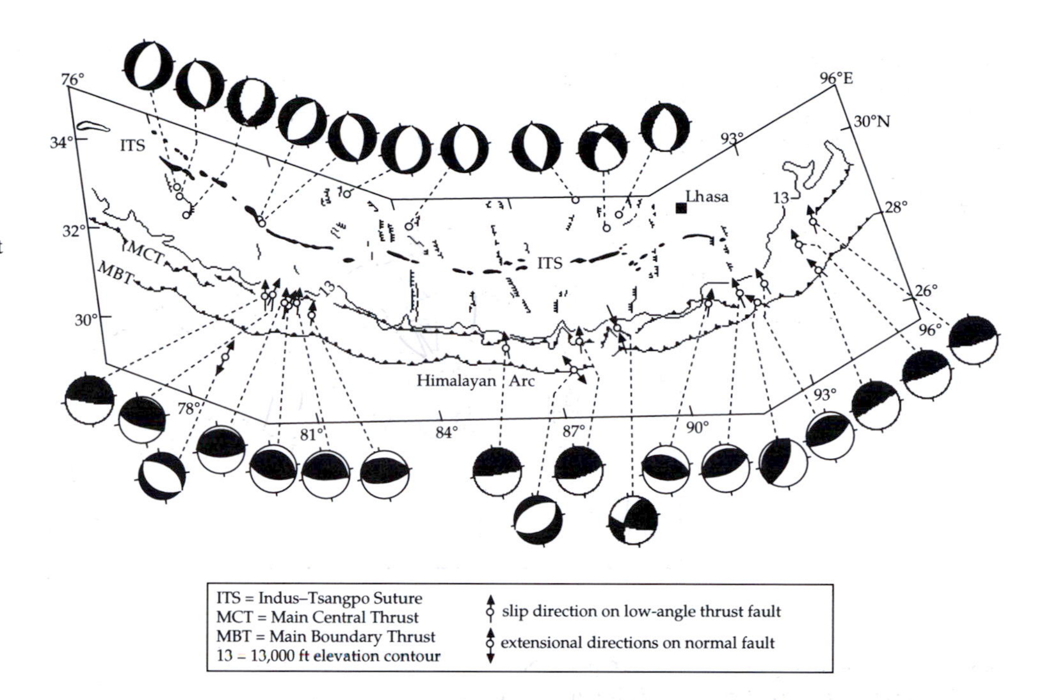

Fig. 3.41 Fault-plane solutions of earthquakes along the arcuate Himalayan mountain belt show normal faulting on north–south oriented fault planes in southern Tibet, and mainly low-angle thrust faults along the Lesser Himalaya mountain chain (after Ni and Barazangi, 1984).

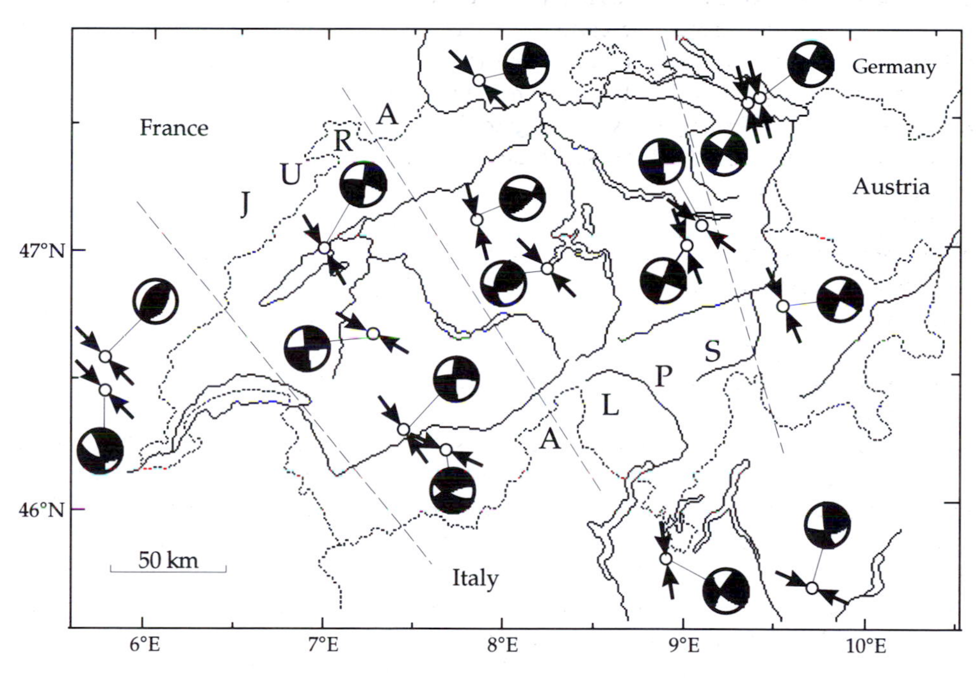

Fig. 3.42 Fault-plane solutions for earthquakes in or near the Swiss Alps in central Europe. The arrows show the horizontal components of the interpreted axes of maximum compression. They are approximately at right angles to the strike of the mountain ranges of the Jura and Alps (after Mayer Rosa and Müller, 1979, and Pavoni, 1977).

directions of the principal stresses to change, so that under southern Tibet the *maximum* compressive stress is vertical.

A different type of collisional tectonics is shown by fault-plane solutions from the Alpine mountain belt in south-central Europe. The Alps were formed during the collision between the African and European plates, starting in the Early Tertiary. The focal mechanisms of many, mostly small earthquakes show that they are predominantly associated with strike–slip faults (Fig. 3.42). The horizontal projections of the compressional axes are oriented almost perpendicular to the strike of the Alps and rotate along the arc. The fault-plane solutions for the modern seismicity indicate that the alpine fold-belt is a region of continuing interplate collision.

3.5.5 Secondary effects of earthquakes: landslides, tsunami, fires and fatalities

Before discussing methods of estimating the size of earthquakes, it is worthwhile to consider some secondary effects that can accompany large earthquakes: landslides, seismic sea waves and conflagrations. These rather exceptional effects cannot be included conveniently in the definition of earthquake *intensity*, because their consequences cannot be easily generalized or quantified. For example, once a major fire has been initiated, other factors not related directly to the size of the earthquake (such as aridity of foliage, combustibility of building materials, availability and efficiency of fire-fighting equipment) determine how it is extinguished.

A major hazard associated with large earthquakes in mountainous areas is the activation of major landslides, which can cause destruction far from the epicenter. In 1970, an earthquake in Peru with magnitude 7.8 and shallow focal depth of 40 km caused widespread damage and a total death toll of about 66,000. High in the Cordillera Blanca mountains above the town of Yungay, about 15 km away, an enormous slide of rock and ice was released by the tremors. Geologists later speculated that a kind of 'air cushion' had been trapped under the mass of rock and mud, enabling it to acquire an estimated speed of 300–400 km h^{-1}, so that it reached Yungay less than five minutes later. Over 90% of the town was buried under the mud and rock, in places to a depth of 14 m, and 20,000 lives were lost.

When a major earthquake occurs under the ocean, it can actuate a *tsunami* (seismic sea wave). This can be triggered by collapse or uplift of part of the ocean floor, by an underwater landslide, or by submarine volcanism. The tsunami propagates throughout the ocean basin as a wave with period T of around 15–30 min. The velocity of propagation of the wave v is dependent on the water-depth d, and the acceleration due to gravity g, and is given by:

$$v=\sqrt{gd} \tag{3.94}$$

Over the open ocean, where the water-depth is of the order of 5 km, the tsunami velocity will be around 220 m s^{-1} (800 km h^{-1}), and the wavelength (equal to the product vT) may measure 200 km. The amplitude of the tsunami over the open ocean is small, of the order of several cm, so that an observer on a ship would scarcely be aware of its passage. However, on approaching shallower water the leading part of the wave slows down and tends to be overridden by the following water mass, so that the height of the wave increases. The wave height may be amplified by the shapes of the sea-bottom and the coastline to several meters. In 1896 a large earthquake raised the ocean-bottom off the southern shore of Japan and initiated a tsunami that raced ashore with an estimated wave height of more than 20 m, causing 26,000 fatalities. One of the best-studied tsunami was set off by a great earthquake in the Aleutian islands in 1946. It travelled across the Pacific and several hours later reached Hilo, Hawaii, where it swept ashore and up river estuaries as a wave 7 m high. A consequence of the devastation by this tsunami around the Pacific basin was the formation of the Tsunami Warning System. When a major earthquake is detected that can produce a tsunami, a warning is issued to alert endangered regions to the imminent threat. The system works well far from the source as there is usually adequate warning time, but casualties still occur near to the generating earthquake.

In addition to causing direct damage to man-made structures, an earthquake can disrupt subterranean supply routes (e.g., telephone, electrical and gas lines) which in turn increases the danger of explosion and fire. Aqueducts and underground water pipelines may be broken, with serious consequences for the inhibition or suppression of fires. The San Francisco earthquake of 1906 was very powerful. The initial shock caused widespread damage, including the disruption of water supply lines. But a great fire followed the earthquake, and because the water supply lines were broken by the tremor, it could not be extinguished. The greatest damage in San Francisco resulted from this conflagration.

3.5.6 Earthquake size

There are two methods of describing how large an earthquake is. The *intensity* of the earthquake is a subjective parameter that is based on an assessment of visible effects. It therefore depends on factors other than the actual size of the earthquake. The *magnitude* of an earthquake is determined instrumentally and is a more objective measure of its size, but it says little directly about the seriousness of the ensuing effects. Illogically, it is usually the magnitude that is reported in news coverage of a major earthquake, whereas the intensity is a more appropriate parameter for describing the severity of its effects on mankind and the environment.

3.5.6.1 *Earthquake intensity*

Large earthquakes produce alterations to the Earth's natural surface features, or severe damage to man-made structures such as buildings, bridges and dams. Even small earthquakes can result in disproportionate damage to these edifices when inferior constructional methods or materials have been utilized. The intensity of an earthquake at a particular place is classified on the basis of the local character of the visible effects it produces. It depends very much on the acuity of the observer, and is in principle subjective. Yet,

Table 3.1 *Abridged and simplified version of the European Macroseismic Scale 1992 (European Seismological Commission, 1993) for earthquake intensity*

The scale focuses especially on the effects on people and buildings. It takes into account classifications of both the vulnerability of a structure (i.e., the materials and method of construction) and the degree of damage.

Intensity	Description of effects
I–IV	*light to moderate earthquakes*
I	**Not felt**: no damage.
II	**Scarcely felt**: felt only by a few people at rest indoors; no damage.
III	**Weak**: felt indoors by a few; hanging objects swing lightly; no damage.
IV	**Largely observed**: felt indoors by many, outdoors by very few; hanging objects swing; windows and dishes rattle; no damage.
V–VIII	*moderate to severe earthquakes*
V	**Strong**: felt indoors by most, outdoors by few; sleepers awakened; hanging objects swing strongly; building shakes or rocks; slight damage to a few buildings.
VI	**Slightly damaging**: frightening to many people; felt by all; glasses and dishware may break; moderate damage to some buildings.
VII	**Damaging**: most people frightened; furniture moved or overturned; moderate to heavy damage to poorly constructed buildings.
VIII	**Heavily damaging**: many people find it hard to stand, even outdoors; furniture overturned; poorly constructed buildings suffer heavy damage or destruction; damage to sturdy buildings.
IX–XII	*severe to destructive earthquakes*
IX	**Destructive**: general panic; monuments and columns fall; waves seen on soft ground; weak structures destroyed; moderate to heavy damage to well built, unreinforced structures; damage to reinforced buildings.
X	**Very destructive**: very heavy damage to masonry and unreinforced structures; moderate damage to reinforced structures.
XI	**Devastating**: unreinforced structures heavily damaged or destroyed; widespread devastation of weak structures.
XII	**Completely devastating**: practically all structures above and below ground are destroyed.

intensity estimates have proved to be a viable method of assessing earthquake size, including historical earthquakes.

The first attempt to grade earthquake severity was made in the late 18th century by Domenico Pignataro, an Italian physician, who classified more than 1000 earthquakes that devastated the southern Italian province of Calabria in the years 1783–1786. His crude analysis classified the earthquakes according to whether they were very strong, strong, moderate or slight. In the mid 19th century an Irish engineer, Robert Mallet, produced a list of 6831 earthquakes and plotted their estimated locations, producing the first map of the world's seismicity and establishing that earthquakes occurred in distinct zones. He also used a four-stage intensity scale to grade earthquake damage, and constructed the first isoseismal maps with lines that outlined areas with broadly equal grades of damage. The Rossi–Forel intensity scale, developed in the late 19th century by the Italian scientist M. S. de Rossi and the Swiss scientist F. Forel, incorporated ten stages describing effects of increasing damage. In 1902 an Italian seismologist, G. Mercalli, proposed a still more extensive, expanded intensity scale which reclassified earthquake severity in twelve stages. A variation, the Modified Mercalli (MM) scale, was developed in 1931 to suit building conditions in the United States, where a later modification is in common use. The Medvedev–Sponheuer–Karnik (MSK) scale, introduced in Europe in 1964, also has twelve stages and differs from the MM scale mainly in details. Modified in 1981, it has proved useful for over 30 years. A new European Macroseismic Scale (EMS) was proposed in 1992; an abridged version is shown in Table 3.1. The new 12-stage EMS scale is based on the MSK scale but takes into account the vulnerability of buildings to earthquake damage and incorporates more rigorous evaluation of the degree of damage to structures with different building standards.

In order to evaluate the active seismicity of a region questionnaires may be distributed to the population, asking for observations that can be used to estimate the intensity experienced. The questionnaires are evaluated with the aid of an intensity scale, and the intensity recorded at the location of each observer is plotted on a map. Continuous lines are then drawn to outline places with the same intensity

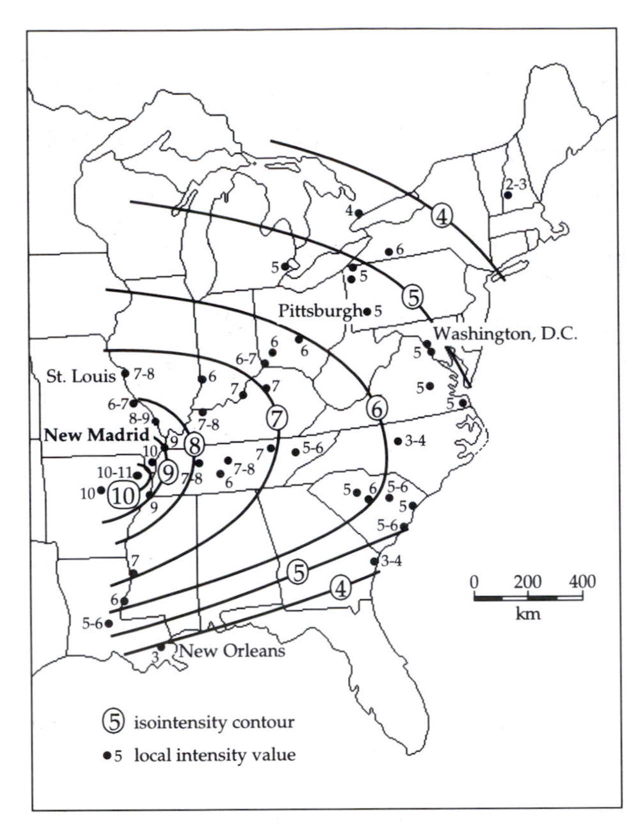

Fig. 3.43 Isoseismal map with contours of equal intensity for the New Madrid, Missouri, earthquake of 1811 (after Nuttli, 1973).

(Fig. 3.43), in the same way that contour lines are used on topographic maps to show elevation. Comparison of the *isoseismal* maps with geological maps helps explain the response of the ground to the shake of an earthquake. This is valuable information for understanding earthquake risk. The foundation on which structures are erected plays a vital role in their survival of an earthquake. For example, soft sediments can amplify the ground motion, enhancing the damage caused. This is even more serious when the sediments have a high water content, in which case liquefaction of the sediments can occur, robbing structures built on them of support and promoting their collapse.

There are numerous examples of this having occurred. In the great Alaskan earthquake in 1964 a section of Anchorage that was built on a headland underlain by a wet clay substratum collapsed and slid downslope to the sea. In 1985 a very large earthquake with magnitude 8.1 struck the Pacific coast of Mexico. About 350 km away in Mexico City, despite the large distance from the epicenter, the damage to buildings erected on the alluvium of a drained lakebed was very severe while buildings set on a hard rock foundation on the surrounding hills suffered only minor damage. In the Loma Prieta earthquake of 1989 severe damage was caused to houses built on landfill in San Francisco's Mission district, and an overhead freeway built on pillars on young alluvium north of Oakland collapsed dramatically. Both regions of destruction were more than 70 km from the epicenter in the Santa Cruz mountains. Similarly, in the San Francisco earthquake of 1906 worse damage occurred to structures built on landfill areas around the shore of the bay than to those with hard rock foundations in the hills of the San Francisco peninsula.

Intensity data play an important role in determining the historic seismicity of a region. An earthquake has dramatic consequences for a population; this was especially the case in the historic past, when real hazards were augmented by superstition. The date (and even the time) of occurrence of strong earthquakes and observations of their local effects have been recorded for centuries in church and civil documents. From such records it is sometimes possible to extract enough information for a given earthquake to estimate the intensity experienced by the observer. If the population density is high enough, it may be possible to construct an isoseismal map from which the epicenter of the tremor may be roughly located. An interesting example of this kind of analysis is the study of the New Madrid earthquakes of 1811–1812, which caused devastation in the Mississippi valley and were felt as far away as the coastlines along the Atlantic and Gulf of Mexico (Fig. 3.43). There were probably three large earthquakes, but the events occurred before the invention of the seismograph so details of what happened are dependent on the subjective reports of observers. Historical records of the era allow development of an intensity map for the settled area east of the Mississippi, but the pioneering population west of the river was at that time too sparse to leave adequate records for intensity interpretation.

Earthquake intensity data are valuable for the construction of seismic risk maps, which portray graphically the estimated earthquake hazard of a region or country. The preparation of a seismic risk map is a lengthy and involved task, which typically combines a study of the present seismicity of a region with analysis of its historic seismicity. A map of the maximum accelerations experienced in recent seismicity helps to identify the areas that are most likely to suffer severe damage in a large earthquake. The likelihood of an earthquake happening in a given interval of time must also be taken into account. Even in a region where large earthquakes occur, they happen at irregular intervals, and so knowledge of local and regional earthquake frequency is important in risk estimation. A seismic risk map of the United States (Fig. 3.44) shows the peak accelerations that are likely to be experienced at least once in a 50-year period.

Devastating earthquakes can occur even in countries

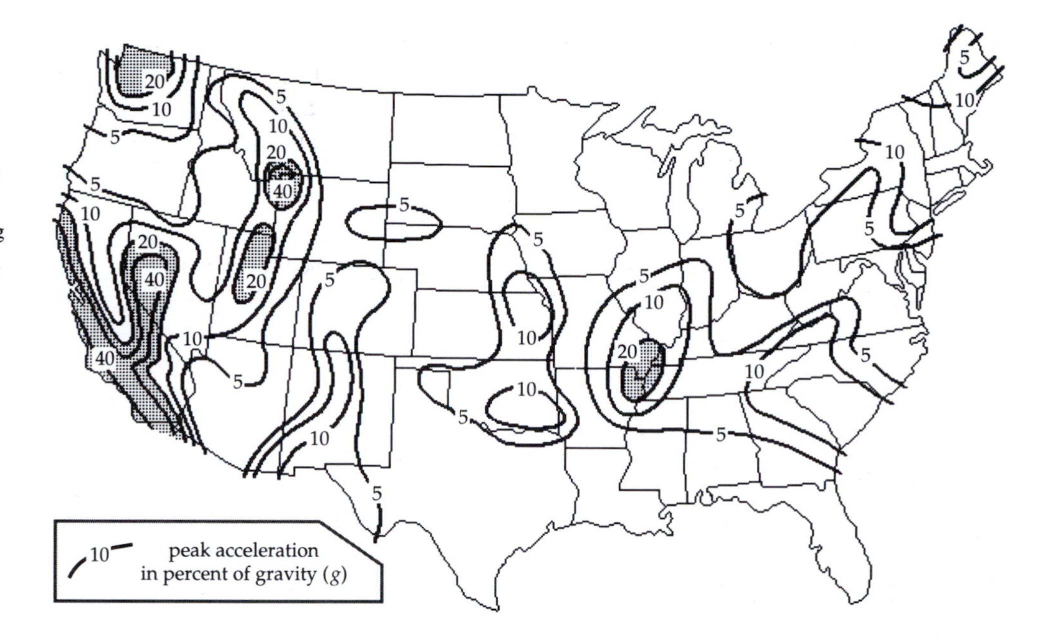

Fig. 3.44 Seismic risk map of the United States. The numbers on contour lines give the maximum acceleration (in percent of g) that might be expected to be exceeded with a probability of 1 in 10 during a 50-year period (after Bolt, 1988).

with relatively low seismic risk. Switzerland, in the center of Europe, is not regarded as a country prone to earthquakes. The seismic activity consists mostly of small earthquakes, mainly in the Alps in the collision zone of the African and European plates. Yet, in 1356 the northern Swiss town of Basel was destroyed by a major earthquake. Seismic risk maps are useful in planning safe sites for important edifices like nuclear power plants or high dams for hydroelectric power, which supply a substantial proportion of Switzerland's energy needs. Risk maps are also valuable to insurance companies, which must know the seismic risk of a region in order to assess the costs of earthquake insurance coverage for private and public buildings.

3.5.6.2 *Earthquake magnitude*

Magnitude is an experimentally determined measure of the size of an earthquake. In 1935 C. F. Richter attempted to grade the sizes of local earthquakes in Southern California on the basis of the amplitude of the ground vibrations they produced at a known distance from the epicenter. The vibrations were recorded by seismographs, which were standardized to have the same response to a given stimulus. Richter's original definition of magnitude was for seismographs at an epicentral distance of 100 km. But, because earthquakes occur at various distances from a seismograph, an extra term was added to compensate for attenuation of the signal with increasing epicentral distance. A commonly used equation for computing the *surface-wave magnitude* (M_s) of a shallow-focus earthquake from seismograph records at epicentral distances greater than 20° is the following one proposed by M. Båth in 1966:

$$M_s = \log_{10}\left(\frac{A_s}{T}\right) + 1.66 \log_{10}\Delta° + 3.3 \qquad (3.95)$$

where A_s is the maximum amplitude of the horizontal ground motion in microns (μm) deduced from the surface-wave amplitude, T is the period of these surface waves (around 20±2 seconds), and $\Delta°$ is the epicentral distance in degrees. The surface-wave magnitudes of some important historical earthquakes are given in Table 3.2.

The depth of the source affects the nature of the seismic wave train, even when the same energy is released. An earthquake with a deep focus generates only a small surface-wave train, while shallow earthquakes cause very strong surface waves. The equation for M_s was derived from the study of shallow earthquakes, observed at a distance of 100 km. Therefore, corrections must be added to the computed value of M_s to compensate for the effects of a focal depth greater than 50 km or epicentral distance less than 20°.

The amplitude of body-waves is not sensitive to the focal depth. As a result, earthquake magnitude scales have also been developed for use with body waves. An equation, proposed by B. Gutenberg in 1945, can be used to calculate a *body-wave magnitude* (m_b) from the maximum amplitude (A_p) of the ground motion associated with P-waves having a period of about 1–5 s:

Table 3.2 *Some important historical earthquakes, with their surface-wave magnitudes M_s, moment magnitudes M_w (where available), and the numbers of fatalities.*

Year	Epicenter	Magnitude M_s	M_w	Fatalities	Comments
1906	San Francisco, California	8.3	7.9	700	San Francisco fire
1908	Messina, Italy	7.5	—	120,000	
1923	Kanto, Japan	8.2	7.9	143,000	Tokyo fire
1960	Chile	8.5	9.5	5,700	
1960	Agadir, Morocco	5.9	—	14,000	
1964	Alaska	8.6	9.2	131	Pacific tsunami
1970	Peru	7.8	—	66,000	Great rock slide
1971	San Fernando, California	6.5	—	65	
1975	Haicheng, China	7.4	—	≈300	Predicted
1976	Tangshan, China	7.8	—	243,000	Not predicted
1980	El Asnam, Algeria	7.7	—	3,500	
1985	Mexico	7.9	8.0	9,500	
1989	Loma Prieta, California	7.1	—	62	

$$m_b=\log_{10}\left(\frac{A_P}{T}\right)+0.01\Delta^\circ+5.9 \quad (3.96)$$

For some earthquakes both M_s and m_b can be calculated. Although variable from one region to another, the magnitudes estimated from different phases of the seismic wave train are roughly related by the following equation (Båth, 1966):

$$m_b=0.56M_s+2.9 \quad (3.97)$$

The values estimated for M_s and m_b do not always correspond well, sometimes because the period of the P-wave is appreciably shorter than the requisite 12 seconds, and the surface-wave magnitude is often felt to give a better representation of the size of the earthquake. However, magnitude estimates based on both body waves and surface waves are dependent on the period of the portion of the wave train with maximum amplitude. Seismologists believe that, for very large earthquakes, M_s and m_b underestimate the energy released. An alternative definition of magnitude, based upon the long-period spectrum of the seismic wave, has been proposed for use with very large earthquakes. It makes use of the physical dimensions of the focus.

As discussed in the elastic rebound model (§3.1), a tectonic earthquake arises from abrupt displacement of a segment of a fault. The area F of the fractured segment and the distance s by which it slipped can be inferred. Together with the rigidity modulus μ of the rocks adjacent to the fault, these quantities define the *seismic moment* M_0 of the earthquake. Assuming that the displacement and rigidity are constant over the area of the rupture:

$$M_0=\mu Fs \quad (3.98)$$

The seismic moment can be used to define a *moment magnitude* (M_w), using an equation of the form given by Aki and Richards (1980):

$$M_w=\tfrac{2}{3}\log_{10}M_0-10.7 \quad (3.99)$$

M_w appears to be more appropriate for describing the magnitudes of very large earthquakes. It has largely replaced M_s in scientific evaluation of earthquake size, although M_s is often quoted in reports in the media.

The magnitude scale is, in principle, open-ended. Negative Richter magnitudes are possible, but the limit of sensitivity of seismographs is around −2. The maximum possible magnitude is limited by the shear strength of the crust and upper mantle, and since the beginning of instrumental recording none has been observed with a surface-wave magnitude M_s as high as 9.

The definition of earthquake magnitude by means of the amplitude of seismic disturbances at a known distance from the epicenter does not take into account the effects that happen in the focus, which, according to the elastic rebound model, involve the displacement of a segment of a fault. This rupture is not always evident at the surface of the Earth, but when it is, the length (L, in kilometers) of the ruptured section can be measured. An approximate empirical relationship is found between magnitude and the length of the surface rupture:

$$M_s=6.1+0.7\log_{10}L \quad (3.100)$$

3.5.6.3 *Relationship between magnitude and intensity*

The intensity and magnitude scales for estimating the size of an earthquake are defined independently but they have some common features. Intensity is a measure of earthquake size based on the degree of local damage it causes at the location of an observer. The definition of magnitude is based on the amplitude of ground motion inferred from the signal recorded by the observer's seismograph, and of course it is the nature of the ground motion – its amplitude, velocity and acceleration – which produce the local damage used to classify intensity. However, in the definition of magnitude the ground-motion amplitude is corrected for epicentral distance and converted to a focal characteristic. Isoseismal maps showing the regional distribution of damage give the maximum intensity (I_{max}) experienced in an earthquake, which, although influenced by the geographic patterns of population and settlement, is usually near to the epicenter. A moderately strong, shallow-focus earthquake under a heavily populated area can result in higher intensities than a large deep-focus earthquake under a wilderness area (compare, for example, the 1960 Agadir and 1964 Alaskan earthquakes in Table 3.2). However, for earthquakes with focal depth $h<50$ km the dependence of I_{max} on the focal depth can be taken into account, and it is possible to relate the maximum intensity to the magnitude with an empirical equation (Karnik, 1969):

$$I_{max}=1.5M_s-1.8\log_{10}h+1.7 \qquad (3.101)$$

This type of equation is useful for estimating quickly the probable damage that an earthquake causes. For example, it predicts that in the epicentral region of an earthquake with magnitude 5 and a shallow focal depth of 10 km, the maximum MSK intensity will be VII (moderately serious damage), whereas, if the focal depth is 100 km, a maximum intensity of only IV–V (minor damage) can be expected.

3.5.7 **Earthquake frequency**

Every year there are many small earthquakes, and only a few large ones. According to a compilation published by Gutenberg and Richter in 1954, the mean annual number of earthquakes in the years 1918–1945 with magnitudes 4–4.9 was around 6000, while there were only an average of about 100 earthquakes per year with magnitudes 6–6.9. The relationship between annual frequency (N) and magnitude (M_s) is logarithmic and is given by an equation of the form

$$\log N=a-bM_s \qquad (3.102)$$

The value of a varies between about 8 and 9 from one region to another, while b is approximately unity for regional and global seismicity. The mean annual numbers of earthquakes in different magnitude ranges are listed in Table 3.3; the frequency decreases with increasing magnitude (Fig. 3.45a), in accordance with Eq. (3.102). The annual number of large earthquakes with magnitude $M_s\geq7$ in the years 1900–1989 has varied between extremes of about 10 and 40, but the long-term average is about 20 per year (Fig. 3.45b).

Table 3.3 *Earthquake frequencies since 1900 (based on data from the US Geological Survey National Earthquake Information Center) and the estimated mean annual energy release computed with the energy–magnitude relation of Båth (1966)*

Earthquake magnitude	Number per year	Annual energy (10^{15} joule yr^{-1})
≥ 8.0	0–1	0–600
7–7.9	18	200
6–6.9	120	43
5–5.9	800	12
4–4.9	6,200	3
3–3.9	49,000	1
2–2.9	≈350,000	0.2
1–1.9	≈3,000,000	0.1

3.5.8 **Energy released in an earthquake**

The definition of earthquake magnitude relates it to the logarithm of the amplitude of a seismic disturbance. Noting that the energy of a wave is proportional to the square of its amplitude it should be no surprise that the magnitude is also related to the logarithm of the energy. Several equations have been proposed for this relationship. An empirical formula worked out by Gutenberg and Richter (Gutenberg, 1956), relates the energy release E to the surface-wave magnitude M_s:

$$\log_{10}E=4.4+1.5M_s \qquad (3.103)$$

where E is in joules. An alternative version of the energy–magnitude relation, suggested by Båth (1966) for magnitudes $M_s>5$, is:

$$\log_{10}E=5.24+1.44M_s \qquad (3.104)$$

For magnitudes 1–9, Båth's formula gives earthquake energies that are two to six times larger than the Gutenberg–Richter formula for the same magnitude; both relations may significantly overestimate the amount of energy released. The logarithmic nature of each formula means that the energy release increases very rapidly with increasing magnitude. For example, when the magnitudes of

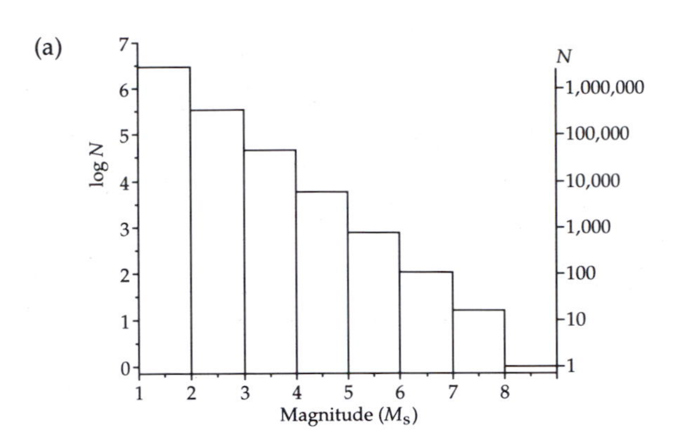

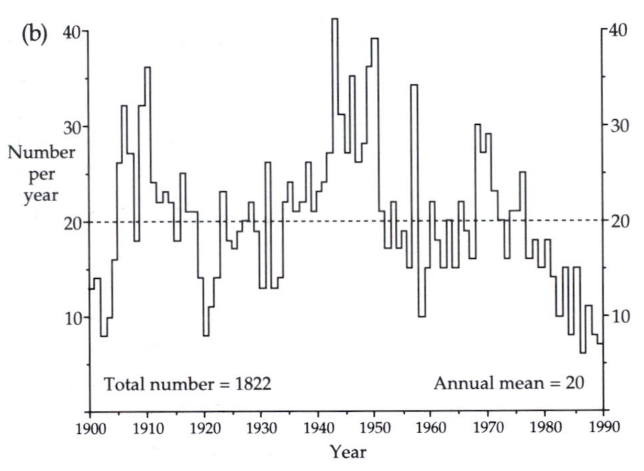

Fig. 3.45 Histograms of (a) the number (N) of earthquakes per year with magnitude M_s, and (b) the annual number of earthquakes with magnitudes $M_s \geq 7$ since 1900 (based on data from the US Geological Survey National Earthquake Information Center).

two earthquakes differ by 1, their corresponding energies differ by a factor 28 ($=10^{1.44}$) according to Båth's equation, or 32 ($=10^{1.5}$) according to the Gutenberg–Richter formula. Hence, a magnitude 7 earthquake releases about 760 ($=10^{2.88}$) to 1000 ($=10^3$) times the energy of a magnitude 5 earthquake. Another way of regarding this observation is that it takes 760–1000 magnitude 5 earthquakes to release the same amount of energy as a single large earthquake with magnitude 7. Multiplying the mean number of earthquakes per year by their estimated energy (using one of the energy–magnitude equations) gives an impression of the importance of very large earthquakes. Table 3.3 shows that the earthquakes with $M_s \geq 7$ are responsible for most of the annual seismic energy. In a year in which a very large earthquake ($M_s \geq 8$) occurs, most of the annual seismic energy is released in that single event.

It is rather difficult to appreciate the amount of energy released in an earthquake from the numerical magnitude alone. A few examples help illustrate the amounts of energy involved. Earthquakes with $M_s = 1$ are so weak that they can only be recorded instrumentally; they are referred to as microearthquakes. The energy associated with one of these events is equivalent to the kinetic energy of a medium sized automobile weighing 1.5 tons which is travelling at 130 km h^{-1} (80 m.p.h.). The energy released by explosives provides another means of comparison, although the conversion of energy into heat, light and shock waves is proportionately different in the two phenomena. One ton of the explosive tri–nitro–toluene (TNT) releases about 4.2×10^9 joules of energy. Båth's equation shows that the 11 kiloton atomic bomb which destroyed Hiroshima released about the same amount of energy as an earthquake with magnitude 5; similarly, one finds that a 1 megaton nuclear bomb releases energy equivalent to an earthquake with magnitude 7.2.

3.5.9 Earthquake prediction.

The problem of earthquake prediction is extremely difficult and is associated with sundry other problems of a sociological nature. To predict an earthquake correctly means deciding, as far in advance as possible, exactly where and when it will occur. It is also necessary to judge how strong it will be, which means realistically that people want to know what the likely damage will be, a feature expressed in the earthquake intensity. In fact the geophysicist is almost helpless in this respect, because at best an estimate of the predicted magnitude can be made. As seen above, even if it is possible to predict accurately the magnitude, the intensity depends on many factors (e.g., local geology, construction standards, secondary effects like fires and floods) which are largely outside the influence of the seismologist who is asked to presage the seriousness of the event. The problem of prediction rapidly assumes sociological and political proportions. Even if the approximate time and place of a major earthquake can be predicted with reasonable certainty, the question then remains of what to do about the situation. Properly, the threatened area should be evacuated, but this would entail economic consequences of possibly enormous dimension.

The difficulties are illustrated by the following possible scenario. Suppose that seismologists conclude that a very large earthquake, with probable magnitude 7 or greater, will take place sometime in a known month of a given year under a specific urban center. No more precise details are possible; in particular, the time of occurrence can not be determined more exactly. Publication of this kind of prediction would cause great consternation, even panic, which for

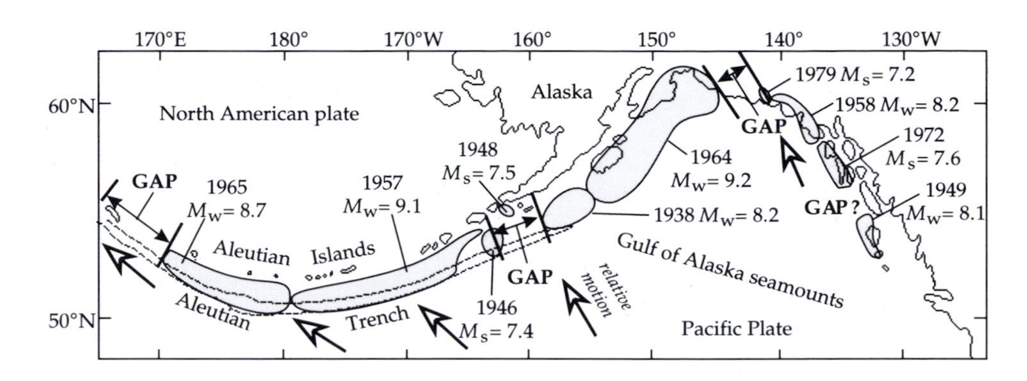

Fig. 3.46 Seismic gaps along the Aleutian island arc. Shaded regions mark the areas of rupture of very large historic earthquakes (M_s or M_w>7.4). Three large gaps in seismicity are potential locations of a future large earthquake (after Sykes *et al.*, 1981).

some of the population could be as disastrous as the earthquake itself. Should the warning be withheld? If the entire area, with its millions of inhabitants, were to be evacuated, the economic dislocation would be enormous. How long should the evacuation last, when every day is economically ruinous? Clearly, earthquake prediction is only useful in a practical sense when it is accurate in both place and time. The responsible authorities must also be provided with a reasonable estimate of the expected magnitude, from which the maximum intensity I_{max} may be gauged with the aid of a relationship like that of Eq. (3.101). The problem then passes out of the domain of the scientist and into that of the politician, with the scientist retaining only a peripheral role as a consultant. But if the earthquake prediction is a failure, the consequences will rebound with certainty on the scientist.

3.5.9.1 *Prediction of the location of an earthquake*

It is easier to predict where a major earthquake is likely to occur than when it will occur. The global seismicity patterns demonstrate that some regions are relatively aseismic. They are not completely free of earthquakes, many of which appear to occur randomly and without warning in these regions. Some intraplate earthquakes have a history of repeated occurrence at a known location, which can sensibly be expected to be the locus of a future shock. However, because most earthquakes occur in the seismically active zones at plate margins, these are the prime areas for trying to predict serious events. Predicting the location of a future earthquake in these zones combines a knowledge of the historical seismicity pattern with the elastic rebound model of what causes an earthquake.

The *seismic gap* theory is based on the simple idea that global plates move under the influence of forces which affect the plates as entities. The interactions at a plate margin therefore act along the entire length of the interplate boundary. Models of plate tectonic reconstructions assume continuity of plate motions on a scale of millions of years. However, the global seismicity patterns show that on the scales of decades or centuries, the process is discontinuous both in time and place. This is because of the way individual earthquakes occur. According to the elastic rebound model, stress accumulates until it exceeds the local strength of the rocks, rupture produces motion on the fault, and an earthquake occurs. During the time of stress accumulation, the area experiences no major earthquake, and the regional pattern of seismicity shows a local gap (Fig. 3.46). This is the potential location for an earthquake that is in the process of accumulating the strain energy necessary to cause rupture.

3.5.9.2 *Prediction of the time and size of an earthquake*

Seismic gap theory holds great promise as a means of determining where an earthquake is likely to occur. Unfortunately, it does not help to predict when it will occur or how large it will be. These factors depend on the local strength of the rocks and the rate at which strain accumulates. There are various ways to observe the effects of the strain accumulation, but the largely unknown factor of local breaking strength of the rocks hinders prediction of the time of an earthquake.

The strain accumulation results in precursory indications of both sociological and scientific nature. The People's Republic of China has suffered terribly from the ravages of earthquakes, partly because of the unavoidable use of low-quality construction materials. In 1966 the highly disciplined society in the People's Republic of China was marshalled to report any strange occurrences associated with earthquake occurrence. They noticed, for example, that wells and ponds bubbled, and sometimes gave off odors. Highly intriguing was the odd behavior of wild and domestic animals prior to many earthquakes. Dogs howled unaccountably, many creatures fell into panic, rats and mice left their holes, snakes abandoned their dens; even fish in ponds behaved in an agitated manner. It is not known how these creatures sense the imminent disaster, and the qualitative reports do not lend themselves to convenient statistical evaluation. However, prediction by scientific methods is still uncertain, so the usefulness of alternative premonitory phenomena cannot be rejected out of hand.

Several scientific methods have been tested as possible ways of predicting the time of earthquake occurrence. They are based on detecting ground changes, or effects related to them, that accompany the progression of strain. For example, a geochemical method which has had some degree of success is the monitoring of radon. Some minerals in the Earth's crust contain discrete amounts of uranium. The gas radon is a natural product of the radioactive decay of uranium. It migrates through pores and cracks and because of its own radioactivity it is a known environmental health hazard in buildings constructed in some geographic areas. Radon gas can become trapped in the Earth's crust, and in many areas it forms a natural radioactive background. Prior to some earthquakes anomalous levels of radon have been detected. The enhanced leakage of radon from the crust may be due to porosity changes at depth in response to the accumulating strains.

The build-up of strain is manifest in horizontal and vertical displacements of the Earth's surface, depending on the type of faulting involved. These displacements can be measured geodetically by triangulation or by modern techniques of trilateration which employ laser-ranging devices to measure the time taken for a laser beam to travel to a reflecting target and back to its source. The travel-time of the beam is measured extremely accurately, and converted into the distance between emitter and reflector. A shift on the fault will change this distance. With laser techniques the constant creep of one side of a horizontal fault relative to the other can be observed accurately. In one method the source and receiver of a laser beam are placed on one side of a fault, with a reflector on the opposite side. In an alternative method a laser source and receiver are placed on each side of the fault. Pulsed signals are beamed from each unit to an orbiting reflecting satellite, the position of which is known accurately. For each ground station, differences in the elapsed time of the pulse are converted by computer into ground movement, and into differential motion on the fault.

Several methods are suited to detecting differential vertical motion. Sensitive gravimeters on opposite sides of a fault can detect vertical displacement of as little as one centimeter of one instrument relative to the other. Distension of the Earth can be monitored with a tiltmeter, which is an instrument designed on the principle of a water-level. It consists of a long tube about 10 m in length, connecting two water-filled containers, in which the difference in water levels is monitored electronically. A tiltmeter is capable of determining tilt changes of the order of 10^{-7} degrees. Tiltmeters and gravimeters installed near to active faults have shown that episodes of ground uplift and tilt precede major earthquakes.

Geodetic and geophysical observations, such as changes in the local geomagnetic field or the electrical resistivity of the ground, are of fundamental interest. However, the most promising methods of predicting the time and size of an imminent earthquake are based on seismic observations. According to the elastic rebound model, the next earthquake on an active fault (or fault segment) will occur when the stress released in the most recent earthquake again builds up to the local breaking point of the rocks. Thus the probability at any time that an earthquake will occur on the fault depends on the magnitude of the latest earthquake, the time elapsed since it occurred and the local rate of accumulation of stress. A symptom of the stress build-up before a major earthquake is an increase of foreshock activity, and this was evidently a key parameter in the successful prediction of a large earthquake in Liaonping province, China, in February, 1975. The frequency of minor shocks increased, at first gradually but eventually dramatically, and then there was an ominous pause in earthquake activity. Chinese seismologists interpreted this 'time gap' as an indication of an impending earthquake. The population was ordered to vacate their homes several hours before the province was struck by a magnitude 7.4 earthquake that destroyed cities and communities. Because of the successful prediction of this earthquake, the epicenter of which was near to the city of Haicheng, the death toll among the 3,000,000 inhabitants of the province was very low.

A technique, no longer in favor, but which at one time looked promising for predicting the time and magnitude of an earthquake is the *dilatancy hypothesis*, based upon systematic variations in the ratio of the travel-times of P-waves and S-waves which originated in the focal volume of larger shocks. Russian seismologists noticed that prior to an earthquake the travel-time ratio t_s/t_p changed systematically: at first it decreased by up to 5%, then it returned to normal values just before the earthquake.

The observations have been attributed to changes in the dilatancy of the ground. Laboratory experiments have shown that, before a rock fractures under stress, it develops minute cracks which cause the rocks to dilate or swell. This dilatancy alters the P-wave velocity, which drops initially (thereby increasing t_p) as, instead of being water-filled, the new volume of the dilated pores at first fills with air. Later, water seeps in under pressure, replaces the air and the t_s/t_p ratio returns to normal values. At this point an earthquake is imminent. The time for which the ratio remains low is a measure of the strain energy that is stored and therefore a guide to the magnitude of the earthquake that is eventually unleashed. Initial success in predicting small earthquakes with the dilatancy model led to a period of optimism that an ultimate solution to earthquake prediction had been found. Unfortunately, it has become apparent that the dilatancy effect is not universal, and its importance appears to be restricted only to certain kinds of earthquakes.

In summary, present seismicity patterns, in conjunction with our knowledge of where historic earthquakes have occurred, permit reasonable judgements of *where* future earthquakes are most likely to be located. However, despite years of effort and the investigation of various scientific methods, it is still not possible to predict reliably *when* an earthquake is likely to happen in an endangered area.

3.5.10 Earthquake control

Earthquakes constitute a serious natural environmental hazard. Despite great efforts by scientists in various countries, successful prediction is not yet generally possible. Consequently, the protection of people against seismic hazard depends currently on the identification of especially perilous areas (such as active faults), the avoidance of these as the sites of constructions, the development and enforcement of appropriate building codes, and the education and training of the population in emergency procedures to be followed during and in the aftermath of a shock. Unfortunately, many densely populated regions are subject to high seismic risk. It is impossible to prevent the cumulation of strain in a region subject to tectonic earthquakes; the efforts of the human race are not likely to have much effect on the processes of plate tectonics! However, it may be possible to influence the manner in which the strain energy is released. The catastrophic earthquakes are those in which a huge amount of strain energy that has accumulated over a long period of time is suddenly released in a single event. If the energy could be released progressively over a longer period of time in many smaller shakes, the violence and disastrous consequences of a major earthquake might be avoided. The intriguing possibility of this type of earthquake control has been investigated in special situations.

In 1962 the U.S. Army began to dispose of liquid toxic waste from the manufacture of chemical weapons by injection into a well, more than 3 km deep, near Denver, Colorado. Although the region had been devoid of earthquake activity for the preceding 80 years, earthquakes began to occur several weeks after pumping started. Until 1965, when waste injection was halted, more than 1000 earthquakes were recorded. They were mostly very small, of the microearthquake category, but some had Richter magnitudes as high as 4.6. When pumping was halted, the seismicity ceased; when pumping was resumed, the earthquake activity started anew. It was conjectured that the liquid waste had seeped into old faults, and by acting as a kind of lubricant, had repeatedly permitted slippage, with an accompanying small earthquake. This incident suggested that it might be possible to control fault motion, either by injecting fluids to lubricate sections of the fault plane, or by pumping out fluid to lock the fault. This opened the intriguing possibility that, by making alternate use of the two processes, the slippage on a fault might be controlled so that it took place by a large number of small earthquakes rather than by a few disastrous earthquakes.

In 1969 the U.S. Geological Survey carried out a test of this lubrication effect in the depleted Rangely oil field in western Colorado. There were many disused wells in the oil field, through which fluids were pumped in and out of the ground over a considerable area. Meanwhile the local seismicity was monitored. This controlled test agreed with the observations at the Denver site, confirming that the earthquake activity correlated with the amount of fluid injected. Moreover, it was established that the earthquake activity increased when the pore pressure exceeded a critical threshold value, and it ceased when the pressure dropped below this value as the fluid was withdrawn. Despite the apparent success of this experiment, it was agreed that further testing was necessary to explore the validity of the method. An obvious difficulty of testing the modification of seismic activity on critical faults is that the tests must be made in remote areas so as to avoid costly damage caused by the testing. The conditions under which the method may be applicable have not been established definitively.

3.5.11 Monitoring nuclear explosions

Since 1963 most nuclear explosions have been conducted underground to prevent the dangerous radioactive fallout that accompanies nuclear explosions conducted underwater or in the atmosphere. The detection and monitoring of such testing activity became important tasks for seismologists. The bilateral Threshold Test Ban Treaty of 1974 between the former Soviet Union and the U.S.A. prohibited underground testing of nuclear devices with a yield greater than 150 kilotons of TNT equivalent, which corresponds roughly to an earthquake with magnitude about 6. Current efforts are underway to establish a global monitoring system to verify compliance with a future Comprehensive Test Ban Treaty. The system should detect nuclear explosions with a yield as low as one kiloton (a well-coupled kiloton explosion has a magnitude of around 4). Detection of these events at distances of several thousand kilometers, and discriminating them from the approximately 7000 earthquakes that occur annually with magnitudes of 4 or above (Table 3.3) poses a monumental challenge to seismologists.

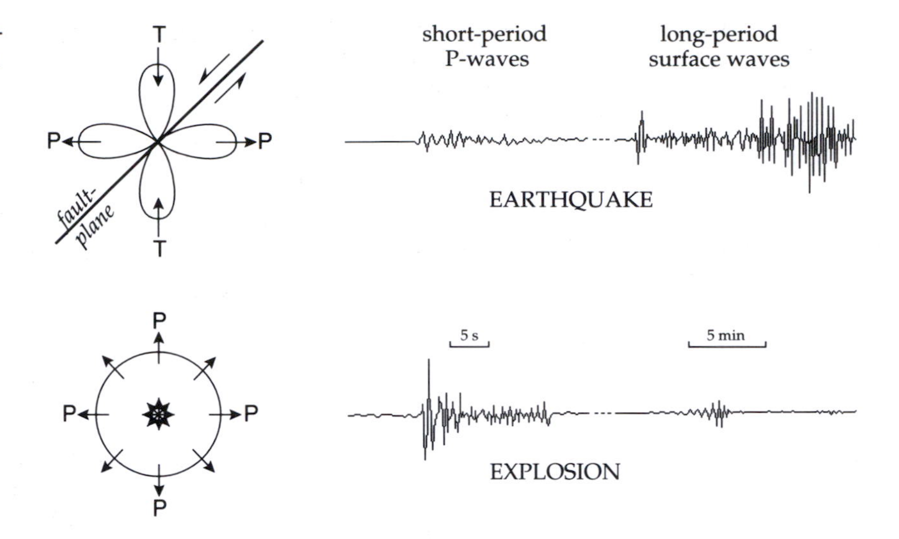

Fig. 3.47 Comparison of P-wave radiation patterns and the relative amplitudes of long-period and short-period surface waves for an earthquake and a nuclear explosion (after Richards, 1989).

In order to achieve the high detection capability needed to monitor underground testing many so-called seismic arrays have been set up. An array consists of several individual seismometers, with spacing on the order of a kilometer or less, that feed their output signals in parallel into a centralized data-processing center. By filtering, delaying and summing the signals of the individual instruments, the incoherent noise is reduced and the coherent signal is increased, thus improving the signal-to-noise ratio significantly over that for a single sensor. A local seismic disturbance arrives at the array on an almost horizontal path and triggers the individual seismometers at successively different times, whereas a teleseismic arrival from a very distant source reaches all seismometers in the array at nearly the same time along a steeply inclined path. The development of seismic arrays permitted the analysis of distant weak events. The enhanced sensitivity led to several advances in seismology. Features of the deep structure of the Earth (e.g., the inner core) could be investigated, earthquake location became more accurate and the analysis of focal mechanisms received a necessary impetus. These improvements were essential because of the need to identify correctly the features that distinguish underground nuclear explosions from small earthquakes.

An earthquake is the result of sudden motion of crustal blocks on opposite sides of a fault-plane. The radiation pattern of P-wave amplitude has four lobes of alternating compression and dilatation (Fig. 3.47). The first motions at the surface of the Earth are either pushes away from the source or tugs toward it, depending on the geometry of the focal mechanism. In contrast, an underground explosion causes outward pressure around the source. The first motions at the surface are all pushes away from the source. Hence, focal mechanism analysis provides an important clue to the nature of the recorded event. Moreover, an explosion produces predominantly P-waves, while earthquakes are much more efficient in also generating surface waves. Consequently, the relative amplitudes of the long-period surface-wave part of the record to the short-period P-wave part are much higher for an earthquake than for an explosion (Fig. 3.47).

Further discrimination criteria are the epicentral location and the focal depth. Intraplate earthquakes are much less common than earthquakes at active plate margins, so an intraplate event might be suspected to be an explosion. If the depth of a suspicious event is determined with a high degree of confidence to be greater than about 15 km, one can virtually exclude that it is an explosion. Deeper holes have not been drilled due to the great technical difficulty, e.g., in dealing with the high temperatures at such depths.

3.6 SEISMIC WAVE PROPAGATION

3.6.1 Introduction

A seismic disturbance is transmitted by periodic elastic displacements of the particles of a material. The progress of the seismic wave through a medium is determined by the advancement of the wavefront. We now have to consider how the wave behaves at the boundary between two media. Historically, two separate ways of handling this problem developed independently in the 17th century. One method, using Huygens' principle, describes the behavior of wavefronts; the other, using Fermat's principle, handles the geometry of ray paths at the interface. The eikonal equation (§3.3.2.5) establishes that these two methods of treating seismic wave propagation are equivalent.

In the Earth's crust the velocities of P- and S-waves are

often proportional to each other. This follows from Eqs. (3.41) and (3.48), which give the body-wave velocities in terms of the Lamé constants λ and μ. For many rocks, Poisson's relation $\lambda=\mu$ applies (see §3.2.4.5), and so

$$\frac{\alpha}{\beta}=\sqrt{\frac{\lambda+2\mu}{\mu}}=\sqrt{3} \quad (3.105)$$

For brevity, the following discussion handles P-waves only, which are assumed to travel with velocities α_1 and α_2 in the two media. However, we can equally apply the analyses to S-waves, by substituting the appropriate shear wave velocities β_1 and β_2 for the media.

3.6.2 **Huygens' principle**

The passage of a wave through a medium and across interfaces between adjacent media was first explained by the 17th century Dutch mathematician and physicist, Christiaan Huygens, who formulated a principle for the propagation of light as a wave, rather than as the stream of particles visualized by his great and influential contemporary, Sir Isaac Newton. Although derived for the laws of optics, Huygens' principle (1678) can be applied equally to any kind of wave phenomenon. The theory is based on simple geometrical constructions and permits the future position of a wavefront to be calculated if its present position is known. Huygens' principle can be stated: *All points on a wavefront can be regarded as point sources for the production of new spherical waves; the new wavefront is the tangential surface (or envelope) of the secondary wavelets.*

This principle can be illustrated simply for a plane wavefront (Fig. 3.48), although the method also applies to curved wavefronts. Let the wavefront initially occupy the position AB and let the open circles represent individual particles of the material in the wavefront. The particles are agitated by the arrival of the wavefront and act as sources of secondary wavelets. If the seismic velocity of the material is V, the distance travelled by each wavelet after time t is Vt and it describes a small sphere around its source particle. If the original wavefront contained numerous closely spaced particles instead of a discrete number, the plane CD tangential to the small wavelets would represent the new position of the wavefront. It is also planar, and lies at a perpendicular distance Vt from the original wavefront. In their turn the particles in the wavefront CD act as sources for new secondary wavelets, and the process is repeated. This principle can be used to derive the laws of reflection and refraction of seismic waves at an interface, and also to describe the process of diffraction by which a wave is deflected at a corner or at the edge of an object in its path.

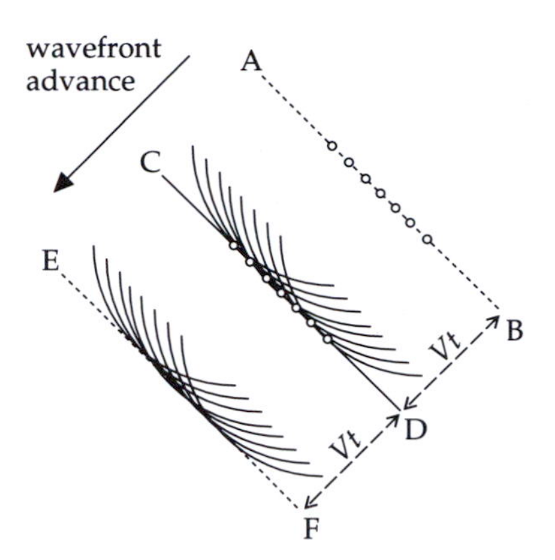

Fig. 3.48 Application of Huygens' principle to explain the advance of a plane wavefront. The wavefront at CD is the envelope of wavelets set up by particle vibrations when the wavefront was at the previous position AB. Similarly, the envelope of wavelets set up by vibrating particles in the wavefront CD forms the wavefront EF.

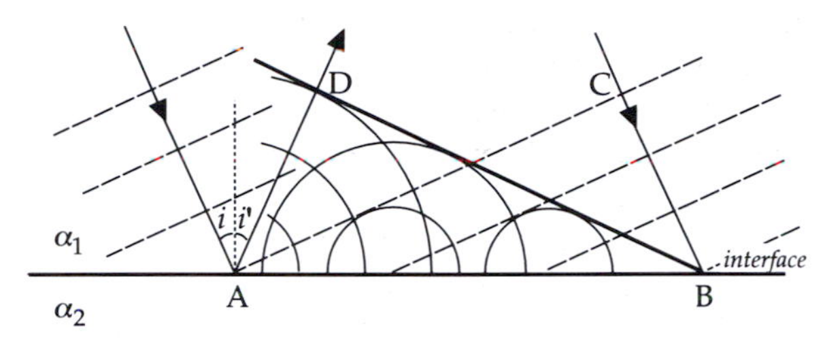

Fig. 3.49 The reflection of a plane P-wave at an interface between two media with different seismic velocities: incident plane waves (e.g., AC); spherical wavelets set up in the upper medium by vibrating particles in the segment AB of the interface; and the reflected plane wave BD, which is the envelope of the wavelets.

3.6.2.1 *The law of reflection using Huygens' principle*

Consider what happens to a plane P-wave travelling in a medium with seismic velocity α_1 when it encounters the boundary to another medium in which the P-wave velocity is α_2 (Fig. 3.49). At the boundary part of the energy of the incident wave is transferred to the second medium, and the remainder is reflected back into the first medium. If the incident wavefront AC first makes contact with the interface at A it agitates particles of the first medium at A and simultaneously the particles of the second medium in contact with the first medium at A. The vibrations of these particles set up secondary waves that travel away from A, back into the first medium as a reflected wave with velocity α_1 (and onward into the second medium as a refracted wave with velocity α_2).

By the time the incident wavefront reaches the interface at B all particles of the wavefront between A and B have been agitated. Applying Huygens' principle, the wavefront

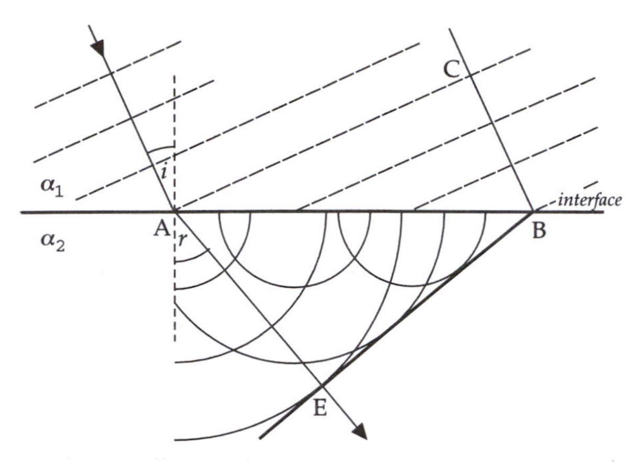

Fig. 3.50 The refraction of a plane P-wave at an interface between two media with different seismic velocities α_1 and α_2 ($>\alpha_1$): incident plane waves (e.g., AC); the spherical wavelets set up in the lower medium by vibrating particles in the segment AB of the interface; and the refracted plane wave BE, which is the envelope of the wavelets. The angles of incidence (i) and refraction (r) are defined between the normal to the interface and the respective rays.

of the reflected disturbance is the tangent plane to the secondary wavelet in the first medium. In Fig. 3.49 this is represented by the tangent BD from B to the circle centered at A, the first point of contact with the boundary. In the time t that elapses between the arrival of the plane wave at A and its arrival at B, the incident wavefront travels a distance CB and the secondary wavelet from A travels the equal distance AD. The triangles ABC and ABD are congruent. It follows that the reflected wavefront makes the same angle with the interface as the incident wave.

It is customary to describe the orientation of a plane by the direction of its normal. The angle between the normal to the interface and the normal to the incident wavefront is called the angle of incidence (i); the angle between the normal to the interface and the normal to the reflected wavefront is called the angle of reflection (i'). This application of Huygens' principle to plane seismic waves shows that the angle of reflection is equal to the angle of incidence ($i=i'$). This is known as the *law of reflection*. Although initially developed for light-beams, it is also valid for the propagation of seismic waves.

3.6.2.2 *The law of refraction using Huygens' principle*

The discussion of the interaction of the incident wave with the boundary can be extended to cover the part of the disturbance that travels into the second medium (Fig. 3.50). This disturbance travels with the velocity α_2 of the second medium. Let t be the time taken for the incident wavefront in the first medium to advance from C to B; then BC$=\alpha_1 t$. In this time all particles of the second medium between A and B have been agitated and now act as sources for new wavelets in the second medium. When the incident wave reaches B, the wavelet from A in the second medium has spread out to the point E, where AE$=\alpha_2 t$. The wavefront in the second medium is the tangent BE from B to the circle centered at A. The angle of incidence (i) is defined as before; the angle between the normal to the interface and the normal to the transmitted wavefront is called the angle of refraction (r). Comparison of the triangles ABC and ABE shows that BC$=$AB sin i, and AE$=$AB sin r. Consequently,

$$\frac{\text{AB}\sin i}{\text{AB}\sin r}=\frac{\text{BC}}{\text{AE}}=\frac{\alpha_1 t}{\alpha_2 t} \tag{3.106}$$

$$\frac{\sin i}{\sin r}=\frac{\alpha_1}{\alpha_2} \tag{3.107}$$

Equation (3.107) is called the *law of refraction* for plane seismic waves. Its equivalent in optics is often called *Snell's law*, in recognition of its discoverer, the Dutch mathematician Willebrod Snellius (or Snell).

3.6.2.3 *Diffraction*

The laws of reflection and refraction derived above with the aid of Huygens' principle apply to the behavior of plane seismic waves at plane boundaries. When a plane or spherical seismic wave encounters a pointed obstacle or discontinuous surface, it experiences *diffraction*. This phenomenon allows the wave to bend around the obstacle, penetrating what otherwise would be a shadow zone for the wave. It is the diffraction of sound waves, for example, that allows us to hear the voices of people who are still invisible to us around a corner, or on the other side of a high wall. Huygens' principle also gives an explanation for diffraction, as illustrated by the following simple case.

Consider the normal incidence of a plane wave on a straight boundary that ends at a sharp corner B (Fig. 3.51). The incident wave is reflected along the entire length AB, with each particle of AB acting as a secondary source according to Huygen's principle. Beyond the edge B the incident wavefronts cannot be reflected. The plane wavefront passes by the edge B, so that the point C should lie in the shadow of AB. However, the corner B also acts as a source of secondary wavelets, part of which contribute to the reflected wavefront and part pass into the shadow zone. The intensity of the wave diffracted into the shadow zone is weaker than in the main wavefront, and it decreases progressively with increasing angle away from the direction of travel of the incident wavefront.

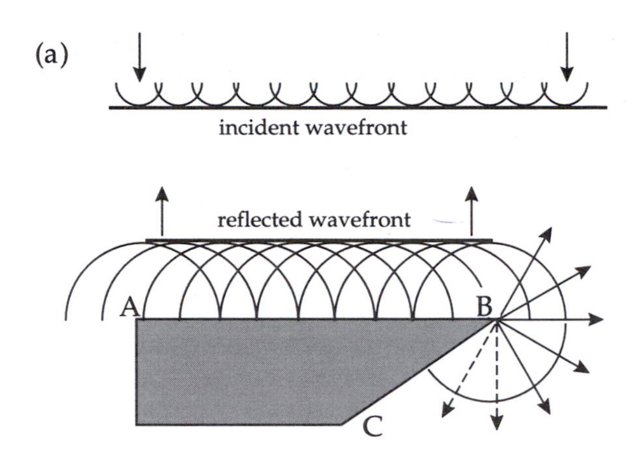

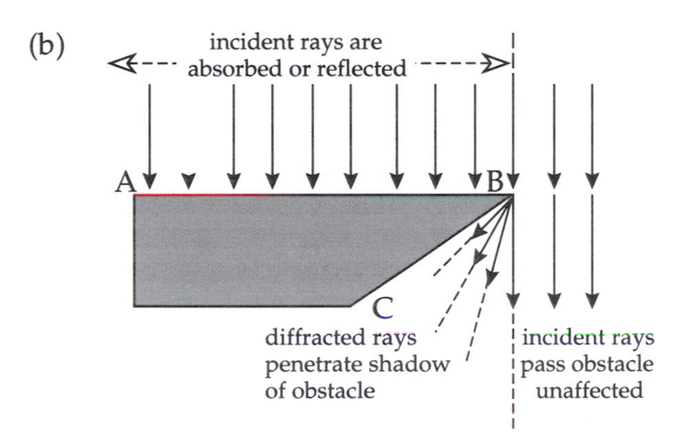

Fig. 3.51 Explanation of diffraction at an edge with the aid of Huygens' principle. (a) The incident and reflected plane wavefronts are the envelopes to Huygens wavelets, which are able to carry the incident disturbance around a sharp corner. (b) The incident rays are absorbed, reflected or pass by the obstacle, but some rays related to the wavelets generated at the point of the obstruction are diffracted into its shadow.

3.6.3 **Fermat's principle**

The behavior of seismic ray paths at an interface is explained by another principle of optics that was formulated – also in the 17th century – by the French mathematician Pierre de Fermat. As applied to seismology, Fermat's principle states that, of the many possible paths between two points A and B, the seismic ray follows the path that gives the shortest travel-time between the points. If ds is the element of distance along a ray path and c is the seismic velocity over this short distance, then the travel-time t between A and B is minimum: thus,

$$t=\int_{A}^{B}\frac{ds}{c}=\text{minimum} \tag{3.108}$$

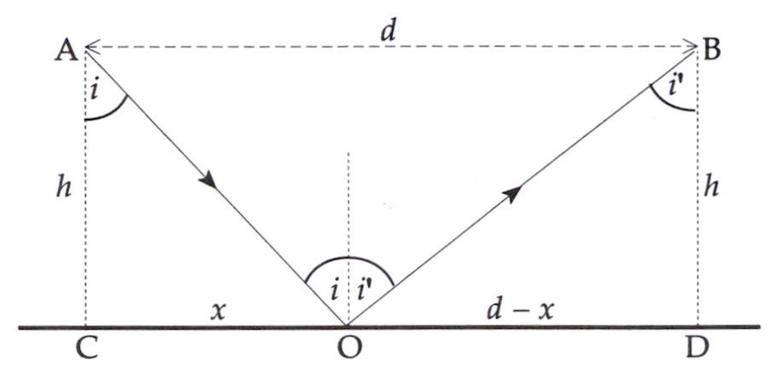

Fig. 3.52 Geometry of incident and reflected rays for derivation of the law of reflection with the aid of Fermat's principle.

Generally, when the velocity varies continuously with position, the determination of the ray path is intricate. In the case of a layered medium, in which the velocity is constant in each layer, Fermat's principle provides us with an independent method for determining the laws of reflection and refraction.

3.6.3.1 *The law of reflection using Fermat's principle*

Consider the reflection of a seismic ray in a medium with constant P-wave velocity α_1 at the boundary to another medium (Fig. 3.52). For convenience we take the boundary to be horizontal. Let A be a point on the incident ray at a vertical distance h from the boundary and let B be the corresponding point on the reflected ray. Let C and D be the nearest points on the boundary to A and B, respectively. Further, let d be the horizontal separation AB, and let O be the point of reflection on the interface at a horizontal distance x from C; than OD is equal to $(d-x)$ and we can write for the travel-time t from A to B:

$$t=\frac{AO}{\alpha_1}+\frac{OB}{\alpha_1}=\frac{1}{\alpha_1}\left[\sqrt{h^2+x^2}+\sqrt{h^2+(d-x)^2}\right] \tag{3.109}$$

According to Fermat's principle the travel-time t must be a minimum. The only variable in Eq. (3.109) is x. To find the condition that gives the minimum travel-time we differentiate t with respect to x and set the result equal to zero:

$$\frac{\partial t}{\partial x}=\frac{1}{\alpha_1}\left[\frac{x}{\sqrt{h^2+x^2}}-\frac{(d-x)}{\sqrt{h^2+(d-x)^2}}\right]=0 \tag{3.110}$$

By inspection of Fig. 3.52 the relationships of these expressions to the angle of incidence (i) and the angle of reflection (i') are evident. The first expression inside the brackets is sin$i$ and the second is sini'. The condition for the minimum travel-time is again $i'=i$; the angle of reflection equals the angle of incidence.

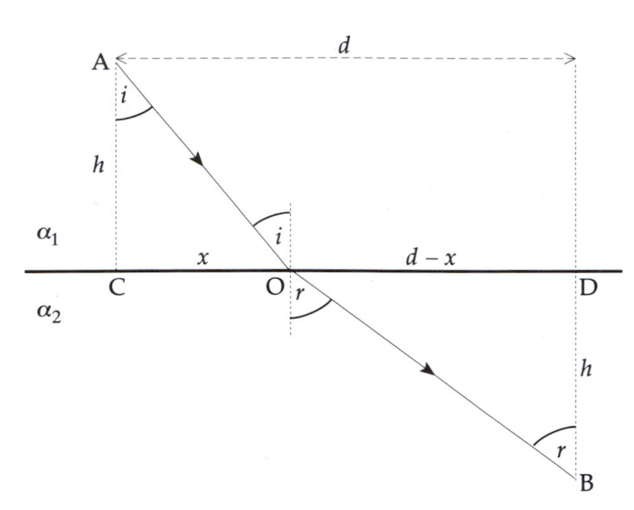

Fig. 3.53 Geometry of incident and refracted rays for derivation of the law of refraction with the aid of Fermat's principle.

3.6.3.2 *The law of refraction using Fermat's principle*

We can use a similar approach to determine the law of refraction. This time we study the passage of the seismic ray from a medium with velocity α_1 into a medium with higher velocity α_2 (Fig. 3.53). Let A again be a point on the incident ray at a vertical distance h from a point C on the interface. The ray traverses the boundary at O, a horizontal distance x from C. Let B now be a point on the ray in the second medium at a distance h from D, the closest point on the interface. The distance CD is d, so that again OD is equal to $(d-x)$. The travel-time t which we have to minimize is given by

$$t=\frac{\mathrm{AO}}{\alpha_1}+\frac{\mathrm{OB}}{\alpha_2}=\frac{\sqrt{h^2+x^2}}{\alpha_1}+\frac{\sqrt{h^2+(d-x)^2}}{\alpha_2} \tag{3.111}$$

Differentiating Eq. (3.111) with respect to x and setting the result equal to zero gives us the condition for the minimum value of t:

$$\frac{x}{\alpha_1\sqrt{h^2+x^2}}-\frac{(d-x)}{\alpha_2\sqrt{h^2+(d-x)^2}}=0 \tag{3.112}$$

By reference to Fig. 3.53 we can write this expression in terms of the sines of the angles of incidence (i) and refraction (r). This application of Fermat's principle to the seismic ray paths gives again the law of refraction that we derived by applying Huygens' principle to the wavefronts (Eq. (3.107)). It can also be stated as

$$\frac{\sin i}{\alpha_1}=\frac{\sin r}{\alpha_2} \tag{3.113}$$

In this example we have assumed that $\alpha_2>\alpha_1$. As it passes from the medium with lower velocity into the medium with higher velocity the refracted ray is bent away from the normal to the boundary, giving an angle of refraction that is greater than the angle of incidence ($r>i$). Under the opposite conditions, if $\alpha_2<\alpha_1$ the refracted ray is bent back toward the normal and the angle of refraction is less than the angle of incidence ($r<i$).

3.6.4 **Partitioning of seismic body waves at a boundary**

The conditions that must be fulfilled at a boundary are that the normal and tangential components of stress, as well as the normal and tangential components of the displacements, must be continuous across the interface. If the normal (or tangential) stress were not continuous, the point of discontinuity would experience infinite acceleration. Similarly, if the normal displacements were not continuous, a gap would develop between the media or parts of both media would overlap to occupy the same space; discontinuous tangential displacements would result in relative motion between the media across the boundary. These anomalies are impossible if the boundary is a fixed surface that clearly separates the media.

As a result of the conditions of continuity, a P-wave incident on a boundary energizes the particles on each side of the boundary at the point of incidence, and sets up four waves. The energy of the incident P-wave is partitioned between P- and S-waves that are reflected from the boundary, and other P- and S-waves that are transmitted into the adjacent layer. The way in which this takes place may be understood by considering the particle motion that is induced at the interface.

The particle motion in the incident P-wave is parallel to the direction of propagation. At the interface the vibration of particles of the lower layer can be resolved into a component perpendicular to the interface and a component parallel to it in the vertical plane containing the incident P-wave. In the second layer each of these motions can in turn be resolved into a component parallel to the direction of propagation (a refracted P-wave) and a component perpendicular to it in the vertical plane (a refracted SV-wave). Because of continuity at the interface, similar vibrations are induced in the upper layer, corresponding to a reflected P-wave and a reflected SV-wave, respectively.

Let the angles between the normal to the interface and the ray paths of the P- and S-waves in medium 1 be i_p and i_s, respectively, and the corresponding angles in medium 2 be r_p and r_s (Fig. 3.54). Applying Snell's law to both the reflected and refracted P- and S-waves gives

$$\frac{\sin i_p}{\alpha_1}=\frac{\sin i_s}{\beta_1}=\frac{\sin r_p}{\alpha_2}=\frac{\sin r_s}{\beta_2} \tag{3.114}$$

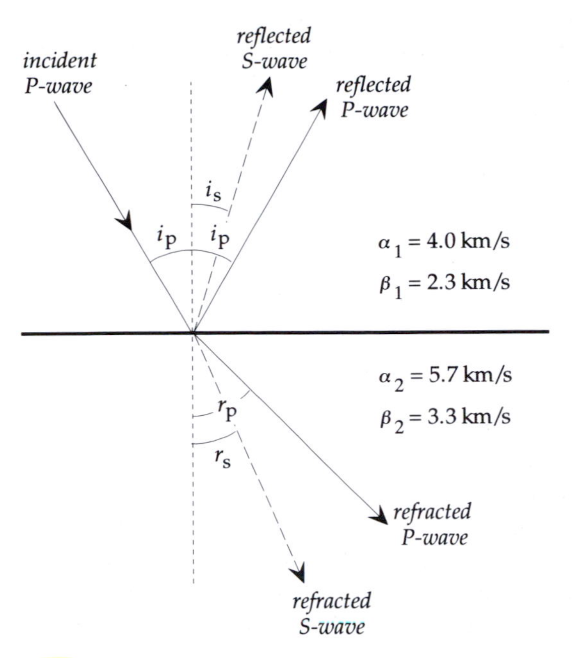

Fig. 3.54 The generation of reflected and refracted P- and S-waves from a P-wave incident on a plane interface.

By similar reasoning it is evident that an incident SV-wave also generates vibrations that have components normal and parallel to the interface, and will set up refracted and reflected P- and SV-waves. The situation is different for an incident SH-wave, which has no component of motion normal to the interface. In this case only refracted and reflected SH-waves are created.

3.6.4.1 *Subcritical and supercritical reflections, and critical refraction*

Let O be a seismic source near the surface of a uniformly thick horizontal layer with P-wave velocity α_1 that lies on top of a layer with higher velocity α_2 (Fig. 3.55). Consider what happens to seismic rays that leave O and arrive at the boundary with all possible angles of incidence. The most simple ray is that which travels vertically to meet the boundary with zero angle of incidence at the point N. This *normally incident ray* is partially reflected back along its track, and partially transmitted vertically into the next medium without change of direction. As the angle of incidence increases, the point of incidence moves from N towards C. The transmitted ray experiences a change of direction according to Snell's law of refraction, and the ray reflected to the surface is termed a *subcritical reflection*.

The ray that is incident on the boundary at C is called the *critical ray* because it experiences *critical refraction*. It encounters the boundary with a *critical angle of incidence*.

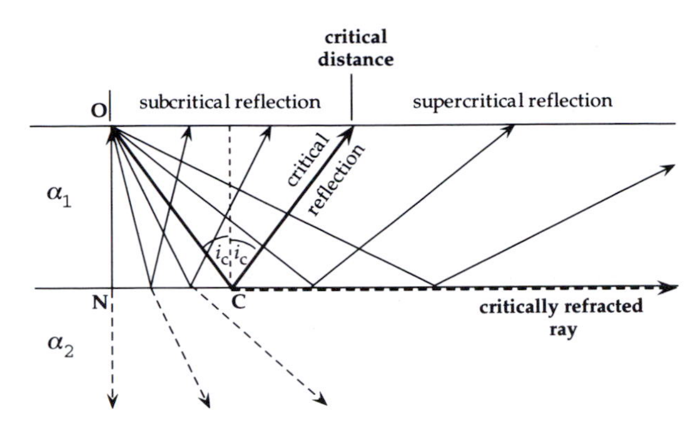

Fig. 3.55 The critical reflection defines two domains, corresponding to regions of subcritical and supercritical reflection, respectively.

The corresponding refracted ray makes an angle of refraction of 90° with the normal to the boundary. As a result, it travels parallel to the boundary in the top of the lower layer with faster velocity α_2. The sine of the angle of refraction of the critical ray is unity, and we can calculate the critical angle, i_c, by applying Snell's law:

$$\sin i_c = \frac{\alpha_1}{\alpha_2} \tag{3.115}$$

The critical ray is accompanied by a *critical reflection*. It reaches the surface at a *critical distance* (x_c) from the source at O. The reflections that arrive inside the critical distance are called *subcritical reflections*. At angles up to the critical angle refracted rays pass into the lower medium, but for rays incident at angles greater than the critical angle refraction is no longer possible. The seismic rays that are incident more obliquely than the critical angle are reflected almost completely. These reflections are termed *supercritical reflections*, or simply *wide-angle reflections*. They lose little energy to refraction, and are thus capable of travelling large distances from the source in the upper medium. Supercritical reflections are recorded with strong amplitudes on seismograms at distant stations.

3.6.5 **Reflection seismology**

Reflection seismology is directed primarily at finding the depths to reflecting surfaces and the seismic velocities of subsurface rock layers. The techniques of acquiring and processing reflection seismology data have been developed and refined to a very high degree of sophistication as a result of the intensive application of this method in the search for petroleum. The principle is simple. A seismic signal (e.g., an explosion) is produced at a known place at a known time, and the echoes reflected from the boundaries between rock layers with different seismic velocities and densities are recorded and analyzed. Compactly designed, robust, electromagnetic seismometers – called 'geophones' in industrial usage – are spread in the region of subcritical reflection, within the critical distance from the shot-point, where no refracted arrivals are possible. Within this distance the only signals received are the wave that travels directly from the shot-point to the geophones and the waves reflected at subsurface interfaces. Surface waves are also recorded and constitute an important disturbing 'noise', because they interfere with the reflected signal. The closer the geophone array is located to the shot-point, the more nearly the paths of the reflected rays travel vertically. Reflection seismic data are most usually acquired along profiles that cross geological structures as nearly as possible normal to the strike of the structure. The travel-times recorded at the geophones along a profile are plotted as a two-dimensional cross-section of the structure. In recent years, three-dimensional surveying, which covers the entire subsurface, has become more important.

Several field procedures are in common use. They are distinguished by different layouts of the geophones relative to the shot-point. The most routine application of reflection seismology is in *continuous profiling*, in which the geophones are laid out at discrete distances along a profile through the shot-point. To reduce seismic noise, each recording point is represented by a group of interconnected geophones. After each shot the geophone layout and shot-point are moved a predetermined distance along the profile, and the procedure is repeated. Broadly speaking, there are two main variations of this method, depending on whether each reflection point on the reflector is sampled only once (*conventional coverage*) or more than once (*redundant coverage*).

The most common form of conventional coverage is a *split-spread* method (Fig. 3.56), in which the geophones are spread symmetrically on either side of the shot-point. If the reflector is flat-lying, the point of reflection of a ray recorded at any geophone is below the point midway between the shot-point and the geophone. For a shot-point at Q the rays QAP and QBR that are reflected to geophones at P and R represent extreme cases. The two-way travel-time of the ray QAP gives the depth of the reflection point A, which is plotted below the mid-point of QP. Similarly, B is plotted below the mid-point of QR. The split-spread layout around the shot-point Q gives the depths of reflection points along AB, which is half the length of the geophone spread PR. The shot-point is now moved to the point R, and the geophones between P and Q are moved to cover the segment RS. From the new shot-point R the positions of reflection points in the segment BC of the reflector are obtained. The ray RBQ from shot-point R to the geophone at Q has the same path as the ray QBR from shot-point Q to the geophone at R. By successively moving the shot-point and half of the split-spread geophone layout a continuous coverage of the subsurface reflector is obtained.

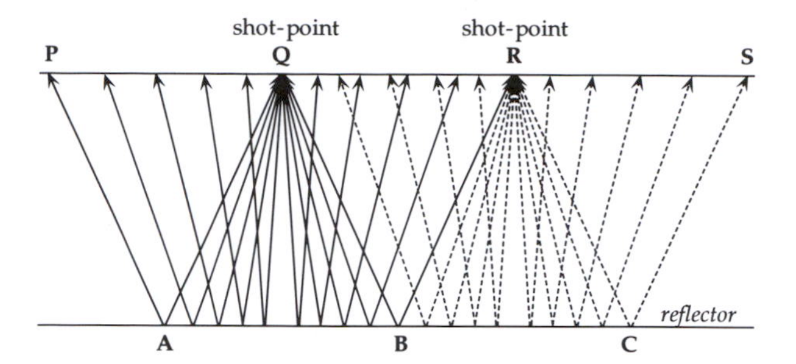

Fig. 3.56 The split-spread method of obtaining continuous subsurface coverage of a seismic reflector.

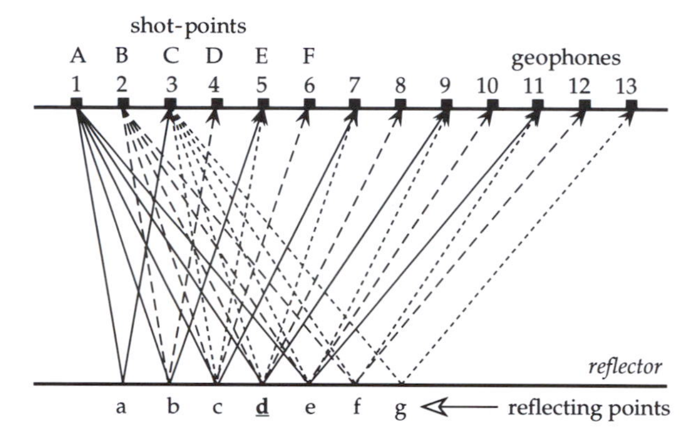

Fig. 3.57 Common-depth-point method of seismic reflection shooting, showing rays from successive shot-points at A, B and C and the repeated sampling of the same point on the reflector (e.g., d) by rays from each shot-point.

Redundant coverage is illustrated by the *common-depth-point* method, which is routinely employed as a means of reducing noise and enhancing the signal-to-noise ratio. Commonly 24 to 96 groups of geophones feed recorded signals into a multi-channel recorder. The principle of common-depth-point coverage is illustrated for a small number of 11 geophone groups in Fig. 3.57. When a shot is fired at A, the signals received at geophones 3–11 give subsurface coverage of the reflector between points a and e. The shot-point is now moved to B, which coincides with the position occupied by geophone 2 for the first shot, and the geophone array is moved forward correspondingly along the direction of the profile to positions 4–12. From shot-point B the subsurface coverage of the reflector is between points b and f. The reflector points b to e are common to

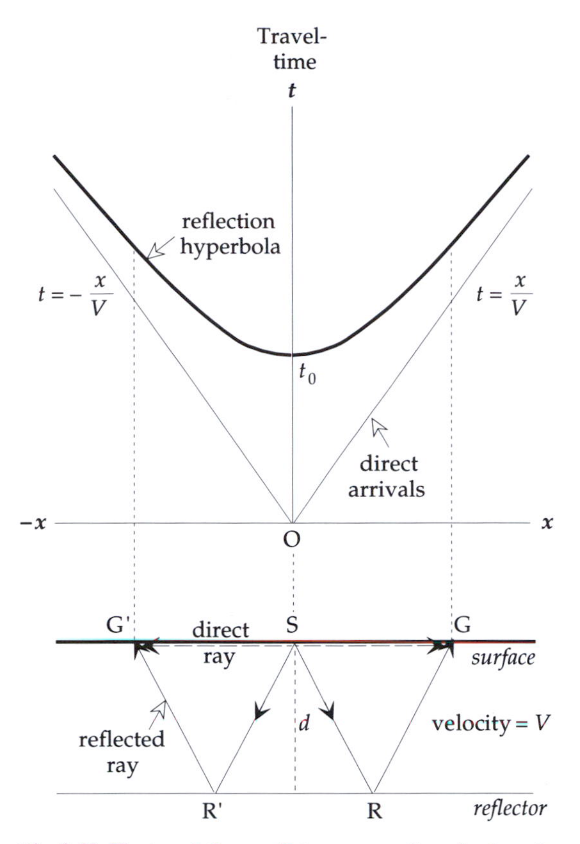

Fig. 3.58 The travel-time vs distance curve for reflections from a horizontal boundary is a hyperbola. The vertical reflection time t_0 is the intercept of the hyperbola with the travel-time axis.

both sets of data. By repeatedly moving the shot-point and geophone array in the described manner, each reflecting point of the interface is sampled multiply. For example, in Fig. 3.57 the reflecting point d is sampled multiply by the rays Ad 9, Bd 8, Cd 7, etc. The lengths of these ray paths are different. During subsequent data-processing the reflection travel-times are corrected for *normal moveout*, which is a geometrical effect related to geophone distance from the shot-point. The records are then *stacked*, which is a procedure for enhancing the signal-to-noise ratio.

3.6.5.1 *Reflection at a horizontal interface*

The simplest case of seismic reflection is the two-dimensional reflection at a horizontal boundary (Fig. 3.58). Let the reflecting bed be at depth d below the shot-point S. The ray that strikes the boundary at R is reflected to the surface and recorded by a geophone at the point G, so that the angles of incidence and reflection are equal. Let G be at a horizontal distance x from the shot-point. If the P-wave velocity is V, the first signal received at G is from the direct wave that travels directly along SG. Its travel-time is given by $t_d = x/V$. It is important to keep in mind that the direct wave is not a surface wave but a body wave that travels parallel to and just below the surface of the top layer. The travel-time t of the reflected ray SRG is (SR+RG)/V. However, SR and RG are equal and therefore

$$t = \frac{2}{V}\sqrt{d^2 + \frac{x^2}{4}} \tag{3.116}$$

$$t = \frac{2d}{V}\sqrt{1 + \frac{x^2}{4d^2}} = t_0\sqrt{1 + \frac{x^2}{4d^2}} \tag{3.117}$$

At $x=0$ the travel-time corresponds to the vertical echo from the reflector; this 'echo-time' is given by $t_0 = 2d/V$. The quantity under the square root in Eq. (3.117) determines the curvature of the t–x curve and is called the *normal moveout factor*. It arises because the ray reaching a geophone at a horizontal distance x from the shot-point has not travelled vertically between it and the reflector. Squaring both sides of Eq. (3.117) and rearranging terms gives

$$\frac{t^2}{t_0^2} - \frac{x^2}{4d^2} = 1 \tag{3.118}$$

This is the equation of a hyperbola (Fig. 3.58) that is symmetrical about the vertical time axis, which it intersects at t_0. For large distances from the shot-point ($x \gg 2d$) the travel-time of the reflected ray approaches the travel-time of the direct ray and the hyperbola is asymptotic to the two lines $t = \pm x/V$.

A principle goal of seismic reflection profiling is usually to find the vertical distance (d) to a reflecting interface. This can be determined from t_0, the two-way reflection travel-time recorded by a geophone at the shot-point, once the velocity V is known. One way of determining the velocity is by comparing t_0 with the travel-time t_x to a geophone at distance x. In reflection seismology the geophones are laid out close to the shot-point and the assumption is made that the geophone distance is much less than the depth of the reflector ($x \ll d$). Eq. (3.117) becomes

$$t_x = t_0\left[1 + \left(\frac{x}{2d}\right)^2\right]^{1/2} = t_0\left[1 + \frac{1}{2}\left(\frac{x}{2d}\right)^2 + \ldots\right]$$

$$= t_0\left[1 + \frac{1}{2}\left(\frac{x}{Vt_0}\right)^2\right] \tag{3.119}$$

The difference between the travel-time t_x and the shot-point travel-time t_0 is the *normal moveout*, $\Delta t_n = t_x - t_0$. By rearranging Eq. (3.119) we get

$$\Delta t_n = \frac{x^2}{2V^2 t_0} \tag{3.120}$$

The echo time t_0 and the normal moveout time Δt_n are found from the reflection data. The distance x of the geo-

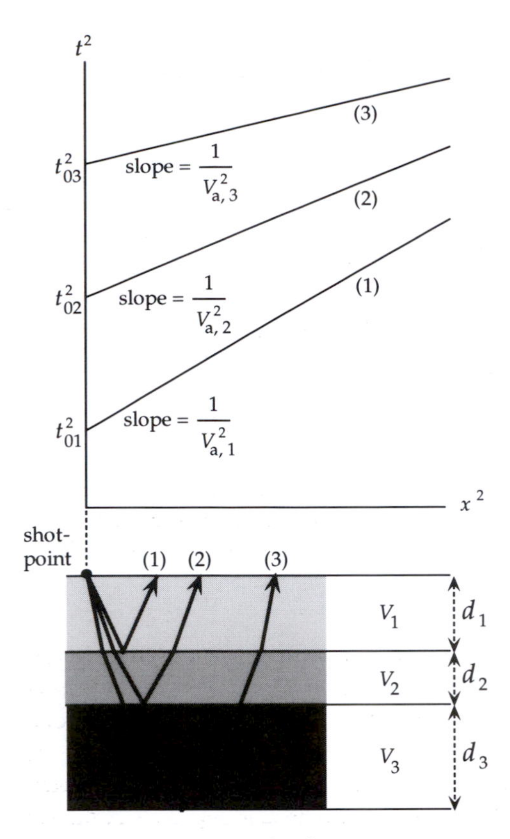

Fig. 3.59 Illustration of a 't^2-x^2 plot' for near-vertical reflections from three horizontal reflectors; $V_{a,1}$ is the true velocity V_1 of layer 1, but $V_{a,2}$ and $V_{a,3}$ are 'average' velocities that depend on the true velocities and the layer thicknesses.

phone from the shot-point is known and therefore the layer velocity V can be determined. The depth d of the reflecting horizon can then be found by using the formula for the echo time.

An alternative way of interpreting reflection arrival times becomes evident when Eq. (3.118) is rearranged in the form

$$t^2 = t_0^2 + \frac{x^2}{V^2} \tag{3.121}$$

A plot of t^2 against x^2 is a straight line that has slope $1/V^2$. Its intercept with the t^2-axis gives the square of the echo time, t_0, from which the depth d to the reflector can be found once the velocity V is known. The record at each geophone will contain reflections from several reflectors. For the first reflector the velocity determined by the t^2–x^2 method is the true *interval velocity* of the uppermost layer, V_1, which, in conjunction with t_{01}, the first echo time, gives the thickness d_1 of the top layer. However, the ray reflected from the second interface has travelled through the first layer with interval velocity V_1 and the second layer with interval velocity V_2. The velocity interpreted in the t^2–x^2 method for this reflection, and for reflections from all deeper interfaces, is an *average velocity*. If the incident and reflected rays travel nearly vertically, the average velocity $V_{a,n}$ for the reflection from the nth reflector is given by

$$V_{a,n} = \frac{d_1 + d_2 + d_3 + \ldots + d_n}{t_1 + t_2 + t_3 + \ldots + t_n} = \frac{\sum_{i=1}^{n} d_i}{\sum_{i=1}^{n} t_i} \tag{3.122}$$

where d_i is the thickness and t_i the interval travel-time for the ith layer.

The t^2–x^2 method is a simple way of estimating layer thicknesses and average velocities for a multi-layered Earth (Fig. 3.59). The slope of the second straight line gives $V_{a,2}$, which is used with the appropriate echo time t_{02} to find the combined depth D_2 to the second interface, given by $D_2 = d_1 + d_2 = (V_{a,2})(t_{02})$; d_1 is known, and so d_2 can be calculated. The two-way travel-time in the second layer is $(t_{02} - t_{01})$ and thus the interval velocity V_2 can be found. In this way the thicknesses and interval velocities of deeper layers can be determined successively.

In fact, of course, the rays do not travel vertically but are bent as they pass from one layer to another (Fig. 3.59). Moreover, the elastic properties of a layer are rarely homogeneous so that the seismic velocity is variable and the ray path in the layer is curved. Exploration seismologists have found it possible to compensate for these effects by replacing the average velocity with the root-mean-square velocity V_{rms} defined by

$$V_{rms}^2 = \frac{\sum_{i=1}^{n} V_i^2 t_i}{\sum_{i=1}^{n} t_i} \tag{3.123}$$

where V_i is the interval velocity and t_i the travel-time for the ith layer.

3.6.5.2 *Reflection at an inclined interface*

When the reflecting interface is inclined at an angle θ to the horizontal, as in Fig. 3.60, the shortest distance d between the shot-point S and the reflector is the perpendicular distance to the inclined plane. The paths of reflected rays on the down-dip side of the shot-point are longer than those on the up-dip side; this has corresponding effects on the travel-times. The rays obey the laws of reflection optics and appear to return to the surface from the point S', which is the image point of the shot-point with respect to the reflector. The travel-time t through the layer with velocity V is readily found with the aid of the image point. For example for the ray SRG, recorded by a geophone on the surface at G, we get

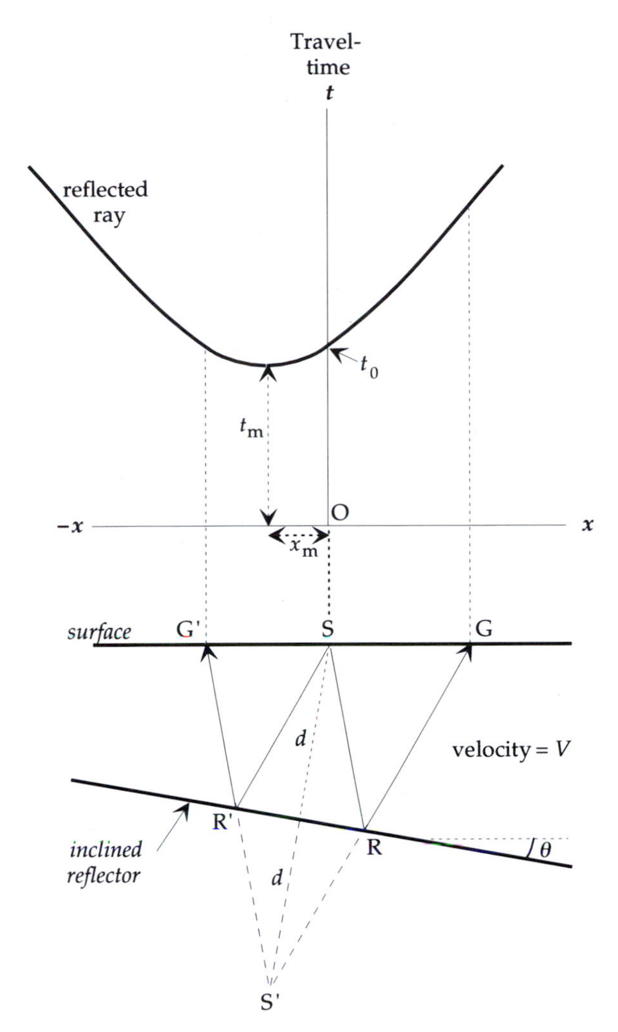

Fig. 3.60 The travel-time vs distance curve for an inclined reflector is also a hyperbola with vertical axis (cf. Fig. 3.58), but the minimum travel-time (t_m) is measured at distance x_m from the shot-point.

$$t=\frac{SR+RG}{V}=\frac{S'R+RG}{V}=\frac{S'G}{V} \tag{3.124}$$

The image point S' is as far behind the reflector as the shot-point is in front: S'S=$2d$. In triangle S'SG the side SG equals the geophone distance x and the obtuse angle S'SG equals (90°+θ). If we apply the law of cosines to the triangle S'SG we can solve for S'G and substitute the answer in Eq. (3.124). This gives the travel-time of the reflection from the inclined boundary:

$$t=\frac{1}{V}\sqrt{(x^2+4xd\sin\theta+4d^2} \tag{3.125}$$

Equation (3.125) is the equation of a hyperbola whose axis of symmetry is vertical, parallel to the t-axis. For a flat reflector the hyperbola was symmetric about the t-axis (Fig. 3.58) and the minimum travel-time (echo time) corresponded to the vertical reflection below the shot-point (x=0). For an inclined reflector the minimum travel-time t_m is no longer the perpendicular path to the reflector, which would give the travel-time t_0 in Fig. 3.60. Although the perpendicular path is the shortest distance from shot-point to *reflector*, it is not the shortest path of a reflected ray between the shot-point and a *geophone*. The shortest travel-time is recorded by the geophone at a horizontal distance x_m on the up-dip side of the shot-point. The coordinates (x_m, t_m) of the minimum point of the travel-time hyperbola are

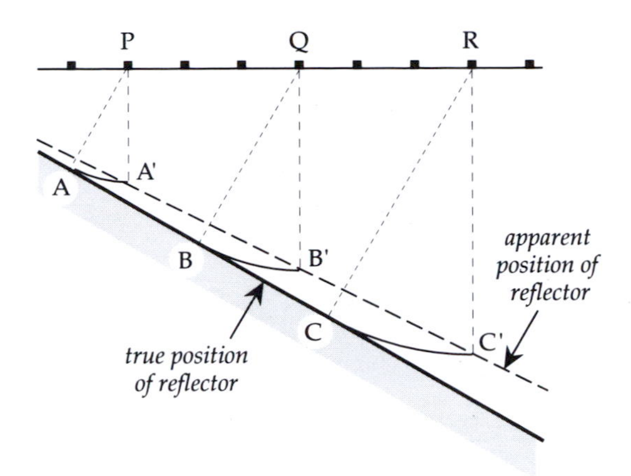

Fig. 3.61 When the reflector is inclined and depths are plotted vertically under geophone positions, the true reflecting points A, B and C are mapped at A', B' and C', falsifying the position of the reflector.

$$x_m=-2d\sin\theta \text{ and } t_m=\frac{2d\cos\theta}{V} \tag{3.126}$$

In practice it is not known a priori whether a reflector is horizontal or inclined. If reflection records are not corrected for the effect of layer-dip, an error results in plotting the positions of dipping beds. The shot-point travel-time t_0 gives the direct distance to a reflector, but the path along which the echo has travelled is not known. Consider the geometry of the inclined boundary in Fig. 3.61. First arrival reflections recorded for shot-points P, Q, and R come from the true reflection points A, B, and C. If the computed reflector depths are plotted directly below the shot-points at A', B' and C', the dipping boundary will appear to lie at a shallower depth than its true position, and the apparent dip of the reflector will be less steep than the true dip. This leads to a distorted picture of the underground structure. For example, an anticline appears broader and less steep-sided than it is. Similarly, if the limbs of a syncline dip steeply

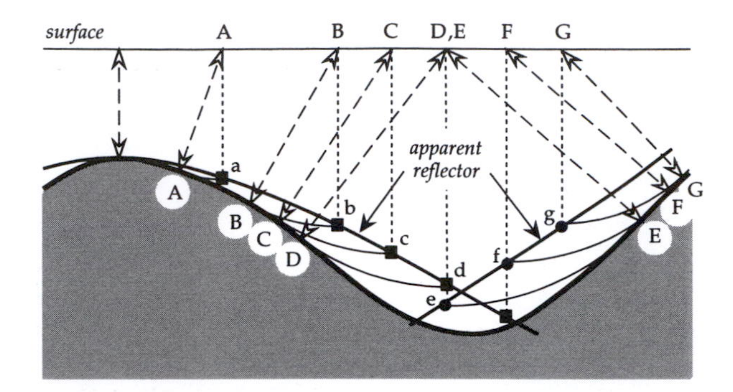

Fig. 3.62 Paths of reflected rays over an anticline and syncline, showing the false apparent depth to the reflecting surface. True reflection points A–G are wrongly mapped at locations a–g beneath the corresponding shot-points.

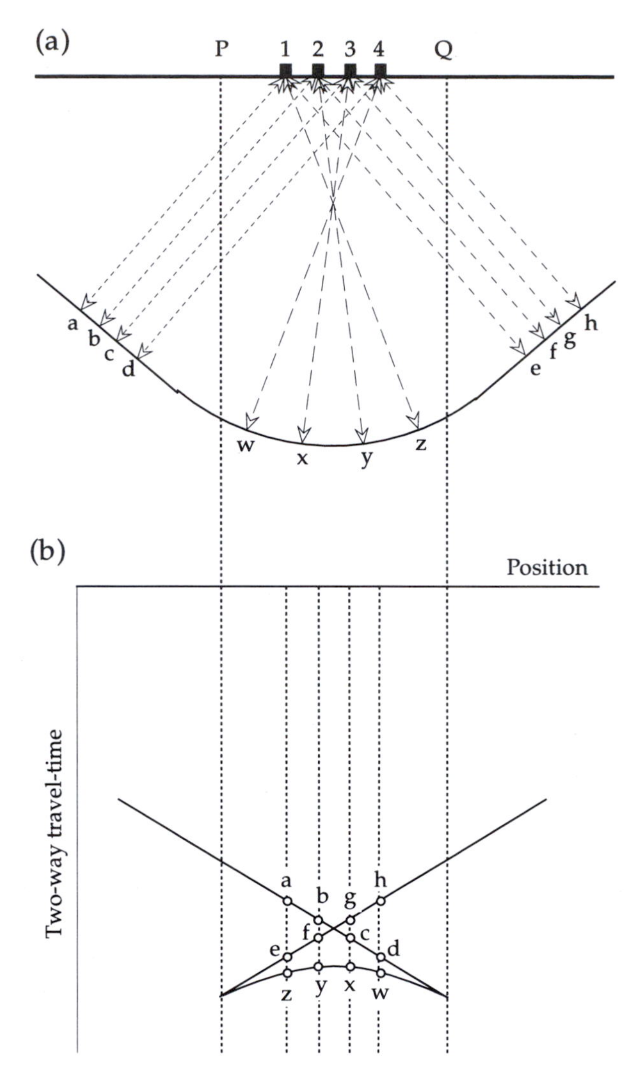

Fig. 3.63 (a) Paths of rays reflected from both flanks and the trough of a tightly curved syncline. (b) Appearance of the corresponding reflection record; the letters on the cusped feature refer to the reflection points in (a).

enough, the first arrivals from the dipping limbs can conceal the true structure (Fig. 3.62).

This happens when the radius of curvature of the bottom of the syncline is less than the subsurface depth of its axis. Over the axis of the syncline, rays reflected from the dipping flanks may be the first to reach the shot-point geophone. The bottom of the syncline is seen as an upwardly convex reflection between two cusps (Fig. 3.63). On an uncorrected reflection record, the appearance of a tight syncline resembles a diffraction.

Reflection seismic records must be corrected for non-vertical reflections. The correctional process is called *migration*. It is an essential part of a reflection seismic study. When the reflection events on seismic cross-sections are plotted vertically below control points on the surface (e.g., as the two-way vertical travel-time to a reflector below the shot-point), the section is said to be *unmigrated*. As discussed above, an unmigrated section misrepresents the depth and dip of inclined reflectors. A *migrated* section is one which has been corrected for non-vertical reflections. It gives a truer picture of the positions of subsurface reflectors.

The process of migration is complex, and requires prior knowledge of the seismic velocity distribution, which in an unexplored or tectonically complicated region is often inadequately known. Several techniques, mostly computer-based, can be used but to treat them adequately is beyond the scope of this text.

3.6.5.3 *Reflection and transmission coefficients*

The partitioning of energy between refractions and reflections at different angles of incidence on a boundary is rather complex. For example, an incident P-wave may be partially reflected and partially refracted, or it may be totally reflected, depending how steeply the incident ray encounters the boundary. The fraction of the incident P-wave energy that is partitioned between reflected and refracted P- and S-waves depends strongly on the angle of incidence (Fig. 3.64). In the case of oblique incidence at less than the critical angle, the amplitudes of the different waves are given by complicated functions of the wave velocities and the angles of incidence, reflection and refraction. The relative amounts of energy in the refracted and reflected P- and S-waves do not change much for angles of incidence up to about 15°. Beyond the critical angle, the refracted P-wave ceases, so that the incident energy is partly reflected as a P-wave and partly converted to refracted and reflected S-waves.

In practice, reflection seismology is carried out at

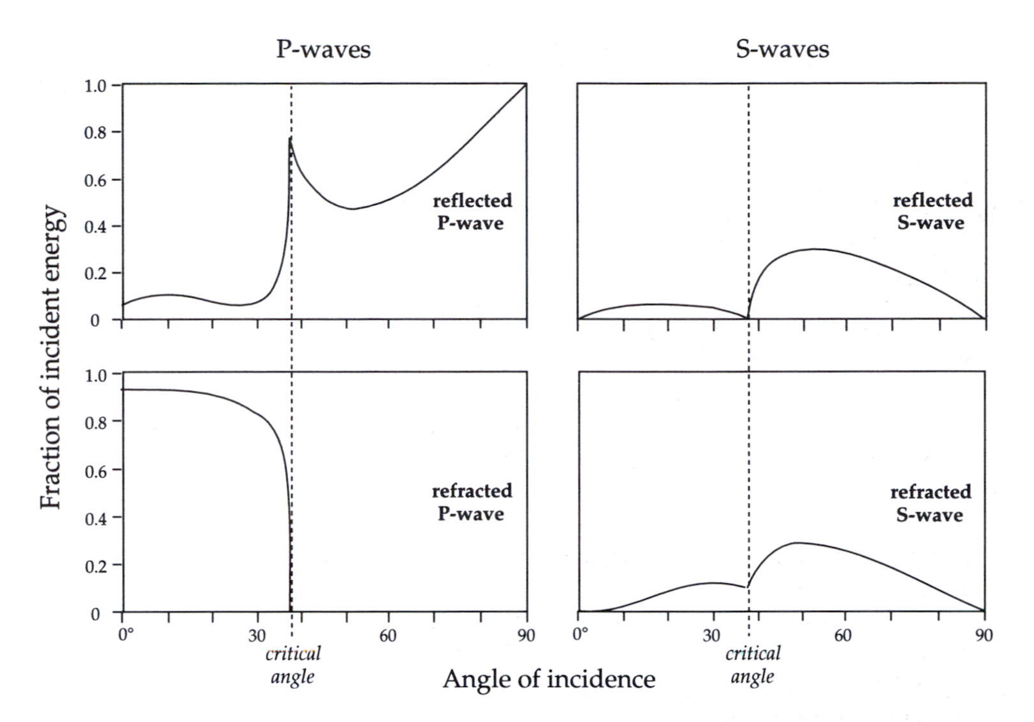

Fig. 3.64 Partitioning of the energy of an incident P-wave between refracted and reflected P- and S-waves (after Dobrin, 1976, and Richards, 1961).

comparatively small angles of incidence. At *normal incidence* on an interface a P-wave excites no tangential stresses or displacements, and no shear waves are induced. The partitioning of energy between the reflected and refracted P-waves then becomes much simpler. It depends on a property of each medium known as its *acoustic impedance*, Z, which is defined as the product of the density ρ of the medium and its P-wave velocity α; thus $Z=\rho\alpha$. The solution of the equations for the amplitudes A_1 and A_2 of the reflected and refracted P-waves, respectively, in terms of the amplitude A_0 of the incident wave are given by:

$$RC=\frac{A_1}{A_0}=\frac{Z_2-Z_1}{Z_2+Z_1}=\frac{\rho_2\alpha_2-\rho_1\alpha_1}{\rho_2\alpha_2+\rho_1\alpha_1}$$

$$TC=\frac{A_2}{A_0}=\frac{2Z_1}{Z_2+Z_1}=\frac{2\rho_1\alpha_1}{\rho_2\alpha_2+\rho_1\alpha_1} \quad (3.127)$$

The amplitude ratios RC and TC are called the reflection coefficient and the transmission coefficient, respectively. As shown earlier (see Eq. (3.75)), the energy of a wave is proportional to the square of its amplitude. The fractions E_R and E_T of the incident energy that are reflected and transmitted are given by the squares of RC and TC.

When the incident wave is reflected at the surface of a medium with higher seismic impedance ($Z_2>Z_1$), the reflection coefficient RC is positive. This means that the reflected wave is in phase with the incident wave. However, if the wave is incident on a medium with lower seismic impedance ($Z_2<Z_1$), the reflection coefficient will be negative. This implies that the reflected wave is 180° out of phase with the incident wave. The fraction of energy reflected from an interface is equal to RC^2, and therefore does not depend on whether the incidence is from the medium of higher or lower seismic impedance.

3.6.5.4 *Synthetic seismograms*

The travel-time of a reflection from a deep boundary in a multi-layered Earth is determined by the thicknesses and seismic velocities of the layers the seismic ray traverses. The amplitude of the recorded reflection is determined by the transmission and reflection coefficients for the subsurface interfaces. If the densities and seismic velocities of subsurface layers are known (for example, from sonic and density logs in conveniently located boreholes), it is possible to reconstruct what the seismogram should look like. Sometimes the density variations are ignored, and reflection and transmission coefficients are calculated simply on the basis of seismic velocities. This approximation can often be very useful, for example in the exploration of the deep structure of the Earth's crust with the seismic reflection method. The densities of deep layers are inaccessible directly, although they can be inferred from seismic velocities. A vertical model of seismic velocities may be available from a related refraction study. These data can be incorporated in a deep seismic reflection study to calculate a *synthetic seismogram*. The comparison of actual and synthetic seismograms is useful for correlating reflection events and for separating real reflections from noise signals such as multiples.

The principle is simple, but the construction is laborious.

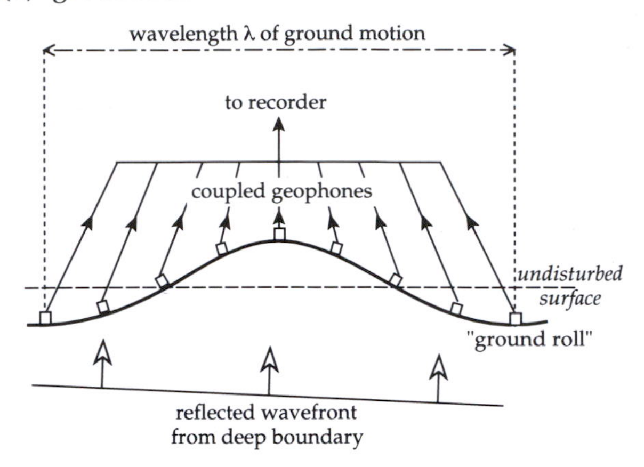

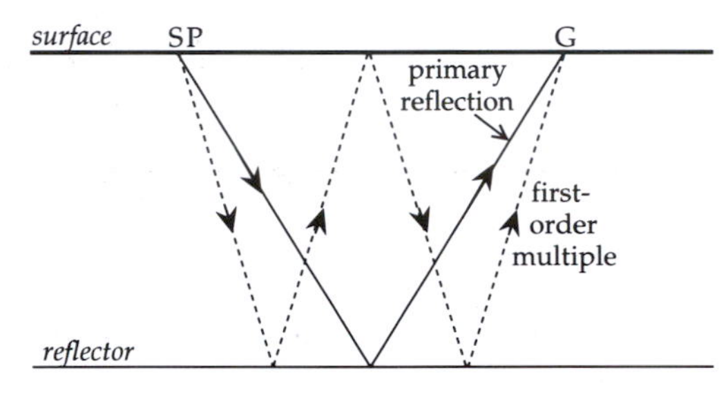

Fig. 3.65 Examples of seismic noise: (a) 'ground roll' due to surface wave, and (b) multiple reflections between a reflector and the free surface.

A vertically incident wave on the first boundary is resolved into reflected and transmitted components, with amplitudes corresponding to the seismic impedances above and below the interface. The transmitted wave is further subdivided at the next deeper interface into other reflected and transmitted components, and this is repeated at each subsequent boundary. Each wave is followed as it is reflected and refracted at subsurface interfaces until it eventually returns to the surface. The theoretical record consists of the superposition of the numerous events, and represents the total travel-time and amplitude of each event. Whether in a high-resolution reflection study of near-surface sedimentary layering for petroleum exploration or in an analysis of deep crustal structure, the construction of a synthetic seismogram is an exacting chore that requires the use of fast modern computers.

3.6.5.5 *Seismic noise*

Controlled-source seismology allows fine resolution of a layered underground through analysis of the seismic travel-times. However, the seismic record contains not only primary reflections, or *signals*, from subsurface interfaces but also spurious secondary events, or *noise*, that interfere with the desired signals. The ratio of the energy in the signal to that in the noise, called the *signal-to-noise ratio*, is a measure of the quality of a seismic record. The higher the signal-to-noise ratio the better the record; records that have a ratio less than unity are unlikely to be usable.

There are many ways in which seismic noise can be excited. They can be divided into *incoherent* (or *random*) *noise* and *coherent noise*. Incoherent noise is local in origin and is caused by shallow inhomogeneities such as boulders, roots or other non-uniformity that can scatter seismic waves locally. As a result of its local nature incoherent noise is different on the records from adjacent geophones unless they are very close. In reflection seismology it is reduced by arranging the geophones in groups or *arrays*, of typically 16 geophones per group, and combining the individual outputs to produce a single record. When n geophones form a group, this practice enhances the signal-to-noise ratio by the factor $\sqrt{n}$.

Coherent noise is present on several adjacent traces of a seismogram. Two common forms result from surface waves and multiple reflections, respectively.

A near-surface explosion excites surface waves (especially Rayleigh waves) that can have strong amplitudes. They travel more slowly than P-waves but often reach the geophones together with the train of subsurface reflections. The resultant 'ground roll' can obscure the reflections, particularly if they are weak. The problem can be minimized by the geometry of the geophone layout (Fig. 3.65a). For example, if the 16 geophones in a group are laid out at equal distances to cover a complete wavelength of the Rayleigh wave, the signals of individual geophones are effectively integrated to a low value. This procedure is only partially effective. Although most of the ground roll is related to Rayleigh waves, part is thought to have more complex near-surface origins. Another method of reducing the effects of ground roll is frequency filtering. The frequency of the ground roll is often lower than that of the reflected P-waves, which allows attenuation of this type of coherent noise by including a high-pass filter in the geophone circuit or in the subsequent processing to cut out low frequencies.

Multiple reflections are a very common source of coherent noise in a layered medium. They can originate in several ways, the most serious of which is the surface multiple (Fig. 3.65b). The reflection coefficient at the free surface of the Earth is high (in principle, $RC \approx -1$), and multiple reflections can occur between the surface and a reflecting interface. The travel-time for a first-order multiple at near-vertical incidence is double that of the primary signal. A copy of the reflector is observed on the seismogram at twice the real depth (or travel-time). Higher-order multiples produce additional apparent reflectors. A further advantage of using the common-depth-point method of reflection profiling is that it is effective in attenuating the surface multiples.

3.6.6 **Refraction seismology**

The method of seismic refraction can be understood by applying Huygens' principle to the critical refraction at the interface between two layers. The seismic disturbance travels immediately below the interface with the higher velocity of the lower medium. It is called a *head wave* (or *Mintrop wave*, after the German seismologist who patented its use in seismic exploration in 1919). The upper and lower media are in contact at the interface and so the upper medium is forced to move in phase with the lower medium. The vibration excited at the boundary by the passage of the head wave acts as a moving source of secondary waves in the upper layer. The secondary waves interfere constructively (in the same way as a reflected wave is formed) to build plane wavefronts; the ray paths return to the surface at the critical angle within the region of supercritical reflections (Fig. 3.66). The doubly refracted waves are especially important for the information they reveal about the layered structure of the deep interior of the Earth.

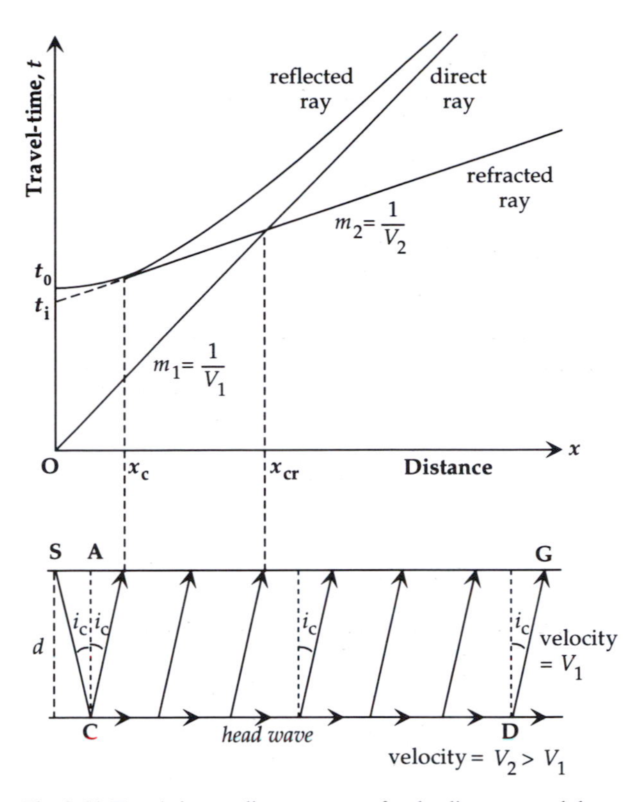

Fig. 3.66 Travel-time vs distance curves for the direct ray and the reflected and refracted rays at a horizontal interface between two layers with seismic velocities V_1 and V_2 ($V_2 > V_1$).

3.6.6.1 *Refraction at a horizontal interface*

The method of refraction seismology is illustrated for the case of the flat interface between two horizontal layers in Fig. 3.66. Let the depth to the interface be d and the seismic velocities of the upper and lower layers be V_1 and V_2 respectively ($V_1 < V_2$). The *direct ray* from the shot-point at S is recorded by a geophone G at distance x on the surface after time x/V_1. The travel-time curve for the direct ray is a straight line through the origin with slope $m_1 = 1/V_1$. The hyperbolic $t-x$ curve for the *reflected ray* intersects the time axis at the two-way vertical reflection ('echo') time t_0. At great distances from the shot-point the reflection hyperbola is asymptotic to the straight line for the direct ray.

The *doubly refracted ray* travels along the path SC with the velocity V_1 of the upper layer, impinges with critical angle i_c on the interface at C, passes along the segment CD with velocity V_2 of the lower layer, and returns to the surface along DG with velocity V_1. The segments SC and DG are equal, CD$=x-2$SA and the travel-time for the path SCDG can be written

$$t = \frac{2\mathrm{SC}}{V_1} + \frac{\mathrm{CD}}{V_2} \tag{3.128}$$

$$\text{i.e., } t = \frac{2d}{V_1 \cos i_c} + \frac{(x - 2d \tan i_c)}{V_2} \tag{3.129}$$

Rearranging terms and using Snell's law, $\sin i_c = V_1/V_2$, we get for the travel-time of the doubly refracted ray

$$t = \frac{x}{V_2} + \frac{2d}{V_1}\cos i_c \tag{3.130}$$

The equation represents a straight line with slope $m_2 = 1/V_2$. The doubly refracted rays are only recorded at distances greater than the critical distance x_c. The first arrival recorded at x_c can be regarded as both a doubly refracted ray and a reflection; the travel-time line for the head wave is tangential to the reflection hyperbola at x_c. By backward extrapolation, the refraction $t-x$ curve is found to intersect the time axis at the *intercept time* t_i, given by

$$t_i = \frac{2d}{V_1}\cos i_c = 2d\frac{\sqrt{V_2^2 - V_1^2}}{V_1 V_2} \tag{3.131}$$

Close to the shot-point the direct ray is the first to be recorded. However, the doubly refracted ray travels part of its path at the faster velocity of the lower layer, so that it eventually overtakes the direct ray and becomes the first arrival. The straight lines for the direct and doubly refracted rays cross each other at this distance, which is accordingly called the *crossover distance*, x_{cr}. It is computed by equating the travel-times for the direct and refracted rays:

$$\frac{x}{V_1}=\frac{x}{V_2}+2d\frac{\sqrt{V_2^2-V_1^2}}{V_1V_2} \tag{3.132}$$

$$x_{cr}=2d\sqrt{\frac{V_2+V_1}{V_2-V_1}} \tag{3.133}$$

Refraction seismology gives the velocities of subsurface layers directly from the reciprocal slopes of the straight lines corresponding to the direct and doubly refracted rays. Once these velocities have been determined it is possible to compute the depth d to the interface by using either the intercept time t_i or the crossover distance x_{cr}, which can be read directly from the $t-x$ plot:

$$d=\frac{1}{2}t_i\frac{V_1V_2}{\sqrt{V_2^2-V_1^2}} \tag{3.134a}$$

$$d=\frac{1}{2}x_{cr}\sqrt{\frac{V_2-V_1}{V_2+V_1}} \tag{3.134b}$$

3.6.6.2 *Refraction at an inclined interface*

In practice, the refracting interface is often not horizontal. The assumption of flat layers then leads to errors in the velocity and depth estimates. When the refractor is suspected to have a dip, the velocities of the beds and the dip of the interface can be obtained by shooting a second complementary profile in the opposite direction. Suppose a refractor dips at an angle θ as in Fig. 3.67. Shot-points A and B are located at the ends of a geophone layout that covers AB. The ray ACDB from the shot-point A strikes the interface at the critical angle i_c at C, runs as a head wave with velocity V_2 along the dipping interface, and the ray emerging at D eventually reaches a geophone at the end of the profile at B. During reverse shooting the ray from the shot-point at B to a geophone at A traverses the same path in the reverse direction. However, the $t-x$ curves are different for the up-dip and down-dip shots. Let d_A and d_B be the perpendicular distances from the shot-points A and B to the interface at P and Q, respectively. For the down-dip shot at A the travel-time to distance x is given by

$$t_d=\frac{AC+DB}{V_1}+\frac{CD}{V_2} \tag{3.135}$$

and, since

$$AC=\frac{d_A}{\cos i_c},\ DB=\frac{d_B}{\cos i_c}$$

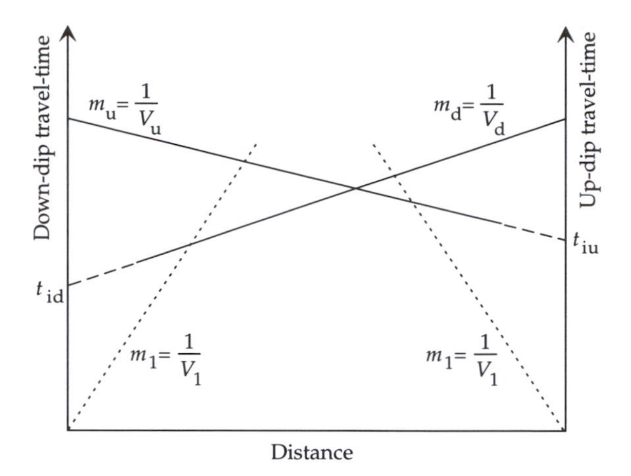

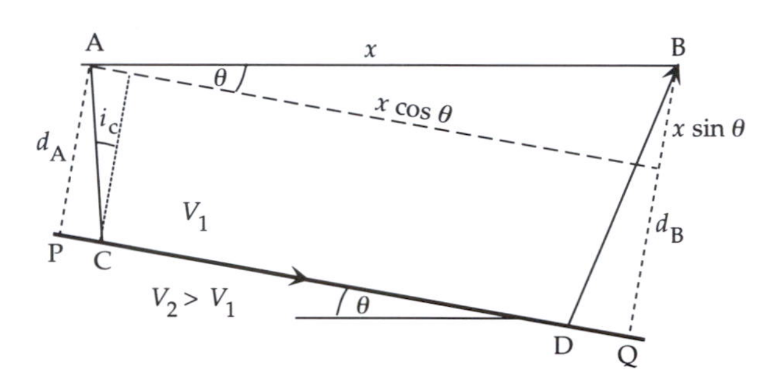

Fig. 3.67 Travel-time vs distance curves of direct and refracted rays for up-dip and down-dip profiles when the refracting boundary dips at angle θ.

$$PC=d_A\tan i_c,\ DQ=d_B\tan i_c,$$

$$CD=x\cos\theta-(PC+DQ)$$

we have

$$t_d=\frac{(d_A+d_B)}{V_1\cos i_c}+\frac{x\cos\theta-(d_A+d_B)\tan i_c}{V_2}$$

$$=\frac{x\cos\theta}{V_2}+\frac{(d_A+d_B)}{V_1}\cos i_c \tag{3.136}$$

Equation (3.136) can be simplified by noting that

$$d_B=d_A+x\sin\theta,\text{ and }\frac{1}{V_2}=\frac{\sin i_c}{V_1}$$

After substitution and gathering terms the down-dip travel-time is given by

$$t_d=\frac{x\sin i_c\cos\theta}{V_1}+\frac{x\cos i_c\sin\theta}{V_1}+\frac{2d_A\cos i_c}{V_1}$$

$$=\frac{x}{V_1}\sin(i_c+\theta)+t_{id} \tag{3.137}$$

where t_{id} is the intercept time for the down-dip shot:

$$t_{id} = \frac{2d_A}{V_1} \cos i_c$$

The analysis for shooting in the up-dip direction is analogous and gives

$$t_u = \frac{x}{V_1} \sin(i_c - \theta) + t_{iu} \tag{3.138}$$

where t_{iu} is the intercept time for the up-dip shot:

$$t_{iu} = \frac{2d_B}{V_1} \cos i_c$$

If the upper layer is homogeneous, the segments for the direct ray will have equal slopes, the reciprocals of which give the velocity V_1 of the upper layer. The segments of the $t-x$ curves corresponding to the doubly refracted ray are different for up-dip and down-dip shooting. The total travel-times in either direction along ACDB must be equal, but the $t-x$ curves have different intercept times. As these are proportional to the perpendicular distances to the refractor below the shot points, the up-dip intercept time t_{iu} is larger than the down-dip intercept time t_{id}. This means that the slope of the up-dip refraction in Fig. 3.67 is flatter than the down-dip slope. If we interpret the reciprocal of the slope as the velocity of the lower medium, we get two *apparent velocities*, V_d and V_u, given by

$$\frac{1}{V_d} = \frac{1}{V_1} \sin(i_c + \theta), \quad \frac{1}{V_u} = \frac{1}{V_1} \sin(i_c - \theta) \tag{3.139}$$

Once the real velocity V_1 and the apparent velocities V_d and V_u have been determined from the $t-x$ curves, the dip of the interface θ and the critical angle i_c (and from it the true velocity V_2 of the lower layer) can be computed:

$$\theta = \frac{1}{2}\left\{\sin^{-1}\left(\frac{V_1}{V_d}\right) + \sin^{-1}\left(\frac{V_1}{V_u}\right)\right\} \tag{3.140}$$

$$i_c = \frac{1}{2}\left\{\sin^{-1}\left(\frac{V_1}{V_d}\right) - \sin^{-1}\left(\frac{V_1}{V_u}\right)\right\} \tag{3.141}$$

If the reciprocal apparent velocities (Eq. (3.139)) are added, a simple approximation for the true velocity of the lower layer is obtained:

$$\frac{1}{V_d} + \frac{1}{V_u} = \frac{1}{V_1}\{\sin(i_c - \theta) + \sin(i_c - \theta)\}$$

$$= \frac{2}{V_1} \sin i_c \cos\theta$$

$$= \frac{2}{V_2} \cos\theta \tag{3.142}$$

If the refractor dip is small, $\cos\theta \approx 1$ (for example, if $\theta < 15°$, $\cos\theta > 0.96$) and an approximate formula for the true velocity of the second layer is

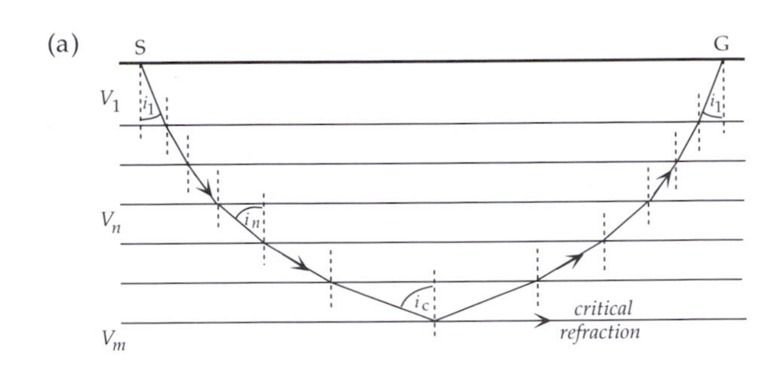

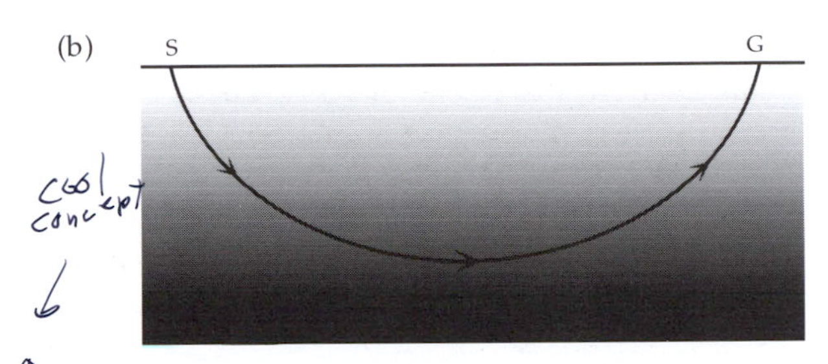

Fig. 3.68 (a) The path of a seismic wave through a horizontally layered medium, in which the seismic velocity is constant in each layer and increases with increasing depth, becomes ever flatter until critical refraction is reached; the return path of each emerging ray mirrors the incident path. (b) When the velocity increases continuously with depth, the ray is a smooth curve that is concave upward.

$$\frac{1}{V_2} \approx \frac{1}{2}\left(\frac{1}{V_d} + \frac{1}{V_u}\right) \tag{3.143}$$

3.6.6.3 *Refraction with continuous change of velocity with depth*

Imagine the Earth to have a multi-layered structure with numerous thin horizontal layers, each characterized by a constant seismic velocity, which increases progressively with increasing depth (Fig. 3.68). A seismic ray that leaves the surface with angle i_1 will be refracted at each interface until it is finally refracted critically. The ray that finally returns to the surface will have an emergence angle equal to i_1. Snell's law applies to each successive refraction (e.g., at the top surface of the nth layer, which has velocity V_n)

$$\frac{\sin i_1}{V_1} = \frac{\sin i_2}{V_2} = \ldots = \frac{\sin i_n}{V_n} = \text{constant} = p \tag{3.144}$$

The constant p is called the *ray parameter*. It is characteristic for a particular ray with emergence angle i_1 and velocity V_1 in the surface layer. If V_m is the velocity of the deepest layer, along whose surface the ray is eventually critically refracted ($\sin i_m = 1$), then the value of p must be equal to $1/V_m$.

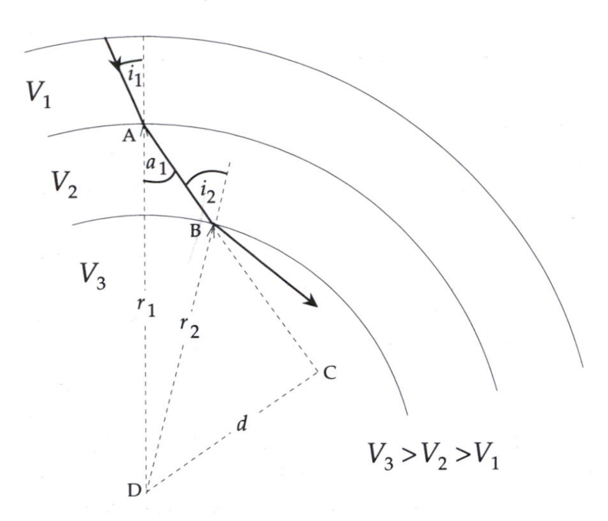

Fig. 3.69 Refraction of a seismic ray in a spherically layered Earth, in which the seismic velocity is constant in each layer and the layer velocity increases with depth.

As the number of layers increases and the thickness of each layer decreases, the situation is approached in which the velocity increases constantly with increasing depth. Each ray then has a smoothly curved path. If the vertical increase of velocity is linear with depth, the curved rays are circular arcs.

In the above we have assumed that the refracting interfaces are horizontal. This type of analysis is common in seismic prospecting, where only local structures and comparatively shallow depths are evaluated. The passage of seismic body waves through a layered spherical Earth can be treated to a first approximation in the same way. We can represent the vertical (radial) velocity structure by subdividing the Earth into concentric shells, each with a faster body-wave velocity than the shell above it (Fig. 3.69). Snell's law of refraction applies to the interface between each pair of shells. For example, at point A we can write

$$\frac{\sin i_1}{V_1}=\frac{\sin a_1}{V_2} \tag{3.145}$$

Multiplying both sides by r_1 gives

$$\frac{r_1 \sin i_1}{V_1}=\frac{r_1 \sin a_1}{V_2} \tag{3.146}$$

In triangles ACD and BCD, respectively, we have

$$d=r_1 \sin a_1=r_2 \sin i_2 \tag{3.147}$$

Combining Eqs. (3.145), (3.146) and (3.147) gives the result

$$\frac{r_1 \sin i_1}{V_1}=\frac{r_2 \sin i_2}{V_2}=\ldots=\frac{r_n \sin i_n}{V_n}=\text{constant}=p \tag{3.148}$$

The constant p is again called the *ray parameter*, although it has different dimensions than in Eq. (3.144) for flat horizontal layers. Here, the seismic ray is a straight line within each spherical layer with constant velocity. If velocity increases continuously with depth, the seismic ray is refracted continuously and its shape is curved concavely upward. It reaches its deepest point when $\sin i=1$, at radius r_0 where the velocity is V_0; these parameters are related by the *Benndorf relationship*:

$$\frac{r \sin i}{V}=\frac{r_0}{V_0}=p \tag{3.149}$$

Determination of the ray parameter is the key to determining the variation of seismic velocity inside the Earth. Access to the Earth's interior is provided by analysis of the travel-times of seismic waves that have traversed the various internal regions and emerge at the surface, where they are recorded. We will see in §3.7.3.1 that the travel-time (t) of a seismic ray to a known epicentral distance (Δ) can be mathematically inverted to give the velocity V_0 at the deepest point of the path. The theory applies for P- and S-waves, the general velocity V being replaced by the appropriate velocity α or β, respectively.

3.7 INTERNAL STRUCTURE OF THE EARTH

3.7.1 Introduction

It is well known that the Earth has a molten core. What is now general knowledge was slow to develop. In order to explain the existence of volcanoes some 19th century scientists postulated that the Earth must consist of a rigid outer crust around a molten interior. It was also known in the last century that the mean density of the Earth is about 5.5 times that of water. This is much larger than the known specific density of surface rocks, which is about 2.5–3. From this it was inferred that density increased towards the Earth's center under the effect of gravitational pressure. The density at the Earth's center was estimated to be comparatively high, greater than 7000 kg m^{-3} and probably in the range 10,000–12,000 kg m^{-3}. It was known that some meteorites had a rock-like composition, while others were much denser, composed largely of iron. In 1897 E. Wiechert, who subsequently became a renowned German seismologist, suggested that the interior of the Earth might consist of a dense metallic core, cloaked in a rocky outer cover. He called this cloak the 'Mantel', which later became anglicized to mantle.

The key to modern understanding of the interior of the Earth – its density, pressure and elasticity – was provided by the invention of the Milne seismograph. The progressive refinement of this instrument and its systematic employment world-wide led to the rapid development of the modern science of seismology. Important results were obtained early in the 20th century. The Earth's fluid core was first detected seismologically in 1906 by R. D. Oldham. He observed that, if the travel-times of P-waves observed at epicentral distances of less than 100° were extrapolated to greater distances, the expected travel-times were less than those observed. This meant that the P-waves arriving at large epicentral distances were delayed in their passage through the Earth. Oldham inferred from this the existence of a central core in which the P-wave velocity was reduced. He predicted that there would be a region of epicentral distances (a 'shadow zone') in which P-waves could not arrive. About this time it was found that P- and S-waves passed through the mantle but that no S-waves arrived beyond an epicentral distance of 105°. In 1914 B. Gutenberg verified the existence of a shadow zone for P-waves in the range of epicentral distances between 105° and 143°. Gutenberg also located the depth of the core–mantle boundary with impressive accuracy at about 2900 km. A modern estimate of the radius of the core is 3485±3 km, giving a mantle 2885 km thick. Gutenberg also predicted that P-waves and S-waves would be reflected from the core–mantle boundary. These waves, known today as PcP and ScS waves, were not observed until many years later. In honor of Gutenberg the core–mantle boundary is known today as the *Gutenberg seismic discontinuity*.

While studying the P-wave arrivals from an earthquake in Croatia in 1909 Andrija Mohorovičić found only a single arrival (P_g) at distances close to the epicenter. Beyond about 200 km there were two arrivals; the P_g event was overtaken by another arrival (P_n) which had evidently travelled at higher speed. Mohorovičić identified P_g as the direct wave from the earthquake and P_n as a doubly refracted wave (equivalent to a head wave) that travelled partly in the upper mantle. Mohorovičić calculated velocities of 5.6 km s^{-1} for P_g and 7.9 km s^{-1} for P_n and estimated that the sudden velocity increase occurred at a depth of 54 km. This seismic discontinuity is now called the *Mohorovičić discontinuity*, or *Moho* for short. It represents the boundary between the crust and mantle. The crustal thickness is known to be very variable. It averages about 33 km, but measures as little as 5 km under oceans and as much as 60–80 km under some mountain ranges.

The *seismological Moho* is commonly defined as the depth at which the P-wave velocity exceeds 7.6 km s^{-1}. This seismic definition is dependent on the density and elastic properties of crustal and mantle rocks, and need not correspond precisely to a change of rock type. An alternative definition of the Moho as the depth where the rock types change is called the *petrological Moho*. For most purposes the two definitions of the crust–mantle boundary are equivalent.

It is now known that the crust is not homogeneous but has a layered structure. In 1925 V. Conrad separated arrivals from a Tauern (Eastern Alps) earthquake of 1923 into P_g and S_g waves in an upper crustal layer and faster P* and S* waves that travelled with velocities 6.29 km s^{-1} and 3.57 km s^{-1}, respectively, in a deeper layer. Because the P* and S* velocities are significantly slower than corresponding upper mantle velocities, Conrad deduced the existence of a lower crustal layer. The interface separating the continental crust into an upper crustal layer and a lower crustal layer is called the *Conrad discontinuity*. Influenced by early petrological models of crustal composition and by comparison with seismic velocities in known materials, seismologists referred to the upper and lower crustal layers as the granitic layer and the basaltic layer, respectively. This petrological separation is now known to be overly simplistic. In contrast to the Moho, which is found everywhere, the Conrad discontinuity is poorly defined or absent in some areas.

The core shadow-zone and its interpretation in terms of a fluid core were well established in 1936 when Inge Lehmann, a Danish seismologist, reported weak P-wave arrivals within the shadow zone. She interpreted these in terms of an inner core with higher seismic velocity. However, the existence of the inner core remained controversial for many years. Improved seismometer design, digital signal treatment and the setting up of seismic arrays have provided corroborating evidence. The existence of a solid inner core is also supported by analyses of the Earth's natural vibrations.

The gross internal structure of the Earth is modelled as a set of concentric shells, corresponding to the inner core, outer core and mantle (Fig. 3.70). An important step in understanding this layered structure has been the development of travel-time curves for seismic rays that passed through the different shells. To facilitate identification of the arrivals of these rays on seismograms a convenient shorthand notation is used. A P- or S-wave that travels from an earthquake directly to a seismometer is labelled with the appropriate letter P or S; until the margin of the core shadow-zone, the P- and S-waves follow identical curved paths. (The curvature, as explained in §3.6.6.3, arises from the increase of seismic velocity with depth.) A wave that reaches the seismometer after being reflected once from the crust is labelled PP (or SS), as its path consists of two identical P- or S-segments.

The energy of an incident P- or S-wave is partitioned at an interface into reflected and refracted P- and S-waves (see

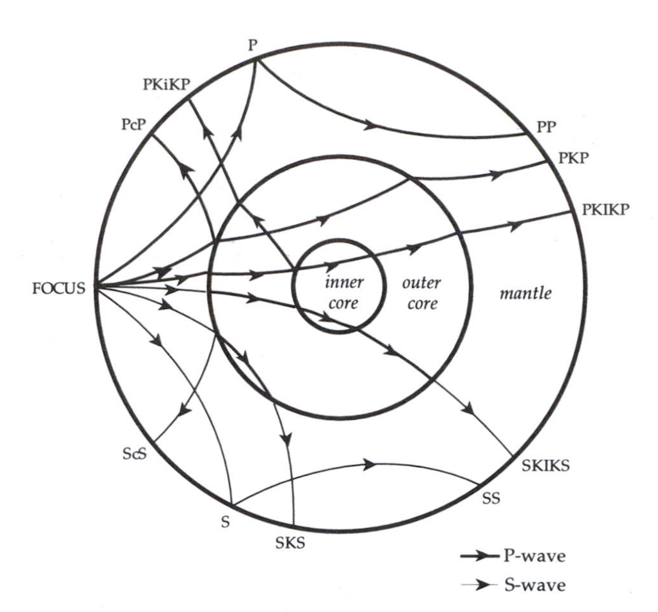

Fig. 3.70 Seismic wave paths of some important refracted and reflected P-wave and S-wave phases from an earthquake with focus at the Earth's surface.

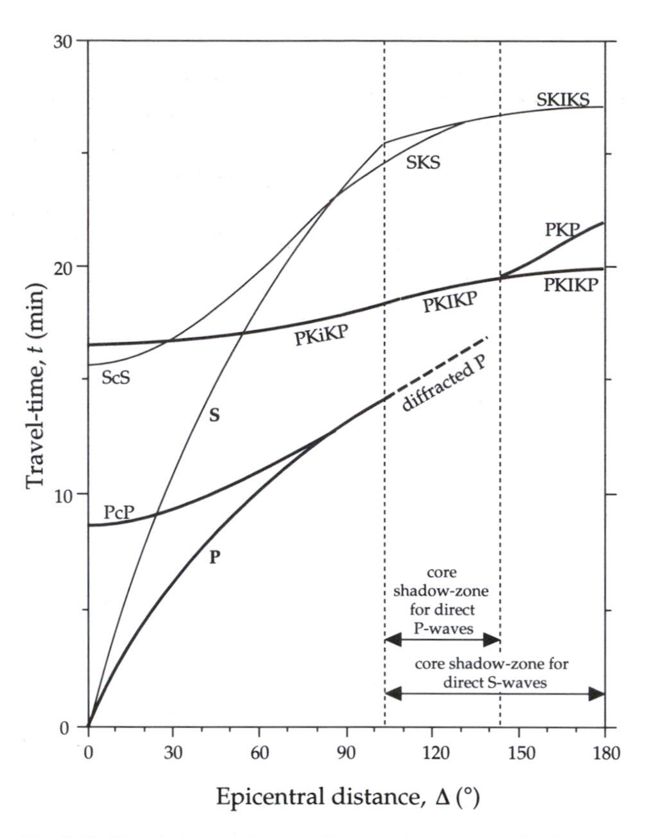

Fig. 3.71 Travel-time vs Δ curves for some important seismic phases (modified from Jeffreys and Bullen, 1940).

§3.6.4). A P-wave incident on the boundary between mantle and fluid outer core is refracted towards the normal to the interface, because the P-wave velocity drops from about 13 km s^{-1} to about 8 km s^{-1} at the boundary. After a second refraction it emerges beyond the shadow zone and is called a PKP wave (the letter K stands for Kern, the German word for core). An S-wave incident at the same point has a lower mantle velocity of about 7 km s^{-1}. Part of the incident energy is converted to a P-wave in the outer core, which has a higher velocity of 8 km s^{-1}. The refraction is away from the normal to the interface. After a further refraction the incident S-wave reaches the surface as an SKS phase. A P-wave that travels through mantle, fluid core and inner core is labelled PKIKP. Each of these rays is refracted at an internal interface. To indicate seismic phases that are reflected at the outer core boundary the letter c is used, giving rise, for example, to PcP and ScS phases (Fig. 3.70). Reflections from the inner core are designated with the letter i, as for example in the phase PKiKP.

Read 3.7.2 – 3.7.4

3.7.2 Refractions and reflections in the Earth's interior

If it possesses sufficient energy, a seismic disturbance may be refracted and reflected – or converted from a P-wave to an S-wave, or vice-versa – many times at the several seismic discontinuities within the Earth and at its free surface. As a result, the seismogram of a large earthquake contains numerous overlapping seismic signals and the identification of individual phases is a difficult task. Late-arriving phases that have been multiply reflected or that have travelled through several regions of the Earth's interior are difficult to resolve from the disturbance caused by earlier arrivals. In the period 1932–1939 H. Jeffreys and K. E. Bullen analyzed a large number of good records of earthquakes registered at a worldwide, though sparse, distribution of seismic stations. In 1940 they published a set of tables giving the travel-times of P- and S-waves through the Earth. A slightly different set of tables was reported by B. Gutenberg and C. F. Richter. The good agreement of the independent analyses confirmed the reliability of the results. The Jeffreys–Bullen seismological tables were used by the international seismic community as the standard of reference for many years.

The travel-time of a seismic wave to a given epicentral distance is affected by the focal depth of the earthquake, which may be as much as several hundred kilometers. The travel-time vs distance curves of some important phases are shown in Fig. 3.71 for an earthquake occurring at the Earth's surface. The figure assumes that the Earth is spherically symmetric, with the same vertical structure under-

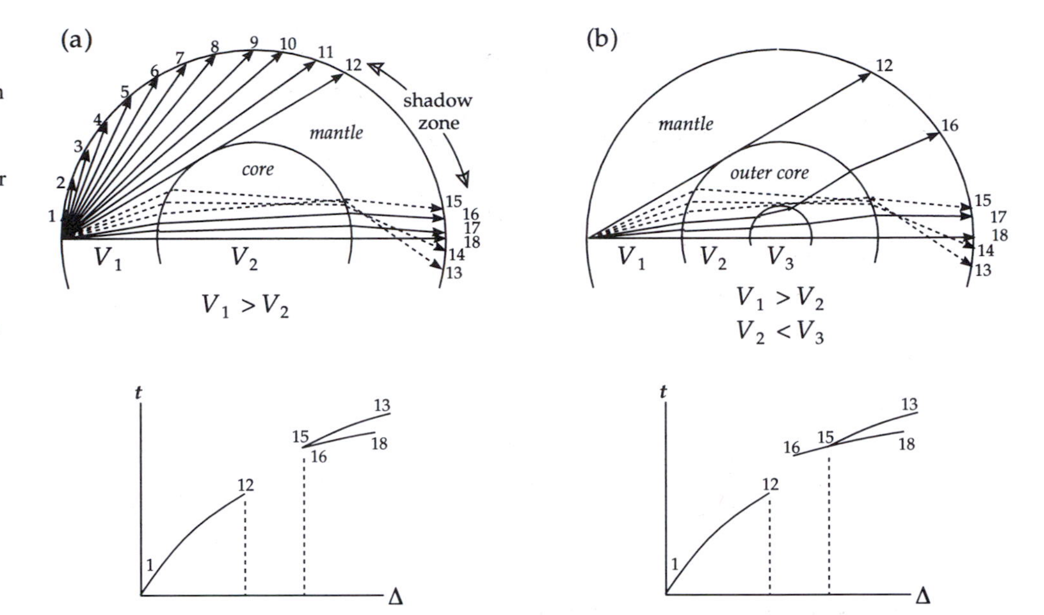

Fig. 3.72 Seismic wave paths and their t–Δ curves for P-waves passing through a spherical Earth with constant velocities in the mantle, outer core and inner core, respectively. (a) Development of a shadow zone when the mantle velocity (V_1) is higher than the outer core velocity (V_2). (b) Penetration of the shadow zone by rays refracted in an inner core with higher velocity than the outer core ($V_3 > V_2$).

neath each place on the surface. This assumption works fairly well, although it is not quite true. Lateral variations of seismic velocity have been found at many depths within the Earth. For example, there are lateral differences in seismic velocity between oceanic and continental crust, and between oceanic and continental lithosphere. At even greater depths significant lateral departures from the spherical model have been detected. These discrepancies form the basis of the branch of seismology called *seismic tomography*, which we will examine later.

3.7.2.1 *Seismic rays in a uniformly layered Earth*

It is important to understand clearly the relationship between the travel-time (t) vs epicentral distance (Δ) curves and the paths of seismic waves in the Earth, like those shown in Fig. 3.70. Consider first an Earth that consists of two concentric shells representing the mantle and core (Fig. 3.72a). The P-wave velocity in each shell is constant, and is faster in the mantle than in the core ($V_1 > V_2$). The figure shows the paths of 18 rays that leave a surface source at angular intervals of 5°. Rays 1–12 travel directly through the mantle as P-waves and emerge at progressively greater epicentral distances. The convex upwards shape of the t–Δ curve is here due to the curved outer surface, the layer velocity being constant. Ray 13 is partially refracted into the core, and partially reflected (not shown in the figure). Because $V_2 < V_1$ the refracted ray is bent towards the Earth's radius, which is normal to the refracting interface. This ray is further bent on leaving the core and reaches the Earth's surface as a PKP phase at an epicentral distance greater than 180°. Rays 14 and 15 impinge more directly on the core and are refracted less severely; their epicentral distances become successively smaller and their travel-times become shorter than ray 13, as indicated by branch 13–15 of the t–Δ curve. This branch is offset in time from the extrapolation of branch 1–12 because of the lower velocity in the core. The paths of rays 16, 17 and 18 (which is a straight line through mantle and core and emerges at an epicentral distance of 180°) become progressively longer, and a second branch 16–18 develops on the t–Δ curve. The two branches meet at a sharp point, or cusp. No P-waves reach the surface in the gap between rays 12 and 15 in this simple model. There is a *shadow zone* between the last P-wave that just touches the core and the PKP-wave with the smallest epicentral distance. The existence of a shadow zone for P-waves is evidence for a core with lower P-wave velocities than the mantle. S-waves in the mantle follow the same ray paths as the P-waves. However, no direct S-waves arrive in the shadow zone, which indicates that the core must be fluid.

Now suppose an inner core with a constant velocity V_3 that is higher than the velocity V_2 in the outer core (Fig. 3.72b). The paths of rays 1–15 are the same as before, through the mantle and outer core. The segments 1–12 and 13–15 of the t–Δ curve are the same as previously. Ray 16 impinges on the inner core and is sharply refracted away from the Earth's radius; on returning to the outer core it is again refracted, back towards the radius. After further refraction at the core–mantle interface this ray emerges at the Earth's surface at a smaller epicentral distance than ray 15, within the P-wave shadow-zone, as a PKIKP event. Successive PKIKP rays are bent less strongly. The PKIKP

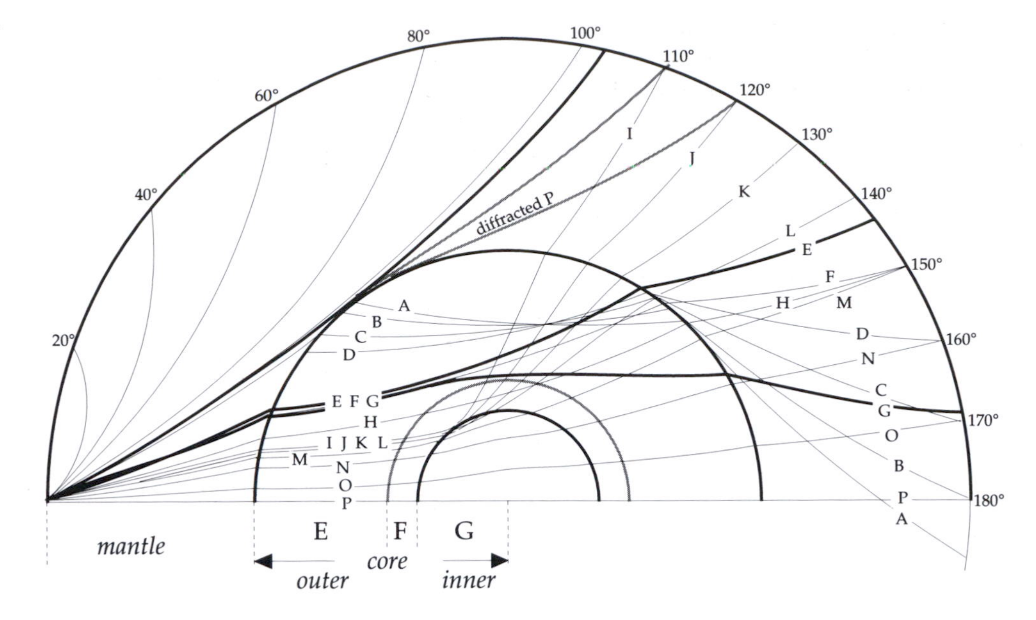

Fig. 3.73 The wave paths of some P, PKP, and PKIKP rays (after Gutenberg, 1959).

rays map out a new branch 16–18 of the t–Δ curve (see also Fig. 3.71).

3.7.2.2 *Travel-time curves for P-, PKP- and PKIKP-waves*

In general the velocities of P-waves and S-waves increase with depth. As described in §3.6.6.3 and illustrated in Fig. 3.68a, the ray paths are curved lines, concave towards the surface. However, the explanation of the paths of P, PKP, and PKIKP phases shown in Fig. 3.73 closely follows the preceding discussion. There is a shadow-zone for direct P-waves between about 103° and 143°, and no direct S-waves are found beyond 103°. The shallowest PKP ray (A in Fig. 3.73) is deviated the furthest, emerging at an epicentral distance greater than 180°. Successively deeper PKP rays (B–E) emerge at ever-smaller epicentral distances until about 143°, after which the epicentral distance increases again to almost 170° (rays F, G). It was long believed that the boundary between inner and outer core was a transitional region (called region F in standard Earth models) with higher P-wave velocity, and that PKP rays traversing this region would again emerge at smaller epicentral distances (ray H). The first rays penetrating the inner core are sharply refracted and emerge in the P-wave shadow-zone. The most strongly deviated (ray I) is observed at an epicentral distance of about 110°; deeper rays (J–P) arrive at ever-greater distances up to 180°. There are at least two branches of the t–Δ curve for $\Delta>143°$, corresponding to the PKP and PKIKP phases, respectively (Fig. 3.71). In fact, depending how the transitional region F is modelled, the t–Δ curve near 143° can have several branches.

The edges of the shadow zone defined by P and PKP phases are not sharp. One reason is the intrusion of PKIKP phases at the 143° edge. Another is the effect of diffraction of P-waves at the 103° edge (Fig. 3.73). The bending of plane waves at an edge into the shadow of an obstruction was described in §3.6.2.3, and explained with the aid of Huygen's principle. The diffraction of plane waves is called *Fraunhofer diffraction*. When their source is not at infinity, waves must be handled as spherical waves. Spherical wavefronts that pass an obstacle are also diffracted. The type of behavior is called *Fresnel diffraction*, and it is also explainable with Huygen's principle as the product of interference between the primary wavefront and secondary waves generated at the obstacle. Wave energy penetrates into the shadow of the obstacle, as though the wavefront were bent around the edge. In this way very deep P-waves are diffracted around the core and into the shadow zone. The intensity of the diffracted rays falls off with increasing angular distance from the diffracting edge, in this case the core–mantle boundary. Modern instrumentation enables detection of long-period diffracted P-waves to large epicentral distances (Figs. 3.71, 3.73). The velocity structure above the core–mantle boundary, in particular in the D"-layer (§3.7.5.3), has a strong influence on the ray paths, travel-times and waveforms of the diffracted waves.

3.7.3 **Radial variations of seismic velocities**

Models of the radial variations of physical parameters inside the Earth implicitly assume spherical symmetry. They are therefore 'average' models of the Earth that do not take into account lateral variations (e.g., of velocity or density) at the same depth. This is a necessary first step in approaching the true distributions as long as lateral variations are relatively small. This appears to be the case; although geophysically significant, the lateral variations in physical properties remain within a few percent of the average value at any depth.

There are two main ways to determine the distributions of body-wave velocities in a spherically symmetric Earth. They are referred to as forward and inverse modelling. Both methods have to employ the same sets of observations, which are the travel-times of different seismic phases to known epicentral distances. The forward technique starts with a known or assumed variation of seismic velocities and calculates the corresponding travel-times. The inversion method starts with the observed t–Δ curves and computes a model of the velocity distributions that could produce the curves. The inversion method is the older one, in use since the early part of this century, and forms an important branch of mathematical theory. Forward modelling is a more recent method that has been successfully employed since the advent of powerful computers.

3.7.3.1 *Inversion of travel-time vs distance curves*

In 1907 the German geophysicist E. Wiechert, building upon an evaluation of the Benndorf problem (Eq. (3.149)) by the mathematician G. Herglotz, developed an analytical method for computing the internal distributions of seismic velocities from observations made at the Earth's surface. The technique is called inversion of travel-times, and it is considered one of the classical methods of geophysics. The observational data consist of the t–Δ curves for important seismic phases (Fig. 3.71). The clues to deciphering the velocity distributions were the Benndorf relationship for the ray parameter p (Eq. (3.149)) and the recognition that the value of p can be obtained from the slope of the travel-time curve at the epicentral distance where the ray returns to the surface.

Consider two rays that leave an earthquake focus at infinitesimally different angles, reaching the surface at points P and P' at epicentral distances Δ and $\Delta+\mathrm{d}\Delta$, respectively (Fig. 3.74). The distance PP' is $R\mathrm{d}\Delta$ (where R is the Earth's radius) and the difference in arc-distances along the adjacent rays is equal to $V\mathrm{d}t$, where V is the velocity in the surface layer and $\mathrm{d}t$ is the difference in travel-times of the two rays. In the small triangle PP'Q the angle QPP' is equal to i, the angle of emergence (incidence). Therefore,

$$\sin i=\frac{V\mathrm{d}t}{R\mathrm{d}\Delta} \tag{3.150}$$

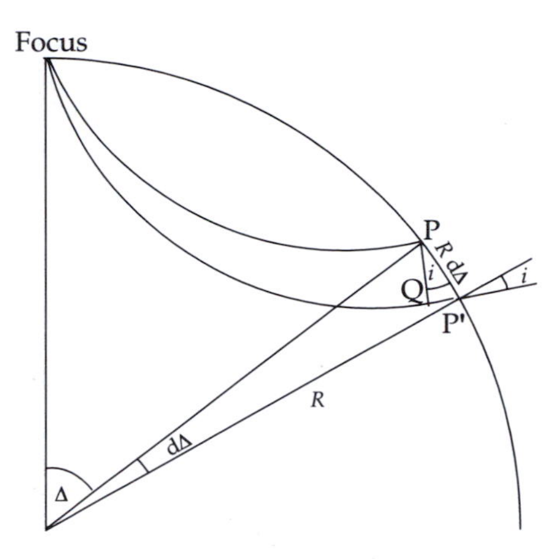

Fig. 3.74 Paths of two rays that leave an earthquake focus at infinitesimally different angles, reaching the surface at points P and P' at epicentral distances Δ and $\Delta+\mathrm{d}\Delta$, respectively.

$$\frac{R\sin i}{V}=p=\frac{\mathrm{d}t}{\mathrm{d}\Delta} \tag{3.151}$$

This means that the value of p for the ray emerging at epicentral distance Δ can be obtained by calculating the slope ($\mathrm{d}t/\mathrm{d}\Delta$) of the travel-time curve for that distance. This is an important step in finding the velocity V_0 at the deepest point of the ray, at radius r_0, because $V_0=r_0/p$. However, before we can find the velocity we need to know the value of r_0. The continuation of the Wiechert–Herglotz analysis is an intricate mathematical procedure, beyond the scope of this text, which fortunately results in a fairly simple formula:

$$\ln\frac{R}{r_0}=\frac{1}{\pi}\int_0^{\Delta_1}\cosh^{-1}\frac{p(\Delta)}{p(\Delta_1)}\mathrm{d}\Delta \tag{3.152}$$

where $p(\Delta)$ is the slope of the t–Δ curve at Δ, the epicentral distance of emergence, and $p(\Delta_1)$ is the slope at any intermediate epicentral distance Δ_1. Eq. (3.152) can be used to integrate numerically along the ray to give the value of r_0 for the ray, and $V_0=r_0/p$.

The Wiechert–Herglotz analysis is valid for regions of the Earth in which p varies monotonically with Δ. It cannot be used in the Earth's crust, because conditions are too inhomogeneous. Seismic velocity distributions in the crust are deduced empirically from long seismic refraction profiles. The Wiechert–Herglotz method cannot be used where a low-velocity zone is present, because the ray does not bottom in the zone. It works well for the Earth's mantle, but care must be taken where a seismic discontinuity is present. In the Earth's core the refraction of P-waves at the core–mantle boundary means that no PKP waves reach their deepest point in the outer layers of the core (Fig. 3.73). However, SKS-waves bottom in the outer core. Inversion of

SKS-wave travel-times complements the inversion of PKP-wave data to give the P-wave velocity distribution in the core.

3.7.3.2 *Forward modelling: polynomial parametrization*

The forward modelling method starts with a presupposed dependence of seismic velocity with depth. The method assumes that the variation of velocity can be expressed by a smooth polynomial function of radial distance within limited depth ranges. This procedure is called polynomial parametrization and in constructing models of the Earth's interior it is applied to the P-wave and S-wave velocities, the seismic attenuation, and the density.

The travel-times of P- and S-waves to any epicentral distance are calculated on the basis of the spherically symmetric, layered model. The computed travel-times are compared with the observed t–Δ curves, and the model is adjusted to account for differences. The procedure is repeated as often as necessary until an acceptable agreement between the computed and real travel-times is achieved. The method requires good travel-time data for many seismic phases and involves intensive computation.

In 1981 A. M. Dziewonski and D. L. Anderson constructed a *Preliminary Reference Earth Model* (acronym: PREM) in which the distributions of body-wave velocities in important layers of the Earth were represented by cubic or quadratic polynomials of normalized radial distance; in thin layers of the upper mantle linear relationships were used. A similar, revised parametrized velocity model (*iasp91*) was proposed by B. L. N. Kennett and E. R. Engdahl in 1991. The variations of P- and S-wave velocities with depth in the Earth according to the *iasp91* model are shown in Fig. 3.75.

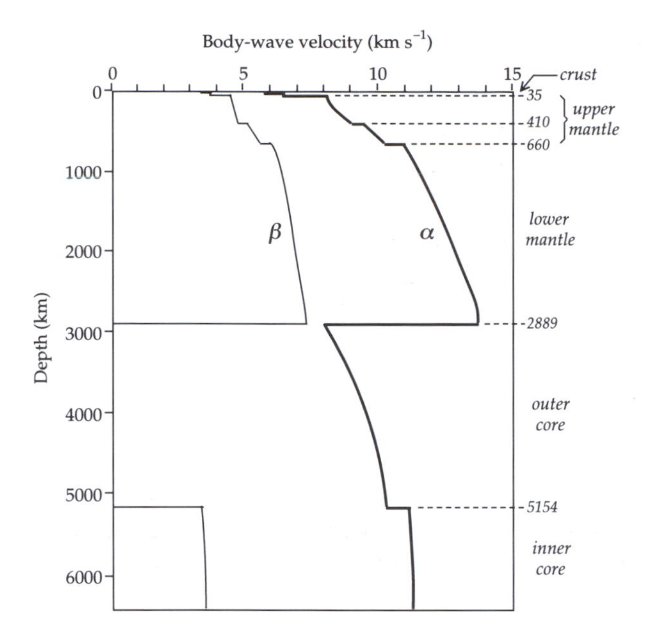

Fig. 3.75 The variations with depth of longitudinal- and shear-wave velocities, α and β, respectively, in the Earth's interior, according to the Earth model *iasp91* (data source: Kennett and Engdahl, 1991).

3.7.4 **Radial variations of density, gravity and pressure**

In order to determine density, gravity and pressure in the Earth's interior several simplifying assumptions must be made, which appear to be warranted by experience. The Earth is assumed to be spherically symmetric and composed of concentric homogeneous shells or layers (e.g., inner core, outer core, mantle, etc.). Possible effects of chemical and phase changes within a shell are not taken into account. Pressure and density are assumed to increase purely hydrostatically. If the distributions of seismic body-wave velocities α and β are known, an important seismic parameter Φ can be defined:

$$\Phi=\alpha^2-\tfrac{4}{3}\beta^2 \tag{3.153}$$

From Eq. (3.49) Φ is equal to K/ρ, where K is the bulk modulus and ρ the density.

3.7.4.1 *Density inside the Earth*

Consider a vertical prism between depths z and dz (Fig. 3.76). The hydrostatic pressure increases from p at depth z to $(p+\mathrm{d}p)$ at depth $(z+\mathrm{d}z)$ because of the extra pressure due to the extra material in the small prism of height dz. The pressure increase dp is equal to the weight w of the prism divided by the area A of the base of the prism, over which the weight is distributed.

$$\mathrm{d}p=\frac{w}{A}=\frac{(\text{volume}\times\rho)g}{A}=\frac{(A\mathrm{d}z\rho)g}{A}=\rho g\mathrm{d}z=-\rho g\mathrm{d}r \tag{3.154}$$

From the definition of bulk modulus, K (Eq. (3.16)) we can write

$$K=-V\frac{\mathrm{d}p}{\mathrm{d}V}=\rho\frac{\mathrm{d}p}{\mathrm{d}\rho} \tag{3.155}$$

Combining Eqs. (3.153), (3.154) and (3.155) gives

$$\frac{-\mathrm{d}\rho}{\rho g\mathrm{d}r}=\frac{\rho}{K}=\frac{1}{\Phi} \tag{3.156}$$

$$\frac{\mathrm{d}\rho}{\mathrm{d}r}=-\frac{\rho(r)g(r)}{\Phi(r)} \tag{3.157}$$

Eq. (3.157) is known as the Adams–Williamson equation. It was first applied to the estimation of density in the Earth in 1923 by E. D. Williamson and L. H. Adams. It yields the density gradient at radius r, when the quantities on the

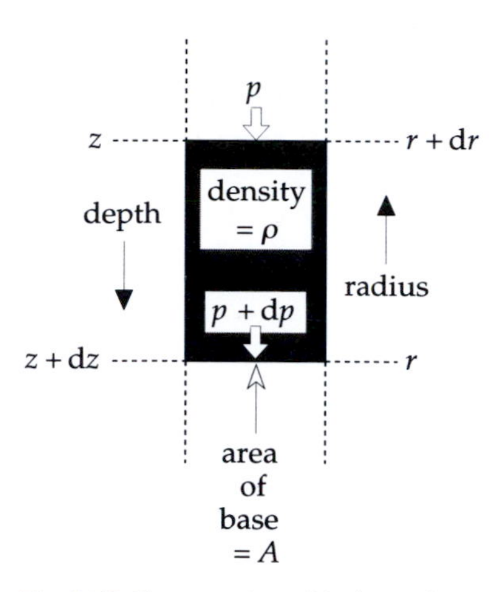

Fig. 3.76 Computation of hydrostatic pressure in the Earth, assuming that a change in pressure dp with depth increase dz is due only to the increase in weight of overlying material.

right-hand side are known. The seismic parameter Φ is known accurately, but the density ρ is unknown; it is in fact the goal of the calculation. The value of gravity g used in the equation must be computed separately for radius r. It is due only to the mass contained within the sphere of radius r, because external (homogeneous) shells of the Earth do not contribute to gravitation inside them. This mass is the total mass E of the Earth minus the cumulative mass of all spherical shells external to r.

The procedure requires that a starting value for ρ be assumed at a known depth. Analyses of crustal and upper mantle structure in isostatically balanced areas give estimates of upper mantle density around 3300 kg m^{-3}. Using this as a starting value at an initial radius r_1 the density gradient in the uppermost mantle can be calculated from Eq. (3.157). Linear extrapolation of this gradient to a chosen greater depth gives the density ρ at radius r_2; the corresponding new value of g can be calculated by subtracting the mass of the shell between the two depths; together with the value of $\Phi(r_2)$ the density gradient can be computed at r_2. This iterative type of calculation gives the variation of density with depth (or radius). Fine steps in the extrapolations give a smooth density distribution (Fig. 3.77). There are two important boundary conditions on the computed density distribution. Integrated over the Earth's radius it must give the correct total mass (E=5.974×10^{24} kg). It must also fulfil the relationship between Earth's moment of inertia (C), mass and radius (R) explained in §2.4.3: C=0.3308 ER^2.

Density changes abruptly at the major seismic discontinuities (Fig. 3.77), showing that it is affected principally by changes in composition. Density is found to be

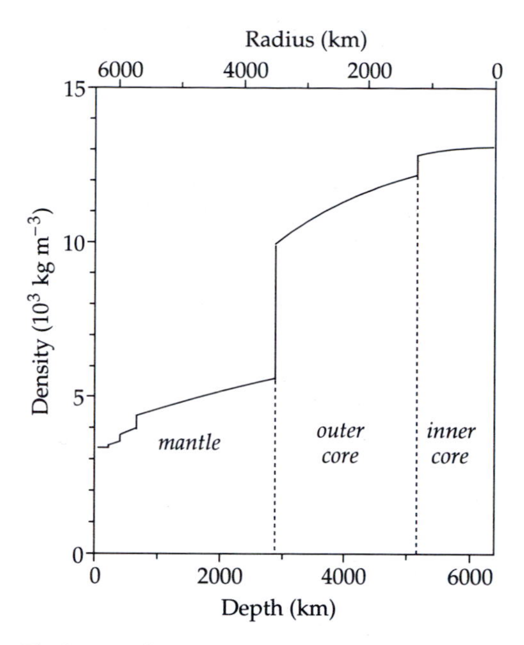

Fig. 3.77 Radial distribution of density within the Earth according to Earth model PREM (data source: Dziewonski, 1989).

around 4500–5500 kg m^{-3} in the mantle, 10,000–12,000 kg m^{-3} in the outer core and around 13,000 kg m^{-3} in the inner core. Between the major discontinuities density increases smoothly as a result of the increases in pressure and temperature.

3.7.4.2 *Gravity and pressure inside the Earth*

The radial variation of gravity can be computed from the density distribution. As stated above, the value of $g(r)$ is due only to the mass $m(r)$ contained within the sphere of radius r. Let the density at radius x ($\leq r$) be $\rho(x)$; the gravity at radius r is then given by

$$g(r) = -G\frac{m(r)}{r^2} = -\frac{G}{r^2}\int_0^r 4\pi x^2 \rho(x)\,dx \qquad (3.158)$$

A remarkable feature of the internal gravity (Fig. 3.78) is that it maintains a value close to 10 m s^{-2} throughout the mantle, rising from 9.8 m s^{-2} at the surface to 10.8 m s^{-2} at the core–mantle boundary. It then decreases almost linearly to zero at the Earth's center.

Hydrostatic pressure is due to the force (N) per unit area (m^2) exerted by overlying material. The SI unit of pressure is the pascal (1 Pa=1 N m^{-2}). In practice this is a small unit. The high pressures in the Earth are commonly quoted in units of gigapascal (1 GPa=10^9 Pa), or alternatively in kilobars or megabars (1 bar=10^5 Pa; 1 kbar=10^8 Pa; 1 Mbar=10^{11} Pa=100 GPa).

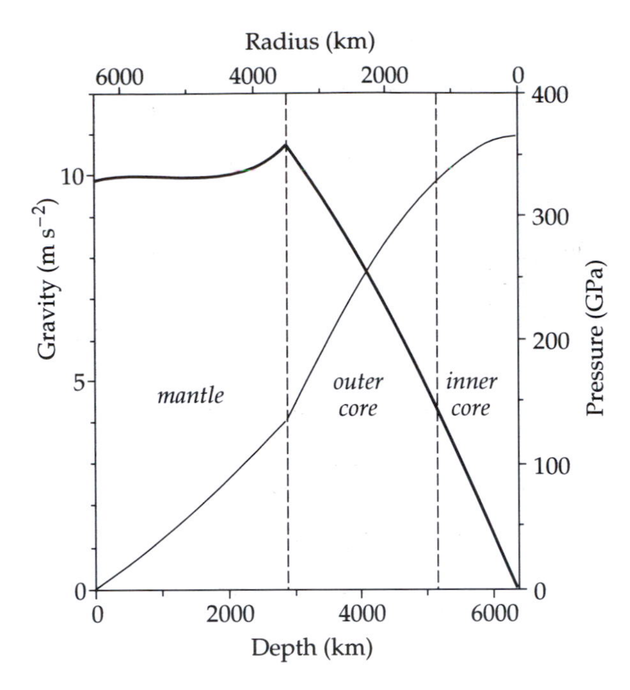

Fig. 3.78 Radial variations of internal gravity (*thick curve*) and pressure (*thin curve*) according to Earth model PREM (data source: Dziewonski, 1989).

Within the Earth the hydrostatic pressure $p(r)$ at radius r is due to the weight of the overlying Earth layers between r and the Earth's surface. It can be computed by integrating Eq. (3.154) using the derived distributions of density and gravity. This gives

$$p(r) = \int_r^R \rho(r) g(r) \mathrm{d}r \tag{3.159}$$

Pressure increases continuously with increasing depth in the Earth (Fig. 3.78). The rate of increase (pressure gradient) changes at the depths of the major seismic discontinuities. The pressure reaches a value close to 380 GPa (3.8 Mbar) at the center of the Earth, which is about 4 million times atmospheric pressure at sea-level.

3.7.5 Models of the Earth's internal structure

Once the velocity distributions of P- and S-waves inside the Earth were known the broad internal structure of the Earth – crust, mantle, inner and outer core – could be further refined. In 1940–1942 K. E. Bullen developed a model of the internal structure consisting of seven concentric shells. The boundaries between adjacent shells were located at sharp changes in the body-wave velocities or the velocity gradients. For ease of identification the layers were labelled A–G (Table 3.4); this nomenclature has been carried over into more modern models. The seismic layering of the Earth is better known than the composition of the layers, which must be inferred from laboratory experiments and petrological modelling.

Table 3.4 *Comparison of Earth's internal divisions according to Model A (Bullen, 1942) and PREM (Dziewonski and Anderson, 1981)*

Model A	PREM		
Region (km)	Layer	Depth range (km)	Comments
A (0–33)	*crust*:		lithosphere 0–80 km
	upper	0–15	
	lower	15–24	
B (33–410)	*upper mantle*:		
	uppermost mantle	24–80	
	low-velocity layer	80–220	asthenosphere
	transition zone	220–400	
C (410–1000)	*upper mantle*:		
	transition zones	400–670	
		670–770	
D (1000–2900)	*lower mantle*:		
	layer D'	770–2740	
	layer D"	2740–2890	
E (2900–4980)	*outer core*	2890–5150	
F (4980–5120)	*transition layer*	----	
G (5120–6370)	*inner core*	5150–6370	

In the original Model A the density distribution was not well constrained. Two different density distributions (A and A') that fitted the known mass and moment of inertia gave disparate central densities of 17,300 kg m^{-3} and 12,300 kg m^{-3}, respectively. In 1950 Bullen presented Earth Model B, in which the bulk modulus (K) and seismic parameter (Φ) were assumed to vary smoothly with pressure below a depth of 1000 km. The model suggested a central density around 18,000 kg m^{-3}.

In the 1950s the development of long-period seismographs made possible the observation of the natural oscillations of the Earth. After very large earthquakes numerous modes of free oscillation are excited with periods up to about one hour (§ 3.3.4). These were first observed unambiguously after the huge Chilean earthquake of 1960.

The free oscillations form an independent constraint on Earth models. The lowest-frequency spheroidal modes involve radial displacements that take place against the restoring force of gravitation and are therefore affected by the density distribution. Starting from a spherically symmetric Earth model with known distributions of density and elastic properties, the *forward problem* consists of calculating how such a model will reverberate. The calculated and observed normal modes of oscillation are compared and the model is adjusted until the required fit is obtained. The *inverse problem* consists of computing the model of density and elastic properties by inverting the frequency spectrum of the free oscillations. The parametrized model PREM, based upon the inversion of body-wave, surface-wave and free-oscillation data, is the current standard model of the Earth's internal structure. It predicts a central density of 13,090 kg m^{-3}.

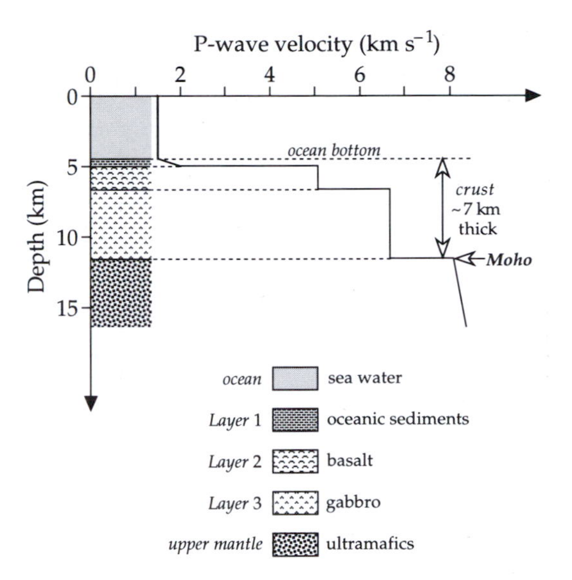

Fig. 3.79 Generalized petrological model and P-wave velocity–depth profile for oceanic crust.

3.7.5.1 *The crust*

The Earth's crust corresponds to Bullen's region A. The structures of the crust and upper mantle are complex and show strong lateral variations. This prohibits using the inversion of body-wave travel-times to get a vertical distribution of seismic velocities. The most reliable information on crustal seismic structure comes from seismic refraction profiles and deep crustal reflection sounding. The variation of seismic velocity with depth in the crust differs according to where the profiles are carried out. Ancient continental shield domains have different vertical velocity profiles than younger continental or oceanic domains. In view of this variability any generalized petrological model of crustal structure is bound to be an oversimplification. However, with this reservation, it is still possible to summarize some general features of crustal structure, and the corresponding petrological layering.

A generalized model of the structure of oceanic crust is shown in Fig. 3.79. Oceanic crust is only 5–10 km thick. Under a mean water-depth of about 4.5 km the top part of the oceanic crust consists of a layer of sediments that increases in thickness away from the oceanic ridges. The igneous oceanic basement consists of a thin (~0.5 km) upper layer of superposed basaltic lava flows underlain by a complex of basaltic intrusions, the sheeted dike complex. Below this the oceanic crust consists of gabbroic rocks.

The vertical structure of continental crust is more complicated than that of oceanic crust, and the structure under ancient shield areas differs from that under younger basins. It is more difficult to generalize a representative model (Fig. 3.80). The most striking difference is that the continental crust is much thicker than oceanic crust. Under stable continental areas the crust is 35–40 km thick and under young mountain ranges it is often 50–60 km thick. The continental Moho is not always a sharp boundary. In some places the transition from crust to mantle may be gradual, with a layered structure. The Conrad seismic discontinuity is believed to separate an upper crustal (granitic) layer from a lower crustal (basaltic) layer. However, the Conrad discontinuity is not found in all places and there is some doubt as to its real nature. Crustal velocity studies have defined two anomalous zones that often disrupt the otherwise progressive increase of velocity with depth. A low-velocity layer within the middle crust is thought to be due to intruded granitic laccoliths; it is called the sialic low-velocity layer. It is underlain by a middle crustal layer composed of migmatites. Below this layer the velocity rises sharply, forming a 'tooth' in the velocity profile. This tooth and the layer beneath it often make up a thinly layered lower crust. Refractions and reflections at the top of the tooth are thought to explain the Conrad discontinuity.

3.7.5.2 *The upper mantle*

In his 1942 model of the Earth's interior (see table 3.4) Bullen made a distinction between the upper mantle (layers B and C) and the lower mantle (layer D). The upper mantle is characterized by several discontinuities of body-wave velocities and steep velocity gradients (Fig. 3.81). The top of the mantle is defined by the Mohorovičić discontinuity (Moho), below which the P-wave velocity exceeds 7.6 km s^{-1}. The Moho depth is very variable, with a global mean value around 30–40 km. A weighted average of oceanic and continental structures equal to 24.4 km is used in model PREM. The assumption of a spherically symmetric Earth does not hold well for the crust and upper mantle. Lateral differences in structure are important down to depths of at least 400 km. The uppermost mantle between the Moho and a depth of 80–120 km is rigid, with increasing P- and S-wave velocities. This layer is sometimes called the *lid* of the underlying low-velocity layer. Together with the crust, the lid forms the *lithosphere*, the rigid outer shell of the Earth that takes part in plate tectonic processes (see §1.2). The lithosphere is subdivided laterally into tectonic plates that may be as large as 10,000 km across (e.g., the Pacific plate) or as small as a few thousand km (e.g., the Philippines plate). The plates are very thin in comparison to their horizontal extent.

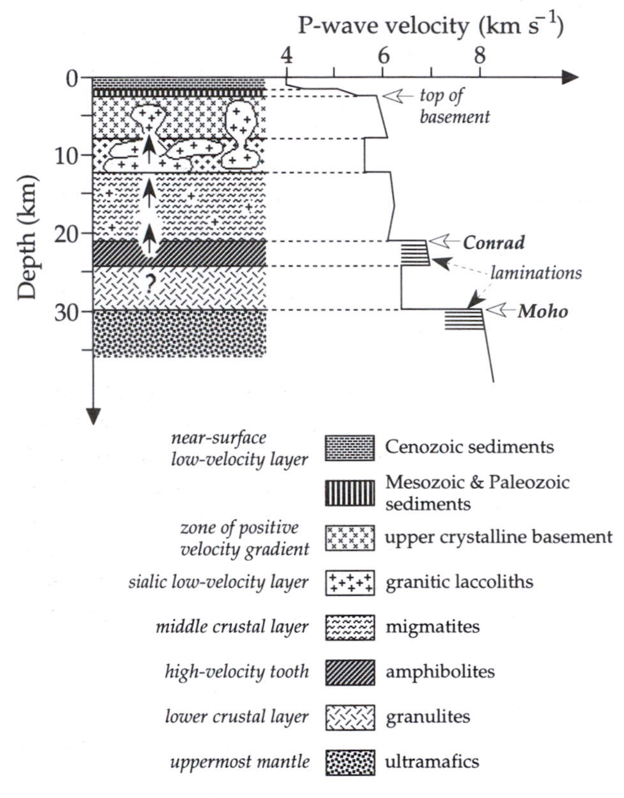

Fig. 3.80 Generalized petrological model and P-wave velocity–depth profile for continental crust (after Mueller, 1977).

An abrupt increase of P- and S-wave velocities by 3–4% has been observed at around 220±30 km depth; it is called the *Lehmann discontinuity*. Like the Conrad discontinuity in the crust it is not found everywhere and its true meaning is in question. Between the lid and the Lehmann discontinuity, in the depth range 100–200 km, body-wave velocity gradients are weakly negative, i.e., the velocities *decrease* with increasing depth. The layer is called the low-velocity layer (LVL). Its nature cannot be evaluated from body waves, because they do not bottom in the layer and its lower boundary is not sharp. The evidence for understanding the LVL comes from the inversion of surface-wave data. Only long-period surface waves with periods longer than about 200 s can penetrate to the depths of the base of the LVL. The data from surface waves are not precise. Their depth resolution is poor and only the S-wave velocity can be determined. Thus, the top and bottom of the LVL are not sharply defined.

The LVL is usually associated with the *asthenosphere*, which also plays an important role in plate tectonic theory. The decreases in seismic velocities are attributed to reduced rigidity in this layer. Over geological time-intervals the mantle reacts like a viscous medium, with a viscosity that depends on temperature and composition. From the point of view of plate tectonics the asthenosphere is a viscous layer that decouples the lithosphere from the deeper mantle; by allowing slow convection, it permits or promotes the relative motions of the global plates. The LVL and asthenosphere reflect changes of rheological properties of the upper mantle material.

The composition of the upper mantle is generally taken

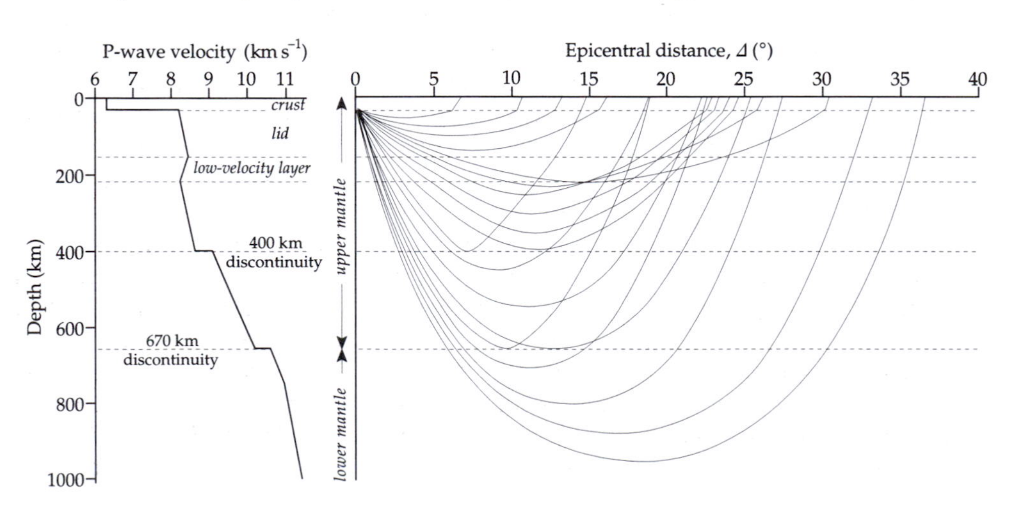

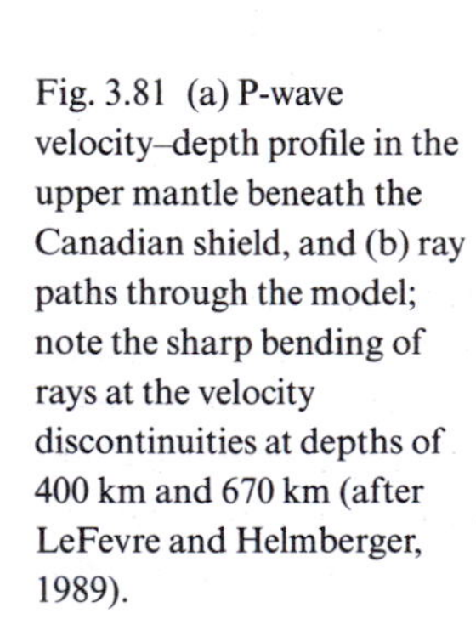

Fig. 3.81 (a) P-wave velocity–depth profile in the upper mantle beneath the Canadian shield, and (b) ray paths through the model; note the sharp bending of rays at the velocity discontinuities at depths of 400 km and 670 km (after LeFevre and Helmberger, 1989).

to be peridotitic, with olivine [$(Mg,Fe)_2SiO_4$] as the dominant mineral. With increasing depth the hydrostatic pressure increases and eventually causes high-pressure transformation of the silicate minerals. This is reflected in the seismic properties. Travel-time–Δ curves of body waves show a distinct change in slope at epicentral distances of about 20°. This is attributed to a discontinuity in mantle velocities at a depth of around 400 km (Fig. 3.81). The 400 km discontinuity (or 20° discontinuity) is interpreted as due to a petrological change from an olivine-type lattice to a more closely packed spinel-type lattice.

The lid, low-velocity layer and the zone down to the 400 km discontinuity together correspond to layer B in Bullen's Earth Model A. A further seismic discontinuity occurs at a depth of 650–670 km. This is a major feature of mantle structure that has been observed world-wide. In the transition zone between the 400 km and 670 km discontinuities there is a further change in structure from β-spinel to γ-spinel, but this is not accompanied by appreciable changes in physical properties. At 670 km the spinel transforms to perovskite. This transition zone corresponds to the upper part of Bullen's layer C.

3.7.5.3 *The lower mantle*

The lower mantle is now classified as the part below the important seismic discontinuity at 670 km. Its composition is rather poorly known, but it is thought to consist of oxides of iron and magnesium as well as iron–magnesium silicates with a perovskite structure. The uppermost part of the lower mantle between 670 and 770 km depth has a high positive velocity gradient and corresponds to the lower part of Bullen's layer C. Beneath it lies Bullen's layer D', which represents a great thickness of normal mantle, characterized by smooth velocity gradients and the absence of seismic discontinuities.

Just above the core–mantle boundary an anomalous layer, approximately 150–200 km thick, has been identified in which body-wave velocity gradients are very small and may even be negative. Although part of the lower mantle, it evidently serves as a boundary layer between the mantle and core. It is labelled D" to distinguish it from the normal mantle above it. The structure and role of the D" layer are not yet known with confidence, but it is the focus of intensive current research. Models of the internal structure of D" have been proposed with positive velocity gradients, others with negative velocity gradients, and some with small velocity discontinuities. The latter possibility is important because it would imply some stratification within D".

The most interesting aspect of D" is the presence – revealed by seismic tomographic imaging (§3.7.6.3) – of velocity variations of several percent that take place over lateral distances comparable in size to the continents and oceans in Earth's crust. The term 'grand structure' has been coined for these regions; the thicker parts have also been termed 'crypto-continents' and the thinner parts 'crypto-oceans' (see Fig. 6.26). Moreover, the seismically fast regions (in which temperatures are presumed to be cooler than normal) lie beneath present subduction zones. This suggests that cold subducted lithosphere may eventually sink to the bottom of the mantle where it is colder and more rigid than the surrounding mantle and hence has higher body-wave velocities. A large low-velocity (hot) region underlies the Pacific basin, in which many centers of volcanism and locally high heat flow ('hotspots', see §6.3.2) are located. The D" layer is suspected of being the source of the mantle plumes that cause these anomalies. Exceptionally hot material from D" rises in thin pipe-like mantle plumes to the 670 km discontinuity, which opposes further upward motion in the same way that it resists the deeper subduction of cold lithospheric slabs. Occasionally a hot plume is able to break through the 670 km barrier, producing a surface hotspot.

If our current understanding of the D" layer is correct, it plays an important role in geodynamic and geothermal behavior. On the one hand, D" serves as the source of material for the mantle plumes that give rise to hotspots, which are important in plate tectonics. On the other hand, the thermal properties of D" could influence the outward transport of heat from the Earth's core; in turn, this could affect the intricate processes that generate the Earth's magnetic field.

3.7.5.4 *The core*

Early in Earth's history, dense metallic elements are thought to have settled towards the Earth's center, forming the core, while lighter silicates ascended and solidified to form the mantle. Studies of the compositions of meteorites and of the behavior of metals at high pressure and temperature give a plausible picture of the composition and formation of the core. It consists mainly of iron, with perhaps up to 10% nickel. The observed pressure–density relationships suggest that some less-dense non-metallic elements (Si, S, O) may be present in the outer core. It is not known if small amounts of the more common radioactive elements (^{40}K, ^{232}Th, ^{235}U and ^{238}U) are present in large enough abundances to contribute to the heat supply of the core.

The core has a radius of 3480 km and consists of a solid inner core (layer G in Bullen's Earth Model A) surrounded by a liquid outer core (Bullen layer E) that is 1220 km thick. The transitional layer (F) was modelled by Bullen as a zone in which the P-wave velocity gradient is negative (i.e., α *decreases* with increasing depth). Not all seismologists agreed with this interpretation of the $t-\Delta$ curves and so the nature of layer F remained controversial for many years. The need for a layer F has now been discarded. Improved seismographic resolution has yielded a large quantity of high-quality data for reflections from the boundaries of the outer core (PcP, ScS) and the inner core (PKiKP) which have helped clear up the nature of these boundaries. The PKiKP phase contains high frequencies; this implies that the inner core boundary is sharp, probably no more than 5 km thick. The seismic events earlier interpreted as due to a layer F are now regarded as rays that have been scattered by small-scale features at the bottom of the mantle.

The inner core transmits P-waves (PKIKP phase) but S-waves in the inner core (PKJKP phase), although in principle possible, have not yet been observed unequivocally. Body-wave travel-times do not constrain the rigidity of the inner core. However, the amplitude spectrum of the frequencies of higher modes of the Earth's free oscillations show that the inner core is likely solid. However, it is possible that it is not completely solid. Rather, it may be a mixture of solid and liquid phases at a temperature close to the solidification temperature. An analogy can be made with the mushy, semi-frozen state that water passes through on freezing to ice.

The outer core is fluid, with a viscosity similar to that of water. It is assumed to be homogeneous and its thermal state is supposed to be adiabatic. These are the conditions to be expected in a fluid that has been well mixed, in the Earth's case by convection and differential rotation. One theory of core dynamics holds that the iron-rich inner core is solidifying from the fluid outer core, leaving behind its lighter elements. These constitute a less-dense, therefore gravitationally buoyant, fluid, which rises through the denser overlying liquid. This compositional type of buoyancy could be an important contributor to convection in the outer core, and therefore to the dynamics of the core and generation of the Earth's magnetic field.

The core–mantle boundary (CMB) is also called the *Gutenberg discontinuity*. It is characterized by very large changes in body-wave velocities and is the most sharply defined seismic discontinuity. Seismic data show that the boundary is not smooth but has a topography of hills and valleys. Anomalies in the travel-times of PKKP phases – which are reflected once internally at the CMB – have been attributed to scattering by topographic features with a relief of a few hundred meters. However, depending on conditions in the hot D" layer of the lower mantle immediately above the CMB some topographic features may be up to 10 km high. Interference between the CMB topography and fluid motions in the outermost core may couple the core and mantle to each other dynamically.

3.7.6 Seismic tomography

A free translation of the term tomography is 'representation in cross-section'. Neighboring two-dimensional cross-sections can be combined to give a three-dimensional model. The use of computer-aided tomography (CAT) in medical diagnosis is well known as a non-invasive method of examining internal organs for abnormal regions. X-rays or ultrasonic rays are absorbed unequally by different materials. CAT consists of studying the attenuation of x-rays or ultrasonic waves that pass through the body in distinctly controlled planar sections. Seismic tomography uses the same principles, with the difference that the travel-times of the signals, rather than their attenuation, are observed. Hence the technique may be described as the three-dimensional modelling of the velocity distribution of seismic waves in the Earth. The technique requires powerful computational facilities and sophisticated programing.

The travel-time of a seismic wave from an earthquake focus to a seismograph is determined by the velocity distribution along its path. For an idealized, spherically symmetric Earth model the radial distributions of velocity are known. The velocities in the model are mean values, which average out lateral fluctuations. If such a velocity model were used to compute travel-times of different phases to any epicentral distance, a set of curves indistinguishable from Fig. 3.71 would result. In reality, the observed travel-times usually show small deviations from the calculated times. These discrepancies are called *travel-time residuals* or *anomalies*, and they can have several causes. An obvious cause is that the focal depth of an earthquake is not zero, as assumed in Fig. 3.71, but may be up to several hundred kilometers. The parametrized Earth model *iasp91* takes this into account and tabulates travel-times for several focal depths. Clearly, precise determination of earthquake focal parameters (epicentral location, depth and time of occurrence) are essential prerequisites for seismic tomography.

Another cause of travel-time residuals arises from the particular local velocity–depth distribution under the observational network. Ideally, the local vertical profile should be known, so as to allow compensation of an observed travel-time for local anomalous structure. In practice, the signals from a selected earthquake are averaged for several stations in a given area of the surface (e.g., about 3° square) to reduce local perturbations.

The most important geophysical reasons for the remaining travel-time discrepancies is that the assumption of spherical symmetry is not perfectly valid, and the ellipticity of the Earth's figure must be taken into account. There are lateral variations in P- and S-wave velocity at any given depth. The fluctuations may amount to several percent of the 'average' velocity for that depth assumed in the reference model. If, at a certain depth, a seismic ray passes through a region in which the velocity is slightly faster than average, the wave will arrive slightly sooner than expected at the receiver; if the anomalous velocity is slower than average, the wave will arrive late. This permits classification of travel-time as 'early' or 'late' depending on whether a ray has traversed a region that is 'fast' or 'slow' with respect to the assumed model. The velocity of a seismic wave is determined by elastic parameters and density, which are affected by temperature. The velocity anomalies obtained from seismic tomography on a global scale are generally interpreted in terms of abnormal temperature and rigidity. A 'slow' region is associated with above-average temperature and lower rigidity, while a 'fast' region is due to lower temperature and higher rigidity.

3.7.6.1 *Travel-time residuals due to anomalous velocity structure*

As a simple example consider the passage of rays in six directions through a square region containing four equal areas with P-wave velocities 5.4 km s^{-1}, 6.0 km s^{-1}, 6.3 km s^{-1} and 6.6 km s^{-1} (Fig. 3.82a). Let the expected velocity in the region be 6.0 km s^{-1}. The velocity anomaly in each area is found by subtracting the reference value and can be expressed as a percentage of the expected value. This gives zones that are 10% and 5% fast, a zone that is 10% slow and a zone with no anomaly (Fig. 3.82b). Suppose that six seismic rays traverse the square region (Fig. 3.82c), and let the expected travel-time be 2.0 s for each ray; some of the observed travel-times are shorter and some are longer than the expected value. The travel-time anomaly is computed by subtracting the observed times from the expected travel-time (Fig. 3.82d); it is positive or negative depending on whether the ray travels faster or slower than average. The *travel-time residual* is obtained by expressing the anomaly as a percentage of the expected travel-time (Fig. 3.82e). A *positive residual* results for an early arrival that has travelled through a 'fast' zone with positive velocity anomaly; a *negative residual* corresponds to passage through a 'slow' zone with negative velocity anomaly. The set of travel-time residuals forms the data base for tomographic inversion, which is designed to yield the original velocity structure (Fig. 3.82a), or, more commonly, the distribution of fast or slow regions (Fig. 3.82b).

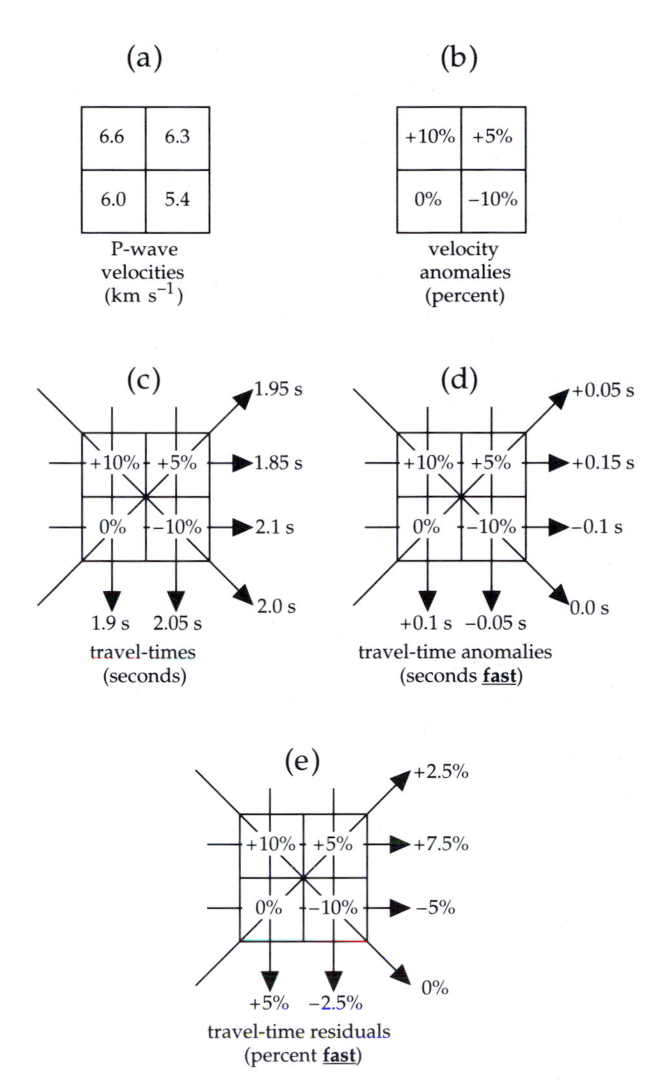

Fig. 3.82 Computation of relative travel-time residuals (in percent) for a simple four-block structure: (a) velocity distribution, (b) velocity anomalies in percent of the reference value (6 km s^{-1}), (c) observed travel-times in different directions through the blocks, and (d) travel-time residuals in percent of the travel-time (2 s) for homogeneous material (after Kissling, 1993).

3.7.6.2 *Velocity anomalies from inversion of travel-time residuals*

The principle of tomographic inversion consists of successively adjusting a model of velocity structure until it gives the observed travel-time residuals along paths that traverse the region. To illustrate how this is done we take as starting point the set of relative travel-time residuals from the previous example. Consider a horizontal ray that traverses the upper two 'fast' blocks. To account for the (early) travel-time anomaly of +7.5% let each block be allocated a (fast) velocity anomaly of +7.5%. Similarly, let each of the bottom two blocks be allocated velocity-anomalies of −5%. This simple velocity distribution (Fig. 3.83a) satisfies the travel-time anomalies for the horizontal rays. However, in the vertical direction it gives travel-time anomalies of +1.25% (the mean of +7.5% and −5%) for each ray

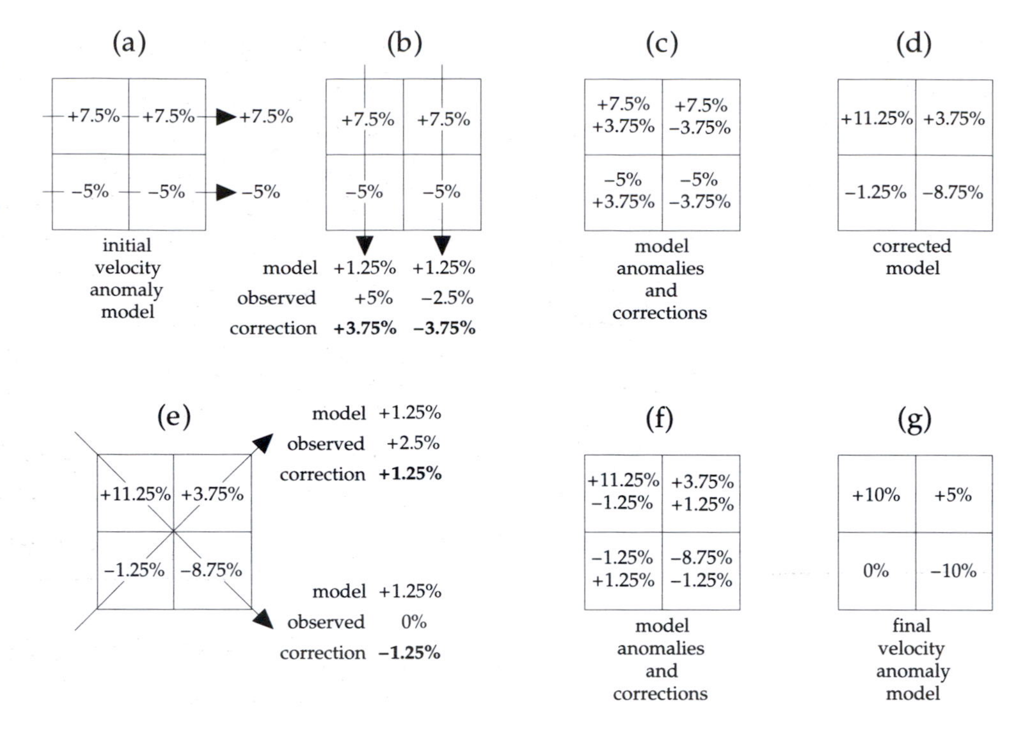

Fig. 3.83 Inversion of relative travel-time residuals to obtain velocity anomalies. In steps (a)–(g) successive corrections are made to a starting model of the velocity anomalies in the four blocks until the travel-time residuals along all six directions in Fig. 3.82 are satisfied (after Kissling, 1993).

(Fig. 3.83b). This does not agree with the observed anomalies for the two vertical rays (+5% and −2.5%, respectively); one vertical anomaly is 3.75% too large, the other is 3.75% too small. The velocities in the blocks can be adjusted accordingly by making a correction of +3.75% to the left-hand blocks and −3.75% to the right-hand blocks (Fig. 3.83c). This gives a new distribution of velocity anomalies which satisfies the horizontal and vertical rays (Fig. 3.83d). The model now gives travel-time anomalies along each of the diagonal rays of +1.25%, compared to observed anomalies of +2.5% and 0%, respectively (Fig. 3.83e). Further corrections of +1.25% are now made to the upper right and lower left blocks, and −1.25% to the lower right and upper left blocks (Fig. 3.83f). The resulting distribution of velocity anomalies (Fig. 3.83g) satisfies all six rays through the anomalous region, and is the same as the original distribution of velocity anomalies (Fig. 3.82b).

This simple example shows how a tomographic image of velocity anomalies in the Earth can be obtained by making successive adjustments to an initial velocity distribution to account for the observed travel-time residuals. In practice this must be done for numerous rays that traverse the volume of interest. In some studies the signals are of local origin, generated by earthquakes or explosions within or near the volume of interest. Other studies are based on *teleseismic* signals, which originate in earthquakes at more than 20° epicentral distance.

3.7.6.3 *Tomographic imaging of the mantle*

Tomographic imaging of the Earth's three-dimensional velocity distribution can be based on either body waves or surface waves. The inversion of body-wave data is the only method that provides the lateral variations in velocity in the deeper interior. The lateral variations at a given depth are equivalent to the variations on the surface of a sphere at that depth, and so can be depicted with the aid of spherical harmonic functions. The contoured velocity anomalies obtained by spherical harmonic analysis for a depth of 2500 km in the lower mantle are dominated by a ring of fast P-wave velocities around the Pacific and a slow-velocity region in the center (Fig. 3.84). The pattern is present at nearly all depths in the lower mantle (i.e., below 1000 km). This is shown clearly by a vertical cross-section along a profile around the equator (Fig. 3.85, *bottom frame*). Deep fast-velocity ("cold") regions under America and Indonesia and slow-velocity ("warm") regions under the Pacific and Atlantic oceans extend from about 1000 km depth to the core–mantle boundary. The lower mantle velocity anomalies show little correlation to plate tectonic elements, which are features of the lithosphere.

Velocity anomalies in the upper mantle and crust can also be modelled by inverting teleseismic body waves. However, these depths can be probed by long-period surface waves, which are more sensitive to variations in rigidity (and thus temperature). The inversion of long-period surface-wave data is a powerful tool for modelling upper mantle S-

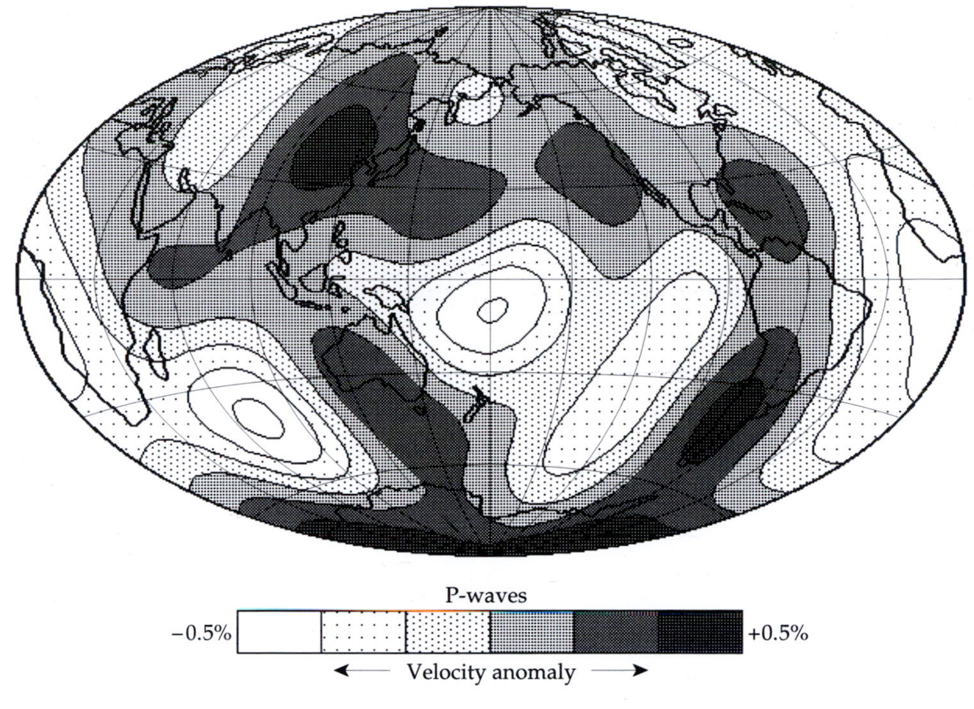

Fig. 3.84 Map of P-wave velocity anomalies in the lower mantle (2500 km depth). The deviations are plotted as percent faster or slower than the reference velocity at this depth (after Dziewonski, 1984, 1989).

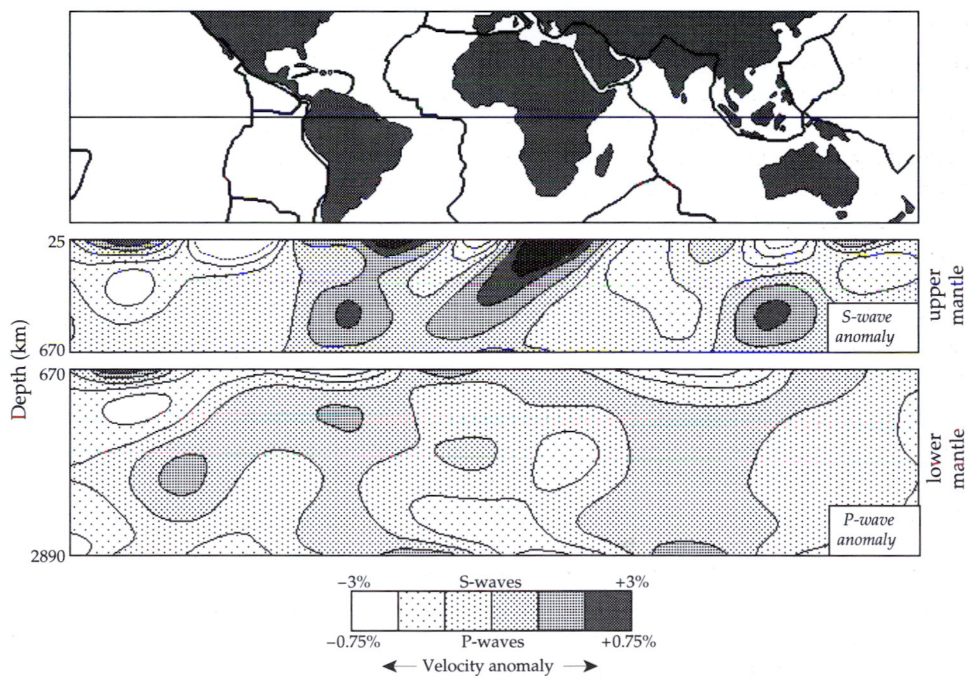

Fig. 3.85 Seismic tomographic section through the mantle along an equatorial profile. *Middle frame*: S-wave anomalies in the upper mantle to a depth of 670 km; *bottom frame*: P-wave anomalies in the lower mantle between 670 and 2890 km (after Woodhouse and Dziewonski, 1984).

wave velocities. The pattern of S-wave velocity anomalies in the upper mantle along the equatorial cross-section (Fig. 3.85, *middle frame*) shows generally increased velocities under the "cold" continents and reduced velocities under the 'warm' oceanic ridge systems. The cold roots of the African and South American continents seem to merge at depths below a few hundred kilometers in the upper mantle. Although some features seem to link up, the transition between the upper mantle and lower mantle cross-sections at 670 km depth is rather poor, because the velocity resolution is poorest near this depth.

3.8 SUGGESTIONS FOR FURTHER READING

INTRODUCTORY LEVEL

Bolt, B. A. 1993. *Earthquakes*, W. H. Freeman, New York.

Kearey, P. and Brooks, M. 1991. *An Introduction to Geophysical Exploration (2nd ed.)*, Blackwell Scientific Publ., Oxford.

Robinson, E. S. and Çoruh, C. 1988. *Basic Exploration Geophysics*, John Wiley & Sons, New York.

Walker, B. S. 1982. *Earthquake*, Time-Life Books Inc., Alexandria, Virginia.

INTERMEDIATE LEVEL

Bott, M. H. P. 1982. *The Interior of the Earth (2nd ed.)*, Edward Arnold, London.

Fowler, C. M. R. 1990. *The Solid Earth: an Introduction to Global Geophysics*, Cambridge Univ. Press, Cambridge.

Lay, T. and Wallace, T. C. 1995. *Modern Global Seismology*, Academic Press, San Diego.

ADVANCED LEVEL

Aki, K. and Richards, P. G. 1980. *Quantitative Seismology: Theory and Methods*, W. H. Freeman, San Francisco.

Bullen, K. E. 1963. *An Introduction to the Theory of Seismology (3rd ed.)*, Cambridge Univ. Press, Cambridge,

Garland, G. D. 1979. *Introduction to Geophysics*, W. B. Saunders., Philadelphia.

Gutenberg, B. and Richter, C. F. 1954. *Seismicity of the Earth and Associated Phenomena*, Princeton Univ. Press, Princeton.

Officer, C. B. 1974. *Introduction to Theoretical Geophysics*, Springer-Verlag, New York.

4 Earth's age, thermal and electrical properties

4.1 GEOCHRONOLOGY

4.1.1 Time

Time is both a philosophical and physical concept. Our awareness of time lies in the ability to determine which of two events occurred before the other. We are conscious of a present in which we live and which replaces continually a past of which we have a memory; we are also conscious of a future, in some aspects predictable, that will replace the present. The progress of time was visualized by Sir Isaac Newton to be like a river that flows involuntarily at a uniform rate. The presumption that time is an independent entity underlies all of classical physics. Although Einstein's *Theory of Relativity* shows that two observers moving relative to each other will have different perceptions of time, physical phenomena are influenced by this relationship only when velocities approach the speed of light. In everyday usage and in non-relativistic science the Newtonian notion of time as an absolute quantity prevails.

The measurement of time is based on counting cycles (and portions of a cycle) of repetitive phenomena. Prehistoric man distinguished the differences between day and night, he observed the phases of the Moon and was aware of the regular repetition of the seasons of the year. From these observations the day, month and year emerged as the units of time. Only after the development of the clock could the day be subdivided into hours, minutes and seconds.

4.1.1.1 *The clock*

The earliest clocks were developed by the Egyptians and were later introduced to Greece and from there to Rome. About 2000 B.C. the Egyptians invented the water clock (or *clepsydra*). In its primitive form this consisted of a container from which water could escape slowly by a small hole. The progress of time could be measured by observing the change of depth of the water in the container (using graduations on its sides) or by collecting and measuring the amount of water that escaped. The Egyptians (or perhaps the Mesopotamians) are also credited with inventing the sundial. Early sundials consisted of devices – poles, upright stones, pyramids or obelisks – that cast a shadow; the passage of time was observed by the changing direction and length of the shadow. After trigonometry was developed dials could be accurately graduated and time precisely measured. Mechanical clocks were invented around 1000 A.D. but reliably accurate pendulum clocks first came into use in the 17th century. Accurate sundials, in which the shadow was cast by a fine wire, were used to check the setting and calibration of mechanical clocks until the 19th century.

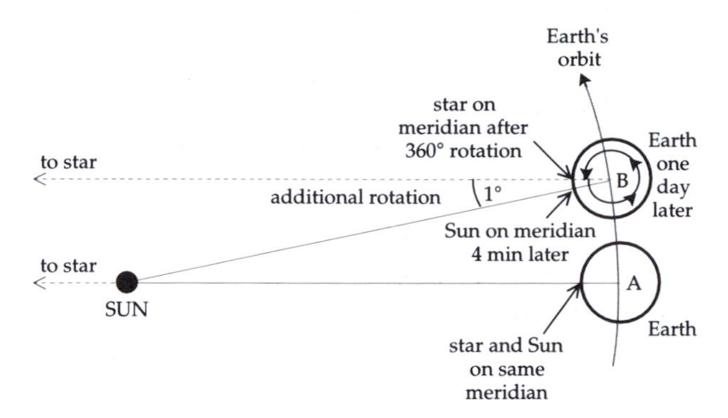

Fig. 4.1 The sidereal day is the time taken for the Earth to rotate through 360° relative to a fixed star; the solar day is the time taken for a rotation between meridians relative to the Sun. This is slightly more than 360° relative to the stars, because the Earth is also orbiting the Sun.

4.1.1.2 *Units of time*

The day is defined by the rotation of the Earth about its axis. The day can be defined relative to the stars or to the Sun (Fig. 4.1). The time required for the Earth to rotate through 360° about its axis and to return to the same meridian relative to a fixed star defines the *sidereal day*. All sidereal days have the same length. Sidereal time must be used in scientific calculations that require rotational velocity relative to the absolute framework of the stars. The time required for the Earth to rotate about its axis and return to the same meridian relative to the Sun defines the *solar day*. While the Earth is rotating about its axis, it is also moving forward along its orbit. The orbital motion about the Sun covers 360° in about 365 days, so that in one day the Earth moves forward in its orbit by approximately 1°. To return to the solar meridian the Earth must rotate this extra degree. The solar day is therefore slightly longer than the sidereal day. Solar days are not equal in length. For example, at perihelion the Earth is moving faster forward in its orbit than at aphelion (see Fig. 1.2). At perihelion the higher angular rate about the Sun means that the Earth has to rotate through a larger than average angle to catch up with the solar meridian.

Thus, at perihelion the solar day is longer than average; at aphelion the opposite is the case. The obliquity of the ecliptic causes a further variation in the length of the solar day. *Mean solar time* is defined in terms of the mean length of the solar day. It is used for most practical purposes on Earth, and is the basis for definition of the hour, minute and second. One *mean solar day* is equal to exactly 86,400 seconds. The length of the *sidereal day* is approximately 86,164 seconds.

The *sidereal month* is defined as the time required for the Moon to circle the Earth and return to the celestial longitude of a given star. It is equal to 27.321 66 (solar) days. To describe the motion of the Moon relative to the Sun, we have to take into account the Earth's motion around its orbit. The time between successive alignments of the Sun, Earth and Moon on the same meridian is the *synodic month*. It is equivalent to 29.530 59 days.

The *sidereal year* is defined as the time that elapses between successive occupations by the Earth of the same point in its orbit with respect to the stars. It is equal to 365.256 mean solar days. Two times per year, in spring and autumn, the Earth occupies positions in its orbit around the Sun where the lengths of day and night are equal at any point on the Earth. The spring occurrence is called the vernal equinox; that in the autumn is the autumnal equinox. The solar year (correctly called the *tropical year*) is defined as the time between successive vernal equinoxes. It equals 365.242 mean solar days, slightly less than the sidereal year. The small difference (0.014 days, about 20 minutes) is due to the *precession of the equinoxes*, which takes place in the retrograde sense (i.e., opposite to the revolution of the Earth about the Sun) and thereby reduces the length of the tropical year.

Unfortunately the lengths of the sidereal and tropical years are not constant but change slowly but measurably. In order to have a world standard the fundamental unit of scientific time was defined in 1956 in terms of the length of the tropical year 1900, which was set equal to 31,556,925.9747 seconds of *ephemeris time*. Even this definition of the second is not constant enough for the needs of modern physics. Highly stable *atomic clocks* have been developed that are capable of exceptional accuracy. For example, the alkali metal cesium has a sharply defined atomic spectral line whose frequency can be determined accurately by resonance with a tuned radio-frequency circuit. This provides the physical definition of the second of *ephemeris time* as the duration of 9,192,631,770 cycles of the cesium atomic clock.

Other units of time are used for specific purposes. Astronomers use a practical unit of time synchronized to the Earth's rotation. This gives a uniform timescale called *universal time* and denoted UT2; it is defined for a particular year by the Royal Observatory at Greenwich, England. The particular ways of defining the basic units of time are important for analyzing some problems in astronomy and satellite geodesy, but for most geophysical applications the minor differences between the different definitions are negligible.

4.1.1.3 *The geological timescale*

Whereas igneous rocks are formed in discrete short-lived eruptions of magma, sequences of sedimentary rocks take very long periods of time to form. Many sedimentary formations contain fossils i.e., relicts of creatures that lived in or near the basin in which the sediments were deposited. Evolution gives a particular identifiable character to the fossils in a given formation, so that it is possible to trace the evolution of different fossil families and correlate their beginnings and extinctions. These characteristics allow correlation of the host rock formation with others that contain part or all of the same fossil assemblage, and permit the development of a scheme for dating sediments relative to other formations. Gradually a *biostratigraphical timescale* has been worked out, which permits the accurate determination of the *relative age* of a sedimentary sequence. This information is intrinsic to any geological timescale.

A geological timescale combines two different types of information. Its basic record is a *chronostratigraphical* scale showing the relationship between rock sequences. These are described in detail in stratigraphical sections that are thought to have continuous sedimentation and to contain complete fossil assemblages. The boundary between two sequences serves as a standard of reference and the section in which it occurs is called a *boundary stratotype*. When ages are associated with the reference points, the scale becomes a geological timescale. Time is divided into major units called *eons*, which are subdivided into *eras*; these in turn are subdivided into *periods* containing several *epochs*. The lengths of time units in early timescales, before ages were determined with radioactive isotopes, were expressed as multiples of the duration of the Eocene epoch. Modern geological timescales are based on isotopic ages, which are calibrated in normal time units.

The geological timescale is constantly being revised and updated. Improved descriptions of important boundaries in global stratotype sections, more refined correlation with other chronostratigraphical units, and better calibration by more accurate isotopic dating lead to frequent revisions. Fig. 4.2 shows an example of a current geological timescale.

4.1.2 **Estimating the Earth's age**

Early estimates of the Earth's age were dominated by religious beliefs. Some oriental philosophers in antiquity

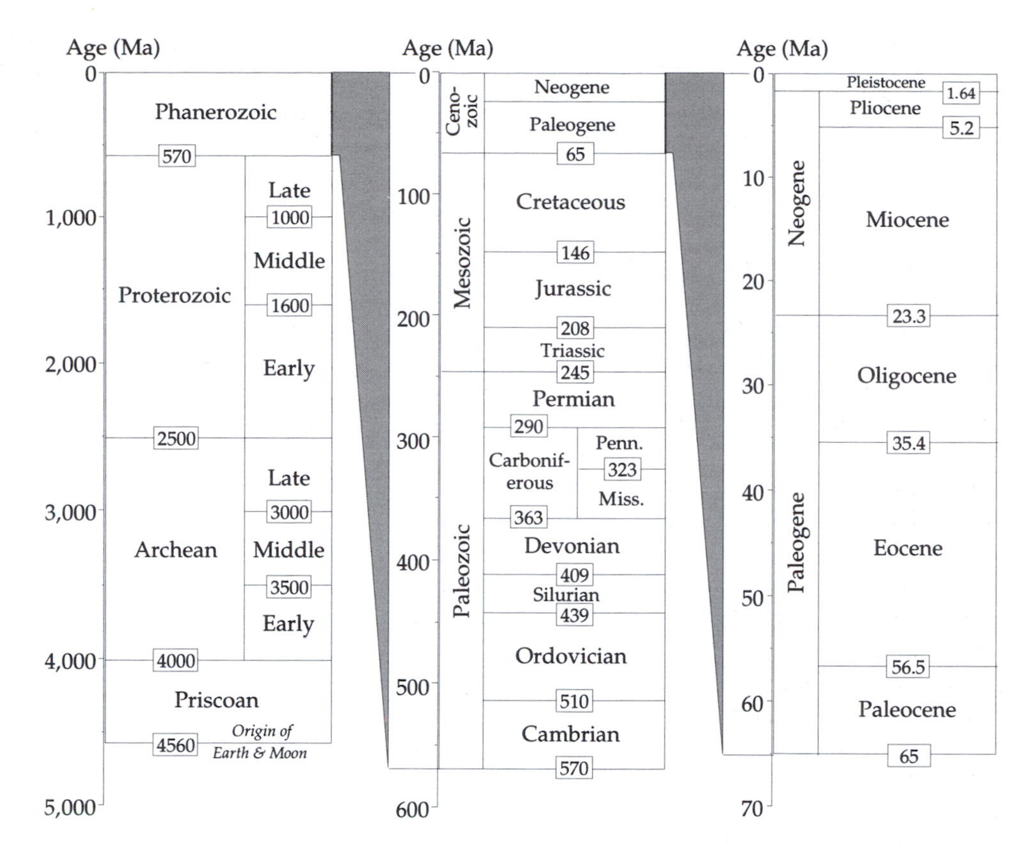

Fig. 4.2 A simplified version of the current geological timescale (*Source*: Harland *et al.*, 1990).

believed that the world had been in existence for millions of years. Yet, western thought on this topic was dominated for centuries by the tenets of the Jewish and Christian faiths. Biblical estimates of the world's age were made on the basis of genealogy by adding up the lengths of lifetimes and generations mentioned in the Old Testament. Some estimates also incorporate information from other ancient scriptures or related sources. The computed age of Creation invariably gave an age of the Earth less than 10,000 yr.

The best-known biblical estimate of the date of Creation was made by James Ussher (1581–1656), an Irish archbishop. His analysis of events recorded in the Old Testament and contemporary ancient scrolls led Ussher to proclaim that the exact time of Creation was at the beginning of night on October 22, in the year 4004 B.C., which makes the age of the Earth approximately 6000 yr. Other biblical scholars inferred similar ages. This type of 'creationist' age-estimate is still favored by many fundamentalist Christians, whose faith is founded on literal interpretation of the Bible. However, biblical scholars with a more broadly based faith recognize age-estimates based upon scientific methodology and measurement.

In the late 19th century the growth of natural philosophy (as physics was then called) fostered calculations of the Earth's age from physical properties of the solar system. Among these were estimates based on the cooling of the Sun, the cooling of the Earth, and the slow increase in the Earth–Moon distance. Chemists tried to date the Earth by establishing the time needed for the seas to acquire their salinity, while geologists conjectured how long it would take for sediments and sedimentary rocks to accumulate.

4.1.2.1 *Cooling of the Sun*

Setting aside the biblical 'sequence' chronicled in the book of Genesis, modern-day philosophers opine that the Earth cannot be older than the Sun. Cooling of the Sun takes place by constant *radiation* of energy into space. The amount of solar energy falling on a square meter per second at the Earth's distance from the Sun (1 AU) is called the solar constant; it equals 1360 W m^{-2}. The amount of energy lost by the Sun per second is obtained by multiplying this value by the surface area of a sphere whose radius is one astronomical unit. This simple calculation shows that the Sun is losing energy at the rate of 3.83×10^{26} W. In the 19th century, before the discovery of radioactivity and nuclear reactions, the source of this energy was not known. A German scientist, H. L. F. von Helmholtz, suggested in 1856 that it might result from the change of potential energy due to gravitational condensation of the Sun from an originally more distended body.

The condensational energy E_s of a mass M_s of uniform density and radius R_s is given by

$$E_s=\frac{3}{5}G\frac{M_s^2}{R_s}=2.28\times10^{41}\,\mathrm{J} \qquad (4.1)$$

In this equation G is the gravitational constant (see §2.2.1). The factor $\frac{3}{5}$ in the result arises from the assumption of a uniform density distribution inside the Sun. Dividing the condensational energy by the rate of energy loss gives an age of 19 million years for the Sun. Allowing for the increase of density towards the center causes an approximately threefold increase in the gravitational energy and the inferred age of the Sun.

4.1.2.2 *Cooling of the Earth*

In 1862 William Thomson, who later became Lord Kelvin, examined the cooling of the Sun in more detail. Unaware (although suspicious) of other sources of solar energy, he concluded that gravitational condensation could supply the Sun's radiant energy for at least 10 million years but not longer than 500 million years (500 Ma). In the same year he investigated the cooling history of the Earth. Measurements in deep wells and mines had shown that temperature increases with depth. The rate of increase, or *temperature gradient*, was known to be variable but seemed to average about 1°F for every 50 ft (0.036 °C m^{-1}). Kelvin inferred that the Earth was slowly losing heat and assumed that it did so by the process of *conduction* alone (§4.2.4). This enabled him to deduce the Earth's age from a solution of the one-dimensional equation of heat conduction, which relates the temperature T at time t and depth z by

$$\rho c\frac{\partial T}{\partial t}=\frac{\partial}{\partial z}\left(k\frac{\partial T}{\partial z}\right) \qquad (4.2)$$

In this equation ρ is the density, c the specific heat and k the thermal conductivity of the cooling body. These parameters are known for particular rock types, but Kelvin had to adopt generalized values for the entire Earth. The model assumes that the Earth initially had a uniform temperature T_0 throughout, and that it cooled from the outside leaving the present internal temperatures higher than the surface temperature. After a time t has elapsed the temperature gradient at the surface of the Earth calculated from Eq. (4.2) is given by

$$\left(\frac{\mathrm{d}T}{\mathrm{d}z}\right)_{z=0}=\sqrt{\frac{\rho c}{\pi k}}\frac{T_0}{\sqrt{t}} \qquad (4.3)$$

Kelvin assumed an initial temperature of 7000 °F (3871 °C, 4144 K) for the hot Earth, and contemporary values for the surface temperature gradient and thermal parameters. The calculation for t yielded an age of about 100 Ma for the cooling Earth.

4.1.2.3 *Increase of the Earth–Moon separation*

The origin of the Moon is still uncertain. George H. Darwin, son of the more famous Charles Darwin and himself a pioneer in tidal theory, speculated that it was torn from the Earth by rapid rotation. Like other classical theories of the Moon's origin (e.g., capture of the Moon from elsewhere in the solar system, or accretion in Earth orbit) the theory is flawed. In 1898 Darwin tried to explain the Earth's age from the effects of lunar tidal friction. The tidal bulges on the Earth interact with the Moon's gravitation to produce a decelerating torque that slows down the Earth's rotation and so causes an increase in the length of the day. The equal and opposite reaction is a torque exerted by the Earth on the Moon's orbit that increases its angular momentum. As explained in §2.3.4.2, this is achieved by an increase in the distance between the Moon from the Earth and a decrease in the rotation rate of the Moon about the Earth, which increases the length of the month. The Earth's rotation decelerates more rapidly than that of the Moon, so that eventually the angular velocities of the Earth and Moon will be equal. In this synchronous state the day and month will each last about 47 of our present days and the Earth–Moon distance will be about 87 Earth radii; the separation is presently 60.3 Earth radii.

Similar reasoning suggests that earlier in Earth's history, when the Moon was much closer to the Earth, both bodies rotated faster so that an earlier synchronous state may be conjectured. The day and month would each have lasted about 5 of our present hours and the Earth–Moon distance would have been about 2.3 Earth radii. However, at this distance the Moon would be inside the *Roche limit*, about 3 Earth radii, at which the Earth's gravitation would tear it apart. Thus, it is unlikely that this condition was ever realized.

Darwin calculated the time needed for the Earth and Moon to progress from an initially unstable close relationship to their present separation and rotation speeds, and concluded that a minimum of 56 Ma would be needed. This provided an independent estimate of the Earth's age, but it is unfortunately as flawed in its underlying assumptions as other models.

4.1.2.4 *Oceanic salinity*

Several determinations of the age of the Earth have been made on the basis of the chemistry of sea water. The reasoning is that salt is picked up by rivers and transported into lakes and seas; evaporation removes water but the vapor is fresh, so the salt is left behind and accumulates with time. If the accumulation rate is measured, and if the initial salt concentration was zero, the age of the sea and, by inference, the Earth can be calculated by dividing the present concentration by the accumulation rate.

Different versions of this method have been investigated. The most noted is that of an Irish geologist, John Joly, in 1899. Instead of measuring salt he used the concentration of a pure element, sodium. Joly determined the total amount of sodium in the oceans and the annual amount brought in by rivers. He concluded that the probable maximum age of the Earth was 89 Ma. Later estimates included corrections for possible sodium losses and non-linear accumulation but gave similar ages less than about 100 Ma.

The principle flaw in the chemical arguments is the assumption that sodium accumulates continuously in the ocean. In fact, all elements are withdrawn from the ocean at about the same rate as they are brought in. As a result, sea water has a chemically stable composition, which is not changing significantly. The chemical methods do not measure the age of the ocean or the Earth, but only the average length of time that sodium resides in the seas before it is removed.

4.1.2.5 *Sedimentary accumulation*

Not to be outdone by the physicists and chemists, late 19th century geologists tried to estimate Earth's age using stratigraphical evidence for the accumulation of sediments. The first step involved determining the thicknesses of sediment deposited during each unit of geological time. The second step was to find the corresponding sedimentation rate. When these parameters are known, the length of time represented in each unit can be calculated. The age of the Earth is the sum of these times.

The geological estimates are fraught with complications. Sediments with a silicate matrix, such as sandstones and shales, are deposited mechanically, but carbonate rocks form by precipitation from sea water. To calculate the mechanical rate of sedimentation the rate of input has to be known. This requires knowing the area of the depositional basin, the area supplying the sediments and its rate of land erosion. Rates used in early studies were largely intuitive. The calculations for carbonate rocks required knowing the rate of solution of calcium carbonate from the land surfaces, but could not correct for the variation of solubility with depth in the depositional basin. The number of unknown or crudely known parameters led to numerous divergent geological estimates for the Earth's age, ranging from tens of millions to hundreds of millions of years.

4.1.3 **Radioactivity**

In 1896 a French physicist, Henri Becquerel, laid a sample of uranium ore on a wrapped, undeveloped photographic plate. After development the film showed the contour of the sample. The exposure was attributed to invisible rays emitted by the uranium sample. The phenomenon, which became known as *radioactivity*, provides the most reliable methods yet known of dating geological processes and calculating the age of the Earth and solar system. To appreciate radioactivity we must consider briefly the structure of the atomic nucleus.

The nucleus of an atom contains positively charged protons and electrically neutral neutrons. The electrostatic *Coulomb force* causes the protons to repel each other. It decreases as the inverse square of the separation (see §4.3.2) and so can act over distances that are large compared to the size of the nucleus. An even more powerful *nuclear force* holds the nucleus together. This attractive force acts between protons, between neutrons, and between proton and neutron. It is only effective at short distances, for a particle separation less than about 3×10^{-15} m.

Suppose that the nucleus of an atom contains Z protons and is surrounded by an equal number of negatively charged electrons, so that the atom is electrically neutral; Z is the *atomic number* of the element and defines its place in the periodic table. The number N of neutrons in the nucleus is its *neutron number*, and the total number A of protons and neutrons is the *mass number* of the atom. Atoms of the same element with different neutron numbers are called *isotopes* of the element. For example, uranium contains 92 protons but may have 142, 143, or 146 neutrons. The different isotopes are distinguished by appending the mass number to the chemical symbol, giving ^{234}U, ^{235}U and ^{238}U.

The Coulomb force of repulsion acts between every pair of protons in a nucleus, while the short-range nuclear force acts only on nearby protons and neutrons. To avoid flying apart due to Coulomb repulsion all nuclei with atomic number Z greater than about 20 have an excess of neutrons ($N>Z$). This helps to dilute the effects of the repulsion-producing protons. However, nuclei with $Z\geq83$ are unstable and disintegrate by *radioactive decay*. This means that they break up spontaneously by emitting elementary particles and other radiation.

At least 28 distinct elementary particles are known to nuclear physics. The most important correspond to the three types of radiation identified by early investigators and called α-, β-, and γ-rays. An α-ray (or α-particle) is a helium nucleus that has been stripped of its surrounding electrons; it is made up of two protons and two neutrons and so has atomic number 2 and mass number 4. A β-particle is an electron. Some reactions emit additional energy in the form of γ-rays, which have a very short wavelength and are similar in character to x-rays.

4.1.3.1 *Radioactive decay*

A common type of radioactive decay is when a neutron n_0 in the nucleus of an atom spontaneously changes to a proton, p^+, and a β-particle (non-orbital electron). The β-particle is at once ejected from the nucleus along with another elementary particle called an antineutrino, ν, which has neither mass nor charge and need not concern us further here. The reaction can be written

$$n_0 \Rightarrow p^+ + \beta^- + \nu \tag{4.4}$$

Radioactive decay is a statistical process. It is customary to call the nucleus that decays the parent and the nucleus after decay the daughter. It is not possible to say in advance which nucleus will spontaneously decay. But the probability that any one will decay per second is a constant, called the *decay constant* or *decay rate*. Statistical behavior is really only applicable to large numbers, but the process of radioactive decay can be illustrated with a simple example. Suppose we start with 1000 nuclei, and the chance per second of a decay is 1 in 10, or 0.1. In the first second 10% of the parent nuclei spontaneously decay (i.e., 100 decays take place); the number of parent nuclei is reduced to 900. The probability that any parent decays in the next second is still 1 in 10, so 90 further decays can be expected. Thus after 2 seconds the number of parent nuclei is reduced to 810, after 3 seconds to 729, and so on. The total number of parent nuclei constantly gets smaller but in principle it never reaches zero, although after a long time it approaches zero asymptotically. The decay is described by an *exponential* curve.

If the decay rate is equal to λ, then in a short time-interval dt the probability that a given nucleus will decay is λdt; if at any time we have P parent nuclei the number that decay in the following interval dt is $P(\lambda dt)$. The change dP in the number of P parent nuclei in a time interval dt due to spontaneous decays is

$$dP = -\lambda P dt$$

$$\frac{dP}{dt} = -\lambda P \tag{4.5}$$

which has the solution

$$P = P_0 e^{-\lambda t} \tag{4.6}$$

Eq. (4.6) describes the exponential decay of the number of parent nuclides, starting from an initial number P_0. While the number of parent nuclides diminishes, the number of daughter nuclides D increases (Fig. 4.3). D is the difference between P and P_0 and so is given by

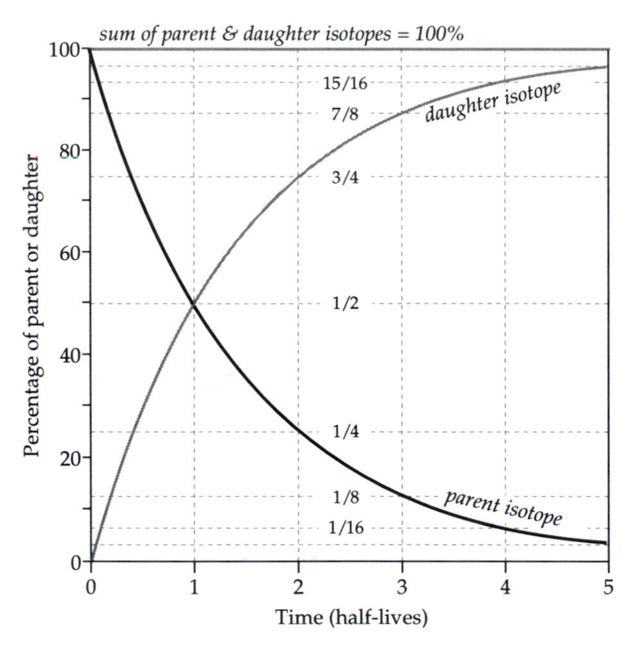

Fig. 4.3 Exponential decrease of the number of parent nuclides and the corresponding growth of the number of daughter nuclides in a typical radioactive decay process.

$$D = P_0 - P = P_0(1 - e^{-\lambda t}) \tag{4.7}$$

The original amount P_0 of the parent nuclide is not known; a rock sample contains a residual amount P of the parent nuclide and an amount D of the daughter product. Eliminating the unknown P_0 from Eq. (4.6) and Eq. (4.7) gives

$$D = P(e^{\lambda t} - 1) \tag{4.8}$$

The experimental description of radioactive decay by Ernest (later Lord) Rutherford and Frederick Soddy in 1902 was based on the observations of times needed for the activity of radioactive materials to decrease by steps of one-half. This time is known as the *half-life* of the decay. In the first half-life the number of parent nuclides decreases to a half, in the second half-life to a quarter, in the third to an eighth, etc. The number of daughter nuclides increases in like measure, so that the sum of parent and daughter nuclides is always equal to the original number P_0. Letting P/P_0 equal 1/2 in Eq. (4.6) we get the relationship between half-life $t_{1/2}$ and decay constant λ:

$$t_{1/2} = \frac{\ln 2}{\lambda} \tag{4.9}$$

The decay rates and half-lives are known for more than 1700 radioactive isotopes. Some are only produced in

Table 4.1 *Decay constants and half-lives of some naturally occurring, radioactive isotopes commonly used in geochronology*

Parent isotope	Daughter isotope	Decay constant (10^{-10} yr^{-1})	Half-life (Ga)
^{40}K	89.5% ^{40}Ca 10.5% ^{40}Ar	5.543	1.25
^{87}Rb	^{87}Sr	0.1420	48.8
^{147}Sm	^{143}Nd	0.0654	106.0
^{232}Th	^{208}Pb	0.4948	14.01
^{235}U	^{207}Pb	9.8485	0.704
^{238}U	^{206}Pb	1.5513	4.468

nuclear explosions and are so short-lived that they last only a fraction of a second. Other short-lived isotopes are produced by collisions between cosmic rays and atoms in the upper atmosphere and have short half-lives lasting minutes or days. A number of naturally occurring isotopes have half-lives of thousands of years (kiloyear, ka), millions of years (megayear, Ma) or billions of years (gigayear, Ga), and can be used to determine the ages of geological events.

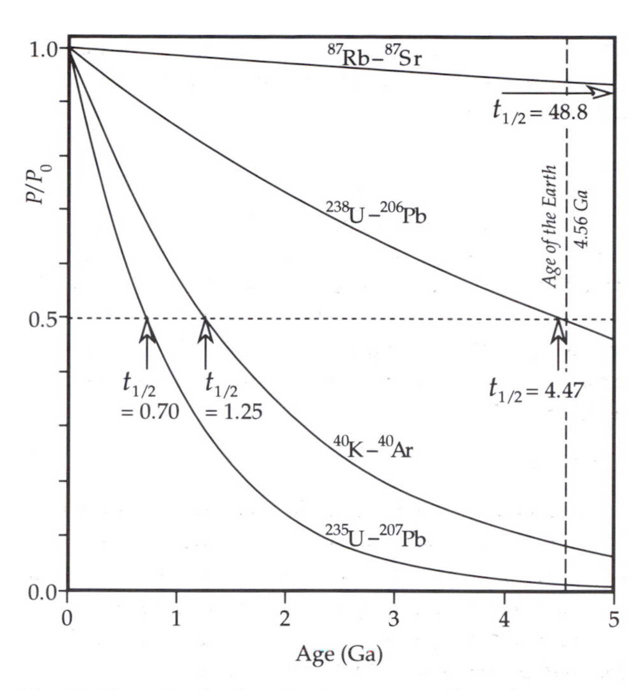

Fig. 4.4 Normalized radioactive decay curves of some important isotopes for dating the Earth and solar system. The arrows indicate the respective half-lives in 10^9 yr (data source: Dalrymple, 1991).

4.1.4 **Radiometric age determination**

Each age-dating scheme involves precise measurement of the concentration of an isotope. This is usually very small. If the radioactive decay has advanced too far, the resolution of the method deteriorates. The best results for a given isotopic decay scheme are obtained for ages less than a few half-lives of the decay. The decay constants and half-lives of some radioactive isotopes commonly used in dating geological events are listed in Table 4.1 and illustrated in Fig. 4.4. Historical and archeological artefacts can be dated by the radioactive carbon method.

4.1.4.1 *Radioactive carbon*

The Earth is constantly being bombarded by cosmic radiation from outer space. Collisions of cosmic particles with atoms of oxygen and nitrogen in the Earth's atmosphere produce high-energy neutrons. These in turn collide with nitrogen nuclei, transforming them into ^{14}C, a radioactive isotope of carbon. ^{14}C decays by β-particle emission to ^{14}N with a half-life of 5730 yr. The production of new ^{14}C is balanced by the loss due to decay, so that a natural equilibrium exists. Photosynthesis in animals and plants replenish living tissue using carbon dioxide, which contains a steady proportion of ^{14}C. When an organism dies, the renewal stops and the residual ^{14}C in the organism decays radioactively.

The radioactive carbon method is a simple decay analysis based on Eq. (4.6). The remaining proportion P of ^{14}C is measured by counting the current rate of β-particle activity, which is proportional to P. This is compared to the original equilibrium concentration P_0. The time since the onset of decay is calculated by solving Eq. (4.6) using the decay rate for ^{14}C ($\lambda = 1.21 \times 10^{-4}$ yr^{-1}).

The radioactive carbon method has been valuable in dating events in the Holocene epoch, which covers the last 10,000 yr of geological time, as well as events related to human prehistory. Unfortunately, human activity has disturbed the natural equilibrium of the replenishment and decay scheme. The concentration of ^{14}C in atmospheric carbon has changed dramatically since the start of the industrial age. Partly this is due to the combustion of fossil fuels like coal and oil as energy sources; they have long lost any ^{14}C and dilute its presence in atmospheric carbon. In the past half-century atmospheric testing of nuclear weapons doubled the concentration of ^{14}C in the atmosphere. Older materials may still be dated by the ^{14}C method, although natural fluctuations in P_0 must be taken into account. These are due to variations in intensity of the geomagnetic field, which acts as a partial shield against the cosmic radiation.

4.1.4.2 *The mass spectrometer*

In the years before World War II, physicists invented the mass spectrometer, an instrument for measuring the mass of

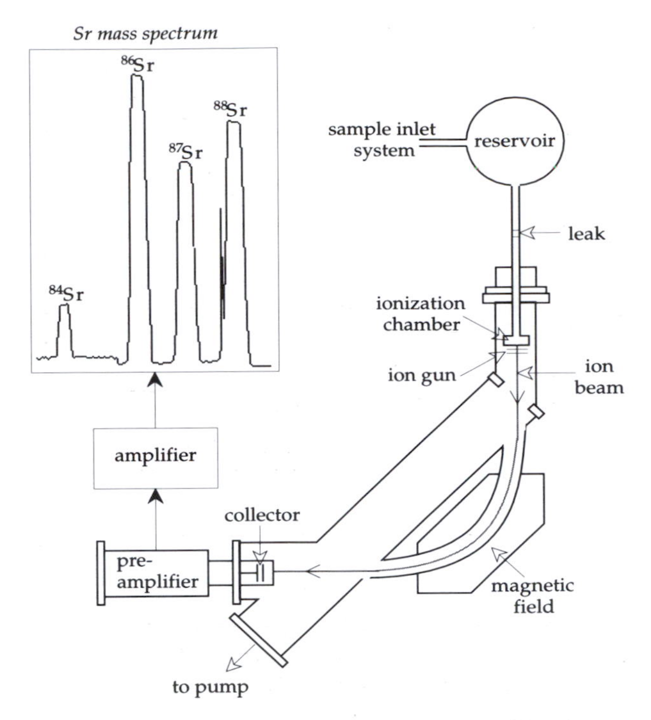

Fig. 4.5 Schematic design of a mass spectrometer, and (*inset*) hypothetical mass spectrum for analysis of Sr isotopes (after York and Farquhar, 1972).

an ion. The instrument was further refined during the development of the atomic bomb. After the war it was adopted into the earth sciences to determine isotopic ratios and became a vital part of the process of isotopic age determination.

The mass spectrometer (Fig. 4.5) utilizes the different effects of electrical and magnetic fields on a charged particle, or ion. First, the element of interest is extracted from selected mineral fractions or from pulverized whole rock. The extract is purified chemically before being introduced into the mass spectrometer, where it is ionized. For the analysis of a gas like argon, bombardment by a stream of electrons may be used. Solid elements, such as potassium, rubidium, strontium or uranium, are vaporized by depositing a sample on an electrically heated filament, or by heating with a high-energy laser beam. The ions then enter an evacuated chamber and pass through an 'ion gun', where they are accelerated by an electrical field and filtered by a velocity selector. This device uses electrical and magnetic fields at right angles to the ion beam to allow only the passage of ions with a selected velocity v. The ion beam is next subjected to a powerful uniform magnetic field B at right angles to its direction of motion. An ion with charge q experiences a Lorentz force (see §5.2.4) equal to (qvB) perpendicular to its velocity and to the magnetic field. Its trajectory is bent to form a circular arc of radius r; the centrifugal force on the particle is equal to (mv^2/r). The curved path focuses the beam on a collector device that measures the intensity of the incident beam. The radius of the circular arc is given by equating the Lorentz and centrifugal forces:

$$r=\frac{m}{B}\frac{v}{q} \tag{4.10}$$

The focal point of the path is determined by the mass of the ion and the strength of the magnetic field B. In the case of a strontium analysis, the ion beam leaving the ion gun contains the four isotopes ^{88}Sr, ^{87}Sr, ^{86}Sr and ^{84}Sr. The beam splits along four paths, each with a different curvature. The magnetic field is adjusted so that only one isotope at a time falls on the collector. The incident current is amplified electronically and recorded. A spectrum is obtained with peaks corresponding to the incidence of individual isotopes (Fig. 4.5, *inset*). The intensity of each peak is proportional to the abundance of the isotope, which can be measured with a precision of about 0.1%. However, the relative peak heights give the relative abundances of the isotopes to a precision of about 0.001%.

An important development in the field of mass spectrometry is the ion microprobe mass analyser. In conventional mass spectrometry the analysis of a particular element is preceded by separating it chemically from a rock sample. The study of individual minerals or the variation of isotopic composition in a grain is very difficult. The ion microprobe avoids contamination problems that can arise during the chemical separation and its high resolution permits isotopic analysis of very small volumes. Before a sample is examined in the instrument its surface is coated with gold or carbon. A narrow beam of negatively charged oxygen ions, about 3–10 μm wide, is focused on a selected grain. Ions are sputtered out of the surface of the mineral grain by the impacting ion beam, accelerated by an electrical field and separated magnetically as in a conventional mass spectrometer. The instrument allows description of isotopic concentrations and distributions in the surface layer of the grain and provides isotopic ratios. It allows isotopic dating of individual mineral grains in a rock.

4.1.4.3 *Rubidium–strontium*

The use of a radioactive decay scheme as given by Eq. (4.7) assumes that the amount of daughter isotope in a sample has been created only by the decay of a parent isotope in a closed system. Usually, however, an unknown initial amount of the daughter isotope is present, so that the amount measured is the sum of the initial concentration D_0 and the fraction derived from decay of the parent P_0. The decay equation is modified to

$$D = D_0 + P = (e^{\lambda t} - 1) \tag{4.11}$$

The need to know the amount of initial daughter isotope D_0 is obviated by the analytical method, which makes use of a third isotope of the daughter element to *normalize* the concentrations of daughter and parent isotopes. The rubidium–strontium method illustrates this technique.

Radioactive rubidium (^{87}Rb) decays by β-particle emission to radiogenic strontium (^{87}Sr). The non-radiogenic, stable isotope ^{86}Sr has approximately the same abundance as the radiogenic product ^{87}Sr, and is chosen for normalization. Writing ^{87}Rb for P, ^{87}Sr for D in Eq. (4.11) and dividing both sides by ^{86}Sr gives

$$\left(\frac{^{87}\mathrm{Sr}}{^{86}\mathrm{Sr}}\right) = \left(\frac{^{87}\mathrm{Sr}}{^{86}\mathrm{Sr}}\right)_0 + \left(\frac{^{87}\mathrm{Rb}}{^{86}\mathrm{Sr}}\right)(e^{\lambda t} - 1) \tag{4.12}$$

In a magmatic rock, the isotopic ratio $(^{87}Sr/^{86}Sr)_0$ is uniform in all minerals precipitated from the melt because the isotopes are chemically identical (i.e., they have the same atomic number). However, the proportion of the different elements Rb and Sr varies from one mineral to another. Eq. (4.12) can be compared with the equation of a straight line, such as

$$y = y_0 + mx \tag{4.13}$$

The ratio $^{87}Sr/^{86}Sr$ is the dependent variable, y, and the ratio $^{87}Rb/^{86}Sr$ is the independent variable, x. If we measure the isotopic ratios in several samples of the rock and plot the ratio $^{87}Sr/^{86}Sr$ against the ratio $^{87}Rb/^{86}Sr$, we get a straight line, called an *isochron* (Fig. 4.6). The intercept with the ordinate axis gives the initial ratio of the daughter isotope. The slope (m) of the line gives the age of the rock, using the known decay constant λ:

$$t = \frac{1}{\lambda}\ln(1+m) = 7.042 \times 10^{10} \ln(1+m) \tag{4.14}$$

Because of its long half-life of 48.8 Ga (Fig. 4.4) the Rb–Sr method is well suited for dating very old events in Earth's history. It has been used to obtain the ages of meteorites and lunar samples, as well as some of the oldest rocks on Earth. For example, the slope of the Rb–Sr isochron in Fig. 4.6 yields an age of 3.55 Ga for the Early Precambrian (Archean) Uivak gneisses from eastern Labrador.

The Rb–Sr and other methods of isotopic dating can be applied to whole rock samples or to individual minerals separated from the rock. The decay equation applies only to a closed system, i.e., to rocks or minerals which have undergone no loss or addition of the parent or daughter isotope since they formed. A change is more likely in a small mineral grain than in the rock as a whole. Individual mineral isochrons may express the age of postformational metamorphism, while the whole rock isochron gives an older age.

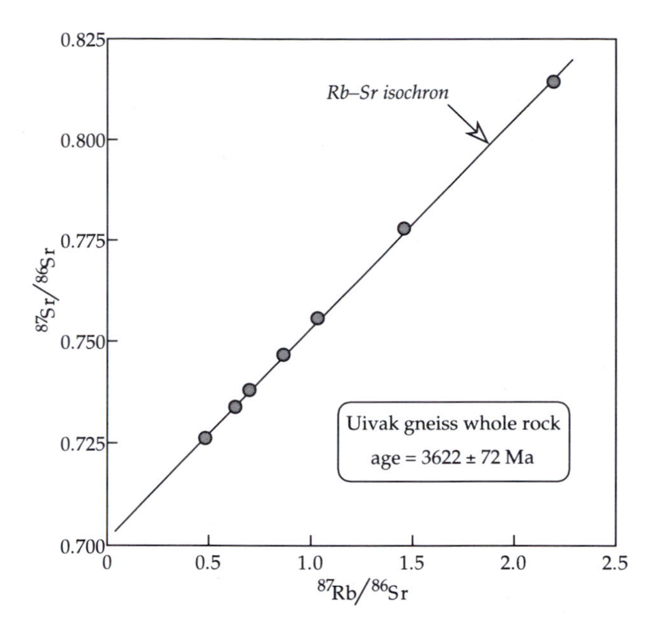

Fig. 4.6 Rb/Sr isochron for the Uivak gneisses from eastern Labrador. The slope of the isochron gives an age of 3.622±0.072 Ga (after Hurst *et al.*, 1975).

On the other hand, spurious ages may result from whole rock analyses if samples of different origin, and hence differing composition, are inadvertently used to construct an isochron.

4.1.4.4 *Potassium–argon*

For several reasons the potassium–argon (K–Ar) method is probably the age-dating technique most commonly used by geologists. The parent isotope, potassium, is common in rocks and minerals, while the daughter isotope, argon, is an inert gas that does not combine with other elements. The half-life of 1250 Ma (1.25 Ga) is very convenient. On the one hand, the Earth's age is equal to only a few half-lives, so radiogenic ^{40}K is still present in the oldest rocks; on the other hand, enough of the daughter isotope ^{40}Ar accumulates in 10^4 yr or so to give fine resolution. In the late 1950s the sensitivity of mass spectrometers was improved by constructing instruments that could be pre-heated at high temperature to drive off contaminating atmospheric argon. This made it possible to use the K–Ar method for dating lavas as young as a few million years.

Radioactive ^{40}K constitutes only 0.01167% of the K in rocks. It decays in two different ways; (a) by β-particle emission to $^{40}Ca_{20}$ with decay rate λ_{Ca}, and (b) by electron capture to $^{40}Ar_{18}$ with decay rate λ_{Ar}. The combined decay constant ($\lambda = \lambda_{Ca} + \lambda_{Ar}$) is equal to 5.543×10^{-10} yr^{-1}. The decay schemes are, respectively:

(a) $^{40}K_{19} \Rightarrow {}^{40}Ca_{20} + \beta^-$

(b) $^{40}K_{19}+e \Rightarrow {}^{40}Ar_{18}$ (4.15)

Electron capture by a nucleus is more difficult and rare than β-particle emission, so the decay of $^{40}K_{19}$ to $^{40}Ca_{20}$ is more common than the formation of $^{40}Ar_{18}$. The ratio of electron capture to β-particle decay is called the *branching ratio*; it equals 0.117. Thus, only the fraction $\lambda_{Ar}/(\lambda_{Ar}+\lambda_{Ca})$, or 10.5%, of the initial radioactive potassium decays to argon. The initial amount of radiogenic ^{40}Ca usually cannot be determined, so the decay to Ca is not used. Allowing for the branching ratio, the K–Ar decay equation is

$$^{40}Ar=0.105\,{}^{40}K(e^{\lambda t-1}) \quad (4.16)$$

The potassium–argon method is an exception to the need to use isochrons. It is sometimes called an accumulation clock, because it is based on the amount of ^{40}Ar that has accumulated. It involves separate measurements of the concentrations of the parent and daughter isotopes. The amount of ^{40}K is a small but constant fraction (0.01167%) of the total amount of K, which can be measured chemically. The ^{40}Ar is determined by mixing with a known amount of another isotope ^{38}Ar before being introduced into the mass spectrometer. The *relative* abundance of the two argon isotopes is measured and the concentration of ^{40}Ar is found using the known amount of ^{38}Ar. By re-ordering Eq. (4.16) and substituting for the decay constant, the age of the rock is obtained from the K–Ar age equation:

$$t=1.804\times10^{9}\ln\left(9.524\frac{^{40}Ar}{^{40}K}+1\right) \quad (4.17)$$

The ^{40}Ar in a molten rock easily escapes from the melt. It may be assumed that all of the radiogenic ^{40}Ar now present in a rock has formed and accumulated since the solidification of the rock. The method works well on igneous rocks that have not been heated since they formed. It cannot be used in sedimentary rocks that consist of the detritus of older rocks. Often it is unsuccessful in metamorphic rocks, which may have complicated thermal histories. A heating phase may drive out the argon, thereby re-setting the accumulation clock. This problem limits the usefulness of the K–Ar method for dating meteorites (which have a fiery entry into Earth's atmosphere) and very old terrestrial rocks (because of their unknown thermal histories). The K–Ar method can be used for dating lunar basalts, as they have not been reheated since their formation.

4.1.4.5 *Argon–argon*

Some uncertainties related to post-formational heating of a rock are overcome in a modification of the K–Ar method that uses the $^{40}Ar/^{39}Ar$ isotopic ratio. The method requires conversion of the ^{39}K in the rock to ^{39}Ar. This is achieved by irradiating the sample with fast neutrons in an atomic reactor.

The terms *slow* and *fast* refer to the energy of the neutron radiation. The energy of slow neutrons is comparable to their thermal energy at room temperature; they are also referred to as *thermal* neutrons. Slow neutrons can be captured and incorporated into a nucleus, changing its size without altering its atomic number. The capture of slow neutrons can increase the size of unstable uranium nuclei beyond a critical value and initiate fission. In contrast, fast neutrons act on a nucleus like projectiles. When a fast neutron collides with a nucleus it may eject a neutron or proton, while itself being captured. If the ejected particle is another neutron, no effective change results. But if the ejected particle is a proton (with a β-particle to conserve charge), the atomic number of the nucleus is changed. For example, bombarding ^{39}K nuclei in a rock sample with fast neutrons converts a fraction of them to ^{39}Ar.

To determine this fraction a control sample of known age is irradiated at the same time. By monitoring the change in isotopic ratios in the control sample the fraction of ^{39}K nuclei converted to ^{39}Ar can be deduced. The age equation is similar to that given in Eq. (4.17) for the K–Ar method. However, ^{39}Ar replaces ^{40}K and an empirical constant J replaces the constant 9.54. The value of J is found from the control sample whose age is known. The $^{40}Ar/^{39}Ar$ age equation is

$$t=1.804\times10^{9}\ln\left(J\frac{^{40}Ar}{^{39}Ar}+1\right) \quad (4.18)$$

In the $^{40}Ar/^{39}Ar$ method the sample is heated progressively to drive out argon at successively higher temperatures. The $^{40}Ar/^{39}Ar$ isotopic ratio of the argon released at each temperature is determined in a mass spectrometer. The age computed for each increment is plotted against the percentage of Ar released. This yields an *age spectrum*. If the rock has not been heated since it was formed, the argon increments given out at each heating stage will yield the same age (Fig. 4.7a). An isochron can be constructed as in the Rb–Sr method by measuring the abundance of a non-radiogenic ^{36}Ar fraction and comparing the isotopic ratios $^{40}Ar/^{36}Ar$ and $^{39}Ar/^{36}Ar$. In an unheated sample all points fall on the same straight line (Fig. 4.7b).

If the rock has undergone postformational heating, the argon formed since the heating is released at lower temperatures than the original argon (Fig. 4.7c). It is not certain why this is the case. The argon probably passes out of the solid rock by diffusion, which is a thermally activated process that depends on both the temperature and the duration of heating. Unless the postformational heating is long and

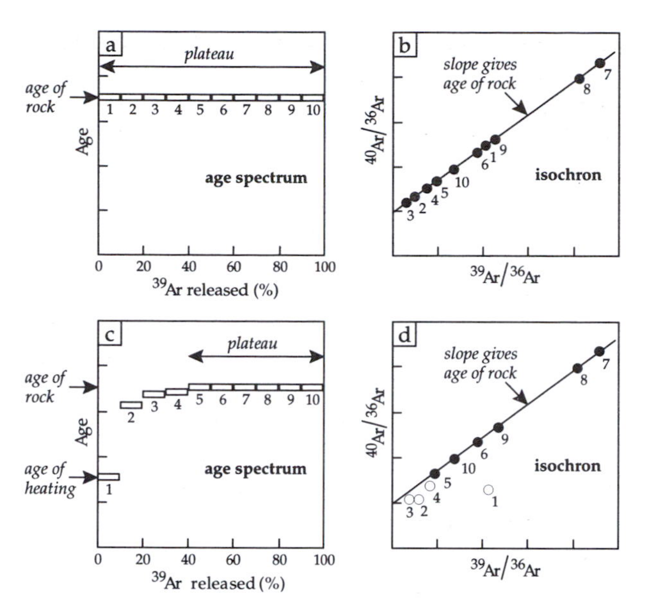

Fig. 4.7 (a) Hypothetical age spectrum and (b) $^{40}Ar/^{39}Ar$ isochron for a sample that has experienced no secondary heating; (c) hypothetical age spectrum and (d) $^{40}Ar/^{39}Ar$ isochron for a sample that was reheated but retains some original argon, (after Dalrymple, 1991).

thorough, only the outside parts of grains may be forced to release trapped argon, while the Ar located in deeper regions of the grains is retained. In a sample that still contains original argon the results obtained at high temperatures form a *plateau* of consistent ages (Fig. 4.7c) from which the optimum age and its uncertainty may be calculated. On an isochron diagram reheated points deviate from the straight line defined by the high-temperature isotopic ratio (Fig. 4.7d).

The high precision of $^{40}Ar/^{39}Ar$ dating is demonstrated by the analysis of samples of melt rock from the Chicxulub impact crater in the Yucatan peninsula of Mexico. The crater is the favored candidate for the impact site of a 10 km diameter meteorite which caused global extinctions at the close of the Cretaceous period. Laser heating was used to release the argon isotopes. The age spectra of three samples show no argon loss (Fig. 4.8). The plateau ages are precisely defined and give a weighted mean age of 64.98±0.05 Ma for the impact. This agrees closely with the 65.01±0.08 Ma age of tektite glass found at a Cretaceous–Tertiary boundary section in Haiti, and so ties the impact age to the faunal extinctions.

The age spectrum obtained in Ar–Ar dating of the Menow meteorite is more complicated and shows the effects of ancient argon loss. Different ages are obtained during incremental heating below about 1200 °C (Fig. 4.9). The plateau ages above this temperature indicate a mean age of 4.48±0.06 Ga. The shape of the age spectrum suggests that about 25% of the Ar was lost much later, about 2.5 Ga ago.

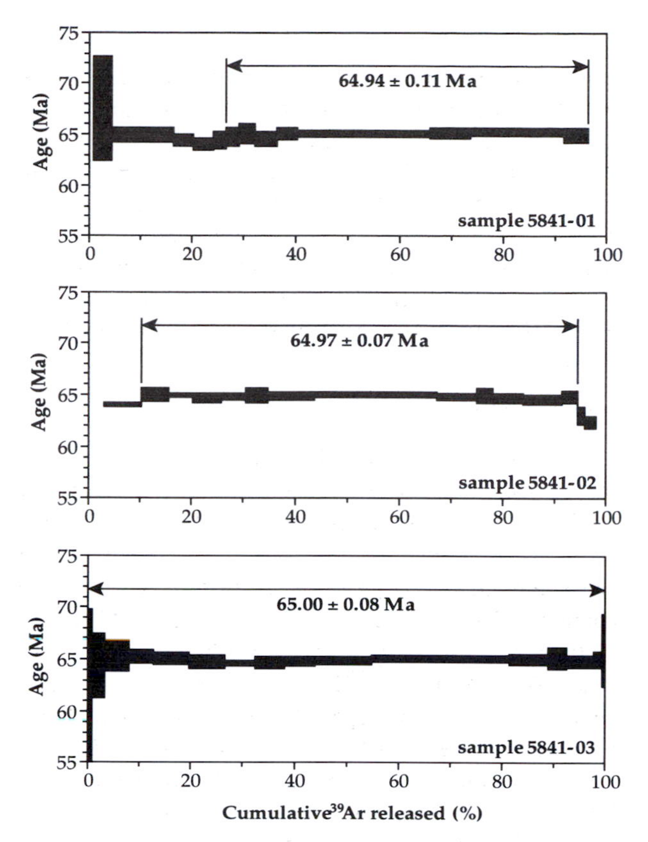

Fig. 4.8 $^{40}Ar/^{39}Ar$ age spectra for three samples of melt rocks from the Chicxulub Cretaceous–Tertiary impact crater (after Swisher *et al.*, 1992).

4.1.4.6 *Uranium–lead: the concordia–discordia diagram*

Uranium isotopes decay through a series of intermediate radioactive daughter products, but eventually they result in stable end-product isotopes of lead. Each of the decays is a multi-stage process but can be described as though it had a single decay constant. We can describe the decay of ^{238}U to ^{206}Pb by

$$\frac{^{206}\mathrm{Pb}}{^{238}\mathrm{U}}=e^{\lambda_{238}t}-1 \tag{4.19}$$

Likewise the decay of ^{235}U to ^{207}Pb can be written

$$\frac{^{207}\mathrm{Pb}}{^{235}\mathrm{U}}=e^{\lambda_{235}t}-1 \tag{4.20}$$

A graph of the $^{206}Pb/^{238}U$ ratio against the $^{207}Pb/^{235}U$ ratio is a curve, called the *concordia* line (Fig. 4.10). All points on this line satisfy Eqs. (4.19) and (4.20) at the same

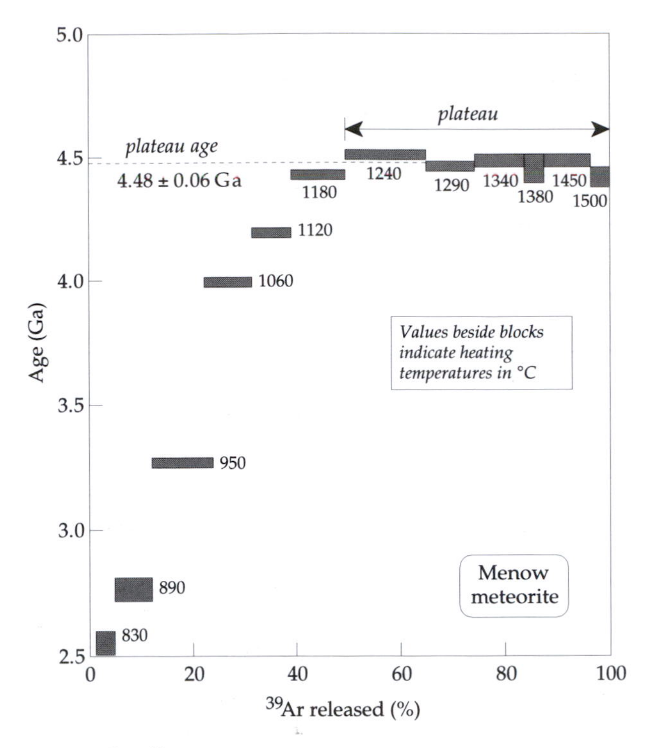

Fig. 4.9 $^{40}Ar/^{39}Ar$ age spectrum for the Menow meteorite, showing a plateau at 4.48±0.06 Ga (after Turner *et al.*, 1978).

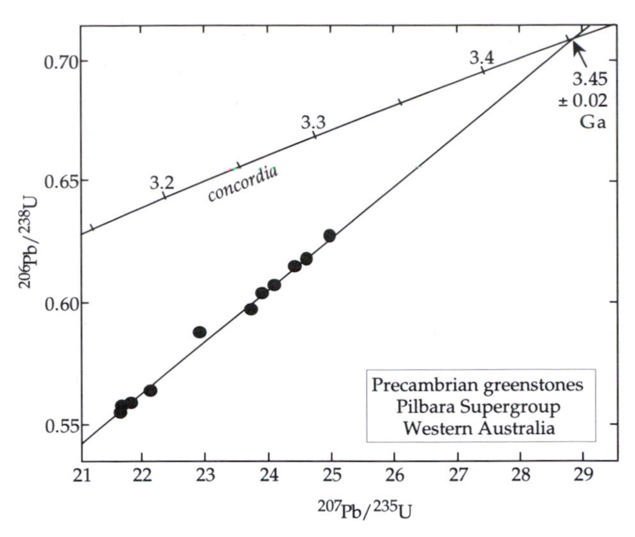

Fig. 4.11 U–Pb concordia–discordia diagram for zircon grains from a volcanic rock in the Precambrian Duffer formation in the Pilbara Block, Western Australia (after Pidgeon, 1978).

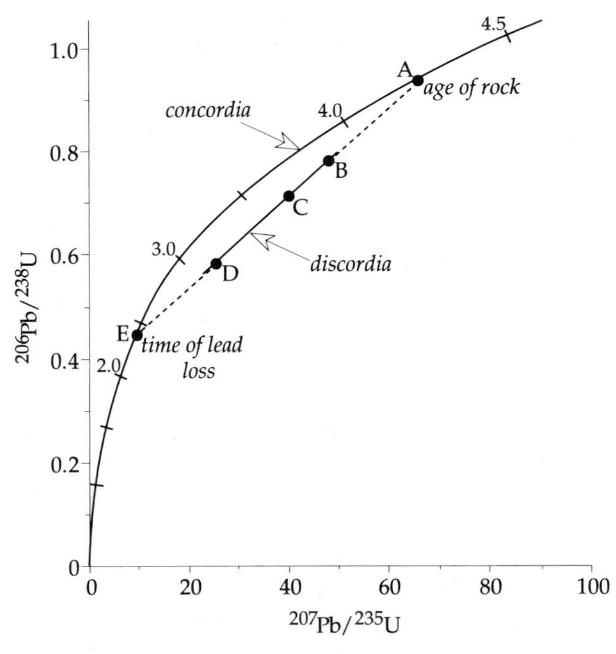

Fig. 4.10 Hypothetical example of a U–Pb concordia–discordia diagram. Lead loss gives points B, C and D on a discordia line. It intersects the concordia curve at A, the age of the rock, and at E, the time since the event that caused lead loss. Marks on the concordia curve indicate age in Ga.

time t. The uranium isotopes decay at different rates, λ_{238} and λ_{235} respectively, which causes the concordia line to have a curved shape.

The amounts of the daughter lead isotopes accumulate at different rates. Lead is a volatile element and is easily lost from minerals, but this does not alter the isotopic ratio of the lead that remains. Loss of lead will cause a point to deviate from the concordia line. However, because the isotopic ratio remains constant the deviant point lies on a straight line between the original age and the time of the Pb-loss. This line corresponds to points that do not agree with the concordia curve; it is called the *discordia* line (Fig. 4.10). Different mineral grains in the same rock experience different amounts of lead loss, so the isotopic ratios in these grains give different points B, C and D on the discordia. The intersection of the concordia and discordia lines at A gives the original age of the rock, or in the case that it has lost all of its original lead, the age of this event. The intersection of the lines at E gives the age of the event that caused the lead loss.

The U–Pb method has been used to date some of the oldest rocks on Earth. The Duffer formation in the Pilbara Supergroup in Western Australia contains Early Precambrian greenstones. U–Pb isotopic ratios for zircon grains separated from a volcanic rock in the Duffer formation define a discordia line that intercepts the theoretical concordia curve at 3.45±0.02 Ga (Fig. 4.11).

4.1.4.7 *Lead–lead isochrons*

It is possible to construct isochron diagrams for the U–Pb decay systems. This is done (as in the Rb/Sr decay scheme) by expressing the radiogenic isotopes ^{206}Pb and ^{207}Pb as ratios of ^{204}Pb, the non-radiogenic isotope of lead. The decay of ^{238}U to ^{206}Pb (or of ^{235}U to ^{207}Pb) can then be described as in Eq. (4.11), where the initial value of the daughter product is unknown. This gives the decay equations

$$\frac{^{207}Pb}{^{204}Pb}-\left(\frac{^{207}Pb}{^{204}Pb}\right)_0=\frac{^{235}U}{^{204}Pb}(e^{\lambda_{235}t}-1) \tag{4.21}$$

$$\frac{^{206}Pb}{^{204}Pb}-\left(\frac{^{206}Pb}{^{204}Pb}\right)_0=\frac{^{238}U}{^{204}Pb}(e^{\lambda_{238}t}-1) \tag{4.22}$$

These equations can be combined into a single isochron equation:

$$\frac{\dfrac{^{207}Pb}{^{204}Pb}-\left(\dfrac{^{207}Pb}{^{204}Pb}\right)_0}{\dfrac{^{206}Pb}{^{204}Pb}-\left(\dfrac{^{206}Pb}{^{204}Pb}\right)_0}=\frac{^{235}U\ (e^{\lambda_{235}t}-1)}{^{238}U\ (e^{\lambda_{238}t}-1)} \tag{4.23}$$

The ratio of ^{235}U to ^{238}U as measured today has been found to have a constant value 1/137.88 in lunar and terrestrial rocks and in meteorites. The decay constants are well known, so for a given age t the right side in Eq. (4.23) is constant. The equation then has the form

$$\frac{y-y_0}{x-x_0}=m \tag{4.24}$$

This is the equation of a straight line of slope m through the point (x_0, y_0). The initial values of the lead isotope ratios are not known, but a plot of the isotopic ratio $^{207}Pb/^{204}Pb$ against the ratio $^{206}Pb/^{204}Pb$ is a straight line. The age of the rock cannot be found algebraically. Values for t must be inserted successively on the right side of Eq. (4.23) until the observed slope is obtained.

4.1.5 Ages of the Earth and solar system

Radiometric dates from three sources – terrestrial rocks, lunar rocks, and meteorites – imply that, if the Earth originated at the same time as meteorites, it is about 4.5–4.6 Ga old (Fig. 4.12). Among the oldest terrestrial rocks are the Isua metasediments from western Greenland, which have been extensively studied and give an age of 3.77 Ga that is consistent with different decay schemes. Even older are the Acasta gneisses in the northwestern part of the Canadian shield, which gave a U–Pb age of 3.96 Ga. Zircon crystals from sedimentary rocks in Australia have yielded U–Pb ages of up to 4.3 Ga. More commonly, however, terrestrial rocks from the oldest Precambrian shield areas in Australia, South Africa, South America and Antarctica give *maximum* ages of 3.4–3.8 Ga.

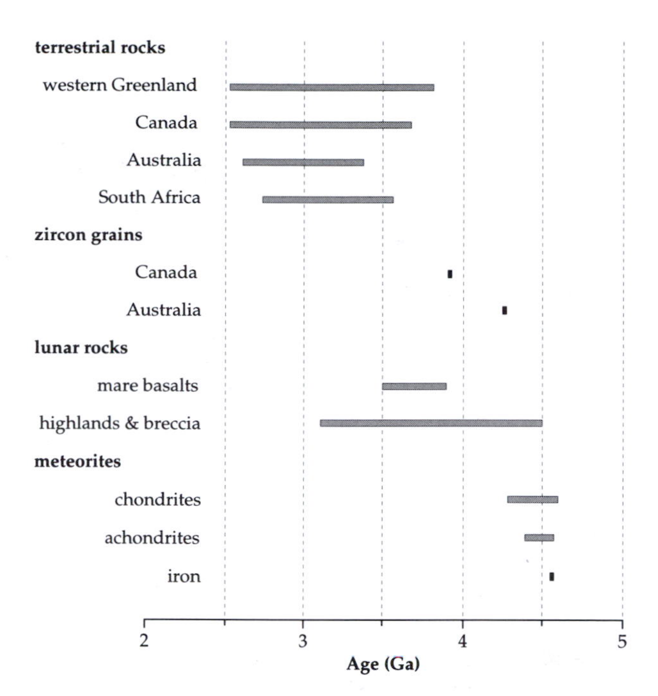

Fig. 4.12 Radiometric age ranges of the oldest terrestrial and lunar rocks and meteorites (compiled from Dalrymple, 1991).

Six American manned missions and three Russian unmanned missions obtained samples of the Moon's surface which have been dated by several isotopic techniques. The ages obtained differ according to the source areas of the rocks. The dark areas of the Moon's surface, the so-called lunar seas or *maria*, were formed when enormous outpourings of basaltic lava filled up low-lying areas and created a flat surface. The lunar volcanism may have persisted until 1 Ga ago. The light areas on the Moon's surface are rough, extensively cratered highlands that reach elevations of 3000–4000 m above the maria. They represent the top part of the lunar crust and are the oldest regions of the Moon. Frequent collisions with large asteroids early in the Moon's history produced numerous craters and pulverized the lunar crust, leaving impact breccia, rock fragments and a dust layer a few meters thick, called the lunar regolith. Age dates from the highland rocks range from about 3.5–4.5 Ga, but the oldest age reported for a lunar rock is 4.51±0.07 Ga, obtained by the Rb–Sr method.

An important source of information concerning the Earth's age is the dating of meteorites. A *meteor* is a piece of solid matter from space that penetrates the Earth's atmosphere at a hypersonic speed of typically 10–20 km/s. Atmospheric friction causes it to become incandescent. Outside the Earth's atmosphere it is known as a *meteoroid*; any part that survives passage through the atmosphere and reaches the Earth's surface is called a *meteorite*. Most meteorites are thought to originate in the asteroid belt between the orbits of Mars and Jupiter (see §1.1.3.3), although tracking of entry paths shows that before colliding with Earth they have highly elliptical counterclockwise orbits about the Sun (in the same sense as the planets). Meteorites are often named after the place on Earth where they are found. They can be roughly divided into three main classes according to their composition. Iron meteorites consist of an alloy of iron and nickel; stony meteorites consist of silicate minerals; and iron–stony meteorites are a mixture of the two. The stony meteorites are further subdivided into chondrites and achondrites. *Chondrites* contain small spherules of high-temperature silicates, and constitute the largest fraction (more than 85%) of recovered meteorites. The *achondrites* range in composition from rocks made up essentially of single minerals like olivine to rocks resembling basaltic lava. Each category is further subdivided on the basis of chemical composition. All main types have been dated isotopically, with most studies being done on the dominant chondrite fraction. There are no obvious age differences between the meteorites of the various groups. Chondrites, achondrites and iron meteorites consistently yield ages of around 4.45–4.50 Ga (Fig. 4.12).

Thus, the oldest isotopically dated terrestrial rocks have ages about 3.6–3.8 Ga, but comparison with the ages of lunar rocks and meteorites suggests that the Earth is about 4.5–4.6 Ga old. This leaves about one-half to three-quarters of a billion years of early Earth history unaccounted for. It is possible that the oldest rocks on Earth have simply not yet been found and dated. Some speculative theories hold that the original crust was destroyed by absorption into the mantle or by intense bombardment by meteorites. The true explanation is not known. What happened in the first half-billion years of its history remains one of Earth's best-kept secrets.

4.2 THE EARTH'S HEAT

4.2.1 Introduction

The radiant energy from the Sun, in conjunction with gravitational energy, determines almost all natural processes that occur at or above the Earth's surface. The hot incandescent Sun emits radiation in a very wide range of wavelengths. The radiation incident on the Earth is largely reflected into space, part enters the atmosphere and is reflected by the clouds or is absorbed and re-radiated into space. A very small part reaches the surface, where it is also partly reflected, especially from the water surfaces that cover three-quarters of the globe. Some is absorbed (e.g., by vegetation) and serves as the source of power for various natural cycles. A small fraction is used to heat up the Earth's surface, but it only penetrates a short distance, some tens of centimeters in the case of the daily cycle and a few tens of meters for the annual changes. As a result, solar energy has negligible influence on internal terrestrial processes. Systems as diverse as the generation of the geomagnetic field and the motion of global lithospheric plates are ultimately powered by the Earth's internal heat.

The Earth is constantly losing heat from its interior. Although diminutive compared to solar energy, the loss of internal heat is many times larger than the energy lost by other means, such as the change in Earth's rotation and the energy released in earthquakes (Table 4.2). Tidal friction slows down the Earth's rotation, and the change can be monitored accurately with modern technology such as very long baseline interferometry (VLBI) and the satellite-based geodetic positioning system (GPS). The associated loss of rotational energy can be computed accurately. The elastic energy released in an earthquake can be estimated reliably, and it is known that most of the energy is released in a few large shocks. However, the annual number of large earthquakes is very variable. The number with magnitude $M_s > 7$ varies between about 10 and 40 (see Fig. 3.45), giving estimates of the annual energy release from about 5×10^{17} J to 4×10^{19} J. The energies of tidal deceleration and earthquakes are small fractions of the geothermal flux, which is the most important form of energy originating in the body of the Earth.

The Earth's internal heat derives from several sources (§4.2.5). For the past 4 Ga or so the Earth's heat has been obtained from two main sources. One source is the cooling of the Earth since its early history, when internal temperatures were much higher than they now are. The other source is the heat produced by the decay of long-lived radioactive isotopes. This is the main source of the Earth's internal heat, which, in turn, powers all geodynamic processes.

4.2.2 Thermodynamic principles

In order to describe thermal energy it is necessary to define clearly some important thermodynamic parameters. The concepts of temperature and heat are easily – and frequently – confused. Temperature – one of the seven fundamental standard parameters of physics – is a quantitative measure of the degree of hotness or coldness of an object relative to some standard. Heat is a form of energy which an object possesses by virtue of its temperature. The difference between temperature and heat is illustrated by a simple example. Imagine a container in which the molecules of a gas move around at a certain speed. Each molecule has a kinetic energy proportional to the square of its velocity. There may be differences from one molecule to the next but

Table 4.2 *Estimates of notable contributions to the Earth's annual energy budget*

Energy source	Annual energy (joule)	Normalized (geothermal flux = 1)
Reflection and re-radiation of solar energy	5.4×10^{24}	$\approx$4000
Geothermal flux from Earth's interior	1.4×10^{21}	1
Rotational deceleration by tidal friction	$\approx10^{20}$	$\approx$0.1
Elastic energy in earthquakes	$\approx10^{19}$	$\approx$0.01

it is possible to determine the mean kinetic energy of a molecule. This quantity is proportional to the temperature of the gas. If we add up the kinetic energies of all molecules in the container we obtain the amount of heat it contains. If heat is added to the container from an external source, the gas molecules speed up, their mean kinetic energy increases and the temperature of the gas rises.

The change of temperature of a gas is accompanied by changes of pressure and volume. If a solid or liquid is heated, the pressure remains constant but the volume increases. Thermal expansion of a suitable solid or liquid forms the principle of the thermometer for measuring temperature. Although Galileo reputedly invented an early and inaccurate 'thermoscope', the first accurate thermometers – and corresponding temperature scales – were developed in the early 18th century by Gabriel Fahrenheit (1686–1736), Ferchaut de Réaumur (1683–1757) and Anders Celsius (1701–1744). Their instruments utilized the thermal expansion of liquids and were calibrated at fixed points such as the melting point of ice and the boiling point of water. The Celsius scale is the most commonly used for general purposes, and it is closely related to the scientific temperature scale.

Temperature apparently has no upper limit. For example, the temperature of the surface of the Sun is less than 10,000 K but the temperature at its center is around 10,000,000 K and temperatures greater than 100,000,000 K have been achieved in physics experiments. But as heat is removed from an object it becomes more and more difficult to lower its temperature further. The limiting low temperature is often called 'absolute zero' and is taken as the zero of the *Kelvin temperature scale*, named in honor of Lord Kelvin. Its divisions are the same as the Celsius scale and the temperature unit is called a *kelvin*. The scale is defined so that the triple point of water – where the solid, liquid and gaseous phases of water can coexist in equilibrium – is equal to 273.16 kelvins, written 273.16 K.

Heat was imagined by early investigators to be exchanged between bodies by the flow of a mystic fluid, called *caloric*. However, in the mid-19th century James Joule, an English brewer, demonstrated in a series of careful experiments that mechanical energy could be converted into heat. In his famous experiment, falling weights drove a paddle wheel in a container of water, raising its temperature. The increase was tiny, less than 0.3 K, yet Joule was able to compute the amount of energy needed to raise the temperature by 1 K. His estimate of this energy – called the *mechanical equivalent of heat* – was within 5% of our modern value. The unit of energy is called the *joule* in recognition of his pioneering efforts. Originally, however, the unit of heat energy was defined as the amount needed to raise the temperature of one gram of water from 14.5 °C to 15.5 °C. This unit, the *calorie* (cal), is equivalent to 4.1868 J.

In physics and engineering it is often important to know the change of heat energy in a unit of time, known as the power. The unit of power is the *watt*, named after James Watt, the Scottish engineer who played an important role in harnessing thermal energy as a source of mechanical power. In geothermal problems we are usually concerned with the loss of heat from the Earth per unit area of its surface. This quantity is called the *heat flux* (or more commonly *heat flow*); it is the amount of heat that flows per second across a square meter of surface. The mean heat flow from the Earth is very small and is measured in units of milliwatt per square meter (mW m^{-2}). Until the adoption of SI units and their slow acceptance into geophysical studies, heat flow results were expressed in 'heat flow units' (HFU), defined so that 1 HFU equals 10^{-6} cal cm^{-2} sec^{-1}. A simple conversion shows that one HFU is equivalent to 41.9 mW m^{-2}.

The addition of a quantity of heat ΔQ raises the temperature by an amount ΔT, which is proportional to ΔQ. The larger the mass m of the body, the smaller is the temperature change, and a given amount of heat produces different temperature changes in different materials. The amount of heat needed to raise the temperature of 1 kg of a material by 1 K is called its *specific heat*, denoted c_p for a process that occurs at constant pressure (and c_v when it happens at constant volume). These observations are summarized in the equation

$$\Delta Q = c_p m \Delta T \tag{4.25}$$

The added heat causes a fractional change of volume that is proportional to the temperature change but which differs from one material to another. The material property is called the *volume coefficient of expansion* α, and is defined by the equation

$$\alpha=\frac{1}{V}\left(\frac{\partial V}{\partial T}\right)_{p} \tag{4.26}$$

When thermal energy is added to a system, part is used to increase the internal energy of the system – i.e., the kinetic energy of the molecules – and part is expended as work, for example, by changing the volume. If the change in total energy ΔQ occurs at constant temperature T, we can define a new thermodynamic parameter, the *entropy* S, which changes by an amount ΔS equal to $\Delta Q/T$. Thus we can write

$$\Delta Q=T\Delta S=\Delta U+\Delta W \tag{4.27}$$

where ΔU is the change of internal energy and ΔW is the work done externally. A thermodynamic process in which heat cannot enter or leave the system is said to be *adiabatic*. The entropy of an adiabatic reaction remains constant: $\Delta S=0$. This is the case when a process occurs so rapidly that there is no time for heat transfer. An example is the passage of a seismic wave in which the compressions and rarefactions occur too rapidly for heat to be exchanged. The adiabatic temperature gradient in the Earth serves as an important reference for estimates of the actual temperature gradient and for determining how heat is transferred.

4.2.3 Temperature inside the Earth

In contrast to the radial distributions of density, seismic velocity and elastic parameters, which are known with a good measure of reliability, our knowledge of the temperature inside the Earth is still imprecise. The temperature can only be measured in the immediate vicinity of the Earth's surface, in boreholes and deep mines. As early as 1530 Georgius Agricola (the latinized name of Georg Bauer, a German physician and pioneer in mineralogy and mining) noted that conditions were warmer in deep mines. In fact, near-surface temperatures increase rapidly with depth by roughly 30 K km^{-1}. At this rate, linear extrapolation would give a temperature around 200,000 K at the center of the Earth. This is greater than the temperature of the surface of the Sun and is unrealistically high.

The conditions of high temperature and pressure in the deep interior can be inferred from experiments, and the adiabatic and melting-point temperatures can be computed with reasonable assumptions. Nevertheless, the temperature–depth profile is poorly known and conjectured temperatures have ranged widely. Limits are placed on the actual temperature by the known physical state of the Earth's interior deduced from seismology. The temperature in the solid inner core must be lower than the melting point, while the temperature of the molten outer core is above the melting point. Similarly the temperature in the solid mantle and crust are below the melting point; the asthenosphere has low rigidity because its temperature comes close to the solidus ("softening point"). The relationship of the actual temperature to the melting point determines how different parts of the Earth's interior behave rheologically (see §6.2).

The experimental approach to estimating the variation of temperature with depth combines knowledge obtained from seismology with laboratory results. The travel-times of seismic body waves show that changes in mineral structure (phase transitions) occur at certain depths (see §3.7.5). Important examples are the olivine–spinel transition at 400 km depth and the spinel–perovskite transition at 670 km depth in the upper mantle. The conditions of temperature and pressure (and hence depth) at which these phase transitions take place can be observed in laboratory experiments, so that the temperatures at the transition depths in the Earth can be determined. Similarly, the depth variation of the melting points of mantle rocks and the iron–nickel core can be inferred from laboratory observations at high pressure and temperature. Seismic velocities in the Earth are now so well known that deviations from normal velocities can be determined by seismic tomography (see §3.7.6) and interpreted in terms of temperature anomalies.

4.2.3.1 *The adiabatic temperature gradient*

An alternative way of estimating temperature inside the Earth is by using physical equations in which the parameters are known from other sources. In the late 19th century James Clerk Maxwell expressed the laws of thermodynamics in four simple equations involving entropy (S), pressure (p), temperature (T) and volume (V). One of these equations is

$$\left(\frac{\partial T}{\partial p}\right)_{S}=\left(\frac{\partial V}{\partial S}\right)_{p} \tag{4.28}$$

The left side is the adiabatic change of temperature with pressure, from which we obtain the adiabatic change of temperature with depth by substituting $dp=\rho g\, dz$, as in §3.7.4. Substituting from Eqs. (4.25) and (4.26) we get

$$\left(\frac{dT}{dp}\right)_{S}=T\frac{\alpha V\Delta T}{c_{p}m\Delta T}=T\frac{\alpha}{\rho c_{p}} \tag{4.29}$$

from which we obtain

$$\left(\frac{dT}{dz}\right)_{\text{adiabatic}}=T\frac{\alpha g}{c_{p}} \tag{4.30}$$

The dependence of density and gravity on depth z are known from seismic travel-times, and the profiles of α and c_p

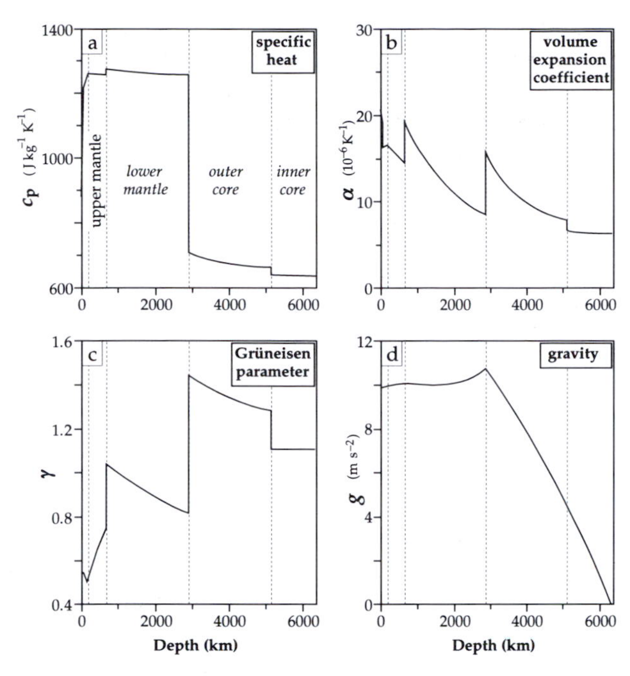

Fig. 4.13 Variations with depth in the Earth of (a) specific heat at constant pressure, (b) volume coefficient of thermal expansion, (c) Grüneisen parameter, and (d) gravity (based upon data from Stacey, 1992).

can be estimated from laboratory observations (Fig. 4.13). For example, in the lower mantle at a depth of 1500 km g=9.9 m s^{-2}, c_p=1200 J kg^{-1} K^{-1}, α=14×10^{-6} K^{-1}, and T=2400 K. This gives an adiabatic temperature gradient of about 0.3 K km^{-1}. In the outer core at about 3300 km depth the corresponding values are: g=10.1 m s^{-2}, c_p=700 J kg^{-1} K^{-1}, α=14×10^{-6} K^{-1}, and T=4000 K and the adiabatic temperature gradient is about 0.8 K km^{-1}.

Approximate estimates of adiabatic temperatures inside the Earth can also be obtained with the aid of the Grüneisen thermodynamic parameter, γ. This is a dimensionless parameter, defined as

$$\gamma=\frac{\alpha K_s}{\rho c_p} \tag{4.31}$$

where K_s is the adiabatic incompressibility or bulk modulus. It is defined in §3.2.4 and Eq. (3.16), which, by writing dp instead of p and dV/V for the dilatation θ, becomes

$$\mathrm{d}p=-K_s\frac{\mathrm{d}V}{V}=K_s\frac{\mathrm{d}\rho}{\rho} \tag{4.32}$$

where ρ is the density. Substituting in Eq. (4.29) gives

$$\frac{\mathrm{d}T}{\mathrm{d}p}=T\frac{\gamma}{K_s}=T\frac{\gamma}{\rho}\frac{\mathrm{d}\rho}{\mathrm{d}p}$$

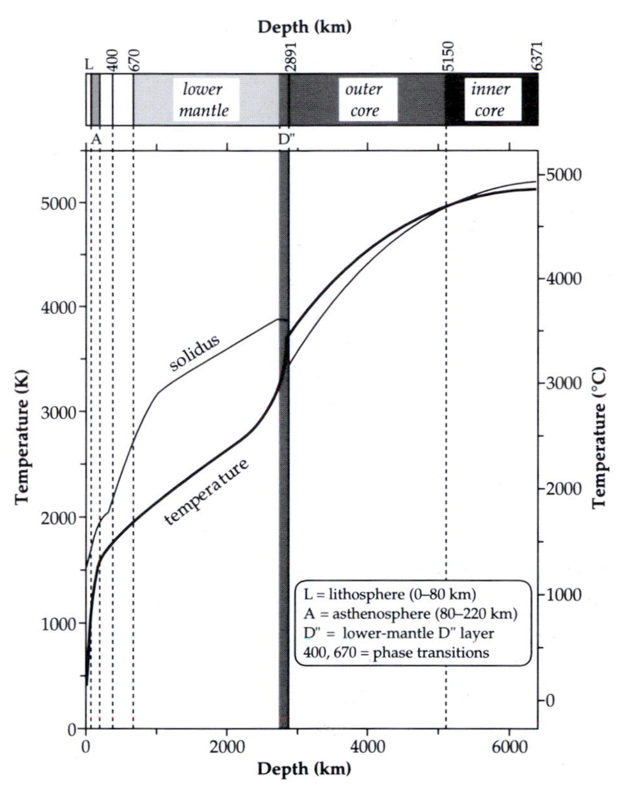

Fig. 4.14 Variations of estimated temperature and melting point with depth in the Earth (based upon data from Stacey, 1992).

$$\frac{\mathrm{d}T}{T}=\gamma\frac{\mathrm{d}\rho}{\rho} \tag{4.33}$$

$$T=T_0\left(\frac{\rho}{\rho_0}\right)^{\gamma} \tag{4.34}$$

With this equation, and knowing the temperature T_0 and density ρ_0 at a given depth, the adiabatic temperature can be computed from the density profile in a region where the Grüneisen parameter γ is known. Fortunately, γ is fairly constant within large regions of the Earth's interior (Fig. 4.13). Clearly, Eq. (4.34) cannot be applied across a boundary between these domains, where γ is discontinuous. If T_0 and ρ_0 are known at calibration points, the adiabatic temperature profile may be computed iteratively within a particular depth interval. A current estimate of the temperature profile in the Earth (Fig. 4.14) has steep gradients in the lithosphere, asthenosphere and in the D" layer above the core–mantle boundary. It indicates a temperature near 3750 K at the core–mantle boundary and a central temperature of about 5100 K.

4.2.3.2 *The melting point gradient*

Another of Maxwell's thermodynamic equations is

$$\left(\frac{\partial S}{\partial p}\right)_T = -\left(\frac{\partial V}{\partial T}\right)_P \tag{4.35}$$

This equation can be applied to the effect of pressure on the melting point of a substance (T_{mp}). The heat required to melt a unit mass of the substance is its latent heat of fusion (L), so the change in entropy on the left hand side of the equation is equal to (mL/T_{mp}). The volume change is the difference between that of the solid phase (V_S) and that of the liquid phase (V_L), so Eq. (4.35) can be rewritten

$$\frac{dT_{mp}}{dp} = \frac{T_{mp}}{mL}(V_S - V_L) \tag{4.36}$$

This is known to physicists as the *Clausius-Clapeyron equation*. It describes the effect of pressure on the melting point, and it is of interest to us because we can easily convert it to give the variation of melting point with depth, assuming that the pressure is hydrostatic so that $dp = \rho g\, dz$ as previously. For a given mass m of the substance we can replace the volumes V_S and V_L with the corresponding densities ρ_S and ρ_L of the solid and liquid phases, respectively, so that

$$\frac{1}{T_{mp}}\frac{dT_{mp}}{dz} = \frac{g}{L}\left(\frac{\rho_S}{\rho_L} - 1\right) \tag{4.37}$$

Again, to obtain the depth distribution of the melting point the variations of density and gravity with depth in the Earth are needed. At outer core pressures the densities of the solid and liquid phases of iron are about 13,000 kg m^{-3} and 11,000 kg m^{-3}, respectively, and the latent heat of fusion of iron is about 7×10^6 J kg^{-1}, so the gradient of the melting point curve in the outer core is about 1 K km^{-1}, i.e., the melting point in the core increases more steeply with depth than the adiabatic temperature. The computations of the adiabatic and melting temperature curves depend on parameters (e.g., L, α, c_p) that are not known with a great degree of reliability in the Earth so the temperature profiles (Fig. 4.14) will undoubtedly change and become more secure as basic knowledge improves.

One factor that must still be evaluated is the role of phase transitions in the mantle. The D" layer just above the core-mantle boundary evidently plays a crucial role in transferring heat from the core to the mantle. It constitutes a thermal boundary layer. Likewise the lithosphere forms a thermal boundary layer that conveys mantle heat to the Earth's surface. It appears unlikely that the phase transition at 400 km constitutes a thermal boundary layer but the phase transition at 670 km depth may do so. In the model used to derive the temperature profiles in Fig. 4.14 the phase transitions do not act as thermal boundary layers. Throughout most of the mantle the temperature gradient is assumed to equal the adiabatic gradient, but the mantle is bounded at top and bottom by thermal boundary layers (the lithosphere and D" layer, respectively) in which the temperature gradient greatly exceeds the adiabatic gradient.

4.2.4 Heat transport in the Earth

Heat can be transported by three processes: conduction, convection and radiation. Conduction and convection require the presence of a material; radiation can pass through space or a vacuum. Conduction is the most significant process of heat transport in solid materials and thus it is very important in the crust and lithosphere. However, it is an inefficient form of heat transport, and when the molecules are free to move, as in a fluid or gas, the process of convection becomes more important. Although the mantle is solid from the standpoint of the rapid passage of seismic waves, the temperature is high enough for the mantle to act as a viscous fluid over long time intervals. Consequently, convection is a more important form of heat transfer than conduction in the mantle. Convection is also the most important form of heat transport in the fluid core, where related changes in the geomagnetic field show that the turnover of core fluid is rapid in geological terms. Radiation is the least important process of heat transport in the Earth. It is only significant in the hottest regions of the core and lower mantle. The absorption of radiant energy by matter increases its temperature and thereby the temperature gradient. Hence, thermal radiation can be taken into account as a modification of the ability of the material to transfer heat by conduction.

4.2.4.1 *Conduction*

Thermal conduction takes place by the transfer of kinetic energy between molecules or atoms. A true understanding of the processes involved would require us to invoke quantum theory and the so-called 'band theory of solids', but a general understanding is possible without resorting to such measures. The electrons in an atom that are most loosely bound – the valence electrons – are essentially free of the ionic cores and can move through a material, so transferring kinetic energy. Hence they are called *conduction electrons*. Because electrons are electrically charged, the net movement of conduction electrons also causes an electrical current. Not surprisingly materials that are good electrical conductors (e.g., silver, copper) also conduct heat well. In this atomic view the conduction electrons move at very high speeds (~ 1000 km s^{-1}) but in random directions so that there is no net energy transfer in any particular direction. In an electrical or temperature field the conduction electrons drift systematically down the slope of the field (i.e., in the direction of the electrical field or temperature gradient). The additional drift velocity is very small (about 0.1

mm s^{-1}) but it passes kinetic energy through the material. This form of conduction is possible in liquids, gases or solids.

An additional mechanism plays an important role in conduction in solids. The atoms in a solid occupy definite positions that form a lattice with a certain symmetry. The atoms are not stationary but vibrate at a frequency that is temperature-dependent. The lattice vibrational energy is quantized, forming units called *phonons*. An increase in temperature at one end of a solid raises the lattice vibrational frequency there. Due to the coupling between atoms the increased vibration is eventually passed through the lattice as an increase in temperature.

The relative importance of electrons and phonons in conducting heat differs from one solid to another. In metals, which contain large numbers of conduction electrons, thermal transport is due largely to the electrons; the lattice conductivity is barely measurable. In an insulator or poor conductor, such as the minerals of the crust and mantle, there are few conduction electrons and thermal conductivity is largely determined by lattice vibrations (phonons).

The transport of heat by conduction in a solid is governed by a simple equation. Consider a solid bar of length L and cross-sectional area A with its ends maintained at temperatures T_1 and T_2, respectively (Fig. 4.15a). Assuming that heat flows only along the bar (i.e., there are no side losses) the net amount of heat (ΔQ) that passes in a given time from the hot end to the cold end depends directly on the temperature difference ($T_2 - T_1$), the area of cross-section (A) and the time of observation (Δt), and inversely on the length of the bar (L). These observations can be summarized in the equation

$$\Delta Q = kA\frac{T_2 - T_1}{L}\Delta t \tag{4.38}$$

The constant of proportionality, k, is the *thermal conductivity*, which is a property of the material of the bar. If the length of the bar is very small or the temperature change across it is uniform, the ratio $(T_2 - T_1)/L$ is the temperature gradient. We can modify the equation to describe the vertical flow of heat out of the Earth by substituting the vertical temperature gradient (dT/dz). Eq. (4.38) can then be rearranged as follows

$$q_z = -\frac{1}{A}\frac{dQ}{dt} = -k\frac{dT}{dz} \tag{4.39}$$

In this equation q_z is the *heat flux*, defined as the flow of heat per unit area per second. The negative sign is needed to account for the direction of the heat flow; if temperature increases in the downward direction of the z-axis, the flow of heat from high to low temperature is upward.

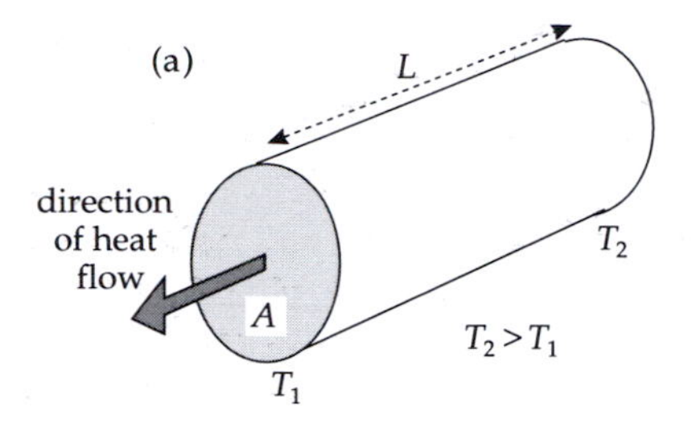

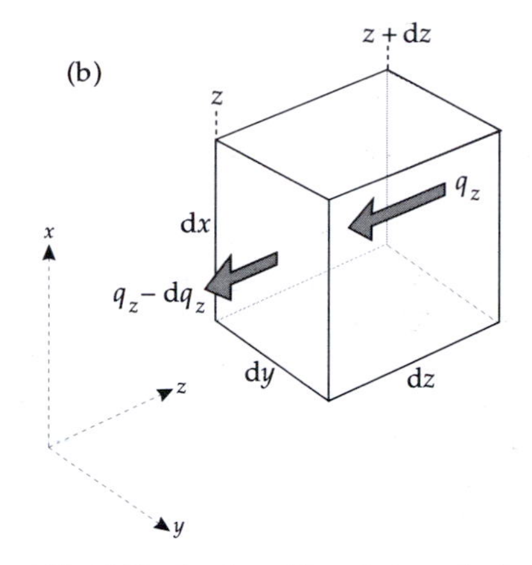

Fig. 4.15 (a) Conduction of heat Q through a bar of length L and cross-sectional area A, with its ends kept at temperatures T_1 and T_2 ($>T_1$). (b) Heat flux entering (q_z) and leaving ($q_z - dq_z$) a short bar of length dz.

The change of temperature within a body is described by the heat conduction equation, which is solved in §4.2.6 for special situations that are of interest for the transfer of thermal energy in the Earth. Conduction is a slow, and less effective means of heat transport than convection. It is important in the rigid crust and lithosphere, where convection cannot take place. However, it cannot be neglected in the fluid core, which is metallic and therefore a good conductor. A significant part of the core's heat is conducted out of the core along the adiabatic temperature gradient. The remainder, in excess of the conductive heat flow, is transported by convection currents.

4.2.4.2 *Convection*

Suppose that a small parcel of material at a certain depth in the Earth is in thermal equilibrium with its surroundings. If the parcel is displaced vertically upward without gaining or losing heat, it experiences a drop in pressure accompanied by a corresponding loss in temperature. If the new temperature of the parcel is the same as that of its surroundings at the new depth, the conditions at each depth are in adiabatic equilibrium. The variation of temperature with depth then defines the adiabatic temperature curve.

Now suppose that the real temperature increases with

depth more rapidly than the adiabatic temperature gradient. The temperature loss of the upwardly displaced parcel is due to the change in pressure, which will be the same as in the previous case. But the real temperature has dropped by a larger amount, so the parcel is now hotter and therefore less dense than its surroundings. Its buoyancy causes it to continue to rise until it reaches a level where it is in equilibrium or can rise no further. Meanwhile, the volume vacated by the displaced parcel is occupied by adjacent material. Conversely, if a parcel of material is displaced downward, it experiences adiabatic increases in pressure and temperature. The temperature increase is less than required by the real temperature gradient, so the parcel remains cooler than its surroundings and sinks further. A pattern of cyclical behavior arises in which material is heated up and rises, while cooler material sinks to take its place, and is in turn heated up and rises, and so on. The process is called *thermal convection* and the physical transportation of material and heat is called a *convection current*.

The difference between the real and adiabatic temperature gradients is the *superadiabatic* temperature gradient, θ. For thermal convection to take place in a fluid, θ must be positive. Suppose that the temperature at a certain depth exceeds the adiabatic temperature by an amount ΔT. The temperature excess causes a volume V of fluid to expand by an amount proportional to the volume coefficient of expansion, α; this causes a mass deficiency of ($V\rho\alpha\ \Delta T$). Archimedes principle applies, so the hot volume V experiences a buoyancy force given by:

$$F_B = V\rho g\alpha\Delta T \tag{4.40}$$

Two effects inhibit the hot volume from rising. First, some of the heat that would contribute to the buoyancy is removed by thermal conduction; the efficacy of this process is expressed by the *thermal diffusivity* κ of the material, which depends on its density ρ, thermal conductivity k and specific heat c_p (see §4.2.6). Second, as soon as the overheated volume of fluid begins to rise, it experiences a resisting drag due to the *viscosity* η of the fluid. The effects combine to produce a force, proportional to $\kappa\eta$, which opposes convection. If the volume V involved in the convection has a typical dimension D, so that $V \sim D^3$, we can define a dimensionless number Ra, the Rayleigh number, which is proportional to the ratio of the buoyancy force to the diffusive–viscous force:

$$\mathrm{Ra} = \frac{g\rho\alpha\Delta T}{\kappa\eta} D^3 \tag{4.41}$$

Initially, heat passes through the material by conduction, but the diffusion takes some time. If the heat flux is large enough, it cannot diffuse entirely. The temperature rises above the adiabatic and buoyancy forces develop. For convection to occur, the buoyancy forces must dominate the resisting forces. This does not happen until the Rayleigh number exceeds a critical value, which is determined additionally by the boundary conditions and the geometry of the convection. For example, the condition for the onset of convection in a thin horizontal fluid layer, heated from below and with the top and bottom surfaces free from stress, was shown by Lord Rayleigh in 1916 to depend on the value of Ra given by

$$\mathrm{Ra} = \frac{g\alpha\theta}{\kappa\nu} D^4 \tag{4.42}$$

Here D is the layer thickness, θ is the superadiabatic temperature gradient and ν (equal to η/ρ) is the *kinematic viscosity*. Convection begins in the flat layer if Ra is greater than $27\pi^4/4 = 658$. In cases with different boundary conditions, or for convection to occur in a spherical shell, the critical Rayleigh number is higher. However, convection generally originates if Ra is of the order of 10^3 and when Ra reaches around 10^5 heat transport is almost entirely by convection with little being transferred by diffusion.

For convection to occur, the real temperature gradient must exceed the adiabatic gradient. However, the loss of heat by convection reduces the difference between the gradients. Accordingly, the adiabatic gradient evolves as a convecting fluid cools. An important effect of convection is to keep the temperature gradient close to the adiabatic gradient. This condition is realized in the Earth's fluid core, where convection is the major mechanism of heat transport. Thermal convection is augmented by *compositional convection* related to the solidification of the inner core. The core fluid is made up of iron, nickel and lower-density elements, e.g., sulfur. Solidification of the inner core separates the dense iron from the lower-density elements at the inner core boundary. Being less dense than the core fluid, the residual materials experience an upward buoyancy force, resulting in a cycle of compositionally driven convection. Thermal and compositional convection in the Earth's core each act as a source of the energy needed to drive the geomagnetic field, with compositional convection the more important type.

Convection is the most important process of thermal transport in the fluid core, but it is also important in the mantle. The material of the Earth's mantle is rigid to the short-lived passage of seismic waves but is believed to yield slowly over long periods of time (§6.2.6) Although the mantle viscosity is high, the timescale of geological processes is so long that long-term flow can take place. The flow patterns are dominated by thermal convection and are influenced by the presence of thermal boundary layers, which

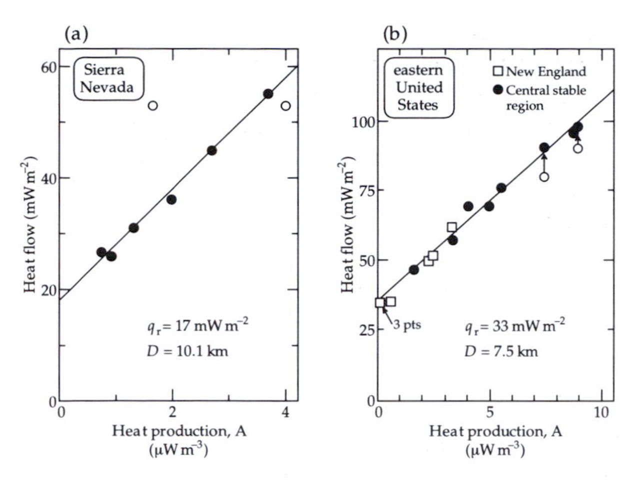

Fig. 4.16 Dependence of surface heat flux on radioactive heat generation in two heat-flow provinces: (a) Sierra Nevada, (b) eastern United States (data source: Roy *et al.*, 1968)

the flowing material cannot cross. However, convection is a more effective mechanism than conduction and it is thought to be the dominant process of heat transfer in the mantle (§6.3).

A further process of heat transfer that involves bodily transport of matter is *advection*. This can be regarded as a form of forced convection. Instead of being conveyed by thermally produced buoyancy, advected heat is transported in a medium that is itself driven by other forces. For example, in a thermal spring the flow of water is due to hydraulic forces and not to density differences in the hot water. Similarly, volcanic eruptions transport advected heat along with the lava flow, but this is propelled by pressure differences rather than by buoyancy.

4.2.4.3 *Radiation*

Atoms can exist in many distinct energy states. The most stable is the ground state, in which the energy is lowest. When an atom changes from an excited state to a lower-energy state, it is said to undergo a transition. Energy corresponding to the difference in energy between the states is emitted as an electromagnetic wave, which we call radiation. Quantum physics teaches that the radiant energy emitted consists of a discrete number of fundamental units, called *quanta*. The particular wavelength of the electromagnetic radiation associated with a transition is proportional to the energy difference between the two states. If several different transitions are taking place simultaneously, the body emits a spectrum of wavelengths. Radio signals, heat, light, and x-rays are examples of electromagnetic radiation that have different wavelengths. The electromagnetic wave consists of fluctuating electric and magnetic fields, which need no medium for their passage. For this reason, radiation can travel through space or a vacuum. In materials it may be scattered or absorbed, depending on its wavelength. Heat radiation corresponds to the *infra-red* part of the electromagnetic spectrum with wavelengths just longer than those of visible light.

The radiation of a commonplace hot object depends on factors that are difficult to assess. Classical physics fails to explain adequately the absorption and emission of radiation. To provide an explanation physicists introduced the concept of a *black body* as a perfect absorber and emitter of radiation. At any temperature it emits a continuous spectrum of radiation; the frequency content of the spectrum does not depend on the material composition of the body but only on its temperature. An ideal black body does not exist in practice, but it can be approximated by a hollow container that has a small hole in its wall. When the container is heated, the radiation escaping through the hole – so-called cavity radiation – is effectively black-body radiation. In 1879 Josef Stefan pointed out that the loss of heat by radiation from a hot object is proportional to the fourth power of the absolute temperature. If R represents the radiant energy per second emitted per unit area of the surface of the body at temperature T, then

$$R=\sigma T^4 \tag{4.43}$$

where σ, known as Stefan's constant or the Stefan–Boltzmann constant, has the value 5.67×10^{-8} W m^{-2} K^{-4}.

In 1900, Max Planck, professor of physics at the university of Berlin, proposed that, in contrast to classical physics, an oscillator could only have discrete amounts of energy. This was the birth of quantum theory. The energy of an oscillator of frequency ν is equal to the product $h\nu$, where the universal constant h (known as Planck's constant) has the value 6.626×10^{-34} J s. The application of quantum principles to black-body radiation provides a satisfactory explanation of Stefan's law and allows Stefan's constant to be expressed in terms of other fundamental physical constants.

Radiation is reflected and refracted in a transparent medium wherever the refractive index n changes; energy is transferred to the medium in each of these interactions. The transparency of the medium is determined by the opacity e, which describes the degree of absorption of electromagnetic radiation. The opacity is wavelength-dependent. In an ionic crystal the absorption of infrared radiation is large. It alters the vibrational frequency, and thereby influences the ability of the crystal lattice to transport heat by conduction. Thus the effect can be taken into account by increasing the conductivity by an extra radiative amount, k_r, given by

$$k_r = \frac{16}{3}\frac{n^2\sigma}{e}T^3 \tag{4.44}$$

The T^3-dependence in this expression suggests that radiation might be more important than lattice conductivity in the hotter regions of the Earth. In fact other arguments lead to the conclusion that this is probably not the case in the upper mantle, because the effect of increasing temperature is partly offset by an increase in the opacity, e. The lower mantle is believed to have a high density of free electrons, which efficiently absorb radiation and raise the opacity. This may greatly reduce the efficacy of heat transfer by radiation in the mantle.

4.2.5 Sources of heat in the Earth

The interior of the Earth is losing heat via geothermal flux at a rate of about 4.4×10^{13} W, which amounts to 1.4×10^{21} J yr^{-1} (see Table 4.2). The heat is brought to the surface in different ways. The creation of new lithosphere at oceanic ridges releases the largest fraction of the thermal energy. A similar mechanism, the spreading of the sea-floor, releases heat in the marginal basins behind island arcs. Rising plumes of magma originating deep in the mantle bring heat to the surface where they break through the oceanic or continental lithosphere at 'hotspots', characterized by intense localized volcanic activity. These important thermal fluxes are superposed on a background consisting of heat flowing into and through the lithosphere from deeper parts of the earth. There are two main sources of the internal heat. Part of it is probably due to the slow cooling of the Earth from an earlier hotter state; part is generated by the decay of long-lived radioactive isotopes.

The early thermal history of the Earth is obscure and a matter of some speculation. According to the cold accretion model of the formation of the planets (see §1.1.4), colliding bodies in a primordial cloud of dust and gas coalesced by self-gravitation. The gravitational collapse released energy that heated up the Earth. When the temperature reached the melting point of iron, a liquid core formed, incorporating also nickel and possibly sulfur or another light element associated with iron. The differentiation of a denser core and lighter mantle from an initially homogeneous fluid must have released further gravitational energy in the form of heat. The dissipation of Earth's initial heat still has an important effect on internal temperatures.

Energy released by short-lived radioactive isotopes may have contributed to the initial heating, but the short-lived isotopes would be consumed quite early. The heat generated by long-lived radioactive isotopes has been an important heat source during most of Earth's history. These isotopes separated into two fractions: some, associated with heavy elements, sank into the core; some, associated with lighter elements, accumulated in the crust. The present distribution of radiogenic sources within the differentiated Earth is uneven. The highest concentrations are in the rocks and minerals of the Earth's crust, while the concentrations in mantle and core materials are low, However, continuing generation of heat by radioactivity in the deep interior, though small, may influence internal temperatures.

4.2.5.1 *Radioactive heat production*

The relationship between observed heat flow and the heat produced by radioactive isotopes in local rocks was first recognized by J. Jolly in 1909. When a radioactive isotope decays, it emits energetic particles and γ-rays. The two particles that are important in radioactive heat production are α-particles and β-particles. The α-particles are equivalent to helium nuclei and are positively charged, while β-particles are electrons. In order to be a significant source of heat a radioactive isotope must have a half-life comparable to the age of the Earth, the energy of its decay must be fully converted to heat, and the isotope must be sufficiently abundant. The main isotopes that fulfil these conditions are ^{238}U, ^{235}U, ^{232}Th and ^{40}K. The isotope ^{235}U has a shorter half-life than ^{238}U (see Table 4.1) and releases more energy in its decay. In natural uranium the proportion of ^{238}U is 99.28%, that of ^{235}U is about 0.71%, and the rest is ^{234}U. The abundance of the radioactive isotope ^{40}K in natural potassium is only 0.01167%, but potassium is a very common element and its heat production is not negligible. The amounts of heat generated per second by these elements (in μW kg^{-1}) are: natural uranium, 95.2; thorium, 25.6; and natural potassium, 0.00348 (Rybach, 1976, 1988). The heat Q_r produced by radioactivity in a rock that has concentrations C_U, C_{Th} and C_K, respectively, of these elements is

$$Q_r = 95.2C_U + 25.6C_{Th} + 0.00348C_k \tag{4.45}$$

Rates of radioactive heat production computed with this equation are shown for some important rock types in Table 4.3. Chondritic meteorites, made up of silicate minerals like olivine and pyroxene, are often taken as a proxy for the initial composition of the mantle; likewise, the olivine-dominated rock dunite represents the ultramafic rocks of the upper mantle. It is apparent that very little heat is produced by radioactivity in the mantle or in the basaltic rocks that dominate the oceanic crust and lower continental crust. The greatest concentration of radiogenic heat sources is in the granitic rocks in the upper continental crust. Multiplying the radioactive heat production values in the last column of Table 4.3 by the rock density gives the radiogenic heat generated in a cubic meter of the rock, A. If we assume that all

Table 4.3 *Estimates of radioactive heat production in selected rock types, based on heat production rates (from Rybach, 1976, 1988) and isotopic concentrations (from Stacey, 1992)*

	Concentration (p.p.m. by weight)			Heat production (10^{-11} W kg^{-1})			
Rock type	U	Th	K	U	Th	K	Total
granite	4.6	18	33,000	43.8	46.1	11.5	101
alkali basalt	0.75	2.5	12,000	7.1	6.4	4.2	18
tholeiitic basalt	0.11	0.4	1,500	1.05	1.02	0.52	2.6
peridotite, dunite	0.006	0.02	100	0.057	0.051	0.035	0.14
chondrites	0.015	0.045	900	0.143	0.115	0.313	0.57
continental crust	1.2	4.5	15,500	11.4	11.5	5.4	28
mantle	0.025	0.087	70	0.238	0.223	0.024	0.49

the heat generated in a rock layer of thickness D meters escapes vertically, the amount crossing a square meter at the surface per second (i.e., the radioactive component of the heat flow) is DA. For example, a one-kilometer thick layer of granite contributes about 3 mW m^{-2} to the continental heat flow. The figures suggest that the 10–20 km thick upper crust produces about one-half of the mean continental heat flow, which is 65 mW m^{-2}.

In fact the relative importance of radiogenic heat in the crust is variable from one region to another. A region in which the heat flow is linearly related to the heat produced by radioactivity is called a *heat-flow province*. Some examples of heat-flow provinces are Western Australia, the Superior Province in the Canadian Shield, and the Basin-and-Range Province in the western United States. As shown in Fig. 4.16 each province is characterized by a different linear relation between q and A, such that

$$q=q_r+DA \tag{4.46}$$

The parameters q_r and D typify the heat-flow province. The intercept of the straight line with the heat-flow axis, q_r, is called the *reduced heat flow*. This is the heat flow that would be observed in the province if there were no radiogenic crustal heat sources. It is due in part to the heat flowing from deeper regions of the Earth into the base of the crustal layer, and partly to cooling of the originally hotter upper crustal layer. Investigations in different heat-flow provinces show that the reduced heat flow averages about 55% of the mean measured heat flow in a province (Fig. 4.17).

The simplest interpretation of D is to regard it as a characteristic thickness of crust involved in radioactive heat production. This assumes that the radiogenic heat sources are distributed uniformly in a crustal slab of constant thickness, which is an unlikely situation. A more likely model is that the radioactive heat generation decreases with depth.

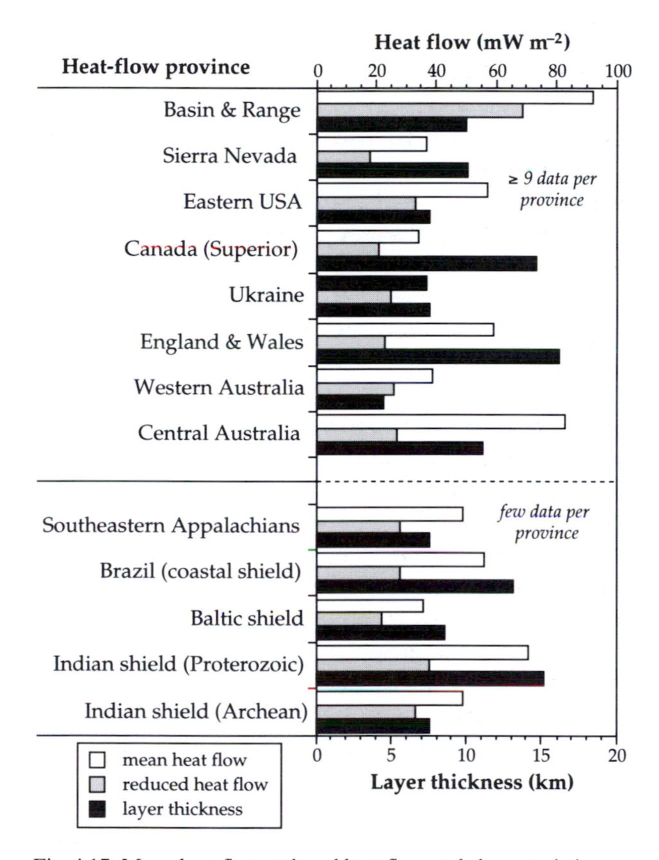

Fig. 4.17 Mean heat flow, reduced heat flow and characteristic thickness of the layer of radioactive heat production in several heat-flow provinces (data source: Vitorello and Pollack, 1980).

Assuming an exponential decrease, the heat production A(z) at depth z is related to the surface heat generation A_0 as

$$A(z)=A_0e^{-z/D} \tag{4.47}$$

where D is a characteristic depth (the depth at which A(z) has decreased to e^{-1} of its surface value). Integrating from

Table 4.4 *Approximate relative contributions (in %) of the main sources of heat flow in oceanic and continental lithosphere (from Bott, 1982)*

Heat source	Contribution to heat flow in: Continents (%)	Oceans (%)
Cooling of the lithosphere	20	85
Heat flow from below the lithosphere	25?	85
Radiogenic heat:	55?	5
upper crust	40	—
rest of lithosphere	15?	—

the surface to infinite depth gives the total radioactive heat production:

$$\int_0^\infty A(z)dz = A_0\int_0^\infty e^{-z/D}dz = DA_0 \tag{4.48}$$

which is the same as for the uniform distribution. The infinite lower limit to the exponential distribution is obviously unrealistic. If the radiogenic sources are distributed in a layer of finite thickness s, the integration becomes

$$A_0\int_0^s e^{-z/D}dz = DA_0(1-e^{-s/D}) \tag{4.49}$$

If s is greater than three times D, this expression differs from DA_0 by less than 5%. The value of D estimated from studies of heat-flow and radioactive heat generation averages about 10 km, but varies from ~4 km to ~16 km (Fig. 4.17).

The three main sources of the Earth's surface heat flow are (i) heat flowing into the base of the lithosphere from the deeper mantle, (ii) heat lost by cooling of the lithosphere with time, and (iii) radiogenic heat production in the crust. The contributions are unequal and different in the oceans and continents (Table 4.4). The most obvious disparity is in the relative importance of lithospheric cooling and radioactivity. The lithosphere is hot when created at oceanic ridges and cools slowly as it ages. The loss of heat by lithospheric cooling is most pronounced in the oceanic crust, which moreover contains few radiogenic heat sources. In contrast, the older continental lithosphere has lost much of the early heat of formation, and the higher concentration of radioactive minerals increases the importance of radiogenic heat production. Regardless of its source, the passage of heat through the rigid outer layers takes place predominantly by conduction, although in special circumstances, such as the flow of magma in the crust or hydrothermal circulation near to oceanic ridges, convection also plays an important role.

4.2.6 The heat conduction equation

Jean Baptiste Joseph Fourier (1768–1830), a noted French mathematician and physicist, developed the theory of heat conduction in 1822. Here we consider the example of one-dimensional heat flow, which typifies many interesting problems involving the flow of heat in a single direction. The equation of heat conduction is most easily developed for this case, from which it can be readily extended to three-dimensional heat flow.

Consider the flow of heat in the negative direction of the z-axis through a small rectangular prism with sides dx, dy and dz (see Fig. 4.15b). We will assume that there are no sources of heat inside the box. Let the amount of heat entering the prism at $(z+dz)$ be Q_z. This is equal to the heat flow q_z multiplied by the area of the surface it flows across (dx dy) and by the duration of the flow (dt). The heat leaving the box at z is $Q_z - dQ_z$, which can be written $Q_z-(dQ_z/dz)\,dz$. The increase in heat in the small box is the difference between these amounts:

$$\frac{dQ_z}{dz}dz = \frac{dq_z}{dz}dz(dxdy)dt = k\frac{d^2T}{dz^2}dV dt \tag{4.50}$$

where dV is the volume of the box (dxdydz). Note that Q_z, q_z and T are all understood to decrease in the direction of flow, so we have substituted for q_z from Eq. (4.39) without using the negative sign. The heat increase in the box causes its temperature to rise by an amount (dT), determined by the specific heat at constant pressure (c_p) and the mass of material (m) in the box. Using Eq. (4.25), we write

$$c_p m dT = c_p \rho\, dV dT \tag{4.51}$$

where ρ is the density of the material in the box. If we equate Eq. (4.50) and Eq. (4.51) for the amount of heat left in the box, we get the equation of heat conduction

$$\frac{dT}{dt} = \frac{k}{\rho c_p}\frac{d^2T}{dz^2}$$

$$\frac{\partial T}{\partial t} = \kappa\frac{\partial^2 T}{\partial z^2} \tag{4.52}$$

where κ $(=k/\rho c_p)$ is called the *thermal diffusivity*; it has the dimensions m^2 s^{-1}. The equation is written with partial differentials because the temperature T is a function of both time and position: $T=T(z, t)$. This just means that, on the one hand, the temperature at a certain position changes with time, and, on the other hand, the temperature at any given time varies with position in the body.

The same arguments can be applied to the components of heat flow through the box in the x- and y-directions. We obtain the three-dimensional heat conduction equation (also called the *diffusion equation*):

$$\frac{\partial T}{\partial t}=\kappa\left(\frac{\partial^2 T}{\partial x^2}+\frac{\partial^2 T}{\partial y^2}+\frac{\partial^2 T}{\partial z^2}\right) \tag{4.53}$$

The equation is solved for any set of boundary conditions by using the method of *separation of variables*. We will apply it to one-dimensional heat flow. Assume that the temperature $T(z, t)$ can be written as the product of $Z(z)$, a function of position only, and $\theta(t)$, a function of time only:

$$T(z,t)=Z(z)\theta(t) \tag{4.54}$$

The actual solution for $T(z, t)$ usually does not have this final form, but once $Z(z)$ and $\theta(t)$ are known they can be combined to fit the boundary conditions. On substituting in Eq. (4.52) we get

$$Z\frac{d\theta}{dt}=\kappa\frac{d^2Z}{dz^2}\theta \tag{4.55}$$

We can use full differentials again because θ depends only on t and Z only on z. Dividing both sides by $Z\theta$ gives

$$\frac{1}{\theta}\frac{d\theta}{dt}=\kappa\frac{1}{Z}\frac{d^2Z}{dz^2} \tag{4.56}$$

Again, note that the left side depends only on t and the right side depends only on z, and that z and t are independent variables. On substituting a particular value for the time t the left side becomes a numerical constant. We have not restricted z, which varies independently. Eq. (4.56) requires that the variable right side remains equal to the numerical constant for any value of z. Conversely, if we substitute a particular value for z, the right side becomes a (new) numerical constant and the variable left side must equal this constant for any t. The identity inherent in Eq. (4.56) implies that each side must be equal to the same *separation constant*. Its value is determined by the boundary conditions of the problem.

A problem of immediate interest for determining the geothermal flux from the Earth's interior is the effect of solar energy that reaches the Earth's surface. The surface rocks heat up during the day and cool down at night. The effect is not restricted to the immediate surface, but affects a volume of rock near the surface. Similarly, the mean surface temperature varies throughout the year with the changing seasons. The heat conduction equation allows us to estimate what depths are affected by these cyclic temperature variations.

4.2.6.1 *Penetration of external heat into the Earth*

Suppose that the surface temperature of the Earth varies cyclically with angular frequency ω, so that at time t it is equal to $T_0\cos\omega t$, where T_0 is the peak temperature during a cycle. It is convenient to handle such problems by using complex numbers. Writing $i=\sqrt{-1}$, some relevant complex identities are

$$e^{i\theta}=\cos\theta+i\sin\theta$$

$$e^{i\pi/2}=i$$

$$\sqrt{i}=e^{i\pi/4}=\frac{1}{\sqrt{2}}(1+i) \tag{4.57}$$

In the first expression, $\cos\theta$ is the real part of $e^{i\theta}$ and $\sin\theta$ is the imaginary part. Thus, the surface-temperature variation $T_0\cos\omega t$ is the real part of $T_0e^{i\omega t}$. Let z represent the depth of a point below the surface. The heat conduction equation can be separated and written as two parts equal to the same constant, which we write (for reasons soon apparent) as $-i\omega$, so that

$$\frac{1}{\theta}\frac{d\theta}{dt}=-i\omega \tag{4.58}$$

$$\kappa\frac{1}{Z}\frac{d^2Z}{dz^2}=-i\omega \tag{4.59}$$

The solution of Eq. (4.58) is simply $\theta=\theta_0\, e^{-i\omega t}$. We can rewrite Eq. (4.59) as

$$\frac{d^2Z}{dz^2}+i\frac{\omega}{\kappa}Z=0 \tag{4.60}$$

This is equivalent to the harmonic equation

$$\frac{d^2Z}{dz^2}+n^2Z=0 \tag{4.61}$$

which has the solution $(Z_0e^{inz}+Z_1e^{-inz})$. In our example temperature decreases downward, away from the surface (i.e., in the z-direction), so we choose $Z_1=0$. The solution of Eq. (4.60) is found by writing

$$n=i\frac{\omega}{\kappa}$$

$$n=\sqrt{i\frac{\omega}{\kappa}}=\sqrt{\frac{\omega}{2\kappa}}(1+i) \tag{4.62}$$

Combining the solutions for θ and Z we get

$$T(z,t)=Z_0e^{i\sqrt{\frac{\omega}{2\kappa}}(1+i)z}\theta_0e^{-i\omega t}$$

$$=Z_0\theta_0e^{-\sqrt{\frac{\omega}{2\kappa}}z}e^{-i(\omega t-\sqrt{\frac{\omega}{2\kappa}}z)} \tag{4.63}$$

The solution of the heat conduction equation at depth z and time t is found by taking the real part and matching it to the boundary conditions on the surface. For a surface temperature variation $T=T_0\cos\omega t$ the solution is

$$T(z,t)=T_0e^{-\frac{z}{d}}\cos(\omega t-\frac{z}{d}) \tag{4.64}$$

where $T_0 = Z_0\theta_0$ and $d = (2\kappa/\omega)^{1/2}$ is the *decay depth* of the temperature. At this depth the amplitude of the temperature fluctuation is attenuated to 1/e of its value on the surface. At a depth of $5d$ the amplitude is less than 1% of the surface value and is effectively zero. Note that d depends inversely on the frequency, so long-period fluctuations penetrate more deeply than rapid fluctuations. This is illustrated by a quick comparison of the decay depths for daily and annual temperature variations in the same ground

$$\frac{d_{\text{annual}}}{d_{\text{daily}}} = \sqrt{\frac{(2\pi/1)}{(2\pi/365}} = \sqrt{365} = 19.1 \tag{4.65}$$

i.e., the annual variation penetrates about 19 times the depth of the daily variation (Fig. 4.18). Moreover, the temperature at any depth varies with the same frequency ω as the surface variation, but it experiences a phase shift or delay, reaching its maximum value progressively later than the surface effect (Fig. 4.18).

As representative values for crustal rocks we take: density, $\rho = 2650$ kg m^{-3}; thermal conductivity, $k = 2.5$ W m^{-1} K^{-1}; and specific heat $c_p = 700$ J kg^{-1} K^{-1}. These give a thermal diffusivity, $\kappa = 1.25 \times 10^{-6}$ m^2 s^{-1}. The daily temperature variation (period 86,400 s) has $\omega = 7 \times 10^{-5}$ s^{-1}, so its penetration depth is about 20 cm. The daily variation has negligible effect deeper than about a meter. Similarly, the annual temperature variation accompanying seasonal changes has a penetration depth of 3.8 m and is negligible deeper than about 19 m. Heat-flow measurements made within 20 m of the surface will be falsified by the daily and annual variations of surface temperature. This is not a problem in the deep oceans, where the Sun's rays never reach the bottom, but it must be taken into account in continental heat-flow measurements. A serious effect is the role of the ice ages, which recur on a timescale of about 10,000 yr and have a penetration depth of about 2 km. The measured temperature gradient must be corrected appropriately.

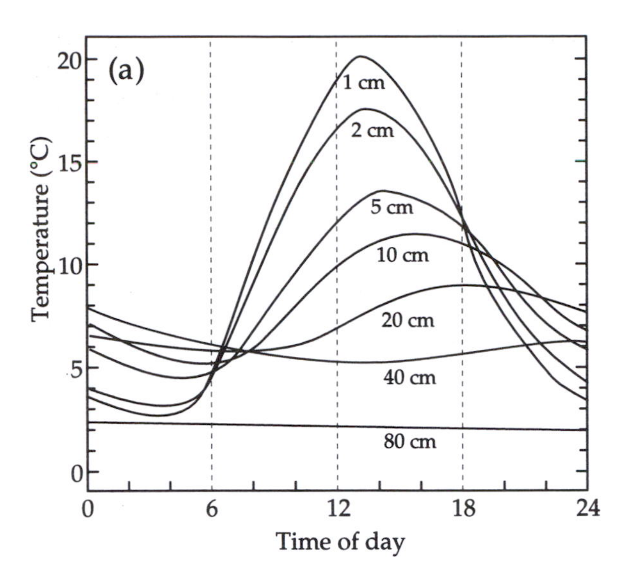

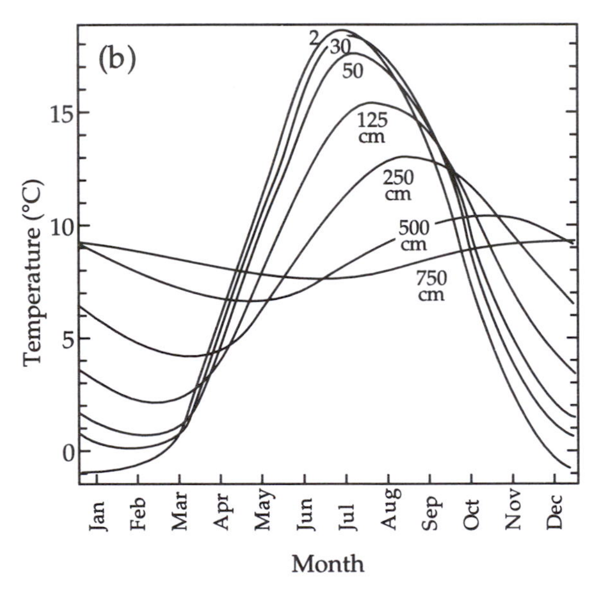

Fig. 4.18 Temperature variations at various depths in a sandy soil: (a) daily fluctuations, (b) annual (seasonal) variations.

4.2.6.2 *Cooling of a semi-infinite half-space*

A thermodynamic problem that is commonly encountered in geology is the sudden heating or cooling of a body. For example, a dike of molten lava intrudes cool host rocks virtually instantaneously, but the heat is conducted to the adjacent rocks over a long period of time until it slowly dissipates. Another case, which we will discuss further here, is when a hot body loses heat at its surface. The simplest example is the one-dimensional cooling of a semi-infinite half-space extending to infinity in the z-direction. Initially at a uniform temperature, the body loses heat through the surface at $z=0$, which has a temperature T_s. This simple model can be used to estimate the temperature distribution inside the cooling lithosphere. The initial temperature corresponds to that of the hot mantle (T_m) and the surface temperature is that of cold ocean-bottom sea-water.

The solution of the one-dimensional heat conduction equation (Eq. (4.52)) with these boundary conditions involves functions called the error function (erf) and complementary error function (erfc), which are defined as

$$\text{erf}(\eta) = \frac{2}{\sqrt{\pi}} \int_0^{\eta} e^{-u^2} du$$

$$\text{erfc}(\eta) = 1 - \text{erf}(\eta) \tag{4.66}$$

The shapes of these functions are shown in Fig. 4.19; their values for any particular value of η are obtained from tables, just like other statistical or trigonometric functions. The complementary error function sinks asymptotically to zero, and is effectively zero for $\eta \geq 2$. The temperature T at depth z and time t after the half-space starts to cool is given by

$$\frac{T-T_s}{T_m-T_s}=\mathrm{erf}\left(\frac{z}{2\sqrt{\kappa t}}\right) \tag{4.67}$$

For large values of z, the temperature is that of the hot mantle, T_m, and at the surface ($z=0$) it equals T_s, the temperature of the ocean floor, which is not far from 0 °C. The temperatures on the left side can be expressed in either K or °C; if we choose the latter we get the simpler equation

$$T=T_m\mathrm{erf}\left(\frac{z}{2\sqrt{\kappa t}}\right) \tag{4.68}$$

The surface heat flow of a semi-infinite half-space is proportional to the temperature gradient, obtained by differentiating Eq. (4.68) with respect to z:

$$\begin{aligned} q_z &= -k\frac{\partial T}{\partial z} = -k\frac{\partial \eta}{\partial z}\frac{\partial T}{\partial \eta} \\ &= -k\frac{1}{2\sqrt{\kappa t}}\frac{\partial}{\partial \eta}\{T_m \mathrm{erf}\,\eta\} \\ &= -\frac{kT_m}{2\sqrt{\kappa t}}\frac{\partial}{\partial \eta}\left\{\frac{2}{\sqrt{\pi}}\int_0^{\eta} e^{-u^2}du\right\} \end{aligned} \tag{4.69}$$

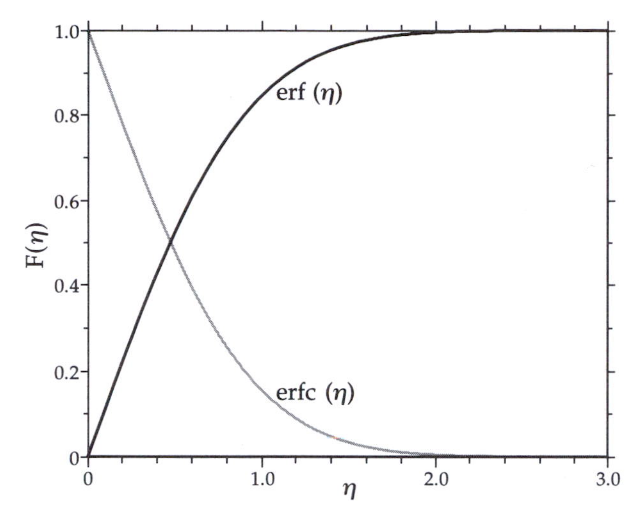

Fig. 4.19 The error function erf(η) and complementary error function erfc(η).

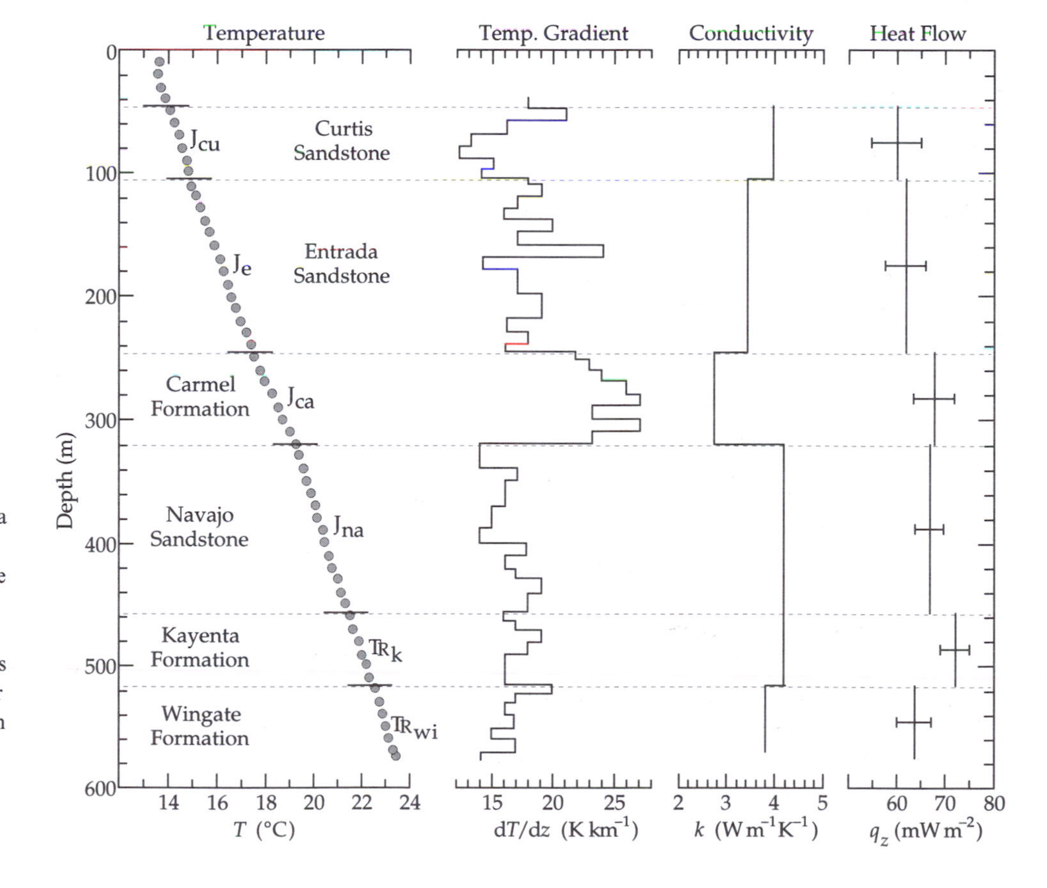

Fig. 4.20 Computation of heat flow by the interval method for geothermal data from drillhole WSR.1 on the Colorado Plateau of the western USA; horizontal bars show standard deviations of measurements in each depth interval (after Powell *et al.*, 1988; based on data from Bodell and Chapman, 1982).

which simplifies to

$$q_z = -\frac{kT_m}{\sqrt{\pi\kappa t}} e^{-\eta^2} \tag{4.70}$$

At the surface, $z=0$, $\eta=0$, and $\exp(-\eta^2)=1$. The surface heat flow at time t is

$$q_z = -\frac{kT_m}{\sqrt{\pi\kappa t}} \tag{4.71}$$

The negative sign here indicates that the heat flows upward, in the direction of decreasing z. The semi-infinite half-space is a good model for the cooling of the oceanic lithosphere, where the heat flow indeed varies as $1/\sqrt{t}$.

4.2.7 Continental heat flow

The computation of heat flow at a locality requires two measurements. The thermal conductivities of a representative suite of samples of the local rocks are measured in the laboratory. The temperature gradient is measured in the field at the investigation site. At continental sites this is usually carried out in a borehole (Fig. 4.20). There are several ways of determining the temperature in the borehole. During commercial drilling the temperature of the drilling fluid can be measured as it returns to the surface. This gives a more or less continuous record, but is influenced strongly by the heat generated during the drilling. At times when drilling is interrupted, the bottom-hole temperature can be measured. Both of these methods give data of possibly commercial interest but they are too inaccurate for heat-flow determination.

In-hole measurements of temperature for heat-flow analyses are made by lowering a temperature-logging tool into the borehole and continuously logging the temperature during its descent. The circulation of drilling fluids redistributes heat in the hole, so it is necessary to allow some time after drilling has ceased for the hole to return to thermal equilibrium with the penetrated formations. The temperature of the water in the hole is taken to be the ambient temperature of the adjacent rocks, provided there are no convection currents.

The most common devices for measuring temperature are the platinum resistance thermometer and the thermistor. A thermistor is a ceramic solid-state device with an electrical resistance that is strongly dependent on temperature. Its resistance depends non-linearly on temperature, requiring accurate calibration, but the sensitivity of the device makes feasible the measurement of temperature differences of 0.001–0.01 deg K. The platinum resistance thermometer and the thermistor are used in two basic ways. In one method the sensor element constitutes an arm of a sensitive Wheatstone bridge, with which its resistance is measured directly. The other common method uses the thermal sensor as the resistive element in a tuned electrical circuit. The tuned frequency depends on the resistance of the sensor element, which is related in a known way to temperature.

Table 4.5 *Computation of heat flow from temperature measurements in WSR.1, a 570 m deep borehole and thermal conductivity measurements on cored samples (after Powell et al., 1988)*

Depth interval (m)	Temperature gradient ($mK\,m^{-1}$)	Thermal conductivity ($W\,m^{-1}\,K^{-1}$)	Interval heat flow ($mW\,m^{-2}$)
45–105	15.0	3.96	60
105–245	18.0	3.43	62
245–320	24.8	2.75	68
320–455	16.0	4.18	67
455–515	17.2	4.20	72
515–575	16.5	3.86	64
		Mean heat flow=65 $mW\,m^{-2}$	

From the measured temperature distribution, the average temperature gradient is computed for a geological unit or a selected depth interval (Fig. 4.20). The gradient is then multiplied by the mean thermal conductivity of the rocks to obtain the interval (or formation) heat flow. Thermal conductivity can show large variations even between adjacent samples, so the *harmonic mean* of at least four samples is usually used. The interval heat-flow values may then be averaged to obtain the mean heat flow for the borehole (Table 4.5).

Several effects can falsify continental heat-flow computations from borehole data. An important assumption is that heat flow is only vertical. Well below the surface the isotherms (surfaces of constant temperature) are flat-lying and the flow of heat (normal to the isotherms) is vertical. However, the surface of the Earth is also presumed to be locally isothermal (i.e., to have constant temperature) near to the borehole. The near-surface isotherms adapt to the topography (Fig. 4.21) so that the direction of heat flow is deflected and acquires a horizontal component, while the vertical temperature gradient is also modified. Consequently, heat flow measured in a borehole must be corrected for the effects of local topography.

The need for a topographic correction was recognized late in the 19th century. Further corrections must be applied for long-term effects such as the penetration of external heat related to cyclical climatic changes, for example the ice ages.

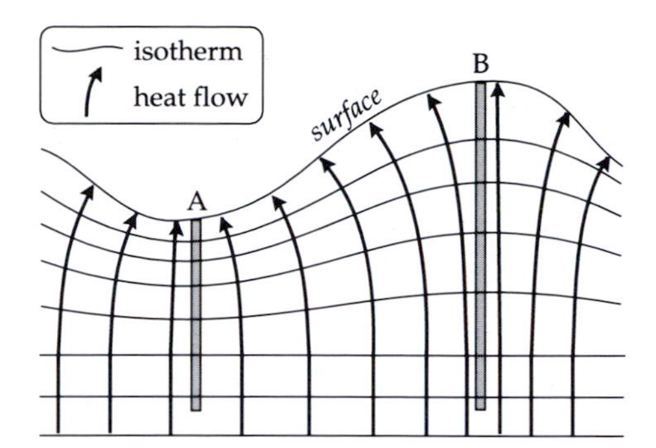

Fig. 4.21 Schematic effect of surface topography on isotherms (*thin lines*) and the direction of heat flow (*thick lines*).

Erosion, sedimentation and changes in the thermal conductivities of surface soils are other long-term effects that may require compensation.

4.2.7.1 *Variation of continental heat flow with age*

Many processes contribute to continental heat flow. Apart from the heat generated by radioactive decay, the most important sources are those related to tectonic events. During an orogenic episode various phenomena may introduce heat into the continental crust. Rocks may be deformed and metamorphosed in areas of continental collision. In extensional regions the crust may be thinned, with intrusion of magma. Uplift and erosion of elevated areas and deposition in sedimentary basins also affect the surface heat flow. After a tectonic event convective cooling takes place efficiently in circulating fluids, while some excess heat is lost by conductive cooling. Consequently, the variation of continental heat flow with time is best understood in terms of the tectonothermal age, which is the age of the last tectonic or magmatic event at a measurement site. The continental heat-flow values comprise a broad spectrum. Even when grouped in broad age categories there is a large degree of overlap (Fig. 4.22). The greatest scatter is seen in the youngest regions. The mean heat flow decreases with increasing crustal or tectonothermal age (Fig. 4.23), falling from 70–80 mW m^{-2} in young provinces to a steady-state value of 40–50 mW m^{-2} in Precambrian regions older than ~800 Ma .

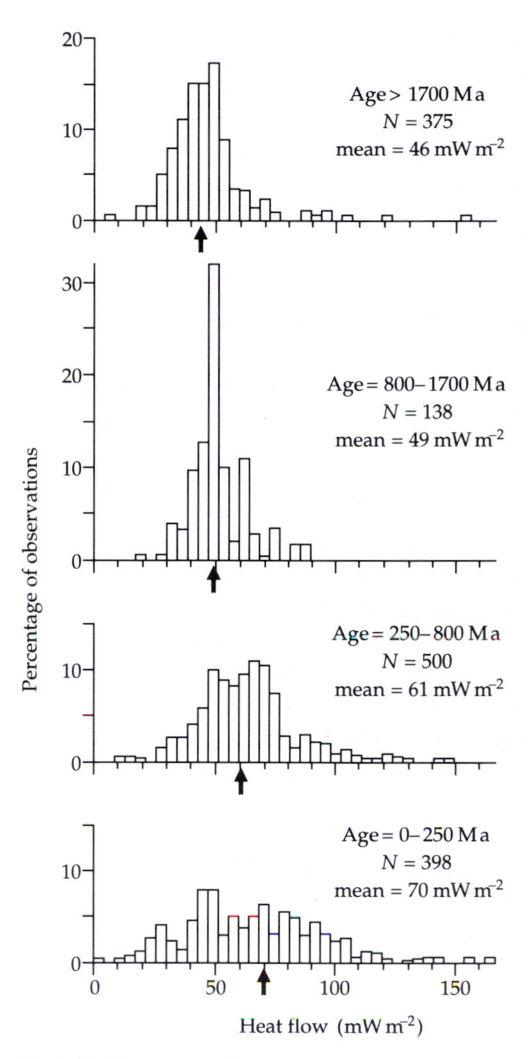

Fig. 4.22 Histograms of continental heat flow for four different age provinces (after Sclater *et al.*, 1981).

4.2.8 Oceanic heat flow

Whereas the mean altitude of the continents is only 840 m above sea-level, the mean depth of the oceans is 3880 m; the abyssal plains in large ocean basins are 5–6 km deep. At these depths extra-terrestrial heat sources have no effect on heat-flow measurements and the flatness of the ocean bottom (except near ridge systems or seamounts) obviates the need for topographic corrections. Measuring the heat flow through the ocean bottom presents technical difficulties that were overcome with the development of the Ewing piston corer. This device, intended for taking long cores of marine sediment from the ocean floor, enables *in-situ* measurement of the temperature gradient. It consists of a heavily weighted, hollow sampling pipe (Fig. 4.24a), commonly about 10 m long although in special cases cores over 20 m in length have been taken (very long coring pipes tend to bend before they reach maximum penetration). A plunger inside the pipe is displaced by sediment during coring and makes a seal with the sediment surface, so that sample loss and core deformation are minimized when the core is withdrawn from the ocean floor. Thermistors are mounted on short arms a few centimeters from the body of the pipe, and the temperatures are recorded in a water-tight

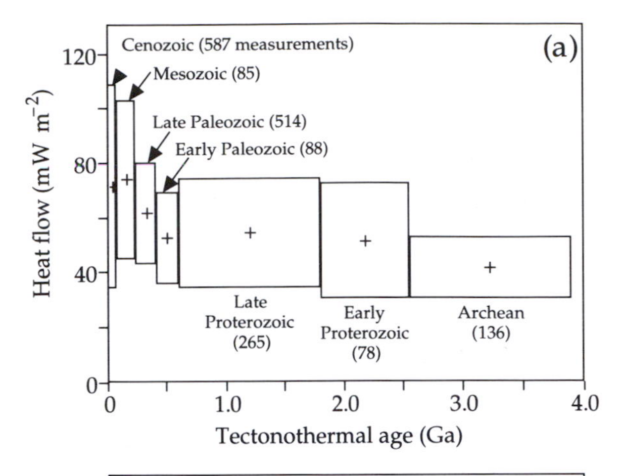

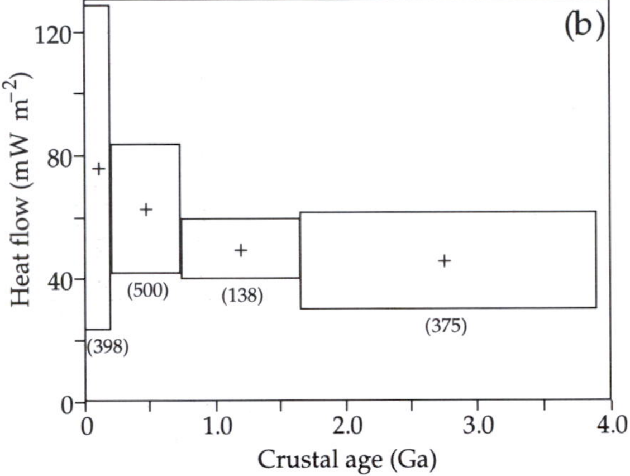

Fig. 4.23 Continental heat-flow data averaged (a) by tectonothermal age, defined as the age of the last major tectonic or magmatic event (based on data from Vitorello and Pollack, 1980), and (b) by radiometric crustal age (after Sclater *et al.*, 1980). The width of each box shows the age range of the data; the height represents one standard deviation on each side of the mean heat flow indicated by the cross at the center of each box. Numbers indicate the quantity of data for each box.

casement. The instrument is lowered from a surface ship until a free-dangling trigger-weight makes contact with the bottom (Fig. 4.24b). This releases the corer, which falls freely and is driven into the sediment by the one-ton lead weight. The friction accompanying this process generates heat, but the ambient temperatures in the sediments can be measured and recorded before the heat reaches the offset sensors (Fig. 4.24c). The sediment-filled corer is hauled back on board the ship, where the thermal conductivity of the sediment can be determined. The recovered core is used for paleontological, sedimentological, geochemical, magnetostratigraphic and other scientific analyses.

Special probes have been devised explicitly for *in-situ* measurement of heat flow. They consist of two parallel tubes about 3–10 m in length and 5 cm apart. One tube is about 5–10 cm in diameter and provides strength; the other, about 1 cm in diameter, is oil-filled and contains arrays of thermistors. After penetration of the ocean-bottom sediments, as described for the Ewing corer, the equilibrium temperature gradient is measured. A known electrical current, either constant in value or pulsed, is then passed along a heating wire and the temperature response is recorded. The observations allow the thermal conductivity of the sediment to be found. In this way a complete determination of heat flow is obtained without having to recover the contents of the corer.

4.2.8.1 *Variation of oceanic heat flow and depth with lithospheric age*

The most striking feature of oceanic heat flow is the strong relationship between the heat flow and distance from the axis of an oceanic ridge. The heat flow is highest near to the ridge axis and decreases with increasing distance from it. For a uniform sea-floor spreading rate the age of the oceanic crust (and lithosphere) is proportional to the distance from the ridge axis, and so the heat flow decreases with increasing age (Fig. 4.25). The lithospheric plate accretes at the spreading center, and as the hot material is transported away from the ridge crest it gradually cools. Model calculations for the temperature in the cooling plate are discussed in the next section: they all predict that the heat flow q caused by cooling of the plate decreases with age t as $1/\sqrt{t}$, when the age of the plate is less than about 55–70 Ma. Older lithosphere cools slightly less rapidly. Currently the decrease of heat flow with age is best explained by a global model called the Global Depth and Heat Flow model (GDH1). The model predicts the following relationships between heat flow (q, mW m^{-2}) and age (t, Ma):

$$q=\frac{510}{\sqrt{t}} \qquad (t\leq 55\text{Ma})$$

$$q=q_s[1+2\exp(-\kappa\pi^2 t/a^2)] \quad (t>55\text{Ma})$$

$$=48+96\exp(-0.0278t) \tag{4.72}$$

Here q_s is the asymptotic heat flow, to which the heat flow decreases over very old oceanic crust ($\approx$48 mW m^{-2}), a is the asymptotic thickness of old oceanic lithosphere ($\approx$95 km), and κ is its thermal diffusivity ($\approx 0.8\times10^{-6}$ m^2 s^{-1}).

Close to a ridge axis the measured heat flow is unpredictable: extremely high values and very low values have been recorded. Over young lithosphere the observed heat flow is systematically less than the values predicted by cooling models (Fig. 4.25). The divergence is related to the process

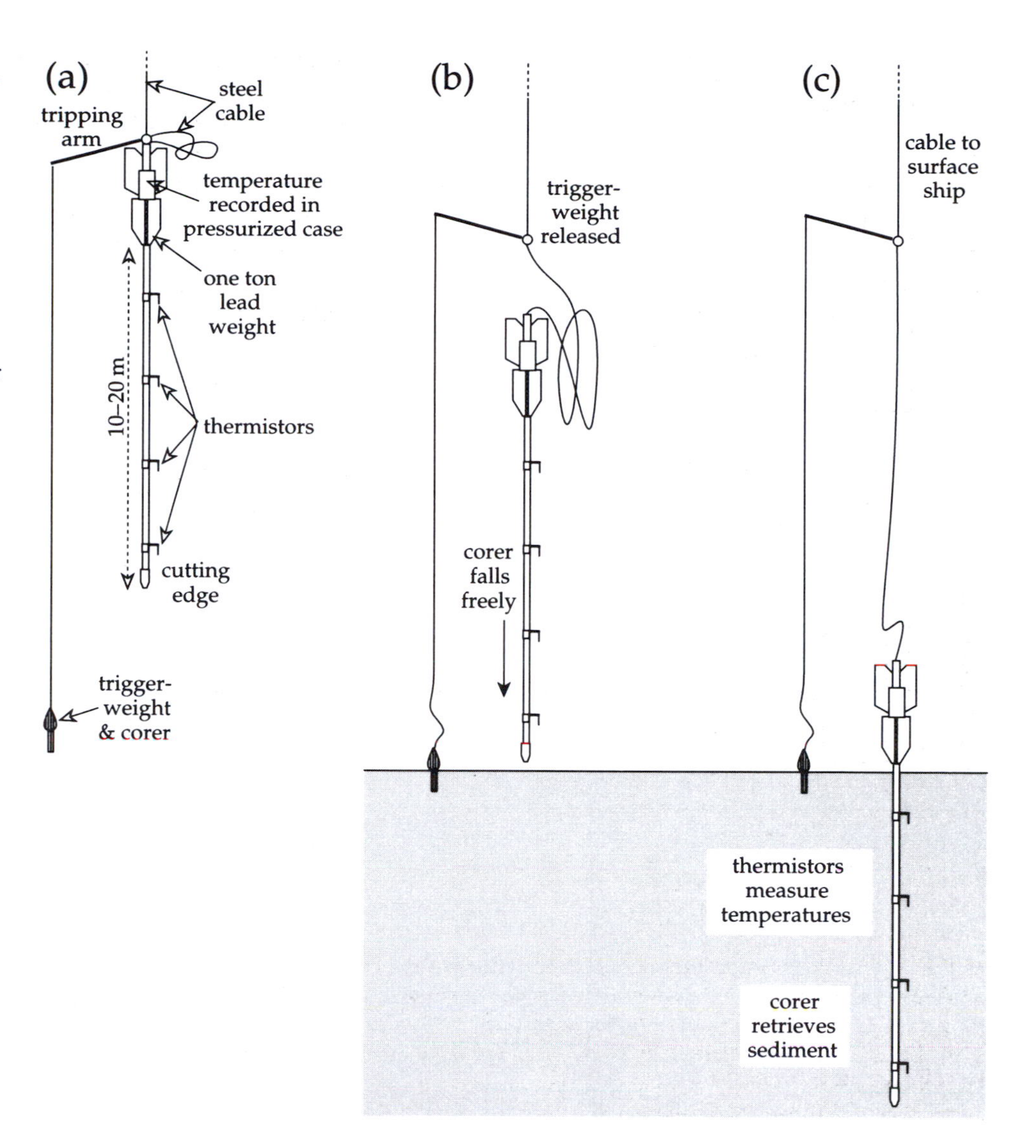

Fig. 4.24 Method of measuring oceanic heat flow and recovering samples of marine sediments: (a) a coring device is lowered by cable to the sea-floor, (b) when a trigger-weight contacts the bottom, the corer falls freely, and (c) temperature measurements are made in the ocean floor and the sediment-filled corer is recovered to the surface ship.

of accretion of the new lithosphere. At a ridge crest magma erupts in a narrow zone through feeder dykes and/or supplies horizontal lava flows. Very hot material is brought in contact with sea-water, which cools and fractures the fresh rock. The water is in turn heated rapidly and a hydrothermal circulation is set up, which transports heat out of the lithosphere by convection. The eruption of hot hydrothermal currents has been observed directly from manned submersibles in the axial zones of oceanic ridges. The expeditions witnessed strong outpourings of mineral-rich hot water (called 'black smokers' and 'white smokers') in the narrow axial rift valley. The heat output of these vents is high: the power associated with a single vent has been estimated to be about 200 MW.

About 30% of the hydrothermal circulation takes place very near to the ridge axis through crust younger than 1 Ma. The rest is due to off-ridge circulation, which is possible because the fractured crust is still permeable to sea-water at large distances from the ridge axis. As it moves away from the ridge sedimentation covers the basement with a progressively thicker layer of low permeability sediments, inhibiting the convective heat loss. The hydrothermal circulation eventually ceases, perhaps because it is sealed by the thick sediment cover, but probably also because the cracks and pore spaces in the crust become closed with increasing age. This is estimated to take place by about 55–70 Ma, because for greater ages the observed decrease of heat flow is close to that predicted by plate cooling models. The hydrothermal circulation in oceanic crust is an important part of the Earth's heat loss. It accounts for about a third of the total oceanic heat flow, and a quarter of the global heat flow.

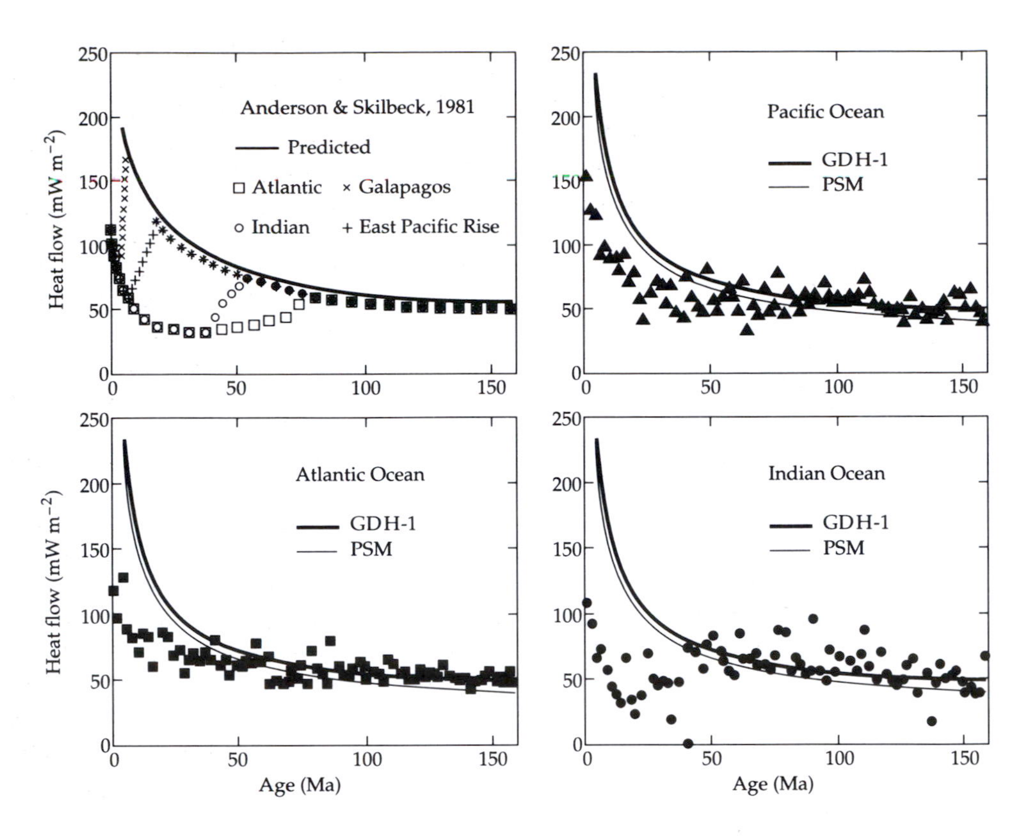

Fig. 4.25 Comparison of observed and predicted heat flow as a function of age of oceanic lithosphere. (a) Schematic summary for all oceans, showing the influence of hydrothermal heat flow at the ocean ridges (after Anderson and Skilbeck, 1981). Comparisons with the reference cooling models PSM (Parsons and Sclater, 1977) and GDH1 (Stein and Stein, 1992) for (b) the Pacific, (c) Atlantic and (d) Indian oceans.

The free-air gravity anomaly over an oceanic ridge system is generally small and related to ocean-bottom topography (§2.6.4.3), which suggests that the ridge system is isostatically compensated. As hot material injected at the ridge crest cools, its volume contracts and its density increases. To maintain isostatic equilibrium a vertical column sinks into the supporting substratum as it cools. Consequently, the depth of the ocean floor (the top surface of the column) is expected to increase with age of the lithosphere. The cooling half-space model predicts an increase in depth proportional to $\sqrt{t}$, where t is the age of the lithosphere, and this is observed up to an age of about 80 Ma (Fig. 4.26). However, the square-root relationship is not the best fit to the observations. Other cooling models fit the observations more satisfactorily, although the differences from one model to another are small. Beyond 20 Ma the data are better fitted by an exponential decay. The optimum relationships between depth (d, m) and age (t, Ma) can be written

$$d=2600+365\sqrt{t} \qquad (t<20\text{Ma})$$

$$d=d_r+d_s[1-(8/\pi^2)\exp(-\kappa\pi^2 t/a^2)] \qquad (t\geq 20\text{Ma})$$

$$=5651-2473\exp(-0.0278t) \qquad (4.73)$$

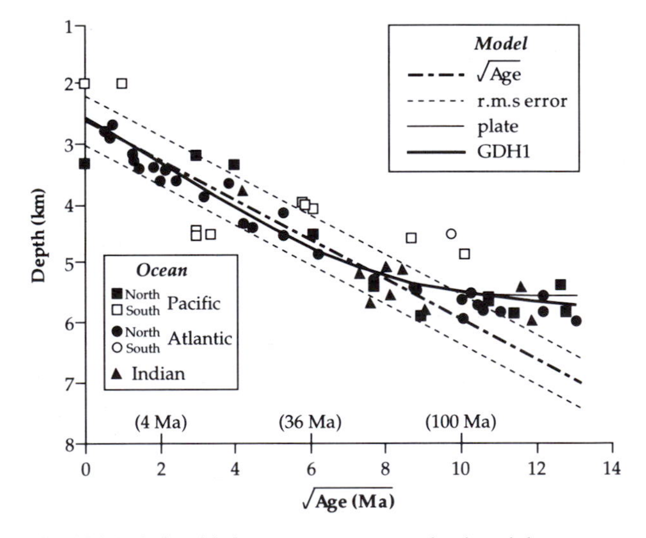

Fig. 4.26 Relationship between mean ocean depth and the square root of age for the Atlantic, Pacific and Indian oceans, compared with theoretical curves for different models of plate structure (after Johnson and Carlson, 1992).

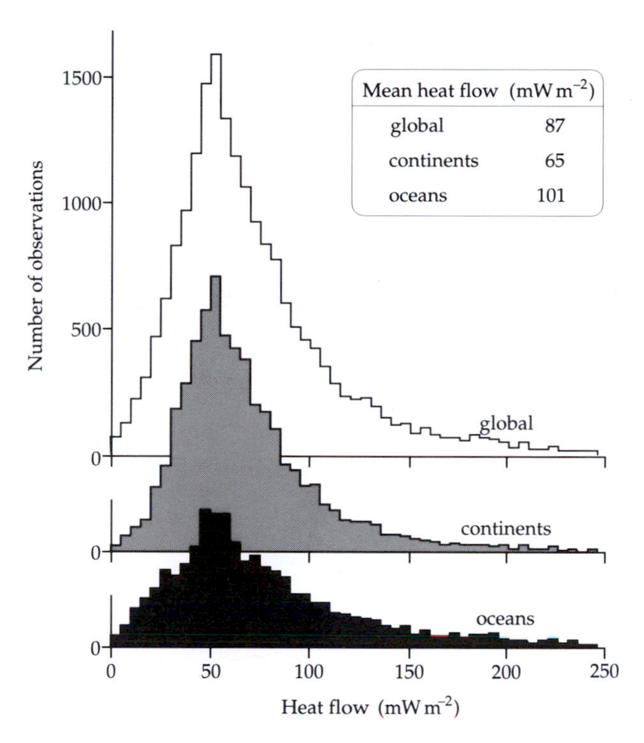

Fig. 4.27 Histograms of continental, oceanic and global heat-flow values (after Pollack *et al.*, 1993).

where d_r is the mean depth of the ocean floor at ridge crests, d_s is the asymptotic subsidence of old lithosphere and the other parameters are as before.

4.2.8.2 *Global heat flow*

Oceanic heat flow has been measured routinely in oceanographic surveys since the 1950s and *in-situ* profiles have been made since the 1970s. In contrast to the measurement of continental heat flow it is not necessary to have an available (and usually expensive) drillhole. However, the areal coverage of the oceans by heat-flow measurements is uneven. A large area in the North Pacific Ocean is still unsurveyed, and most of the oceanic areas south of about latitude 35°S (the approximate latitude of Cape Town or Buenos Aires) are unsurveyed or only sparsely covered. The uneven data distribution is dense along the tracks of research vessels and absent or meager between them. The sites of measured continental heat flow are even more irregularly distributed. Antarctica, most of the interiors of Africa and South America, and large expanses of Asia are either devoid of heat-flow data or are represented by only a few sites.

In recent years a global data set of heat-flow values has been assembled, representing 20,201 heat-flow sites. The data set is almost equally divided between observations on land (10,337 sites) and in the oceans (9,864 sites). Histograms of the heat-flow values are spread over a wide range for each domain (Fig. 4.27). The distributions have similar characteristics, extending from very low, almost zero values to more than 200 mW m^{-2}. The high values on the continents are from volcanic and tectonically active regions, while the highest values in the oceans are found near to the axes of oceanic ridges. Both on the continents (Fig. 4.23) and in the oceans (Fig. 4.25), heat flow varies with crustal age. To determine global heat-flow statistics, the fraction of the Earth's surface area having a given age is multiplied by the mean heat flow measured for that age domain. The weighted sum gives a mean heat flow of 65 mW m^{-2} for the continental data set. The oceanic data must be corrected for hydrothermal circulation in young crust; the areally weighted mean heat flow is then 101 mW m^{-2} for the oceanic data set. The oceans cover 60.6% and the continents 39.4% of the Earth's surface, the latter figure including 9.1% for the continental shelves and other submerged continental crust. The weighted global mean heat flow is 87 mW m^{-2}. Multiplying by the Earth's surface area, the estimated global heat loss is found to be 4.42×10^{13} W (equivalent to an annual heat loss of 1.4×10^{21} J). About 70% of the heat is lost through the oceans and 30% through the continents.

The heat-flow values in both continental and oceanic domains are found to depend on crustal age and geological characteristics. These relationships make it possible to create a map of global heat flow that allows for the uneven distribution of actual measurements (Fig. 4.28). The procedure in creating this map was as follows. First, the Earth's surface was divided into 21 geological domains, of which 12 are in the oceans and 9 on the continents. Next, relationships between heat flow and age were used to associate a representative heat flow with each domain (Table 4.6). This made it possible to estimate heat flow for regions that have no measurement sites. Allowance was also made for the loss of heat by hydrothermal circulation near to ridge systems. The surface of the globe was next divided into a grid of 1° elements (i.e., each element measures $1°\times1°$), and the mean heat flow through each element was estimated. This gave a complete data set (partly observed and partly synthesized) covering the entire globe. The gridded data were fitted by spherical harmonic functions (as in the representation of the geoid, §2.4.5) up to degree and order 12. The results of the analysis were used to compute smooth contours of equal heat flow, which were than plotted as the global heat-flow map (Fig. 4.28). If the global mean value is subtracted, the Earth's surface can be divided into regions with above-average and below-average heat flow, respectively (Fig. 4.29). The regions with above-average heat flow are notably associated with the oceanic ridge systems. About half of the Earth's heat is lost by the cooling of oceanic lithosphere of Cenozoic age (younger than 65 Ma).

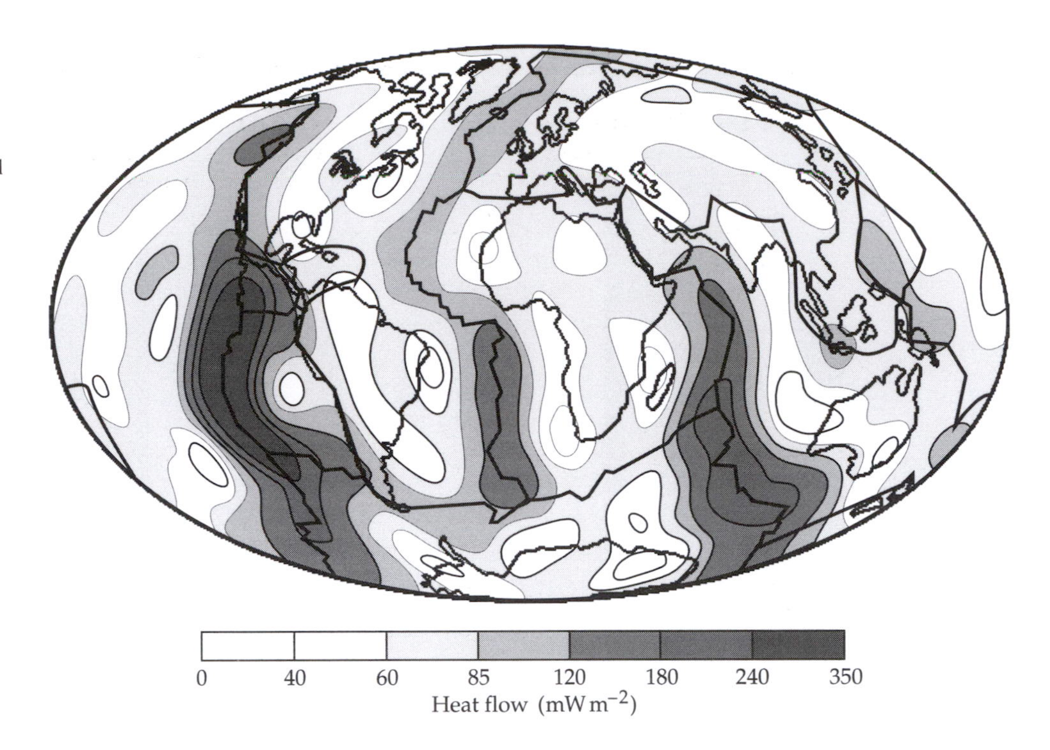

Fig. 4.28 Global distribution of heat flow. The contours show a 12th-degree spherical harmonic representation of the global heat flow based on direct measurements and empirical estimators for regions without data (after Pollack *et al.*, 1993).

One must keep in mind that this global model is based on a mixture of actual heat-flow measurements in regions where they are available, and estimated values in inaccessible regions. Moreover, the measured data near ocean ridges are replaced with values predicted by cooling models to compensate for the known loss of heat by hydrothermal circulation. Nevertheless, these global heat-flow maps are the best available representations of the geographical pattern and flux of the heat flowing out of the Earth's interior. Although details may eventually need modification, the main features are not in doubt.

4.2.8.3 *Models for the cooling of oceanic lithosphere*

The variations of heat flow and ocean depth with time constrain the possible thermal models for cooling of the lithosphere in different ways. The predicted heat flow is contingent on the temperature gradient in a model, but the oceanic depth is defined by the vertical distribution of density, which, in turn, depends on the volume coefficient of expansion and the temperature profile in the plate. Thus, oceanic bathymetry depends on the temperature integrated over depth.

Several cooling models have been proposed, all of which satisfy the decrease in heat flow and increase in ocean depth with age. The simplest model represents the cooling lithosphere as a semi-infinite half-space (Fig. 4.30a). Initially, the temperature inside the half-space is uniform and higher than on its upper surface, which is maintained at the temperature of cold ocean-bottom sea-water. As long as the lithosphere is thin – which it is near the ridge – horizontal heat conduction can be neglected. The heat flow in the uniform half-space is vertical, along the z-axis, and is equivalent to the one-dimensional flow in a thin vertical column (Fig. 4.31a). The spreading process can be envisioned as transporting the column away from the ridge axis, during which conductive cooling takes place and the temperature distribution in the column changes.

This model allows us to compute the temperature distribution in the oceanic lithosphere. We need to compute the depth z at which a given temperature T is reached after time t, when the vertical column has moved at velocity v to a distance vt from the ridge. First, the desired temperature T is expressed as a fraction of the mantle temperature T_m. Using Eq. (4.68) and the appropriate table, the argument η_0 is found which gives an error function equal to (T/T_m). Setting the numerical value η_0 equal to $z/2\sqrt{kt}$ gives the shape of the isotherm for the temperature T:

$$\frac{z}{2\sqrt{\kappa t}}=\eta_0$$

$$z=(2\eta_0\sqrt{\kappa})\sqrt{t} \tag{4.74}$$

The isotherms in the cooling lithosphere have a parabolic shape with respect to the time (or horizontal distance) axis, of which only the part for $z>0$ is of interest. The surface heat flow for this model is given by Eq. (4.71), and so is inversely proportional to $\sqrt{t}$.

Table 4.6 *Mean heat-flow values for the oceans and continents, based on measurements at 20,201 sites (after Pollack et al., 1993)*

The oceanic heat-flow values in italics are corrected for hydrothermal circulation according to the model of Stein and Stein (1992).

Description	Number of sites	Area of Earth (%)	Heat flow ($mW\ m^{-2}$)
OCEANS			
Quaternary	415	1.2	*806*
Pliocene	712	2.4	*286*
Miocene	1,211	9.2	*142*
Oligocene	593	7.7	*93*
Eocene	691	7.8	*75*
Paleocene	205	3.9	*65*
Late Cretaceous	359	6.9	60
Middle Cretaceous	695	11.2	54
Early Cretaceous	331	4.3	51
Late Jurassic	295	3.8	49
Cenozoic undifferentiated	846	2.2	89
Mesozoic undifferentiated	599	0.2	45
All oceanic data	6,952	60.6	101
CONTINENTS			
Continental shelf regions	295	9.1	78
Cenozoic: igneous	3,705	1.1	97
sedimentary and metamorphic	2,912	8.1	64
Mesozoic: igneous	1,591	1.6	64
sedimentary and metamorphic	1,310	4.5	64
Paleozoic: igneous	1,810	0.4	61
sedimentary and metamorphic	403	5.9	58
Proterozoic	260	6.2	58
Archean	963	2.5	52
All continental data	13,249	39.4	65

The half-space model has some unrealistic aspects. It predicts infinitely large heat flow at the ridge axis, and the initial mantle temperature T_m is approached asymptotically and is only reached at infinite depth. The distances between successive isotherms for equal increments in temperature get progressively larger. The near-surface layer in which the temperature changes are significant has been called a *thermal boundary layer*. Its base is defined arbitrarily as the depth at which the temperature reaches a chosen fraction of T_m. The layer can be regarded as a thermal model of the lithosphere (Fig. 4.30b). Instead of being defined mechanically as the depth where seismic shear waves are attenuated, the base of the lithosphere in the thermal model is an isotherm (Fig. 4.31b). The model predicts that the lithosphere becomes thicker with increasing age, as also inferred from seismic data, and the thickness is proportional to $\sqrt{t}$.

Parker and Oldenburg (1973) proposed a modification of the boundary-layer model in which a solid lithosphere overlies a fluid asthenosphere. The base of the lithosphere is taken to be the solid–liquid phase boundary of the material. It is defined by the melting-point isotherm, and denotes a phase change. This is probably a closer representation of the real situation, although, by treating the asthenosphere as a fluid, it exaggerates the change in rheology. The temperature of the asthenosphere lies close to the solidus temperature, but its condition is only partially molten (perhaps about 5%).

The half-space and boundary-layer models fit the observed variations of heat flow and ocean depth with age for young lithosphere. For ages greater than about 70 Ma the ocean depths in particular are less than predicted by the $\sqrt{t}$ relationship (Fig. 4.26). This suggests that the source of heat from below the lithosphere may be shallower than in the half-space model at large ages. As an alternative to the half-space models the oceanic lithosphere has been modelled as a flat layer or plate of finite thickness, bounded

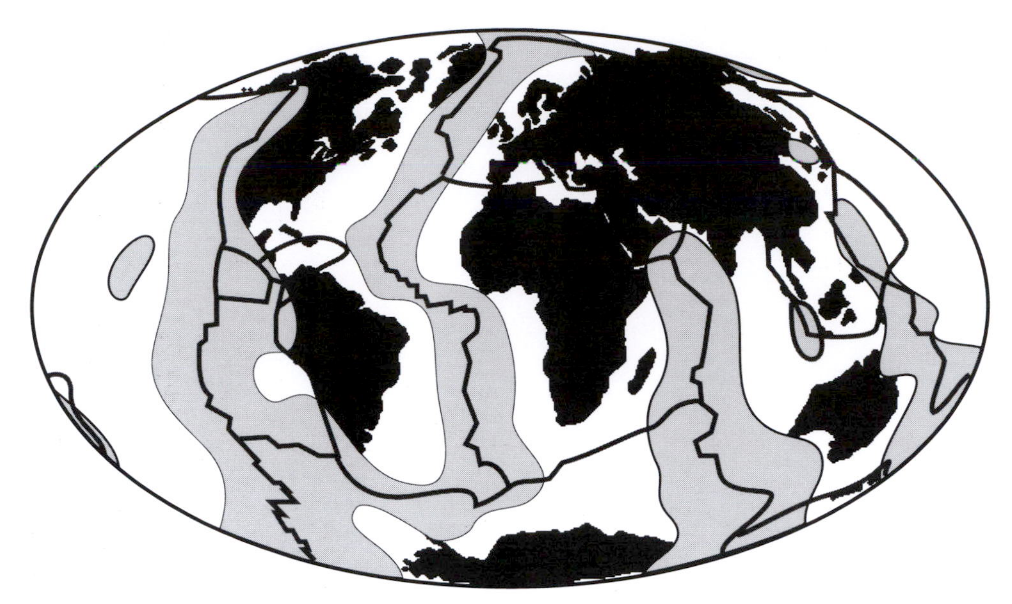

Fig. 4.29 Geographic regions where the heat flow is higher (*lighter-shaded*) and lower (*unshaded*) than the global mean heat flow; lines (*darker shaded*) mark positions of plate boundaries (after Pollack *et al.*, 1993).

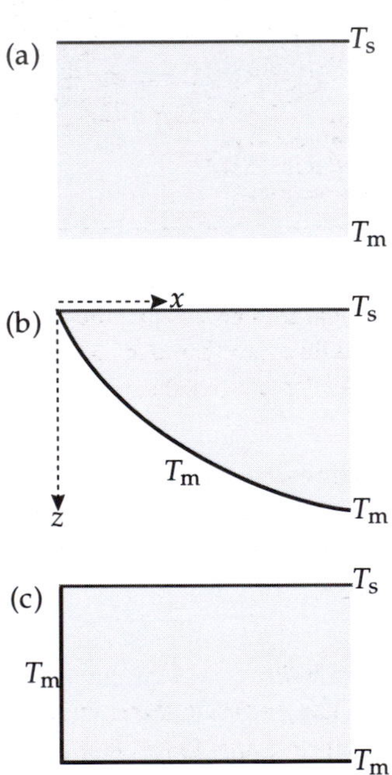

Fig. 4.30 (a) Semi-infinite half-space, (b) thermal boundary layer, and (c) plate models for the cooling of oceanic lithosphere. T_s and T_m are surface and mantle temperature, respectively.

above by cold sea-water and with a constant temperature on its lower surface. Far from a ridge axis this model brings hot mantle temperatures nearer to the surface than in the half-space model. Below the ridge axis the vertical edge of the new plate has the same high temperature of its lower surface (Fig. 4.30c), which results in heat being conducted horizontally through the plate. This is not a serious problem as long as the plate is much thinner than its horizontal extent away from a ridge. This condition is clearly met for the main lithospheric plates, which are several thousand kilometers across and only of the order of a hundred kilometers thick.

The plate model is not intended to model the vertical mechanical structure of the plate, but only to explain in a phenomenological way the typical age-dependence of both ocean depth and heat flow. The plate thickness in the model is the asymptotic thermal thickness of old oceanic lithosphere and reflects the combined effects of temperature and rheology. Its horizontal isothermal base requires additional deep heat sources that prevent the lithosphere from cooling as a half-space at great ages. The model allows simple computation of the thermal cooling history. The best-known version, proposed by Parsons and Sclater (1977), assumed a plate thickness of 125 km and a basal temperature of 1350 °C. At large distances from each spreading center it gives very good fits to both the observed heat flow (Fig. 4.25) and the ocean depth (Fig. 4.26). The most recent update, the GDH1 plate model, has a thickness of 95 km and a basal temperature of 1450 °C. It fits the observations even better.

4.2.8.4 *Thermal structure of oceanic lithosphere*

The plate models explain observed thermal data better than the boundary-layer models. The boundary-layer model is most appropriate near to a ridge axis, and agrees better with other geophysical data, which show that the lithosphere thickens with distance from a ridge. However, the plate model is needed at great distances to explain heat flux and ocean depths over old lithosphere. To reconcile these contrasting attributes a two-layered model of the lithosphere has been proposed (Fig. 4.32). The upper layer is rigid and

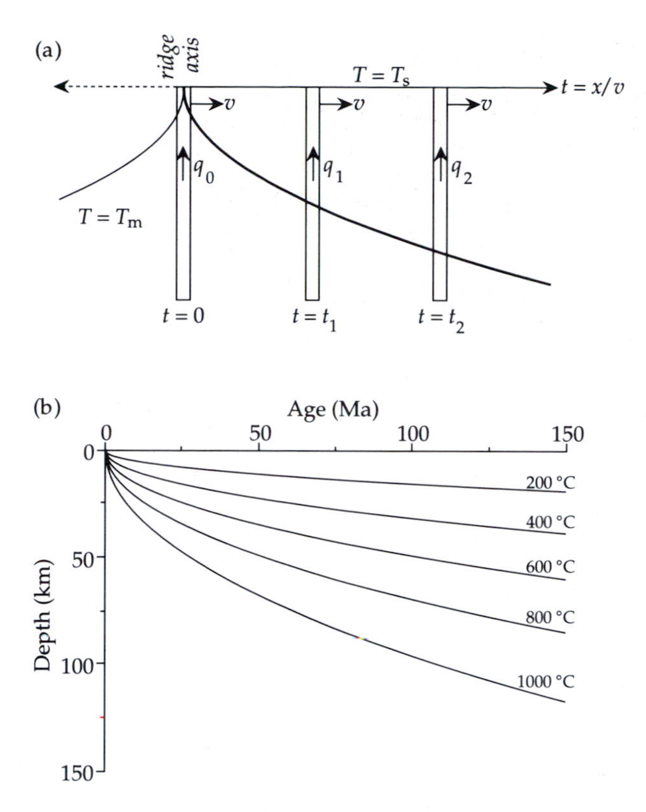

Fig. 4.31 Application of the infinite half-space model to explain the cooling of oceanic lithosphere: (a) vertical heat flow in narrow columns that move away from the ridge crest, and (b) predicted thermal structure in the cooling plate (after Turcotte and Schubert, 1982).

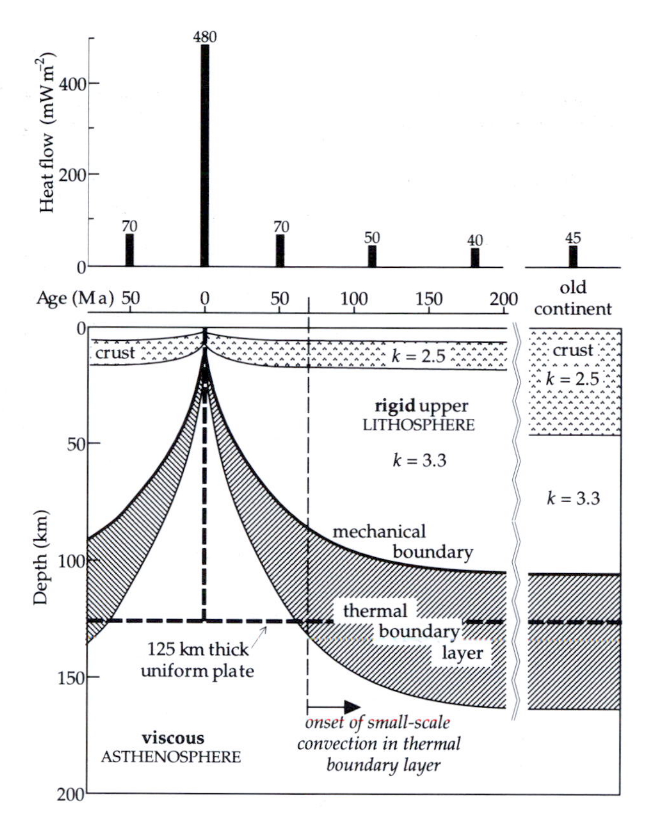

Fig. 4.32 Schematic diagram of lithospheric plate structure beneath oceans and continents. The dashed line indicates the approximation as a plate of constant thickness (based upon Parsons and McKenzie, 1978, and Sclater *et al.*, 1981).

has a mechanically defined lower boundary, above which heat transfer is by conduction. Below this level the increasing temperature causes a change in mechanical properties. The lower lithosphere is plastic enough to permit material movement, and so behaves like a viscous solid. The base of the upper layer is an isotherm, representing the temperature at which rigidity is lost. The base of the lower lithosphere is a thermally defined boundary, and is also an isotherm. Several suggestions have been made as to how this structure may approximate the plate model for old lithosphere. They include additional heat sources such as radiogenic heating, frictional heating as a result of shear at the base of the lithosphere, and reheating of old lithosphere due to the intrusion of mantle plumes at hotspots. It has also been postulated that, at lithosphere ages greater than about 70 Ma, small-scale convection currents in the lower lithosphere may augment thermal conduction. This would bolster the transfer of heat from the convecting asthenosphere into the lithosphere, effectively giving a thinner lithosphere than in the half-space models. Analysis of the dispersion of seismic surface waves indicates that there are differences in structure between continents and oceans down to about 200 km. This is compatible with the thermal model of a rigid mechanical layer underlain by a convecting thermal boundary layer extending to about 150–200 km.

4.2.8.5 *Heat flow at subduction zones*

The oceanic lithosphere is bent sharply downward beneath the overriding plate in a subduction zone. It extends as an inclined slab deep into the upper mantle, which it penetrates at a rate of a few cm per year. The old lithosphere is cold, having lost much of its original heat of formation at the ridge axis. By the time it reaches an ocean trench the isotherms in the plate are far apart and the temperature gradient is small. The separation of the isotherms is increased by the downward bending of the plate. Since the heat flow is proportional to the temperature gradient, very low heat-flow values ($\approx$ 35 mW m^{-2}) are measured in oceanic trenches. After bending downward the plate is subducted to great depths, subjecting it to increases in pressure and temperature. Heat is conducted into the plate from the adjacent mantle. This process is so slow that the interior of the sub-

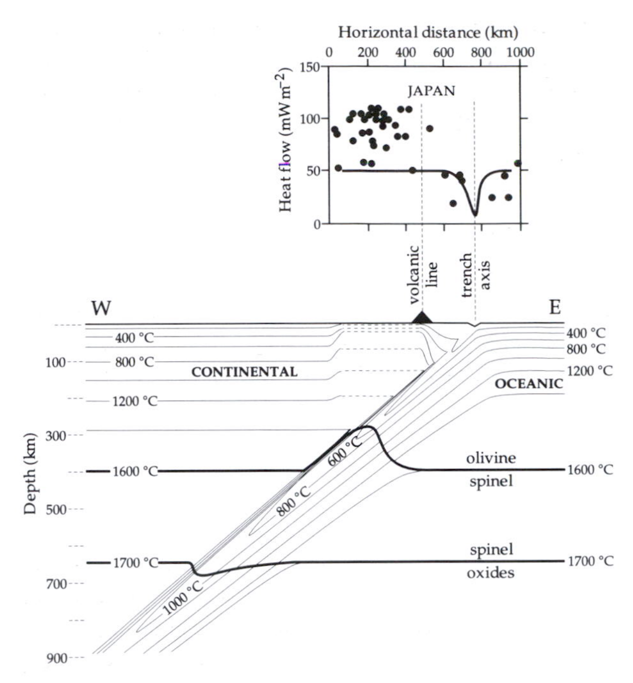

Fig. 4.33 (*Bottom*): The thermal structure of a subduction zone and back-arc region (the model of Schubert *et al.*, 1975, inverted horizontally), showing the possible isotherms in the cold subducting plate and the thermal effects of the olivine–spinel and spinel–oxide phase changes. (*Top*): comparison of heat-flow measurements across the Japanese trench with the theoretical heat flow (solid curve) computed by Toksöz *et al.* (1971).

ducting slab remains colder than its environment (Fig. 4.33). A temperature of 800 °C is normally reached at about 70 km depth in the oceanic plate but in the descending slab this temperature exists to deeper than 500 km. Above this depth the coldest part of the slab has a horizontal temperature deficit of 800–1000 K.

Heat conducted from the mantle is not the only heat source that must be taken into account in modelling the thermal structure of the subducting slab. An important additional source is the frictional heating that results from shear deformation at the surfaces of the slab where it is in contact with the mantle. In the upper part of a subduction zone the shear heating melts the basaltic layer of the oceanic lithosphere and forms a layer of eclogite in the top of the slab. The high density of the eclogite causes a positive gravity anomaly (see Fig. 2.59), and adds to the forces propelling the slab downward. The phase transition in which the open structure of olivine-type minerals converts to a denser spinel-type structure normally takes place at a depth of 400 km. The phase transition depends on temperature and pressure. Laboratory experiments indicate that it takes place at lower pressure at low temperature than at high temperature. Consequently it occurs at shallower depths within the cold plate than in the adjacent mantle. As a result the transition depth is deflected upward by about 100 km. The transition is exothermic and the latent heat given out in the transition is an additional heat source that contributes to the thermal structure of the subduction zone. The transition also results in a density increase, which adds to the forces driving the plate downward.

The deeper transition at 670 km is less well understood. High temperature apparently causes it to take place at higher pressure, and so the depth of occurrence is deflected downward inside the subducting slab. It is uncertain if the transition is endothermic, absorbing heat from the environment, or exothermic as assumed in the model in Fig. 4.33. An endothermic phase change has the effect of reducing the density, and acts against the other downward forces on the slab.

Although other, slightly different models have been derived for the temperature distribution in the descending slab, they all have in common the downward deflection of isotherms in the cold descending slab. The heat flow can be computed for a given thermal model. When compared with the observed heat flow on a profile across the subduction zone, the models fail to explain adequately the high heat flow observed on the overriding plate (Fig. 4.33). Volcanic activity is partly responsible, fed by magmas produced by partial melting of oceanic crustal material in the descending slab and of the upper mantle in the overriding plate. Shallow melting is promoted by water from the subducting plate and generates basaltic magma; deeper melting involves less water and results in andesitic magma. When the overriding plate is continental, volcanic chains form along the continental margin parallel to the deep oceanic trench. The volcanicity is typified by the eruption of both basaltic and andesitic lavas. The lavas are more felsic than those formed when two oceanic plates collide, which may imply that they include melted material from the upper mantle of the overriding continental plate.

When two oceanic plates converge, a volcanic arc is formed on the overriding plate. Behind the arc, high heat flow on the overriding plate is related to back-arc spreading, in which new oceanic crust is generated by the intrusion of basaltic magma from partial melting in the upper mantle. This form of sea-floor spreading produces a marginal basin behind the island arc. The intrusion of magma is not confined to a single location, as at a ridge axis, but is spread diffusely in the basin. Consequently, the stripes of lineated oceanic magnetic anomalies characteristic of sea-floor spreading at ridge systems are missing or at best weakly defined in a marginal basin.

4.3 GEOELECTRICITY

4.3.1 Introduction

Electric charge – together with mass, length and time – is a fundamental property of nature. The name *electric* derives from the Greek word for amber ('elektron'), the naturally occurring fossilized resin of coniferous trees that has been used since antiquity in the making of jewelry. The Greek philosopher Thales of Miletus (ca. 600 B.C.) is credited with first reporting the power of amber, when rubbed with a cloth, to attract light objects. The ancient sages could not understand this behavior in terms of their everyday world, and so, together with the power of *magnetism* possessed by natural lodestone (see §5.1.1), electricity remained a wonderful but unknown phenomenon for more than two millennia. In 1600 A.D. the English physician William Gilbert summarized previous investigations and extant knowledge in the first systematic study of these phenomena.

In the following century it was established that there were two types of electric charge, now referred to as positive and negative. Objects that carried like types of charge were observed to repel each other, and those that carried opposite types were attracted to each other. In 1752 the American statesman, diplomat and scientist Benjamin Franklin performed a celebrated experiment; by flying a kite during a thunderstorm, he established that lightning is an electrical phenomenon. Having survived this risky endeavour Franklin developed the far-sighted theory that electricity consisted of an omnipresent fluid, and that the different types of charge represented surplus and scarcity of this fluid. This view strikingly resembles modern theory, in which the 'fluid' consists of electrons.

The laws of electrostatic attraction and repulsion were established in 1785 as a result of careful experiments by a French scientist, Charles Augustin de Coulomb (1736–1806), who also established the laws of magnetostatic force (§5.1.3). Coulomb invented a sensitive torsion balance, with which he could measure accurately the force between electrically charged spheres. His results represent the culmination of knowledge of electrostatic phenomena.

The 18th century concept of electricity as a fluid finds further expression in electrical nomenclature. Electricity is said to flow between charged objects when they are brought in contact, and the rate of flow is called an electric current. The study of the properties and effects of electric currents became possible around 1800, when an Italian physicist, Alessandro Volta (later elevated by Napoleon to the rank of Count), invented a primitive electric battery, called a voltaic pile, in which electricity was produced by chemical action. The relationship between the electric current in a conductor and the voltage of the battery was established in 1827 by Georg Ohm, a German physicist. The magnetic effects produced by electric currents were established in the early 19th century by Oersted, Ampère, Faraday and Lenz. Their contributions are discussed in more detail in a later chapter (§5.1.3) on the physical origins of magnetism.

4.3.2 Electrical principles

Coulomb established that the force of attraction or repulsion between two charged spheres was proportional to the product of the individual electric charges and inversely proportional to the square of the distance between the centers of the spheres. His law can be written as the following equation:

$$F = K\frac{Q_1 Q_2}{r^2} \tag{4.75}$$

where Q_1 and Q_2 are the electric charges, r is their separation and K is a constant. This inverse-square law strongly resembles the law of universal gravitation (Eq. (2.2)), formulated by Newton more than a century before Coulomb's law. However, in gravitation the force is always attractive, whereas in electricity it may be attractive or repulsive, depending on the nature of the charges. In the law of gravitation the units of mass, distance and force are already defined, so that the gravitational constant is predetermined; only its numerical value needed to be measured. In Coulomb's law, F and r are defined from mechanics (as the newton and meter, respectively), but the units of Q and K are undefined. The value of K was originally set equal to unity, thereby defining the unit of electric charge. This definition led to unfortunate complications when the magnetic effects of electric currents were analyzed. The alternative is to define independently the unit of charge, thereby fixing the meaning of the constant K.

The unit of charge is the coulomb (C), defined as the amount of charge that passes a point in an electrical circuit when an electric current of one ampère (A) flows for one second (i.e., 1 C=1 A s). In turn, the ampère is defined from the magnetic effects of a current (see §5.2.4). When a current flows in the same direction through two parallel long straight conductors, magnetic fields are produced around the conductors, which cause them to attract each other. If the current flows through the conductors in opposite directions they repel each other. The ampère is defined as the current that produces a force of 2×10^{-7} N per meter of length between infinitely long thin conductors that are one meter apart in vacuum. Thus, the unit of charge is defined precisely, if rather indirectly. In the Système Internationale (SI) units K is written as $(4\pi\epsilon_0)^{-1}$, so that Coulomb's law becomes

$$F = \frac{1}{4\pi\epsilon_0}\frac{Q_1 Q_2}{r^2} \tag{4.76}$$

where the constant ϵ_0 is called the *permittivity constant*. It is approximately equal to $8.854\,19\times10^{-12}\,C^2\,N^{-1}\,m^{-2}$.

Modern electrical theory descends from the discovery in 1897 by the English physicist Joseph J. Thomson of the *electron* as the basic elementary unit of electric charge. It has a *negative charge* of 1.602×10^{-19} C. A proton in the nucleus of an atom has an equal *positive charge*. Normally an atom contains as many electrons as it has protons in its nucleus and is electrically neutral. If an atom or molecule loses one or more electrons, it has a net positive charge and is called a *positive ion*; similarly, a *negative ion* is an atom or molecule with a surplus of electrons.

In metals, some electrons are only loosely bound to the atoms. They can move with relative ease through the material, which is called an electrical *conductor*. Metals like copper and silver are good conductors. In other materials, called *insulators*, the electrons are tightly bound to the atoms. Glass, rubber, and dry wood are typical insulators. A perfect insulator does not allow electrons to move through it, whereas a perfect conductor offers no opposition to the passage of electrons. Real conductors offer different degrees of opposition.

A flow of charge, or *electric current*, results when the free electrons in a conductor move in a common direction. A current of one ampère corresponds to a flow of about 6,250,000,000,000,000,000 electrons per second past any point of a circuit! The direction of an electric current is defined to be the direction of flow of positive charge, which is opposite to the direction of motion of the electrons.

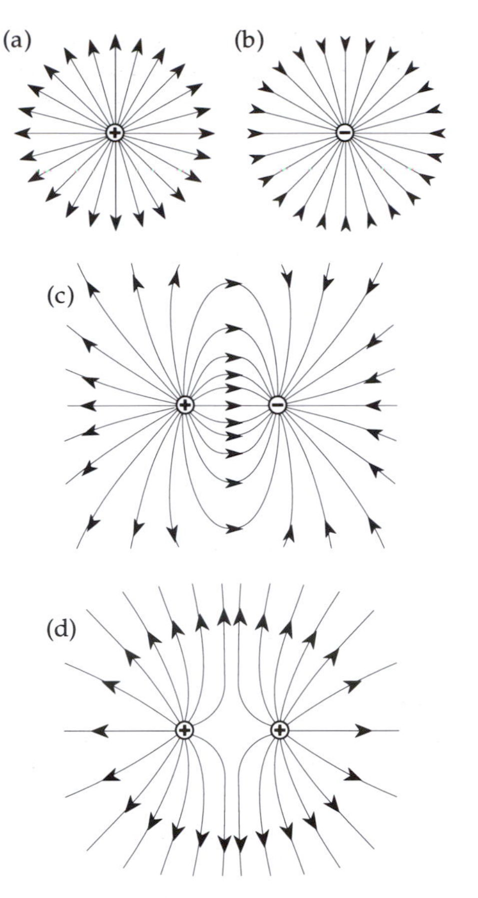

Fig. 4.34 Planar cross-sections of electric field lines around point charges: (a) single positive, (b) single negative, (c) two equal and opposite, and (d) two equal positive charges.

4.3.2.1 *Electric field and potential*

The force exerted on a unit electric charge by another charge Q is called the *electric field* of the charge Q. Thus, if we let $Q_1 = Q$ and $Q_2 = 1$ in Eq. (4.76), we obtain the equation for the electric field E at distance r from a charge Q

$$E = \frac{Q}{4\pi\epsilon_0 r^2} \tag{4.77}$$

According to this definition E has the dimensions of newton/coulomb ($N\,C^{-1}$).

The term 'field' also has another connotation, introduced by Michael Faraday (1791–1867) to refer to the geometry of the lines of force near a charge. Around a positive point charge the field lines are radially outward, describing the (divergent) direction along which a free positive charge would move (Fig. 4.34a); around a negative point charge they are radially inward (convergent) (Fig. 4.34b). The field lines of a pair of opposite point charges diverge from the positive charge, spread apart and converge on the negative charge (Fig. 4.34c); they give the appearance of drawing the opposite charges together. The combined field of two positive point charges is characterized by field lines that leave each charge and diverge in the space between (Fig. 4.34d); the field lines appear visibly to push the like charges apart. The direction of the electric field at any point is tangential to the electric field line. The strength of the field is represented by the spatial concentration of the field lines. Close to either electrical charge the field is strong and it weakens with increasing distance from the charge. Consequently, work is required to move a charged particle from one point in the field to another. This work contributes to the potential energy of the system.

For example, at an infinite distance from a positive charge Q the repulsive force on a unit positive charge is zero, but at a distance r it is given by Eq. (4.77). The potential energy of the unit charge at r is called the *electric potential* at r; we will denote it U. The units of U are energy per unit charge, i.e. joules/coulomb. If we move a distance dr against the field E, the potential changes by an amount dU equal to the work done against E, which is ($-E$ dr). i.e., d$U=-E$ dr, so that

$$E=-\frac{dU}{dr} \tag{4.78}$$

We can readily compute the electric potential U at r by integration:

$$U=-\int_{\infty}^{r} E dr=-\int_{\infty}^{r} \frac{Q}{4\pi\epsilon_0 r^2} dr \tag{4.79}$$

from which

$$U=\frac{Q}{4\pi\epsilon_0 r} \tag{4.80}$$

The energy needed to move a unit charge from one point to another in the electric field of Q is the *potential difference* between the two points. The unit of potential difference is the same as that of U (i.e., joules/coulomb) and is called a *volt*. From Eq. (4.78) we obtain the more common alternative units of volt/meter (V m^{-1}) for the electric field E.

Electric charge flows from a point with higher potential to a point with lower potential. The situation is analogous to the flow of water through a pipe from one level to a lower level. The rate of flow of water through the pipe is determined by the difference in gravitational potential between the two levels. Likewise, the electric current in a circuit depends on the potential difference in the circuit.

4.3.2.2 *Ohm's law*

The German scientist Georg Simon Ohm established in 1827 that the electric current I in a conducting wire is proportional to the potential difference V across it. The linear relationship is expressed by the equation

$$V=IR \tag{4.81}$$

where R is the *resistance* of the conductor. The unit of resistance is the *ohm* (Ω). The inverse of resistance is called the *conductance* of a circuit; its unit is the reciprocal ohm (Ω^{-1}), variously also called a *mho* or *siemens* (S).

Experimental observations on different wires of the same material showed that a long wire has a larger resistance than a short wire, and a thin wire has a larger resistance than a thick wire. Formulated more precisely, for a given material the resistance is proportional to the length L and inversely proportional to the cross-sectional area A of the conductor (Fig. 4.35). These relationships are expressed in the equation

$$R=\rho\frac{L}{A} \tag{4.82}$$

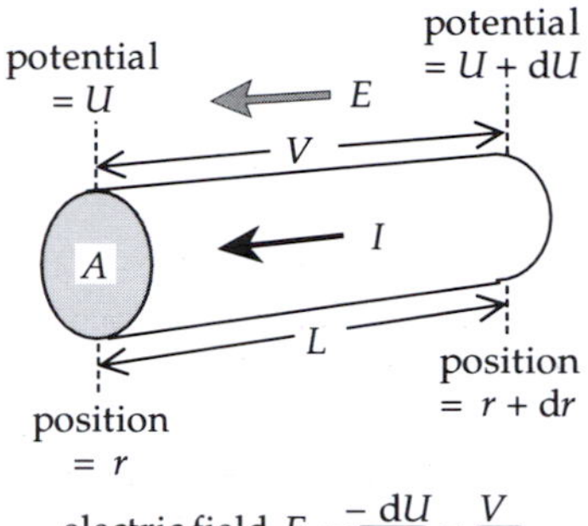

Fig. 4.35 Parameters used to define Ohm's law for a straight conductor.

The proportionality constant ρ is the *resistivity* of the conductor. It is a physical property of the material of the conductor, which expresses its ability to oppose a flow of charge. The inverse of ρ is called the *conductivity* σ of the material. The unit of resistivity is the ohm-meter (Ω m); the unit of conductivity is the reciprocal ohm-meter (Ω^{-1} m^{-1}).

If we substitute Eq. (4.82) for R in Eq. (4.81) and rearrange the terms we get the following expression:

$$\frac{V}{L}=\rho\frac{I}{A} \tag{4.83}$$

The ratio V/L on the left side of this equation is, by comparison with Eq. (4.78), the electric field E (assuming the potential gradient to be constant along the length of the conductor). The ratio I/A is the current per unit cross-sectional area of the conductor; it is called the *current density* and denoted J (Fig. 4.35). We can now rewrite Ohm's law as

$$E=\rho J \tag{4.84}$$

This form is useful for calculating the formulas used in resistivity methods of electrical surveying. However, the quantities that are measured are V and I.

4.3.2.3 *Types of electrical conduction*

Electric current passes through a material by one of three different modes: by electronic, dielectric, or electrolytic conduction. Electronic (or ohmic) conduction occurs in metals and crystals, dielectric conduction in insulators, and electrolytic conduction in liquids.

Electronic conduction is typical of a metal. The free electrons in a metal have a high average speed (about 1.6×10^6 ms^{-1} in copper). They collide with the atoms of the metal, which occupy fixed lattice sites, and bounce off in random directions. When an electric field is applied, the electrons acquire a common drift velocity, which is superposed on their random motions, so that they move at a much smaller speed (about 4×10^{-5} ms^{-1} in copper) in the direction of the field. The resistivity is determined by the mean free time between collisions. If the atomic arrangement causes frequent collisions, the resistivity is high, whereas a long mean free time between collisions results in low resistivity. The energy lost in the collisions appears in the form of heat.

A form of *semi-conduction* is important in some crystals, such as the silicate minerals. The resistivity of the mineral is higher than a conductor but lower than an insulator, and it is called a *semiconductor*. Different types of semi-conduction are possible. Silicates contain fewer conduction electrons than a metal, but the electrons are not rigidly bound to atoms as in an insulator. The energy needed to liberate additional electrons from their atoms is not large, and thermal excitation is enough to allow them to take part in *electronic semi-conduction*. The liberated electron leaves a vacancy or *hole* in the valence level of the atomic structure, which behaves as a positive charge. Natural crystals also contain impurity atoms, which may have a different valency than that required by the lattice for charge balance. The impurity is a source of holes or excess electrons which take part in *impurity semi-conduction*. At high temperature ions may detach from the lattice; they behave like ions in an electrolyte and give rise to electrical currents by *ionic semi-conduction*. A potential difference across a semi-conductor produces an electric current made up of opposite flows of negative electrons and positive holes. If most of the current is carried by the negative electrons, the semiconductor is called n-type; if the positive holes predominate, the semiconductor is said to be p-type.

Dielectric conduction occurs in insulators, which contain no free electrons. Normally, the electrons are distributed symmetrically about a nucleus. However, an electric field displaces the electrons in the direction opposite to that of the field, while the heavy nucleus shifts slightly in the direction of the field. The atom or ion acquires an electric polarization and acts like an electric dipole. The net effect is to change the permittivity of the material from ϵ_0 to a different value ϵ, given by

$$\epsilon=\kappa\epsilon_0 \tag{4.85}$$

Here, κ is the *dielectric constant* of the material; it is dimensionless, and has a value commonly in the range 3–81. Examples of κ for some natural materials are: sandstone, 5–12; granite, 3–19; diorite 6; basalt, 12; water 81. Dielectric effects are unimportant in constant current situations. However, in an alternating electric field the polarization changes with the frequency of the field. The fluctuating polarization of the electric charge contributes to the alternating current, and so modifies the effective conductivity or resistivity. In practice, this effect depends strongly on the frequency of the inducing alternating field. The higher the frequency, the greater is the effect of dielectric conduction. Some geoelectric methods utilize signals in the audio-frequency range, where dielectric conduction is insignificant, but ground-penetrating radar uses frequencies in the MHz to GHz range and depends on dielectric contrasts.

Electrolytic conduction occurs in aqueous solutions that contain free ions. The water molecule is polar (i.e., it has a permanent electric dipole moment) with a strong electric field which breaks down molecules of dissolved salts into positively and negatively charged ions. For example, in a saline solution the molecule of sodium chloride (NaCl) dissociates into separate Na^+ and Cl^- ions. The solution is called an *electrolyte*. The ions in the electrolyte are mobilized by an electric field, which causes a current to flow. Electric charge is transported by positive ions in the direction of the field and negative ions in the opposite direction. The resistivity of an electrolyte may be understood by analogy with the flow of water through a partially blocked pipe. The electric current in the electrolyte involves the physical transport of material (ions), which results in collisions with the molecules of the medium (electrolyte), causing resistance to the flow. Ionic conduction is consequently slower than electronic conduction.

4.3.3 Electrical properties of the Earth

In our daily lives we experience frequent reminders of the Earth's gravity field. It is less obvious that the Earth also has an electric field. Its presence mainly becomes evident during thunderstorms, when electrical discharges take place as lightning. The Earth's electric field acts radially inward, so that the Earth behaves like a negatively charged sphere. At its surface the vertically downward electric field amounts to about 200 V m^{-1}. The atmosphere has a net positive charge, equal and opposite to that of the Earth, resulting from the distribution of positively and negatively ionized air molecules. The charges originate from the continual bombardment of the Earth by cosmic rays.

Cosmic rays are subatomic particles with very high energy. Primary cosmic rays reach the Earth from outer space, travelling at velocities close to that of light. They consist largely of protons (hydrogen nuclei) and α-particles (helium nuclei), with lesser amounts of other ions. Their origin is still unknown. Some are emitted by the Sun at the time of solar flares, but these occur too infrequently to be the main source. This source lies elsewhere in our galaxy. It is thought that a large proportion of the galactic cosmic rays are accelerated to high speed by supernova explosions. The path of a cosmic ray is easily deflected by a magnetic field.

Even the weak interstellar magnetic field is enough to disperse fast-moving cosmic rays, so that they reach the Earth equally from all directions. The incoming particles collide with nuclei in the upper atmosphere, producing showers of secondary cosmic rays, consisting of protons, neutrons, electrons and other elementary particles. Consequently, at any given time a fraction of the molecules of the atmosphere are electrically charged. The Earth's electric field accelerates positive particles downward to the Earth's surface, where they neutralize negative surface charges. This would rapidly eliminate the negative surface charge, which is maintained by thunderstorm activity.

Thunderstorms, and the causes of lightning, are not yet fully understood. A possible scenario is the following. In a storm-cloud droplets of water vapor become electrically charged. The Earth's downward electric field may cause polarization within a droplet, with positive charge on the bottom and negative on the top. When the drop is heavy enough to fall, negative charges from molecules pushed out of its path are attracted to the bottom while fewer positive charges gather on the top. As a result the droplet becomes negatively charged. The wind action in storm-clouds causes the negative charge to accumulate at the base of the cloud, while a corresponding positive charge gathers in its upper extent. When the potential difference between the two charges exceeds the break-down voltage of the atmosphere, a brief but powerful electric current flows. Most lightning strokes occur within the storm-cloud. However, the negative charge on the base of the cloud repels the negative charge on the ground surface beneath it. Once again, if the potential difference between the cloud and the ground becomes large enough to overcome the break-down voltage of the air, a lightning stroke ensues. This carries negative charge to the ground. In this way, the numerous daily lightning storms that occur worldwide maintain the negative charge of the Earth.

To a first approximation the Earth may be regarded as a uniform electrical conductor. Electric charges on the surface of a conductor disperse so that the electric potential is the same at all points on the surface, i.e., it is an electrical equipotential surface. The surface potential is commonly used as the reference level for electrical potential energy and is defined to be zero. Thus a positively charged body has a positive potential difference (voltage) with respect to ground, while a negatively charged body has a negative voltage.

4.3.3.1 *Electrical surveying*

As with other physical parameters, geoelectrical properties are utilized in both applied and general geophysics. They are exploited commercially in the search for valuable orebodies, which may be located by their anomalous electrical conductivities. Deep electrical sounding provides valuable information about the internal structure of the Earth's crust and mantle. Electrical surveys may be based on natural sources of potential and current. More commonly, they involve the detection of signals induced in subsurface conducting bodies by electric and magnetic fields generated above ground. Investigations in this category include resistivity and electromagnetic methods. These techniques have long been used in commercial geophysical surveying. In recent years they have also become important in the scientific investigation of environmental problems. The electrical techniques require the measurement of potential differences in the ground between suitably implanted electrodes. The electromagnetic techniques detect subsurface conductivity anomalies remotely; they do not need contact with the ground. As well as being employed in surface surveys they are especially suited to airborne use.

The important physical properties of rocks for electrical surveying are the permittivity (for georadar) and the resistivity (or conductivity), on which several techniques are based. Anomalies arise, for example, when a good conductor (such as a mineralized dike or orebody) is present in rocks that have higher resistivities. The resistivity contrast between orebody and host rock is often large, because the resistivities of different rocks and minerals vary widely (Fig. 4.36). In metallic ores the resistivity can be very low, but igneous rocks that contain no water can have a very high resistivity. For example, in a high-grade pyrrhotite ore ρ is of the order of 10^{-5} Ω m, while in dry marble it is around 10^8 Ω m. The range between these extremes spans 13 orders of magnitude. Moreover, the resistivity range of any given rock type is wide and overlaps with other rock types (Fig. 4.36).

The resistivity of rocks is strongly influenced by the presence of groundwater, which acts as an electrolyte. This is especially important in porous sediments and sedimentary rocks. The minerals that form the matrix of a rock are generally poorer conductors than groundwater, so the conductivity of a sediment increases with the amount of groundwater it contains. This depends on the fraction of the rock that consists of pore spaces (the *porosity*, ϕ), and the fraction of this pore volume that is water-filled (the *water saturation*, S). The conductivity of the rock is proportional to the conductivity of the groundwater, which is quite variable because it depends on the concentration and type of dissolved minerals and salts it contains. These observations are summarized in an empirical formula, called Archie's law, for the resistivity ρ of the rock

$$\rho=\frac{a}{\phi^m S^n}\rho_w \qquad (4.86)$$

By definition ϕ and S are fractions between 0 and 1, ρ_w is the resistivity of the groundwater, and the parameters a, m and

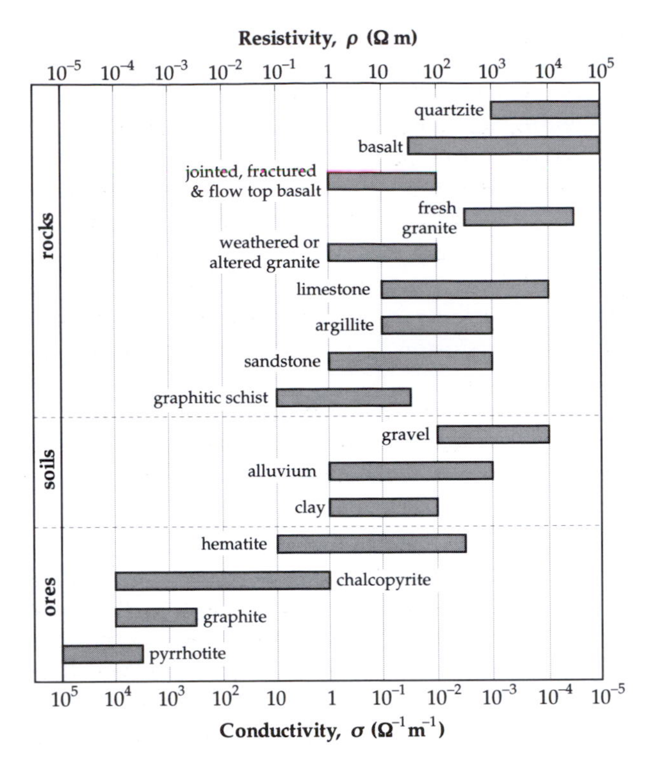

Fig. 4.36 Ranges of electrical resistivity for some common rocks, soils and ores (data source: Ward, 1990; augmented by data from Telford *et al.*, 1990).

n are empirical constants that have to be determined for each case. Generally, $0.5 \le a \le 2.5$, $1.3 \le m \le 2.5$ and $n \approx 2$.

4.3.4 Natural potentials and currents

Electrical investigations of natural electrical properties are based on the measurement of the voltage between a pair of electrodes implanted in the ground. Natural differences in potential occur in relation to subsurface bodies that create their own electric fields. The bodies act like simple voltaic cells; their potential arises from electrochemical action. Natural currents (called *telluric currents*) flow in the crust and mantle of the Earth. They are induced electromagnetically by electric currents in the *ionosphere* (described in §5.4.3.2). In studying natural potentials and currents the scientist has no control over the source of the signal. This restricts the interpretation, which is mostly only qualitative. The natural methods are not as useful as controlled induction methods, such as resistivity and electromagnetic techniques, but they are inexpensive and fast.

4.3.4.1 *Self-potential (spontaneous potential)*

A potential that originates spontaneously in the ground is called a *self-potential* (or *spontaneous potential*). Some self-potentials are due to man-made disturbances of the environment, such as buried electrical cables, drainage pipes or waste disposal sites. They are important in the study of environmental problems. Other self-potentials are natural effects due to mechanical or electrochemical action. In every case the groundwater plays a key role by acting as an electrolyte.

Some self-potentials have a mechanical origin. When an electrolyte is forced to flow through a narrow pipe, a potential difference (voltage) may arise between the ends of the pipe. Its amplitude depends on the electrical resistivity and viscosity of the electrolyte, and on the pressure difference that causes the flow. The voltage is due to differences in the *electrokinetic* or *streaming potential*, which in turn is influenced by the interaction between the liquid and the surface of the solid (an effect called the *zeta-potential*). The voltage can be positive or negative and may amount to some hundreds of millivolts. This type of self-potential can be observed in conjunction with seepage of water from dams, or the flow of groundwater through different lithological units.

Most self-potentials have an electrochemical origin. For example, if the ionic concentration in an electrolyte varies with location, the ions tend to diffuse through the electrolyte so as to equalize the concentration. The diffusion is driven by an electric *diffusion potential*, which depends on the temperature as well as the difference in ionic concentration. When a metallic electrode is inserted in the ground, the metal reacts electrochemically with the electrolyte (i.e., groundwater), causing a contact potential. If two identical electrodes are inserted in the ground, variations in concentration of the electrolyte cause different electrochemical reactions at each electrode. A potential difference arises, called the *Nernst potential*. The combined diffusion and Nernst potentials are called the *electrochemical* self-potential. It is temperature sensitive and may be either positive or negative, amounting to at most a few tens of millivolts.

The self-potentials that originate by the above mechanisms are attracting increased attention in environmental and engineering situations. However, in the exploration for subsurface regions of mineralization they are often smaller than the potentials associated with orebodies and are classified accordingly as 'background potentials'. The self-potential associated with an orebody is called its 'mineralization potential'. Self-potential (SP) anomalies across orebodies are invariably negative, amounting usually to a few hundred millivolts. They are most commonly associated with sulfide ores, such as pyrite, pyrrhotite, and chalcopyrite, but also with graphite and some metallic oxides.

The origin of the mineralization type of self-potential is still obscure, despite decades of applied investigations. At

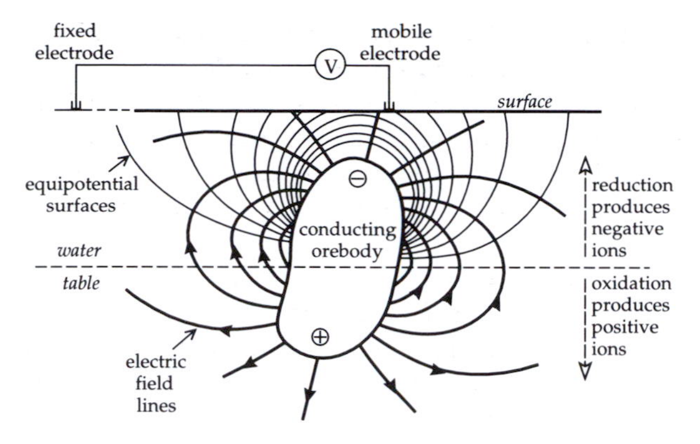

Fig. 4.37 A schematic model of the origin of the self-potential anomaly of an orebody. The mechanism depends on differences in oxidation potential above and below the water table.

one time it was thought that the effect arose from galvanic action. This occurs when dissimilar metal electrodes are placed in an electrolyte. Unequal contact potentials are formed between the metals and the electrolyte, giving rise to a potential difference between the electrodes. According to this model an orebody behaves like a simple voltaic cell, with groundwater acting as the electrolyte. It was believed that oxidation of the part of the orebody above the water table produced a potential difference between the upper and lower parts, causing a spontaneous electric polarization of the body. Oxidation involves the addition of electrons, so the top of the orebody becomes negatively charged, explaining the observed negative anomalies. Unfortunately, this simple model does not explain many of the observed features of self-potential anomalies and has proved to be untenable.

Another mechanism for self-potential depends on variations in oxidation (redox) potential with depth (Fig. 4.37). The ground above the water table is more accessible to oxygen than the submersed part, so moisture above the water table contains more oxidized ions than that below it. An electrochemical reaction takes place at the surface between the orebody and the host rock above the water table. It results in reduction of the oxidized ions in the adjacent solution. An excess of negative ions appears above the water table. A simultaneous reaction between the submersed part of the orebody and the groundwater causes oxidation of the reduced ions present in the groundwater. This produces excess positive ions in the solution and liberates electrons at the surface of the orebody, which acts as a conductor connecting the two half-cells. Electrons flow from the deep part to the shallow part of the orebody. Outside the orebody, positive ions move from bottom to top along the electric field lines. The equipotential surfaces are normal to the field lines. The self-potential is measured where they intersect the ground surface (Fig. 4.37).

The redox model is inadequate for the same reason as the galvanic model; it fails to account for many of the observed features of self-potential anomalies. In particular, the association of self-potential models with the water table has been cast in doubt. Moreover, sulfide orebodies appear to persist for geological lengths of time, so that a mechanism involving permanent flow of charge appears unlikely. Self-potential is a feature of a stable system that is perturbed by making an electrical connection between the host rock and the sulfide conductor through the inserted electrodes and their connecting wire. The observed potential difference appears to be due to the difference in oxidation potential between the locations of the measurement electrodes, one inside and the other outside the zone of mineralization.

4.3.4.2 *SP surveying*

The equipment needed for an SP survey is very simple. It consists of a sensitive high-impedance digital voltmeter to measure the natural potential difference between two electrodes implanted in the ground. Simple metal stakes are inadequate as electrodes. Electrochemical reactions take place between the metal and moisture in the ground, causing the build-up of spurious charges on the electrodes, which can falsify or obscure the small natural self-potentials. To avoid or minimize this effect *non-polarizable electrodes* are used. Each electrode consists of a metal rod submersed in a saturated solution of its own salt; a common arrangement is a copper rod in copper sulfate solution. The combination is contained in a ceramic pot which allows the electrolyte to leak slowly through its porous walls, thereby making electrical contact with the ground.

Two field methods are in common use (Fig. 4.38). The gradient method employs a fixed separation between the electrodes, of the order of 10 m apart. The potential difference is measured between the electrodes, then the pair is moved forward along the survey line until the trailing electrode occupies the location previously occupied by the leading electrode. The total potential at a measurement station relative to a starting point outside the study area is found by summing the incremental potential differences. Some electrode polarization is unavoidable, even with non-polarizable electrodes. This gives rise to a small error in each measurement; these add up to a cumulative error in the total potential. The polarization effects can sometimes be reduced by interchanging the leading and trailing electrodes. In this ‘leapfrog’ technique the leading electrode for one measurement is kept in place and becomes the trailing electrode for the next measurement; meanwhile the previous trailing electrode is moved ahead to become the leading electrode. Cumulative error is the most serious disadvantage

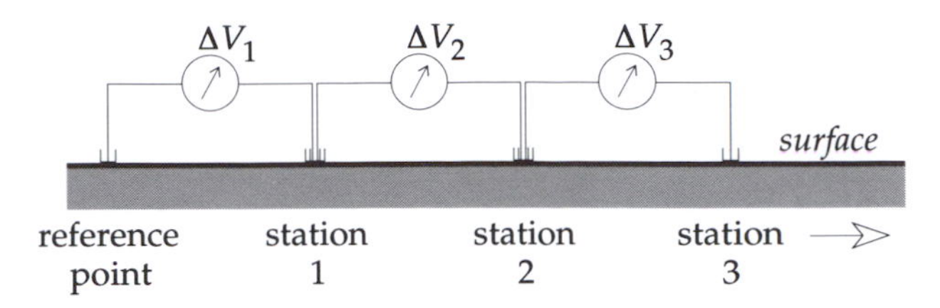

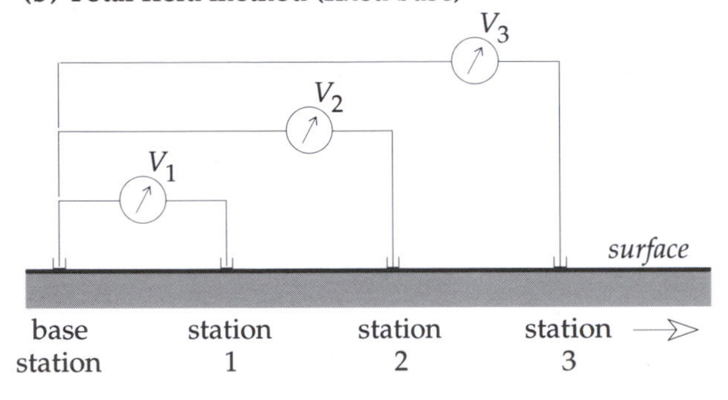

Fig. 4.38 The field techniques of measuring self-potential by (a) the gradient method and (b) the total field method. The total potential V at a station in the gradient method is found by summing the previous potential differences ΔV; in the total field method V is measured directly.

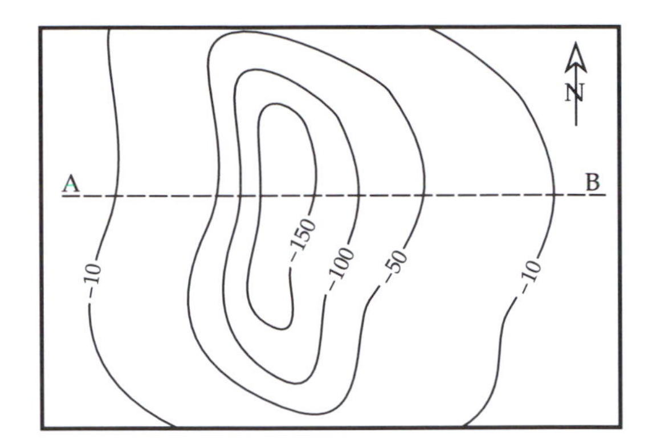

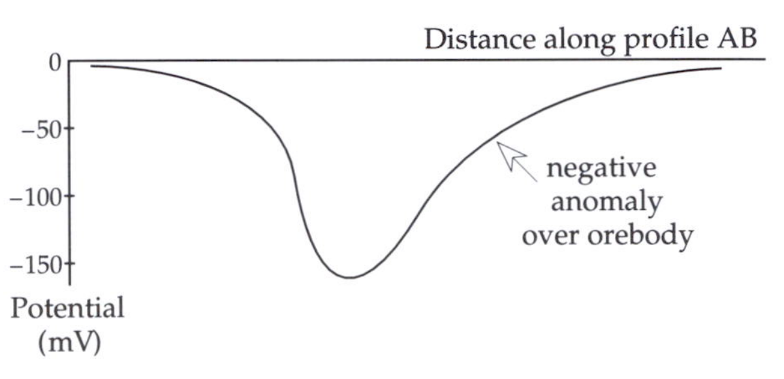

Fig. 4.39 Hypothetical contour lines of a negative self-potential anomaly over an orebody; the asymmetry of the anomaly along the profile AB suggests that the orebody dips toward A.

of the fixed electrode configuration. A practical advantage of the technique is that only a short length of connecting wire must be moved along with the electrodes.

The total field method utilizes a fixed electrode at a base station outside the area of exploration and a mobile measuring electrode. With this method the total potential is measured directly at each station. The wire connecting the electrodes has to be long enough to allow good coverage of the area of interest. This necessitates a long wire that must be wound or unwound on a reel for each measurement station. However, the total field method results in smaller cumulative error than the gradient method. It allows more flexibility in placing the mobile electrode and usually gives data of better quality. Hence, the total field method is usually preferred except in difficult terrain.

The surveying procedure with each technique consists of measuring potential at discrete stations along a profile. As in gravity and magnetic surveys, the data are mapped (Fig. 4.39) and interpretations of anomalies are based on their geometry. Methods used to interpret self-potential anomalies are often qualitative or are based on simple geometric models. Visual inspection of mapped anomalies may reveal trends related to elongation of the orebody; crowding of contour lines can indicate its orientation. Profiles plotted in known directions across the anomaly can be compared with curves generated from simple models of the source. For example, a polarized sphere may be used to model the source of approximately circular anomalies, while a horizontal line source (or polarized cylinder) may be used to model an elongate anomaly. A common and effective method is to model SP anomalies with point sources; complex anomalies are modelled with combinations of sources and sinks.

4.3.4.3 *Telluric currents*

Ultraviolet radiation from the Sun ionizes molecules of air in the thin upper atmosphere of the Earth. The ions accumulate in several layers, forming the ionosphere (see §5.4.3.2) at altitudes between about 80 km and 1500 km above the Earth's surface. Electric currents in the ionosphere arise from systematic motions of the ions, which are affected by various factors such as the daily and monthly tides, seasonal variations in insolation and the periodic fluctuation in ionization related to the 11-yr sunspot cycle. The currents produce varying magnetic fields with the same frequencies, which are observed at the surface of the Earth and can be analyzed from long-term continuous records of the geomagnetic field. The ionospheric effects show up in the energy spectrum of the geomagnetic field as distinct

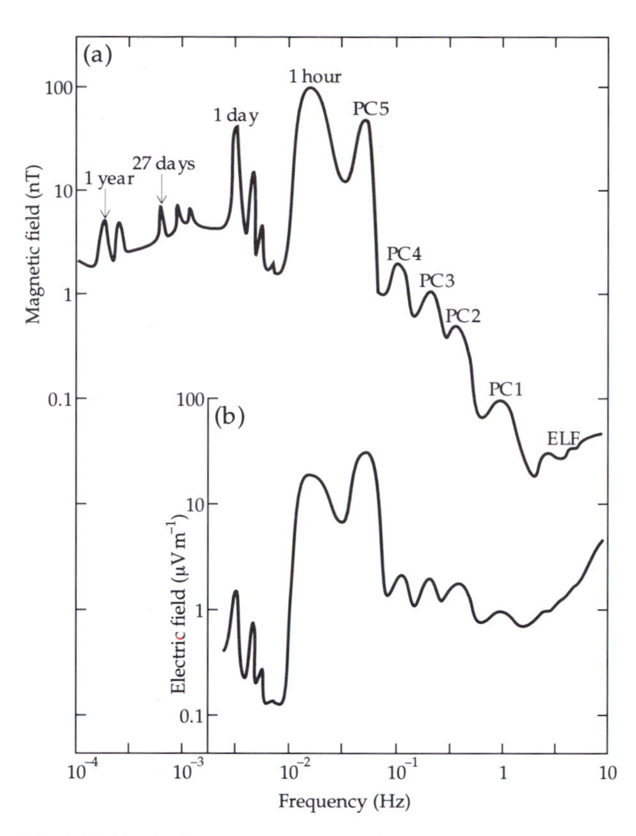

Fig. 4.40 (a) The frequency spectrum of natural variations in the horizontal intensity of the geomagnetic field, and (b) the corresponding spectrum of induced electric field fluctuations, computed for a model Earth with uniform resistivity 20 Ω m (after Serson, 1973). Peaks PC1–5 correspond to geomagnetic pulsations.

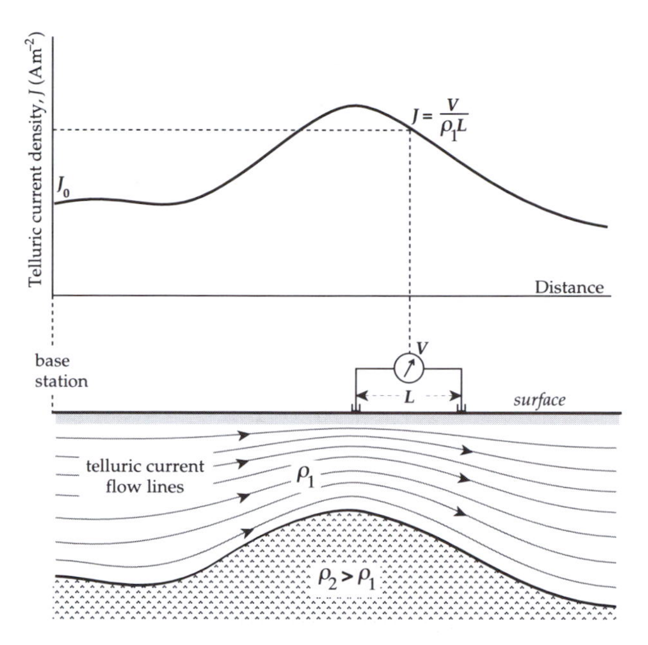

Fig. 4.41 Telluric current lines are deflected by changes in thickness of a conducting layer over a more resistive structure (*bottom*). The telluric current density (*top*) is obtained from the voltage measured between a pair of fixed-separation electrodes at the surface (after Robinson and Çoruh, 1988).

peaks representing periods that range from fractions of a second (geomagnetic *pulsations*) to several years (Fig. 4.40). The magnetic fields induce fluctuating electric currents, called *telluric currents*, that flow in horizontal layers in the crust and mantle. The current pattern consists of several huge whorls, thousands of kilometers across, which remain fixed with respect to the sun and thus move around the Earth as it rotates.

The distribution of telluric current density depends on the variation of resistivity in the horizontal conducting layers. At shallow crustal depths the lines of current flow are disturbed by subsurface structures which cause contrasts in resistivity. These could arise from geological structures or the presence of mineralized zones. Consider, for example, a buried anticline which has a highly resistive rock (such as granite) as its core and is overlain by a conducting layer of porous sedimentary rocks saturated with groundwater. The horizontal flow of telluric current across the anticline chooses the less-resistive path through the conducting sediments. The current lines bunch together over the axis of the anticline, increasing the horizontal current density (Fig. 4.41). The equipotential surfaces normal to the current lines intersect the ground surface, where potential differences can be measured with a high-impedance voltmeter.

The field equipment for measuring telluric current density is simple. The sensors are a pair of non-polarizable electrodes with a fixed separation L of the order of 10–100 m. The potential difference V between the electrodes is measured with a high-impedance voltmeter. The electric field E at a point mid-way between the electrodes can be assumed to be V/L. Using Ohm's law (Eq. (4.84)) and assuming that the telluric current flows in conducting rock layer with resistivity ρ_1, the telluric current density J at each measurement station along a profile is given by

$$J=\frac{V}{\rho_1 L} \tag{4.87}$$

The direction of the telluric current is not known, so two pairs of electrodes oriented perpendicular to each other are used. One pair is aligned north–south, the other east–west. Telluric currents vary unpredictably with time, but they change only slowly within a homogeneous region. To keep track of the temporal changes an orthogonal pair of electrodes is set up at a fixed base station outside the area to be explored. Another orthogonal pair is moved across the survey area. The potential differences across each electrode pair in the mobile and base arrays are recorded simultaneously for several minutes at each measurement station. Correlation of the records allows removal of the temporal changes in direction and intensity of the telluric currents.

The deflection of telluric current by a resistive subsurface structure as shown in Fig. 4.41 is greatly idealized. It assumes an infinite resistivity ρ_2 in the core of the anticline. In practice, the current is not completely diverted through the better-conducting layer; part flows through the more resistive layer as well. Thus we cannot assume that the resistivity ρ_1 in Eq. (4.87) corresponds to the good conductor. Rather, it represents some undefined mixture of the values ρ_1 and ρ_2. It is not the true resistivity of either layer, but the *apparent resistivity* of the measurement.

4.3.5 **Resistivity surveying**

The large contrast in resistivity between orebodies and their host rocks (see Fig. 4.36) is exploited in electrical resistivity prospecting, especially for minerals that occur as good conductors. Representative examples are the sulfide ores of iron, copper and nickel. Electrical resistivity surveying is also an important geophysical technique in environmental applications. For example, due to the good electrical conductivity of groundwater the resistivity of a sedimentary rock is much lower when it is waterlogged than in the dry state.

Instead of relying on natural currents, two electrodes are used to supply a controlled electrical current to the ground. As in the telluric method, the lines of current flow adapt to the subsurface resistivity pattern so that the potential difference between equipotential surfaces can be measured where they intersect the ground surface, using a second pair of electrodes. A simple direct current can cause charges to accumulate on the potential electrodes, which results in spurious signals. A common practice is to commutate the direct current so that its direction is reversed every few seconds; alternatively a low-frequency alternating current may be used. In multi-electrode investigations the current electrode-pair and potential electrode-pair are usually interchangeable.

4.3.5.1 *Potential of a single electrode*

Consider the flow of current around an electrode that introduces a current I at the surface of a uniform half-space (Fig. 4.42a). The point of contact acts as a current source, from which the current disperses outward. The electric field lines are parallel to the current flow and normal to the equipotential surfaces, which are hemispherical in shape. The current density J is equal to I divided by the surface area, which is $2\pi r^2$ for a hemisphere of radius r. The electric field

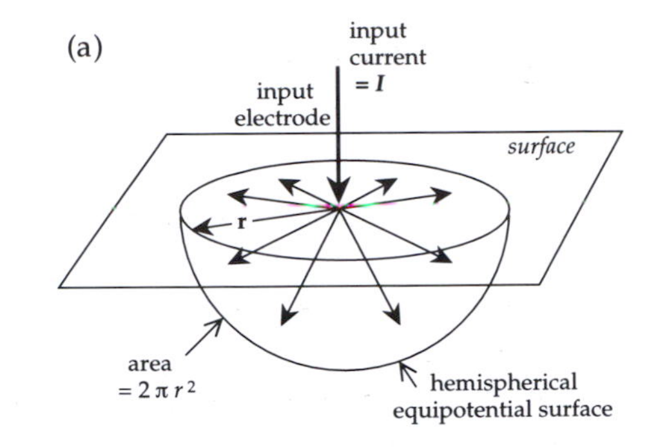

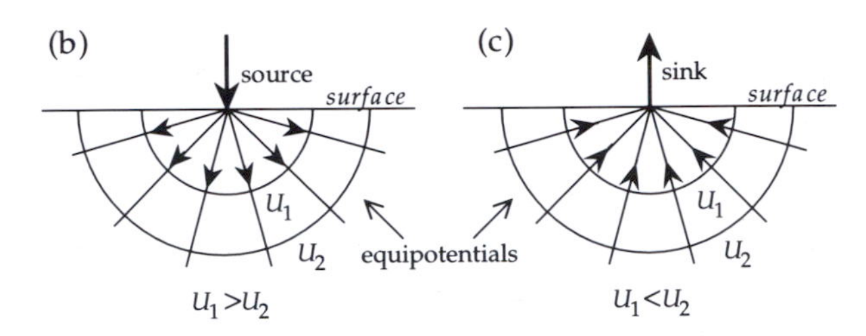

Fig. 4.42 Electric field lines and equipotential surfaces around a single electrode at the surface of a uniform half-space: (a) hemispherical equipotential surfaces, (b) radially outward field lines around a source, and (c) radially inward field lines around a sink.

E at distance r from the input electrode is obtained from Ohm's law (Eq. (4.84))

$$E=\rho J=\rho\frac{I}{2\pi r^2} \tag{4.88}$$

Putting this expression in Eq. (4.78) yields the electric potential U at distance r from the input electrode:

$$\frac{dU}{dr}=-\rho\frac{I}{2\pi r^2}$$

$$U=\rho\frac{I}{2\pi r} \tag{4.89}$$

If the ground is a uniform half-space, the electric field lines around a *source* electrode, which supplies current to the ground, are radially outward (Fig. 4.42b). Around a *sink* electrode, where current flows out of the ground, the field lines are radially inward (Fig. 4.42c). The equipotential surfaces around a source or sink electrode are hemispheres, if we regard the electrode in isolation. The potential around a source is positive and diminishes as $1/r$ with increasing distance. The sign of I is negative at a sink, where the current flows out of the ground. Thus, around a sink the potential is negative and increases (becomes less negative) as $1/r$ with increasing distance from the sink. We can use these observations to calculate the potential difference between a second

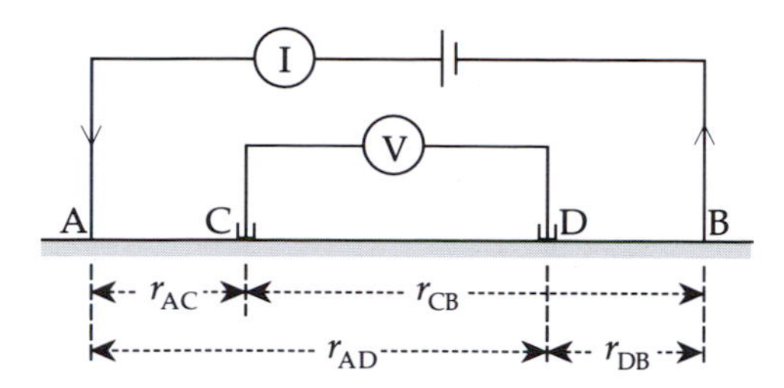

Fig. 4.43 General four-electrode configuration for resistivity measurement, consisting of a pair of current electrodes (A, B) and a pair of potential electrodes (C, D).

pair of electrodes at known distances from the source and sink.

4.3.5.2 *The general four-electrode method*

Consider an arrangement consisting of a pair of current electrodes and a pair of potential electrodes (Fig. 4.43). The current electrodes A and B act as source and sink, respectively. At the detection electrode C the potential due to the source A is $+\rho I/(2\pi r_{AC})$, while the potential due to the sink B is $-\rho I/(2\pi r_{CB})$. The combined potential at C is

$$U_C = \frac{\rho I}{2\pi}\left(\frac{1}{r_{AC}} - \frac{1}{r_{CB}}\right) \tag{4.90}$$

Similarly, the resultant potential at D is

$$U_D = \frac{\rho I}{2\pi}\left(\frac{1}{r_{AD}} - \frac{1}{r_{DB}}\right) \tag{4.91}$$

The potential difference measured by a voltmeter connected between C and D is

$$V = \frac{\rho I}{2\pi}\left[\left(\frac{1}{r_{AC}} - \frac{1}{r_{CB}}\right) - \left(\frac{1}{r_{AD}} - \frac{1}{r_{DB}}\right)\right] \tag{4.92}$$

All quantities in this equation can be measured at the ground surface except the resistivity, which is given by

$$\rho = 2\pi\frac{V}{I}\left[\frac{1}{\left(\frac{1}{r_{AC}} - \frac{1}{r_{CB}}\right) - \left(\frac{1}{r_{AD}} - \frac{1}{r_{DB}}\right)}\right] \tag{4.93}$$

4.3.5.3 *Special electrode configurations*

The general formula for the resistivity measured by a four-electrode method is simpler for some special geometries of the current and potential electrodes. The most commonly used configurations are the Wenner, Schlumberger and double-dipole arrangements. In each configuration the four electrodes are collinear but their geometries and spacings are different.

In the *Wenner configuration* (Fig. 4.44a) the current and potential electrode pairs have a common mid-point and the

(a) Wenner

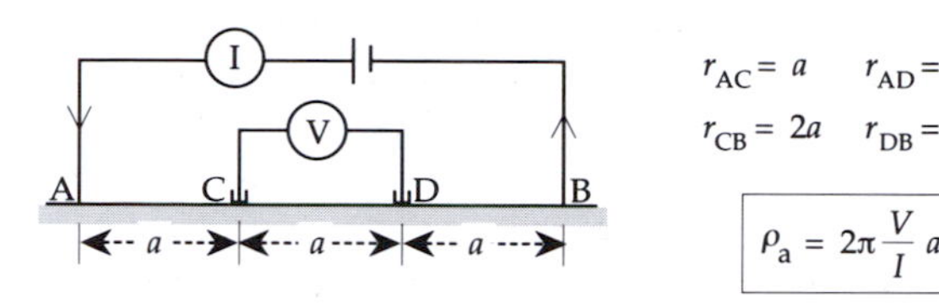

$r_{AC} = a$ $r_{AD} = 2a$
$r_{CB} = 2a$ $r_{DB} = a$

$\rho_a = 2\pi\frac{V}{I}a$

(b) Schlumberger

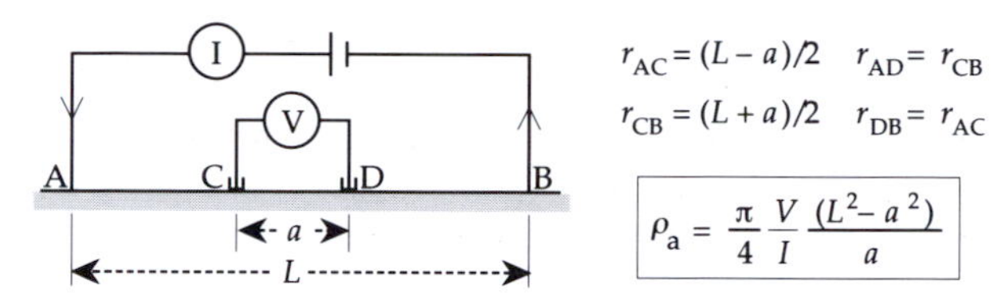

$r_{AC} = (L-a)/2$ $r_{AD} = r_{CB}$
$r_{CB} = (L+a)/2$ $r_{DB} = r_{AC}$

$\rho_a = \frac{\pi}{4}\frac{V}{I}\frac{(L^2-a^2)}{a}$

(c) Double-dipole

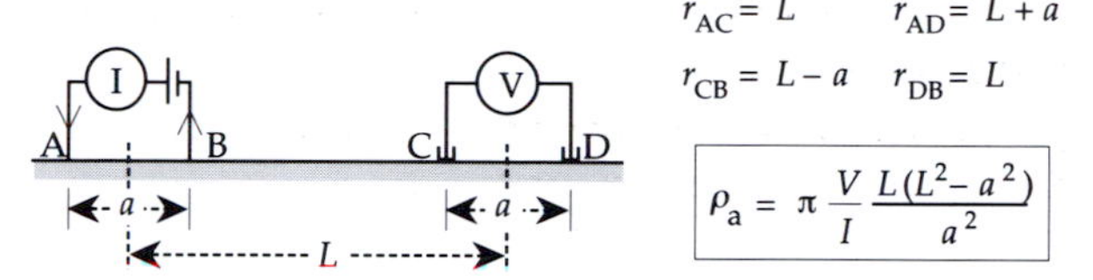

$r_{AC} = L$ $r_{AD} = L+a$
$r_{CB} = L-a$ $r_{DB} = L$

$\rho_a = \pi\frac{V}{I}\frac{L(L^2-a^2)}{a^2}$

Fig. 4.44 Special geometries of current and potential electrodes for (a) Wenner, (b) Schlumberger and (c) double-dipole configurations.

distances between adjacent electrodes are equal, so that $r_{AC}=r_{DB}=a$, and $r_{CB}=r_{AD}=2a$. Inserting these values in Eq. (4.93) gives

$$\rho = 2\pi\frac{V}{I}\left[\frac{1}{\left(\frac{1}{a} - \frac{1}{2a}\right) - \left(\frac{1}{2a} - \frac{1}{a}\right)}\right] \tag{4.94}$$

$$\rho = 2\pi a\frac{V}{I} \tag{4.95}$$

In the *Schlumberger configuration* (Fig. 4.44b) the current and potential pairs of electrodes also have a common mid-point, but the distances between adjacent electrodes differ. Let the separations of the current and potential electrodes be L and a, respectively. Then $r_{AC}=r_{DB}=(L-a)/2$ and $r_{AD}=r_{CB}=(L+a)/2$. Substituting in the general formula, we get

$$\rho = 2\pi\frac{V}{I}\left[\frac{1}{\left(\frac{2}{L-a} - \frac{2}{L+a}\right) - \left(\frac{2}{L+a} - \frac{2}{L-a}\right)}\right] \tag{4.96}$$

$$\rho = \frac{\pi}{4}\frac{V}{I}\left(\frac{L^2-a^2}{a}\right) \tag{4.97}$$

In the *double-dipole configuration* (Fig. 4.44c) the spacing of the electrodes in each pair is a, while the distance between their mid-points is L. In this case $r_{AC}=r_{DB}=L$, $r_{CB}=L-a$, and $r_{AD}=L+a$. The measured resistivity is

$$\rho=2\pi\frac{V}{I}\left[\frac{1}{\left(\frac{1}{L}-\frac{1}{L-a}\right)-\left(\frac{1}{L+a}-\frac{1}{L}\right)}\right] \tag{4.98}$$

$$\rho=\pi\frac{V}{I}\left(\frac{L(L^2-a^2)}{a^2}\right) \tag{4.99}$$

Two modes of investigation can be used with each electrode configuration. The Wenner configuration is best adapted to *lateral profiling*. The assemblage of four electrodes is displaced stepwise along a profile while maintaining constant values of the inter-electrode distances corresponding to the configuration employed. The separation of the current electrodes is chosen so that the current flow is maximized in depths where lateral resistivity contrasts are expected. Results from a number of profiles may be compiled in a resistivity map of the region of interest. The regional survey reveals the horizontal variations in resistivity within an area at a particular depth. It is best suited to locating steeply dipping contacts between rocks with a strong resistivity contrast and good conductors such as mineralized dikes, which may be potential orebodies.

In *vertical electrical sounding (VES)* the goal is to observe the variation of resistivity with depth. The technique is best adapted to determining depth and resistivity for flat-lying layered rock structures, such as sedimentary beds, or the depth to the water table. The Schlumberger configuration is most commonly used for VES-investigations. The mid-point of the array is kept fixed while the distance between the current electrodes is progressively increased. This causes the current lines to penetrate to ever greater depths, depending on the vertical distribution of conductivity.

4.3.5.4 *Current distribution*

The current pattern in a uniform half-space extends laterally on either side of the profile line. Viewed from above, the current lines bulge outward between source and sink with a geometry similar to that shown in Fig. 4.34c. In a vertical section the current lines resemble half of a dipole geometry. In three dimensions the current can be visualized as flowing through tubes that fatten as they leave the source and narrow as they converge towards the sink. Fig. 4.45 shows the flow pattern of the current in a vertical section through the 'tubes' in a uniform half-space.

In order to evaluate the depth penetration of current in a uniform half-space we define orthogonal Cartesian coordinates with the x-axis parallel to the profile and the z-axis

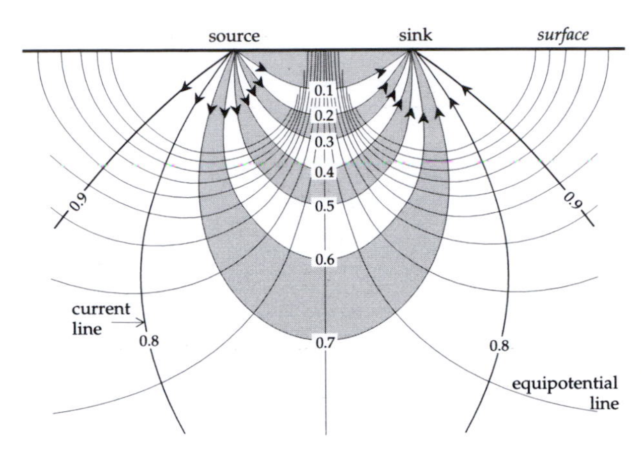

Fig. 4.45 Cross-section of current 'tubes' and equipotential surfaces between a source and sink; numbers on the current lines indicate the fraction of current flowing above the line (after Robinson and Çoruh, 1988; based upon Van Nostrand and Cook, 1966).

vertical (Fig. 4.46a). Let the spacing of the current electrodes be L and the resistivity of the half-space be ρ. The horizontal electric field E_x at (x, y, z) is

$$E_x=-\frac{\partial U}{\partial x}=-\frac{\partial}{\partial x}\left\{\frac{\rho I}{2\pi}\left(\frac{1}{r_1}-\frac{1}{r_2}\right)\right\} \tag{4.100}$$

where $r_1=(x^2+y^2+z^2)^{1/2}$ and $r_2=((L-x)^2+y^2+z^2)^{1/2}$. Differentiating and using Ohm's law (Eq. (4.84)) gives the horizontal current density J_x at (x, y, z):

$$J_x=\frac{I}{2\pi}\left\{\frac{x}{r_1^3}+\frac{(L-x)}{r_2^3}\right\} \tag{4.101}$$

If (x, y, z) is on the vertical plane mid-way between the current electrodes, $x=L/2$, $r_1=r_2$ and the current density is given by

$$J_x=\frac{IL}{2\pi}\frac{1}{\{(L/2)^2+y^2+z^2\}^{3/2}} \tag{4.102}$$

The horizontal current dI_x across an element of area $(dy\,dz)$ in the median vertical plane is $dI_x=J_x\,dy\,dz$. The fraction of the input current I that flows across the median plane above a depth z is obtained by integration:

$$\frac{I_x}{I}=\frac{L}{2\pi}\int_0^z dz\int_{-\infty}^{+\infty}\frac{dy}{\{(L/2)^2+y^2+z^2\}^{3/2}} \tag{4.103}$$

$$\frac{I_x}{I}=\frac{L}{\pi}\int_0^z\frac{dz}{(L/2)^2+z^2} \tag{4.104}$$

$$\frac{I_x}{I}=\frac{2}{\pi}\tan^{-1}\frac{2z}{L} \tag{4.105}$$

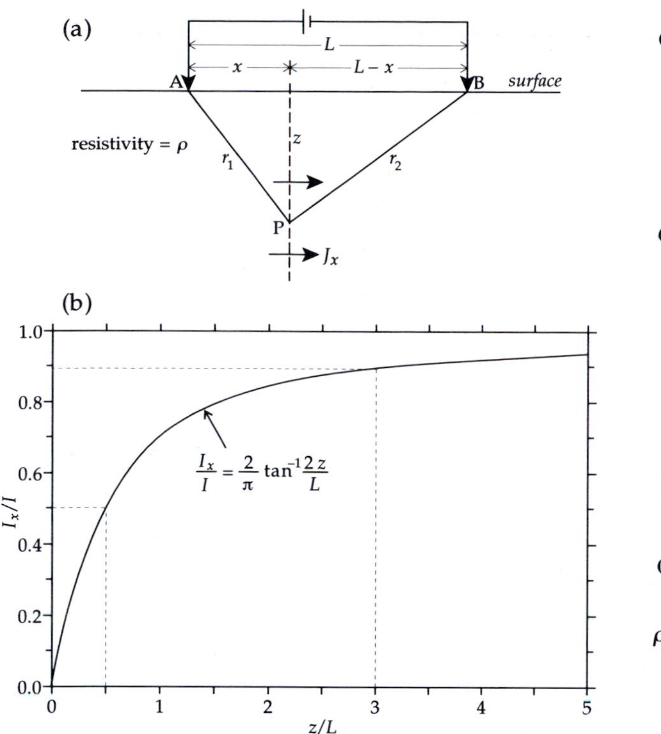

Fig. 4.46 (a) Geometry for determining current density in uniform ground below two electrodes, and (b) fraction of current (I_x/I) that flows above depth z across the median plane between current electrodes with spacing L (after Telford *et al.*, 1990).

Equation (4.105) shows that I_x depends upon the current-electrode spacing L (Fig. 4.46b). Half the current crosses the plane above a depth $z=L/2$, and almost 90% passes above the depth $z=3L$. The fraction of current between any two depths is found from the difference in the fractions above each depth calculated with Eq. (4.105).

4.3.5.5 *Apparent resistivity*

In the idealized case of a perfectly uniform conducting half-space the current flow lines resemble a dipole pattern (Fig. 4.45), and the resistivity determined with a four-electrode configuration is the true resistivity of the half-space. But in real situations the resistivity is determined by different lithologies and geological structures and so may be very inhomogeneous. This complexity is not taken into account when measuring resistivity with a four-electrode method, which assumes that the ground is uniform. The result of such a measurement is the *apparent resistivity* of an equivalent uniform half-space and generally does not represent the true resistivity of any part of the ground.

Consider a horizontally layered structure in which a layer of thickness d and resistivity ρ_1 overlies a conducting half-space with a lower resistivity ρ_2 (Fig. 4.47). If the

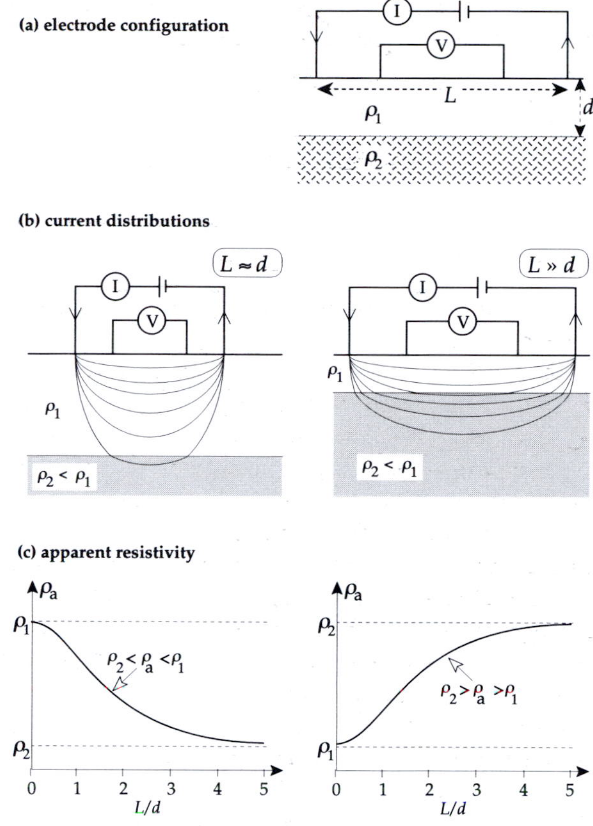

Fig. 4.47 (a) Parameters of the four-electrode arrangement, (b) distribution of current lines in a two-layer ground with resistivities ρ_1 and ρ_2 ($\rho_1>\rho_2$) and (c) the variation of apparent resistivity as the current electrode spacing is varied for the two cases of $\rho_1>\rho_2$ and $\rho_1<\rho_2$.

current electrodes are close together, so that $L \ll d$, all or most of the current flows in the more resistive upper layer, so that the measured resistivity is close to the true value of the upper layer, ρ_1. With increasing separation of the current electrodes the depth reached by the current lines increases. Proportionally more current flows in the less resistive layer, so the measured resistivity decreases. Conversely, if the upper layer is a better conductor than the lower layer, the apparent resistivity increases with increasing electrode spacing. When the electrode separation is much larger than the thickness of the upper layer ($L \gg d$) the measured resistivity is close to the value ρ_2 of the bottom layer. Between the extreme situations the apparent resistivity determined from the measured current and voltage is not related simply to the true resistivity of either layer.

4.3.5.6 *Vertical electrical sounding*

A two-layer situation is encountered often in electrical prospecting, for example when a conducting overburden overlies a resistive basement. It is also common in environmental applications, when the conducting water table lies under drier, more resistive soil or rocks. Before the advent of portable computers two-layer cases were interpreted with the aid of *characteristic curves*. These theoretical curves, calculated for a particular four-electrode array, take into account the change of depth penetration when current lines cross the boundary to a layer with different resistivity. The electrical boundary conditions require continuity of the component of current density J normal to the interface and of the component of electric field E tangential to the interface. At a boundary the current lines behave like optical or seismic rays, and are guided by similar laws of reflection and refraction. For example, if θ is the angle between a current line and the normal to the interface, the electrical 'law of refraction' is

$$\frac{\tan\theta_1}{\tan\theta_2}=\frac{\rho_2}{\rho_1} \tag{4.106}$$

In a set of characteristic curves the apparent resistivity ρ_a is normalized by the resistivity ρ_1 of the upper layer and the electrode spacing is expressed as a multiple of the layer thickness. The shape of the curve of apparent resistivity versus electrode spacing depends on the resistivity contrast between the two layers, and a family of characteristic curves is calculated for different ratios of ρ_2/ρ_1 (Fig.4.48). The resistivity contrast is conveniently expressed by a k-factor defined as

$$k=\frac{\rho_2-\rho_1}{\rho_2+\rho_1} \tag{4.107}$$

The k-factor ranges between -1 and $+1$ as the resistivity ratio ρ_2/ρ_1 varies between 0 and ∞. The characteristic curves, drawn as full logarithmic plots on a transparent overlay, are compared graphically with the field data to find the best-fitting characteristic curve. The comparison yields the resistivities ρ_1 and ρ_2 of the upper and lower layers, respectively, and the layer thickness, d.

Although characteristic curves can also be computed for the interpretation of structures with multiple horizontal layers, modern VES analyses take advantage of the flexibility offered by small computers with graphic outputs on which the apparent resistivity curves can be visually assessed. The first step in the analysis consists of classifying the shape of the vertical sounding profile.

The apparent resistivity curve for a three-layer structure generally has one of four typical shapes, determined by the vertical sequence of resistivities in the layers (Fig. 4.49). The type K curve rises to a maximum then decreases, indicating that the intermediate layer has higher resistivity than the top and bottom layers. The type H curve shows the opposite effect; it falls to a minimum then increases again due to an intermediate layer that is a better conductor than the top and bottom layers. The type A curve may show some changes of gradient but the apparent resistivity generally increases continuously with increasing electrode separation, indicating that the true resistivities increase with depth from layer to layer. The type Q curve exhibits the opposite effect; it decreases continuously along with a progressive decrease of resistivity with depth.

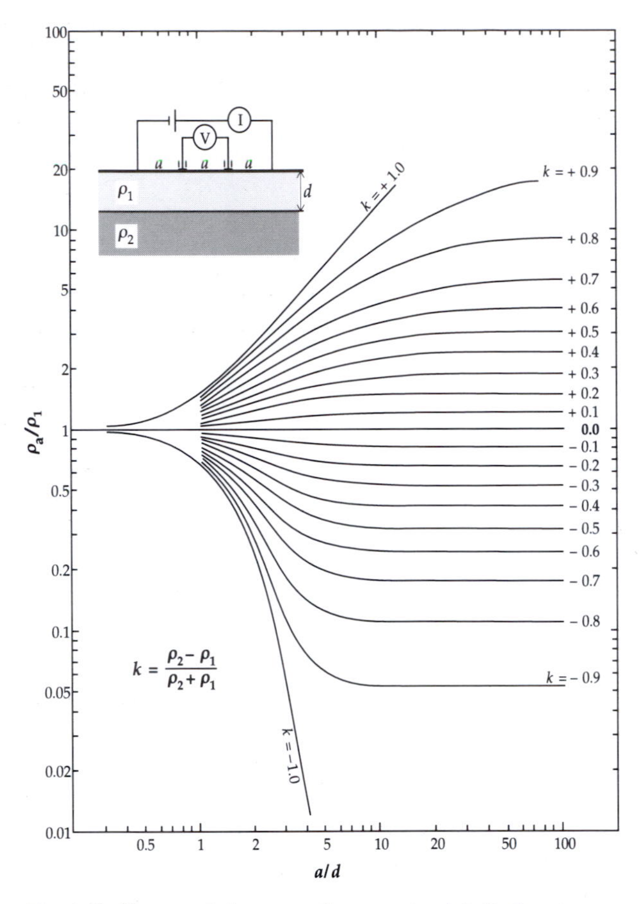

Fig. 4.48 Characteristic curves of apparent resistivity for a two-layer structure using the Wenner array; parameters are defined in the inset.

Once the observed resistivity profile has been identified as of K, H, A or Q type, the next step is equivalent to one-dimensional *inversion* of the field data. The technique involves iterative procedures that would be very time-consuming without a fast computer. The method assumes the equations for the theoretical response of a multi-layered ground. Each layer is characterized by its thickness and resistivity, each of which must be determined. A first estimate of these parameters is made for each layer and the predicted curve of apparent resistivity versus electrode spacing

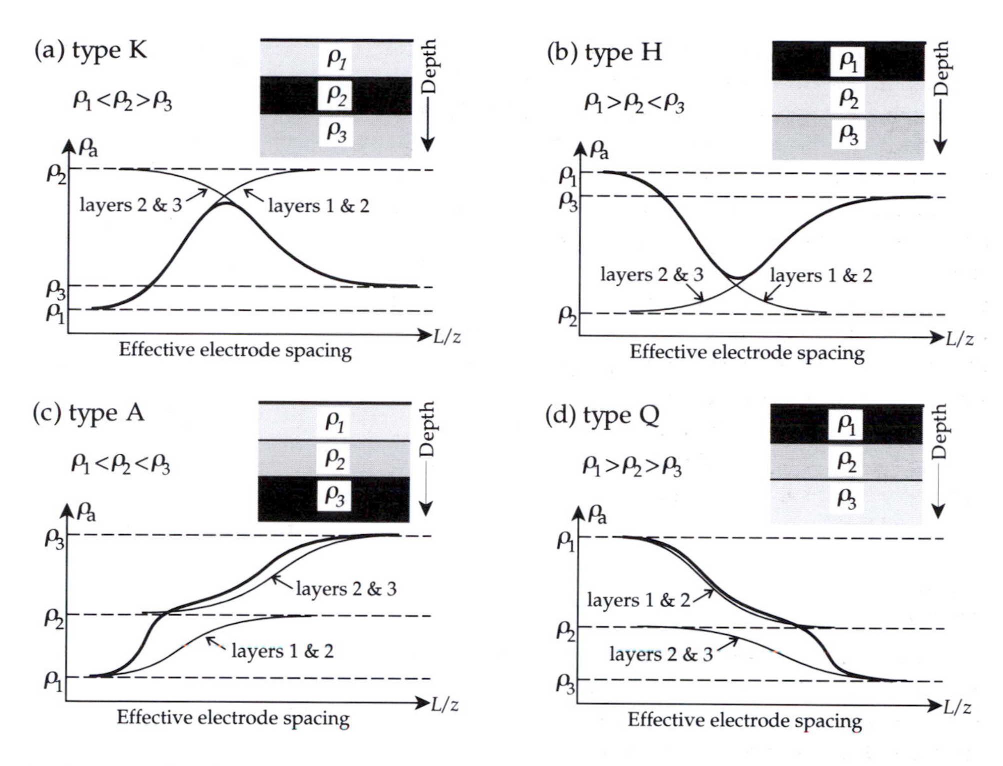

Fig. 4.49 The four common shapes of apparent resistivity curves for a layered structure consisting of three horizontal layers.

is computed. The discrepancies between the observed and theoretical curves are then determined point by point. The layer parameters used in the governing equations are next adjusted, and the calculation is repeated with the corrected values, giving a new predicted curve to compare with the field data. Using modern computers the procedure can be reiterated rapidly until the discrepancies are smaller than a pre-determined value.

The inversion method is equivalent to matching automatically the observed and theoretical curves. A one-dimensional analysis accommodates only the variations of resistivity and layer thickness with depth. The response of a vertically layered structure has an analytical solution, so efficient inversion algorithms can be established. In recent years, procedures have been proposed that also take into account lateral heterogeneities. The response of two- or three-dimensional structures must be approximated by a numerical solution, based on the finite-difference or finite-element techniques. The number of unknown quantities increases, as do the computational difficulties of the inversion.

4.3.5.7 *Induced polarization*

If commutated direct current is used in a four-electrode resistivity survey, the sequence of positive and negative flow may be interspersed with periods when the current is off. The inducing current then has a box-like appearance (Fig. 4.50a). When the current is interrupted, the voltage across the potential electrodes does not drop immediately to zero. After an initial abrupt drop to a fraction of its steady-state value it decays slowly for several seconds (Fig. 4.50b). Conversely, when the current is switched on, the potential rises suddenly at first and then gradually approaches the steady-state value. The slow decay and growth of part of the signal are due to *induced polarization*, which results from two similar effects related to the rock structure: membrane polarization and electrode polarization.

Membrane polarization is a feature of electrolytic conduction. It arises from differences in the ability of ions in pore fluids to migrate through a porous rock. The minerals in a rock generally have a negative surface charge and thus attract positive ions in the pore fluid. They accumulate on the grain surface and extend into the adjacent pores, partially blocking them. When an external voltage is applied, positive ions can pass through the 'cloud' of positive charge but negative ions accumulate, unless the pore size is large enough to allow them to bypass the blockage. The effect is like a membrane, which selectively allows the passage of one type of ion. It causes temporary accumulations of negative ions, giving a polarized ionic distribution in the rock. The effect is most pronounced in rocks that contain clay minerals; firstly, because the grain and pore sizes are small, and, secondly, because clay grains are relatively strongly charged and adsorb ions on their surfaces. The ionic build-up takes a short time after the voltage is switched on; when the current is switched off, the ions drift back to their original positions.

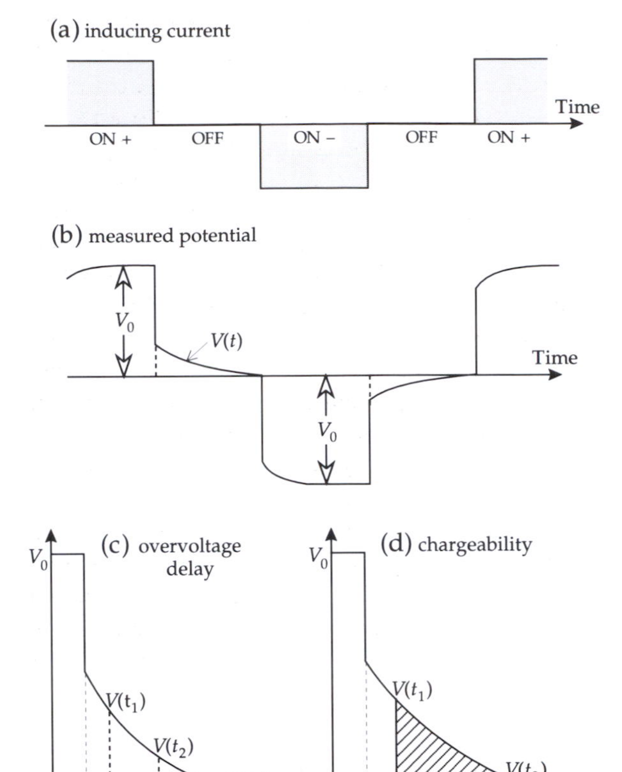

Fig. 4.50 (a) Illustration of the IP-related decay of potential after interruption of the primary current. (b) Effect of the IP decay time on the potential waveform for a square-wave input current.

Electrode polarization is a similar effect that occurs when ore minerals are present. The metallic grains conduct charge by electronic conduction, while electrolytic conduction takes place around them. However, the flow of electrons through the metal is much faster than the flow of ions in the electrolyte, so opposite charges accumulate on facing surfaces of a metallic grain that blocks the path of ionic flow through the pore fluid. An *overvoltage* builds up for some time after the external current is switched on. The size of the effect is commensurate with the metallic concentration. After the current is switched off, the accumulated ions disperse and the overvoltage decays slowly.

The two effects responsible for induced polarization are indistinguishable at measurement level. The field method for an induced polarization (IP) survey is most often based on the double-dipole array. The current electrodes form a transmitter pair, while the potential electrodes form a receiver pair. The steady-state voltage V_0 is recorded and compared with the amplitude of the decaying residual voltage $V(t)$ at time t after the current is interrupted (Fig. 4.50c). The ratio $V(t)/V_0$ is expressed as a percentage, which decays during the 0.1–10 s between switching the current on and off. If the decay curve is sampled at many points, its shape and the area under the curve may be obtained (Fig. 4.50d). The area under the decay curve, expressed as a fraction of the steady-state voltage, is called the *chargeability M*, defined as

$$M=\frac{1}{V_0}\int_{t_1}^{t_2} V(t)\mathrm{d}t \tag{4.108}$$

M has the dimensions of time and is expressed in seconds or milliseconds. It is the most commonly used parameter in IP studies.

The induced polarization determines the length of the potential decay time. If it is shorter than the time when the inducing current is off, successive half-cycles of the potential will not interfere. However, if a disseminated conductor is present, the decay time increases, causing overlap and distortion of the half-cycles. The higher the signal frequency the more pronounced is the effect. It increases the ratio $V(t)/V_0$, giving the impression of a better conductor than is really present (i.e., the apparent resistivity decreases with increasing frequency). Clearly, IP and resistivity surveys with alternating current are also influenced. The frequency dependence of the IP effect is exploited by measuring apparent resistivity at two low frequencies. Let these be f and F ($>f$). Commonly $f\approx 0.05-0.5$ Hz and $F\approx 1-10$ Hz. Then $\rho_f>\rho_F$ and we can define a *frequency effect* as

$$\mathrm{FE}=\frac{\rho_f-\rho_F}{\rho_F} \tag{4.109}$$

The ratio FE is often multiplied by 100 to express it as a percentage (PFE). If no IP effect is present the resistivity will be the same at both frequencies. The larger the value of FE or PFE, the greater is the induced polarization in the ground. At frequencies above 10 Hz mutual inductance effects between the cables of the primary and detection circuits can produce troublesome potentials, which must be avoided by the field procedure (such as restricting F) or minimized analytically.

The presence of metallic conductors is expressed by a similar parameter to FE, the metallic factor (MF). This is proportional to the difference in conductivities at the two measurement frequencies.

$$\begin{aligned}\mathrm{MF}&=A(\sigma_F-\sigma_f)=A(\rho_F^{-1}-\rho_f^{-1})\\&=A\left(\frac{\rho_f-\rho_F}{\rho_f\rho_F}\right)\end{aligned} \tag{4.110}$$

The constant A is equal to $2\pi\times 10^5$; the units of MF are those of conductivity (i.e., $\Omega^{-1}\,\mathrm{m}^{-1}$ or $\mathrm{S\,m^{-1}}$).

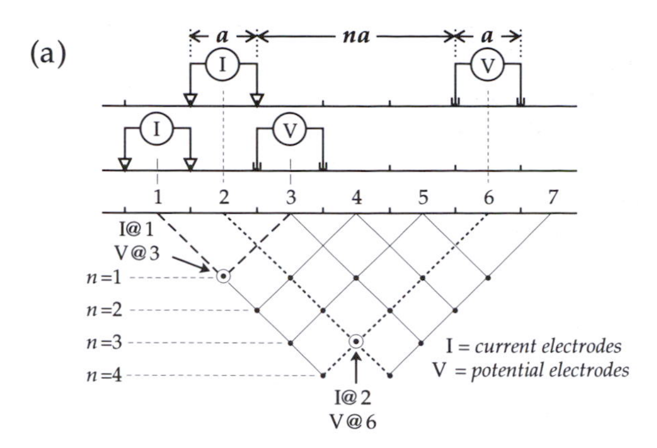

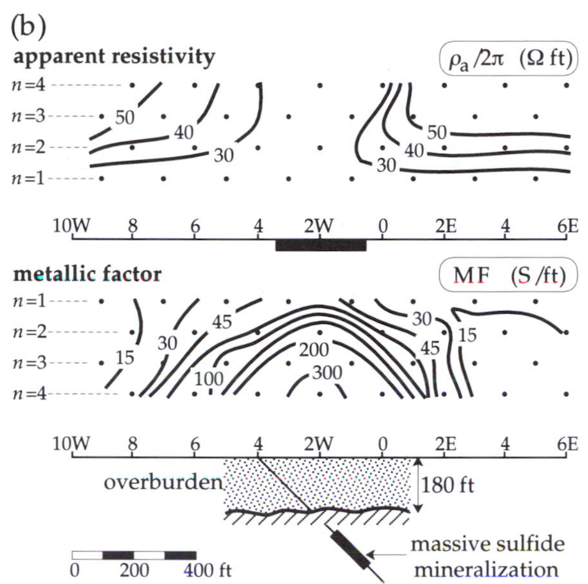

Fig. 4.51 (a) Construction of a pseudo-section for a double-dipole IP survey: the measured parameter is plotted at the intersection of 45° lines extending from the mid-points of the transmitter and receiver pairs. (b) Pseudo-sections of apparent resistivity and metallic factor for an IP survey over a sulfide orebody (redrawn from Telford *et al.*, 1990).

An IP survey includes both lateral profiling and vertical sounding with the expanding spread method. Using a double-dipole array the distance between nearest electrodes of the transmitter and receiver pairs is a multiple (na) of the electrode spacing a in each pair. Measurements are made at several discrete positions as the receiver pair is moved incrementally away from the fixed transmitter pair. The transmitter pair is then moved by one increment along the profile and the procedure is repeated. The value of ρ_a, FE or MF obtained in each measurement is plotted below the midpoint of the array at the intersection of two lines inclined at 45° (Fig. 4.51a). Information is obtained from increasingly greater depths as the transmitter–receiver array expands (i.e., as n increases). The plotted value is, however, not the real value of the parameter at the indicated depth. (Recall, for example, that a measurement of apparent resistivity represents an equivalent half-space beneath the array.) A two-dimensional picture of the variation of the IP parameter beneath the profile is synthesized by contouring the results (Fig. 4.51b). The plot is called a *pseudo-section*; it provides a convenient (though artificial) image of the presence of anomalous conductors, but does not represent their true lateral or vertical extent. The presence of anomalous regions may be investigated further by exploratory drilling.

Resistivity anomalies depend on the presence of continuous conductors, such as groundwater or massive orebodies. If mineralization is disseminated through a rock it may not cause a significant resistivity anomaly. The good response of the IP method for disseminated concentrations of conducting ore minerals led to its development in base-metal exploration, where large low-grade orebodies may be commercially important. However, the IP effect also depends on the porosity and saturation of the rock. As a result, it can also be used in the search for groundwater and in other environmental applications.

4.3.6 **Electromagnetic surveying**

The pioneering observations of electrical and magnetic phenomena early in the 19th century by Coulomb, Oersted, Ampère, Gauss and Faraday were unified in 1873 by the Scottish mathematical physicist James Clerk Maxwell (1831–1879). His achievement is of similar stature to that of Newton in gravitation or Einstein in relativity. Like Newton, Maxwell gathered existing knowledge and unified it in a way that allowed the prediction of other phenomena. His book *A Treatise on Electricity and Magnetism* was as important as Newton's *Principia* to the further development of physics. Just as all discussions in dynamics start with Newton's laws, all arguments in electromagnetism begin with Maxwell's equations. In particular, he proposed the theory of the *electromagnetic field*, which classifies light as an electromagnetic phenomenon in the same sense as electricity and magnetism. This ultimately led to the recognition of the wave nature of matter. Unfortunately, Maxwell died while still in the prime of his career, before his theoretical predictions were verified. The German physicist Heinrich Hertz established the existence of electromagnetic waves experimentally in 1887, eight years after Maxwell's untimely death.

Coulomb's law shows that an electric charge is surrounded by an electric field, which exerts forces on other charges, causing them to move, if they are free to do so. Ampère's law shows that an electric charge (or current)

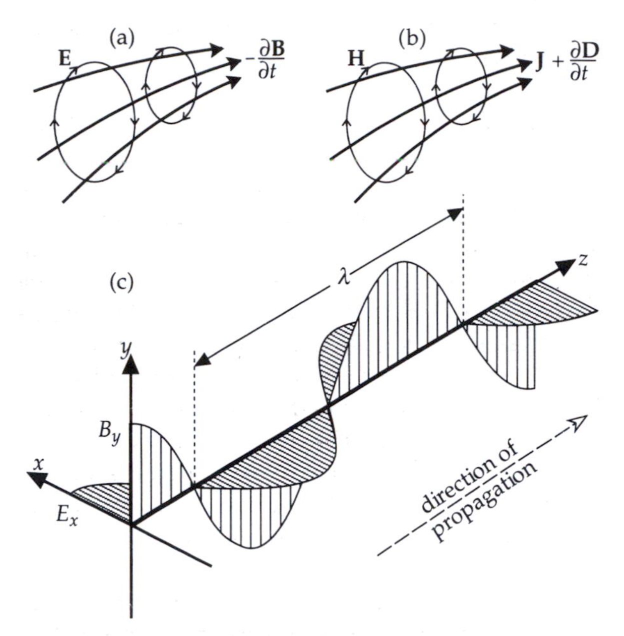

Fig. 4.52 (a) An electric field **E** is generated by a changing magnetic field ($\partial\mathbf{B}/\partial t$), while (b) a magnetic field **B** is produced by the current density **J** and the changing displacement-current density ($\partial\mathbf{D}/\partial t$); (c) in an electromagnetic wave an electric field E_x and a magnetic field B_y fluctuate normal to each other in the plane normal to the propagation direction (z-axis).

moving in a conductor produces a magnetic field proportional to the speed of the charge. If the electric field increases, so that the charge is accelerated, its changing velocity produces a changing magnetic field, which in turn induces another electric field in the conductor (Faraday's law) and thereby influences the movement of the accelerated charge. The coupling of the electric and magnetic fields is called electromagnetism. If two straight conductors are laid end-to-end and connected in series, they act as an electrical dipole. An alternating electric field applied to the conductors causes the dipole to oscillate, acting as an antenna for the emission of an electromagnetic wave. This consists of a magnetic field B and an electric field E, which vary with the frequency of the oscillator, and are oriented at right angles to each other in the plane perpendicular to the direction of propagation (Fig. 4.52). In a vacuum all electromagnetic waves travel at the speed of light (c=2.997 924 58×10^8 m s^{-1}, about 300,000 km s^{-1}), which is one of the fundamental constants of nature.

The derivation of electromagnetic field equations from Maxwell's equations is beyond the level of this textbook, but their meaning can be readily understood. Two equations, identical in form, are obtained. They describe the propagation of the **B** and **E** field vectors, respectively, and are written as:

$$\nabla^2\mathbf{B}=\mu_0\sigma\frac{\partial \mathbf{B}}{\partial t}+\mu_0\epsilon\frac{\partial^2\mathbf{B}}{\partial t^2} \tag{4.111}$$

$$\nabla^2\mathbf{E}=\mu_0\sigma\frac{\partial \mathbf{E}}{\partial t}+\mu_0\epsilon\frac{\partial^2\mathbf{E}}{\partial t^2} \tag{4.112}$$

$$\text{where } \nabla^2=\frac{\partial^2}{\partial x^2}+\frac{\partial^2}{\partial y^2}+\frac{\partial^2}{\partial z^2} \tag{4.113}$$

In these equations σ is the electrical *conductivity* and μ is the magnetic *permeability*, which in most materials (unless they are ferromagnetic) is close to the value μ_0 for free space, ($\mu_0=4\pi\times10^{-7}$ N A^{-2}). The parameter ϵ is the electrical *permittivity* of the material, which is related to the value for free space ($\epsilon_0=8.8542\times10^{-12}$ C^2 N^{-1} m^{-2}) by the *dielectric constant*, κ, where $\epsilon=\kappa\epsilon_0$. The value of κ is ~5–20 in most rocks and minerals and ~80 in water, so that usually $\epsilon>\epsilon_0$ (see also §4.3.2.3).

Electromagnetic radiation encompasses a wide frequency spectrum. It extends from very high-frequency (short-wavelength) γ-rays and x-rays to low-frequency (long-wavelength) radio signals (Fig. 4.53). Visible light constitutes a narrow part of the spectrum. Two ranges of electromagnetic radiation are of particular importance in solid Earth geophysics: a high-frequency range in the radar part of the spectrum, and a broad range of low frequencies extending from audio frequencies to signals with periods of hours, days or years. The electromagnetic equations reduce to simpler forms for these two particular frequency ranges.

The left side of Eq. (4.112) describes the variation of the E-component of the electromagnetic wave in space. The right side describes its variation with time and so its frequency dependence. The first term is related to the familiar *conduction* of electricity in a conductor. Maxwell introduced the second term, which is related to a *displacement* current. It originates when charges are displaced but not separated from their atoms, causing an electric polarization; fluctuations in the polarization have the effect of an alternating displacement current. Suppose that the electric dipole emitting E and B oscillates sinusoidally with angular frequency ω. Then $|\partial E/\partial t|\sim\omega E$, and $|\partial^2E/\partial t^2|\sim\omega^2E$, so the *magnitude ratio* (MR) of the second (*displacement*) term to the first (*conduction*) term on the right side of Eq. (4.112) is

$$\mathrm{MR}=\frac{\left|\mu_0\epsilon\dfrac{\partial^2E}{\partial t^2}\right|}{\left|\mu_0\epsilon\dfrac{\partial E}{\partial t}\right|}=\frac{\epsilon\omega^2E}{\sigma\omega E}=2\pi f\frac{\epsilon}{\sigma} \tag{4.114}$$

where f is the frequency of the signal. The conductivity σ of rocks and soils is generally in the range 10^{-5} to 10^{-1} Ω^{-1} m^{-1}; in orebodies σ may be as large as 10^3 to 10^5 Ω^{-1} m^{-1} (see Fig. 4.36). Electromagnetic induction sur-

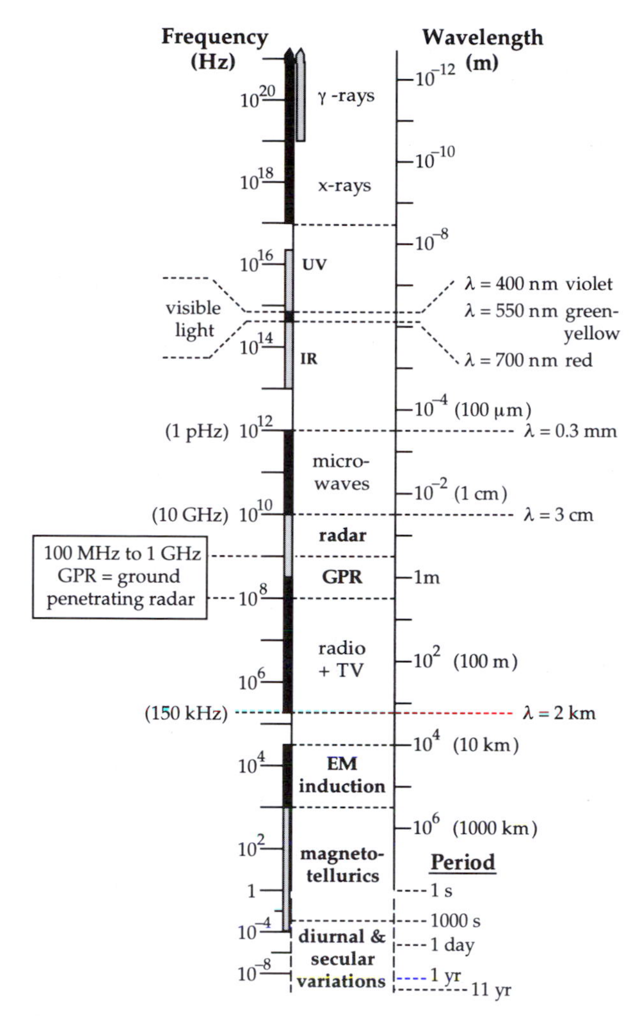

Fig. 4.53 The electromagnetic spectrum, showing the frequency and wavelength ranges of some common phenomena and the frequencies and periods used in electromagnetic surveying.

veying is usually carried out at frequencies below 10^4 Hz, for which the magnitude ratio MR is much less than unity in both good and bad conductors. At these frequencies the electromagnetic signal passes through the ground in a diffusive manner, by conversion of the changing magnetic fields to electric currents and vice versa. High-frequency surveying employs radar signals with frequencies around 10^8 Hz, for which the magnitude ratio MR is very small in an orebody but can be much greater than unity in rocks and soils. Under these conditions the electromagnetic signal propagates like a wave, and so is subject to diffraction, refraction and reflection.

4.3.6.1 *Ground-penetrating radar*

At high frequencies in poorly conducting media the conduction term in the electromagnetic equations is negligible compared to the displacement term. The electric field equation then becomes

$$\nabla^2\mathbf{E}=\mu_0\epsilon\frac{\partial^2 E}{\partial t^2} \tag{4.115}$$

with a similar equation for the magnetic field. This has the familiar form of the wave equation, which describes the propagation of an elastic disturbance (Eqs. (3.54) and (3.55)). Analogously, Eq. (4.115) describes the propagation of the electric part of an electromagnetic wave. By comparing with the seismic wave equations we see that the **E** and **B** fields in an electromagnetic wave have the same velocity v, where $v^2=1/\mu_0\epsilon$. In a vacuum the wave velocity is equal to the velocity of light c, given by $c^2=1/\mu_0\epsilon$. Using the relationship $\epsilon=\kappa\epsilon_0$ we get $v^2=c^2/k$, and taking into account that the dielectric constant κ is ~5–20 in earth materials, the velocity of an electromagnetic wave in the ground is found to be about 0.2–0.6 c.

To simplify further discussion, suppose that the electromagnetic disturbance propagates along the z-axis (i.e., $\partial/\partial x=\partial/\partial y=0$), so that $\nabla^2=\partial^2/\partial z^2$ in Eq. (4.113). For this one-dimensional case

$$\frac{\partial^2 E}{\partial z^2}=\mu_0\epsilon\frac{\partial^2 E}{\partial t^2}=\frac{1}{v^2}\frac{\partial^2 E}{\partial t^2} \tag{4.116}$$

If we compare this equation with Eq. (3.57) for a seismic wave, we see that the solution for a component E_i of the electric field is

$$E_i=E_0\sin 2\pi\left(\frac{z}{\lambda}-ft\right) \tag{4.117}$$

where λ is the wavelength, f the frequency and $f\lambda=v$, the velocity of the wave.

The above considerations suggest that high-frequency electromagnetic waves travel in the ground in an analogous manner to seismic waves. Instead of being determined by the elastic parameters the propagation of radar signals is dependent on the dielectric properties of the ground. A comparatively young branch of geophysical exploration has been developed to investigate underground structures with ground-penetrating radar (GPR, or *georadar*). GPR makes use of the familiar 'echo-principle' used in reflection seismology. A very short radar pulse, lasting only several nanoseconds (i.e., ~10^{-8} s) is emitted by a mobile antenna on the ground surface. The path of the radar signal through the ground can be traced as a ray, which experiences refractions, reflections and diffractions at boundaries where the dielectric constant changes. A second antenna, the receiver, is located close to the transmitter, as in the case of seismic reflection, so as to receive near-vertical reflections from underground discontinuities. The signal-processing tech-

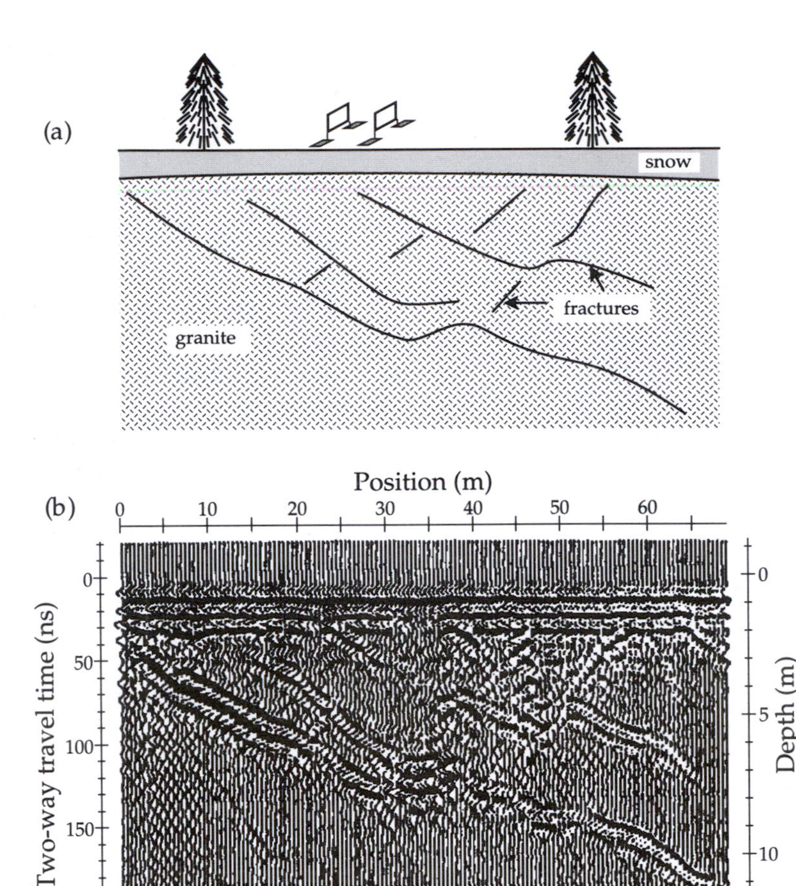

Fig. 4.54 Fracture pattern in a granitic bedrock revealed by ground-penetrating radar: (a) geological cross-section, (b) processed georadar reflection section (courtesy of A. G. Green).

niques of reflection seismology can also be applied to the georadar signal to help minimize the effects of diffractions and other noise. Consequently, georadar provides a detailed picture of the shallow subsurface structure (Fig. 4.54). It has become an important tool in environmental studies of near-surface features, such as buried and forgotten waste deposits, fracture patterns in otherwise uniform rock bodies, or the investigation of groundwater resources.

High-frequency signals are rapidly attenuated with depth. On the one hand, geometrical spreading of the signal outward from its source (spherical divergence) causes a decrease of intensity with distance. More important is absorption of the signal by ground materials, which is a function of their conductivity. Depending on the composition of the soil or rocks (e.g., the presence of clay-rich layers or groundwater), the nature of subsurface structures and the frequency of the radar signal, the effective penetration may be up to 10 m, although conditions commonly restrict it to only a few meters. However, at a radar frequency of 10^8–10^9 Hz and with a velocity of $\sim 10^8$ m s^{-1} the resolution is in the range 0.1–1 m. Thus, despite its limited depth penetration, the high resolution of georadar makes it a powerful tool for near-surface geophysical exploration.

4.3.6.2 *Electromagnetic induction*

Electromagnetic (EM) surveys carried out at frequencies below 50 kHz are based on the principle of electromagnetic induction. An alternating magnetic field in a coil or cable induces electric currents in a conductor. The conductivity of rocks and soils is too poor to permit significant induction currents, but when a good conductor is present a system of *eddy-currents* is set up. In turn, the eddy-currents produce secondary magnetic fields that are superposed on the primary field and can be measured at the ground surface (Fig. 4.55a).

Suppose that a low-frequency plane wave propagates along the vertical z-axis. The displacement current is now negligible compared to the conduction current and Eq. (4.111) becomes

$$\frac{\partial^2 B}{\partial z^2} = \mu_0 \sigma \frac{\partial B}{\partial t} \tag{4.118}$$

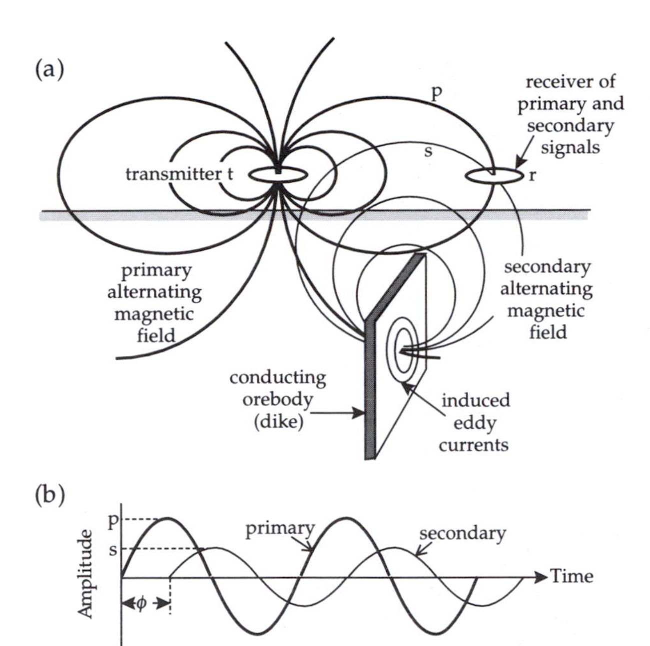

Fig. 4.55 (a) Illustration of primary and secondary fields in the horizontal loop induction method of electromagnetic exploration for shallow orebodies. (b) Amplitudes and phases of the primary (p) and secondary (s) fields.

where the magnetic field has components B_x and B_y. The form of this equation is reminiscent of the one-dimensional equation of *diffusion* or *heat conduction* (Eq. (4.52)), whose solution (Eq. (4.64)) describes how the temperature changes with time and position when a fluctuating temperature acts on the surface. By analogy, the solution of Eq. (4.118) for the components B_x or B_y of an alternating magnetic field with angular frequency ω (= $2\pi f$) in a conductor with conductivity σ is

$$B_{x,y}(z,t)=B_0 e^{-z/d}\cos\left(\omega t-\frac{z}{d}\right) \tag{4.119}$$

$$\text{where } d=\sqrt{\frac{2}{\mu_0\sigma\omega}} \tag{4.120}$$

Here, d is called the *skin depth*. At this depth the magnetic field is attenuated to e^{-1} (~37%) of its value outside the conductor. The skin depth is dependent on the conductivity of the body and the frequency of the field. The skin depth in normal ground (σ~10^{-3} Ω^{-1} m^{-1}) for a low-frequency alternating magnetic field (f~10^3 Hz) is about 500 m but in an orebody (σ~10^4 Ω^{-1} m^{-1}) it is only ~16 cm. The comparable figures for a high-frequency radar signal (f~10^9 Hz) are 50 cm and 0.16 mm, respectively. Note that the skin depth is not the maximum depth of penetration of the magnetic field. It helps to indicate how rapidly the field is attenuated, but the magnetic field is effective at depths that are many times the skin depth. However, it decays to ~1% at a depth $z=5d$ and to ~0.1% at $z=7d$, effectively limiting the practical depth of exploration with the induction method.

The many field methods of EM induction have a common principle. A coil or cable is used as transmitter of the primary alternating magnetic field, while another coil serves as receiver of both the primary signal and a secondary signal from the eddy-currents induced in a conductor (Fig. 4.55a). The magnetic field in the conductor experiences a phase shift (equal to z/d, Eq. (4.119)) due to the conductivity. This results in a phase difference ϕ between the secondary and primary signals in the receiver (Fig. 4.55b). The exact theory of EM induction is complicated, even in a simple situation, but we can obtain a simple qualitative appreciation by applying some concepts from electrical circuit theory. As in §4.2.6.1 we will use complex numbers involving $i=\sqrt{-1}$.

Let the current systems in transmitter, receiver and conductor be represented by simple loops carrying currents I_t, I_r and I_c, respectively. If the currents are sinusoidal, each has the form $I=I_0 e^{i\omega t}$, so that $dI/dt=i\omega I$. Let the resistance of the conductor be R and its self-inductance be L. The voltage V_c in the conductor is composed of two parts. A resistive part due to the current I_c in the resistance R is equal to I_cR. An inductive part due to the change of current is equal to $L\ dI_c/dt$. The complete voltage in the conductor is then

$$V_c=I_cR+L\frac{dI_c}{dt}=I_c(R+i\omega L) \tag{4.121}$$

The voltage V_c is induced in the conductor by the changing current I_t in the transmitter circuit. Let the mutual inductance between transmitter and conductor be M_{tc}; then $V_c=-M_{tc}\ dI_t/dt$. Similarly, the transmitter current induces a primary voltage $V_p=-M_{tr}\ dI_t/dt$ in the receiver, in which the eddy currents in the conductor also induce a secondary voltage $V_s=-M_{cr}\ dI_c/dt$. Here M_{tr} and M_{cr} are the mutual inductances between transmitter and receiver, and conductor and receiver, respectively. The following relationships exist between the different voltages and currents:

$$V_p=-M_{tr}\frac{dI_t}{dt}=-i\omega M_{tr}I_t \tag{4.122}$$

$$V_s=-M_{cr}\frac{dI_c}{dt}=-i\omega M_{cr}I_c \tag{4.123}$$

$$V_c=-M_{tc}\frac{dI_t}{dt}=-i\omega M_{tc}I_t \tag{4.124}$$

Combining Eq. (4.121) and Eq. (4.124) we get

$$\frac{I_c}{I_t} = \frac{-i\omega M_{tc}}{R+i\omega L} = \frac{-i\omega M_{tc}}{R^2+\omega^2 L^2}(R-i\omega L) \qquad (4.125)$$

From Eq. (4.121) and Eq. (4.124) the ratio of V_s to V_p in the receiver is

$$\frac{V_s}{V_p} = \frac{M_{cr}}{M_{tr}}\frac{I_c}{I_t} = -\frac{M_{tc}M_{cr}}{M_{tr}}\frac{(\omega^2 L+i\omega R)}{R^2+\omega^2 L^2} \qquad (4.126)$$

which can be written in the form

$$\frac{V_s}{V_p} = -\frac{M_{tc}M_{cr}}{M_{tr}L}\left(\frac{\beta^2+i\beta}{1+\beta^2}\right) \qquad (4.127)$$

where $\beta=\omega L/R$ is the *response parameter* of the conductor. The function in parentheses in Eq. (4.127) is a complex number, so the voltage ratio (or *response* of the measuring system) can be written $P+iQ$. The real part P has the same phase as the primary signal and is called the *in-phase component* of the response. The imaginary part Q is 90° out of phase with the primary signal (i.e., if the primary signal is $\sim\cos\omega t$, the imaginary part is $\sim\sin\omega t=\cos[\omega t-\pi/2]$); it is called the *quadrature component.*

4.3.6.3 *EM induction surveying*

The propagation of ground penetrating radar is described by a wave equation. Just as the seismic reflection and refraction methods are highly sensitive to elastic and density changes, GPR is capable of high resolution of differences in electrical properties. In contrast, the resemblance of the basic equations of EM induction to the diffusion equation (Eq. (4.53)) classifies the method as a *diffusive* one. Diffusive techniques – for example, the gravity, magnetic field, geothermal and seismic surface-wave methods – respond to a volumetric average of the specific physical parameter and do not show fine detail of its distribution. Hence, EM induction yields an average value of the electrical conductivity in a particular volume, but the resolution is better than that of potential field methods.

The EM induction method is very suitable for airborne surveys. These were carried out originally with fixed wing aircraft but they now more commonly use helicopters, which can adapt better to terrain roughness while flying close to the ground. The transmitter and receiver coils are mounted (usually with their axes coaxial or parallel to the line of flight) in fixed positions in the aircraft or in a towed 'bird', using similar configurations as in airborne magnetometer surveys (see Fig. 5.44). Alternatively, the transmitter may be in the airplane and the receiver in the bird or in another airplane. The increased separation of transmitter and receiver in this configuration gives greater depth penetration. However, in flight the bird yaws and pitches, altering the separation and parallelism of the coils, so that normally only the quadrature component is usable. The flight patterns consist of parallel profiles traversing the terrain. Lines are flown at about 100 m above ground level with fixed-wing aircraft and 30 m with helicopters.

When a potential conductivity anomaly has been located from the air, it is usual to investigate it further with a ground-based EM induction method. The transmitter may be a long cable or large horizontal loop on the ground surface, or it may be a small coil (diameter ~1 m) with its axis vertical or horizontal. The receiver is usually a similar small coil. It can be used to detect the direction, intensity or phase of the secondary signal. In its most simple application the tilt of the coil about a horizontal axis measures the dip-angle of the combined primary and secondary fields at the receiver. The method allows location, outlining and, to some extent, depth determination of a conductor. However, the dip-angle method registers only part of the available information in the secondary signal and does not describe the electrical properties of the conductor. As shown by Eq. (4.127) these properties affect the phase and relative amplitudes of the in-phase and quadrature components of the secondary signal relative to the primary. Phase-component EM measurement methods, therefore, allow more detailed interpretation.

The methods are illustrated by the horizontal loop electromagnetic method (HLEM), popularly known also by the commercial names *Slingram* or *Ronka*. The receiver and transmitter are coupled by a fixed cable about 30 to 100 m in length, and kept at a constant separation while the pair is moved along a traverse of a suspected conductor (Fig. 4.56a). The cable supplies a direct signal that exactly cancels the primary signal at the receiver, leaving only the secondary field of the conductor. This is separated into in-phase and quadrature components, which are expressed as percentages of the primary field and plotted against the position of the mid-point of the pair of coils (Fig. 4.56b). The in-phase and quadrature signals are zero far from the conductor and at the places where either the transmitter or receiver passes over the conductor. This enables the outline of a buried conductor to be charted. The signal rises to a positive peak on either side and falls to a negative peak over the middle of the conductor. The peak-to-peak responses of the in-phase and quadrature components depend on the quality of the conductor, which is expressed by a response parameter such as β in Eq. (4.127). A suitable function is the dimensionless parameter $\alpha=\mu_0\sigma\omega sl$, which contains the conductivity σ and width s of the conductor as well as the coil spacing l and frequency ω of the EM system. The systematic variation of the response curves with the value of α (Fig. 4.56b) allows interpreta-

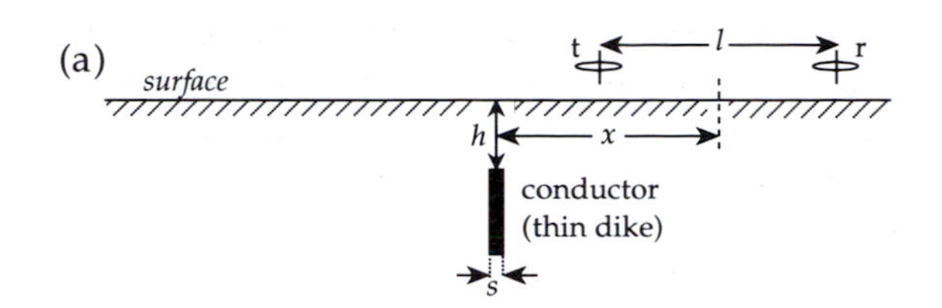

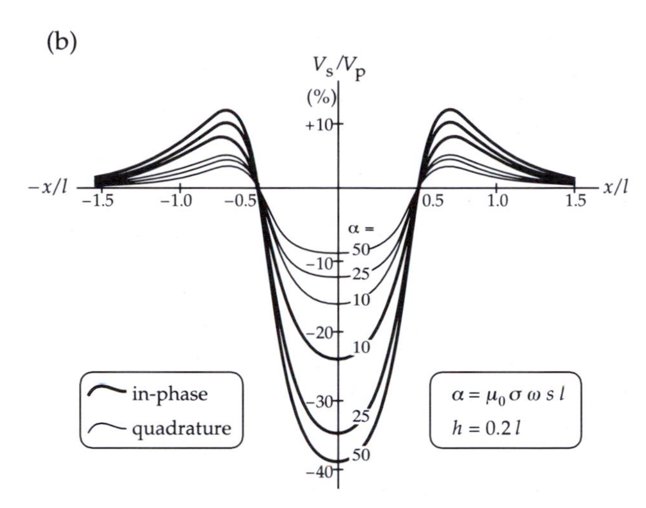

Fig. 4.56 (a) Geometry of an HLEM profile across a thin vertical dike. (b) In-phase and quadrature profiles over a dike at depth $h/l = 0.2$ for some values of the response parameter α.

tion of the quality of the conductor. A simple way of doing this is with the aid of model response curves. The variation of in-phase and quadrature signals over a conducting orebody can be modelled experimentally on a smaller scale in the laboratory. The smaller values of s and l are compensated by larger values of σ and ω to give the same response parameter α. The model response curves for different α are then directly applicable to the interpretation of real conductors measured in the field.

The most common use of EM induction methods is in lateral profiling, usually on traverses at right angles to the geological strike of dikes or other suspected conducting bodies. In environmental applications it is useful for locating buried pipes that may carry fluids or gases. Ground-based EM methods may also be used for vertical sounding, applying the same principles as in resistivity methods to obtain the conductivities of horizontal layers. The greater the separation of transmitter and receiver, the deeper is the maximum depth at which conductors may be analyzed. An important form of vertical EM sounding is the magnetotelluric method, which takes advantage of the penetrative ability of low-frequency signals from natural sources in the external geomagnetic field.

4.3.6.4 *Magnetotelluric sounding*

Magnetotelluric (MT) sounding is a natural-source electromagnetic method. The fluctuating electromagnetic fields that originate in the ionosphere are partly reflected at the Earth's surface; the returning fields are again reflected off the conducting ionosphere. This happens repeatedly, so that the fields eventually have a strong vertical component and may be regarded as vertically propagating plane waves with a wide spectrum of frequencies. These fields penetrate into the ground and induce telluric electric currents (§4.3.4.3), which in turn generate secondary magnetic fields. The telluric currents are detected with two pairs of electrodes, usually oriented NS and EW. Three components of the magnetic fields are measured: the vertical component and a horizontal component parallel to each of the telluric components. The method yields conductivity information from much greater depths than artificial-source induction methods. It has been applied in the search for petroleum and deep zones of mineralization in the upper crust. Utilizing long periods in the range 10–1000 s it is an important method for the investigation of the structure of the crust and upper mantle.

Consider a plane electromagnetic wave propagating in the z-direction (Fig. 4.52). Let the electric component E_x be along the x-axis, so that the magnetic field B_y (being normal to E_x) is along the y-axis. Ampère's law, as summarized in Maxwell's equations, relates E_x to the gradient of B_y in the z-direction. Because B has only a y-component, Ampère's law simplifies to

$$E_x = -\frac{1}{\mu_0 \sigma}\frac{\partial By}{\partial z} \tag{4.128}$$

where B_y has the form of Eq. (4.119). Differentiating B_y by parts gives

$$E_x = -\frac{B_0}{\mu_0 \sigma}\left\{\left(-\frac{e^{-z/d}}{d}\right)\cos(\omega t - \frac{z}{d}) + e^{-z/d}\left(\frac{-1}{d}\right)\left(-\sin(\omega t - \frac{z}{d})\right)\right\} \tag{4.129}$$

$$= \frac{B_0}{\mu_0 \sigma d} e^{-z/d}\left\{\cos(\omega t - \frac{z}{d}) - \sin(\omega t - \frac{z}{d})\right\} \tag{4.130}$$

$$= \frac{B_0}{\mu_0 \sigma d} e^{-z/d}\sqrt{2}\cos(\omega t - \frac{z}{d} + \frac{\pi}{4}) \tag{4.131}$$

Comparison of Eq. (4.119) and Eq. (4.131) shows a phase shift of 45° ($\pi/4$) between E_x and B_y. However, the ratio of the maximum amplitudes of the two components is

$$\frac{|E_x|}{|B_y|} = \frac{\sqrt{2}}{\mu_0 \sigma d} \tag{4.132}$$

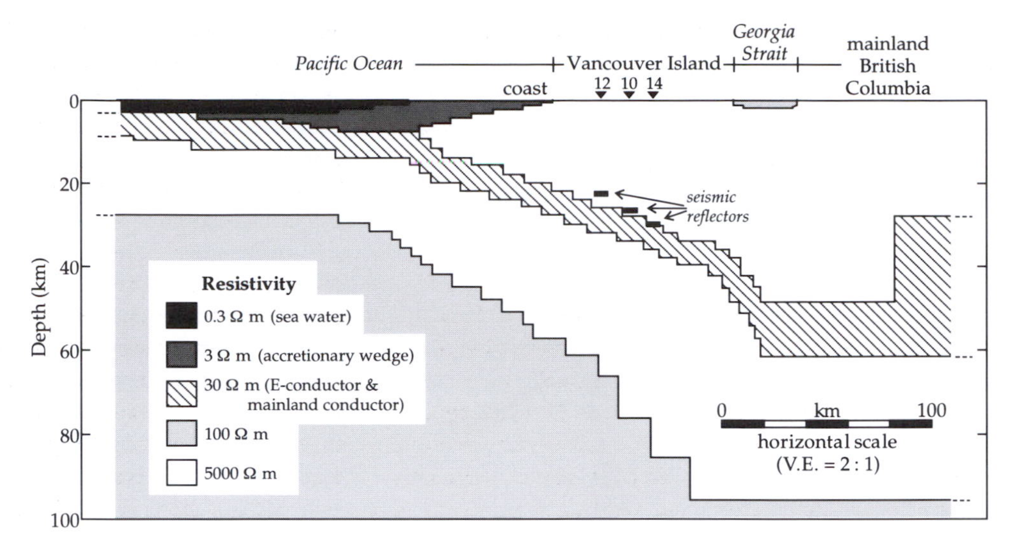

Fig. 4.57 Two-dimensional resistivity model of the crust and upper mantle beneath Vancouver Island and the adjacent mainland derived from magnetotelluric results (redrawn from Kurtz *et al.*, 1986).

If we now substitute for d from Eq. (4.120) and write $\rho=1/\sigma$, we get

$$\rho=\frac{\mu_0}{\omega}\frac{|E_x|^2}{|B_y|^2} \tag{4.133}$$

$$d=\frac{2}{\omega}\frac{|E_x|}{|B_y|} \tag{4.134}$$

for the effective resistivity ρ at depth d. The analysis gives similar results for an electric field along the y-axis and a corresponding magnetic field along the x-axis. In this case the ratio of the field amplitudes is $|E_y|/|B_x|$.

In addition to the horizontal magnetic fields, B_x and B_y, the vertical component B_z is also recorded for use in the interpretation of two-dimensional structures. Thus the data set from an MT site consist of two electrical components and three magnetic field components recorded continuously during a lengthy observation interval covering some hours or days. The recorded magnetic fields consist of an external part from the ionosphere and an internal part related to the induced current distribution. These components must be separated analytically. The electric and magnetic records contain numerous frequencies, of which some are simply noise and some are of geophysical interest. As a result, sophisticated data-processing is required, involving power-spectrum analysis and filtering.

The interpretation of MT data is based on either modelling or inversion. The modelling method is a direct approach to solving the conductivity distribution. It assumes a conductivity model for which a theoretical response is calculated and compared with the real response. The parameters of the model are adjusted in turn repeatedly to obtain the most favorable fit to the observations. As in the case of vertical electrical sounding with direct currents (§4.3.5.6), the inversion method seeks a solution to the EM induction problem by using the frequency spectrum of the observations to establish the causal conductivity distribution.

Although MT sounding can be carried out in the sub-audio to audio range ($f\sim 10-10^4$ Hz), its main application is in determining the electrical conductivity at great depths using very low frequencies ($f \ll 1$ Hz). The investigation of resistivity in the crust and upper mantle using MT sounding is illustrated by a profile across Vancouver Island (Fig. 4.57). Twenty-seven MT sounding stations were located along a NW–SE reflection seismic profile. One-dimensional analysis of the vertical distribution of resistivity beneath three stations (10, 12 and 14 in Fig. 4.57) showed an electrical discontinuity at virtually the same depth as a major seismic reflector observed in the associated seismic reflection profile. The information acquired about the mean resistivity above this depth was then used in a two-dimensional inversion of the MT records at all the stations. The resistivity pattern was interpreted down to depths of 100 km. It shows a north-eastward dipping zone of high conductivity ($\rho\approx 30$ Ω m), referred to as the E-conductor, surrounded by much more resistive material ($\rho\geq 5000$ Ω m). The E-conductor was interpreted as the top of the descending Juan de Fuca plate, where it subducts under the North American plate. The anomalously high conductivity in the top of the plate was attributed to conducting fluids in sediments derived from the accretionary wedge.

4.3.7 Electrical conductivity in the Earth

The complicated structure of the crust and upper mantle results in large lateral variations in electrical conductivity. Apart from the oceans, sediments and individual anomalous conductors, the outer carapace of the Earth is generally a poor electrical conductor. The physical mechanism of conductivity in silicate rocks is by semi-conduction, which can take place in three different ways (§4.3.2.3). Each type of semi-conduction is governed by a thermally activated process, in which the conductivity σ at temperature T is given by

$$\sigma=\sigma_0 e^{-E_a/kT} \tag{4.135}$$

where k is Boltzmann's constant ($k=1.38065\times10^{-23}$ J K^{-1}). The constant σ_0 is the hypothetical maximum value of the conductivity, reached asymptotically at very high temperatures. E_a is the activation energy of the particular type of semi-conduction. Its value determines the temperature range in which the thermally activated process becomes effective as a mechanism for σ. In the crust and upper mantle (i.e., in the lithosphere) impurity semi-conduction is likely the main mechanism in dry rocks. Electronic semi-conduction is probably dominant in the asthenosphere and deeper regions of the mantle. Ionic semi-conduction is an important mechanism at high temperature, but is unlikely to be significant below about 400 km, because it is suppressed by the high pressure at greater depth.

The electrical conductivity in the Earth at great depths is inferred from four sources: deep electrical sounding, geomagnetic variations, secular variations and extrapolation from laboratory experiments. The first two methods are based on induction effects arising from changes in the external part of the geomagnetic field; these encompass a broad spectrum with peaks of energy at several periods (see Fig. 4.40). Electrical and magnetotelluric sounding use the components with periods from milliseconds to one or two days. The inversion of MT data gives a conductivity pattern that is generally concordant with seismic data and related to the broad geological structure of the crust and upper mantle.

The time spectrum of external geomagnetic field variations contains some prominent periods that are longer than a day (see Fig. 4.40). The study of the longer-period geomagnetic variations provides information about conductivity in the Earth down to about 2000 km. The longer the period of the variation the deeper its penetration depth. The daily (or diurnal) variation (§5.4.3.3) yields conductivity information to about 900 km. Magnetic storms last several days or weeks and have a strong 48 hr component, which is used to extend conductivity information to about 1000 km. In addition to the spectrum of geomagnetic variation shown in Fig. 4.40 there is a longer-period component related to the 11-yr sunspot cycle. It results from increased solar activity and is accompanied by solar flares (see §5.4.8.1) and emissions of charged particles that augment the solar wind and excite ionospheric activity. Analysis of the 11-yr component allows the model of mantle conductivity to be extended to about 2000 km depth. Our knowledge of the electrical conductivity in the mantle at depths greater than 2000 km cannot be obtained from analysis of effects related to the external magnetic field.

The secular variation of the internal geomagnetic field (§5.4.5) originates in the upper part of the fluid outer core. It consists of fluctuations in intensity and direction with periods of the order of 10–10^4 yr. If the secular variation could be observed at the core–mantle boundary it would be possible to determine conductivity throughout the mantle. Unfortunately, secular variation must be observed at the Earth's surface, after it has passed through the conducting mantle, which acts as a filter. The signal is attenuated by the skin effect, preferentially affecting the highest frequencies. Thus, observations of high-frequency changes in secular variation place an upper limit on the average conductivity of the mantle, because a greater conductivity would block them out. From time to time, abrupt changes in the rate of secular variation take place for unknown reasons. A conspicuous example of these 'geomagnetic jerks' occurred in 1969–70, when a pulse in secular variation occurred with an estimated duration of less than two years. Although not all analysts concur, the effect is widely believed to be of internal origin. Analysis of the propagation of a secular-variation pulse provides an estimate of the mean conductivity of the whole mantle, which, integrated with data from other sources, gives the conductivity in the lower mantle.

Some of the different models of mantle conductivity that have been proposed are shown in Fig. 4.58. The differences between the models reflect increases in quantity and improvements in quality of geomagnetic data as well as advances in the techniques of data-processing, especially the development of inversion methods. Although the models diverge in many respects, they have some features in common. The conductivity averages about $10^{-2}\,\Omega^{-1}\,\mathrm{m}^{-1}$ in the lithosphere and increases with greater depth. Sharper rates of increase are found at depths of 400 and 670 km, where the olivine–spinel and spinel–perovskite phase changes occur, respectively (see §3.7.5.2). At about 700 km depth each model gives a conductivity of about $1\,\Omega^{-1}\,\mathrm{m}^{-1}$ which rises in the lower mantle to be about 10–200 $\Omega^{-1}\,\mathrm{m}^{-1}$ at the core–mantle boundary. The secular variation is not uniform over the Earth's surface. Large areas of continental size (the Central Pacific is the best studied) are characterized by slow rates of variation. This is possibly due to the additional screening effect of features in the D″-layer above the core–mantle boundary (§3.7.5.3), so-called 'crypto-continents' (see Fig. 6.26) in which the conductivity may be 1000 times higher than in the overlying mantle (Stacey, 1992).

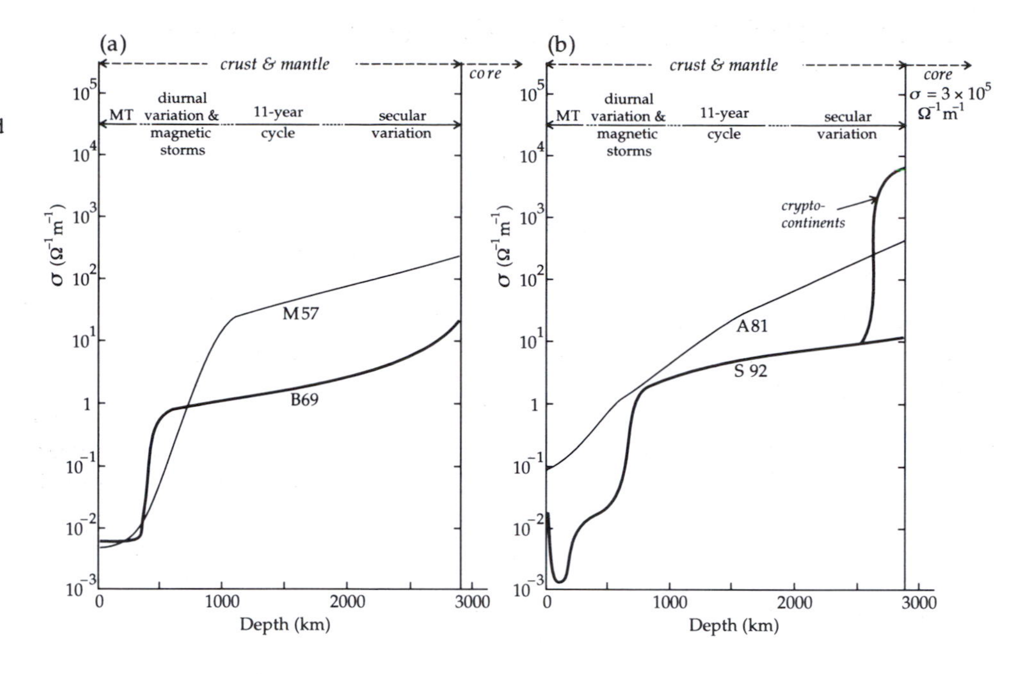

Fig. 4.58 Models of electrical conductivity (σ) at depth in the mantle proposed by (a) MacDonald (1957; M57) and Banks (1969; B69), (b) Achache *et al.*, (1981; A81) and Stacey (1992; S92).

Conductivity in the outer core is estimated by extrapolation from laboratory experiments. The core has the composition of an iron alloy, with an iron content of ~ 83% and a concentration of the alloying elements of ~ 17%. The effect of pressure on the conductivity of the alloy is not large at this concentration. Measurements of resistivity at atmospheric pressure and different temperatures lead to an extrapolated resistivity of $\rho=3.3\times10^{-6}\,\Omega$ m at the temperature of the outer core, with a corresponding conductivity $\sigma=3\times10^{5}\,\Omega^{-1}\,\mathrm{m}^{-1}$.

4.4 SUGGESTIONS FOR FURTHER READING

INTRODUCTORY LEVEL

Kearey, P. and Brooks, M. 1991. *An Introduction to Geophysical Exploration (2nd ed.)*, Blackwell Scientific Publ., Oxford.

Robinson, E. S. and Çoruh, C. 1988. *Basic Exploration Geophysics*, John Wiley & Sons, New York.

INTERMEDIATE LEVEL

Bott, M. H. P. 1982. *The Interior of the Earth (2nd ed.)*, Edward Arnold, London.

Dobrin, M. B. and Savit, C. H. 1988. *Introduction to Geophysical Prospecting (4th ed.)*, McGraw-Hill, New York.

Faure, G. 1986, *Principles of Isotope Geology (2nd ed.)*, John Wiley & Sons, New York.

Fowler, C. M. R. 1990. *The Solid Earth: an Introduction to Global Geophysics*, Cambridge Univ. Press, Cambridge.

Jessop, A.M. 1990. *Thermal Geophysics*, Elsevier, Amsterdam.

Sheriff, R. E. 1989. *Geophysical Methods*, Prentice Hall, New Jersey.

Telford, W. M., Geldart, L. P. and Sheriff, R. E. 1990. *Applied Geophysics*, Cambridge Univ. Press, Cambridge.

ADVANCED LEVEL

Dalrymple, G. B. 1991. *The Age of the Earth*, Stanford Univ. Press, Stanford, California.

Garland, G. D. 1979. *Introduction to Geophysics*, W. B. Saunders., Philadelphia.

Grant, F. S. and West, G. F. 1965. *Interpretation Theory in Applied Geophysics*, McGraw-Hill, New York.

Officer, C. B. 1974. *Introduction to Theoretical Geophysics*, Springer-Verlag, New York.

Stacey, F. D. 1992. *Physics of the Earth* (*3rd ed.*), Brookfield Press, Brisbane, Australia

York, D. and Farquhar, R. M. 1972. *The Earth's Age and Geochronology*, Pergamon Press, Oxford.

5 Geomagnetism and paleomagnetism

5.1 HISTORICAL INTRODUCTION

5.1.1 The discovery of magnetism

Mankind's interest in magnetism began as a fascination with the curious attractive properties of the mineral lodestone, a naturally occurring form of magnetite. Called loadstone in early usage, the name derives from the old English word load, meaning 'way' or 'course'; the loadstone was literally a stone which showed a traveller the way.

The earliest observations of magnetism were made before accurate records of discoveries were kept, so that it is impossible to be sure of historical precedents Nevertheless, Greek philosophers wrote about lodestone around 800 B.C. and its properties were known to the Chinese by 300 B.C. To the ancient Greeks science was equated with knowledge, and was considered an element of philosophy. As a result, the attractive forces of lodestone were ascribed to metaphysical powers. Some early animistic philosophers even believed lodestone to possess a soul. Contemporary mechanistic schools of thought were equally superstitious and gave rise to false conceptions that persisted for centuries. Foremost among these was the view that electrical and magnetic forces were related to invisible fluids. This view persisted well into the 19th century. The power of a magnet seemed to flow from one pole to the other along lines of induction that could be made visible by sprinkling iron filings on a paper held over the magnet. The term 'flux' (synonymous with flow) is still found in 'magnetic flux density', which is regularly used as an alternative to 'magnetic induction' for the fundamental magnetic field vector **B**.

One of the greatest and wealthiest of the ancient Greek city-colonies in Asia Minor was the seaport of Ephesus, at the mouth of the river Meander (modern Küçük Menderes) in the Persian province of Caria, in what is now the Turkish province of western Anatolia. In the fifth century B.C. the Greek state of Thessaly founded a colony on the Meander close to Ephesus called *Magnesia*, which after 133 B.C. was incorporated into the Roman empire as *Magnesia ad Maeandrum*. In the vicinity of Magnesia the Greeks found a ready supply of lodestone, pieces of which subsequently became known by the Latin word *magneta* from which the term magnetism derives.

It is not known when the directive power of the magnet – its ability to align consistently north–south – was first recognized. Early in the Han dynasty, between 300 and 200 B.C., the Chinese fashioned a rudimentary compass out of lodestone. It consisted of a shaped spoon object, whose bowl balanced and could rotate on a flat polished surface. This compass may have been used in the search for gems and in the selection of sites for houses. Before 1000 A.D. the Chinese had developed suspended and pivoted-needle compasses. Their directive power led to the use of compasses for navigation long before the origin of the aligning forces was understood. As late as the 12th century, it was supposed in Europe that the alignment of the compass arose from its attempt to follow the pole star. It was later shown that the compass alignment was produced by a property of the Earth itself. Subsequently, the characteristics of terrestrial magnetism played an important role in advancing the understanding of magnetism.

5.1.2 Pioneering studies in terrestrial magnetism

In 1269 the medieval scholar Pierre Pélerin de Maricourt, who took the Latin nom-de-plume of Petrus Peregrinus, wrote the earliest known treatise of experimental physics (*Epistola de Magnete*). In it he described simple laws of magnetic attraction. He experimented with a spherical magnet made of lodestone, placing it on a flat slab of iron and tracing the lines of direction which it assumed. These lines circled the lodestone sphere like geographical meridians and converged at two antipodal points, which Peregrinus called the *poles* of the magnet, by analogy to the geographical poles. He called his magnetic sphere a *terrella*, for 'little Earth'.

It was known to the Chinese around 500 A.D., in the Tang dynasty, that magnetic compasses did not point exactly to geographical north, as defined by the stars. The local deviation of the magnetic meridian from the geographical meridian is called the magnetic *declination*. By the 14th century, the ships of the British navy were equipped with a mariner's compass, which became an essential tool for navigation. It was used in conjunction with celestial methods, and gradually it became apparent that the declination changed with position on the globe. During the 15th and 16th centuries the world-wide pattern of declination was established. By the end of the 16th century, Mercator recognized that declination was the principal cause of error in contemporary map-making.

Georg Hartmann, a German cleric, discovered in 1544 that a magnetized needle assumed a non-horizontal attitude in the vertical plane. The deviation from the horizontal is now called the magnetic *inclination*. He reported his discovery in a letter to his superior, Duke Albrecht of Prussia, who evidently was not impressed. The letter lay unknown to the world in the royal archives until its discovery in 1831. Meanwhile, an English scientist, Robert Norman, rediscovered the inclination of the Earth's magnetic field independently in 1576.

In 1600 William Gilbert (1544–1603), an English scientist and physician to Queen Elizabeth, published *De Magnete*, a landmark treatise in which he summarized all that was then known about magnetism, including the results of about seventeen years of his own research. His studies extended also to the electrostatic effects seen when some materials were rubbed, for which he coined the name 'electricity' from the Greek word for amber. Gilbert was the first to distinguish clearly between electrical and magnetic phenomena. His magnetic studies followed the work of Peregrinus three centuries earlier. Using small magnetic needles placed on the surface of a sphere of lodestone to study its magnetic field, he recognized the poles, where the needles stood on end, and the equator, where they lay parallel to the surface. Gilbert achieved the leap of imagination that was necessary to see the analogy between the attraction of the lodestone sphere and the known magnetic properties of the Earth. He recognized that the Earth itself behaved like a large magnet. This was the first unequivocal recognition of a geophysical property, preceding the laws of gravitation in Newton's *Principia* by almost a century. Although founded largely on qualitative observations, *De Magnete* was the most important work on magnetism until the 19th century.

The discovery that the declination of the geomagnetic field changed with time was made by Henry Gellibrand (1597–1637), an English mathematician and astronomer, in 1634. He noted, on the basis of just three measurements made by William Borough in 1580, Edmund Gunter in 1622 and himself in 1634, that the declination had decreased by about 7° in this time. From these few observations he deduced what is now called the *secular variation* of the field.

Gradually the variation of the terrestrial magnetic field over the surface of the Earth was established. In 1698–1700 Edmund Halley, the English astronomer and mathematician, carried out an important oceanographic survey with the prime purpose of studying compass variations in the Atlantic ocean. In 1702 this resulted in the publication of the first declination chart.

5.1.3 The physical origins of magnetism

By the end of the 18th century many characteristics of terrestrial magnetism were known. The *qualitative* properties of magnets (e.g. the concentration of their powers at their poles) had been established, but the accumulated observations were unable to provide a more fundamental understanding of the phenomena because they were not *quantitative*. A major advance was achieved by Charles Augustin de Coulomb (1736–1806), the son of a noted French family, who in 1784 invented a torsion balance that enabled him to make quantitative measurements of electrostatic and magnetic properties. In 1785 he published the results of his intensive studies. He established the inverse-square law of attraction and repulsion between small electrically charged balls. Using thin, magnetized steel needles about 24 inches (61 cm) in length, he also established that the attraction or repulsion between their poles varied as the inverse square of their separation.

Alessandro Volta (1745–1827) invented the voltaic cell with which electrical currents could be produced. The relationship between electrical currents and magnetism was detected in 1820 by Hans Christian Oersted (1777–1851), a Danish physicist. During experiments with a battery of voltaic cells he observed that a magnetic needle is deflected at right angles to a conductor carrying a current, thus establishing that an electrical current produces a magnetic force.

Oersted's result was met with great enthusiasm and was followed at once by other notable discoveries in the same year. The law for the direction and strength of the magnetic force near a current-carrying wire was soon formulated by the French physicists Jean-Baptiste Biot (1774–1862) and Felix Savart (1791–1841). Their compatriot André Marie Ampère (1775–1836) quickly undertook a systematic set of experiments. He showed that a force existed between two parallel straight current-carrying wires, and that it was of a type different from the known electrical forces. Ampère experimented with the magnetic forces produced by current loops and proposed that internal electrical currents were responsible for the existence of magnetism in iron objects (i.e., ferromagnetism). This idea of permanent magnetism due to constantly flowing currents was audacious for its time.

At this stage, the ability of electrical currents to generate magnetic fields was known, but it fell to the English scientist Michael Faraday (1791–1867), to demonstrate in 1831 what he called 'magneto-electric' induction. Faraday came from a humble background and had little mathematical training. Yet he was a gifted experimenter, and his results demonstrated that the change of magnetic flux in a coil (whether produced by introducing a magnet or by the change in current in another coil) induced an electric current in the coil. The rule that governs the direction of the induced current was formulated three years later by a Russian physicist, Heinrich Lenz (1804–1865). Unhampered by mathematical equations, Faraday made fundamental contributions to understanding magnetic processes. Instead of regarding magnetic and electrical phenomena as the effects of centers of force acting at a distance, he saw in his mind's eye fictional lines of force traversing space. This image emphasized the role of the medium and led eventually to the concept of *magnetic field*, which Faraday first used in 1845.

Although much had been established by the early 1830s, it was still necessary to interpret the strengths of magnetic forces by relating magnetic units to mechanical units. This was achieved in 1832 by the German scientist and mathematician, Carl Friedrich Gauss (1777–1855), who assumed that static magnetism was carried by magnetic 'charges', analogous to the carriers of static electricity. Experiment had shown that, in contrast to electric charge, magnetic poles always occur as oppositely signed pairs, and so the basic unit of magnetic properties corresponds to the dipole. Together with Wilhelm Weber (1804–1891), Gauss developed a method of absolute determination of the intensity of the Earth's magnetic field. They founded a geomagnetic observatory at Göttingen where the Earth's magnetic field was observed at regular intervals. Subsequently some 50 observatories were founded in different countries. By 1837 global charts of the total intensity, inclination and declination were in existence, although the data had been measured at different times and their areal coverage was incomplete. To analyse the data-set Gauss applied the mathematical techniques of spherical harmonic analysis and the separation of variables, which he had invented. In 1839 he established that the main part of the Earth's magnetic field was a dipole field that originated inside the Earth.

The fundamental physical laws governing magnetic effects were now firmly established. In 1872 James Clerk Maxwell (1831–1879), a Scottish physicist, derived a set of equations that quantified all known relationships between electrical and magnetic phenomena: Coulomb's laws of force between electric charges and magnetic poles; Oersted's and Ampère's laws governing the magnetic effects of electric currents; Faraday's and Lenz's laws of electromagnetic induction and Ohm's law relating current to electromotive force. Maxwell's mathematical studies predicted the propagation of electric waves in space, and concluded that light is also an electromagnetic phenomenon transmitted through a medium called the *luminiferous ether*. The need for this light-transmitting medium was eliminated by the theory of relativity. By putting the theory of the electromagnetic field on a mathematical basis, Maxwell enabled a greater understanding of electromagnetic phenomena before the discovery of the electron.

A further notable discovery was made in 1879 by Heinrich Lorentz (1853–1928), a Dutch physicist. In experiments with vacuum tubes he observed the deflection of a beam of moving electrical charge by a magnetic field. The deflecting force acted in a direction perpendicular to the magnetic field and to the velocity of the charged particles, and was proportional to both the field and the velocity. This result now serves to define the unit of magnetic induction.

Since the time of man's first awareness of magnetic behavior, students of terrestrial magnetism have made important contributions to the understanding of magnetism as a physical phenomenon. In turn, advances in the physics of magnetism have helped geophysicists to understand the morphology and origin of the Earth's magnetic field, and to apply this knowledge to geological processes, such as global tectonics. The physical basis of magnetism is fundamental to the geophysical topics of geomagnetism, rock magnetism and paleomagnetism.

5.2 THE PHYSICS OF MAGNETISM

5.2.1 **Introduction**

Early investigators conceptualized gravitational, electrical and magnetic forces between objects as instantaneous effects that took place through direct action-at-a-distance. Faraday introduced the concept of the *field* of a force as a property of the space in which the force acts. The force-field plays an intermediary role in the interaction between objects. For example, an electric charge is surrounded by an electrical field that acts to produce a force on a second charge. The pattern of a field is portrayed by *field lines*. At any point in a field the direction of the force is tangential to the field line and the intensity of the force to the number of field lines per unit cross-sectional area.

Problems in magnetism are often more complicated for the student than those in gravitation and electrostatics. For one thing, gravitational and electrostatic fields act centrally to the source of force, which varies in each case as the inverse square of distance. Magnetic fields are not central; they vary with azimuth. Moreover, even in the simplest case (that of a magnetic dipole or a small current loop) the field strength falls off inversely as the cube of distance. To make matters more complicated, the student has to take account of *two* magnetic fields (denoted by **B** and **H**).

The confusion about the **B**-field and the **H**-field may be removed by recalling that all magnetic fields originate with electrical currents. This is the case even for permanent

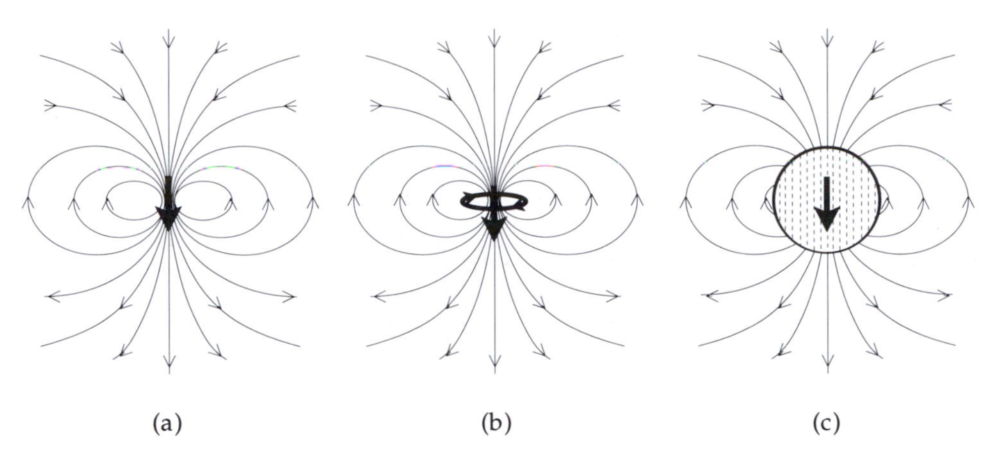

Fig. 5.1 The characteristic field lines of a magnetic dipole are found around (a) a short bar magnet, (b) a small loop carrying an electric current, and (c) a uniformly magnetized sphere.

magnets, as Ampère astutely recognized in 1820. We now know that these currents are associated with the motions of electrons about atomic nuclei in the permanent magnets. The fundamental magnetic field associated with currents in any medium is **B**. The quantity **H** should be regarded as a computational parameter proportional to **B** in non-magnetizable materials. Inside a magnetizable material, **H** describes how **B** is modified by the magnetic polarization (or magnetization, **M**) of the material. The magnetic **B**-field is alternatively called the magnetic induction or magnetic flux density.

Historically, the laws of magnetism were established by relating the **B**-field to fictitious centers of magnetic force called *magnetic poles*, defined by comparison with the properties of a bar magnet. Gauss showed that, in contrast to electro static charges, free magnetic poles cannot exist; each positive pole must be paired with a corresponding negative pole. The most important type of magnetic field – and also the dominant component of the geomagnetic field – is that of a *magnetic dipole* (Fig. 5.1a). This is the field of two magnetic poles of opposite sense that are infinitesimally close to each other. The geometry of the field lines shows the paths along which a free magnetic pole would move in the vicinity of the dipole. A tiny current loop (Fig. 5.1b) and a uniformly magnetized sphere (Fig. 5.1c) also have dipole-type magnetic fields around them. Although magnetic poles do not exist physically, many problems that arise in geophysical situations can be readily solved in terms of surface distributions of poles or dipoles. So we will first examine these concepts.

5.2.2 Coulomb's law for magnetic poles

Coulomb's experiments in 1785 established that the force between the ends of long thin magnets was inversely proportional to the square of their separation. Gauss expanded Coulomb's observations and attributed the forces of attraction and repulsion to fictitious magnetic charges, or poles. An inverse square law for the force F between magnetic poles with strengths p_1 and p_2 at distance r from each other can be formulated as

$$F(r)=K\frac{p_1p_2}{r^2} \tag{5.1}$$

The proportionality constant K was originally defined to be dimensionless and equal to unity, analogously to the law of electrostatic force. This gave the dimensions of pole strength in the centimeter-gram-second (c.g.s.) system as dyne$^{1/2}$ cm.

5.2.2.1 *The field of a magnetic pole*

The gravitational field of a given mass is defined as the force it exerts on a unit mass (§2.2.2). Similarly, the electric field of a given charge is the force it exerts on a unit charge. These ideas cannot be transferred directly to magnetism, because magnetic poles do not really exist. Nevertheless, many magnetic properties can be described and magnetic problems solved in terms of fictitious poles. For example, we can define a magnetic field B as the force exerted by a pole of strength p on a unit pole at distance r. From Eq. (5.1) we get

$$B(r)=K\frac{p}{r^2} \tag{5.2}$$

Setting $K=1$, the unit of the magnetic **B**-field has dimensions dyne$^{1/2}$ cm^{-1} in c.g.s. units and is called a *gauss*. Geophysicists employ a smaller unit, the *gamma* (γ), to describe the geomagnetic field and to chart magnetic anomalies ($1\gamma=10^{-5}$ gauss).

Unfortunately, the c.g.s. system required units of electrical charge that had different dimensions and size in electrostatic and electromagnetic situations. By international agreement the units were harmonized and rationalized. In the modern Système Internationale (SI) units the proportionality constant K is not dimensionless. It has the value $\mu_0/4\pi$, where μ_0 is called the *permeability constant* and is equal to $4\pi\times10^{-7}$ N A^{-2} (or henry/meter, H m^{-1}, which are equivalent units to N A^{-2}).

5.2.2.2 *The potential of a magnetic pole*

In studying gravitation we also used the concept of a field to describe the region around a mass in which its attraction could be felt by another test mass. In order to move the test mass away from the attracting mass, work had to be done against the attractive force and this was found to be equal to the gain of potential energy of the test mass. When the test mass was a unit mass, the attractive force was called the gravitational field and the gain in potential energy was called the change of potential. We calculated the gravitational potential at distance r from an attracting point mass by computing the work that would have to be expended *against* the field to move the unit mass from r to infinity.

We can define the magnetic potential W at a distance r from a pole of strength p in exactly the same way. The magnetic field of the pole is given by Eq. (5.2). Using the value $\mu_0/4\pi$ for K and expressing the pole strength p in SI units, the magnetic potential at r is given by

$$W=-\int_r^\infty B\mathrm{d}r=\frac{\mu_0 p}{4\pi r} \tag{5.3}$$

5.2.3 The magnetic dipole

In Fig. 5.1 the line joining the positive and negative poles (or the normal to the plane of the loop, or the direction of magnetization of the sphere) defines an axis, about which the field has rotational symmetry. Let two equal and opposite poles, $+p$ and $-p$, be located a distance d apart (Fig. 5.2). The potential W at a distance r from the mid-point of the pair of poles, in a direction that makes an angle θ to the axis, is the sum of the potentials of the positive and negative poles. At the point (r, θ) the distances from the respective poles are r_+ and r_- and we get for the magnetic potential of the pair

$$W=\frac{\mu_0 p}{4\pi}\left(\frac{1}{r_+}-\frac{1}{r_-}\right) \tag{5.4}$$

$$W=\frac{\mu_0 p}{4\pi}\left(\frac{r_- - r_+}{r_+ r_-}\right) \tag{5.5}$$

The pair of opposite poles are considered to form a *dipole* when their separation becomes infinitesimally small compared to the distance to the point of observation (i.e., $d \ll r$). In this case, we get the approximate relations

$$r_+ \approx r-\frac{d}{2}\cos\theta$$

$$r_- \approx r+\frac{d}{2}\cos\theta' \tag{5.6}$$

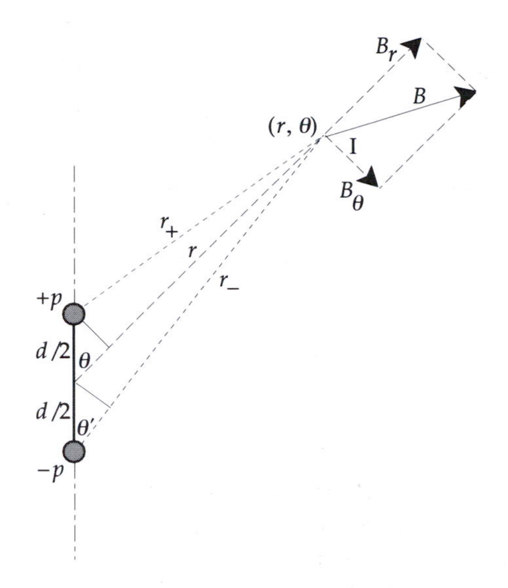

Fig. 5.2 Geometry for the calculation of the potential of a pair of magnetic poles.

When $d \ll r$, we can write $\theta \approx \theta'$ and terms of order $(d/r)^2$ and higher can be neglected. This leads to the further simplifications

$$r_- - r_+ \approx \frac{d}{2}(\cos\theta' + \cos\theta) \approx d\cos\theta$$

$$r_+ r_- \approx r^2 - \frac{d^2}{4}\cos^2\theta \approx r^2 \tag{5.7}$$

Substituting Eq. (5.7) in Eq. (5.5) gives the dipole potential at the point (r, θ):

$$W=\frac{\mu_0}{4\pi}\frac{(dp)\cos\theta}{r^2}=\frac{\mu_0 m\cos\theta}{4\pi r^2} \tag{5.8}$$

The quantity $m=(dp)$ is called the *magnetic moment* of the dipole. This definition derives from observations on bar magnets. The torque exerted by a magnetic field to turn the magnet parallel to the field direction is proportional to m. This applies even when the separation of the poles becomes very small, as in the case of the dipole.

The torque can be calculated by considering the forces exerted by a uniform magnetic field B on a pair of magnetic poles of strength p separated by a distance d (Fig. 5.3). A force equal to (Bp) acts on the positive pole and an equal and opposite force acts on the negative pole. If the magnetic axis is oriented at angle θ to the field, the perpendicular distance between the lines of action of the forces is $d\sin\theta$. The torque τ felt by the magnet is equal to $B(pd)\sin\theta$ (i.e., $\tau=mB\sin\theta$). Taking into account the direction of the torque and using the conventional notation for the cross product of two vectors this gives for the magnetic torque

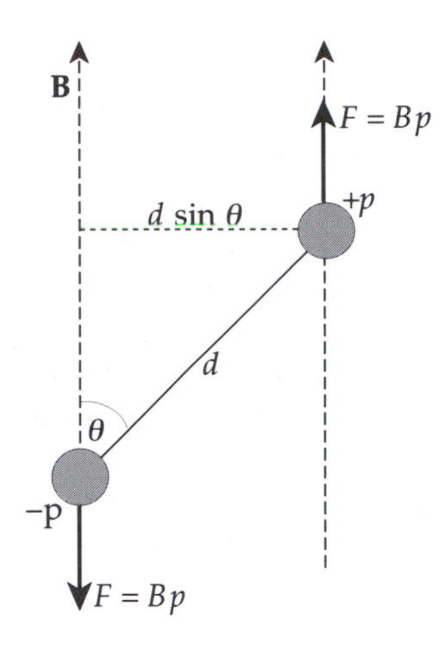

Fig. 5.3 Definition of the magnetic moment **m** of a pair of magnetic poles.

$$\boldsymbol{\tau}=\mathbf{m}\times\mathbf{B} \tag{5.9}$$

5.2.4 The magnetic field of an electrical current

The equation used to define the magnetic **B**-field was formulated by Lorentz in 1879. Let q be an electrical charge that moves with velocity v through a magnetic field **B** (Fig. 5.4a). The charged particle experiences a deflecting force **F** given by Lorentz's law, which in SI units is:

$$\mathbf{F}=q(\mathbf{v}\times\mathbf{B}) \tag{5.10}$$

The SI unit of the magnetic **B**-field defined by this equation is called a *tesla*; it has the dimensions N A^{-1} m^{-1}.

Imagine the moving charge to be confined to move along a conductor of length dl and cross-section A (Fig. 5.4b). Let the number of charges per unit volume be N. The number inside the element dl is then NA d**l**. Each charge experiences a deflecting force given by Eq. (5.10). Thus the total force transferred to the element **dl** is

$$\mathrm{d}\mathbf{F}=NA\mathrm{d}l\, q(\mathbf{v}\times\mathbf{B})=NAvq(\mathrm{d}\mathbf{l}\times\mathbf{B}) \tag{5.11}$$

The electrical current I along the conductor is the total charge that crosses A per second, and is given by $I=NAvq$. From Eq. (5.11) we get the law of Biot and Savart for the force experienced by the element **dl** of a conductor carrying a current I in a magnetic field **B**:

$$\mathrm{d}\mathbf{F}=I(\mathrm{d}\mathbf{l}\times\mathbf{B}) \tag{5.12}$$

The orienting effect of an electrical current on magnetic compass needles, reported by Oersted and Ampère in 1820,

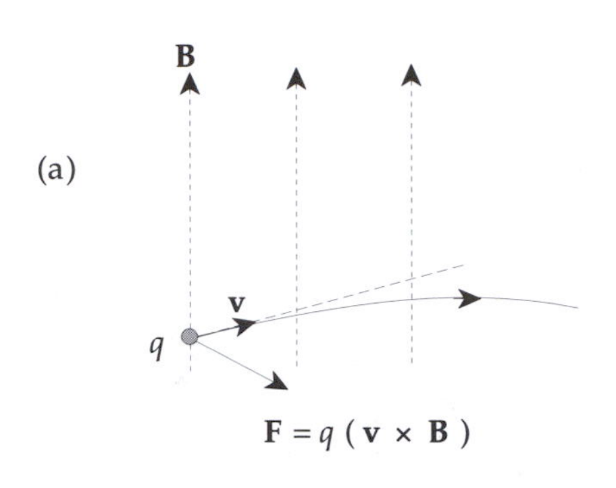

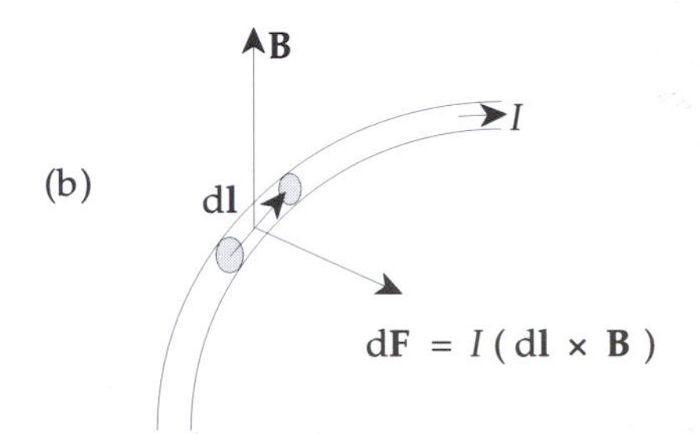

Fig. 5.4 Illustrations of (a) Lorentz's law for the deflecting force **F** experienced by an electrical charge that moves with velocity **v** through a magnetic field **B**, and (b) the law of Biot and Savart for the force experienced by an element **dl** of a conductor carrying a current I in a magnetic field **B**.

is illustrated in Fig. 5.5. The magnetic field lines around an infinitely long straight wire form concentric circles in the plane normal to the wire. The strength of the **B**-field around the wire is

$$B=\frac{\mu_0 I}{2\pi r} \tag{5.13}$$

The Biot–Savart law can be applied to determine the torque exerted on a small rectangular loop PQRS in a magnetic field (Fig. 5.6a). Let the lengths of the sides of the loop be a and b, respectively, and define the x-axis parallel to the sides of length a. The area of the loop can be expressed as a vector with magnitude $A=ab$, and direction **n** normal to the plane of the loop. Suppose that a current I flows in the loop and that a magnetic field **B** acts normal to the x-axis, making an angle θ with the normal to the plane of the loop. Applying Eq. (5.12), a force F_x equal to $(IbB\cos\theta)$ acts on the side PQ in the direction of $+x$; its effect is cancelled by an equal and opposite force F_x acting on side RS in the direction of $-x$. Forces equal to (IaB) act in opposite directions on the sides QR and SP (Fig. 5.6b). The perpendicular distance between their lines of action is $d \sin\theta$, so the torque τ experienced by the current loop is

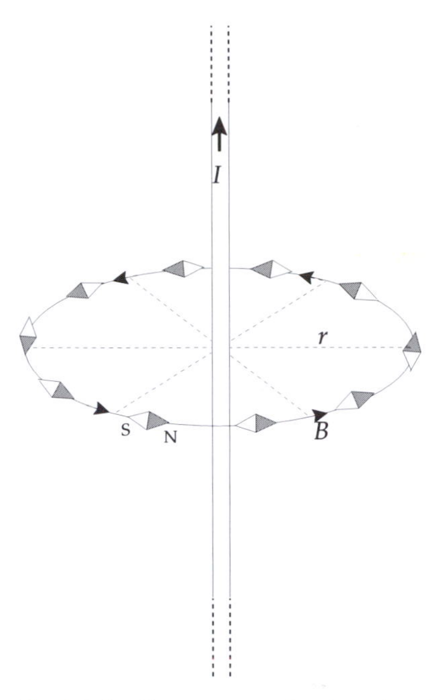

Fig. 5.5 Small compass needles show that the magnetic field lines around an infinitely long straight wire carrying an electrical current form concentric circles in a plane normal to the wire.

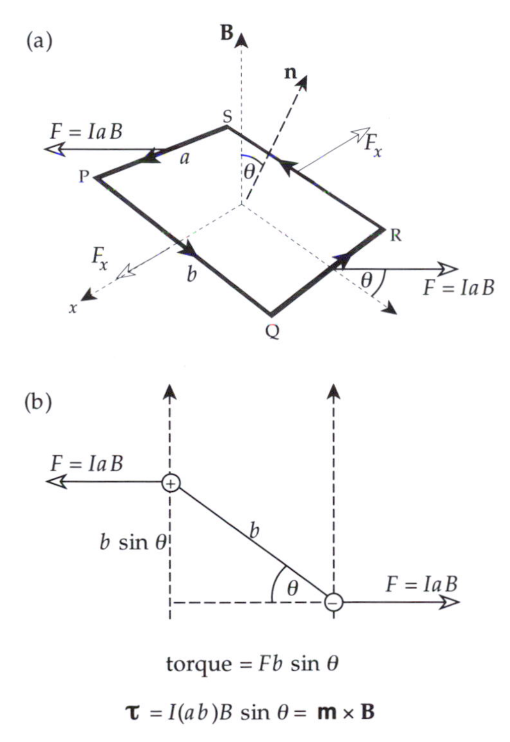

Fig. 5.6 (a) Rectangular loop carrying a current I in a uniform magnetic field B; (b) derivation of the torque $\boldsymbol{\tau}$ experienced by the loop.

$$\boldsymbol{\tau}=(Iab)B\sin\theta=(IA)B\sin\theta$$

$$\boldsymbol{\tau}=\mathbf{m}\times\mathbf{B} \tag{5.14}$$

The quantity $\mathbf{m}=IA$ is a vector with direction parallel to the normal to the plane of the current loop. This expression is valid for an arbitrary small loop of area A, regardless of its shape. By comparing Eqs. (5.14) and (5.9) for the torque on a dipole, it is evident that $\mathbf{m}$ corresponds to the magnetic moment of the current loop. At distances greater than the dimensions of the loop, the magnetic field is that of a dipole at the center of the loop (see Fig. 5.1b). The definition of $\mathbf{m}$ in terms of a current-carrying loop shows that magnetic moment has the units of current times area, or ampère meter^2 (A m^2).

5.2.5 Magnetization and the magnetic field inside a material

A true picture of magnetic behavior requires a quantum-mechanical analysis. Fortunately, a working understanding of the magnetic behavior of materials can be acquired without getting involved in the quantum-mechanical details. The simplified concept of atomic structure introduced by Ernest Rutherford in 1911 gives a readily understandable model for the magnetic behavior of materials. The motion of an electron around an atomic nucleus is treated like the orbital motion of a planet about the Sun. The orbiting charge forms an electrical current with which an *orbital magnetic moment* is associated. A planet also rotates about its axis; likewise each electron can be visualized as having a spin motion about an axis. The spinning electrical charge produces a *spin magnetic moment*. Each magnetic moment is directly related to the corresponding *angular momentum*. In quantum theory each type of angular momentum of an electron is quantized. Thus the spin and orbital magnetic moments are restricted to having discrete values. The spin magnetic moment is usually more important than the orbital moment in the rock-forming minerals (see §5.2.6).

A simplified picture of the magnetic moments inside a material is shown in Fig. 5.7. The magnetic moment $\mathbf{m}$ of each atom is associated with a current loop as illustrated in Fig. 5.1b and described in the previous section. The net magnetic moment of a volume V of the material depends on the degree of alignment of the individual atomic magnetic moments. It is the vector sum of all the atomic magnetic moments in the material. The magnetic moment per unit volume of the material is called its *magnetization*, denoted $\mathbf{M}$:

$$\mathbf{M}=\sum_i \mathbf{m}_i/V \tag{5.15}$$

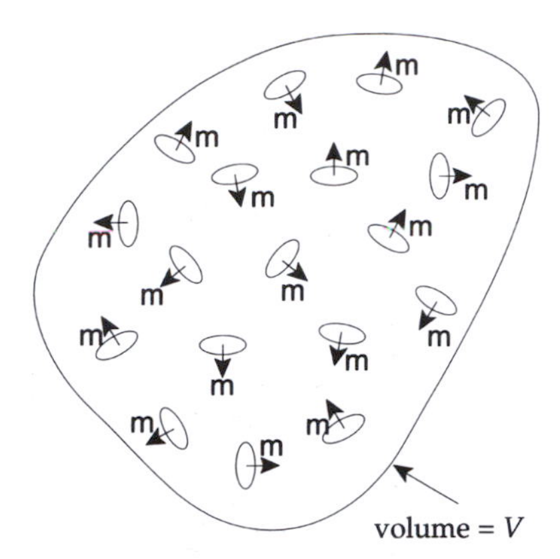

Fig. 5.7 Schematic representation of the magnetic moments inside a material; each magnetic moment m is associated with a current loop on an atomic scale.

Magnetization has the dimensions of magnetic moment (A m^2) divided by volume (m^3), so that the SI units of **M** are A m^{-1}. The dimensions of **B** are N A^{-1} m^{-1} and those of μ_0 are N A^{-2}; consequently the dimensions of B/μ_0 are also A m^{-1}. In general, the magnetization **M** inside a magnetic material will not be exactly equal to B/μ_0 ; let the difference be **H**, so that

$$\mathbf{H}=\mathbf{B}/\mu_0-\mathbf{M} \tag{5.16}$$

In the earlier c.g.s. system **H** was defined by the vector equation $\mathbf{H}=\mathbf{B}-4\pi\mathbf{M}$, and the dimensions of **H** and **B** were the same. For this reason **H** became known as the *magnetizing field* (or **H**-field). It is a readily computed quantity that is useful in determining the value of the true magnetic field **B** in a medium. The fundamental difference between the **B**-field and the **H**-field can be understood by inspection of the configurations of their respective field lines. The field lines of **B** always form closed loops (see Fig. 5.1). The field lines of **H** are discontinuous at surfaces where the magnetization **M** changes in strength or direction. Magnetic methods of geophysical exploration take advantage of surface effects that arise where the magnetization is interrupted.

Anomalous magnetic fields arise over geological structures that cause a magnetization contrast between adjacent rock types. Many magnetic anomalies can be analyzed by replacing the change in magnetization at a surface by an appropriate surface distribution of fictitious magnetic poles. The methodology, though based on a fundamentally false concept, is quite practical for modelling anomaly shapes and is often much simpler than a physically correct analysis in terms of current distributions. For example, in a uniformly magnetized rod, the N-poles of the elementary magnetic moments are considered to be exposed on one end of the rod, with a corresponding distribution of S-poles on the opposite end; inside the material the N-poles and S-poles cancel each other (Fig. 5.8a). The **H**-field inside the material arises from these pole distributions and acts in the *opposite* direction to the magnetization **M**. Outside the magnet the **B**-field and **H**-field are parallel; the **H**-field is discontinuous at the ends of the magnet.

The same situation can be portrayed in terms of current loops. The physical source of every **B**-field is an electrical current, even in a permanent magnet (Fig. 5.8b). Atomic current loops give a continuous **B**-field that emerges from the magnet at one end, re-enters at the other end and is closed inside the magnet. The aligned magnetic moments of the elementary current loops cancel out inside the body of the magnet, but the currents in the loops adjacent to the sides of the magnet combine to form a surface 'current' that maintains the magnetization **M**.

In a vacuum there is no magnetization ($\mathbf{M}=0$); the vectors **B** and **H** are parallel and proportional ($\mathbf{B}=\mu_0\mathbf{H}$). Inside a magnetizable material the magnetic **B**-field has two sources. One is the external system of real currents that produce the magnetizing field **H**; the other is the set of internal atomic currents that cause the atomic magnetic moments whose net alignment is expressed as the magnetization **M**. In a general, anisotropic magnetic material **B**, **M** and **H** are not parallel. However, many magnetic materials are not strongly anisotropic and the elementary atomic magnetic moments align in a statistical fashion with the magnetizing field. In this case **M** and **H** are parallel and proportional to each other

$$\mathbf{M}=k\mathbf{H} \tag{5.17}$$

The proportionality factor k is a physical property of the material, called the *magnetic susceptibility*. It is a measure of the ease with which the material can be magnetized. Because **M** and **H** have the same units (A m^{-1}), k is a dimensionless quantity. The susceptibility of most materials is temperature-dependent, and in some materials (ferromagnets and ferrites) k depends on **H** in a complicated fashion. In general, Eq. (5.16) can be rewritten

$$\mathbf{B}=\mu_0(\mathbf{H}+\mathbf{M})=\mu_0\mathbf{H}(1+k)=\mu_0\mu\mathbf{H} \tag{5.18}$$

The quantity $\mu=(1+k)$ is called the *magnetic permeability* of the material. The term 'permeability' recalls the early 19th century association of magnetic powers with an invisible fluid. For example, the permeability of a material expresses the ability of the material to allow a fluid to pass through it. Likewise, the magnetic permeability is a measure of the ability of a material to convey a magnetic flux. Ferromagnetic metals have high permeabilities; in contrast, minerals and rocks have low susceptibilities and permeabilities $\mu\approx1$.

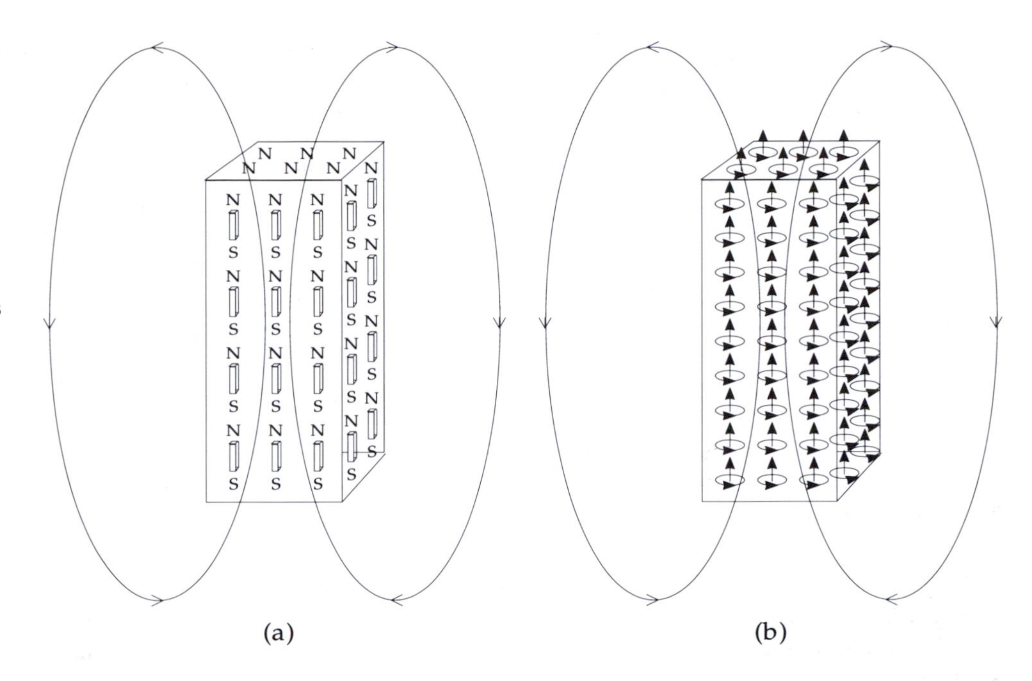

Fig. 5.8 The magnetization of a material may be envisaged as due to an alignment of (a) small dipoles or (b) equivalent current loops; even in a permanent magnet the physical source of the **B**-field of the material is a system of electrical currents on an atomic scale.

5.2.6 The magnetic properties of materials

The magnetic behavior of a solid depends on the magnetic moments of the atoms or ions it contains. As discussed above, atomic and ionic magnetic moments are proportional to the quantized angular momenta associated with the orbital motion of electrons about the nucleus and with the spins of the electrons about their own axes of rotation. In quantum theory the *exclusion principle* of Wolfgang Pauli states that no two electrons in a given system can have the same set of quantum numbers. When applied to an atom or ion, Pauli's principle stipulates that each possible electron orbit can be occupied by up to two electrons with opposite spins. The orbits are arranged in shells around the nucleus. The magnetic moments of paired opposite spins cancel each other out. Consequently, the net angular momentum and the net magnetic moment of a filled shell must be zero. The net magnetic moment of an atom or ion arises from incompletely filled shells that contain unpaired spins. The atoms or ions in a solid are not randomly distributed but occupy fixed positions in a regular lattice, which reflects the symmetry of the crystalline structure and which controls interactions between the ions. Hence, the different types of magnetic behavior observed in solids depend not only on the presence of ions with unpaired spins, but also on the lattice symmetry and cell size.

Three main classes of magnetic behavior can be distinguished on the basis of magnetic susceptibility: *diamagnetism*, *paramagnetism* and *ferromagnetism*. In diamagnetic materials the susceptibility is low and negative, i.e., a magnetization develops in the opposite direction to the applied field. Paramagnetic materials have low, positive susceptibilities. Ferromagnetic materials can be subdivided into three categories. True ferromagnetism is a cooperative phenomenon observed in metals like iron, nickel and cobalt, in which the lattice geometry and spacing allows the exchange of electrons between neighboring atoms. This gives rise to a *molecular field* by means of which the magnetic moments of adjacent atoms reinforce their mutual alignment parallel to a common direction. Ferromagnetic behavior is characterized by high positive susceptibilities and strong magnetic properties. The crystal structures of certain minerals permit an indirect cooperative interaction between atomic magnetic moments. This *indirect exchange* confers magnetic properties that are similar to ferromagnetism. The mineral may display *antiferromagnetism* or *ferrimagnetism*. The small group of ferrimagnetic minerals is geophysically important, especially in connection with the analysis of the Earth's paleomagnetic field.

5.2.6.1 *Diamagnetism*

All magnetic materials show a diamagnetic reaction in a magnetic field. The diamagnetism is often masked, because it is overlain by stronger paramagnetic or ferromagnetic properties. It is characteristically observable in materials in which all electron spins are paired.

The Lorentz law (Eq. (5.10)) shows that a change in the **B**-field alters the force experienced by an orbiting electron. The plane of the electron orbit is compelled to precess around the field direction; the phenomenon is called *Larmor*

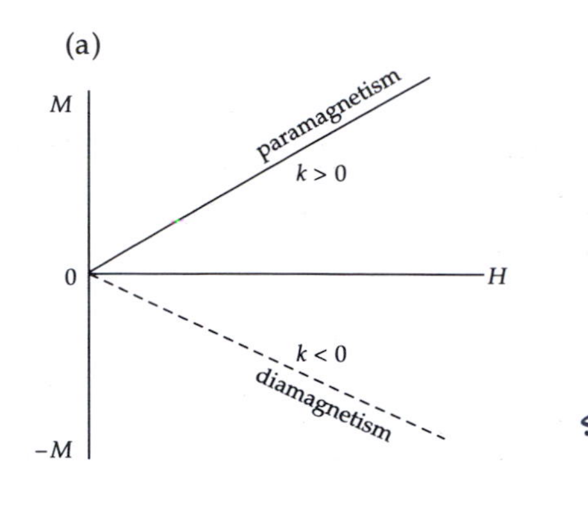

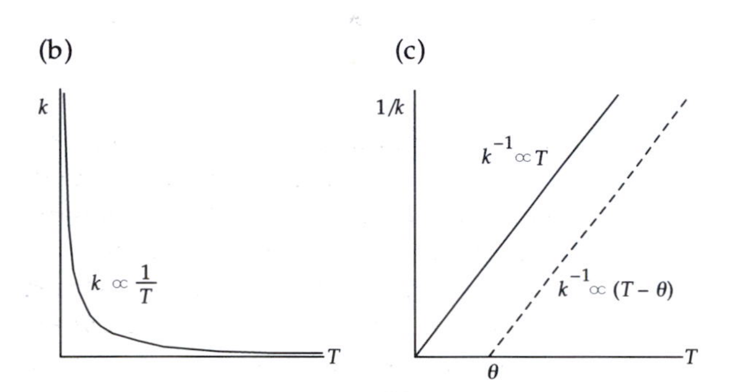

Fig. 5.9 (a) Variations of magnetization **M** with applied magnetic field **H** in paramagnetic and diamagnetic materials; (b) the variation of paramagnetic susceptibility with temperature, and (c) the linear plot of the inverse of paramagnetic susceptibility against temperature.

precession. It represents an additional component of rotation and angular momentum. The sense of the rotation is opposite to that of the orbital rotation about the nucleus. Hence, the magnetic moment associated with the Larmor precession opposes the applied field. As a result a weak magnetization proportional to the field strength is induced in the opposite direction to the field. The magnetization vanishes when the applied magnetic field is removed. Diamagnetic susceptibility is reversible, weak and negative (Fig. 5.9a); it is independent of temperature. Many important rock-forming minerals belong to this class, amongst them quartz and calcite. They have susceptibilities around -10^{-6} in SI units.

5.2.6.2 *Paramagnetism*

Paramagnetism is a statistical phenomenon. When one or more electron spins is unpaired, the net magnetic moment of an atom or ion is no longer zero. The resultant magnetic moment can align with a magnetic field. The alignment is opposed by thermal energy which favors chaotic orientations of the spin magnetic moments. The magnetic energy is small compared to the thermal energy, and in the absence of a magnetic field the magnetic moments are oriented randomly. When a magnetic field is applied, the chaotic alignment of magnetic moments is biassed towards the field direction. A magnetization is induced proportional to the strength of the applied field and parallel to its direction. The susceptibility is reversible, small and positive (Fig. 5.9a). An important paramagnetic characteristic is that the susceptibility k varies inversely with temperature (Fig. 5.9b) as given by the *Curie law*

$$k=\frac{C}{T} \tag{5.19}$$

where the constant C is characteristic of the material. Thus, a plot of $1/k$ against temperature is a straight line (Fig. 5.9c). In solids and liquids mutual interactions between ions may be quite strong and paramagnetic behavior is only displayed when the thermal energy exceeds a threshold value. The temperature above which a solid is paramagnetic is called the *paramagnetic Curie temperature* or *Weiss constant* of the material, denoted by θ; it is close to zero kelvins in *paramagnetic* solids. At temperatures $T>\theta$ the paramagnetic susceptibility k is given by the *Curie–Weiss law*

$$k=\frac{C}{T-\theta} \tag{5.20}$$

For a solid the plot of $1/k$ against $(T-\theta)$ is a straight line (Fig. 5.9c). Many clay minerals and other rock-forming minerals (e.g., chlorite, amphibole, pyroxene, olivine) are paramagnetic at room temperature, with susceptibilities commonly around 10^{-5}–10^{-4} in SI units.

5.2.6.3 *Ferromagnetism*

In paramagnetic and diamagnetic materials the interactions between individual atomic magnetic moments are small and often negligible. However, in some metals (e.g., iron, nickel, cobalt) the atoms occupy lattice positions that are close enough to allow the exchange of electrons between neighboring atoms. The exchange is a quantum-mechanical effect that involves a large amount of energy, called the *exchange energy* of the metal. The exchange interaction produces a very strong *molecular field* within the metal, which aligns the atomic magnetic moments (Fig. 5.10a) exactly parallel and produces a *spontaneous magnetization* (M_s). The magnetic moments react in unison to a magnetic field, giving rise to a class of strong magnetic behavior known as *ferromagnetism*.

A rock sample may contain thousands of tiny ferromagnetic mineral grains. The magnetization loop of a rock sample shows the effects of *magnetic hysteresis* (Fig. 5.11). In strong fields the magnetization reaches a saturation value

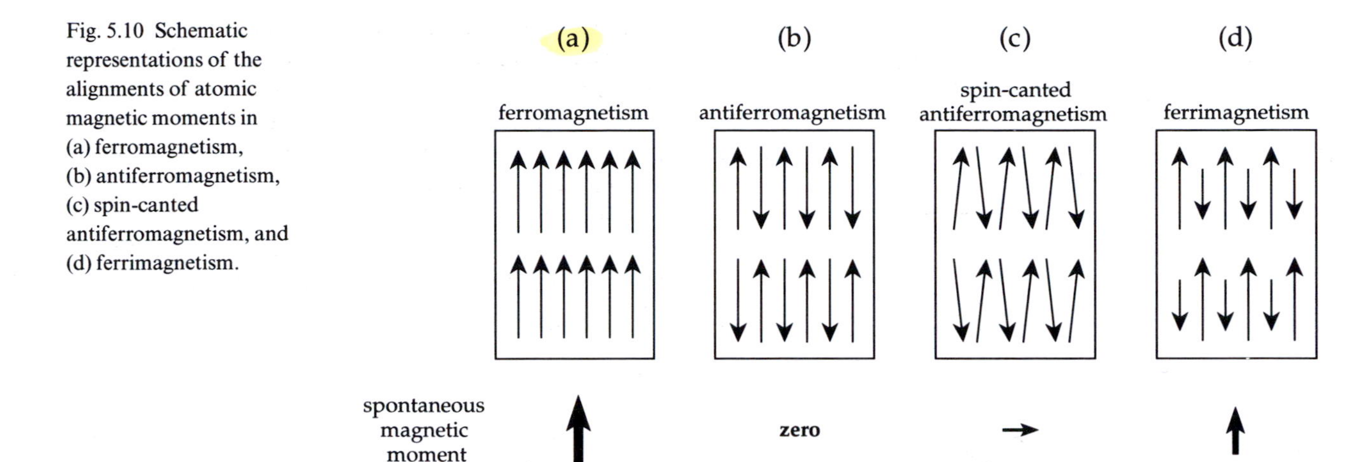

Fig. 5.10 Schematic representations of the alignments of atomic magnetic moments in (a) ferromagnetism, (b) antiferromagnetism, (c) spin-canted antiferromagnetism, and (d) ferrimagnetism.

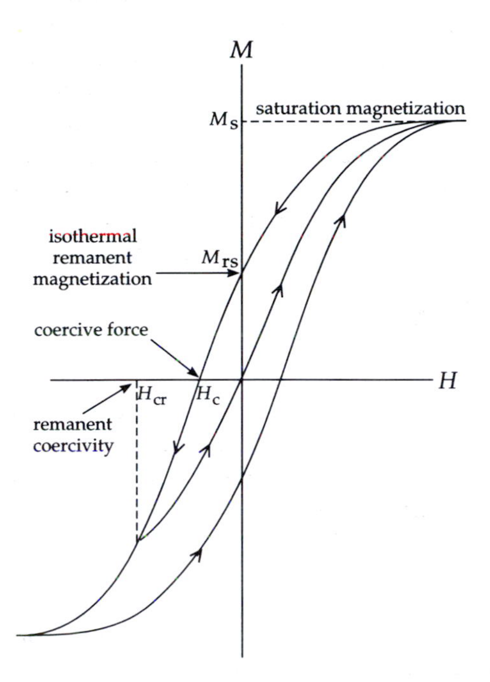

Fig. 5.11 The magnetization loop of an arbitrary ferromagnetic material.

(equal to M_s), at which the individual magnetic moments are aligned with the applied field. If the magnetizing field is reduced to zero, a ferromagnetic material retains part of the induced magnetization. The residual magnetization is called the *remanence*, or *isothermal remanent magnetization* (IRM); if the sample was magnetized to saturation, the remanence is a saturation IRM (M_{rs}). For a given ferromagnetic mineral, the ratio M_{rs}/M_s depends on grain size. If a magnetic field is applied in the opposite direction to the IRM, it remagnetizes part of the material in the antiparallel direction. For a particular value H_c of the reverse field (called the *coercive force*) the induced reverse magnetization exactly cancels the original remanence and the net magnetization is zero. If the reverse field is removed at this stage, the residual remanence is smaller than the original IRM. By repeating the process in ever stronger reverse fields a back-field H_{cr} (called the *coercivity of remanence*) is found which gives a reverse remanence that exactly cancels the IRM, so that the residual remanence is zero. The ratio H_{cr}/H_c also depends on grain-size. Rock-forming magnetic minerals often have natural remanences with very high coercive properties.

When a ferromagnetic material is heated, its spontaneous magnetization disappears at the *ferromagnetic Curie temperature* (T_c). At temperatures higher than the *paramagnetic Curie temperature* (θ) the susceptibility k becomes the paramagnetic susceptibility, so that $1/k$ is proportional to $(T-\theta)$ as given by the Curie–Weiss law (Eq. (5.20)). The paramagnetic Curie temperature for a ferromagnetic solid is several degrees higher than the ferromagnetic Curie temperature, T_c. The gradual transition from ferromagnetic to paramagnetic behavior is explained by persistence of the molecular field due to short-range magnetic order above T_c.

5.2.6.4 *Antiferromagnetism*

In oxide crystals the oxygen ions usually keep the metal ions far apart, so that direct exchange of electrons between the metal ions is not possible. However, in certain minerals interaction between magnetic spins becomes possible by the exchange of electrons from one metal ion to the other through the electron 'cloud' of the oxygen ion. This *indirect exchange* (or *superexchange*) process results in antiparallel directions of adjacent atomic magnetic moments (Fig. 5.10b), giving two sublattices with equal and opposite intrinsic magnetic moments. As a result, the susceptibility of an antiferromagnetic crystal is weak and positive, and remanent magnetization is not possible. The antiferromagnetic alignment breaks down at the *Néel temperature*, above which paramagnetic behavior is shown. The Néel temperature T_N of many antiferromagnetic substances is lower than room temperature, at which they are paramagnetic. A common example of an antiferromagnetic mineral is ilmenite ($FeTiO_3$), which has a Néel temperature of 50 K.

5.2.6.5 *Parasitic ferromagnetism*

When an antiferromagnetic crystal contains defects, vacancies or impurities, some of the antiparallel spins are unpaired. A weak 'defect moment' can result due to these lattice imperfections. Also if the spins are not exactly antiparallel but are inclined at a small angle, they do not cancel out completely and again a ferromagnetic type of magnetization can result (Fig. 5.10c). Materials that exhibit this form of *parasitic ferromagnetism* have the typical characteristics of a true ferromagnetic metal, including hysteresis, a spontaneous magnetization and a Curie temperature. An important geological example is the common iron mineral hematite (α-Fe_2O_3), in which both the spin-canted and defect moments contribute to the ferromagnetic properties. Hematite has a variable, weak spontaneous magnetization of about 2000 A m^{-1}, very high coercivity and a Curie temperature around 675 °C. The variable magnetic properties are due to variation in the relative importances of the defect and spin-canted moments.

5.2.6.6 *Ferrimagnetism*

The metallic ions in an antiferromagnet occupy the voids between the oxygen ions. In certain crystal structures, of which the most important geological example is the spinel structure, the sites of the metal ions differ from each other in the coordination of the surrounding oxygen ions. Tetrahedral sites have four oxygen ions as nearest neighbors and octahedral sites have six. The tetrahedal and octahedral sites form two sublattices. In a normal spinel the tetrahedral sites are occupied by divalent ions and the octahedral sites by Fe^{3+} ions. The most common iron oxide minerals have an *inverse spinel* structure. Each sublattice has an equal number of Fe^{3+} ions. The same number of divalent ions (e.g. Fe^{2+}) occupy other octahedral sites, while the corresponding number of tetrahedral sites is empty.

When the indirect exchange process involves antiparallel and unequal magnetizations of the sublattices (Fig. 5.10d), resulting in a net spontaneous magnetization, the phenomenon is called *ferrimagnetism*. Ferrimagnetic materials (called *ferrites*) exhibit magnetic hysteresis and retain a remanent magnetization when they are removed from a magnetizing field. Above a given temperature – sometimes called the ferrimagnetic Néel temperature but more commonly the Curie temperature – the long-range molecular order breaks down and the mineral behaves paramagnetically. The most important ferrimagnetic mineral is magnetite (Fe_3O_4), but maghemite, pyrrhotite and goethite are also significant contributors to the magnetic properties of rocks.

5.2.7 **Magnetic anisotropy**

Anisotropy is the directional dependency of a property. The magnetism of metals and crystals is determined by the strengths of the magnetic moments associated with atoms or ions, and the distances between neighbors. Here the symmetry of the lattice plays an important role, and so the magnetic properties of most ferromagnetic materials depend on direction. Magnetic anisotropy is an important factor in the grain size-dependence of the magnetic behavior of rocks and minerals. There are three important types: magnetocrystalline, magnetostatic and magnetostrictive anisotropies.

5.2.7.1 *Magnetocrystalline anisotropy*

The direction of the spontaneous magnetization (M_s) in a ferromagnetic metal is not arbitrary. The molecular field that produce M_s originates in the direct exchange of electron spins between neighboring atoms in a metal. The symmetry of the lattice structure of the metal affects the exchange process and gives rise to a *magnetocrystalline anisotropy energy*, which has a minimum value when M_s is parallel to a favored direction referred to as the *easy axis* (or *easy direction*) of magnetization. The simplest form of magnetic anisotropy is uniaxial anisotropy, when a metal has only a single easy axis. For example, cobalt has a hexagonal structure and the easy direction is parallel to the *c*-axis at room temperature. Iron and nickel have cubic unit cells; at room temperature, the easy axes in iron are the edges of the cube, but the easy axes in nickel are the body diagonal directions.

Magnetocrystalline anisotropy is also exhibited by ferrites, including the geologically important ferrimagnetic minerals. The exchange process in a ferrite is indirect, but energetically preferred easy axes of magnetization arise that reflect the symmetry of the crystal structure. This gives rise to different forms of the anisotropy in hematite and magnetite.

Hematite has a rhombohedral or hexagonal structure and a *uniaxial anisotropy* with regard to the *c*-axis of symmetry. Oxygen ions form a close-packed hexagonal lattice in which two-thirds of the octahedral interstices are occupied by ferric (Fe^{3+}) ions. When the spontaneous magnetization makes an angle ϕ with the *c*-axis of the crystal, the uniaxial anisotropy energy density can be written to first order

$$E_a = K_u \sin^2\phi \tag{5.21}$$

K_u is called the uniaxial magnetocrystalline anisotropy constant. Its value in hematite is around -10^3 J m^{-3} at room temperature. The negative value of K_u in Eq. (5.21) means that E_a decreases as the angle ϕ increases, and is minimum when $\sin^2\phi$ is maximum, i.e., when ϕ is 90°. As a result, the spontaneous magnetization lies in the basal plane of the hematite crystal at room temperature.

Because of its inverse spinel structure, magnetite has *cubic anisotropy*. Let the direction of the spontaneous magnetization be given by direction cosines α_1, α_2 and α_3 relative to the edges of the cubic unit cell. The magnetocrystalline anisotropy energy density is then given by

$$E_a = K_1(\alpha_1^2\alpha_2^2 + \alpha_2^2\alpha_3^2 + \alpha_3^2\alpha_1^2) + K_2\alpha_1^2\alpha_2^2\alpha_3^2 \tag{5.22}$$

The anisotropy constants K_1 and K_2 of magnetite are equal to -1.36×10^4 J m^{-3} and -0.44×10^4 J m^{-3}, respectively, at room temperature. Because these constants are negative, the anisotropy energy density E_a is minimum when the spontaneous magnetization is along a [111] body diagonal, which is the magnetocrystalline easy axis of magnetization at room temperature.

5.2.7.2 *Magnetostatic (shape) anisotropy*

In strongly magnetic materials the shape of the magnetized object causes a *magnetostatic anisotropy*. In rocks this effect is associated with the shapes of the individual grains of ferrimagnetic mineral in the rock, and to a lesser extent with the shape of the rock sample. The anisotropy is magnetostatic in origin and can be conveniently explained with the aid of the concept of magnetic poles.

The spontaneous magnetization of a uniformly magnetized material can be pictured as giving rise to a distribution of poles on the free end surfaces (Fig. 5.12). As noted above, a property of the magnetic **B**-field is that its field lines form closed loops, whereas the **H**-field begins and ends on boundary surfaces, at which it is discontinuous. The field lines of the magnetic field **H** outside a magnet are parallel to the B-field and are directed from the surface distribution of N-poles to the distribution of S-poles. In the absence of an externally applied field the **H**-field inside the magnet is also directed from the distribution of N-poles on one end to the S-poles on the other end. It forms a *demagnetizing field* ($\mathbf{H}_d$) that opposes the magnetization. The strength of the demagnetizing field varies directly with the surface density of the magnetic pole distribution on the end surfaces of the magnet, and inversely with the distance between these surfaces. $\mathbf{H}_d$ thus depends on the shape of the magnet and the intensity of magnetization; it can be written

$$\mathbf{H}_d = -N\mathbf{M} \tag{5.23}$$

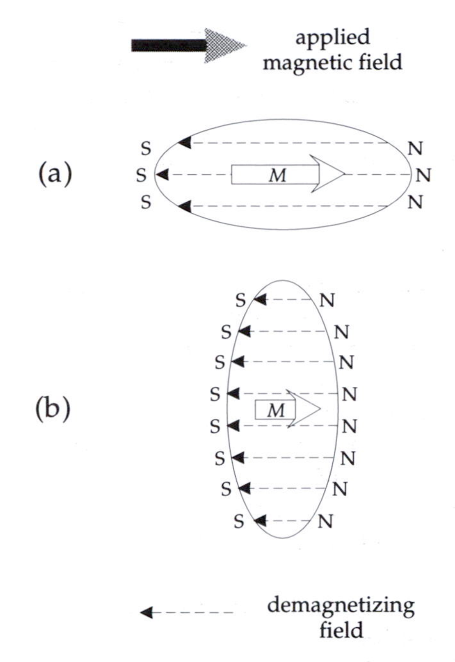

Fig. 5.12 The origin of shape anisotropy: the distributions of magnetic poles on surfaces that intersect the magnetization of a uniformly magnetized prolate ellipsoid produce internal demagnetizing fields; these are weak parallel to the long axis (a) and strong parallel to the short axis (b). As a result, the net magnetization is stronger parallel to the long axis than parallel to the short axis.

N is called the *demagnetizing factor*. It is a dimensionless constant determined by the shape of the magnetic grain. It can be computed for a geometrical shape, such as a triaxial ellipsoid. The demagnetizing factors N_1, N_2 and N_3 parallel to the symmetry axes of an ellipsoid satisfy the relationship

$$N_1 + N_2 + N_3 = 1 \tag{5.24}$$

The magnetostatic energy of the interaction of the grain magnetization with the demagnetizing field is called the demagnetizing energy (E_d). For a grain with uniform magnetization M in a direction with demagnetizing factor N,

$$E_d = \frac{\mu_0}{2}NM^2 \tag{5.25}$$

Consider the shape anisotropy of a small grain shaped like a prolate ellipsoid. When the spontaneous magnetization M_s is along the long axis of the ellipsoid (Fig. 5.12a), the opposing pole distributions are further away from each other and their surface density is lower than when M_s is parallel to the short axis (Fig. 5.12b). The demagnetizing field and energy are smallest when M_s is parallel to the long axis, which is the energetically favored direction of magnetization. The demagnetizing energy is larger in any other direction, giving a *shape anisotropy*. If N_1 is the demagnetizing factor for the long axis and N_2 that for the short axis ($N_1 < N_2$), the difference in energy between the two directions of magnetization defines a magnetostatic anisotropy energy density given by

$$E_a = \frac{\mu_0}{2}(N_2 - N_1)M_s^2 \sin^2\theta \tag{5.26}$$

which is minimum when M_s is parallel to the longest dimension of the grain ($\theta = 0$).

Shape-dependent magnetic anisotropy is important in minerals that have a high spontaneous magnetization. The more elongate the grain is, the higher the shape anisotropy will be. It is the predominant form of anisotropy in very fine grains of magnetite (and maghemite) if the longest axis exceeds the shortest axis by only about 20%.

5.2.7.3 *Magnetostrictive anisotropy*

The process of magnetizing some materials causes them to change shape. Within the crystal lattice the interaction energy between atomic magnetic moments depends on their separations (called the bond length) and on their orientations, i.e., on the direction of magnetization. If an applied field changes the orientations of the atomic magnetic moments so that the interaction energy is increased, the bond lengths adjust to reduce the total energy. This produces strains which result in a shape change of the ferromagnetic specimen. This phenomenon is called *magnetostriction*. A material which elongates in the direction of magnetization exhibits positive magnetostriction, while a material that shortens parallel to the magnetization shows negative magnetostriction. The maximum difference in magnetoelastic strain, which occurs between the demagnetized state and that of saturation magnetization, is called the *saturation magnetostriction*, denoted λ_s.

The inverse effect is also possible. For example, if pressure is applied to one face of a cubic crystal, it will shorten elastically along the direction of the applied stress and will expand in directions perpendicular to it. These strains alter the separations of atomic magnetic moments, thereby perturbing the effects that give rise to magnetocrystalline anisotropy. Thus the application of stress to a magnetic material can change its magnetization; the effect is called *piezomagnetism*. On a submicroscopic scale the stress field that surrounds a vacancy, defect or dislocation in the crystal structure can locally affect the orientations of ionic magnetic spins.

Magnetostriction is a further source of anisotropy in magnetic minerals. The magnetostrictive (or magnetoelastic) anisotropy energy density E_a depends on the amount and direction of the stress σ. If the saturation magnetization makes an angle θ to the stress, E_a is given for a uniaxial magnetic mineral by

$$E_a = \tfrac{3}{2}\lambda_s \sigma \cos^2\theta \tag{5.27}$$

This is the simplest expression for magnetostrictive energy. It assumes that the magnetostriction is isotropic i.e., that it has the same value in all directions. This condition is fulfilled if the magnetocrystalline axes of the ferrimagnetic minerals in a rock are randomly distributed. The magnetoelastic energy of a cubic mineral is more complicated. Instead of a single magnetostriction constant λ_s separate constants λ_{100} and λ_{111} are required for the saturation magnetostriction along the [100] and [111] directions, respectively, corresponding to the edge and body diagonal directions of the cubic unit cell.

In magnetite the magnetoelastic energy is more than an order of magnitude less than the magnetocrystalline energy at room temperature. Consequently, magnetostriction plays only a secondary role in determining the direction of magnetization of magnetite grains. However, in minerals that have high magnetostriction (e.g. titanomagnetites (see §5.3.2.1) with a compositional factor $x \approx 0.65$) the magnetoelastic energy may be significant in determining easy directions of magnetization, and the magnetization may be sensitive to modification by deformation.

5.3 ROCK MAGNETISM

5.3.1 The magnetic properties of rocks

A rock may be regarded as a heterogeneous assemblage of minerals. The matrix minerals are mainly silicates or carbonates, which are diamagnetic in character. Interspersed in this matrix is a lesser quantity of secondary minerals (such as the clay minerals) that have paramagnetic properties. The bulk of the constituent minerals in a rock contribute to the magnetic susceptibility but are incapable of any contribution to the remanent magnetic properties, which are due to a dilute dispersion of ferrimagnetic minerals (e.g., commonly less than 0.01% in a limestone). The variable concentrations of ferrimagnetic and matrix minerals result in a wide range of susceptibilities in rocks (Fig. 5.13).

The weak and variable concentration of ferrimagnetic minerals plays a key role in determining the magnetic properties of the rock that are significant geologically and geophysically. The most important factors influencing rock magnetism are the type of ferrimagnetic mineral, its grain size, and the manner in which it acquires a remanent magnetization.

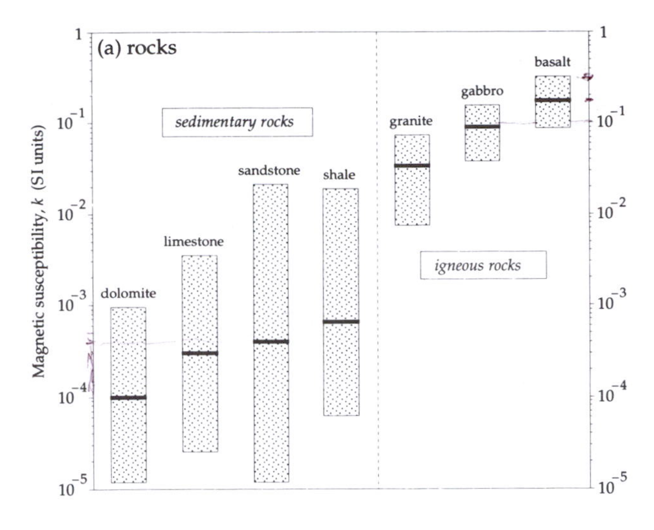

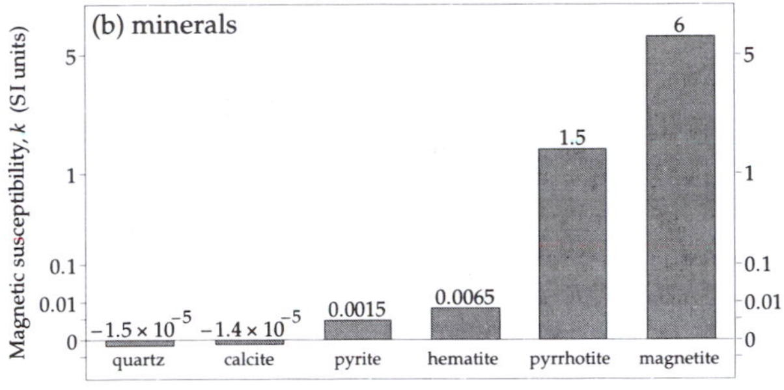

Fig. 5.13 (a) Median values and ranges of the magnetic susceptibility of some common rock types, and (b) the susceptibilities of some important minerals.

5.3.2 The ternary oxide system of magnetic minerals

The most important magnetic minerals are iron–titanium oxides, which are naturally occurring ferrites. The mineral structure consists of a close-packed lattice of oxygen ions, in which some of the interstitial spaces are occupied by regular arrays of ferrous (Fe^{2+}) and ferric (Fe^{3+}) iron ions and titanium (Ti^{4+}) ions. The relative proportions of these three ions determine the ferrimagnetic properties of the mineral. The composition of an iron–titanium oxide mineral can be illustrated graphically on the ternary oxide diagram (Fig. 5.14), the corners of which represent the minerals rutile (TiO_2), wustite (FeO), and hematite (Fe_2O_3). The proportions of these three oxides in a mineral define a point on the ternary diagram. The vertical distance of the point above the FeO–Fe_2O_3 baseline reflects the amount of titanium in the lattice. Hematite is in a higher state of oxidation than wustite; hence the horizontal position along the FeO–Fe_2O_3 axis expresses the degree of oxidation.

The most important magnetic minerals belong to two solid-solution series: (a) the titanomagnetite, and (b) the

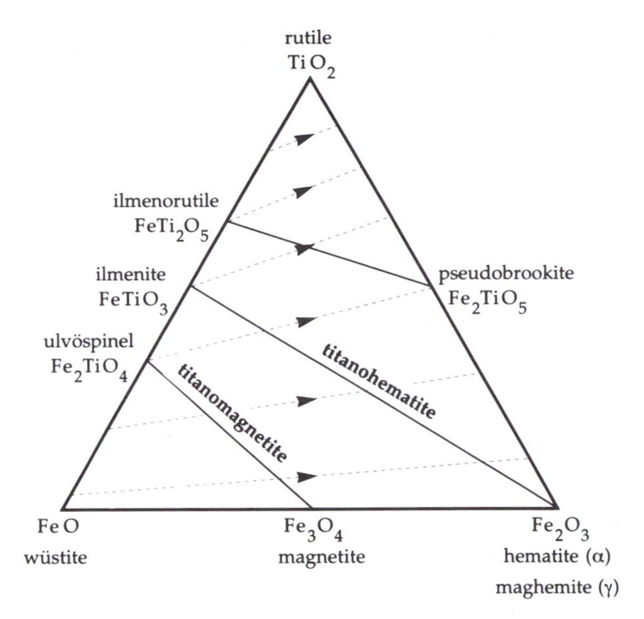

Fig. 5.14 Ternary compositional diagram of the iron–titanium oxide solid solution magnetic minerals.

titanohematite series. The minerals of a third series, pseudobrookite, are paramagnetic at room temperature. They are quite rare and are of minor importance in rock magnetism. The compositions of naturally occurring forms of titanomagnetite and titanohematite usually plot as points on the ternary diagram that are displaced from the ideal lines towards the TiO_2–Fe_2O_3 axis, which indicates that they are partly oxidized.

5.3.2.1 *The titanomagnetite series*

Titanomagnetite is the name of the family of iron oxide minerals described by the general formula $Fe_{3-x}Ti_xO_4$ ($0 \le x \le 1$). These minerals have an inverse spinel structure and exemplify a solid-solution series in which ionic replacement of two Fe^{3+} ions by one Fe^{2+} and one Ti^{4+} ion can take place. The compositional parameter x expresses the relative proportion of titanium in the unit cell. The end members of the solid-solution series are magnetite (Fe_3O_4), which is a typical strongly magnetic ferrite, and ulvöspinel (Fe_2TiO_4), which is antiferromagnetic at very low temperature but is paramagnetic at room temperature. An alternative form of the general formula is $x\,Fe_2TiO_4 \cdot (1-x)\,Fe_3O_4$. Written in this way, it is apparent that the compositional parameter x describes the molecular fraction of ulvöspinel. As the amount of titanium increases, the cell size increases and the Curie temperature θ and spontaneous magnetization M_s of the titanomagnetite decrease (Fig. 5.15).

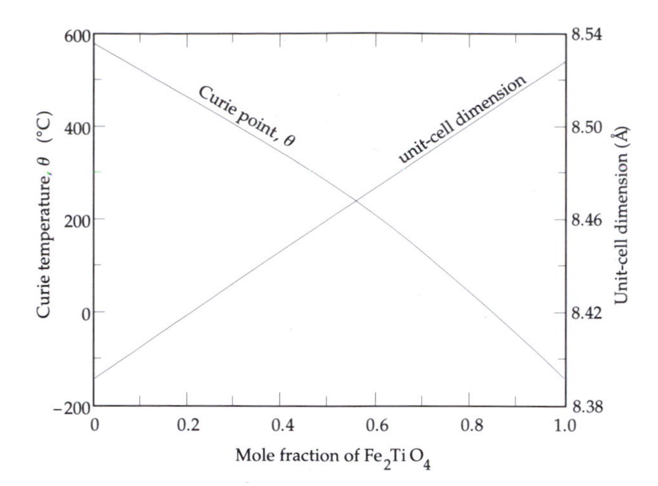

Fig. 5.15 Variations of Curie temperature and unit-cell size with composition in titanomagnetite (after Nagata, 1961).

Magnetite is one of the most important ferrimagnetic minerals. It has a strong spontaneous magnetization (M_s=4.8×10^5 A m^{-1}) and a Curie temperature of 578 °C. Because of the high value of M_s, magnetite grains can have a strong shape anisotropy. The magnetic susceptibility is the strongest of any naturally occurring mineral (k≈1–10 SI). For many sedimentary and igneous rocks the magnetic susceptibility is proportional to the magnetite content.

Maghemite (γ-Fe_2O_3) can be produced by low-temperature oxidation of magnetite. It is a strongly magnetic mineral (M_s≈4.5×10^5 A m^{-1}). Experiments on maghemite doped with small amounts of foreign ions indicate that it has a Curie temperature of 675 °C. However, it is metastable and reverts to hematite (α-Fe_2O_3) when heated above 300–350 °C. The low temperature oxidation of titanomagnetite leads to a 'titanomaghemite' solid-solution series.

Titanomagnetite is responsible for the magnetic properties of oceanic basalts. The basaltic layer of the oceanic crust is the main origin of the marine magnetic anomalies that are of vital importance to modern plate tectonic theory. The magnetic properties of the 0.5 km thick basaltic layer are due to the presence of very fine grained titanomagnetite (or titanomaghemite, depending on the degree of ocean-floor weathering). The molecular fraction (x) of Fe_2TiO_4 in titanomagnetite in oceanic basalts is commonly around 0.6.

5.3.2.2 *The titanohematite series*

The minerals of the titanohematite solid-solution series are also variously referred to as 'hemoilmenite', 'hematite-ilmenite' or 'ilmenohematite'. They have the general formula $Fe_{2-x}Ti_xO_3$. The unit cell has rhombohedral symmetry. Ionic substitution is the same as for titanomagnetite, and the compositional parameter x has the same implications for the titanium content of the unit cell. The end members of the solid-solution series are hematite (Fe_2O_3) and ilmenite ($FeTiO_3$). The chemical formula can be written in the alternative form x $FeTiO_3 \cdot (1-x)$ Fe_2O_3, where x represents the molecular fraction of ilmenite. As in the case of titanomagnetite, the cell size increases and the Curie point decreases as the titanium content increases. The Curie point of hematite is 675 °C, while ilmenite is antiferromagnetic at low temperature and paramagnetic at room temperature. For titanium contents $0.5<x<0.95$ titanohematite is ferrimagnetic and for $x<0.5$ it exhibits parasitic ferromagnetism.

The end member hematite (α-Fe_2O_3) is an extremely important magnetic mineral. Its magnetic properties arise from parasitic ferromagnetism due to the spin-canted magnetic moment and the possible defect moment of its otherwise antiferromagnetic lattice. Hematite has a weak spontaneous magnetization (M_s≈2.2×10^3 A m^{-1}) and a strong magnetocrystalline anisotropy (K_u≈10^3 J m^{-3}). Hematite is paleomagnetically important because of its common occurrence and its high magnetic and chemical stability. It often occurs as a secondary mineral, formed by oxidation of a precursor mineral, such as magnetite, or by precipitation from fluids passing through a rock.

5.3.3 **Other ferrimagnetic minerals**

Although the iron–titanium oxides are the dominant magnetic minerals, rocks frequently contain other minerals with ferromagnetic properties. Although pyrite (FeS_2) is a very common sulfide mineral, especially in sedimentary rocks, it is paramagnetic and therefore cannot carry a remanent magnetization. As a result it does not contribute directly to the paleomagnetic properties of rocks, but it may act as a source for the formation of goethite or secondary magnetite.

Pyrrhotite is a common sulfide mineral which can form authigenically or during diagenesis in sediments, and which can be ferrimagnetic in certain compositional ranges. It is non-stoichiometric (i.e., the numbers of anions and cations in the unit cell are unequal) and has the formula $Fe_{1-x}S$. The parameter x refers to the proportion of vacancies among the cation lattice sites and is limited to the range $0<x<0.14$. Pyrrhotite has a pseudohexagonal crystal structure and would be antiferromagnetic but for the presence of the cation vacancies. The Néel temperature at which the fundamental antiferromagnetism disappears is around 320 °C. Pyrrhotite with the formula Fe_7S_8 is ferrimagnetic with a Curie temperature close to the Néel temperature and a strong spontaneous magnetization of about 10^5 A m^{-1} at room temperature. The magnetocrystalline anisotropy restricts the easy axis of magnetization to the hexagonal basal plane at room temperature.

The iron oxyhydroxide goethite (FeOOH) is another common authigenic mineral in sediments. Like hematite, goethite is antiferromagnetic, but has a weak parasitic ferromagnetism. It has a very high coercivity (with maximum values in excess of 5 T) and a low Curie point around 100 °C or less. It is thermally unstable relative to hematite under most natural conditions, and decomposes on heating above about 350 °C. It is a common (and paleomagnetically undesirable) secondary mineral in limestones and other sedimentary rocks.

5.3.4 Identification of ferrimagnetic minerals

It is often difficult to identify the ferrimagnetic minerals in a rock, because their concentration is so low, especially in sedimentary rocks. If the rock is coarse grained, ferrimagnetic minerals may be identified optically among the opaque grains by studying polished sections in reflected light. However, in many rocks that are paleomagnetically important (e.g., basaltic lava, pelagic limestone) optical examination may be unable to resolve the very fine grain size of the ferrimagnetic mineral. Identification of the ferrimagnetic mineral fraction may then be made by means of their properties of Curie temperature and coercivity.

The Curie temperature is measured using a form of balance in which a strong magnetic field gradient exerts a force on the sample that is proportional to its magnetization. The field (usually 0.4–1 T) is strong enough to saturate the magnetization of many minerals. The sample is heated and the change of magnetic force (i.e., sample magnetization) is observed with increasing temperature. When the Curie point is reached, the ferromagnetic behavior disappears; at higher temperatures the sample is paramagnetic. The Curie point is diagnostic of many minerals. For example, an extract of magnetic minerals from a pelagic limestone (Fig. 5.16) shows the presence of goethite (~100 °C Curie point, sample SR 3A), magnetite (~570 °C Curie point, all samples) and hematite (~650 °C Curie point, sample SR 11). Some Curie balances are sensitive enough to analyse whole rock samples, but this is generally only possible in strongly magnetized igneous rocks. For most rocks it is necessary to extract the ferrimagnetic minerals, or to concentrate them. The extraction is sometimes difficult, and often it is not certain that the extract is representative of the rock as a whole. This is also a drawback of optical methods, which only allow description of large grains.

To avoid the difficulties and uncertainties associated with making a special extract or concentrate of the ferrimagnetic

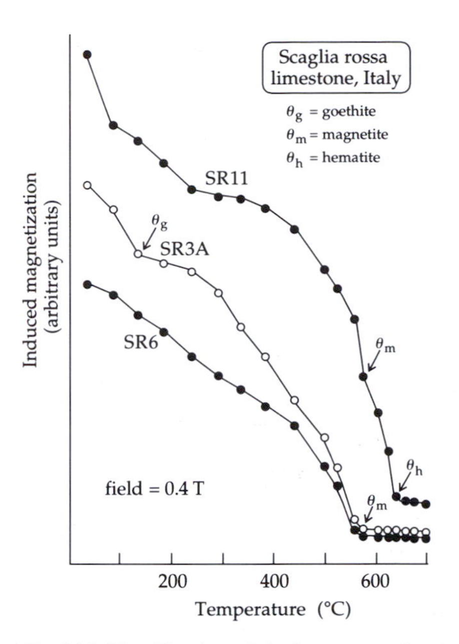

Fig. 5.16 Identification of the ferromagnetic minerals in a pelagic limestone by determination of their Curie temperatures in concentrated extracts (after Lowrie and Alvarez, 1975).

fraction, alternative methods of ferrimagnetic mineral identification, based on bulk magnetic properties, are more widely used. One simple method makes use of the coercivities and Curie temperatures, as expressed in the thermal demagnetization of the isothermal remanent magnetization (see §5.3.6.4). Another method, useful for pure magnetite and hematite, takes advantage of the low-temperature variations of the magnetocrystalline anisotropy constants of these minerals.

5.3.5 Grain size dependence of ferrimagnetic properties

The ferromagnetic properties of metals and ferrites vary sensitively with grain size. Consider an assemblage of uniformly magnetized grains of a ferrimagnetic mineral characterized by a spontaneous magnetization M_s and uniform grain volume V. Let the spontaneous magnetization be oriented parallel to an easy direction of magnetization (crystalline or shape-determined), defined by the anisotropy energy K_u per unit volume. The energy that keeps the magnetization parallel to the easy direction is equal to VK_u. Thermal energy, proportional to the temperature, has the effect of disturbing this alignment. At temperature T the thermal energy of a grain is equal to kT, where k is Boltzmann's constant (k=1.381×10^{-23} J K^{-1}). At any instant in time there is a chance that thermal energy will deflect the magnetic moment of a grain away from its easy direction. Progressively, the net magnetization of the material (the sum of all the magnetic moments of the numerous magnetic grains) will be randomized by the thermal energy, and the magnetization will be observed to decay. If the initial magnetization of the assemblage is M_{r0}, after time t it will decrease exponentially to $M_r(t)$, according to

$$M_r(t) = M_{r0} \exp\left(-\frac{t}{\tau}\right) \tag{5.28}$$

In this equation τ is known as the *relaxation time* of the grain. If the relaxation time is long, the exponential decrease in Eq. (5.28) is slow and the magnetization is stable. The parameter τ depends on properties of the grain and is given by the equation

$$\tau = \frac{1}{\nu_0} \exp\left(\frac{VK_u}{kT}\right) \tag{5.29}$$

The constant ν_0 is related to the lattice vibrational frequency and has a very large value ($\approx 10^8 - 10^{10}$ s^{-1}). The value of K_u depends on whether the easy direction of the magnetic mineral is determined by the magnetocrystalline anisotropy or the magnetostatic (shape) anisotropy. For example, in hematite the magnetocrystalline anisotropy prevails because the spontaneous magnetization is very weak, and K_u is equal to the magnetocrystalline anisotropy. In grains of magnetite the value of K_u is equal to $K_1/3$, if magnetocrystalline anisotropy K_1 controls the magnetization (as in an equidimensional grain); if the magnetite grain is elongate with demagnetizing factors N_1 and N_2, shape anisotropy determines K_u, which is then equal to the energy density E_a given by Eq. (5.26).

This theory applies only to very small grains that are uniformly magnetized. Fine grained ferrimagnetic minerals are, however, very important in paleomagnetism and rock magnetism. The very finest grains, smaller than a critical size, exhibit an unstable type of magnetic behavior called *superparamagnetism*, with relaxation times typically less than 100 s. Above the critical size the uniformly magnetized grain is very stable and is called a single domain grain.

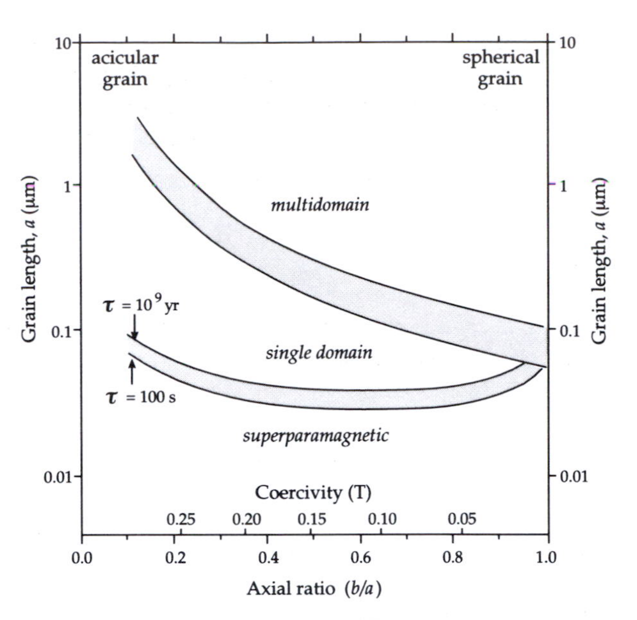

Fig. 5.17 Ranges of grain sizes and shapes for superparamagnetic, single domain and multi-domain magnetic behavior in ellipsoidal magnetite grains (after Evans and McElhinny, 1969).

5.3.5.1 *Superparamagnetism*

In a ferromagnetic material the strong molecular fields keep the atomic spin magnetic moments uniformly aligned with each other, and the grain anisotropy requires this spontaneous magnetization to lie parallel to an 'easy' direction. If the temperature is too high, thermal energy (kT) may exceed the anisotropy energy (VK_u) but still be too small to break up the spontaneous magnetization. The thermal energy causes the entire magnetic moment of the grain to fluctuate coherently in a manner similar to paramagnetism (the theory of which applies to individual atomic magnetic moments). The grain magnetization has no stable direction, and the behavior is said to be *superparamagnetic*. It is important to note that superparamagnetic grains themselves are immobile; only their uniform magnetization fluctuates relative to the grain. Whether the ferrimagnetic grain exists in a stable or superparamagnetic state depends on the grain size, the grain shape (if the origin of K_u is magnetostatic) and the temperature. If the grain volume V is very small, unstable magnetic behavior due to superparamagnetism becomes likely. Magnetite and hematite grains finer than about 0.03 μm in diameter are superparamagnetic at room temperature.

5.3.5.2 *Single domain particles*

When the anisotropic magnetic energy (VK_u) of a grain is greater than the thermal energy (kT), the spontaneous magnetization direction favors one of the easy directions. The entire grain is uniformly magnetized as a *single domain*. This situation occurs in very fine grains of ferrimagnetic minerals.

In magnetite K_u is the magnetostatic energy related to the particle shape. The theoretical range of single domain sizes in magnetite is narrow, from about 0.03 to 0.1 μm in equant grains and up to about 1 μm in elongate grains (Fig. 5.17). In hematite K_u is the large magnetocrystalline anisotropy energy, and the range of single domain sizes is larger, from about 0.03 to 15 μm.

The magnetization of a single domain particle is very

stable, because to change it requires rotating the entire uniform spontaneous magnetization of the grain against the grain anisotropy, which requires a very strong magnetic field. The magnetic field required to reverse the direction of magnetization of a single domain grain is called its *coercivity* B_c and is given by:

$$B_c = \frac{2K_u}{M_s} \tag{5.30}$$

The maximum coercivity of single domain magnetite is around 0.3 T for needle-shaped elongate grains. The magnetocrystalline anisotropy of hematite gives it higher maximum coercivities, in excess of 0.5 T. However, the magnetic properties of hematite are very variable and its maximum coercivity commonly exceeds 2T. Because of their stable remanent magnetizations, single domain particles play a very important role in paleomagnetism.

5.3.5.3 *Multidomain particles*

Single domain behavior is restricted to a limited range of grain sizes. When a grain is large enough, the magnetic energy associated with its magnetization becomes too large for the magnetization to remain uniform. This is because the demagnetizing field of a uniformly magnetized grain (Fig. 5.18a) interacts with the spontaneous magnetization and generates a magnetostatic (or self-demagnetizing) energy. To reduce this energy, the magnetization subdivides into smaller, uniformly magnetized units, called *Weiss domains* after P. Weiss, who theoretically predicted domain structure in 1907. In the simplest case the magnetization divides into two, oppositely magnetized domains (Fig. 5.18b). The net magnetization is reduced to zero, and the magnetostatic energy is reduced by about a half. Further subdivision (Fig. 5.18c) reduces the magnetostatic energy correspondingly. In a grain with n domains of alternately opposed spontaneous magnetizations the magnetostatic energy is reduced by $1/n$. The domains are separated from one another by thin regions, about 0.1 μm thick, that are usually much thinner than the domains they divide. These regions are called *Bloch domain walls* in recognition of F. Bloch, who in 1932 proposed a theory for the structure of the domain wall on an atomic scale. Within the domain wall the magnetization undergoes small progressive changes in direction from each atom to its neighbor. The crystalline magnetic anisotropy of the material attempts to keep the atomic magnetic spins parallel to favored crystalline directions, while the exchange energy resists any change of direction from parallel alignment by the molecular field. The energy of these competing effects is expressed as a *domain wall energy* associated with each unit area of the wall.

The magnetization of a *multi-domain* grain can be changed by moving the position of a domain wall, which causes some domains to increase in size and others to decrease. A large multi-domain grain may contain many easily movable domain walls. Consequently, it is much easier to change the magnetization of a multi-domain grain than of a single domain grain. As a result multi-domain grains are less stable carriers of remanent magnetization than single domain grains.

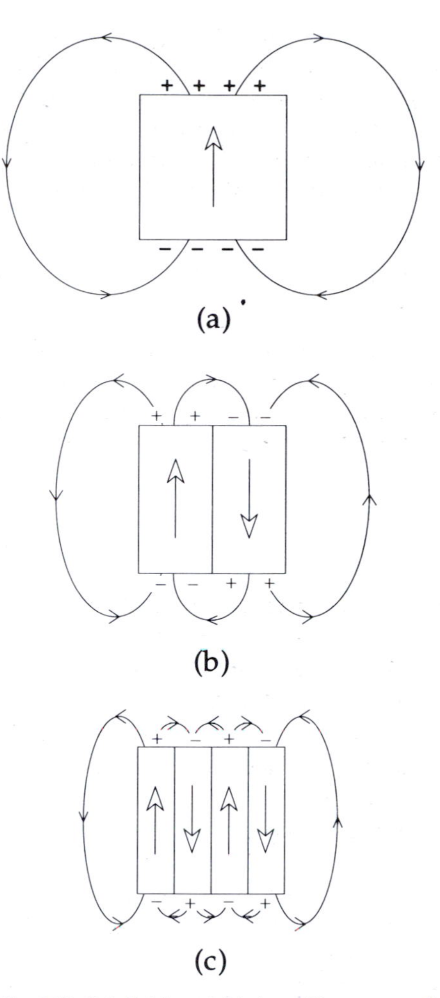

Fig. 5.18 Subdivision of (a) the uniform magnetization of a large grain into (b) two oppositely magnetized magnetic domains and (c) four alternately magnetized domains.

The transition between single domain and multi-domain behavior occurs when the reduction of magnetostatic energy is balanced by the energy associated with the domain wall that has been added. If magnetite grains are elongate, they can persist as single domain grains up to about 1 μm (Fig. 5.17). Magnetite grains larger than a few μm in diameter are probably multi-domain. In equidimensional grains of magnetite the transition should occur in grains of about 0.05–0.1 μm diameter. However, at this point the grain is not physically large enough to contain a wall, which has a thickness of about 0.1 μm. Grains in the intermediate range of sizes, and those large enough to contain only a few walls, are said to carry a *pseudo-single domain* magnetization. True multi-domain behavior in magnetite is observed when the grain size exceeds 15–20 μm.

In a pseudo-single domain grain that is large enough to contain only two domains, the domain wall separating them is not able to move freely. Its freedom of movement is restricted by interactions with the grain surface. A small grain in this size range has more stable magnetic properties than a multi-domain particle but is not as stable as a true single domain grain. Magnetite grains between about 0.1 and several μm in diameter have pseudo-single domain properties.

5.3.6 **Remanent magnetizations in rocks**

The small concentration of ferrimagnetic minerals in a rock gives it the properties of magnetic hysteresis. Most important of these is the ability to acquire a remanent magnetization (or *remanence*). The untreated remanence of a rock is called its natural remanent magnetization (NRM). It may be made up of several components acquired in different ways and at different times. The geologically important types of remanence are acquired at known times in the rock's history, such as at the time of its formation or subsequent alteration. The remanence of a rock can be very stable against change; the high coercivity (especially of the fine grains) of the ferrimagnetic mineral assures preservation of the magnetic signal during long geological epochs.

A remanence acquired at or close to the time of formation of the rock is called a *primary* magnetization; a remanence acquired at a later time is called a *secondary* magnetization. Examples of primary remanence are thermoremanent magnetization, which an igneous rock acquires during cooling, and the remanent magnetizations acquired by a sediment during or closely after deposition. Secondary remanences may be caused by chemical change of the rock during diagenesis or weathering, or by sampling and laboratory procedures.

5.3.6.1 *Thermoremanent magnetization*

The most important type of remanent magnetization in igneous (and high-grade metamorphic) rocks is *thermoremanent magnetization* (TRM). Igneous rocks solidify at temperatures well above 1000 °C. At this temperature the grains are solid and fixed in a rigid matrix. The grains of a ferrimagnetic mineral are well above their Curie temperature, which in magnetite is 578 °C and in hematite is 675 °C. There is no molecular field and the individual atomic magnetic moments are free to fluctuate chaotically; the magnetization is paramagnetic (Fig. 5.19).

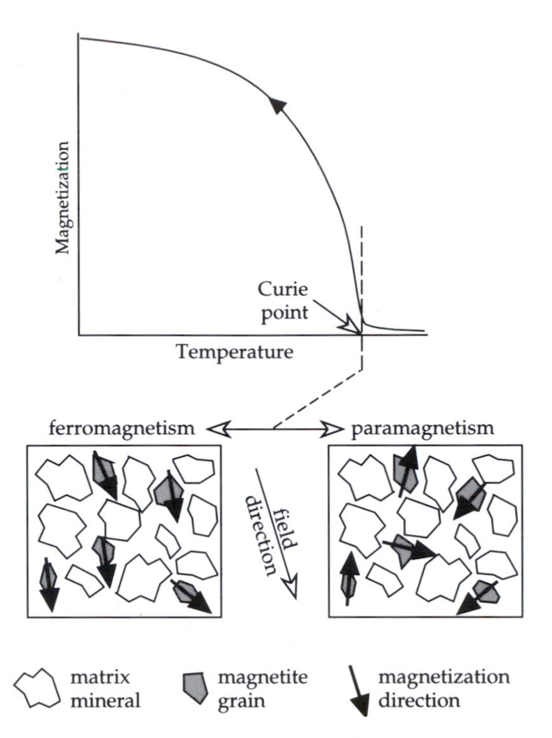

Fig. 5.19 On cooling through the Curie temperature the magnetic state of magnetite grains changes from paramagnetism to ferromagnetism. On cooling further the magnetizations in the magnetite grains become blocked along easy directions of magnetization close to the field direction. The resultant thermoremanent magnetization is parallel to the field direction.

As the rock cools, the temperature eventually passes below the Curie temperature of the ferrimagnetic grains and a spontaneous magnetization appears. In single domain grains the relaxation time of the grain magnetization is governed by Eq. (5.29), which can be modified by writing the anisotropy energy density K_u in terms of the spontaneous magnetization M_s and coercivity B_c of the grain. This gives the relaxation time as follows

$$\tau = \frac{1}{\nu_0} \exp\left(\frac{VM_sB_c}{2kT}\right) \tag{5.31}$$

At high temperature the thermal energy (kT) is larger than the magnetic energy ($VM_sB_c/2$) and the magnetization is unstable. Although the individual atomic magnetic moments are forced by the molecular field to act as coherent units, the grain magnetizations are superparamagnetic. As the rock cools further, the spontaneous magnetization and the magnetic anisotropy energy K_u of the grain increase. Eventually the temperature passes below a value at which the thermal energy, whose effect is to randomize the grain magnetic moments, is no longer greater than the magnetic anisotropy energy. The spontaneous magnetization then becomes 'blocked' along an easy direction of magnetization of the grain. In the absence of an external magnetic field the grain magnetic moments will be randomly oriented (assuming the easy axes to be randomly distributed). If the grain cools below its *blocking temperature* in a magnetic field, the grain magnetic moment is 'blocked' along the easy axis that is closest to the direction of the field at that time (Fig. 5.19). The alignment of grain magnetic moments with the field is neither perfect nor complete; it represents a statistical preference. This means that in an assemblage of grains more of the grains have their magnetic moments aligned close to the field direction than any other direction. The degree of alignment depends on the strength of the field.

It is important to note that the mineral grains themselves are immobile throughout the process of acquisition of TRM. Only the internal magnetizations of the grains can change direction and eventually become blocked. The blocking temperatures of TRM are dependent on the grain size, grain shape, spontaneous magnetization and magnetic anisotropy of the ferrimagnetic mineral. If the rock contains a wide range of grain sizes and perhaps more than a single magnetic mineral, there may be a broad spectrum of blocking temperatures. Maximum blocking temperatures may range as high as the Curie point, and the spectrum can extend to below ambient temperature. If the magnetic field is applied only while the rock is cooling through a limited temperature range, only grains with blocking temperatures in this range are activated, and a partial TRM (or pTRM) results.

TRM is a very stable magnetization which can exist unchanged for long intervals of geological time. The ability of TRM to record accurately the field direction is demonstrated by the results from a lava that erupted on Mt. Etna at a time for which a record of the magnetic field direction is available from observatory data. The directions of the TRM in the lava samples are the same as the direction of the ambient field (Fig. 5.20).

Mt. Etna lavas
(*Chevallier, 1925*)

360°/0°, 30°, 60°, 90°, 120°, 150°, 180°, 210°, 240°, 270°, 300°, 330°
90° 60° 30°

■ direction of magnetic field during eruption
○ direction of magnetization of lava sample

Fig. 5.20 Agreement of directions of thermoremanent magnetization in a basaltic lava flow on Mt. Etna (Sicily) with the direction of the geomagnetic field during eruption of the lava (based upon data from Chevallier, 1925).

5.3.6.2 *Sedimentary remanent magnetizations*

The acquisition of *depositional remanent magnetization* (DRM) during deposition of a sediment takes place at constant temperature. Magnetic and mechanical forces compete to produce a physical alignment of detrital ferrimagnetic particles. During settling through still water these particles are oriented by the ambient magnetic field in the same way that it orients a compass needle. The particles become aligned statistically with the Earth's magnetic field (Fig. 5.21). The action of mechanical forces may at times spoil this alignment. Water currents cause hydromechanical

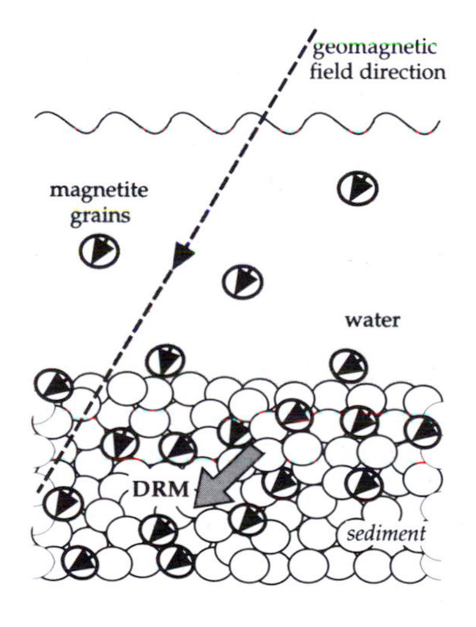

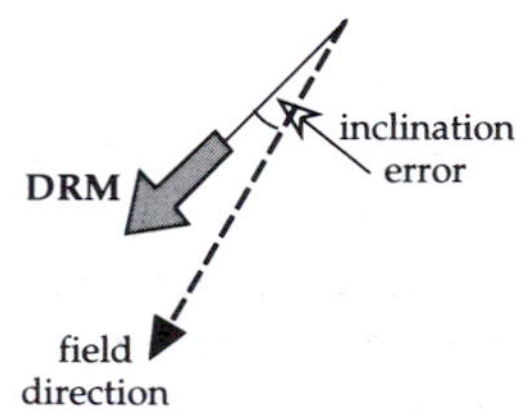

Fig. 5.21 Acquisition of depositional remanent magnetization (DRM) in a sediment; gravity causes an inclination error between the magnetization and field directions.

(a)

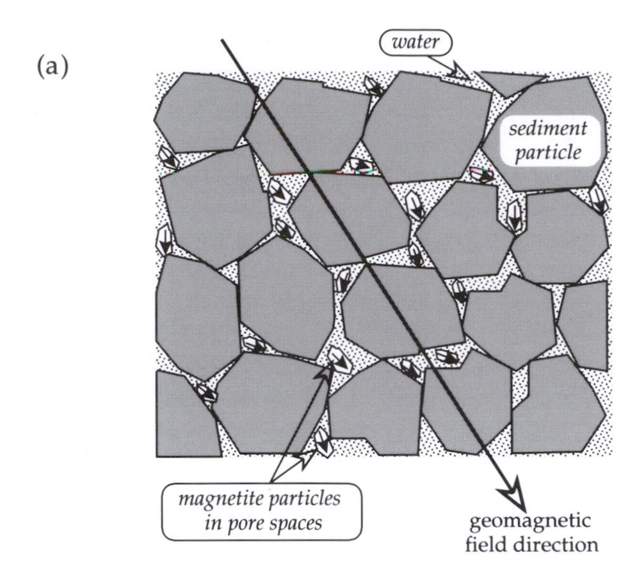

(b)

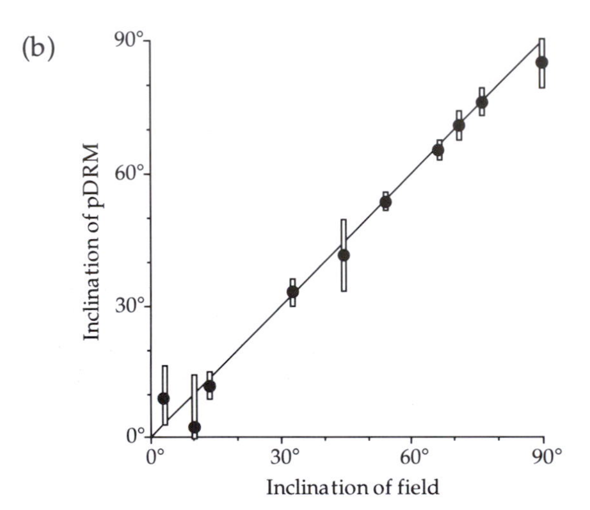

Fig. 5.22 (a) Post-depositional remanent magnetization (pDRM) is acquired by reorientation of ferromagnetic grains in the pore spaces of a deposited sediment. (b) Comparison of the pDRM inclination with the field inclination in a redeposited deep-sea sediment (after Irving and Major, 1964).

forces that disturb the alignment during settling, giving rise to a *declination error*. On contact with the bottom of the sedimentary basin, the mechanical force of gravity rolls the particle into a stable attitude, causing an *inclination error*. The pressure of overlying sediment during deep burial results in compaction, which can produce further directional errors. The DRM is finally fixed in sedimentary rocks during diagenesis.

A modified form of *post-depositional remanence* (pDRM) is important in fine grained sediments. A water-logged slurry forms at the sediment–water interface. Fine grained magnetic minerals in the water-filled pore spaces in the sediment are partially in suspension and may be reoriented by the magnetic field, if they are free enough to move (Fig. 5.22a). The energy for this motion of the particles is obtained from the Brownian motion of the water molecules, which continuously and randomly collide with the particles in the pore spaces. Large particles are probably in contact with the surrounding grains and are unaffected by collisions with the water molecules. Very fine particles that are virtually floating in the pore spaces may acquire enough freedom from the Brownian agitation to align statistically with the Earth's magnetic field. Laboratory experiments have established that the direction of pDRM is an accurate record of the depositional field, without inclination error (Fig. 5.22b). The pDRM is acquired later than the actual time of sedimentation, and is fixed in the sediment during compaction and de-watering at a depth of ~10 cm. This may represent a lock-in time delay of 100 yr in lacustrine sediments or 10,000 yr in pelagic marine sediments, where it is no more important geologically than the errors involved in locating paleontological stage boundaries. The pDRM process is particularly effective in fine grained sediments containing strongly magnetic magnetite grains. For example, pDRM is the most important mechanism of primary magnetization in pelagic limestones. Compaction may cause a flattening of the inclination under some conditions.

Bioturbation may mix the sediment, typically to a depth of about 10 cm in pelagic sediments. This affects the positions of stratigraphic marker levels. First occurrence datum levels of fossils are carried deeper, and last occurrences are carried higher than the true stratigraphic levels. Agitation of bioturbated sediment by the burrowing organisms assists the Brownian motion of the magnetic particles. Under these conditions the pDRM is acquired at the base of the bioturbated zone.

5.3.6.3 *Chemical remanent magnetization*

Chemical remanent magnetization (CRM) is usually a secondary form of remanence in a rock. It occurs when the magnetic minerals in a rock suffer chemical alteration or when new minerals form authigenically. An example is the precipitation of hematite from a goethite precursor or from iron-saturated fluids that pass through the rock. The magnetic minerals may also experience diagenetic modification or oxidation by weathering, which usually happens on the grain surface and along cracks (Fig. 5.23). The growth of a new mineral (or the alteration of an existing one) involves changes in grain volume V, spontaneous magnetization M_s and coercivity B_c. The chemical change affects the relaxation time of the grain magnetization, according to Eq. (5.31). The grains eventually grow through a critical volume, at which the grain magnetization becomes blocked. The new CRM is acquired in the direction of the ambient

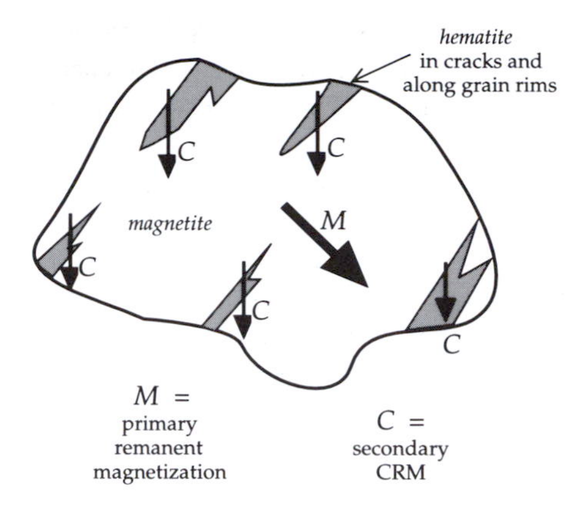

Fig. 5.23 Acquisition of chemical remanent magnetization (CRM) accompanies the diagenetic modification or oxidation by weathering of magnetic minerals; this often happens on the grain surface and along cracks.

field during the chemical change, and so it is usually a secondary remanence. It has stable magnetic properties similar to TRM. A common example is the formation of hematite during diagenesis or weathering. Hematite that originates in this way typically carries a secondary remanent magnetization.

5.3.6.4 *Isothermal remanent magnetization*

Isothermal remanent magnetization (IRM) is induced in a rock sample by placing it in a magnetic field at constant temperature. For example, rock samples are exposed to the magnetic fields of the sampling equipment and to other magnetic fields during transport to the laboratory. A common technique of rock magnetic analysis consists of deliberately inducing IRM in a known magnetic field produced in a large coil or between the poles of an electromagnet. The magnetic moments within each grain are partially aligned by the applied field. The degree of alignment depends on the field strength and on the resistance of the magnetic mineral to being magnetized, that is, its coercivity. After removing the sample from the applied field an IRM remains in the rock (see Fig. 5.11). If the rock sample is placed in progressively stronger fields the IRM increases to a maximum value called the saturation IRM, which is determined by the type and concentration of the magnetic mineral. The shape of the progressive acquisition curve and the field needed to reach saturation IRM depend on the coercivities of the magnetic minerals in the rock (Fig. 5.24).

The maximum coercivities of the most common ferromagnetic minerals in rocks are fairly well known (Table 5.1). These minerals also have distinctive maximum blocking temperatures (§5.3.6.1). The combination of these properties provides a method of identification of the predominant magnetic minerals in rocks. Starting with the strongest field available, IRM is imparted in successively smaller fields, chosen to remagnetize different coercivity fractions, along two or three orthogonal directions. The compound IRM is then subjected to progressive thermal demagnetization. The demagnetization characteristics of the different coercivity fractions help to identify the magnetic mineralogy (Fig. 5.24).

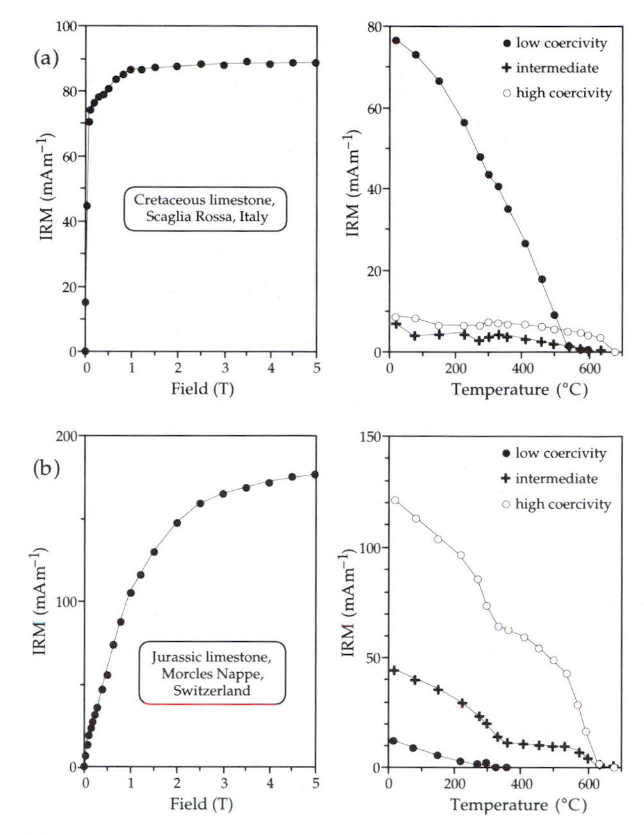

Fig. 5.24 Examples of the identification of magnetic minerals by acquisition and subsequent thermal demagnetization of IRM. Hematite is present in both (a) and (b), because saturation IRM requires fields>1 T and thermal demagnetization of the hard fraction persists to T≈675 °C. In (a) the soft fraction that demagnetizes at T≈575 °C is magnetite, while in (b) no magnetite is indicated but pyrrhotite is present in all three fractions, shown by thermal unblocking at T≈300–330 °C (after Lowrie, 1990).

5.3.6.5 *Other remanent magnetizations*

If a rock containing magnetic minerals with unstable magnetic moments experiences a magnetic field, there is a finite probability that some magnetic moments opposite to the field may switch direction to be parallel to the field. As time goes on, the number of magnetic moments in the direction of the field increases and the magnetization grows logarith-

Table 5.1 *Maximum coercivities and blocking temperatures for some common ferromagnetic minerals*

Ferromagnetic mineral	Maximum coercivity (T)	Maximum blocking temperature (° C)
magnetite	0.3	575
maghemite	0.3	≈350
titanomagnetite ($Fe_{3-x}Ti_xO_4$):		
x=0.3	0.2	350
x=0.6	0.1	150
pyrrhotite	0.5–1	325
hematite	1.5–5	675
goethite	>5	80–120

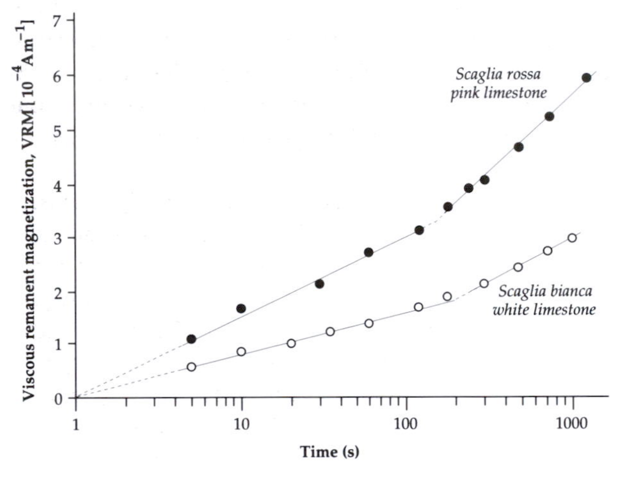

Fig. 5.25 Viscous remanent magnetization (VRM) in pelagic limestone samples, showing logarithmic growth with increasing time (after Lowrie and Heller, 1982).

mically with time (Fig. 5.25). The time-dependent remanence acquired in this way is called a *viscous remanent magnetization* (VRM). The VRM also decreases logarithmically when the field is removed or changed and is often identifiable during laboratory analysis as an unstable time-dependent change in remanence. The direction of a VRM is often parallel to the present-day field direction, which can be useful in its identification. However, it is always a secondary remanence and it can mask the presence of geologically interesting stable components. Techniques of progressive demagnetization have been designed to remove VRM and IRM, effectively 'cleaning' the remanence of a rock of undesirable components.

An important form of remanent magnetization can be produced in a rock sample by placing it in a coil that carries an alternating magnetic field, whose amplitude is then slowly reduced to zero. In the absence of another field, this procedure randomizes the orientations of grain magnetic moments with coercivities less than the peak field. If, however, the rock sample is exposed to a small constant magnetic field while the amplitude of the alternating magnetic field is decreasing to zero, the magnetic moments are not randomized. Their distribution is biassed with a statistical preference for the direction of the constant field. This produces an *anhysteretic remanent magnetization* (ARM) in the sample. The intensity of ARM increases with the amplitude of the alternating field, and also with the strength of the constant bias field. ARM may be produced deliberately, as described, and it is commonly observed as a spurious effect during progressive alternating field demagnetization of rock samples when the shielding from external fields is imperfect (§5.6.3.2).

5.4 GEOMAGNETISM

5.4.1 **Introduction**

The magnetic field of the Earth is a vector; that is, it has both magnitude and direction. The magnitude, or intensity F, of the field is measured in the same units as other **B**-fields, namely in *tesla* (see Eq. (5.10)). However, a tesla is an extremely strong magnetic field, such as one would observe between the poles of a powerful electromagnet. The Earth's magnetic field is much weaker; its maximum intensity is reached near to the magnetic poles, where it amounts to about 6×10^{-5} T. Modern instruments for measuring magnetic fields (called magnetometers) have a sensitivity of about 10^{-9} T; this unit is called a *nanotesla* (nT) and has been adopted in geophysics as the practical unit for expressing the intensity of geomagnetic field intensity. There is a further reason for adopting this unit. Most geomagnetic surveys carried out until the 1970s used the now abandoned c.g.s. system of units, in which the **B**-field was measured in *gauss*, equivalent to 10^{-4} T. The practical unit of geophysical exploration was then 10^{-5} gauss, called a *gamma*(γ). Thus, the former unit (γ) is conveniently equal to 10^{-9} T, which is the new unit (nT).

The magnetic vector can be expressed as Cartesian components parallel to any three orthogonal axes. The *geomagnetic elements* are taken to be components parallel to the geographic north and east directions and the vertically downward direction (Fig. 5.26). Alternatively, the geomagnetic elements can be expressed in spherical polar coordinates. The magnitude of the magnetic vector is given by the field-strength F; its direction is specified by two angles. The *declination* D is the angle between the magnetic merid-

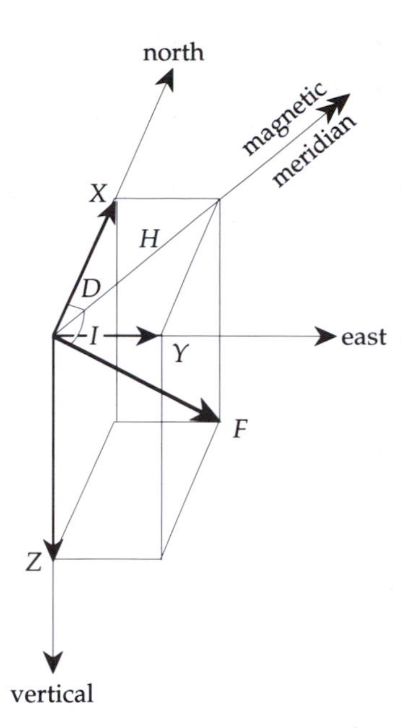

Fig. 5.26 Definition of the geomagnetic elements. The geomagnetic field can be described by north (X), east (Y) and vertically downward (Z) Cartesian components, or by the angles of declination (D) and inclination (I) together with the total field intensity (F).

ian and the geographic meridian; the *inclination* I is the angle at which the magnetic vector dips below the horizontal (Fig. 5.26). The Cartesian (X, Y, Z) and spherical polar (F, D, I) sets of geomagnetic elements are related to each other as follows:

$$X = F\cos I\cos D; \quad Y = F\cos I\sin D; \quad Z = F\sin I$$

$$F^2 = X^2 + Y^2 + Z^2$$

$$D = \arctan\left(\frac{Y}{X}\right); \quad I = \arctan\left(\frac{Z}{\sqrt{X^2+Y^2}}\right) \tag{5.32}$$

5.4.2 Separation of the magnetic fields of external and internal origin

The magnetic field B and its potential W at any point can be expressed in terms of the spherical polar coordinates (r, θ, ϕ) of the point of observation. Gauss expressed the potential of the geomagnetic field as an infinite series of terms involving these coordinates. Essentially, his method divides the field into separate components that decrease at different rates with increasing distance from the center of the Earth. The detailed analysis is complicated and beyond the scope of this text. The magnitude of the potential is given by

$$W = R\sum_{n=1}^{n=\infty}\left(A_n r^n + \frac{B_n}{r^{n+1}}\right)\sum_{l=0}^{l=n} Y_n^l(\theta,\phi) \tag{5.33}$$

where R is the Earth's radius.

This is a rather formidable expression, but fortunately the most useful terms are quite simple. The summation signs indicate that the total potential is made up of an infinite number of terms with different values of n and l. We will only pay attention here to the few terms for which $n=1$. The expression in parentheses describes the variation of the potential with distance r. For each value of n there will be different dependencies (e.g., on r, r^2, r^3, r^{-2}, r^{-3}, etc.). The function $Y_n^l(\theta, \phi)$ describes the variation of the potential when r is constant, i.e., on the surface of a sphere. It is called a *spherical harmonic function*, because it has the same value when θ or ϕ is increased by an integral multiple of 2π. For observations made on the spherical surface of the Earth, the constants A_n describe parts of the potential that arise from magnetic field sources outside the Earth, which are called the *geomagnetic field of external origin*. The constants B_n describe contributions to the magnetic potential from sources inside the Earth. This part is called the *geomagnetic field of internal origin*.

The potential itself is not measured directly. The geomagnetic elements X, Y and Z (Fig. 5.26) are recorded at magnetic observatories. Ideally these should be distributed uniformly over the Earth's surface but in fact they are predominantly in the northern hemisphere. The geomagnetic field components are directional derivatives of the magnetic potential and depend on the same coefficients A_n and B_n. Observations of magnetic field elements at a large number of measurement stations with a world-wide distribution allows the relative importance of A_n and B_n to be assessed. From the sparse data set available in 1838 Gauss was able to show that the coefficients A_n are very much smaller than B_n. He concluded that the field of external origin was insignificant and that the field of internal origin was predominantly that of a dipole.

5.4.3 The magnetic field of external origin

The magnetic field of the Earth in space has been measured from satellites and spacecraft. The external field has a quite complicated appearance (Fig. 5.27). It is strongly affected by the *solar wind*, a stream of electrically charged particles (consisting mainly of electrons, protons and helium nuclei) that is constantly emitted by the Sun. The solar wind is a *plasma*. This is the physical term for an ionized gas of low particle density made up of nearly equal concentrations of oppositely charged ions. At the distance of the Earth from the Sun (1 AU) the density of the solar wind is about 7 ions

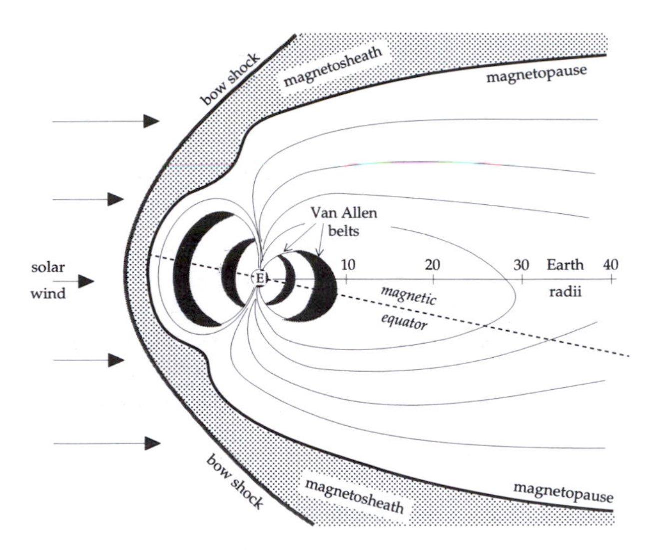

Fig. 5.27 Schematic cross-section through the magnetosphere, showing various regions of interaction of the Earth's magnetic field with the solar wind.

per cm^3, and it produces a magnetic field of about 6 nT. The solar wind interacts with the magnetic field of the Earth to form a region called the *magnetosphere*. At distances greater than a few Earth radii the interaction greatly alters the magnetic field from that of a simple dipole.

The velocity of the solar wind relative to the Earth is about 450 km s^{-1}. At a great distance (about 15 Earth radii) from the Earth, on the day side, the supersonic solar wind collides with the thin upper atmosphere. This produces an effect similar to the build-up of a shock wave in front of a supersonic aircraft. The shock front is called the *bow-shock* region (Fig. 5.27); it marks the outer boundary of the magnetosphere. Within the bow-shock region the solar wind is slowed down and heated up. After passing through the shock front the solar wind is diverted around the Earth in a region of turbulent motion called the *magnetosheath*. The moving charged particles of the solar wind constitute electrical currents. They produce an *interplanetary magnetic field*, which reinforces and compresses the geomagnetic field on the day side and weakens and stretches it out on the night side of the Earth. This results in a geomagnetic tail, or *magnetotail*, which extends to great distances 'downwind' from the Earth. The Moon's distance from the Earth is about 60 Earth radii and so its monthly orbit about the Earth brings it in and out of the magnetotail on each circuit. The transition between the deformed magnetic field and the magnetosheath is called the *magnetopause*.

5.4.3.1 *The Van Allen radiation belts*

Charged particles that penetrate the magnetopause are trapped by the geomagnetic field lines and form the *Van Allen radiation belts*. These constitute two doughnut-shaped regions coaxial with the geomagnetic axis (Fig. 5.28). The inner belt contains mainly protons, the outer belt energetic electrons. Within each belt the charged particles move in helical fashion around the geomagnetic field lines (Fig. 5.29). The pitch of the spiraling motion gets smaller as the particle comes ever closer to the Earth and the field intensity increases; eventually it reaches zero and reverses sense. This compels the particles to shuttle rapidly from one polar region to the other along the field lines. The inner Van Allen belt starts about 1000 km above the Earth's surface and extends to an altitude of about 3000 km (Fig. 5.28); the outer belt occupies a doughnut shaped region at distances between about 3 and 4 Earth radii (20,000–30,000 km) from the center of the Earth.

5.4.3.2 *The ionosphere*

The effects described above illustrate how the Earth's magnetic field acts as a shield against much of the extra-terrestrial radiation. The atmosphere acts as a protective blanket against the remainder. Most of the very short-wavelength fraction of the solar radiation that penetrates the atmosphere does not reach the Earth's surface. Energetic γ- and x-rays and ultraviolet radiation cause ionization of the molecules of nitrogen and oxygen in the thin upper atmosphere in altitudes between about 50 km and 1500 km, forming an ionized region called the *ionosphere*. It is formed of five layers, labelled the D, E, F_1, F_2 and G layers from the base to the top. Each layer can reflect radio waves. The thicknesses and ionizations of the layers change during the course of a day; all but one or two layers on the night side of the Earth disappear while they thicken and strengthen on the day side (Fig. 5.30). A radio transmitter on the day side can bounce signals off the ionosphere that then travel around the world by multiple reflections between the ground surface and the ionosphere. Consequently, radio reception of distant stations from far across the globe is best during the local night hours. The D layer is closest to the Earth at an altitude of about 80–100 km. It was first discovered in 1902, before the nature of the ionosphere was known, because of its ability to reflect long-wavelength radio waves, and is named the *Kennelly–Heaviside layer* in honor of its discoverers. The E layer is used by short-wave amateur radio enthusiasts. The F layers are the most intensely ionized.

5.4.3.3 *Diurnal variation and magnetic storms*

The ionized molecules in the ionosphere release swarms of electrons that form powerful, horizontal, ring-like electrical currents. These act as sources of external magnetic fields that are detected at the surface of the Earth. The ionization

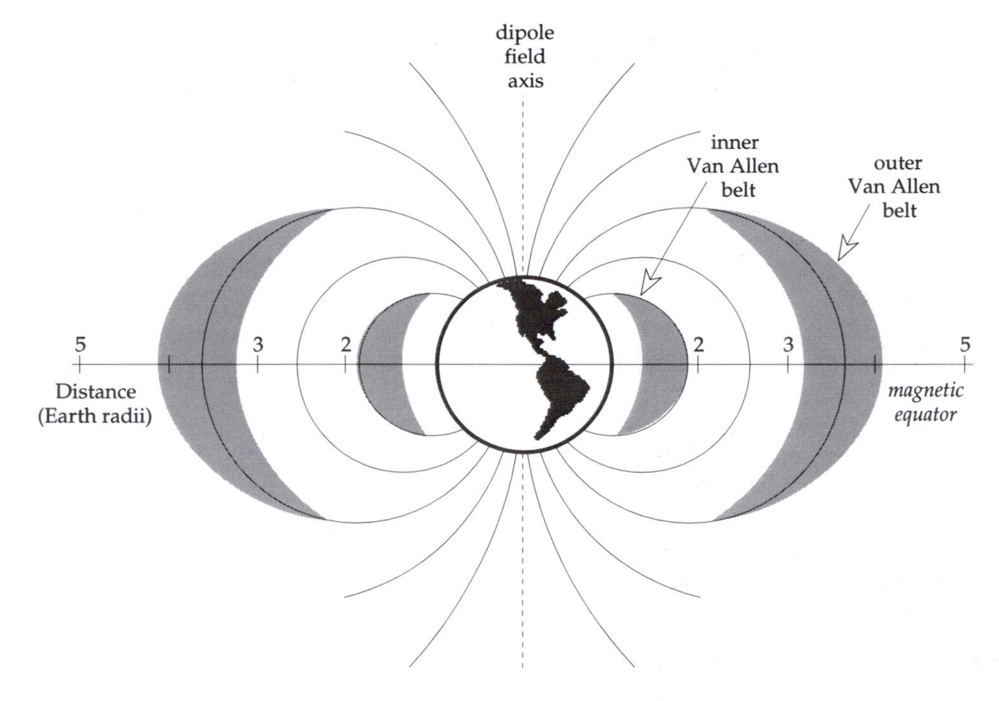

Fig. 5.28 Schematic representation of the inner and outer Van Allen belts of charged particles trapped by the magnetic field of the Earth.

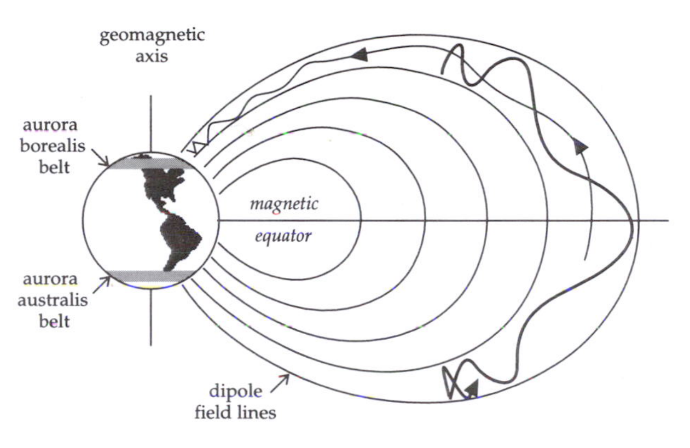

Fig. 5.29 Charged particles from the solar wind are constrained to move in a helical fashion about the geomagnetic field lines (after Vestine, 1962).

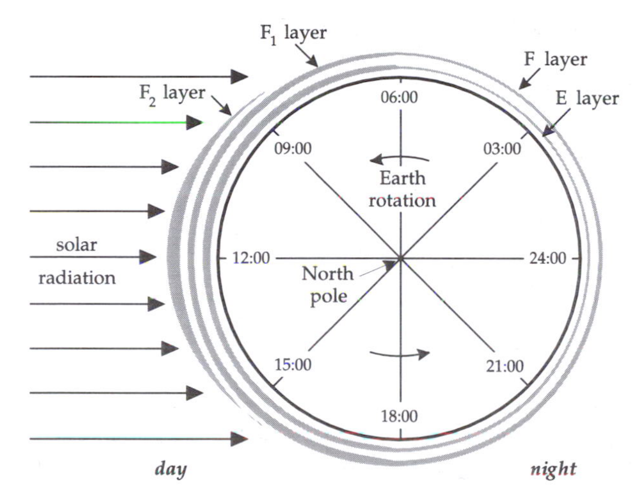

Fig. 5.30 Cross-section through the Earth showing the layered structure of the ionosphere (after Strahler, 1963).

is most intense on the day side of the Earth, where extra layers develop. The Sun also causes atmospheric tides in the ionosphere, partly due to gravitational attraction but mainly because the side facing the Sun is heated up during the day. The motions of the charged particles through the Earth's magnetic field produce an electrical field, according to Lorentz's law (Eq. (5.10)), which drives electrical currents in the ionosphere. In particular, the horizontal component of particle velocity interacts with the vertical component of the geomagnetic field to produce horizontal electrical current loops in the ionosphere. These currents cause a magnetic field at the Earth's surface. As the Earth rotates beneath the ionosphere the observed intensity of the geomagnetic field fluctuates with a range of amplitude of about 10–30 nT at the Earth's surface and a period of one day (Fig. 5.31a). This time-dependent change of geomagnetic field intensity is called the *diurnal* (or *daily*) *variation*.

The magnitude of the diurnal variation depends on the latitude at which it is observed. Because it greatly exceeds the accuracy with which magnetic fields are measured during surveys, the diurnal variation must be compensated by correcting field measurements accordingly. The intensity of the effect depends on the degree of ionization of the ionosphere and is therefore determined by the state of solar

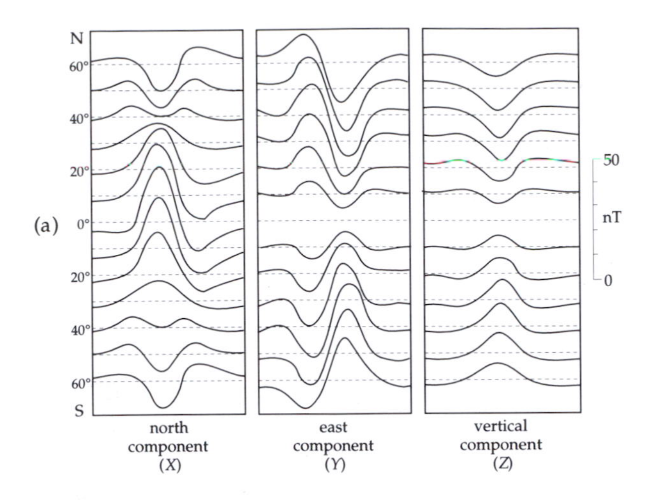

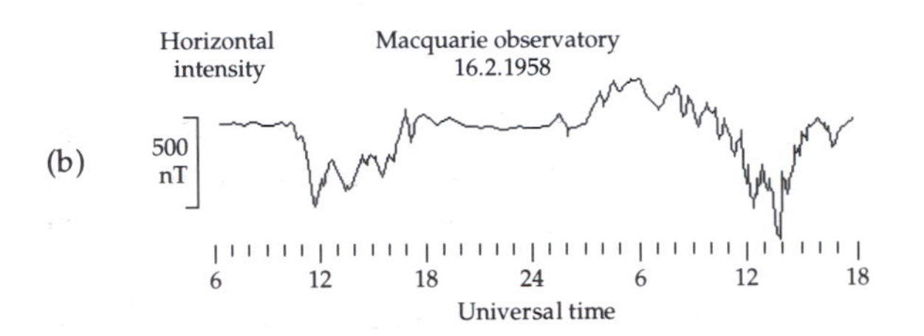

Fig. 5.31 (a) The time-dependent daily (or diurnal) variation of the components of geomagnetic field intensity at different latitudes (after Chapman and Bartels, 1940), and (b) the variation of horizontal field intensity during a magnetic storm (after Ondoh and Maeda, 1962).

activity. As described in §5.4.8.1, the solar activity is not constant. On days when the activity of the Sun is especially low, the diurnal variation is said to be of solar quiet (S_q) type. On normal days, or when the solar activity is high, the S_q variation is overlaid by a solar disturbance (S_D) variation. Solar activity changes periodically with the 11-year cycle of sunspots and solar flares. The enhanced emission of radiation associated with these solar phenomena increases the ionospheric currents. These give rise to rapidly varying, anomalously strong magnetic fields (called *magnetic storms*) with amplitudes of up to 1000 nT at the Earth's surface (Fig. 5.31b). The ionospheric disturbance also disrupts short-wave to long-wave radio transmissions. Magnetic surveying must be suspended temporarily while a magnetic storm is in progress, which can last for hours or days, depending on the duration of the solar activity.

5.4.4 The magnetic field of internal origin

To understand how geophysicists describe the geomagnetic field of internal origin mathematically we return to Eq. (5.33). First, we follow Gauss and omit the coefficients A_n of the external field. The spherical harmonic functions $Y_n^l(\theta, \phi)$ that describe the variation of potential on a spherical surface are then written in expanded form in terms of some new polynomial functions. We need not delve deeply into their meaning to understand what they represent. The potential W of the field of internal origin becomes

$$W = R\sum_{n=1}^{n=\infty}\sum_{l=0}^{l=n}\left(\frac{R}{r}\right)^{n+1}\left(g_n^l\cos l\phi + h_n^l\sin l\phi\right)P_n^l(\cos\theta) \qquad (5.34)$$

Here, R is the Earth's radius, as before, and $P_n^l(\cos\theta)$ are called Schmidt polynomials, which are related to more widely used functions known as associated Legendre polynomials. The Schmidt polynomials with $l=0$ are identical to the corresponding Legendre polynomials. The difference between higher-order Schmidt and Legendre polynomials results from the methods used to normalize them. When the square of an associated Legendre polynomial is integrated over the surface of a sphere the value of the integral is unity. The function is said to be *fully normalized.* However, the Schmidt functions with $l\neq 0$ do not integrate to unity over the surface of the sphere; for order n they integrate to $1/(2n+1)$. The Schmidt functions are said to be *partially normalized.* For our present discussion this difference is unimportant.

Eq. (5.34) is a *multipole* expression of the geomagnetic potential. It allows the very complex geometry of the total field to be broken down into contributions from a number of fields with simple geometries (Fig. 5.32). For example, Eq. (5.34) contains three terms for which $n=1$. Each corresponds to a dipole field. One dipole is parallel to the axis of rotation of the Earth and the other two are at right angles to each other in the equatorial plane. The terms with $n=2$ describe a geometrically more complex field, known as a quadrupole field. Just as a dipole field results when two single poles are brought infinitesimally close to each other (see §5.2.3), a quadrupole field results when two opposing coaxial dipoles are brought infinitesimally close end-to-end. As its name implies, the quadrupole field derives from four magnetic poles of which two are 'north' poles and two are 'south' poles. The terms with $n=3$ describe an octupole field, arising from a combination of eight (2^3) poles; the terms with $n=N$ describe a field that arises from a combination of 2^N poles. By superposing many terms corresponding to these simple geometries a field of great complexity can be generated.

The constants g_n^l and h_n^l in the geomagnetic potential are called the Gauss coefficients of degree n and order l. Inspection of Eq. (5.34) shows that they have the same dimensions (nT) as the **B**-field. Their values are computed from analysis of measurements of the geomagnetic field.

Two main types of data are integrated in modern analy-

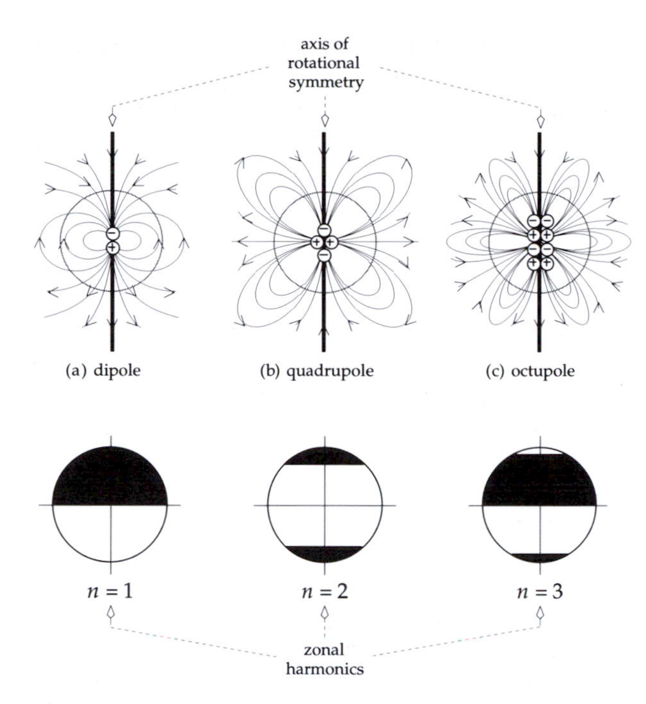

Fig. 5.32 Axial cross-sections showing the field-line geometries of (a) dipole, (b) quadrupole and (c) octupole fields; each field is rotationally symmetrical about the axis of the configuration. The corresponding zonal spherical harmonics are illustrated symbolically by shading the alternate zones in which magnetic field lines leave or return to the surface of a sphere.

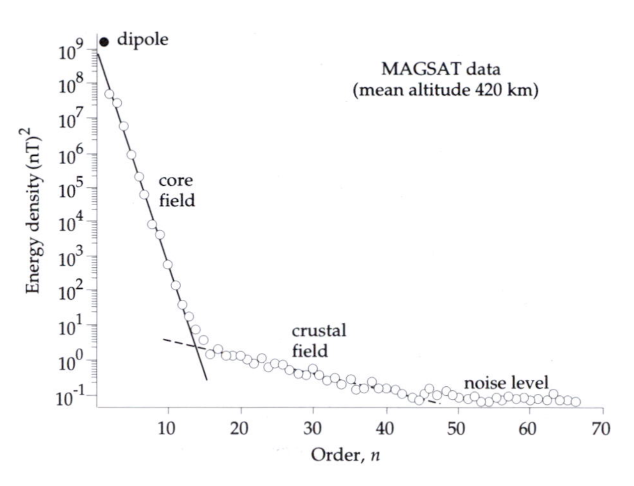

Fig. 5.33 The energy density spectrum derived from measurements of the geomagnetic field made by the MAGSAT Earth-orbiting satellite (after Cain, 1989).

ses of the field. A vast amount of survey data is obtained from satellites orbiting the Earth in low near-polar orbits. Continuous records at magnetic observatories yield average values of the geomagnetic elements from which optimum values for the Gauss coefficients are derived. In principle, an infinite number of Gauss coefficients would be needed to define the field completely. The coefficients of degree 8 and higher are very small and the calculation of Gauss coefficients must usually be truncated. A global model of the field is provided by the International Geomagnetic Reference Field (IGRF), which is based on coefficients up to $n=10$, although analyses of higher degree have been made. It is updated at regular intervals. The IGRF also gives the rate of change of each of the Gauss coefficients (its *secular variation*), which permits correction of the current values between update years.

The Gauss coefficients get smaller with increasing degree n; this decrease provides a way of estimating the origin of the internal field. The analysis involves a technique called *power spectral analysis*. The distance across a feature of the magnetic field (for example, a region where the field is stronger than average) is called the wavelength of the feature. As in the case of gravity anomalies, deep-seated magnetic sources produce broad (long-wavelength) magnetic anomalies, while shallow sources result in narrow (short-wavelength) anomalies. Spectral analysis consists of calculating the power (alternatively called the energy density) associated with each 'frequency' in the signal. This is obtained by computing the sum of the squares of all coefficients with the same order. In the case of the geomagnetic field, the spectral analysis is based on the values of the Gauss coefficients. The spatial frequency of any part of the observed field is contained in the order n of the coefficients. Low-order terms (those with small values of n) correspond to long-wavelength features, high-order terms are related to short-wavelength features.

Measurements of the geomagnetic field from the MAGSAT Earth-orbiting satellite at a mean altitude of 420 km above the Earth's surface have been analysed to yield Gauss coefficients to degree $n=66$, special techniques being invoked for degrees $n>29$. A plot of the energy density associated with each degree n of the geomagnetic field shows three distinct segments (Fig. 5.33). The high frequency terms of degree $n>40$ are uncertain and the terms with $n>50$ are in the 'noise level' of the analysis and cannot be attributed any geophysical importance. The terms $15<n\leq 40$ are due to short-wavelength magnetic anomalies associated with the magnetization of the Earth's crust. The terms $n\leq 14$ dominate the Earth's magnetic field and are due to much deeper sources in the fluid core.

5.4.4.1 *The dipole field*

The most important part of the Earth's magnetic field at the surface of the Earth is the dipole field, given by the Gauss coefficients for which $n=1$. If we write only the first term of Eq. (5.34) we get the potential

$$W = \frac{R^3 g_1^0 \cos\theta}{r^2} \tag{5.35}$$

Note that the spatial variation of this potential depends on ($\cos\theta/r^2$) in the same way as the potential of a dipole, which was found (Eq. (5.8)) to be

$$W = \frac{\mu_0 m \cos\theta}{4\pi r^2} \tag{5.36}$$

Comparison of the coefficients of ($\cos\theta/r^2$) in Eq. (5.35) and Eq. (5.36) gives the dipole moment of the Earth's axial dipole in terms of the first Gauss coefficient:

$$m = \frac{4\pi}{\mu_0} R^3 g_1^0 \tag{5.37}$$

The term g_1^0 ($n=1$, $l=0$) is the strongest component of the field. It describes a magnetic dipole at the center of the Earth and aligned with the Earth's rotation axis. This is called the *geocentric axial magnetic dipole*.

The magnetic field **B** of a dipole is symmetrical about the axis of the dipole. At any point at distance r from the center of a dipole with moment m on a radius that makes an angle θ to the dipole axis the field of the dipole has a radial component B_r and a tangential component B_θ, which can be obtained by differentiating the potential with respect to r and θ, respectively:

$$B_r = -\frac{\partial W}{\partial r} = \frac{\mu_0}{4\pi}\frac{2m\cos\theta}{r^3} \tag{5.38}$$

$$B_\theta = -\frac{1}{r}\frac{\partial W}{\partial \theta} = \frac{\mu_0}{4\pi}\frac{m\sin\theta}{r^3} \tag{5.39}$$

Note that B_r vanishes at the equator ($\theta=90°$) and the field is horizontal; comparing Eqs. (5.37) and (5.39) we get that the horizontal equatorial field B_q is equal to g_1^0. At a point (r, θ) on the surface of a uniformly magnetized sphere the magnetic field line is inclined to the surface at an angle I, which is given by

$$\tan I = \frac{B_r}{B_\theta} = 2\cot\theta = 2\tan\lambda \tag{5.40}$$

The angle I is called the *inclination* of the field, and θ is the angular distance (or *polar angle*) of the point of observation from the magnetic axis. The polar angle is the complement of the magnetic latitude, λ (i.e., $\theta=90°-\lambda$). Eq. (5.40) has an important application in paleomagnetism, as will be seen later.

The terms g_1^1 and h_1^1 are the next strongest in the potential expansion. They describe contributions to the potential from additional dipoles with their axes in the equatorial plane. The total dipole moment of the Earth is then obtained from the vector sum of all three components:

$$m = \frac{4\pi}{\mu_0} R^3 \sqrt{(g_1^0)^2 + (g_1^1)^2 + (h_1^1)^2} \tag{5.41}$$

The analysis of the geomagnetic field for the year 1995 gave the following values for the dipole coefficients: $g_1^0=-29{,}682$ nT; $g_1^1=-1{,}789$ nT; $h_1^1=5{,}318$ nT. Note that the sign of g_1^0 is negative. This means that the axial dipole points opposite to the direction of rotation. Taken together, the three dipole components describe a geocentric dipole inclined at about 10.7° to the Earth's rotation axis. This tilted geocentric dipole accounts for more than 90% of the geomagnetic field at the Earth's surface. Its axis cuts the surface at the north and south *geomagnetic poles*. For epoch 1995 the respective poles were located at 79.3°N, 71.4°W (i.e., 288.6°E) and 79.3°S, 108.6°E. The geomagnetic poles are antipodal (i.e., exactly opposite) to each other.

The *magnetic poles* of the Earth are defined as the locations where the inclination of the magnetic field is ±90° (i.e., where the field is vertically upward or downward). An *isoclinal* map (showing constant inclination values) for the year 1980 shows that the location of the north magnetic pole was at 77.3°N, 258.2°E while the south magnetic pole was at 65.6°S, 139.4°E (Fig. 5.34). These poles are not exactly opposite one another. The discrepancy between the magnetic poles and the geomagnetic poles arises because the terrestrial magnetic field is somewhat more complex than that of a perfect dipole.

5.4.4.2 *The non-dipole field*

The part of the field of internal origin (about 5% of the total field), obtained by subtracting the field of the inclined geocentric dipole from the total field, is collectively called the *non-dipole field*. A map of the non-dipole field consists of a system of irregularly sized, long-wavelength magnetic anomalies (Fig. 5.35). To describe this field requires all the terms in the potential expansion of degree $n \geq 2$.

As related earlier, the terms of degree $n=2$ in the expansion of the potential correspond to magnetic quadrupoles, the next-higher terms of degree $n=3$ to magnetic octupoles, etc. The configurations of axial quadrupole and octupole fields are illustrated in Fig. 5.32, for the case where the order l of the Gauss coefficients is zero. If $l=0$ in Eq. (5.34), the potential does not vary around a circle of 'latitude' defined by a chosen combination of θ and r. This part of the field is said to have *zonal symmetry*. The meaning of this expression is most easily illustrated for a quadrupole field, as follows.

The field of a quadrupole is rotationally symmetric about its axis; the same is the case for an octupole or higher-order constellation of poles. In the northern hemisphere the field of an axial quadrupole is horizontal at a particular latitude. North of this latitude the field lines of the quadrupole

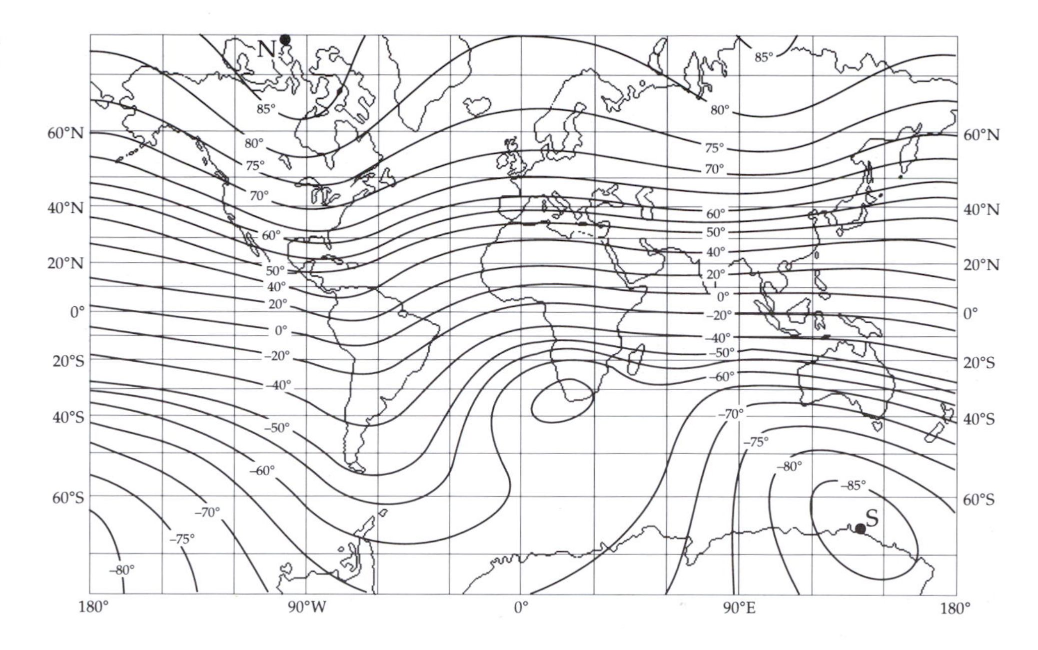

Fig. 5.34 The isoclinal map of the geomagnetic field for the year 1980 A.D. (after Merrill and McElhinny, 1983).

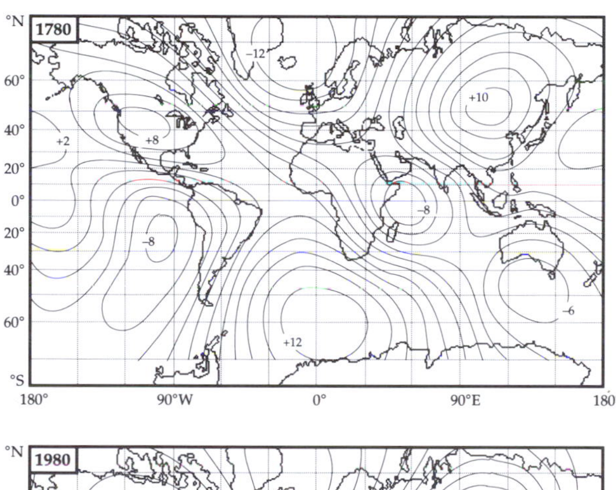

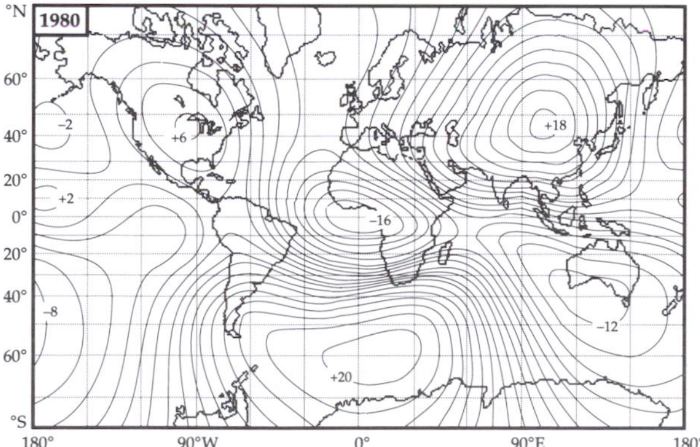

Fig. 5.35 The non-dipole magnetic field for the years 1780 A.D. (after Yukutake and Tachinaka, 1968) and 1980 A.D. (after Barton, 1989).

leave the Earth. A similar circle of latitude is located in the southern hemisphere; south of it the quadrupole field lines also leave the Earth. The field lines re-enter the Earth in the band of latitudes between those where the field is horizontal. The symmetry of the field is described by these three *zones* around the axis (see Fig. 5.32b). The axial octupole field also exhibits zonal symmetry; two zones in which the field leaves the Earth alternate with two zones in which it re-enters it.

Terms in the potential expansion for which the order l of the Gauss coefficients is equal to the degree n (e.g., g_1^1, h_1^1, g_2^2, h_2^2) are called *sectorial harmonics*. Their symmetry relative to the Earth's axis is characterized by an even number of sectors around the equator in which the field lines alternately leave and re-enter the Earth. The potential terms for which $l \neq n$ are known as *tesseral harmonics*. Their pattern of symmetry is defined by the intersections of circles of latitude and longitude, which outline alternating domains in which the field lines leave and re-enter the Earth, respectively.

5.4.5 Secular variation

At any particular place on the Earth the geomagnetic field is not constant in time. When the Gauss coefficients of the internal field are compared from one epoch to another, slow but significant changes in their values are observed. The slow changes of the field only become appreciable over decades or centuries of observation and so they are called *secular variations* (from the Latin word *saeculum* for a long age). They are manifest as variations of both the dipole and non-dipole components of the field.

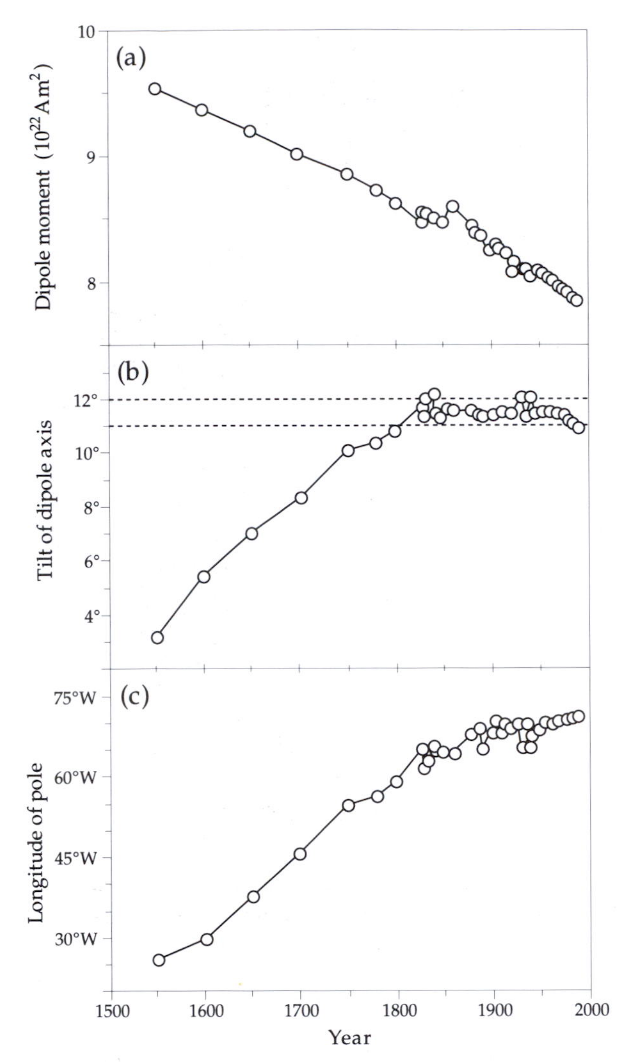

Fig. 5.36 Secular variations of the tilted geomagnetic centered dipole from 1550 A.D. to 1990 A.D. (a) Decrease of dipole moment; (b) slow changes of the tilt of the dipole axis relative to the rotation axis, and (c) longitude variation indicating westward drift of the geomagnetic poles (after Barton, 1989).

5.4.5.1 *Secular variation of the dipole field*

The dipole field exhibits secular variations of intensity and direction. Calculations of the Gauss coefficients for different historical epochs show a near-linear decay of the strength of the dipole moment at a rate of about 3.2% per century between about 1550 A.D. and 1900 A.D. At the start of the 20th century the decay became even faster and has averaged about 5.8% per century during the last 80 yr (Fig. 5.36a). If it continues at the same almost linear rate, the field intensity would reach zero in about another 2000 yr. The cause of the quite rapid intensity decay is not known; it may simply be part of a longer-term fluctuation. However, another possibility is that the dipole moment may be decreasing preparatory to the next reversal of geomagnetic field polarity.

The position of the dipole axis also shows secular variation. The changes can be traced by plotting the colatitude (the angle between the dipole axis and the rotation axis) and longitude of the geomagnetic pole as a function of time. Data are only sufficiently abundant for spherical harmonic analysis since the early 19th century. Less reliable data, enlarged by archeomagnetic results (see §5.6.2.1), allow estimates of the secular variation of the dipole axis since the middle of the 16th century. The earlier data suggest that in the 16th century the dipole axis was tilted at only about 3° to the rotation axis; a gradual increase in tilt took place between the 16th and 19th centuries. During the last 200 yr the dipole axis has maintained an almost constant tilt of about 11–12° to the rotation axis (Fig. 5.36b).

For the past 400 yr the longitude of the geomagnetic pole has drifted steadily westward (Fig. 5.36c). Before the 19th century the pole moved westward at about 0.14 deg yr^{-1}; this corresponds to a pseudo-period of 2600 yr for a complete circle about the geographic pole. However, since the early 19th century the westward motion of the pole has been slower, at an average rate of 0.044 deg yr^{-1}, which corresponds to a pseudo-period of 8200 yr.

5.4.5.2 *Secular variation of the non-dipole field*

Comparison of maps of the non-dipole field for different epochs (Fig. 5.35) show two types of secular variation. Some anomalies (e.g., over Mongolia, the South Atlantic and North America) appear to be stationary but they change in intensity. Other anomalies (e.g., over Africa) slowly change position with time. The secular variation of the non-dipole field therefore consists of a standing part and a drifting part.

Although some foci may have a north–south component of motion the most striking feature of the secular variation of the recent non-dipole field is a slow westward drift. This is superposed on the westward drift of the dipole, but can be separated readily by spherical harmonic analysis. The rate of drift of the non-dipole field can be estimated from longitudinal changes in selected features plotted for different epochs. The mean rate of westward drift of the non-dipole field in the first half of this century has been estimated to be 0.18 deg yr^{-1}, corresponding to a period of about 2000 yr. However, some foci drift at up to about 0.7 deg yr^{-1}, much faster than the average rate. Results from several geomagnetic observatories show that the rate of drift is dependent on latitude (Fig. 5.37).

Westward drift is an important factor in theories of the origin of the geomagnetic field. It is considered to be a

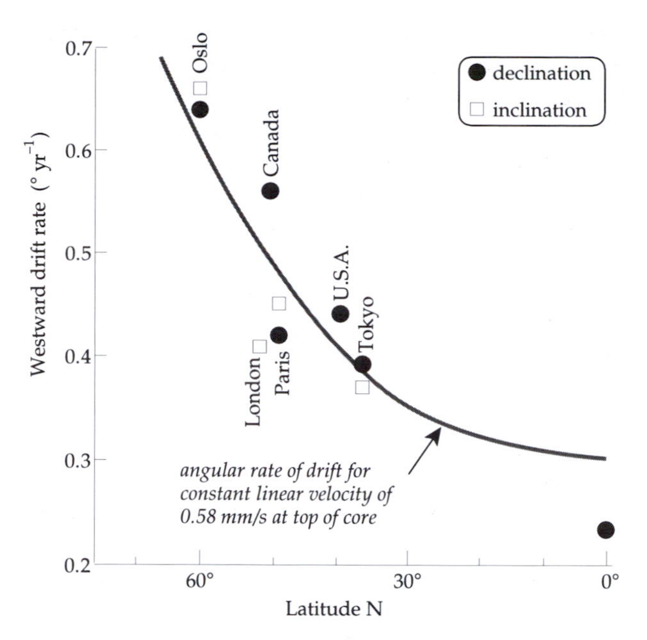

Fig. 5.37 The variation with latitude of average westward drift rates estimated from inclination and declination observations at geomagnetic observatories in the northern hemisphere. The curve gives the angular rotation rate at the surface of the core for a linear velocity of 0.058 cm s^{-1} (after Yukutake, 1967).

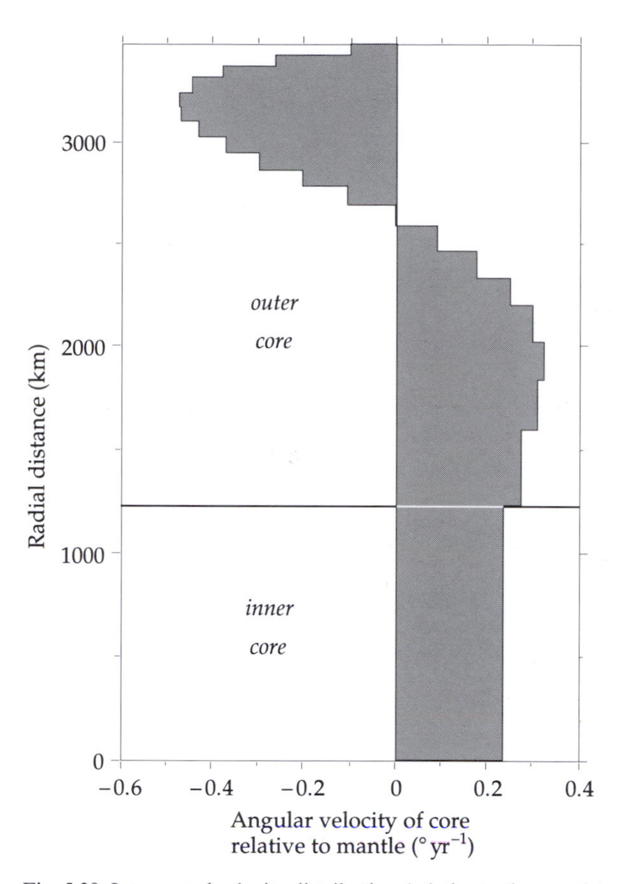

Fig. 5.38 Interpreted velocity distribution (relative to the mantle) for a multi-layered core model in which the change of angular momentum of each layer due to convectional fluid motion is balanced by electromagnetic forces (after Watanabe and Yukutake, 1975).

manifestation of rotation of the outer layers of the core relative to the lower mantle. Theoretical models of the geomagnetic field (discussed in the next section) presume conservation of angular momentum of the fluid core. To maintain the angular momentum of a particle of fluid that moves radially inwards (decreasing the distance from the rotational axis) its angular rate of rotation must speed up. This results in a layered structure for the radial profile of angular rate of rotation relative to the mantle (Fig. 5.38). The outer layers of the core probably rotate more slowly than the solid mantle, imparting a westward drift to features of the magnetic field rooted in the fluid motion.

5.4.6 Origin of the dipole field

Analysis of the Gauss coefficients and the wavelengths of features of the non-dipole field indicate that the main field is produced in the fluid outer core of the Earth. The composition of the fluid core has been estimated from seismic and geochemical data. The major constituent is liquid iron, with smaller amounts of other less-dense elements. Geochemical analyses of iron meteorites suggest that the core composition may have a few percent of nickel, while shock-wave experiments require 6–10% of non-metallic light elements such as silica, sulfur or oxygen. The solid inner core is inferred from seismic and shock-wave data to consist of almost pure iron.

For the generation of the magnetic field the important parameters of the core are its temperature, viscosity and electrical conductivity. Temperature is known very poorly inside the Earth, but probably exceeds 3000 °C in the liquid core. The electrical conductivity of iron at 20 °C is 10^7 $\Omega^{-1}\,m^{-1}$ and decreases with increasing temperature. At the high temperatures and pressures in the core the electrical conductivity is estimated to be around 3–5$\times10^5$ $\Omega^{-1}\,m^{-1}$, which corresponds to a good electrical conductor. For comparison, the conductivity of carbon (used for many electrical contacts) is 3×10^4 $\Omega^{-1}\,m^{-1}$ at 20 °C.

5.4.6.1 *Magnetostatic and electromagnetic models*

As observed by Gilbert in 1600, the dipole field of the Earth resembles that of a uniformly magnetized sphere. However, permanent magnetization is an inadequate explanation for the geomagnetic dipole. The mean magnetization of the Earth would need to be many times greater than that of the most strongly magnetic crustal rocks. The Curie point of the most important minerals at atmospheric pressure is less than 700 °C, which is reached at depths of about 25 km so that only a thin outer crustal shell could be permanently magnetized. The necessary magnetization of this shell is even greater than values observed in crustal rocks. Moreover, a magnetostatic origin cannot account for the temporal changes observed in the internal field, such as its secular variation.

The main magnetic field of the Earth is thought to be produced by electrical currents in the conductive core. Although the core is a good conductor, an electrical current system in the core continually loses energy through ohmic dissipation. The lost electrical energy is converted to heat and contributes to the thermal balance of the core. The equations of electromagnetism applied to the core show that an electrical current in the core would decay to zero in times of the order 10^4 yr unless it is sustained. Paleomagnetic evidence supports the existence of a geomagnetic field since Precambrian time, at least 10^9 yr, which implies that it must be continuously maintained or regenerated. The driving action for the main field is called the *dynamo process*, by analogy to the production of electrical power in a conductor that rotates in a magnetic field.

5.4.6.2 *The geomagnetic dynamo*

When a charged particle moves through a magnetic field, it experiences a deflecting electrical field (called the Lorentz field) proportional to the magnetic flux density B and the particle velocity v, and acting in the direction normal to both B and v. The Lorentz field acts as an additional source of electrical current in the core. Its strength is dependent on the velocity of motion of the conducting fluid relative to the magnetic field lines. When this term is included in the Maxwell electromagnetic equations, a *magnetohydrodynamic* equation relating the magnetic field $\mathbf{B}$ to the fluid flow $\mathbf{v}$ and conductivity σ in the core is obtained. It is written

$$\frac{\partial \mathbf{B}}{\partial t}=\frac{1}{\mu_0\sigma}\nabla^2\mathbf{B}+\nabla\times(\mathbf{v}\times\mathbf{B}) \tag{5.42}$$

This vector equation, although complicated, has immediate consequences. The left side gives the rate of change of magnetic flux in the core; it is determined by two terms on the right side. The first is inversely dependent on the electrical conductivity, and determines the decay of the field in the absence of a driving potential; the better the conductor, the smaller is this *diffusion term*. The second, *dynamo term* depends on the Lorentz electrical field, which is determined by the velocity field of the fluid motions in the core. The conductivity of the outer core ($(3-5)\times10^5$ $\Omega^{-1}\,\mathrm{m}^{-1}$) is high and for a fluid velocity of about 1 mm s^{-1} the dynamo term greatly exceeds the diffusion term. Under these conditions the lines of magnetic flux in the core are dragged along by the fluid flow. This concept is called the *frozen-flux theorem*, and it is fundamental to dynamo theory. The diffusion term is only zero if the electrical conductivity is infinite. There is probably some diffusion of the field through the fluid, because it is not a perfect conductor. However, the frozen-flux theorem appears to approximate well the conditions in the fluid outer core.

The derivation of a solution of the dynamo theory is difficult. In addition to Maxwell's equations with the addition of a term for the Lorentz field, the Navier–Stokes equation for the fluid flow, Poisson's equation for the gravitational potential and the generalized equation of heat transfer must be simultaneously satisfied. The fluid flow consists of a radial component and a rotational component. The energy for the radial flow is released at the boundary between the inner and outer core, where the freezing of the pure iron inner core depletes the fluid of the outer core of its heaviest component. The remaining lighter elements rise through the liquid outer core, causing a buoyancy-driven convection.

The rotational component of the fluid flow is the result of a radial velocity gradient in the liquid core, with inner layers rotating faster than outer layers (Fig. 5.38). The relative rotation of the conducting fluid drags magnetic field lines around the rotational axis to form ring-like, *toroidal* configurations. The toroidal field lines are parallel to the flow and therefore to the surface of the core. This means that the toroidal fields cannot escape from the core and cannot be measured. Their interactions with the upwelling and descending branches of convective currents create electrical current systems that produce poloidal magnetic fields. These, in their turn, escape from the core and can be measured at the surface of the Earth. The fluid motions are subject to the effects of Coriolis forces, which prove to be strong enough to dominate the resultant flow patterns.

5.4.7 **Origin of the non-dipole field**

The non-dipole field is believed to originate from irregularities in the flow pattern of the fluid core. These have been ascribed to interaction between the fluid flow and the rough topography of the core–mantle boundary. The nature of the interaction is not known, although models have been proposed whereby turbulence in the fluid flow is caused by mantle protuberances.

The distribution of positive and negative non-dipole field anomalies (see Fig. 5.35) suggested an alternative representation of the non-dipole field to the conventional portrayal as the superposition of quadrupole, octupole and higher-order terms in a spherical harmonic expansion. The positive and negative anomalies have been modelled by inward or outward oriented radial dipoles in the core at about one-quarter the Earth's radius. Each dipole is presumed to be caused by a toroidal current loop parallel to the surface of the core. A single centered axial dipole and eight auxiliary radial dipoles are adequate to represent the field observed at the Earth's surface. The secular variation at a site is explained as the passage of one of the auxiliary dipoles under the site.

5.4.8 Magnetic fields of the Sun, Moon and planets

Our knowledge of the magnetic fields of the Sun and planets derives from two types of observation. Indirect observations utilize spectroscopic effects. All atoms emit energy related to the orbital and spin motions. The energy is quantized so that the atom possesses a characteristic spectrum of energy levels. The lowest of these is called the ground state. When an atom happens to have been excited to a higher energy level, it is unstable and eventually returns to the ground state. In the process the energy corresponding to the difference between the elevated and ground states is emitted as light. For example atomic hydrogen gas in the Sun and galaxies emits radiation at 1420 MHz, with corresponding wavelength of 21 cm. This frequency is in the microwave range and can be detected by radio telescopes. If the hydrogen gas is in motion the frequency is shifted by the Doppler effect, from which the velocity of the motion can be deduced.

In the presence of a magnetic field a spectral line can split into several lines. This 'hyperfine splitting' is called the *Zeeman effect*. When a hydrogen atom is in the ground state its hyperfine structure has only one line. However, if the atom is in a magnetic field, it acquires additional energy from the interactions between the atom and the magnetic field and hence it can exist in several energy states. As a result the spectral lines of energy emitted by the atom are split into closely spaced lines, representing transitions between the different energy states. The energy differences between these states are dependent on the strength of the magnetic field, which can be estimated from the observations.

Direct observations of extra-terrestrial magnetic fields have been carried out since the 1960s by space probes. Magnetometers mounted in these spacecraft have recorded directly the intensity of the interplanetary magnetic field as well as the magnetic fields around several planets. The manned Apollo missions to the Moon resulted in a large amount of data obtained from the orbiting spacecraft. The materials collected on the Moon's surface and brought back to Earth by the astronauts have provided valuable information about lunar magnetic properties.

5.4.8.1 *Magnetic field of the Sun*

The Sun has nearly 99.9% of the mass of the solar system. About 99% of this mass is concentrated in a massive central core that reaches out to about 80% of the Sun's radius. The remainder of the Sun consists of an outer conducting shell, which contains only 1% of the Sun's mass but has a thickness equivalent to about 20% of the solar radius. Thermonuclear conversion of hydrogen into helium in the dense core produces temperatures of the order of 15,000,000 K. The visible solar disk is called the *photosphere*. Its diameter is about 2,240,000 km (about 175 times the diameter of the Earth) and its surface temperature is about 6000 K. The lower solar atmosphere is called the *chromosphere*; the outer atmosphere is called the *corona*. The chromosphere includes spike-like emissions of hydrogen gas called solar prominences that sometimes can reach far out into the corona.

The heat from the core is radiated out to the outer conducting shell, where it sets up convection currents. Some of these convection currents are small-scale (about 1000 km across) and last only a few minutes; others are larger-scale (30,000 km across) and persist for about a (terrestrial) day. The convection is affected by the Sun's rotation and is turbulent.

The rotation of the Sun has been estimated spectroscopically by measuring the Doppler shift of spectral lines. Independent estimates come from observing the motion of solar features like sunspots, etc. The rotational axis is tilted slightly at 7° to the pole to the ecliptic. Near the Sun's equator the rotational period is about 25 Earth days; the polar regions rotate more slowly with a period of 35–40 days. The Sun's core may rotate more rapidly than the outer regions. Turbulent convection and velocity shear in the outer conducting shell are conducive to a dynamo origin for the Sun's magnetic field. The surface field is dipolar in higher latitudes but has a more complicated pattern in equatorial latitudes.

Temperatures in the outer solar atmosphere (corona) are very high, around 1,000,000 K. The constituent particles achieve velocities in excess of the escape velocity of the Sun's gravitational field. The supersonic stream of mainly protons (H^+ ions), electrons and α-particles (He^{2+} ions) escaping from the Sun forms the solar wind. The flow of electric charge produces an interplanetary magnetic field (IMF) of varying intensity. At the distance of the Earth from the Sun the IMF measures about 6 nT.

Sunspots have been observed since the invention of the telescope (ca. 1610 A.D.). A sunspot is a dark fleck on the surface of the Sun, measuring roughly 1,000 to 100,000 km in diameter. It has a lower temperature than the surrounding photosphere and represents a strong disturbance extending far into the Sun's interior. Sunspots last for several days or weeks and move with the Sun's rotation, providing a means of estimating the rotational speed. The frequency of sunspots changes cyclically with a period of 11 years.

Intense magnetic fields are associated with the sunspots. These often occur in unequally sized pairs of opposite polarity. The predominating magnetic polarity of sunspots changes from one period of maximum sunspot activity to the next, implying that the period is in fact 22 years. The magnetic field of each sunspot is toroidal. It can be imagined to resemble a vortex, or tornado, in which the magnetic field lines leave the solar surface in one sunspot and return to it in the other member of the pair. The polarity of the Sun's dipole field also reverses with the change of polarity of the sunspots, indicating that the features are related.

Associated with the sunspots and their strong magnetic fields are emissions of hydrogen gas, called *solar flares*. The charged particles ejected in the flares contribute to the solar wind, which consequently transfers the sunspot cyclicity to fluctuations in the Earth's magnetic field of external origin and to ancillary terrestrial phenomena such as magnetic storms, brilliant aurora and interference with radio transmissions.

5.4.8.2 *Lunar magnetism*

Classical hypotheses for the origin of the Moon – rotational fission, capture or binary accretion – appear to be flawed. According to a recent 'giant impact' hypothesis, the Earth experienced a catastrophic collision with a Mars-sized protoplanet early in its development. Some debris from the collision remained in orbit around the Earth, where it re-accumulated as the Moon about 4.6 Ga ago. The early Moon was probably covered by an ocean of molten magma at least a hundred kilometers thick, the relicts of which are now the major constituents of the lunar highlands. After cooling and solidification the lunar surface was bombarded by planetesimals and meteoroids until about 4 Ga ago. The huge craters left by the impacts were later filled by molten basalt to form the lunar maria. Rock samples recovered from the maria in the manned Apollo missions have been dated radiometrically at 3.1–3.9 Ga. The volcanism may have continued after this time but probably ceased about 2.5 Ga ago. Since then the lunar surface has been pulverized by the constant bombardment by meteorites, micrometeorites and elementary particles of the solar wind and cosmic radiation. Except for the lunar highlands the surface is now covered by a layer of shocked debris several meters thick called the lunar *regolith*.

Our knowledge of lunar magnetism derives from magnetic field measurements made from orbiting spacecraft, magnetometers set up on the Moon and rock samples collected from the lunar surface and brought back to Earth. An interesting aspect of the satellite field measurements is the use of the electron reflection method to determine the lunar surface field from that measured in orbit. The principle is basically the same as used to explain the origin of the Van Allen belts in the Earth's magnetosphere (see §5.4.3). Electrons rain abundantly upon the lunar surface, in part from the solar wind and in part from the charges trapped in the Earth's magnetotail, which is crossed by the monthly orbit of the Moon about the Earth. In the absence of a lunar magnetic field the electrons would be absorbed by the lunar surface. As in the Earth's magnetosphere an electron incident on the Moon is forced by the Lorentz force to spiral about a magnetic field line (see Fig. 5.29). As the electron approaches the lunar surface the magnetic field strength increases and the pitch of the helix decreases to zero. At the point of closest approach (also called the *mirroring point*) the electron motion is completely rotational about the field line. The electron path then spirals with increasing pitch back along the field-line, away from the Moon. The effectiveness of the electron reflection is directional; for a given electron velocity there is a critical angle of incidence below which reflection ceases. By counting the number of electrons with a known energy that are reflected past a satellite in known directions the surface field B_s can be estimated from the field B_0 measured at the altitude of the satellite.

Measurements of the lunar magnetic field from orbiting spacecraft show that the Moon now has no detectable dipole moment. It has only a very weak non-dipolar magnetic field due to local magnetization of crustal rocks. Estimates of surface anomalies suggest that the largest are of the order of 100 nT. From records of meteoritic impacts and moonquakes made by seismometers left on the Moon for several years it has been inferred that if the Moon has a core (necessary for a lunar dynamo) it must be smaller than 400–500 km in radius, representing less than 2% of the Moon's volume. In contrast, the Earth's core occupies about 16% of the Earth's volume. Electromagnetic experiments have also been unable to establish whether a conducting lunar core exists or not.

On the other hand, samples of lunar rocks recovered in the manned Apollo missions possessed quite strong natural remanent magnetizations. Rock magnetic studies suggest that the lunar samples were magnetized in fields of the order of 10–100 μT (0.1–1 gauss). This implies that the Moon had a stronger magnetic field in the first 1–2 Ga of its history than it does now. Lunar paleointensity estimates suggest a decrease in strength between 4.0 and 3.1 Ga ago. The nature

Table 5.2 *Magnetic characteristics of the planets (data from Russell, 1980).*

PLANET	Mean orbital radius (AU)	Mean radius of planet (km)	Period of rotation (days)	Magnetic dipole moment (m_E)	Equivalent equatorial magnetic field (nT)	Dipole tilt to rotation axis (°)
Mercury	0.387	2440	58.6	$(2\text{–}7)\times10^{-4}$	100–400	—
Venus	0.723	6051	243.7	$<10^{-5}$	<1	—
Earth	1	6371	1	1	30,400	11.4
Moon	0.00257	1738	27.3	—	—	—
Mars	1.524	3370	1.029	$(1\text{–}30)\times10^{-5}$	2–60	(15–20)?
Jupiter	5.203	69,910	0.415	19,400	450,000	9.7
Saturn	9.539	58,230	0.445	575	23,000	<1
Uranus	19.18	25,460	0.718	50	24,000	60

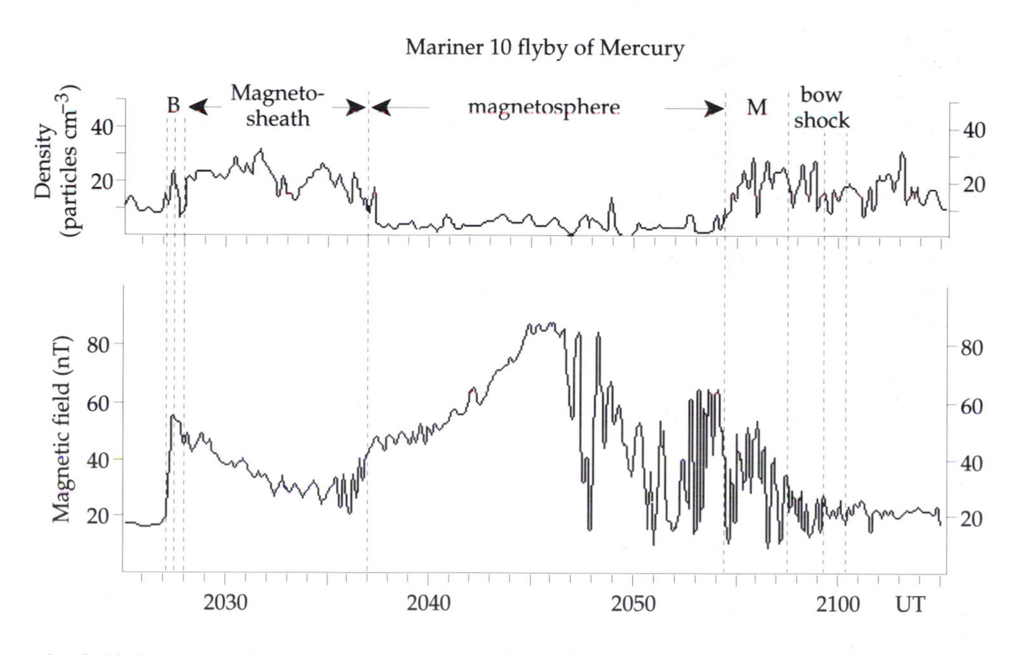

Fig. 5.39 Changes with time of particle density and magnetic field along the path of the Mariner 10 spacecraft during its flight past the planet Mercury (data from Ogilvie *et al.*, 1977).

of the ancient lunar magnetic field is not understood. Possibly it was caused by an early internal dynamo that is now extinct, but the dynamo action would require a liquid core for which there is no evidence. Alternatively, the interior of the Moon may have acquired a primordial remanent magnetization in the external field of the Sun or Earth. In turn this may have provided the fields for acquisition of the observed crustal magnetizations.

5.4.8.3 *Extra-terrestrial magnetic exploration*

Jupiter and Saturn, like the Earth, have magnetic fields that are strong enough to trap charged particles. The motions of these charged particles generate electromagnetic radiation which is detectable as radio waves far from the planet. The magnetic field of Jupiter was first detected in this way. Moreover, the radio emissions are modulated by the rotation of the planet. Analysis of the periodicity of their modulated radio emissions provide the best estimates of the rotational rates of Jupiter and Saturn.

Most data concerning the magnetic fields of the planets (see Table 5.2) have been obtained with flux-gate magnetometers (see §5.5.2.1) installed in passing or orbiting spacecraft. When the spacecraft traverses the magnetosphere of a planet, the magnetometer registers the passage through the bow shock and magnetopause (Fig. 5.39). A bow shock results from the supersonic collision of the solar wind with the atmosphere of a planet, just as it does for the Earth (see Fig. 5.27). Counters of energetic particles register a sudden increase in frequency and the magnetometer shows a change in magnetic field strength during passage through the bow shock into the magnetosheath. When the spacecraft leaves the magnetosheath and crosses the magnetopause it enters the region that is shielded from the solar wind by the magnetic field of the planet. The magnetopause is where the kinetic energy of the plasma is equal to the potential energy of the planetary magnetic field. The existence of a bow shock may be regarded as evidence for a planetary atmosphere, while the magnetopause is evidence that the planet has a magnetic field.

5.4.8.4 *The magnetic fields of the planets*

Mercury was visited by the *Mariner 10* spacecraft, which made three passes of the planet in 1974 and 1975. The on-board magnetometer detected a bow shock and a magnetopause (Fig. 5.39), which imply that the planet has a magnetic field. On the first encounter a magnetic field of about 100 nT was measured at an altitude of 700 km (radial distance of 3100 km), which was the point of closest approach. In estimating the magnetic field of the planet a compensation must be made for the magnetic field of the solar wind. Different models have given estimates of the strength of the dipole moment in the range $(2–7)\times10^{-4}$ of the Earth's magnetic moment ($m_E \approx 7.835\times10^{22}$ A m^{-2} in 1990), which would give a surface equatorial magnetic field of about 100–400 nT. The source of the magnetic field is uncertain. It is possible, but unlikely, that Mercury has a global magnetic moment due to crustal remanent magnetization; it would be difficult to explain the uniformity and duration of the external field needed to produce this. More probably, Mercury has a small active internal dynamo.

Venus was investigated by several American and Russian spacecraft in the 1960s. The instruments on *Mariner 5* in 1967 clearly detected a bow shock from the collision of the solar wind with the planetary atmosphere. Magnetometer data from later spacecraft found no evidence for a planetary magnetic field. If a magnetopause exists, it must lie very close to the planet and may wrap around it. Data from the *Pioneer* Venus orbiter in 1978 set an upper limit of $10^{-5}\,m_E$ for a planetary dipole moment, which would give a surface equatorial field <1 nT. The absence of a detectable magnetic field was not expected. As shown by Eq. (5.37) the dipole magnetic moment is proportional to the equatorial field times the cube of the planetary radius. The strength of a dynamo field is expected to be proportional to the core radius and to the rotation rate. These considerations give a scaling law whereby the dipole magnetic moment of a planet is proportional to its rotation rate and the fourth power of the core radius. The rotation of the planet is very slow compared to that of the Earth; one sidereal day on Venus lasts 243 Earth days, but the size of the planet is close to that of the Earth. It was therefore expected that Venus might have an internal dynamo with a dipole moment about 0.2% that of the Earth and an equatorial surface field of about 86 nT. It seems likely that the slow rotation does not provide enough energy for an active dynamo.

Mars was expected to have a magnetic moment between that of Earth and Mercury, because of its size and rotation rate. *Mariner 4* in 1965 was the only American spacecraft carrying magnetometers that has visited Mars; it detected a bow shock but no conclusive evidence for a magnetopause. Data from orbiting Russian spacecraft have been interpreted ambiguously, and there is no agreement on an accepted model. At present the available data establish only upper limits on the Martian magnetic field. Depending on the model used for interpretation the maximum dipole moment is estimated to be only $(1–30)\times10^{-5}\,m_E$.

Jupiter has been known since 1955 to possess a strong magnetic field, because of the associated polarized radio emissions. The spacecraft *Pioneer 10* and *11* in 1973–4 and *Voyager 1* and *2* in 1979 established that the planet has a bow shock and magnetopause. The huge magnetosphere encounters the solar wind about 5,000,000 km 'upwind' from the planet; its magnetotail may extend all the way to Saturn. Two reasons account for the great size of the magnetosphere compared to that of Earth. First, the solar wind pressure on the Jovian atmosphere is weaker due to the greater distance from the Sun; secondly, Jupiter's magnetic field is much stronger than that of the Earth. The dipole moment is almost 20,000 m_E which gives a powerful equatorial magnetic field of more than 400,000 nT at Jupiter's surface. The quadrupole and octupole parts of the non-dipole magnetic field have been found to be proportionately much larger relative to the dipole field than on Earth. The dipole axis is tilted at 9.7° to the rotation axis. The magnetic field of Jupiter results from an active dynamo in the metallic hydrogen core of the planet. The core is probably very large, with a radius up to 75% of the planet's radius. This would explain the high harmonic content of the magnetic field near the planet.

Saturn was reached by *Pioneer 11* in 1979 and the *Voyager 1* and *2* spacecraft in 1980 and 1981, respectively. The on-board magnetometers detected a bow shock and a magnetopause. Saturn's dipole moment is smaller than expected, but is estimated to be around 500 m_E. This gives an equatorial field of 55,000 nT, almost double that of Earth. The magnetic field has a purer dipole character (i.e., the non-dipole components are weaker) than the fields of Jupiter or Earth. The simplest explanation for this is that the field is generated by an active dynamo in a conducting core that is smaller relative to the size of the planet. The axis of the dipole magnetic field lies only about 1° away from the rotation axis, in contrast to 10.7° on Earth and 9.7° on Jupiter.

Uranus is unusual in that its spin axis has an obliquity of 97.9°. This means that the rotation axis lies very close to the ecliptic plane, and the orbital planes of its satellites are almost orthogonal to the ecliptic plane. Uranus was visited by *Voyager 2* in January 1986. The spacecraft encountered a bow shock and magnetopause and subsequently entered the magnetosphere of the planet. Uranus has a dipole moment about 50 times stronger than Earth's, giving a surface field of 24,000 nT, comparable to that on Earth. Intriguingly, the axis of the dipole field has a large tilt of about 60° to the rotation axis; there is no explanation for this tilt.

It is not yet known whether **Neptune** and **Pluto** have magnetic fields. Because of its similarity in size and composition to Uranus, it is thought that Neptune may also have a magnetic field. Pluto, on the other hand, is probably too small to have a magnetic field resulting from dynamo action.

There is reasonable confidence that Mercury, Earth, Jupiter, Saturn, and Uranus have active planetary dynamos today. The magnetic data for Mars and Neptune are inconclusive. All available data indicate that Venus and the Moon do not have active dynamos now, but possibly each might have had one earlier in its history.

5.5 MAGNETIC SURVEYING

5.5.1 The magnetization of the Earth's crust

The high-order terms in the energy density spectrum of the geomagnetic field (Fig. 5.33) are related to the magnetization of crustal rocks. Magnetic investigations can therefore yield important data about geological structures. By analogy with gravity anomalies we define a magnetic anomaly as the difference between the measured (and suitably corrected) magnetic field of the Earth and that which would be expected from the International Geomagnetic Reference Field (§5.4.4). The magnetic anomaly results from the contrast in magnetization when rocks with different magnetic properties are adjacent to each other, as, for example, when a strongly magnetic basaltic dike intrudes a less magnetic host rock. The stray magnetic fields surrounding the dike disturb the geomagnetic field locally and can be measured with sensitive instruments called magnetometers.

As discussed in §5.3.1, each grain of mineral in a rock can be classified as having diamagnetic, paramagnetic or ferromagnetic properties. When the rock is in a magnetic field, the alignment of magnetic moments by the field produces an induced magnetization ($\mathbf{M}_i$) proportional to the field, the proportionality constant being the magnetic susceptibility, which can have a wide range of values in rocks (see Fig. 5.13). The geomagnetic field is able to produce a correspondingly wide range of induced magnetizations in ordinary crustal rocks. The direction of the induced magnetization is parallel to the Earth's magnetic field in the rock.

Each rock usually contains a tiny quantity of ferromagnetic minerals. As we have seen, these grains can become magnetized permanently during the formation of the rock or by a later mechanism. The remanent magnetization ($\mathbf{M}_r$) of the rock is not related to the present-day geomagnetic field, but is related to the Earth's magnetic field in the geological past. Its direction is usually different from that of the present-day field. As a result the directions of $\mathbf{M}_r$ and $\mathbf{M}_i$ are generally not parallel. The direction of $\mathbf{M}_i$ is the same as that of the present field but the direction of $\mathbf{M}_r$ is often not known unless it can be measured in rock samples.

The total magnetization of a rock is the sum of the remanent and induced magnetizations. As these have different directions they must be combined as vectors (Fig. 5.40a). The direction of the resultant magnetization of the rock is not parallel to the geomagnetic field. If the intensities of $\mathbf{M}_r$ and $\mathbf{M}_i$ are similar, it is difficult to interpret the total magnetization. Fortunately, in many important situations $\mathbf{M}_r$ and $\mathbf{M}_i$ are sufficiently different to permit some simplifying assumptions. The relative importance of the remanent and induced parts of the magnetization is expressed in the *Königsberger ratio* (Q_n), defined as the ratio of the intensity of the remanent magnetization to that of the induced magnetization (i.e., $Q_n = M_r/M_i$).

Two situations are of particular interest. The first is when Q_n is very large (i.e., $Q_n \gg 1$). In this case (Fig. 5.40b), the total magnetization is dominated by the remanent component and its direction is essentially parallel to $\mathbf{M}_r$. Oceanic basalts, formed by extrusion and rapid underwater cooling at oceanic ridges, are an example of rocks with high Q_n ratios. Due to the rapid quenching of the molten lava titanomagnetite grains form with skeletal structures and very fine grain sizes. The oceanic basalts carry a strong thermoremanent magnetization and often have Q_n values of 100 or greater. This facilitates the interpretation of oceanic magnetic anomalies, because in many cases the induced component can be neglected and the crustal magnetization can be interpreted as if it were entirely remanent.

The other important situation is when Q_n is very small (i.e., $Q_n \ll 1$). This requires the remanent magnetization to be negligible in comparison to the induced magnetization. For example, coarse magnetite grains carry multidomain magnetizations (§5.3.5.3). The domain walls are easily moved around by a magnetic field. The susceptibility is high and the Earth's magnetic field can induce a strong

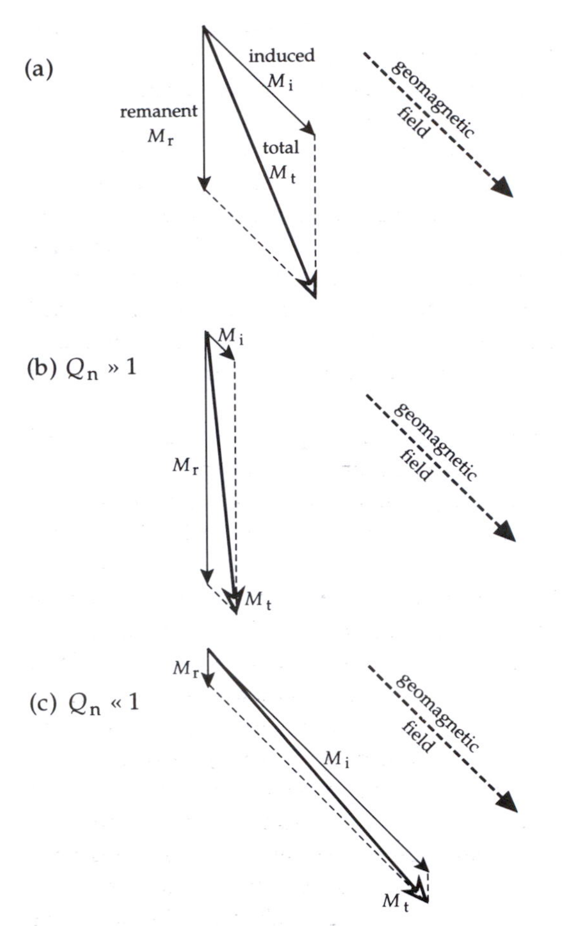

Fig. 5.40 The remanent (M_r), induced (M_i), and total (M_t) magnetizations in a rock. (a) For an arbitrary case M_t lies between M_t and M_r, (b) for a very large Königsberger ratio ($Q_n \gg 1$) M_t is close to M_r, and (c) for a very small Königsberger ratio ($Q_n \ll 1$) M_t is almost the same as M_i.

magnetization. Any remanent magnetization is usually weak, because it has been subdivided into antiparallel domains. These two factors yield a low value for Q_n. Magnetic investigations of continental crustal rocks for commercial exploitation (e.g., in ancient shield areas) can often be interpreted as cases with $Q_n \ll 1$. The magnetization can then be assumed to be entirely induced (Fig. 5.40c) and oriented parallel to the direction of the present-day geomagnetic field at the measurement site, which is usually known. This makes it easier to design a model to interpret the feature responsible for the magnetic anomaly.

5.5.2 Magnetometers

The instrument used to measure magnetic fields is called a magnetometer. Until the 1940s magnetometers were mechanical instruments that balanced the torque of the magnetic field on a finely balanced compass needle against a restoring force provided by gravity or by the torsion in a suspension fiber. The balance types were cumbersome, delicate and slow to operate. For optimum sensitivity they were designed to measure changes in a selected component of the magnetic field, most commonly the vertical field. This type of magnetometer has now been superseded by more sensitive, robust electronic instruments. The most important of these are the flux-gate, proton-precession and optically pumped magnetometers.

5.5.2.1 *The flux-gate magnetometer*

Some special nickel–iron alloys have very high magnetic susceptibility and very low remanent magnetization. Common examples are *Permalloy* (78.5% Ni, 21.5% Fe) and *Mumetal* (77% Ni, 16% Fe, 5% Cu, 2% Cr). The preparation of these alloys involves annealing at very high temperature (1100–1200 °C) to remove lattice defects around which internal stress could produce magnetostrictive energy. After this treatment the coercivity of the alloy is very low (i.e., its magnetization can be changed by a very weak field) and its susceptibility is so high that the Earth's field can induce a magnetization in it that is a considerable proportion of the saturation value.

The sensor of a flux-gate magnetometer consists of two parallel strips of the special alloy (Fig. 5.41a). They are wound in opposite directions with primary energizing coils. When a current flows in the primary coils, the parallel strips become magnetized in opposite directions. A secondary coil wound about the primary pair detects the change of magnetic flux in the cores (Fig. 5.41b), which is zero as soon as the cores saturate. While the primary current is rising or falling, the magnetic flux in each strip changes and a voltage is induced in the secondary coil. If there is no external magnetic field, the signals due to the changing flux are equal and opposite and no output signal is recorded. When the axis of the sensor is aligned with the Earth's magnetic field, the latter is added to the primary field in one strip and subtracted from it in the other. The phases of the magnetic flux in the alloy strips are now different; one saturates before the other. The flux-changes in the two alloy strips are no longer equal and opposite. An output voltage is produced in the secondary coil that is proportional to the strength of the component of the Earth's magnetic field along the axis of the sensor.

The flux-gate magnetometer is a *vector magnetometer*, because it measures the strength of the magnetic field in a particular direction, namely along the axis of the sensor. This requires that the sensor must be accurately oriented along the direction of the field component to be measured.

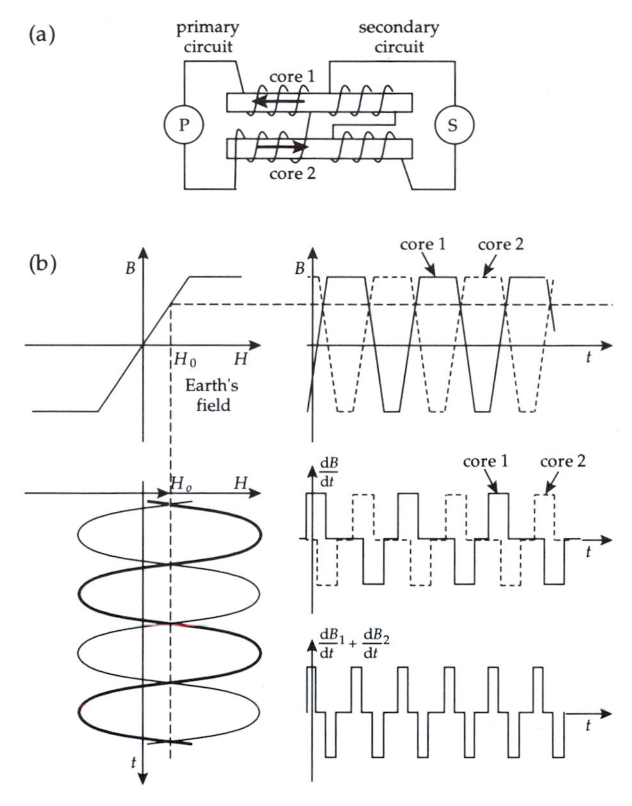

Fig. 5.41 Simplified principle of the fluxgate magnetometer. (a) Primary and secondary electrical circuits include coils wrapped around parallel strips of *Mumetal* in opposite and similar senses, respectively. (b) The output signal in a magnetic field is proportional to the net rate of change of magnetic flux in the *Mumetal* strips (after Militzer *et al.*, 1984).

For total field measurements three sensors are employed. These are fixed at right angles to each other and connected with a feedback system which rotates the entire unit so that two of the sensors detect zero field. The magnetic field to be measured is then aligned with the axis of the third sensor.

The flux-gate magnetometer does not yield absolute field values. The output is a voltage, which must be calibrated in terms of magnetic field. However, the instrument provides a continuous record of field strength. Its sensitivity of about 1 nT makes it capable of measuring most magnetic anomalies of geophysical interest. It is robust and adaptable to being mounted in an airplane, or towed behind it. The instrument was developed during World War II as a submarine detector. After the war it was used extensively in airborne magnetic surveying.

5.5.2.2 *The proton-precession magnetometer*

Since World War II sensitive magnetometers have been designed around quantum-mechanical properties. The proton-precession magnetometer depends on the fact that the nucleus of the hydrogen atom, a proton, has a magnetic moment proportional to the angular momentum of its spin. Because the angular momentum is quantized, the proton magnetic moment can only have specified values, which are multiples of a fundamental unit called the nuclear magneton. The situation is analogous to the quantization of magnetic moment associated with electron spin, for which the fundamental unit is the Bohr magneton. The ratio of the magnetic moment to the spin angular momentum is called the *gyromagnetic ratio* (γ_p) of the proton. It is an accurately known fundamental constant with the value $\gamma_p = 2.675\,13 \times 10^8\ s^{-1}\ T^{-1}$.

The proton-precession magnetometer is simple and robust in design. The sensor of the instrument consists of a flask containing a proton-rich liquid, such as water. Around the flask are wound a magnetizing solenoid and a detector coil (Fig. 5.42); some designs use the same solenoid alternately for magnetizing and detection. When the current in the magnetizing solenoid is switched on, it creates a magnetic field of the order of 10 mT, which is about 2000 times stronger than the Earth's field. The magnetizing field aligns the magnetic moments of the protons along the axis of the solenoid, which is oriented approximately east-west at right angles to the Earth's field. After the magnetizing field is interrupted, the magnetic moments of the proton spins react to the couple exerted on them by the Earth's magnetic field. Like a child's top spinning in the field of gravity, the proton magnetic moments precess about the direction of the ambient magnetic field. They do so at a rate known as the *Larmor precessional frequency*. The motion of the magnetic moments induces a signal in the detector coil. The induced signal is amplified electronically and the precessional frequency is accurately measured by counting cycles for a few seconds. The strength B_t of the measured magnetic field is directly proportional to the *frequency* of the signal (f), and is given by

$$B_t = \frac{2\pi}{\gamma_p} f \tag{5.43}$$

The intensity of the Earth's magnetic field is in the range 30,000–60,000 nT. The corresponding precessional frequency is approximately 1250–2500 Hz, which is in the audio-frequency range. Accurate measurement of the signal frequency gives an instrumental sensitivity of about 1 nT, but requires a few seconds of observation. Although it gives an absolute value of the field, the proton-precession magnetometer does not give a continuous record. Its portability and simplicity give it advantages for field use.

The flux-gate and proton-precession magnetometers are widely used in magnetic surveying. The two instruments

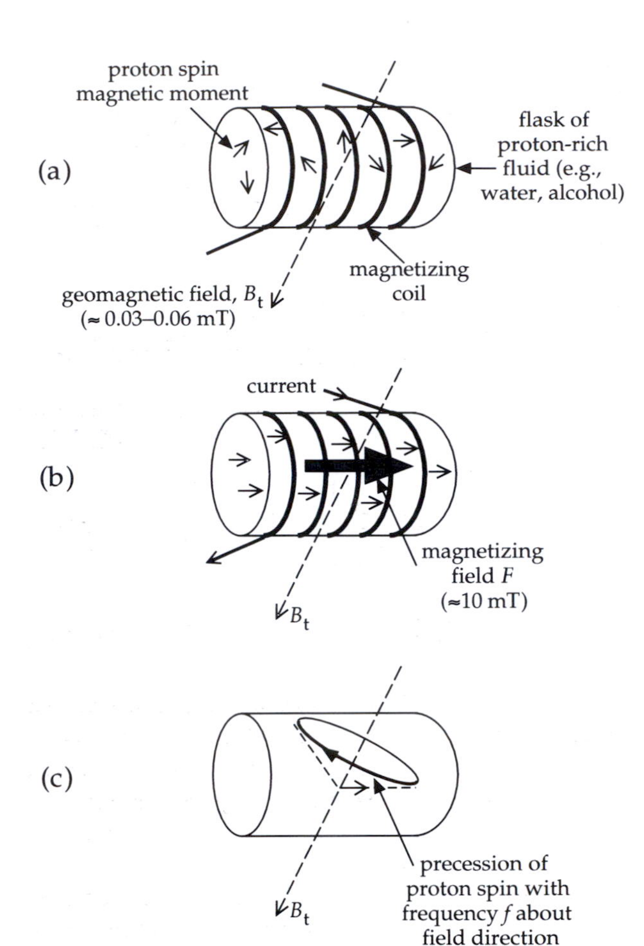

Fig. 5.42 (a) The elements of a proton-precession magnetometer. (b) Current in the magnetizing coil produces a strong field F that aligns the magnetic moments ('spins') of the protons. (c) When the field F is switched off, the proton spins precess about the geomagnetic field B_t, inducing an alternating current in the coil with the Larmor precessional frequency f.

have comparable sensitivities of 0.1–1 nT. In contrast to the flux-gate instrument, which measures the component of the field along its axis, the proton-precession magnetometer cannot measure field components; it is a total-field magnetometer. The total field $\mathbf{B}_t$ is the vector sum of the Earth's magnetic field $\mathbf{B}_E$ and the stray magnetic field $\Delta\mathbf{B}$ of, say, an orebody. Generally, $\Delta B \ll B_E$, so that the direction of the total field does not deviate far from the Earth's field. In some applications it is often adequate to regard the measured total field anomaly as the projection of $\Delta\mathbf{B}$ along the Earth's field direction.

5.5.2.3 *The absorption-cell magnetometer*

The absorption-cell magnetometer is also referred to as the alkali-vapor or optically pumped magnetometer. The principle of its operation is based on the quantum-mechanical model of the atom. According to their quantum numbers the electrons of an atom occupy concentric shells about the nucleus with different energy levels. The lowest energy level of an electron is its ground state. The magnetic moment associated with the spin of an electron can be either parallel or antiparallel to an external magnetic field. The energy of the electron is different in each case. This results in the ground state splitting into two sublevels with slightly different energies. The energy difference is proportional to the strength of the magnetic field. The splitting of energy levels in the presence of a magnetic field is called the *Zeeman effect*.

Absorption-cell magnetometers utilize the Zeeman effect in vapors of alkali elements such as rubidium or cesium, which have only a single valence electron in the outermost energy shell. Consider the schematic representation of an alkali-vapor magnetometer in Fig. 5.43. A polarized light-beam is passed through an absorption cell containing rubidium or cesium vapor and falls on a photoelectric cell, which measures the intensity of the light-beam. In the presence of a magnetic field the ground state of the rubidium or cesium is split into two sublevels, G_1 and G_2. If the exact amount of energy is added to the vapor, the electrons may be raised from their ground state to a higher-energy level, H. Suppose that we irradiate the cell with light from which we have filtered out the spectral line corresponding to the energy needed for the transition G_2H. The energy for the transition G_1H has not been removed, so the electrons in ground state G_1 will receive energy that excites them to level H, whereas those in ground state G_2 will remain in this state. The energy for these transitions comes from the incident light-beam, which is absorbed in the cell. In due course, the excited electrons will fall back to one of the more stable ground states. If an electron in excited state H falls back to sublevel G_1 it will be re-excited into level H; but if it falls back to sublevel G_2 it will remain there. In time, this process – called 'optical pumping' – will empty sublevel G_1 and fill level G_2. At this stage no more energy can be absorbed from the polarized light-beam and the absorption cell becomes transparent. If we now supply electromagnetic energy to the system in the form of a radio-frequency signal with just the right amount of energy to permit transitions between the populated G_2 and unpopulated G_1 ground sublevels, the balance will be disturbed. The optical pumping will start up again and will continue until the electrons have been expelled from the G_1 level. During this time energy is absorbed from the light-beam and it ceases to be transparent.

In the rubidium- and cesium-vapor magnetometers a polarized light-beam is shone at approximately 45° to the magnetic field direction. In the presence of the Earth's mag-

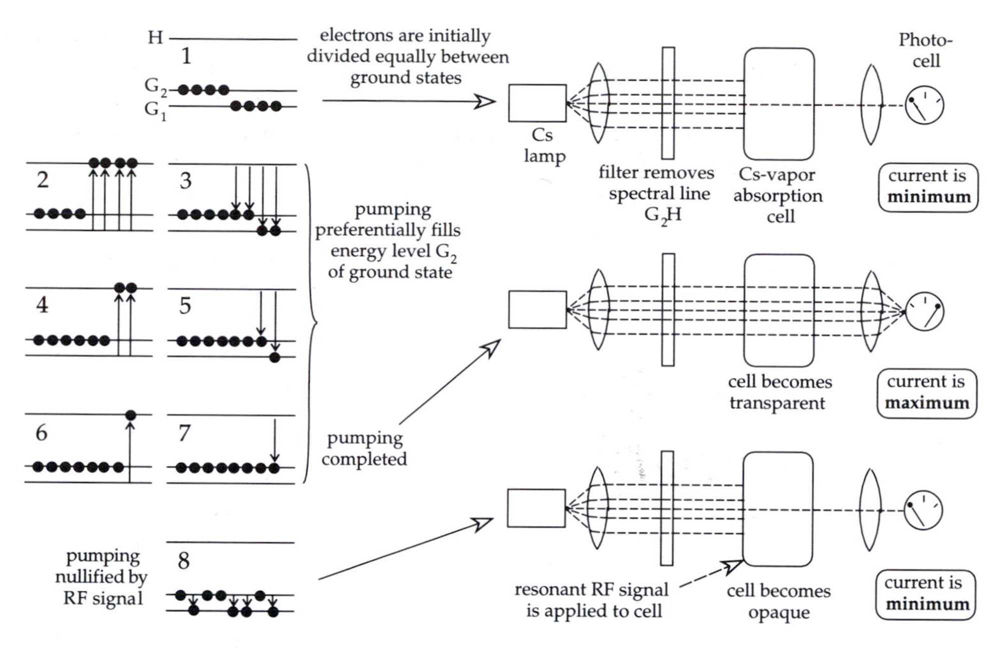

Fig. 5.43 The principle of operation of the optically pumped magnetometer (after Telford *et al.*, 1990).

netic field the electrons precess about the field direction at the Larmor precessional frequency. At one part of the precessional cycle an electron spin is almost parallel to the field direction, and one half-cycle later it is nearly antiparallel. The varying absorption causes a fluctuation of intensity of the light-beam at the Larmor frequency. This is detected by the photocell and converted to an alternating current. By means of a feedback circuit the signal is supplied to a coil around the container of rubidium gas and a radio-frequency resonant circuit is created. The ambient geomagnetic field B_t that causes the splitting of the ground state is proportional to the Larmor frequency, and is given by

$$B_t = \frac{2\pi}{\gamma_e} f \qquad (5.44)$$

Here, γ_e is the gyromagnetic ratio of the electron, which is known with an accuracy of about 1 part in 10^7. It is about 1800 times larger than γ_p, the gyromagnetic ratio of the proton. The precessional frequency is correspondingly higher and easier to measure precisely. The sensitivity of an optically pumped magnetometer is very high, about 0.01 nT, which is an order of magnitude more sensitive than the fluxgate or proton-precession magnetometer.

5.5.3 Magnetic surveying

The purpose of magnetic surveying is to identify and describe regions of the Earth's crust that have unusual (anomalous) magnetizations. In the realm of applied geophysics the anomalous magnetizations might be associated with local mineralization that is potentially of commercial interest, or they could be due to subsurface structures that have a bearing on the location of oil deposits. In global geophysics, magnetic surveying over oceanic ridges provided vital clues that led to the theory of plate tectonics and revealed the polarity history of the Earth's magnetic field since the Early Jurassic.

Magnetic surveying consists of (1) measuring the terrestrial magnetic field at predetermined points, (2) correcting the measurements for known changes, and (3) comparing the resultant value of the field with the expected value at each measurement station. The expected value of the field at any place is taken to be that of the International Geomagnetic Reference Field (IGRF), described in §5.4.4. The difference between the observed and expected values is a *magnetic anomaly*.

5.5.3.1 *Measurement methods*

The surveying of magnetic anomalies can be carried out on land, at sea and in the air. In a simple land survey an operator might use a portable magnetometer to measure the field at the surface of the Earth at selected points that form a grid over a suspected geological structure. This method is slow but it yields a detailed pattern of the magnetic field anomaly over the structure, because the measurements are made close to the source of the anomaly.

In practice, the surveying of magnetic anomalies is most efficiently carried out from an aircraft. The magnetometer must be removed as far as possible from the magnetic environment of the aircraft. This may be achieved by mounting the instrument on a fixed boom, A, several meters long (Fig. 5.44a). Alternatively, the device may be towed

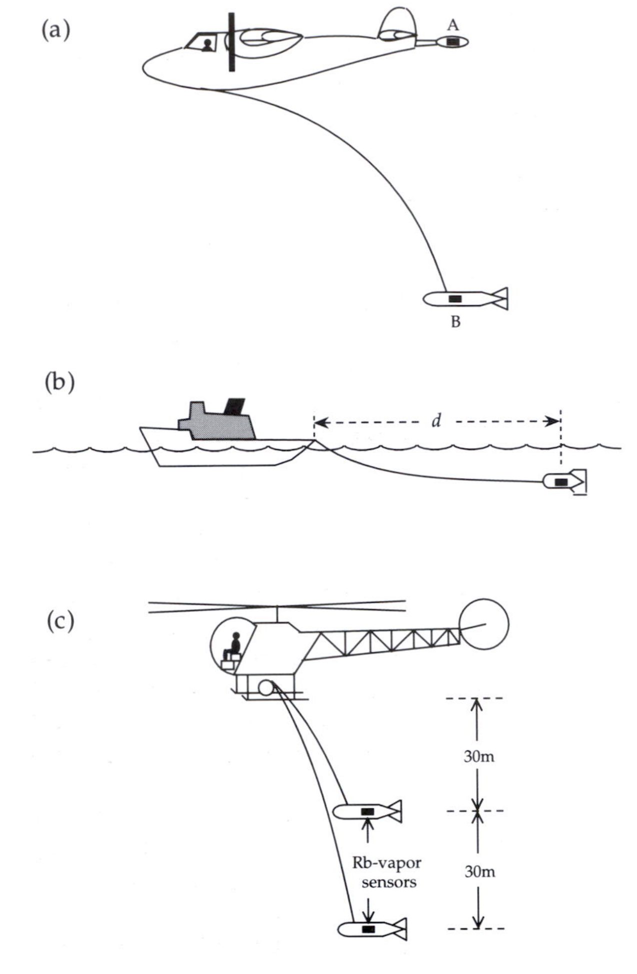

Fig. 5.44 (a) In airborne magnetic surveying the magnetometer may be mounted rigidly on the airplane at the end of a boom (A), or towed in an aerodynamic housing behind the plane (B). (b) In marine studies the magnetometer must be towed some distance *d* behind the ship to escape its magnetic field. (c) A pair of sensitive magnetometers in the same vertical plane act as a magnetic gradiometer (after Slack *et al.*, 1967).

behind the aircraft in an aerodynamic housing, B, at the end of a cable 30 to 150 m long. The 'bird' containing the magnetometer then flies behind and below the aircraft. The flight environment is comparatively stable. Airborne magnetometers generally have higher sensitivity ($\approx$0.01 nT) than those used in ground-based surveying (sensitivity $\approx$1 nT). This compensates for the loss in resolution due to the increased distance between the magnetometer and the source of the anomaly. Airborne magnetic surveying is an economical way to reconnoitre a large territory in a short time. It has become a routine part of the initial phases of the geophysical exploration of an uncharted territory.

The magnetic field over the oceans may also be surveyed from the air. However, most of the marine magnetic record has been obtained by shipborne surveying. In the marine application a proton-precession magnetometer mounted in a waterproof 'fish' is towed behind the ship at the end of a long cable (Fig. 5.44b). Considering that most research vessels consist of several hundred to several thousand tons of steel, the ship causes a large magnetic disturbance. For example, a research ship of about 1000 tons deadweight causes an anomaly of about 10 nT at a distance of 150 m. To minimize the disturbance of the ship the tow-cable must be about 100–300 m in length. At this distance the 'fish' in fact 'swims' well below the water surface. Its depth is dependent on the length of the tow-cable and the speed of the ship. At a typical survey speed of 10 km hr^{-1} its operational depth is about 10–20m.

5.5.3.2 *Magnetic gradiometers*

The magnetic gradiometer consists of a pair of alkali-vapor magnetometers maintained at a fixed distance from each other. In ground-based surveying the instruments are mounted at opposite ends of a rigid vertical bar. In airborne usage two magnetometers are flown at a vertical spacing of about 30 m (Fig. 5.44c). The difference in outputs of the two instruments is recorded. If no anomalous body is present, both magnetometers register the Earth's field equally strongly and the difference in output signals is zero. If a magnetic contrast is present in the subsurface rocks, the magnetometer closest to the structure will detect a stronger signal than the more remote instrument, and there will be a difference in the combined output signals.

The gradiometer emphasizes anomalies from local shallow sources at the expense of large-scale regional variation due to deep-seated sources. Moreover, because the gradiometer registers the *difference* in signals from the individual magnetometers, there is no need to compensate the measurements for diurnal variation, which affect each individual magnetometer equally. Proton-precession magnetometers are most commonly used in ground-based magnetic gradiometers, while optically pumped magnetometers are favored in airborne gradiometers.

5.5.3.3 *The survey pattern*

In a systematic regional airborne (or marine) magnetic survey the measurements are usually made according to a predetermined pattern. In surveys made with fixed-wing aircraft the survey is usually flown at a constant flight elevation above sea-level (Fig. 5.45a). This is the procedure favored for regional or national surveys, or for the investigation of areas with dramatic topographic relief. The survey focuses on the depth to the magnetic basement, which often underlies less-magnetic sedimentary surface rocks at considerable

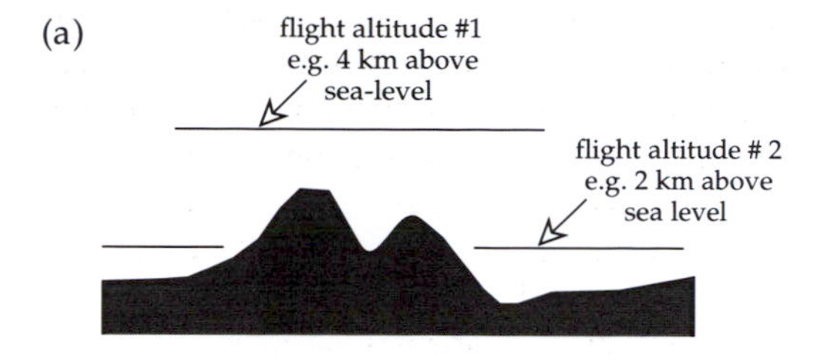

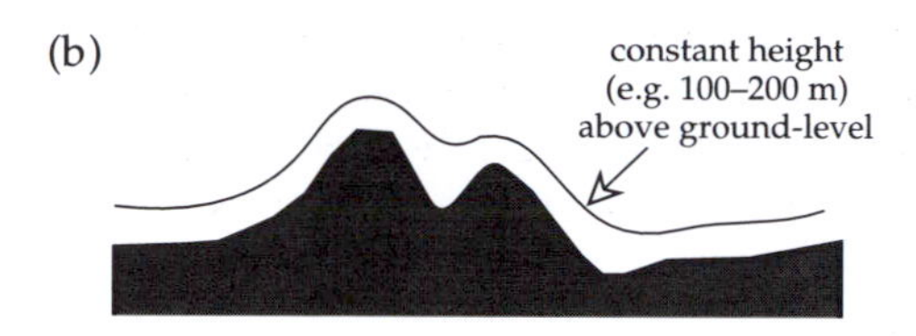

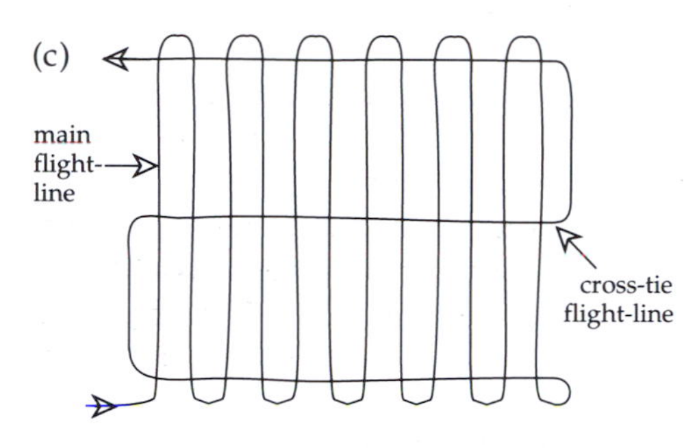

Fig. 5.45 In airborne magnetic surveying the flight-lines may be flown at (a) constant altitude above sea-level, or (b) constant height above ground-level. The flight pattern (c) includes parallel measurement lines and orthogonal cross-tie lines.

depth. In regions that are flat or that do not have dramatic topography, it may be possible to fly a survey at low altitude, as close as possible to the magnetic sources. This method would be suitable over ancient shield areas, where the goal of the survey is to detect local mineralizations with potential commercial value. If a helicopter is being employed, the distance from the magnetic sources may be kept as small as possible by flying at a constant height above the ground surface (Fig. 5.45b).

The usual method is to survey a region along parallel flight-lines (Fig. 5.45c), which may be spaced anywhere from 100 m to a few kilometers apart, depending on the flight elevation used, the intensity of coverage, and the quality of detail desired. The orientation of the flight-lines is selected to be more or less normal to the trend of suspected or known subsurface features. Additional tie-lines are flown at right angles to the main pattern. Their separation is about 5–6 times that of the main flight-lines. The repeatability of the measurements at the intersections of the tie-lines and the main flight-lines provides a check on the reliability of the survey. If the differences (called closure errors) are large, an area may need to be re-surveyed. Alternatively, the differences may be distributed mathematically among all the observations until the closure errors are minimum.

5.5.4 Reduction of magnetic field measurements

In comparison to the reduction of gravity data, magnetic survey data require very few corrections. One effect that must be compensated is the variation in intensity of the geomagnetic field at the Earth's surface during the course of a day. As explained in more detail in §5.4.3.3 this *diurnal variation* is due to the part of the Earth's magnetic field that originates in the ionosphere. At any point on the Earth's surface the external field varies during the day as the Earth rotates beneath different parts of the ionosphere. The effect is much greater than the precision with which the field can be measured. The diurnal variation may be corrected by installing a constantly recording magnetometer at a fixed base station within the survey area. Alternatively, the records from a geomagnetic observatory may be used, provided it is not too far from the survey area. The time is noted at which each field measurement is made during the actual survey and the appropriate correction is made from the control record.

The variations of magnetic field with *altitude*, *latitude* and *longitude* are dominated by the vertical and horizontal variations of the dipole field. The total intensity B_t of the field is obtained by computing the resultant of the radial component B_r (Eq.(5.38)) and the tangential component B_θ (Eq. (5.39)):

$$B_t = \sqrt{B_r^2 + B_\theta^2} = \frac{\mu_0 m}{4\pi} \frac{\sqrt{1+3\cos^2\theta}}{r^3} \tag{5.45}$$

The *altitude* correction is given by the vertical gradient of the magnetic field, obtained by differentiating the intensity B_t with respect to radius, r. This gives

$$\frac{\partial B_t}{\partial r} = -3\frac{\mu_0 m}{4\pi} \frac{\sqrt{1+3\cos^2\theta}}{r^4} = -\frac{3}{\mathrm{r}} B_t \tag{5.46}$$

The vertical gradient of the field is found by substituting $r=R=6371$ km and an appropriate value for B_t. It clearly depends on the latitude of the measurement site. At the magnetic equator ($B_t \approx 30{,}000$ nT) the altitude correction amounts to about 0.015 nT m^{-1}; near the magnetic poles ($B_t \approx 60{,}000$ nT) it is about 0.030 nT m^{-1}. The correction is so small that it is often ignored.

In regional studies the corrections for *latitude* and *longitude* are inherent in the reference field that is subtracted. In a survey of a small region, the latitude correction is given by the north-south horizontal gradient of the magnetic field, obtained by differentiating B_t with respect to polar angle, θ. This gives for the northward increase in B_t (i.e., with increasing latitude)

$$-\frac{1}{r}\frac{\partial B_t}{\partial \theta}=\frac{\mu_0 m}{4\pi}\frac{1}{r^4}\frac{\partial}{\partial \theta}\sqrt{1+3\cos^2\theta}=\frac{3B_t\sin\theta\cos\theta}{r(1+3\cos^2\theta)} \quad (5.47)$$

The latitude correction is zero at the magnetic pole (θ=0°) and magnetic equator (θ=90°) and reaches a maximum value of about 5 nT per *kilometer* (0.005 nT m^{-1}) at intermediate latitudes. It is insignificant in small-scale surveys.

In some land-based surveys of highly magnetic terrains (e.g., over lava flows or mineralized intrusions), the disturbing effect of the magnetized topography may be serious enough to require additional topographic corrections.

5.5.5 Magnetic anomalies

The gravity anomaly of a body is caused by the density contrast ($\Delta\rho$) between the body and its surroundings. Similarly, a magnetic anomaly originates in the magnetization contrast (ΔM) between rocks with different magnetic properties. Geophysical prospecting is carried out largely in continental crustal rocks, for which the Königsberger ratio is much less than unity (i.e., $Q_n \ll 1$) and the magnetization may be assumed to be induced by the present geomagnetic field. The magnetization contrast is then due to susceptibility contrast in the crustal rocks. If k represents the susceptibility of an orebody, k_0 the susceptibility of the host rocks and T the strength of the inducing magnetic field, Eq. (5.17) allows us to write the magnetization contrast as

$$\Delta M=(k-k_0)T \quad (5.48)$$

The physical processes that give rise to a magnetic anomaly can be illustrated for the special case of a vertical dike that strikes north–south and is magnetized by a vertical magnetic field. This is a simplified situation because in practice both the dike and the field will be inclined, probably at different angles. However, it allows us to make a few observations that are generally applicable. Let the dike be infinitely long to avoid perturbing effects that can occur near its ends; consider a west–east section through it normal to its length (Fig. 5.46b). Two scenarios are of particular interest. The first is when the dike has a large vertical extent, such that its bottom surface is at a great depth (Fig. 5.47a); the other is when the dike has a limited vertical extent (Fig. 5.47b). In both cases the vertical field will magnetize the dike parallel to its vertical sides, but the resulting anomalies have different shapes. To understand the anomaly shapes we will use the concept of magnetic pole distributions.

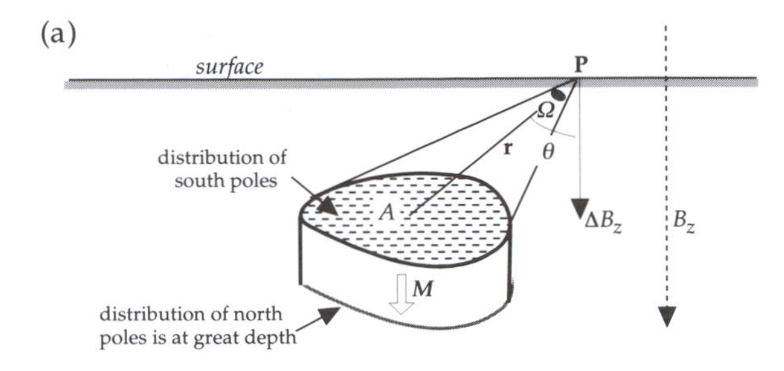

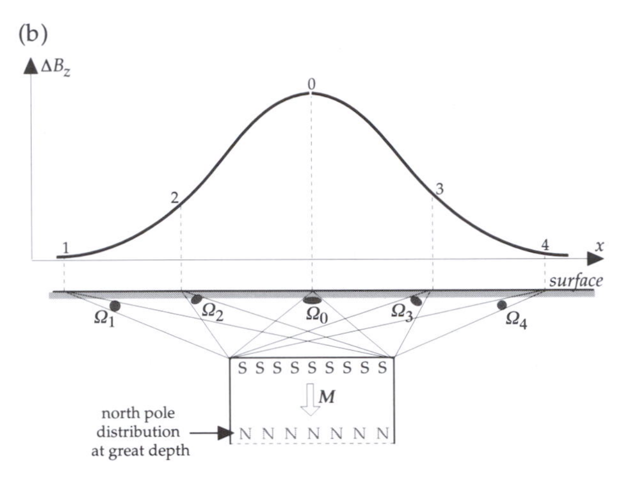

Fig. 5.46 Calculation of the magnetic anomaly of a vertical prism with infinite depth extent. For simplicity the magnetization M and inducing field T are both assumed to be vertical. (a) The distribution of magnetic poles on the top surface of the prism subtends an angle Ω at the point of measurement. (b) The magnetic anomaly ΔT varies along a profile across the prism with the value of the subtended angle Ω.

5.5.5.1 *Magnetic anomaly of a surface distribution of magnetic poles*

Although the magnetic pole is a fictive concept (see §5.2.2.1), it is widely used in magnetic prospecting as a convenient means of describing and computing magnetic fields. If a slice is made through a uniformly magnetized object, simple logic tells us that there will be as many south poles per unit of area on one side of the slice as north poles on the opposite side; these will cancel each other and the net sum of poles per unit area of the surface of the slice is zero. This is no longer the case if the magnetization changes across the intersecting surface. On each unit area of the surface there will be more poles of the stronger magnetization than poles of the weaker one. A quantitative derivation shows that the resultant number of poles per unit area (called the surface density of poles) is proportional to the magnetization contrast. As the example shows, magnetic pole distributions only appear on surfaces that intersect a magnetization.

Consider now the magnetic anomaly caused by such a

distribution of poles. The equation for the anomaly can be calculated exactly for many cases, but a qualitative understanding can be obtained from a simple example, which requires a slight acquaintance with spherical (conical) geometry. Euclidian geometry (and the rules of trigonometry) apply to two-dimensional situations. Spherical geometry handles the geometrical relationships of three-dimensional solid bodies. We will later encounter some of the trigonometrical relationships on the surface of a sphere in the calculation of paleomagnetic pole positions. At present, we only need to define what is meant by a *solid angle* (Fig. 5.46a). A surface of area A at a distance r from a point is said to subtend a solid angle W at the point given by $\Omega=A/r^2$. The minimum value of a solid angle is zero, when the surface is infinitesimally small. The maximum value of a solid angle is when the surface completely surrounds the point. For example, the area of the surface of a sphere of radius r is $A=4\pi r^2$ and the solid angle at its center has the maximum possible value, which is 4π. If the surface A is not normal, but is inclined at an angle θ to the radius r, the projection of A normal to r is used, which is $A\cos\theta$. In this case $\Omega=(A\cos\theta)/r^2$.

Now consider the distribution of poles on the upper surface of area A of a vertical prism with magnetization M induced by a vertical field B_z, as illustrated in Fig. 5.46a. At the surface of the Earth, distant r from the distribution of poles, the strength of their anomalous (or 'stray') magnetic field is proportional to the total number of poles on the surface, which is the product of A and the surface density σ of poles. Eq. (5.2) tells us that the intensity of the field of a pole decreases as the inverse square of distance r. If the direction of r makes an angle θ with the vertical field that induces the magnetization M, the vertical component of the anomalous field at P is found by multiplying by $\cos\theta$. The anomalous **B**-field (magnetic anomaly) ΔT of the surface distribution of poles is

$$\Delta T \propto \frac{\sigma A\cos\theta}{r^2} \propto \sigma\Omega \tag{5.49}$$

A more rigorous derivation leads to essentially the same result. At any point on a measurement profile, the magnetic anomaly ΔT of a distribution of poles is proportional to the solid angle Ω subtended by the distribution at the point. The solid angle changes progressively along a profile (Fig. 5.46b). At the extreme left and right ends, the radius from the observation point is very oblique to the surface distribution of poles and the subtended angles Ω_1 and Ω_4 are very small; the anomaly distant from the body is nearly zero. Over the center of the distribution, the subtended angle reaches its largest value Ω_0 and the anomaly reaches a maximum. The anomaly falls smoothly on each side of its crest corresponding to the values of the subtended angles Ω_2 and Ω_3 at the intermediate positions. A measurement profile across an equal distribution of 'north' poles would be exactly inverted. The north poles create a field of repulsion that acts everywhere to oppose the Earth's magnetic field, so the combined field is less than it would be if the 'north' poles were not there. The magnetic anomaly over 'north' poles is negative.

5.5.5.2 *Magnetic anomaly of a vertical dike*

We can now apply these ideas to the magnetic anomaly of a vertical dike. In this and all following examples we will assume a two-dimensional situation, where the horizontal length of the dike (imagined to be into the page) is infinite. This avoids possible complications related to 'end effects'. Let us first assume that the dike extends to very great depths (Fig. 5.47a), so that we can ignore the small effects associated with its remote lower termination. The vertical sides of the dike are parallel to the magnetization and no magnetic poles are distributed on these faces. However, the horizontal top face is normal to the magnetization and a distribution of magnetic poles can be imagined on this surface. The direction of magnetization is parallel to the field, so the pole distribution will consist of 'south' poles. The magnetized dike behaves like a magnetic *monopole*. The magnetometer measures the sum of the two 'stray fields'. At any point above the dike we measure both the inducing field and the *anomalous* 'stray field', which is directed toward the top of the dike. The anomalous field has a component parallel to the Earth's field and so the total magnetic field will be everywhere stronger than if the dike were not present. The magnetic anomaly is everywhere positive, increasing from zero far from the the dike to a maximum value directly over it (Fig. 5.46b).

If the vertical extent of the dike is finite, the distribution of north poles on the bottom of the dike is close enough to the ground surface to produce a measurable stray field. The upper distribution of south poles causes a positive magnetic anomaly, as in the previous example. The lower distribution of north poles causes a negative anomaly (Fig. 5.47b). The north poles are further from the magnetometer than the south poles, so their negative anomaly over the dike is weaker. However, farther along the profile the deeper distribution of poles subtends a larger angle than the upper one does. As a result, the strength of the weaker negative anomaly does not fall off as rapidly along the profile as the positive anomaly does. Beyond a certain lateral distance from the dike (to the left of L and to the right of R in Fig. 5.47b) the negative anomaly of the lower pole distribution is stronger than the positive anomaly of the upper one. This causes the magnetic anomaly to have negative side

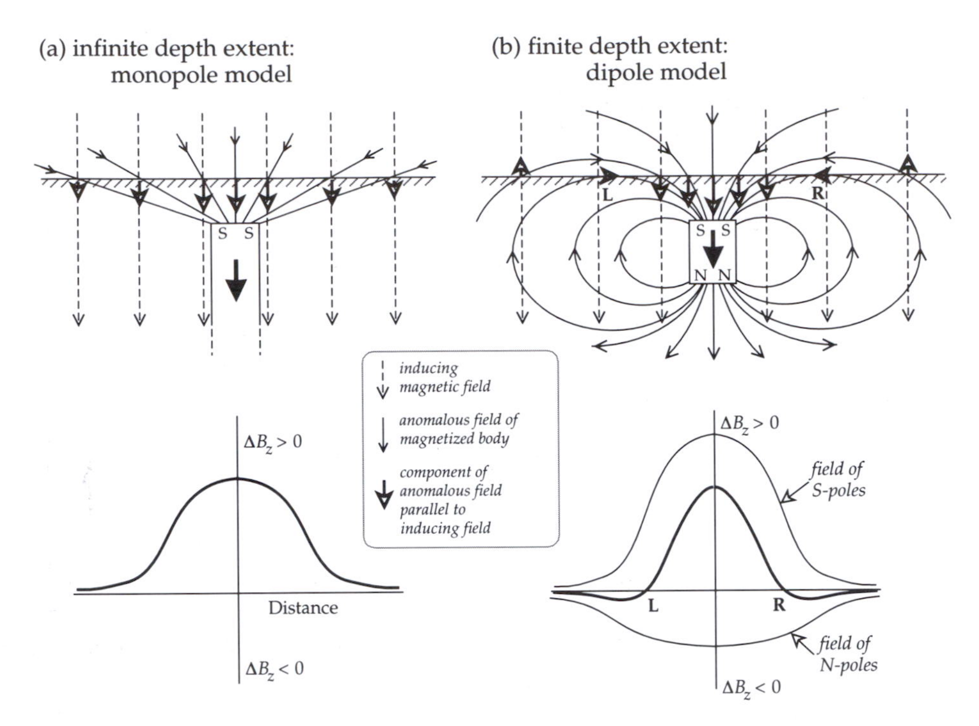

Fig. 5.47 (a) The total-field magnetic anomaly over a vertically magnetized block with infinite depth extent is due only to the distribution of poles on the top surface. (b) If the block has finite depth extent, the pole distributions on the top and bottom surfaces cause the anomaly.

lobes, which asymptotically approach zero with increasing distance from the dike.

The magnetized dike in this example resembles a bar magnet and can be modelled crudely by a *dipole*. Far from the dike, along a lateral profile, the dipole field-lines have a component opposed to the inducing field, which results in the weak negative side lobes of the anomaly. Closer to the dike, the dipole field has a component that reinforces the inducing field, causing a positive central anomaly.

5.5.5.3 *Magnetic anomaly of an inclined magnetization*

When an infinitely long dike is magnetized obliquely rather than vertically, its anomaly can be modelled either by an inclined dipole or by pole distributions (Fig. 5.48). The magnetization has both horizontal and vertical components, which produce magnetic pole distributions on the vertical sides of the dike as well as on its top and bottom. The symmetry of the anomaly is changed so that the negative lobe of the anomaly is enhanced on the side towards which the horizontal component of magnetization points; the other negative lobe decreases and may disappear.

The shape of a magnetic anomaly also depends on the angle at which the measurement profile crosses the dike, and on the strike and dip of the dike. The geometry, magnetization and orientation of a body may be taken into account in forward-modelling of an anomaly. However, as in other potential field methods, the inverse problem of determining these factors from the measured anomaly is not unique.

The asymmetry (or *skewness*) of a magnetic anomaly can be compensated by the method of *reduction to the pole*. This consists of recalculating the observed anomaly for the case that the magnetization is vertical. The method involves sophisticated data-processing beyond the scope of this text. The observed anomaly map is first converted to a matrix of values at the intersections of a rectangular grid overlying the map. The Fourier transform of the matrix is then computed and convolved with a filter function to correct for the orientations of the body and its magnetization. The reduction to the pole removes the asymmetry of an anomaly (Fig. 5.49) and allows a better location of the margins of the disturbing body. Among other applications, the procedure has proved to be important for detailed interpretation of the oceanic crustal magnetizations responsible for lineated oceanic magnetic anomalies.

5.5.5.4 *Effect of block width on anomaly shape*

When the width of an infinitely long dike increases appreciably, it is more appropriate to treat the magnetized body as a crustal block. The effect of block width on anomaly shape can be understood by reverting to the discussion of magnetic pole distributions. If it is magnetized in the direction of the present field, a dike-like crustal block that is rather narrow compared to its depth extent gives a fairly sharp positive anomaly with negative side lobes (Fig. 5.50a), for the reasons given in §5.5.5.2. However, consider the anomaly that would be observed over the middle of a crustal block that is very wide in comparison to its depth extent (Fig. 5.50d). Over the center of the block the angle

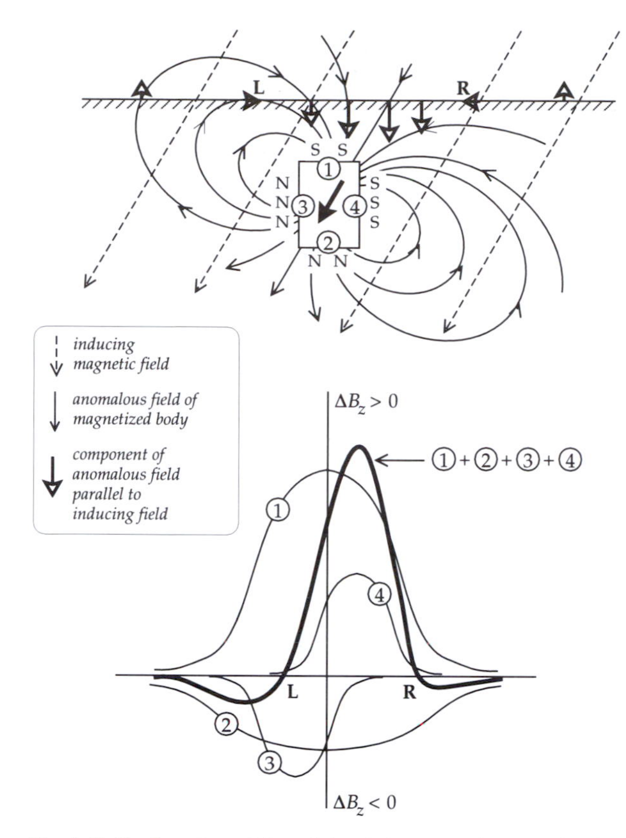

Fig. 5.48 Explanation of the origin of the magnetic anomaly of an infinitely long vertical prism in terms of the pole distributions on top, bottom and side surfaces, when the magnetic field (or magnetization) is inclined.

subtended by the lower distribution of north poles is almost (but not quite) as large as the angle subtended by the upper distribution of south poles. As a result, the positive anomaly over the center of the block is almost zero. Over the block but close to its edges, the angle subtended by the upper layer of south poles is larger and the anomaly is positive; off the edges of the block the angle subtended by the lower layer of north poles is larger and the anomaly is negative. The pronounced dip in the middle of the anomaly is present because of the limited depth extent of the block; if it were infinitely deep, the competing effect of the north poles would be missing and the shape of the anomaly would be flat-topped. The central dip develops progressively as the width of the block increases with respect to its depth extent (Fig. 5.50b, c). Examination of Fig. 5.46 shows that the subtended angle and, consequently, the anomaly shape also depend on the height of the measurement above the surface of the block. A low-level measurement profile over the block will show a large central dip, while a high profile over the same block will show a smaller dip or none at all.

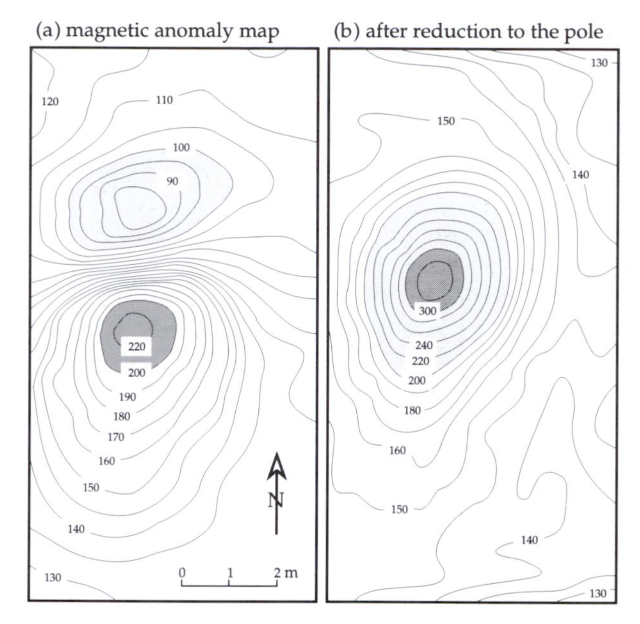

Fig. 5.49 Effect of data-processing by reduction to the pole on the magnetic anomaly of a small vertical prism with an inclined magnetization. In (a) the contour lines define a dipole-type of anomaly with regions of maximum and minimum intensity (in nT); in (b) the anomaly after reduction to the pole is much simpler and constrains better the location of the center of the prism (after Lindner *et al.*, 1984).

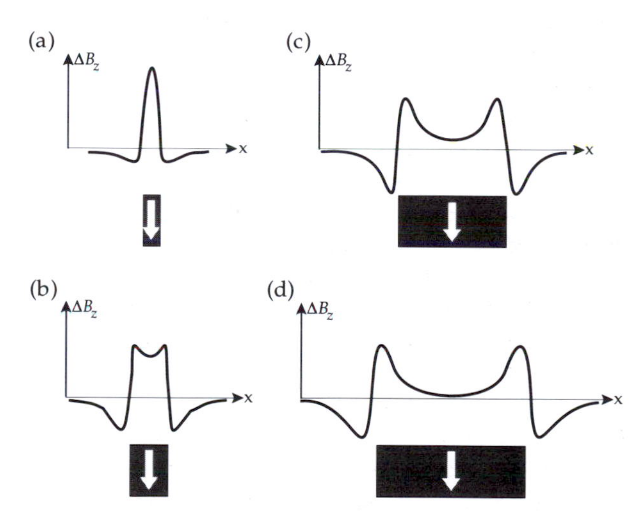

Fig. 5.50 The effect of block width on the shape of the magnetic anomaly over a vertically magnetized crustal block.

5.5.6 Oceanic magnetic anomalies

In the late 1950s marine geophysicists conducting magnetic surveys of the Pacific ocean basin off the west coast of North America discovered that large areas of oceanic crust are characterized by long stripes of alternating positive and

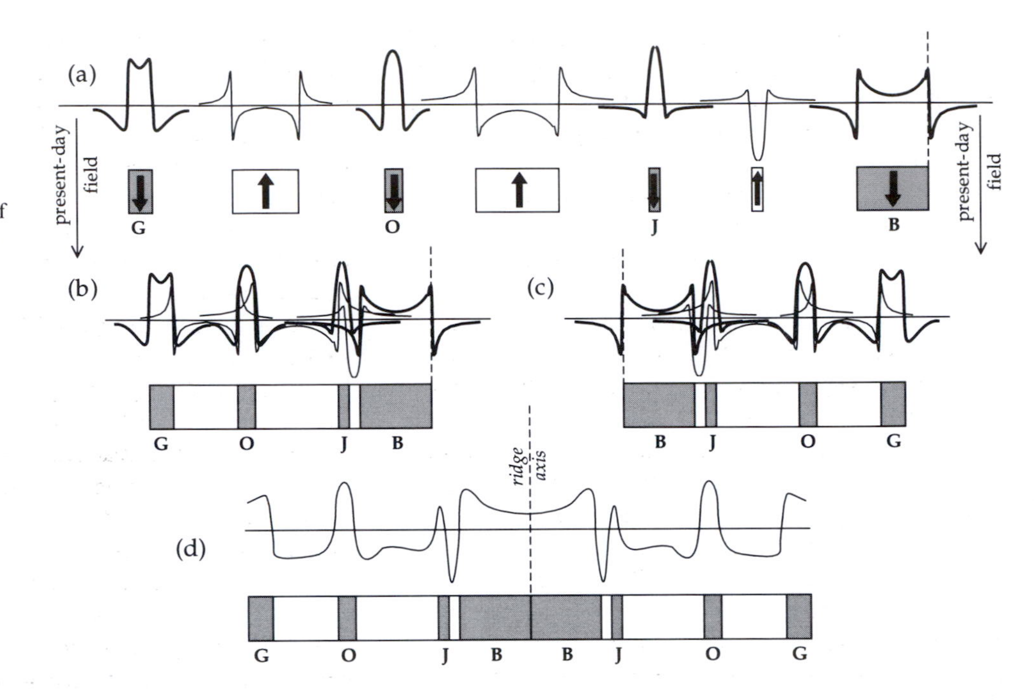

Fig. 5.51 Explanation of the shape of a magnetic profile across an oceanic spreading center: (a) the anomalies of individual oppositely magnetized crustal blocks on one side of the ridge, (b) overlap of the individual anomalies, (c) the effect for the opposite sequence of blocks on the other side of the ridge, and (d) the complete anomaly profile.

negative magnetic anomalies. The striped pattern is best known from studies carried out across oceanic ridge systems (see Fig. 1.13). The striped anomalies are hundreds of kilometers in length parallel to the ridge axis, 10–50 km in width, and their amplitudes amount to several hundreds of nanotesla. On magnetic profiles perpendicular to a ridge axis the anomaly pattern is found to exhibit a remarkable symmetry about the axis of the ridge. The origin of the symmetric lineated anomaly pattern cannot be explained by conventional methods of interpretation based on susceptibility contrast.

Seismic studies indicate a layered structure for the oceanic crust. The floor of the ocean lies at water depths of 2–5 km, and is underlain by a layer of sediment of variable thickness, called seismic Layer 1. Under the sediments lie a complex of basaltic extrusions and shallow intrusions, about 0.5 km thick, forming seismic Layer 2A, under which are found the deeper layers of the oceanic crust consisting of a complex of sheeted dikes (Layer 2B) and gabbro (Layer 3). The magnetic properties of these rocks were first obtained by studying samples dredged from exposed crests and ridges of submarine topography. The rocks of Layers 2B and 3 are much less magnetic than those of Layer 2A. Samples of pillow basalt dredged near to oceanic ridges have been found to have moderate susceptibilities for igneous rocks, but their remanent magnetizations are intense. Their Königsberger ratios are commonly in the range 5–50 and frequently exceed 100. Recognition of these properties provided the key to understanding the origin of the lineated magnetic anomalies. In 1963 the English geophysicists F. J. Vine and D. H. Matthews proposed that the *remanent magnetizations* (and not the susceptibility contrast) of oceanic basaltic Layer 2 were responsible for the striking lineated anomaly pattern. This hypothesis soon became integrated into a working model for understanding the mechanism of sea-floor spreading (see §1.2.5 and Fig. 1.14).

The oceanic crust formed at a spreading ridge acquires a thermoremanent magnetization (TRM) in the geomagnetic field. The basalts in Layer 2A are sufficiently strongly magnetized to account for most of the anomaly measured at the ocean surface. For a lengthy period of time (measuring several tens of thousands to millions of years) the polarity of the field remains constant; crust formed during this time carries the same polarity as the field. After a polarity reversal, freshly formed basalts acquire a TRM parallel to the new field direction, i.e., opposite to the previous TRM. Adjacent oceanic crustal blocks of different widths, determined by the variable time between reversals, carry antiparallel remanent magnetizations.

The oceanic crust is magnetized in long blocks parallel to the spreading axis, so the anomaly calculated for a profile perpendicular to the axis is two-dimensional, as in the previous examples. Consider the case where the anomalies on a profile have been reduced to the pole, so that their magnetizations can be taken to be vertical. We can apply the concept of magnetic pole distributions to each block individually to determine the shape of its magnetic anomaly (Fig. 5.51a). If the blocks are contiguous, as is the case when they form by a continuous process such as sea-floor spreading, their individual anomalies will overlap (Fig. 5.51b). The spreading process is symmetric with respect to the ridge axis, so a mirror-image of the sequence of polarized blocks is formed on the other side of the axis (Fig. 5.51c). If the two sets of crustal blocks are brought together at the spreading axis, a magnetic anomaly sequence ensues that exhibits a symmetric pattern with respect to the ridge axis (Fig. 5.51d).

This description of the origin of oceanic magnetic anomalies is over-simplified, because the crustal magnetization is more complicated than assumed in the block model. For example, the direction of the remanent magnetization, acquired at the time of formation of the ocean crust, is generally not the same as the direction of the magnetization induced by the present-day field. However, the induced magnetization has uniformly the same direction in the magnetized layer, which thus behaves like a uniformly magnetized thin horizontal sheet and does not contribute to the magnetic anomaly. Moreover, oceanic rocks have high Königsberger ratios, and so the induced magnetization component is usually negligible in comparison to the remanent magnetization. An exception is when a magnetic survey is made close to the magnetized basalt layer, in which case a topographic correction may be needed.

Unless the strike of a ridge is north–south, the magnetization inclination must be taken into account. Skewness is corrected by reducing the magnetic anomaly profile to the pole (§5.5.5.3). A possible complication may arise if the oceanic magnetic anomalies have two sources. The strongest anomaly source is doubtless basaltic Layer 2A, but, at least in some cases, an appreciable part of the anomaly may arise in the deeper gabbroic Layer 3. The two contributions are slightly out of phase spatially, because of the curved depth profiles of cooling isotherms in the oceanic crust. This causes a magnetized block in the deeper gabbroic layer to lie slightly further from the ridge than the corresponding block with the same polarity in the basaltic layer above it. The net effect is an asymmetry of inclined magnetization directions on opposite sides of a ridge, so that the magnetic anomalies over blocks of the same age have different skewnesses.

5.6 PALEOMAGNETISM

5.6.1 **Introduction**

A mountain walker using a compass to find his way in the Swiss Alps above the high mountain valley of the Engadine would notice that in certain regions (for example, south of the Septimer Pass) the compass-needle shows very large deviations from the north direction. The deflection is due to the local presence of strongly magnetized serpentinites and ultramafic rocks. Early compasses were more primitive than modern versions, but the falsification of a compass direction near strongly magnetic outcrops was known by at least the early 19th century. In 1797 Alexander von Humboldt proposed that the rocks in these unusual outcrops had been magnetized by lightning strikes. The first systematic observations of rock magnetic properties are usually attributed to A. Delesse (1849) and M. Melloni (1853), who concluded that volcanic rocks acquired a remanent magnetization during cooling. After a more extensive series of studies in 1894 and 1895 of the origin of magnetism in lavas G. Folgerhaiter reached the same conclusion and suggested that the direction of remanent magnetization was that of the geomagnetic field during cooling. By 1899 he had extended his work to the record of the secular variation of inclination in ancient potteries. Folgerhaiter noted that some rocks have a remanent magnetization opposite to the direction of the present-day field. Reversals of polarity of the geomagnetic field were established decisively early in the 20th century.

In 1922 Alfred Wegener proposed his concept of continental drift, based on years of study of paleoclimatic indicators such as the geographic distribution of coal deposits. At the time, there was no way of explaining the mechanism by which the continents drifted. Only motions of the crust were considered, and the idea of rigid continents ploughing through rigid oceanic crust was unacceptable to geophysicists. There was as yet no way to reconstruct the positions of the continents in earlier eras or to trace their relative motions. Subsequently, paleomagnetism was to make important contributions to understanding continental drift by providing the means to trace past continental motions quantitatively.

A major impetus to these studies was the invention of a very sensitive *astatic magnetometer*. The apparatus consists of two identical small magnets mounted horizontally at opposite ends of a short rigid vertical bar so that the magnets are oriented exactly antiparallel to each other. The assembly is suspended on an elastic fiber. In this configuration the Earth's magnetic field has equal and opposite effects on each magnet. If a magnetized rock is brought close to one magnet, the magnetic field of the rock produces a stronger twisting effect on the closer magnet than on the distant one and the assembly rotates to a new position of equilibrium. The rotation is detected by a light beam reflected off a small mirror mounted on the rigid bar. The device was introduced in 1952 by P. M. S. Blackett to test a theory that related the geomagnetic field to the Earth's rotation. The experiment did not support the postulated effect.

However, the astatic magnetometer became the basic tool of paleomagnetism and fostered its development as a scientific discipline. Hitherto it had only been possible to measure magnetizations of strongly magnetic rocks. The astatic magnetometer enabled the accurate measurement of weak remanent magnetizations in rocks that previously had been unmeasurable.

In the 1950s, several small research groups were engaged in determining and interpreting the directions of magnetization of rocks of different ages in Europe, Africa, North and South America and Australia. In 1956 S. K. Runcorn put forward the first clear geophysical evidence in support of continental drift. Runcorn compared the directions of magnetization of Permian and Triassic rocks from Great Britain and North America. He found that the paleomagnetic results from the different continents could be brought into harmony for the time before 200 Ma ago by closing the Atlantic ocean. The evaluation of the scientific data was statistical and at first was regarded as controversial. However, Mesozoic paleomagnetic data were soon obtained from the southern hemisphere that also argued strongly in favor of the continental drift hypothesis. In 1957 E. Irving showed that paleomagnetic data conformed better with geological reconstructions of earlier positions of the continents than with their present-day distribution. Subsequently, numerous studies have documented the importance of paleomagnetism as a chronicle of past motions of global plates and as a record of the polarity history of the Earth's magnetic field.

5.6.2 The time-averaged geomagnetic field

A fundamental assumption of paleomagnetism is that the time-averaged geomagnetic field corresponds to that of an axial geocentric dipole. The data in support of this important hypothesis come partly from studies of secular variation and partly from paleomagnetic observations in young rocks and sediments.

The dipole and non-dipole parts of the historic geomagnetic field are known to change slowly with time. Spherical harmonic analysis of the geomagnetic field (see §5.4.5.1) shows that the axis of the geocentric inclined dipole has drifted slowly westward at about 0.044–0.14 deg yr^{-1} in the last 400 yr (see Fig. 5.36). This would correspond to a complete circuit about the rotation axis in 2500–8000 yr. The rates of westward drift of the historic non-dipole field are around 0.22–0.66 deg yr^{-1} (see Fig. 5.37), giving a periodicity of 550–1650 yr. However, it is important to keep in mind that the historic records of the secular variation of the geomagnetic field cover only a fragment of a complete circuit, which is not enough to confirm cyclicity or to estimate a period. The record of earlier magnetic field intensity and direction must be inferred from *archeomagnetism*.

5.6.2.1 *Archeomagnetic records of secular variation*

Paleomagnetism is the study of the geomagnetic field recorded in rock magnetizations; archeomagnetism is the study of the geomagnetic field recorded in dateable historic artefacts. The age of an archeological relict such as a pot or vase, can often be determined with reliable precision. The pot, and the oven in which it was fired, may carry a thermoremanent magnetization (TRM) acquired during cooling. The direction of the TRM can be measured easily, and, if the attitude of the pot during firing is known or can be assumed, the inclination of the ancient magnetic field in which the artefact was made can be deduced. The same considerations apply to lava flows that can be dated from historic records.

The secular record of paleoinclination during the past 2000 yr is available for two regions in which many archeomagnetic studies have been carried out: southeastern Europe and southwestern Japan. These regions are 110° apart in longitude but lie in similar latitude ranges, 35–40°N. Smoothed curves through the observations show pseudo-cyclical changes with several maxima and minima (Fig. 5.52). The shapes of the curves are not distinctive, so correlation of individual extreme values is risky; however, comparison of the four numbered maxima and minima in the last 1400 yr suggests that the extreme values appear to occur about 400 yr earlier in Japan than in Europe. The equivalent period for a full circuit of the globe is 1300 yr. The pseudo-cyclicity is interpreted as the effect of westward drift of foci of the non-dipole field past the sampling site. More detailed analysis of the archeomagnetic data shows that the drift rates vary with the latitude of the observation site (see Fig. 5.37). The mean drift rate is 0.38° longitude per year, which is faster than the rate deduced from recent secular variation.

A subtle magnetic technique devised by Thellier in 1937 permits the intensity of the field in which the objects acquired TRM to be determined. Assuming this to be due to a dipole field, the strength of the dipole can be inferred. Of course, as the archeomagnetic record of paleoinclination shows, the assumption of a dipole field is not really justifiable; the archeological relict cools rapidly and records the total field at the time of cooling, which contains a substantial non-dipole component. For this reason paleointensity values are usually averaged over intervals of 500–1000 years, in the hope of averaging out the non-dipole contributions. The record of paleointensity of the dipole moment is rather ragged, with a rather broad spread of values for each 500 yr interval, but the

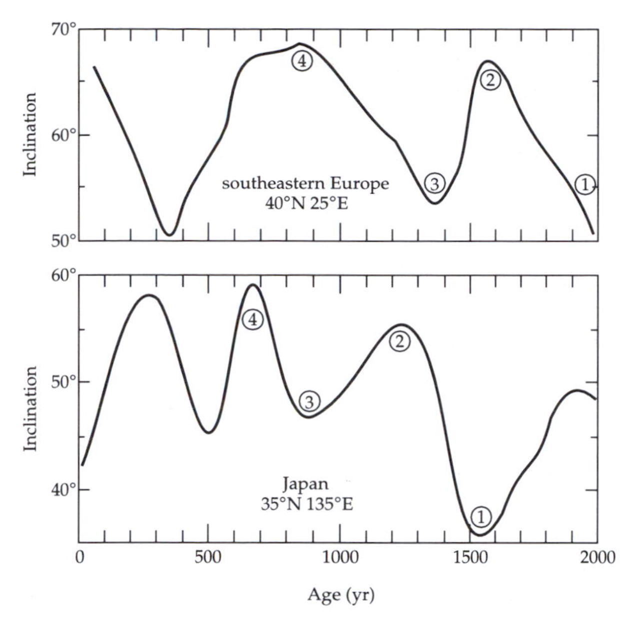

Fig. 5.52 Secular variation of geomagnetic inclination from archeomagnetic studies in southeastern Europe and southwestern Japan (after Merrill and McElhinny, 1983).

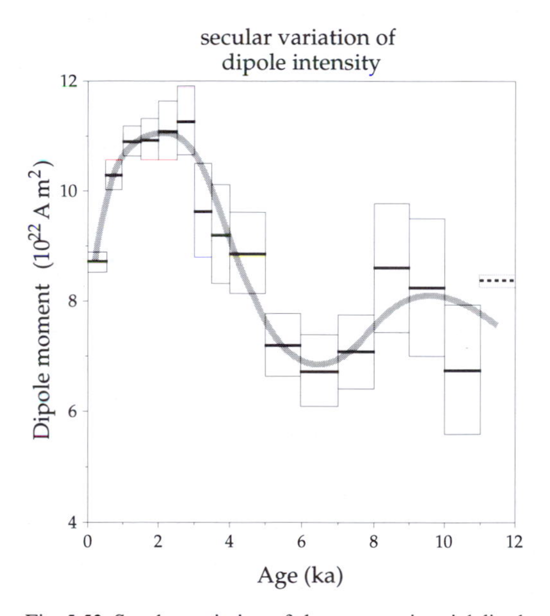

Fig. 5.53 Secular variation of the geocentric axial dipole moment derived from paleointensity measurements (data source: Merrill and McElhinny, 1983).

mean values for the past 10,000 yr show apparently cyclic changes with a pseudo-period of about 8000 yr (Fig. 5.53). However, prior to 10,000 yr ago the data are few and the error ranges are large. Unfortunately, European archeomagnetic intensity data in the age range 8000–12,000 yr have high values, but results of the same age from the rest of the world show low intensities. Thus, the apparent cyclicity in intensity may not be real.

Westward drift of the dipole field implies systematic changes in the equatorial components of the dipole. If the changes are approximately cyclical, the mean long-term strength of the equatorial dipole measured at any give site would average to zero within a few multiples of 10^4 yr. Similarly, if the secular variation of the non-dipole field can be assumed to be roughly periodic its mean value should average to zero within a few multiples of 10^3 yr. According to these arguments, the only long-term component of the geomagnetic field that persists and is not averaged to zero within a few tens of thousands of years is the axial dipole component. The long-term equivalence of the Earth's magnetic field with that of a dipole located at the center of the Earth and oriented along the rotation axis is a fundamental tenet of paleomagnetism; it is called the *axial geocentric dipole hypothesis*.

5.6.2.2 *The axial geocentric dipole hypothesis*

The evidence in support of the axial geocentric dipole hypothesis comes from paleomagnetic studies in modern deep-sea sediments and young igneous and sedimentary rocks. Pelagic sediments are deposited extremely slowly in the deep ocean basins. Sedimentation rates of 1–10 m Ma^{-1} are common. The sediments acquire a post-depositional remanent magnetization (pDRM), which is an accurate record of the depositional field direction. The slow deposition of deep-sea sediments ensures thorough averaging of the magnetic field recorded. For example, at pelagic sedimentation rates a typical one-inch thick sample of deep-sea sediment averages paleomagnetic directions acquired during 2500-25,000 yr of deposition. The test of the axial geocentric dipole hypothesis in modern deep-sea sediments consists of comparing the inclination observed in sediment samples with the inclination expected for the latitude of the site where the sediment was sampled. The relationship between field inclination I, magnetic co-latitude θ and latitude λ was developed in Eq. (5.40) and is shown in Fig. 5.54a. The mean inclinations of remanent magnetization were measured in fifty-two deep-sea sediment cores of Plio-Pleistocene age taken from sites at different latitudes in the northern and southern hemisphere. The observed inclinations agree well with the values predicted by the theoretical curve for the axial geocentric dipole hypothesis (Fig. 5.54b).

Assuming the direction of the magnetic field recorded at a given site to be that of a dipole field, it is possible to calculate where the geomagnetic pole would need to be in order to produce the observed declination and inclination. This

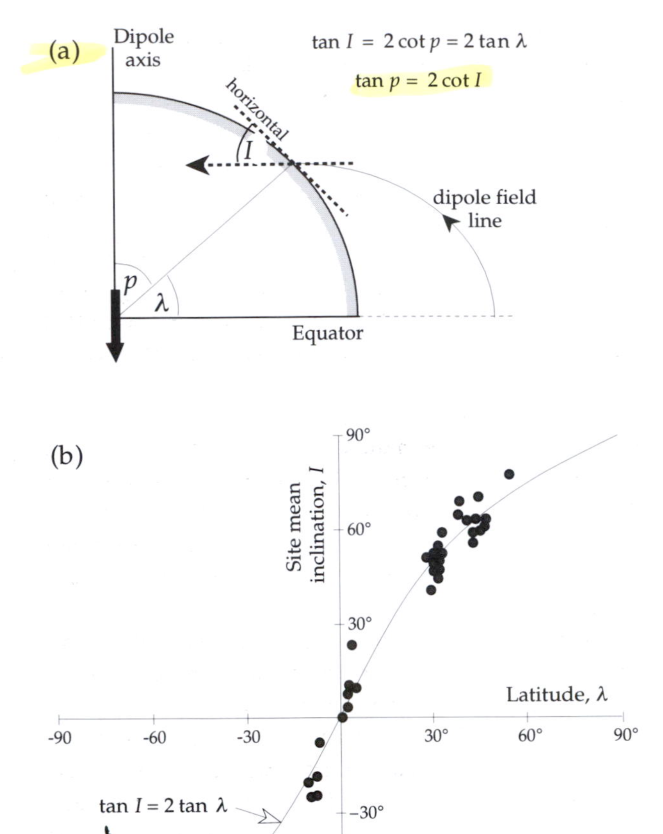

Fig. 5.54 (a) The geocentric axial dipole hypothesis predicts the relationship $\tan I = 2\tan\lambda$ between the inclination I of a dipole field and the magnetic latitude λ. (b) The inclinations measured in modern deep-sea sediment cores agree well with the theoretical curve (based on data from Opdyke and Henry, 1969).

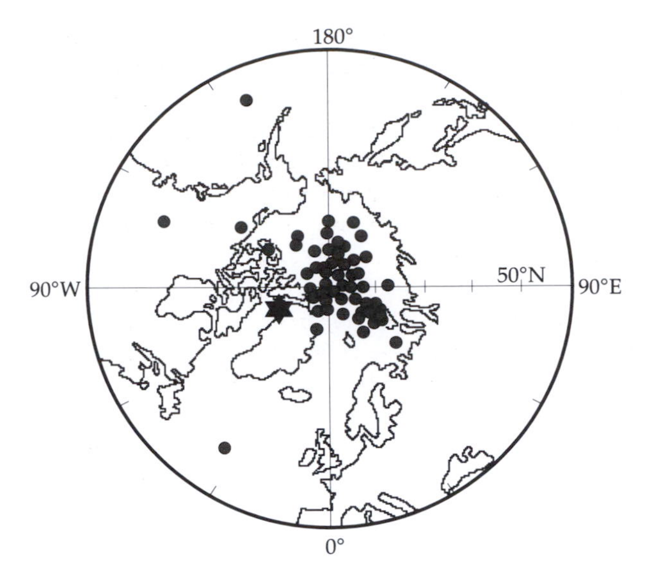

Fig. 5.55 Paleomagnetic pole positions for rocks of Plio-Pleistocene to Recent age (after McElhinny, 1973).

location is called the *virtual geomagnetic pole* (VGP) position. It is useful in computing where the pole lay in ancient times, the so-called *paleomagnetic pole*. The difference between a VGP and a paleomagnetic pole is illustrated by the following example for a recently extruded lava. Each sample from the lava formed in a short interval of time, and the field direction it records will be that of the total geomagnetic field at the site, combining axial dipole, non-dipole and non-axial dipole components. The VGP will therefore not coincide with the rotation axis. If data are collected from several flows of different ages, each will carry a slightly different record of the field. The computed VGP position will be different from flow to flow, and so the distribution of VGP will be scattered. If samples are measured from a large number of recent lava flows covering a long enough period of time to average the non-dipole and non-axial dipole parts to zero, the mean direction of the collection will correspond to the field of an axial geocentric dipole. The pole position calculated from the mean direction of the collection of flows will agree with the rotation axis. This pole, representing an averaged value of the field, is called a *paleomagnetic pole*. The VGP represents a spot estimate of the field, including non-axial dipole components; the paleomagnetic pole represents an averaged field, corresponding to the axial geocentric dipole.

The paleomagnetic pole positions determined in studies of Plio-Pleistocene to Recent volcanic and sedimentary rocks covering the past 5 Ma lend further support to the axial geocentric dipole hypothesis. The distribution of paleomagnetic poles is clustered around the geographic pole and not around the present-day geomagnetic pole (Fig. 5.55). Statistical analysis shows that the mean of the paleomagnetic poles does not differ significantly from the geographic pole.

The axial geocentric dipole hypothesis maintains that, if data are averaged over a long enough interval of time, the mean paleomagnetic pole position will coincide with the axis of rotation of the Earth. In fact, detailed analysis of Late Tertiary paleomagnetic poles has shown that this hypothesis does not hold exactly. This is because the mean pole position calculated for any field that is symmetric about the rotation axis will lie on the axis, provided the directions are obtained at sites covering a wide range of longitudes. When young paleomagnetic data of the same age are averaged for a particular region, they give a paleomagnetic pole position on the far side of the present-day rotation axis. This 'far-sidedness' of paleomagnetic directions is caused by the presence of a small axial geocentric quadrupole, amounting to a few percent of the axial geocentric dipole. The superposition of the axial dipole and quadrupole is equivalent to displacing the center of the dipole about 300 km northward along the rotation axis away from the center of the Earth. This is a second-order effect; to a first approximation the time-averaged paleomagnetic field may be considered to be that of an axial geocentric dipole, and the paleomagnetic pole lies within a few degrees of the rotational pole.

5.6.3 Methods of paleomagnetism

The requirement that the mean paleomagnetic pole position derived for a collection of rocks should represent the axial geocentric dipole is taken into account in the methodology of paleomagnetic analysis. This begins with the sampling of a rock formation on a hierarchical scheme designed to eliminate or minimize non-systematic errors and to average out the effects of secular variation of the paleomagnetic field. At each hierarchical level, averaging and statistical analysis are carried out on the remanent magnetization vectors. Ideally, a paleomagnetic collection should contain a large number of samples per site. In practice, about 6–10 samples are enough to define the mean direction for a site; the mean values of typically 10–20 sites from the same formation are averaged to get a mean paleomagnetic direction for a formation or region.

A further assumption of paleomagnetism is that the natural remanent magnetization (NRM) of a rock was acquired at the time of formation of the rock (or at a known time in its history), and has since remained unaltered. In fact, the NRM is usually made up of several components acquired at different times, including during the procedures of sampling and preparation. Laboratory techniques must be applied that eliminate the undesirable components and isolate the primary magnetization. This process is loosely called 'magnetic cleaning'.

The presentation of paleomagnetic directions measured in rock samples is made with the help of stereographic projection. This is a way of plotting three-dimensional directions by projecting them onto a plane. These plots have already been encountered in the analysis of first motion studies of earthquakes (see §3.5.4.2). A direction is identified by the point where it intersects a unit sphere centered at the observation site. This converts a set of directions to a set of points on the surface of a sphere. The intersection point is then projected onto the horizontal plane to give a stereographic plot. This can be done in different ways. The Lambert equal-area projection is usually preferred in paleomagnetism as it avoids visually distorting the dispersion of directions. In geology, all directions are plotted on a stereogram as projections on the lower hemisphere. In paleomagnetic stereograms directions with positive (downward) inclinations are plotted as lower hemisphere projections; directions with negative (upward) inclinations are plotted with a different symbol as upper hemisphere projections.

5.6.3.1 *Measurement of remanent magnetization*

Measurements of the natural remanent magnetization of rocks with an astatic magnetometer were laborious and time-consuming and the instrument has now fallen into disuse. In modern paleomagnetic laboratories more efficient spinner magnetometers and cryogenic magnetometers are in common use.

Spinner magnetometers originally consisted of a large sensor coil containing many turns of wire in which an alternating signal was induced by rotating the sample at high frequency (around 100 Hz) within the coil. Rapid rotation was needed because the voltage induced was proportional to the rate of change of flux in the coil. After phase-lock detection and electronic amplification of the signal, the calibrated output yielded two components of remanence in the plane normal to the rotational axis. The instrument was susceptible to electrostatic build-up but was capable of measuring magnetizations of around 10^{-3} A m^{-1} in 10–15 minutes.

The flux-gate spinner magnetometer is a subsequent refinement in which the sensor coil is replaced with flux-gate sensors. These detect directly the external magnetic fields of the sample. The signal strength is not dependent on rotational speed which could be reduced to about 5–10 Hz. The rotation of the sample gives a sinusoidal output. A large number of cycles can be averaged to reduce noise. The output is commonly digitized and stored in memory in a small on-line computer. The components of magnetization in the plane normal to the rotational axis are then determined by Fourier analysis. A computer-controlled flux-gate spinner magnetometer is capable of measuring a rock magnetization around 5×10^{-5} A m^{-1} in standard samples under optimum conditions. The complete measurement of a sample takes only a few minutes.

The cryogenic magnetometer is the most sensitive and rapid instrument in current use. Its sensor consists of a coil immersed in liquid helium. At this temperature (4 K) the coil is superconducting. A small change of magnetic field induces a comparatively large current, which because of the superconducting condition is persistent until the sample is removed. In line with the coil is a Josephson junction, which is a quantum-mechanical device consisting of a very thin element that allows the passage of current in distinct units proportional to a quantum of magnetic flux. By counting the number of flux jumps electronically the external magnetic field of the rock specimen can be inferred, and from this its magnetization computed. Most cryogenic magnetometers contain orthogonal sets of coils and can measure two or three axes of magnetization simultaneously within a few seconds. The sensitivity of the instrument corresponds to a rock magnetization of 5×10^{-6} A m^{-1} in standard samples.

5.6.3.2 *Stepwise progressive demagnetization*

The natural remanent magnetization (NRM) of a rock may contain several components, some related to the geological history of the rock and others to the sampling and handling procedures. It is necessary to 'magnetically clean' the natural magnetization so that the structure of the NRM can be analyzed and stable components isolated. This is done in a stepwise procedure, in which progressively more and more of the original magnetization is removed. There are two main methods of doing this.

The first method is progressive *alternating field (AF) demagnetization.* An alternating magnetic field can be produced in a coil by passing an alternating current through it. The field fluctuates between equal and opposite peak values. When a rock sample is placed in the alternating magnetic field, the grain magnetic moments with coercivities less than the peak value of the field are remagnetized in a new direction; the field cannot affect a magnetization component with coercivity higher than the peak field. The intensity of the alternating field is reduced slowly and uniformly to zero. This randomizes the part of the rock magnetization that has coercivities less than the peak value of the alternating magnetic field. The AF demagnetizing coil must be surrounded by magnetic shields or special additional coils to cancel out the Earth's magnetic field; otherwise an anhysteretic remanent magnetization (ARM) is induced along the direction of this field.

The part of the remanence that remains after a demagnetization treatment has been 'magnetically cleaned'. The direction and intensity of the remanent magnetization are affected. The demagnetization procedure is repeated using successively higher values of the peak alternating field, remeasuring the remaining magnetization after each step, until the magnetization is reduced to zero. Suppose that the NRM of a sample consists of two components AB and BC with different directions and different coercivity spectra (Fig. 5.56a, b). In the early stages of progressive demagnetization (steps 1–3) the 'soft' component BC is first

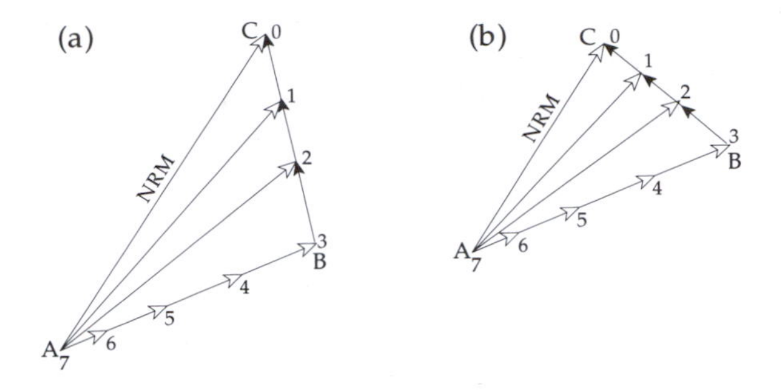

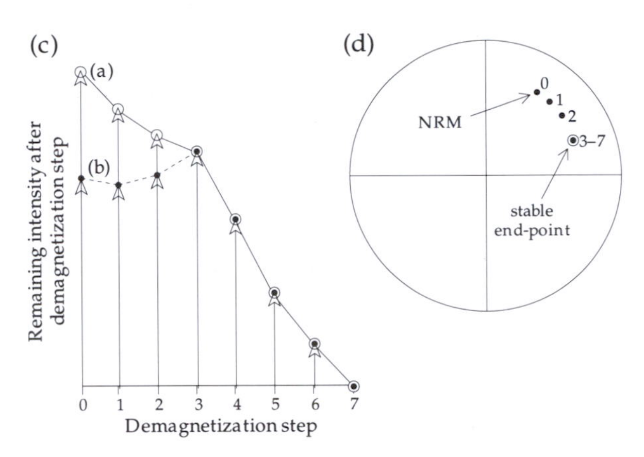

Fig. 5.56 Stepwise demagnetization of a remanent magnetization consisting of two components with different ranges of stability: (a) low stability vector BC demagnetizes before stable stable vector AB; (b) same situation with different angle between AB and BC; (c) variation of intensity for cases (a) and (b); and (d) directional changes on a stereogram. Numbers on points indicate successive demagnetization steps.

reduced to zero. The vector measured after each step in the progressive demagnetization is the sum of the 'hard' component, which has not yet been affected by the field used, and the residual part of the soft component. If the direction of the soft component is within 90°of the hard component the intensity decreases during this demagnetization interval; otherwise it may increase (Fig. 5.56c). The direction of the resultant vector changes continually in steps 1–3 (Fig. 5.56d). After removal of the soft component in step 3, higher alternating fields (steps 4–7) progressively reduce the hard component AB. During this stage the intensity decreases constantly but the direction remains consistent with little scatter, defining the 'stable end-point' or 'characteristic' direction of the magnetization.

The effectiveness of the AF demagnetization method is limited by the strongest peak field that can be produced in the demagnetizing coil. This is commonly 0.1 T, although some equipment can reach around 0.3 T. These peak fields are well below the maximum coercivity of pyrrhotite and far below the coercivities of hematite or goethite. Thus AF demagnetization is not effective in demagnetizing components carried by these minerals. The method is most commonly used for rocks that contain magnetite as the main magnetic mineral.

An alternative method of 'magnetic cleaning' is progressive *thermal demagnetization*. When a rock sample is heated to a given temperature T, magnetic components that have lower blocking temperatures than T are thermally randomized. If the sample is now cooled in field-free space, this part of the NRM remains demagnetized. In stepwise thermal demagnetization the heating and cooling cycle is repeated with progressively higher maximum temperatures. The progressive destruction of the magnetization reveals the components present in the NRM as in Fig. 5.56. This method is often more effective than AF demagnetization, because it is only necessary to heat a sample above the highest Curie temperature of its constituent minerals to destroy all of the NRM. However, if the rock contains thermally unstable magnetic minerals, irreversible changes may complicate the thermal demagnetization method.

5.6.3.3 *Analysis of magnetization components*

The stability of a remanent magnetization during stepwise demagnetization can be demonstrated by plotting the remaining intensity after each step against the corresponding temperature or AF field, as in Fig. 5.56c; the directional stability can be controlled simultaneously by plotting the direction after each step on a stereogram (Fig. 5.56d). However, the analysis of magnetization with an intensity plot and stereogram is outmoded. More sophisticated methods treat the magnetization as a vector and analyse the stability of its individual components.

The most powerful method of analysis of the structure and stability of a remanent magnetization is by constructing a vector diagram. The method was introduced by J. D. A. Zijderveld, a Dutch paleomagnetist, in the early 1960s. The magnetization at each stage of demagnetization is resolved into north (N), east (E) and vertical (V) components. Plots are then made of the north component against the east component, and of a horizontal component (north or east) against the vertical component. This is equivalent to projecting the vector on to the horizontal plane and the north–south (or east–west) vertical plane (Fig. 5.57a).

Components of NRM that have distinct spectra of coercivities or blocking temperatures show as linear segments on a vector demagnetization diagram. The example in Fig. 5.57b shows three distinct linear segments on the horizontal and vertical projections. The slopes of the straight lines represent NRM components with different directions. A component removed below 150 °C is directed downward to the north, and so has normal polarity. The component is probably a soft overprint (perhaps a VRM, see §5.3.6.5) acquired in the present-day or a recent field. A vector removed between 150 °C and 300 °C may represent a more ancient overprint; it is directed in a southerly, upward direction and thus has a reversed polarity. If a stable vector is left after demagnetization of less-stable fractions, it is indicated by a straight line to the origin of each half of the vector diagram. This is the case for the component removed from 300 °C to 580 °C. It is interpreted as a stable primary component acquired when the field had reversed polarity.

If more than one magnetization component is present, it is possible that the spectra of coercivity or blocking temperature of the components may overlap partially. During demagnetization of the overlapping components the vector diagram exhibits a curved trajectory. If the spectra overlap completely, no straight segment can be determined. In this case the sample does not have a single stable magnetization component.

5.6.3.4 *Statistical analysis of paleomagnetic directions*

For the purposes of statistical analysis each paleomagnetic direction in a collection of samples is considered to have equal value and may be regarded as a unit vector. Each vector has unit length but a different direction. The end points of the vectors lie on the surface of a unit sphere and form a distribution of points. The statistical methods for evaluating paleomagnetic directions (or the distribution of points on a sphere) were developed in 1953 by Sir Ronald Fisher. He found that the best estimate of the mean direction of a population of N unit vectors is their vector mean, R. To illustrate this point, consider five paleomagnetic directions, each represented by a unit vector (Fig. 5.58a). Usually the unit vectors are not parallel and when added vectorially their resultant has length $R \leq 5$ (Fig. 5.58b); its direction is the best estimate of the mean of the five paleomagnetic directions.

Fisher proposed that the probability density $P(\theta, \kappa)$ of the angle θ between an individual sample direction and the mean direction of the distribution is:

$$P(\theta, \kappa) = \frac{\kappa}{4\pi \sinh\kappa} \exp(\kappa \cos\theta) \tag{5.50}$$

The parameter κ is called the 'precision parameter' or 'concentration parameter'. It describes the dispersion of the directions, and is akin to the inverse of the variance of the distribution. Strictly speaking, κ is a property of an infinitely large population of directions. However, in paleomagnetic investigations only a small number of directions are usually sampled; it is assumed that they are representative for an infinite population. The parameters that are

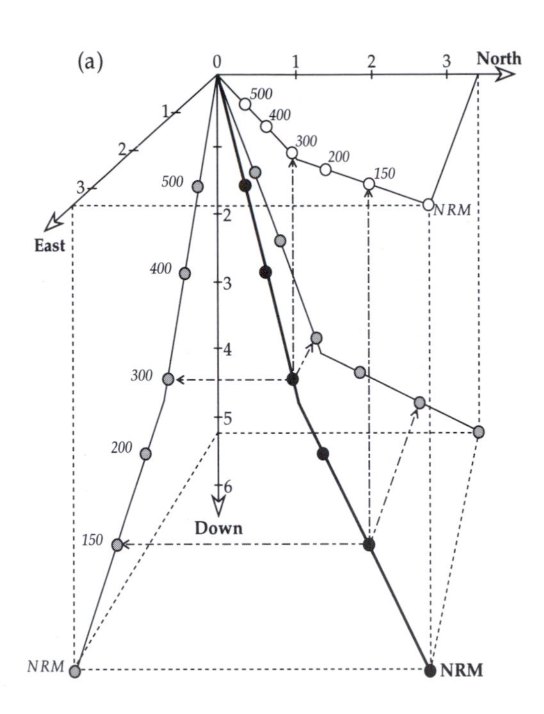

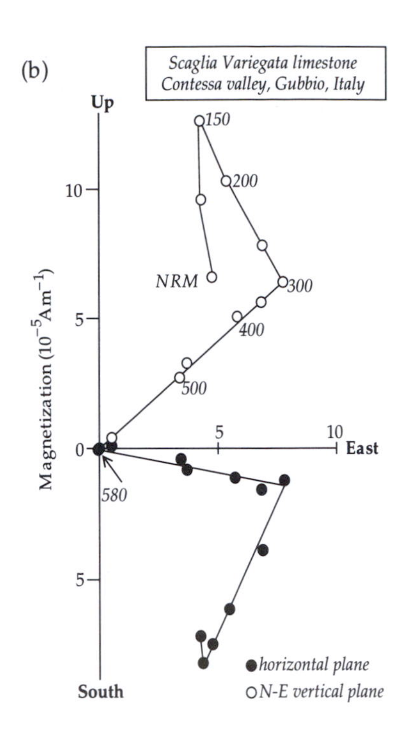

Fig. 5.57 The vector diagram method (Zijderveld, 1967) for analyzing progressive AF or thermal demagnetization. (a) Schematic showing how the components of the vector remaining at each stage of demagnetization are projected as points on three orthogonal planes (horizontal, vertical N–S and vertical E–W). (b) Vector diagram for the thermal demagnetization of a limestone sample. Magnetization components with non-overlapping spectra of thermal blocking temperatures show as linear segments.

computed are approximate estimates of the true parameters of the population. Fisher showed that the best estimate (k) of the precision parameter κ (valid for $k>3$) is given by

$$k=\frac{N-1}{N-R} \tag{5.51}$$

where R is the vector sum of the N unit vectors, computed as in Fig. 5.58b. When k (or κ) is zero, the directions are uniformly or randomly distributed. A badly scattered set of directions has a small value of k; large values of k apply to tightly grouped directions (Fig. 5.58c).

As in other statistical situations, the scatter is a property of the distribution of directions. It is described by the angular standard deviation, which is proportional to $1/\sqrt{k}$. However, it is usually more important to describe how well the mean direction is defined. Keep in mind that we do not know the true mean direction; we have only made an estimate of it, based on the available data. The true mean may differ by several degrees from our estimate. However, if we know that with 95% certainty the true mean lies within, say, 7° of our estimate, we can draw a cone with semi-angle of 7° about the estimated mean direction. The cone is said to define the confidence limits of the mean at the 95% probability level. The size of the confidence limit depends on the number of directions N in the distribution and their dispersion parameter k. The semi-angle of the cone of confidence is denoted a_{95} and is given approximately by

$$\alpha_{95}=\frac{140}{\sqrt{Nk}} \tag{5.52}$$

We could select any level of confidence to describe how well the mean is defined. However, two levels are common in statistics; the 95% (significant) and the 99% (highly significant) levels. In paleomagnetism the level of 95% confidence is used. This means that there is a 95% probability that the true mean of the distribution lies within this cone about the estimated mean direction.

5.6.3.5 *Field tests of magnetization stability*

If possible, paleomagnetic sampling includes a field test that can establish the stability of the magnetization stability over geological time. This was especially important in the early days of paleomagnetism when the laboratory techniques of 'magnetic cleaning' were not yet available. Pioneering researchers devised some ingenious tests of paleomagnetic stability based on field observations. The fold test and reversals test still serve as the best ways to demonstrate the stability of a remanent magnetization through the aeons of geological time and to verify the timing of its acquisition.

The fold test is perhaps the most important paleomagnetic field test. It is applied to samples taken from beds that were originally horizontal and have been tilted by later tectonic effects. If the paleomagnetic direction in the rock is stable, it will experience the same rigid-body rotation as the tilted strata; its direction will vary around the fold (Fig. 5.59a, layer A). This is called a pre-folding magnetiza-

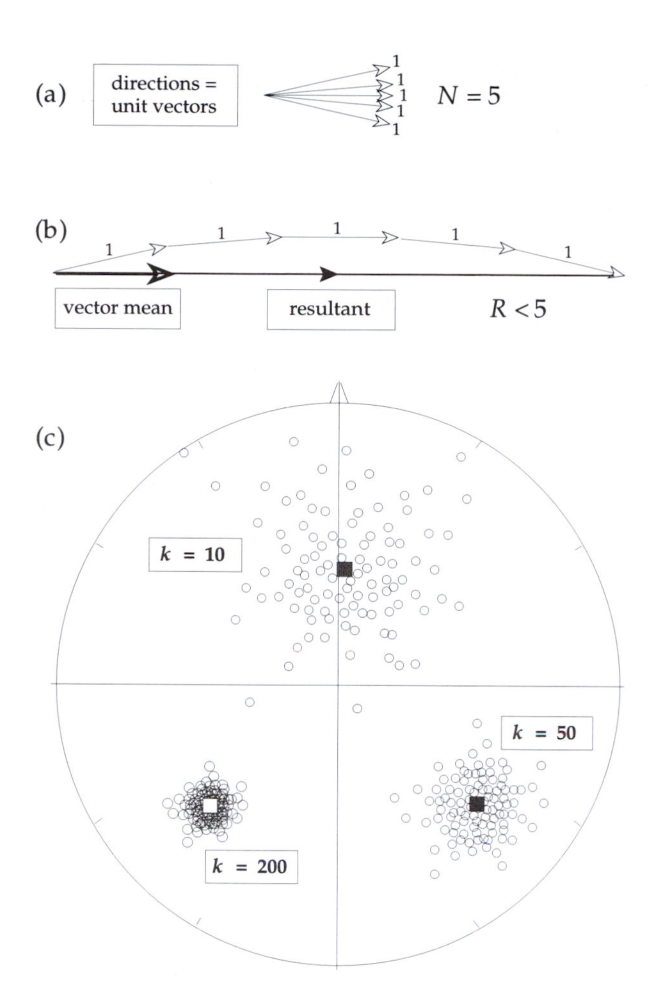

Fig. 5.58 (a) Representation of five magnetization directions as unit vectors. (b) The vector mean direction is that of the resultant vector **R**. (c) Stereograms of some distributions of paleomagnetic directions: the tighter the grouping, the larger the concentration parameter k.

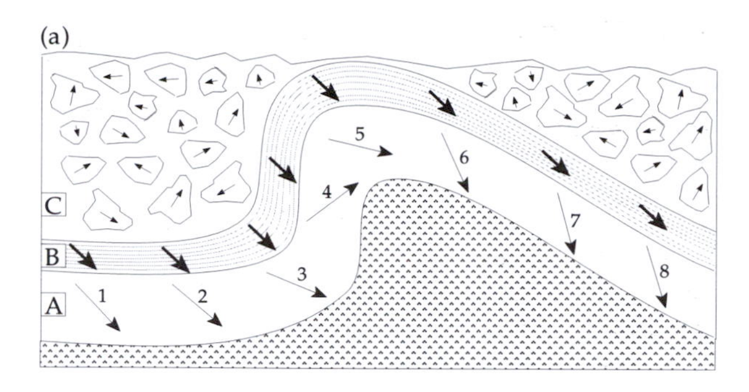

(b)

	layer A	layer B
directions before unfolding		
directions after unfolding		

Fig. 5.59 (a) Magnetization directions (arrows) around a fold in stable (A) and unstable layers (B), and in stably magnetized conglomerate cobbles (C). (b) Comparison of directions in the stable layer (A) and unstable layer (B) before and after unfolding.

tion. On the other hand, if the magnetization was acquired by the rock after it was folded, it will have a uniform direction at all points of the fold (Fig. 5.59a, layer B). This is called a post-folding magnetization. A third situation is common, in which the magnetization is acquired during the tectonic event; in this case the direction of magnetization changes around the fold but by a smaller amount than the folding. This is called a synfolding magnetization.

In practice, the fold test consists of comparing the directions before applying tilt corrections with the directions after unfolding the tilted beds. If samples with a stable magnetization are taken from all parts of the fold, their uncorrected directions should be smeared out. After correction for bedding tilt, the dispersion of directions should be reduced, and the directions should group around a common direction, which is the pre-folding direction of the formation (Fig. 5.59b, layer A). This is called a positive fold test. If the magnetization is unstable or is due to post-folding remagnetization, the tilt corrections will increase the scatter of the distribution of directions (Fig. 5.59b, layer B); this is a negative fold test.

An application of the fold test is shown in Fig. 5.60 for 12 sites of the Scaglia Rossa limestone from the central Apennines in southern Umbria (Italy). The sites were collected on different limbs of long anticlinal structures. Mean directions were computed for about 10–12 'magnetically cleaned' samples at each site. The uncorrected directions are quite scattered, with a confidence cone (α_{95}) equal to 11°. After correcting each site for the local tilt of the bedding the data are much better grouped, and the confidence cone is reduced to 6°. The concentration parameter increases significantly, from 14.6 to 46.3, indicating that the magnetization was acquired before the folding.

The conglomerate test is a field test of stability that is rather seldom used. Suppose that we are investigating a limestone formation and that we discover a conglomerate containing cobbles of the limestone (Fig. 5.59a, layer C). Assuming that the cobbles have been randomly re-oriented by the processes of erosion, transport and re-deposition,

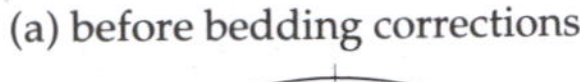

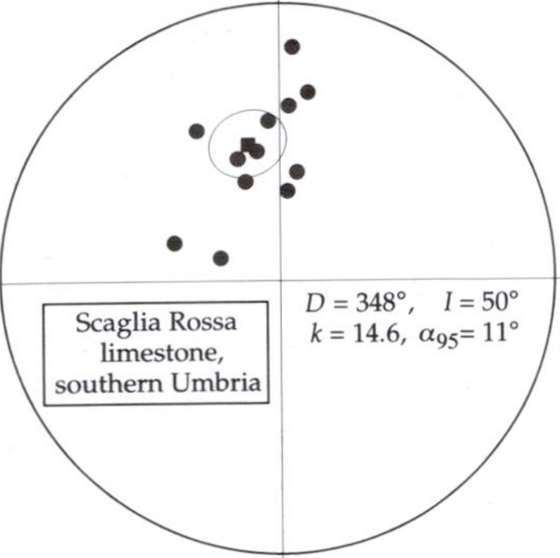

(b) after bedding corrections

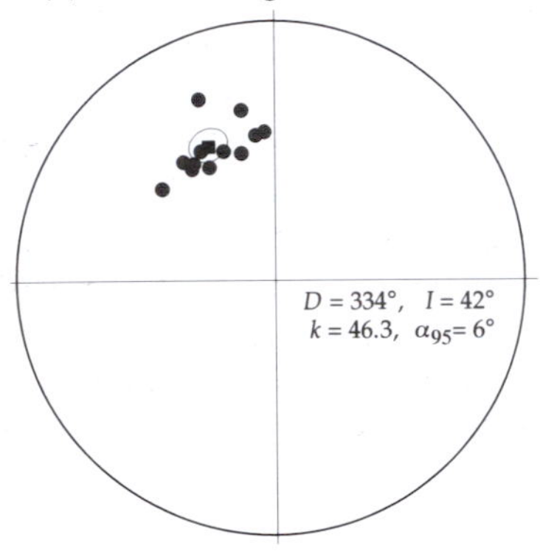

Fig. 5.60 Example of a positive fold test in 12 sites of the Scaglia Rossa limestone from southern Umbria. The directions (a) before correcting for local bedding tilt are more scattered than (b) the corrected directions.

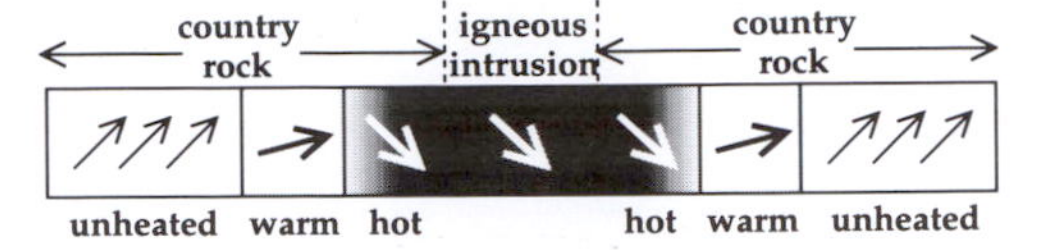

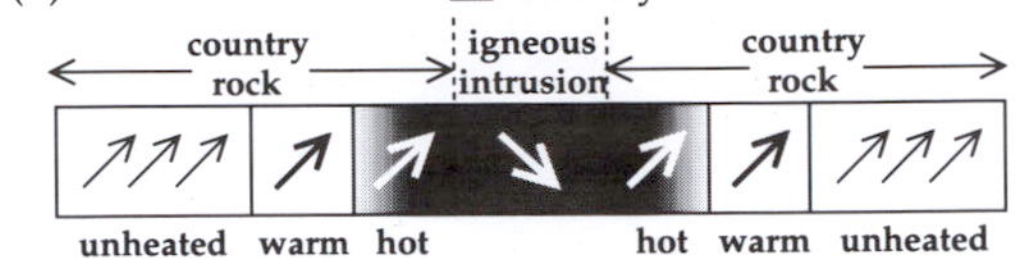

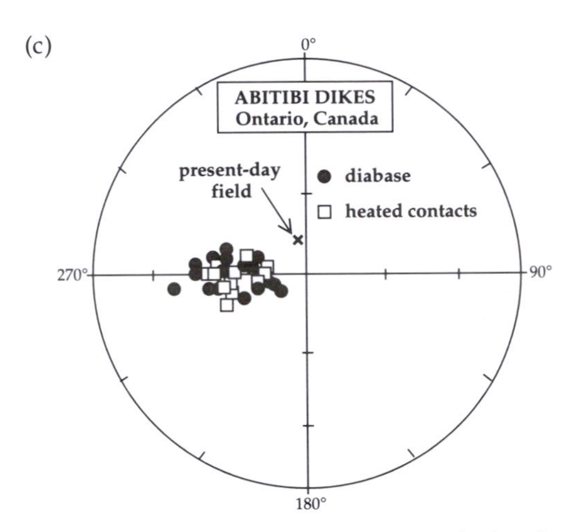

Fig. 5.61 The baked contact test. (a) Magnetization directions of intrusion and country rock when both are stable and (b) when one is unstable. (c) Example of a stable baked contact test for the Abitibi dikes, Ontario, Canada: the directions in the dikes are the same as in the baked country rock and are different from the present-day field direction (after Irving and Naldrett, 1977).

their paleomagnetic directions, if stable, should be randomly distributed. If a systematic direction is found, the magnetization of the limestone may have a large secondary component.

The baked contact test is important in igneous rocks. During intrusion of a dike or sill the adjacent layers of the host rock are baked by contact with the hot lava and acquire a TRM when they cool. In general the magnetic minerals in the lava will differ in composition and grain size from those in the host rock. If samples taken from the lava and contact zone of the host rock have the same magnetization direction (Fig. 5.61a), the lava carries a stable paleomagnetic vector. If they are different (Fig. 5.61b), one of the magnetizations is unstable; alternatively, either the lava or the baked zone was re-magnetized at a later time. An example of stable magnetizations in Precambrian rocks from the Canadian shield is shown by the agreement between the directions in the Abitibi diabase dikes and the heated contact zone (Fig. 5.61c).

The reversals test can be applied when the paleomagnetic samples represent a large enough time interval (>10 ka) to have recorded normal and reversed polarities of the magnetic field. Remanent magnetizations acquired within successive intervals of constant polarity of the Earth's magnetic field should be exactly antiparallel. Let the normal magnetization be represented by the vector **N** and the reversed magnetization by the vector **R** (Fig. 5.62a). The presence of an unremoved secondary component, represented by the vector **S**, will give resultant normal and reversed directions that are no longer antiparallel. If it is possible to clean the directions magnetically, the antipodal normal and reversed directions (**N** and **R**) should be recovered. If magnetic cleaning is inadequate, a residual part of the unremoved secondary component may spoil the antiparallelism.

An example in which the reversals test shows successful 'cleaning' is shown in Fig. 5.62b for samples from a single

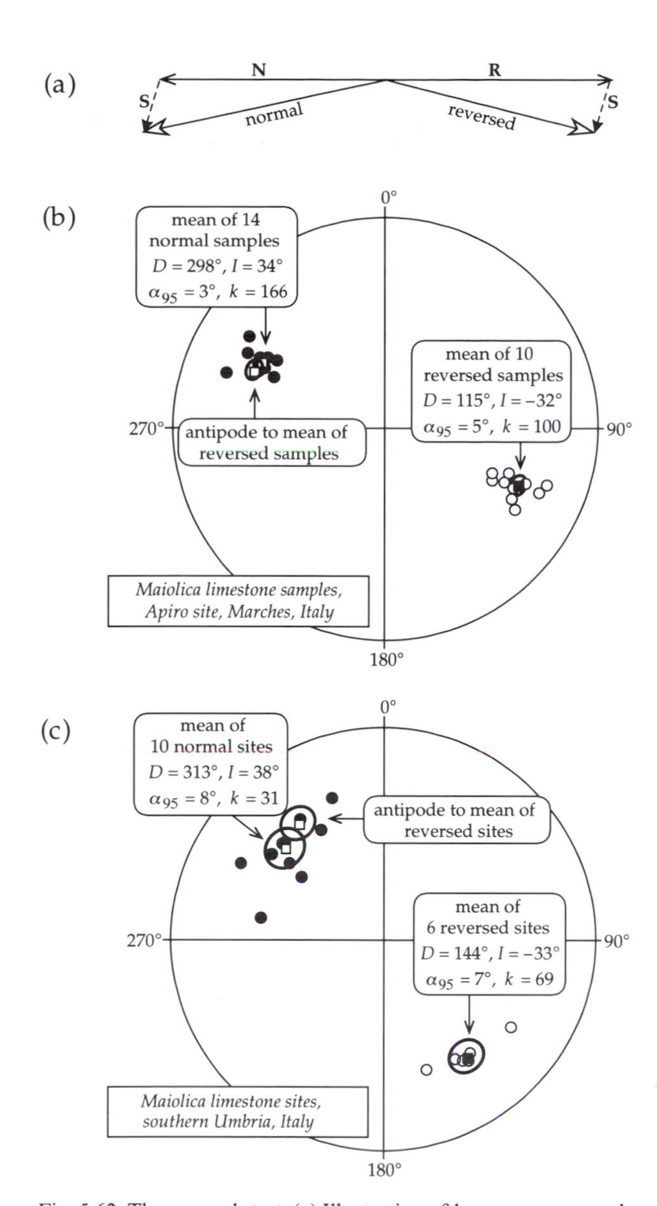

Fig. 5.62 The reversals test. (a) Illustration of how an unremoved secondary component S can spoil the anti-parallel directions of normal and reverse magnetizations. (b) A positive fold test for samples of the Maiolica limestone in a site at Apiro (Marches, Italy): the normal and reverse mean directions are almost exactly opposite. (c) Mean directions of normal and reverse polarity Maiolica sites in southern Umbria are not exactly opposite due to local tectonic rotations about vertical axes.

site in the Early Cretaceous Maiolica limestone in central Italy. The vector mean of 14 normal samples has D_N=298°, I_N=34°, and α_{95}=3°; the mean of 10 reversed polarity samples has D_R=115°, I_R=−32°, and α_{95}=5°. A simple way to compare how well the sets of normal and reversed polarity directions agree is to invert the mean direction and confidence circle for the reversed group of samples through the origin. The common polarity mean directions differ by only 3°. The mean of each group lies within the confidence limits of the other, so there is no significant difference between the normal and reversed directions.

When the site-mean directions from several sites of the Maiolica limestone are compared throughout a large region of the Umbrian Apennine mountain belt, the antiparallelism of sites with normal and reversed polarities no longer holds. The vector mean of 10 normal sites has D_N=313°, I_N=38°, and α_{95}=8°; the mean of 6 reversed polarity sites has D_R=144°, I_R=−33°, and α_{95}=7° (Fig. 5.62c). In this case the common polarity mean directions differ by 10°. The mean of each group lies outside the confidence limits of the other, so there is now a significant difference between the normal and reversed directions. Closer examination shows that the mean *inclinations I* of the normal and reversed groups are equivalent, but the *declinations D* are dispersed along a small circle about a vertical pole. The smeared declinations reflect small rotations of each site about a vertical axis, the result of regional tectonism in the area of investigation. This illustrates an important application of paleomagnetism; the description of tectonic rotations that would otherwise be difficult or impossible to observe in the field.

5.6.4 Paleomagnetism and tectonics

Paleomagnetism has made important contributions in documenting local and regional tectonic motions as well as the motions of lithospheric plates. The reason for the failure of the reversals test in Fig. 5.62c was ascribed to local tectonic disturbances within a region. To make use of paleomagnetic data on a larger scale the observed directions must be compared to suitable reference directions. A reference direction can be computed, if it is known where the paleomagnetic pole was in the geological past. The history of paleomagnetic pole positions can be established on a continental scale.

Paleomagnetic results from central and southern Europe document the effects of large-scale tectonics (Fig. 5.63). Data of Late Paleozoic to Late Cretaceous age are represented in this analysis. During this long time interval large amounts of motion of the European and African plate have taken place. As a result the inclinations measured at the indicated sites show large variations. However, the reference declinations for sites in Europe do not vary much during this time; this is also true for the African reference declinations. There is a large difference between the north–northeast pointing European declinations and the northwest-directed African declinations (Fig. 5.63). The

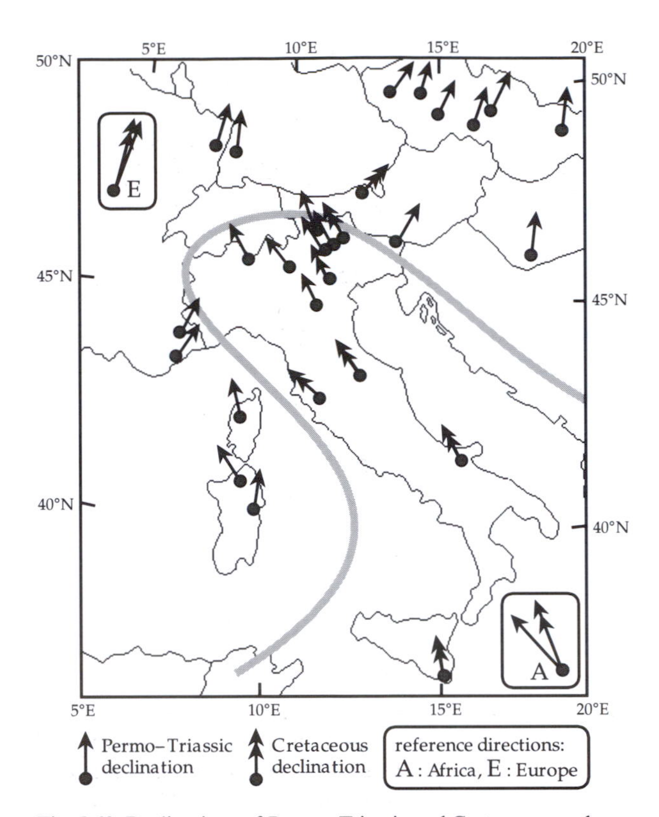

Fig. 5.63 Declinations of Permo–Triassic and Cretaceous rocks from Italy differ systematically from those of Europe north and west of the Alps, but agree well with directions predicted for the African plate. A hypothetical outline of the Adriatic promontory to the African plate is suggested by the shaded line.

paleomagnetic declinations observed at sites in central and southern Europe show distinct affinities. North and west of a crude line through the alpine chain the paleomagnetic declinations agree with those expected for the European continent. The declinations observed south of the Alps, on the Italian peninsula and Sicily are oriented toward the northwest, in agreement with directions expected for the African continent. The pattern of paleomagnetic data support a tectonic interpretation of the Italian peninsula and adjacent regions of the Adriatic as a northern promontory of the African plate. Although the differences in the declination pattern are striking, there is a certain amount of leeway in the interpretation. This is mainly because the reference directions of Africa are derived from paleomagnetic pole locations that are not very well defined for some time intervals.

5.6.4.1 *Location of the virtual geomagnetic pole*

Paleomagnetic poles are computed as the average of virtual geomagnetic pole (VGP) positions calculated for a number of samples at a site. The VGP position is where the pole of a geocentric magnetic dipole would need to be in order to give the observed declination D and inclination I of the remanent magnetization measured in the sample. The method of computation of the VGP position is illustrated in Fig. 5.64. First, from the inclination of the magnetization (i.e., the paleofield) we can calculate how far away the VGP was at the time the rock magnetization was acquired. The angular distance to the pole p, assuming a dipole magnetic field, is obtained by using the relationship between inclination and polar angle (Fig. 5.54a).

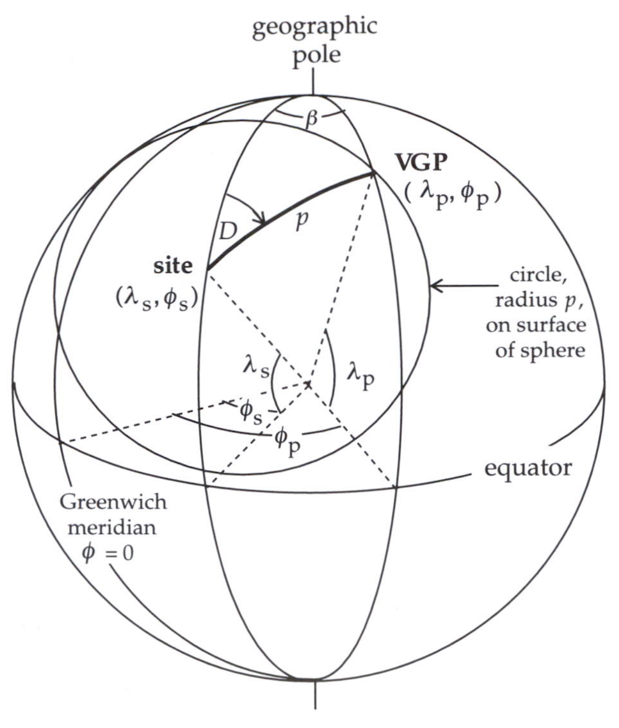

Fig. 5.64 Method of locating the virtual geomagnetic pole (VGP) from the declination D and inclination I measured at a site (after Nagata, 1961).

The value of p determines the radius of a small circle centered on the paleomagnetic sampling site at latitude λ_s and longitude ϕ_s. The circle is the locus of all possible VGP positions that could give the observed inclination I at the site. We next have to decide which point on the small circle is the VGP position. The declination of the remanent magnetization is the angle between geographic north and the horizontal direction to the ancient magnetic pole. In this case the declination defines a meridian (or great circle) which passes through the sampling site and makes an angle D with the north–south meridian (Fig. 5.64). The place where this great circle intersects the small circle with radius p is the location of the virtual geomagnetic pole. Its latitude (λ_p) and longitude (ϕ_p) can be computed exactly from trigonometric formulas:

$$\sin \lambda_p = \sin \lambda_s \cos p + \cos \lambda_s \sin p \cos D \tag{5.53}$$

$$\phi_p = \phi_s + \beta, \text{ for } \cos p \geq \sin \lambda_s \sin \lambda_p$$

$$\phi_p = \phi_s + 180 - \beta, \text{ for } \cos p \leq \sin \lambda_s \sin \lambda_p$$

where

$$\sin \beta = \frac{\sin p \sin D}{\cos \lambda_p} \tag{5.54}$$

The longitude of the paleomagnetic pole is here given relative to a fixed meridian in present-day geographic coordinates. The key paleomagnetic parameter is the distance p of the investigated site from the Earth's rotation axis, which was the position of the paleomagnetic pole at the time of formation of the rocks under investigation. At the time of magnetization all locations on the same latitude (i.e., at the same distance p from the rotation axis) were magnetized with zero declination, because the axial dipole field lines through the site lead to the rotation axis. The longitude of the site (its position on the circle of latitude) remains indeterminate. The declination measured later at the site is the expression of any change of azimuthal orientation, which can result, for example, from local tectonic motion or from large-scale continental displacement.

5.6.4.2 *Apparent polar wander paths*

The observation that paleomagnetic poles obtained from rocks of Pleistocene and Pliocene age are closely grouped about the geographic pole (see Fig. 5.55) is in agreement with the axial geocentric dipole hypothesis. However, when paleomagnetic pole positions are calculated for old rocks from the same continent, they group far away from the geographic pole. This is illustrated by the positions of paleomagnetic poles from the stable European craton. Pliocene and Pleistocene poles group close to the geographic pole but Permian poles are located about 45° away (Fig. 5.65). If the axial dipole hypothesis is valid for rocks of all ages, the pole distributions imply that the geographic pole for Europe in the Permian period (about 250–290 Ma ago) lay far from its present position. An alternative interpretation is that the geographic pole has not changed, but the European continent has moved relative to the pole. This suggests that the position about which the Permian poles now cluster was on the rotation axis in the Permian period. The European continent has subsequently moved to its present-day position with regard to the rotation axis.

Paleomagnetic data allow us to resolve the ambiguity. If paleomagnetic pole positions are computed for rocks of different ages from the same continent, they plot systematically along an irregular, curved path. It appears as though the paleomagnetic pole has moved slowly along this path towards the present rotation axis. The apparent motion of the paleomagnetic pole is called *apparent polar wander* (APW) and the path is called an *apparent polar wander path*. The paleomagnetic data from a particular continent define a unique APW path for that continent, and each continent has a different APW path. Thus, we have a European APW path, African APW path, North American APW path, and so on.

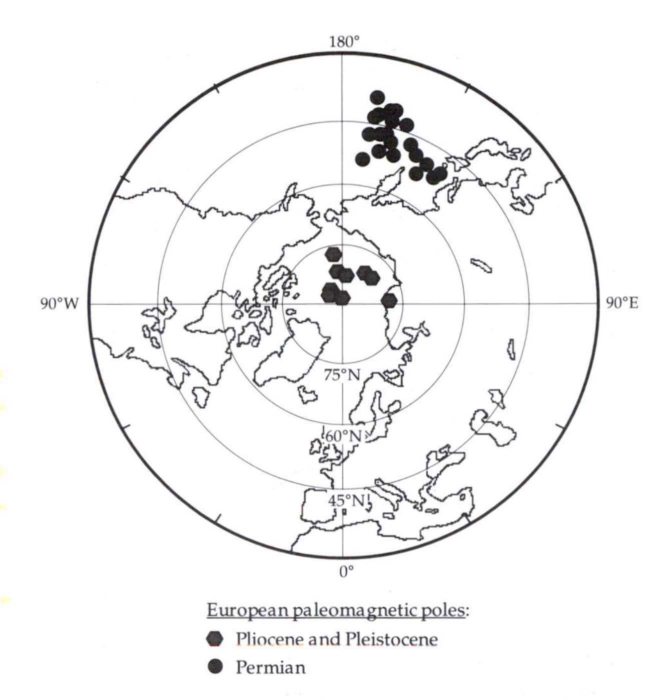

Fig. 5.65 Locations of European paleomagnetic poles. Pliocene and Pleistocene poles (data source: McElhinny, 1973) lie close to the present-day geographic pole, while Permian poles (data source: Van der Voo, 1993) are located at about 45°N in the NW Pacific Ocean.

A schematic plot of the European and North American APW paths since the Late Paleozoic shows clearly distinct curves (Fig. 5.66). Each APW path lies on the opposite side of the geographic pole from the continent to which it belongs. Keeping the axial geocentric dipole hypothesis in mind, it is obviously impossible that the paleomagnetic pole (i.e., the Earth's rotation axis) could have moved simultaneously along two different APW paths. The two APW paths evidently represent the separate motions of the European and North American continents relative to the rotation axis. They constitute paleomagnetic evidence for 'continental drift'.

5.6.4.3 *Paleogeographic reconstructions using APW paths*

A more detailed plot of the two APW paths for the time before the Late Jurassic (Fig. 5.67a) shows strong similarities in their shapes, particularly for the time from the Upper Carboniferous to the Upper Triassic. It is possible to overlay these two segments of the APW paths by moving Europe (including Russia west of the Ural mountains) and North America into different positions relative to each other (Fig. 5.67b). For the time represented by the overlap of the APW paths the two continents formed part of a larger 'supercontinent', called *Euramerica*. When the adjacent part of Asia east of the Urals is included, the continents in the northern hemisphere form an earlier landmass called *Laurasia*. The present separation of the APW paths (Figs. 5.66, 5.67a) is interpreted as evidence for relative plate tectonic motion between Europe and North America that has taken place since the end of the interval for which the APW paths overlap well, i.e., since the Early Jurassic.

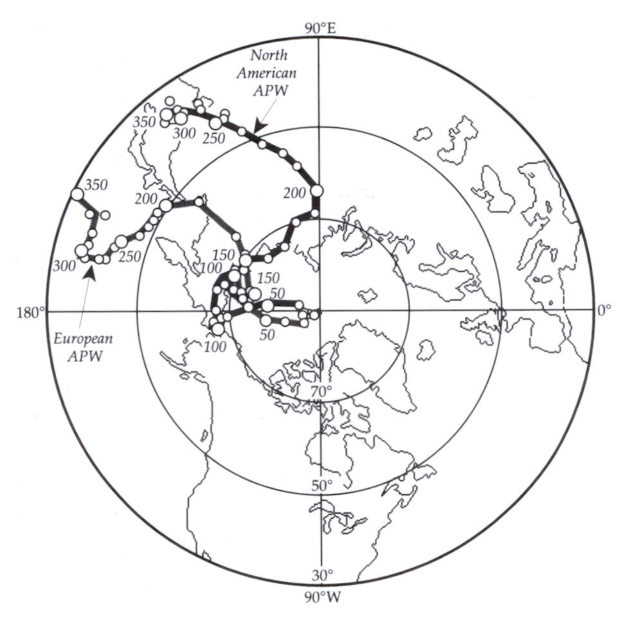

Fig. 5.66 Average apparent polar wander paths for North America and Europe in the past 350 Ma (after Irving, 1977). Numbers on paths are age in Ma.

In order to bring the two APW paths into coincidence we have to move Europe relative to North America (or vice versa) so as to close the present gap between the continents. As shown in §6.3.3, the relative motion of plates on the surface of the spherical Earth is equivalent to a relative rotation about an *Euler pole of rotation*. The computer-generated 'Bullard' fit of the 500 fathom contour lines on opposites sides of the North Atlantic ocean (see §1.2.2.2) can be obtained by displacing Europe toward North America by a clockwise rotation through 38° about the Euler pole located at 88.5°N 27.7°E, which by coincidence is very close to the present-day geographical pole (Fig. 5.67b). The APW path of a continent is constrained to move with the continent. If the European APW path is also rotated by 38° clockwise about the same Euler pole, the observed overlap of the Upper Carboniferous to Upper Triassic sections of the European and North American APW paths is obtained. Later segments of the paths diverge, indicating relative motion between the continents. The Late Jurassic pole position corresponds to the Earth's rotation axis at that time. Circles of paleolatitude about the North American paleomagnetic pole emphasize the paleogeographic reconstruction of the relative positions of Europe and America in the Late Jurassic.

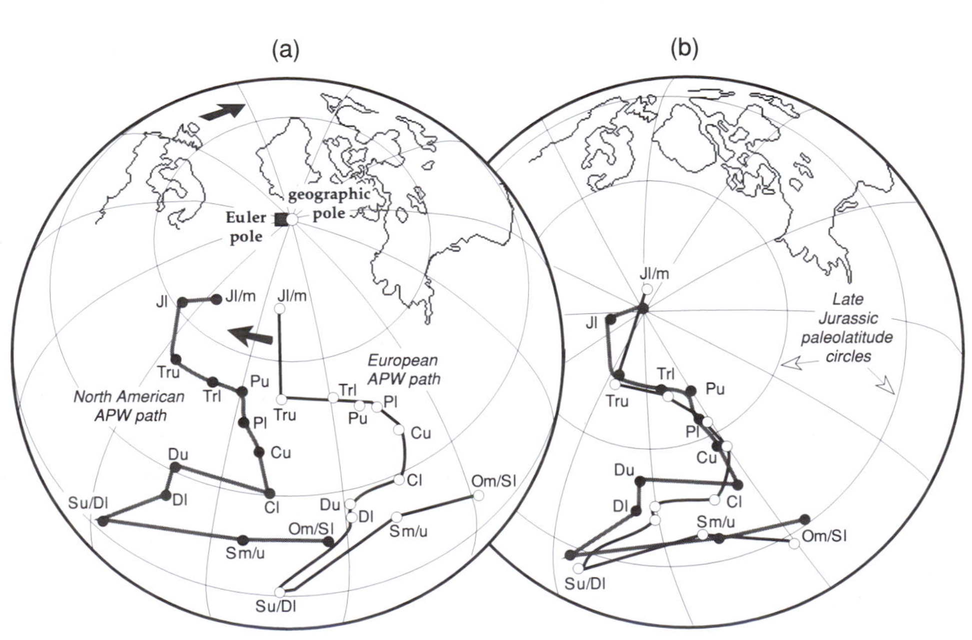

Fig. 5.67 (a) The Ordovician to Jurassic segments of the North American and European APW paths. (b) The same APW paths after rotating Europe by 38° clockwise about the Euler rotation pole at 88.5°N 27.7°E, marked by the square symbol in (a) (after Van der Voo, 1990).

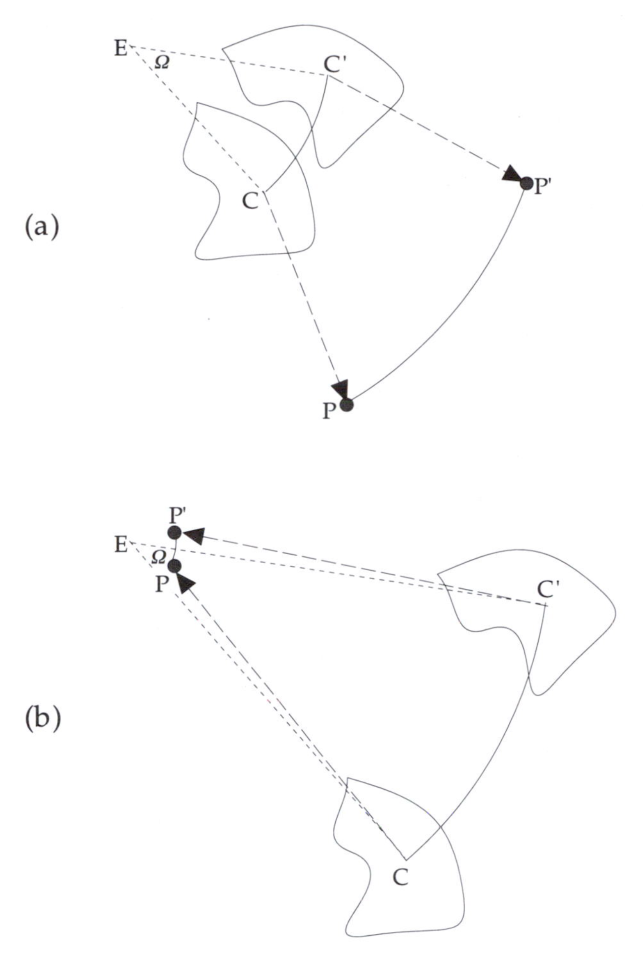

Fig. 5.68 Rotation of a continental plate C about an Euler pole E displaces the paleomagnetic pole P (a) by a large amount, if P is further from E than the continent C, and (b) by only a small amount when P lies close to E.

The interpretation of APW paths is not always as clear-cut as in this example. Consider the situation of the paleomagnetic pole P for continental plate C, which is rotated through an angle Ω about the Euler pole of rotation, E. First, let the plate lie between the paleomagnetic pole and the rotation pole (Fig. 5.68a). The rotation of the plate from C to C′ causes the paleomagnetic pole P to move to P′. The arc PP′ of the polar motion is longer than the arc CC′ of the true plate motion. Next, consider what happens when the paleomagnetic pole lies between the plate and the Euler pole (Fig. 5.68b). In this case the plate moves through a large distance but the paleomagnetic pole moves only a small distance. In the extreme case where the paleomagnetic and Euler poles coincide, the plate rotation does not move the paleomagnetic pole at all. Under these special conditions plate motion leaves no trace in the APW path of the plate.

Clearly, the interpretation of an APW path as a record of plate motion relative to the geographic axis must be made with caution. The rate of motion of the pole along an APW path cannot be simply equated with the rate of motion of the parent continent or global plate. It follows that similarity of APW paths does not imply a unique solution for former relative plate positions. However, if two continents once belonged to the same plate for some length of time, they should have acquired the same APW path for this time. Matching the present APW paths of the continents for the time they were on the same plate should give a unique reconstruction of the earlier positions of the continents relative to each other. To avoid ambiguities, additional independent evidence (such as paleoclimatic data, or computer-matching of coastlines) must be utilized in conjunction with paleomagnetic data for making such reconstructions.

5.6.4.4 *Paleomagnetism and continental drift*

The 19th century geologist Eduard Suess deduced the existence of a great Late Paleozoic continent, which he called Gondwanaland (§1.2.1). It was composed of Africa, Antarctica, Arabia, Australia, India and South America. In 1912, Wegener went a step further by postulating that all the present continents lay close together during the Late Paleozoic, forming a single great continent that he called Pangaea. Wegener's concept was based on paleoclimatic evidence, the matching of Carboniferous coal belts and of regions of Paleozoic glaciation in the different continents. Subsequently, additional geological evidence for the existence of Gondwanaland and Pangaea during the Late Paleozoic and Early Mesozoic accumulated from the fields of sedimentology, paleontology, and tectonics. The earlier great continents were presumed to have dispersed to their present-day location by the process of *continental drift*. Unfortunately, Wegener was unable to offer a satisfactory driving mechanism for continental drift, and some of his ideas were found to be extreme. Scepticism among geophysicists and geologists brought Wegener's theories into disrepute.

Interest in continental drift was re-awakened in the 1950s by the development of paleomagnetism. Soon thereafter some of the most convincing paleomagnetic evidence for continental drift was obtained from the 'southern continents'. Researchers found that Mesozoic paleomagnetic pole positions of the same age from these continents were very dissimilar. In landmark contributions E. Irving showed that the paleomagnetic poles of the southern continents were incompatible with the present-day arrangement of these continents, but that they agreed much better when the continents were rearranged to conform with a

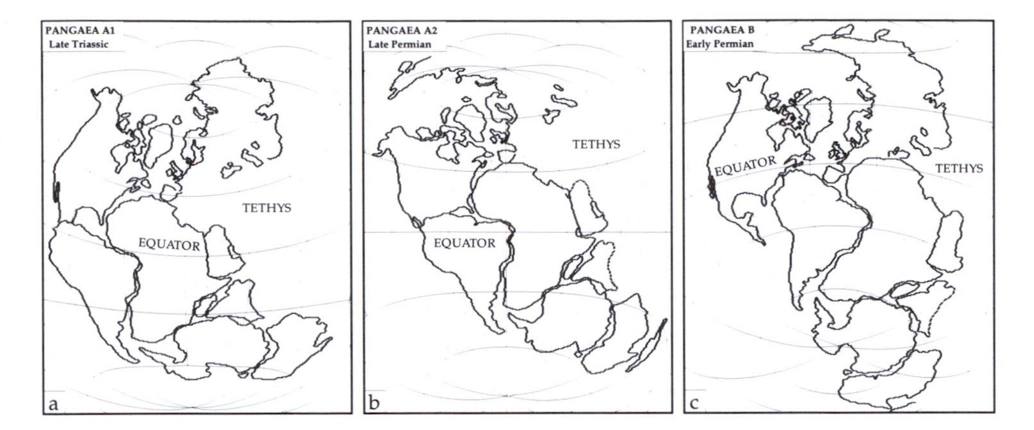

Fig. 5.69 The configurations of Pangaea models A1, A2 and B based on the matching of coastlines and the optimizing of paleomagnetic data (after Morel and Irving, 1981).

Gondwanaland reconstruction. Numerous paleomagnetic investigations have subsequently provided a rich data-base that can be used to test the validity of reconstructions of former great continents at different times in their history.

The reconstructions are generally not made on paleomagnetic evidence alone. Usually, a model is proposed, based on geometrical or geological grounds. The congruity of paleomagnetic pole positions from the separate continents is then evaluated in their reconstructed positions. The model is adjusted iteratively until a configuration of the continents is obtained that gives minimum dispersion of the paleomagnetic poles.

The evaluation of paleomagnetic data from the Gondwanic continents, North America and Europe lends convincing support to the reconstructions and thereby to the continental drift hypothesis. From the Carboniferous to the Triassic, contemporary paleomagnetic poles from the individual continents do not agree when the continents are in their present positions, but are more consistent when the great continent is reconstructed. In fact, the paleomagnetic data are of high enough quality to suggest refinements to the purely geometric reconstructions. The Pangaea model in Fig. 5.69a corresponds closely to computer-assisted matches of the continental coastlines (see §1.2.2.2); it is referred to as Pangaea A1. It places the east coast of North America adjacent to the coast of northwest Africa. This configuration is supported well by Late Triassic and Early Jurassic paleomagnetic poles. It is the generally accepted model of Pangaea immediately prior to its breakup in the Early Jurassic. However, older paleomagnetic data of Permian and Carboniferous age (around 280 Ma) are less compatible with the Pangaea A1 model. Results of Late Permian to Middle Triassic age agree much better with a configuration referred to as Pangaea A2 (Fig. 5.69b), first proposed by R. Van der Voo and R. French in 1974. In this model, North America is much closer to South America and its eastern coast is opposite to western Africa. The transition from from the Late Permian Pangaea A2 to the Late Triassic Pangaea A1 configuration requires a large dextral shear between the continents in this time interval.

Pangaea may have had yet another configuration earlier in its history. In 1977 E. Irving showed that results of Carboniferous and Early Permian age agree better for a Pangaea configuration in which the east coast of North America is adjacent to the west coast of South America (Fig. 5.69c). This model, Pangaea B, is possible because paleomagnetic longitudes are much more poorly constrained than paleolatitudes. The change from Early Permian Pangaea B to Late Permian Pangaea A2 requires a large dextral megashear between Laurasia and Gondwana.

The models Pangaea A1, A2 and B are each consistent with paleomagnetic data for the different times of the reconstructions, which span about 100 Ma. None of the models accounts for the apparent polar wander paths of the individual continents over the entire interval of time from Early Permian to Late Triassic or Early Jurassic. Instead, the differences between the models imply that Pangaea was not a static great continent for this entire time interval, but that internal motions took place between the constituent continents.

Paleomagnetic data can be used to reconstruct the relative positions of continents during any time interval with enough good paleomagnetic data. APW paths (e.g., Fig. 5.66) can be determined fairly precisely by averaging the best available pole positions in 20–40 Ma time windows. Optimum fitting of APW paths of different continents allows reconstructions to be made for the time represented by the matching segments (Fig. 5.67). When this procedure is applied to paleomagnetic data covering the last 375 Ma, a picture of continental drift since the Middle Devonian is obtained (§1.2.2.3). According to this scenario, the supercontinents Laurasia and Gondwana, which were still separated by the Hercynian Ocean (Fig. 5.70) in the Devonian, collided in the Carboniferous to form Pangaea. The paleo-

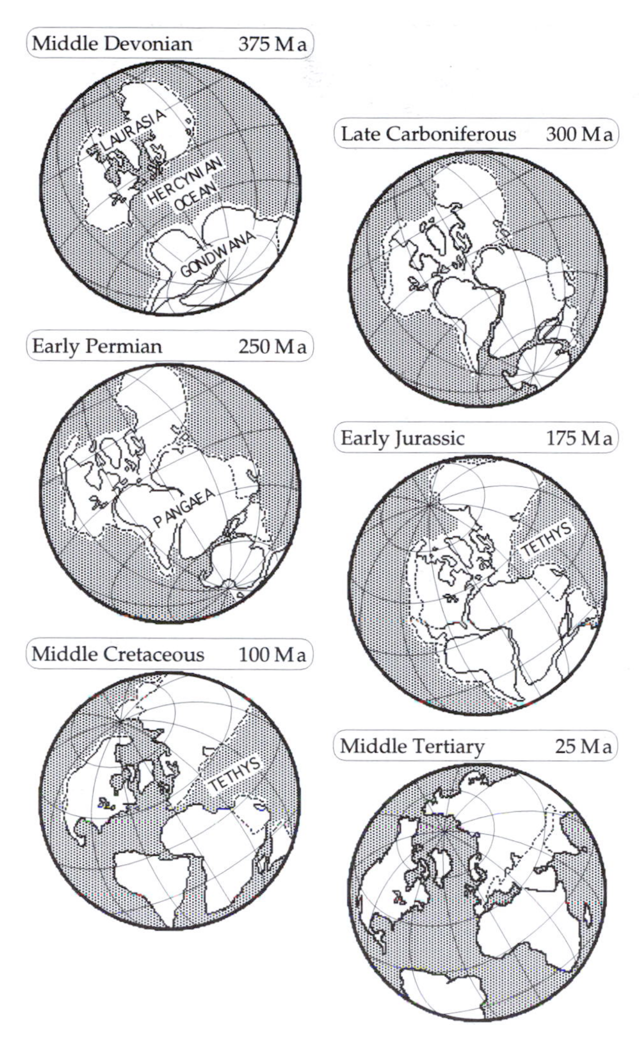

Fig. 5.70 Continental drift since the Devonian is illustrated by reconstructions of the positions of the major continental blocks at different times based on paleomagnetic data (after Irving, 1977).

magnetic reconstruction of continental drift for older epochs becomes tenuous because reliable paleomagnetic data become scarcer. The derivation of durable reconstructions for the Early Paleozoic and Precambrian will be a long and painstaking process.

The positions of continents since the breakup of Pangaea can also be obtained from analysis of APW paths. However, more precise reconstructions can be made by using a different form of paleomagnetic data, namely the record of geomagnetic polarity. Sea-floor spreading has imprinted this record in the oceanic crust, creating lineated magnetic anomalies. Matching coeval anomalies allows us to trace the motions of the lithospheric plates since the Middle Jurassic and describe the drift of the continents which they transport.

5.7 GEOMAGNETIC POLARITY

5.7.1 Introduction

The earliest demonstration that the geomagnetic field has changed polarity in the past was made by the French scientists P. David and B. Brunhes. In 1904–6 they described the magnetic properties of young lava flows in the Massif Central region of France. They found that clays baked by the lava flows had the same direction of remanent magnetization as the lavas. Moreover, when the magnetization direction in the lava was opposite to that of the present-day field, the same was the case in the baked clay. The opposite polarities were interpreted as evidence that the geomagnetic field can reverse its polarity.

A Japanese scientist, M. Matuyama, was the first to associate the polarity of remanent magnetization in lavas with their age, determined stratigraphically. In 1929 he reported finding young Quaternary lavas with magnetization directions close to the present-day field direction, whereas the directions in older Quaternary and Pleistocene lavas were clustered about an antipodal direction. He also found that one of three samples of Miocene basalt was magnetized oppositely to the other two. Matuyama's interpretation was that geomagnetic polarity had changed several times during Late Tertiary time.

The idea that geomagnetic polarity could change was controversial, and for many years sceptics sought alternative interpretations. Scientists realised that the observed reversed polarities might have a mineralogical explanation. Indeed, some ferromagnetic minerals, because of their composition and structure, can acquire a thermoremanent magnetization exactly opposite to the field direction. This mechanism is called self-reversal of magnetization. It has been described in lavas in which the ferromagnetic minerals are particular forms of titanohematite. Fortunately, it is a rather rare phenomenon. Most records of polarity reversals have been found to be a feature of the geomagnetic field.

To envisage what a reversal of geomagnetic polarity means, imagine that the geomagnetic dipole inverts its direction. At present the axial geocentric dipole points from the northern hemisphere towards the southern hemisphere; a polarity reversal would orient the dipole in the opposite direction. At each point on the surface the magnetic inclination I changes sign and the declination D changes by 180°; for example, a normal direction $\{I=40°, D=30°\}$ might change to a reverse direction $\{I=-40°, D=210°\}$. A polarity reversal is a global event, experienced simultaneously all over the Earth. Thus, geomagnetic reversals provide a convenient means of stratigraphic correlation and dating.

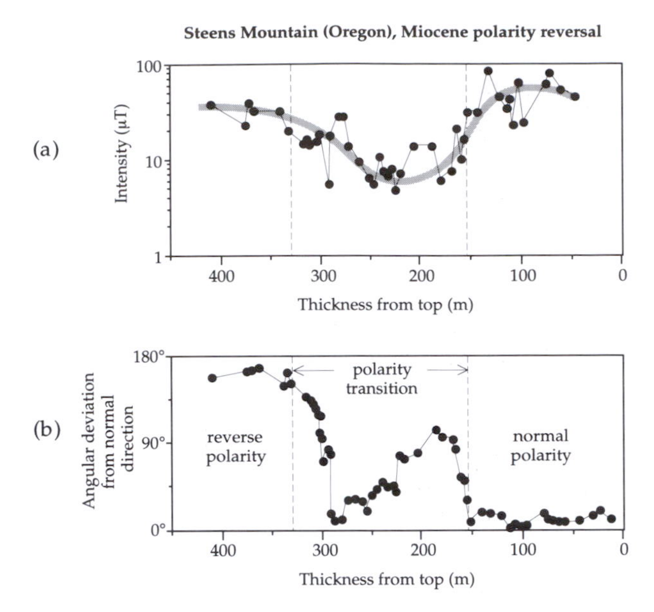

Fig. 5.71 The record of a reversed-to-normal Miocene polarity transition at Steens Mountain, Oregon. (a) The paleointensity record during the transition and (b) the directional record, shown as the angular deviation from the normal paleomagnetic direction outside the transition (after Prévot *et al.*, 1985).

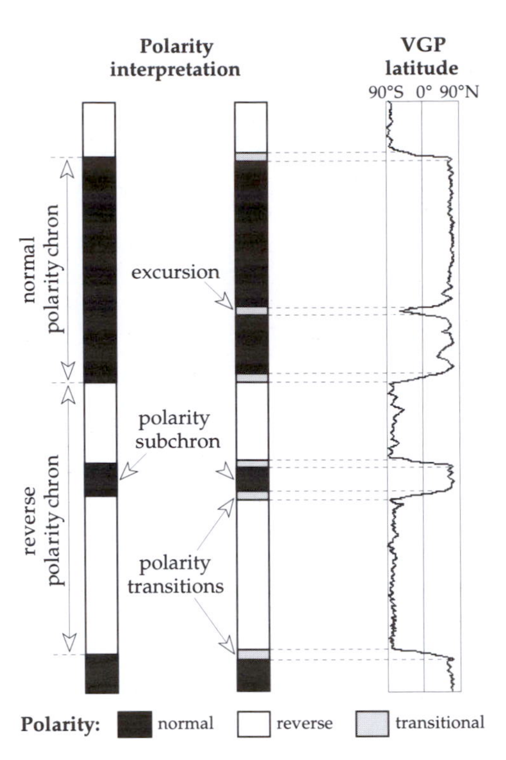

Fig. 5.72 Definition of polarity chrons, subchrons and transitions (modified after Cox, 1982, and Harland *et al.*, 1990).

5.7.1.1 *Geomagnetic polarity transitions*

The change of polarity from one sense to the opposite one is called a *polarity transition*. Paleomagnetic records of polarity transitions have been observed in radiometrically dated lava sequences and in deep-sea sediments with known deposition rates. These records indicate that the duration of a polarity transition is about 3.5–5 ka. This is much shorter than the length of the interval of constant polarity before or after the transition, which may last for hundreds of thousands or millions of years.

It is not yet known for sure how the geomagnetic field behaves during a polarity transition. The dipole field is dominant before and after a transition, but it is not certain that this is the case during the transition. Detailed analyses of field behavior during a polarity transition usually show a notable decrease in field intensity (Fig. 5.71); this is observed in volcanic and sedimentary records of reversals. Possibly the dipole component disappears, granting more importance to higher-order quadrupole or octupole field configurations. There seems to be stronger evidence that, even though its intensity decreases, the transitional field is still dominantly that of a dipole. If this can be assumed, the position of the virtual geomagnetic pole (VGP) of the transitional dipole field can be calculated. During a polarity transition the VGP position changes progressively. It appears to move systematically relative to the Earth's rotation axis, defining a path from one polar region to the opposite one. The transitional paths of many reversals appear to define two longitudinal belts, one over the Americas and an antipodal belt over Southeast Asia. However, many other transitions do not pass over these two belts. It has not been established conclusively that a path over the Americas or Southeast Asia is a preferred feature of polarity transitions.

5.7.1.2 *Geomagnetic polarity intervals*

Long intervals of constant normal or reversed polarity, originally called *polarity epochs*, are referred to as *polarity chrons* (Fig. 5.72); they last typically from 50 ka to 5 Ma. The polarity chrons are interrupted at irregular intervals by shorter *polarity subchrons* (originally called *events*) lasting for 20–50 ka. At times the polarity record shows large departures of the magnetic pole from normal or reversed polarity, but the polarity does not change completely; the pole wanders into equatorial latitudes but returns to its initial location on the rotation axis. The departure is short-lived, lasting less than 10 ka, and the phenomenon is called a magnetic *excursion*. The irregular pattern of polarity intervals in any sequence provides a kind of geological fingerprint which can be used under favorable circumstances to date and correlate some types of sedimentary rocks. This procedure is called magnetic polarity stratigraphy, or *magnetostratigraphy*.

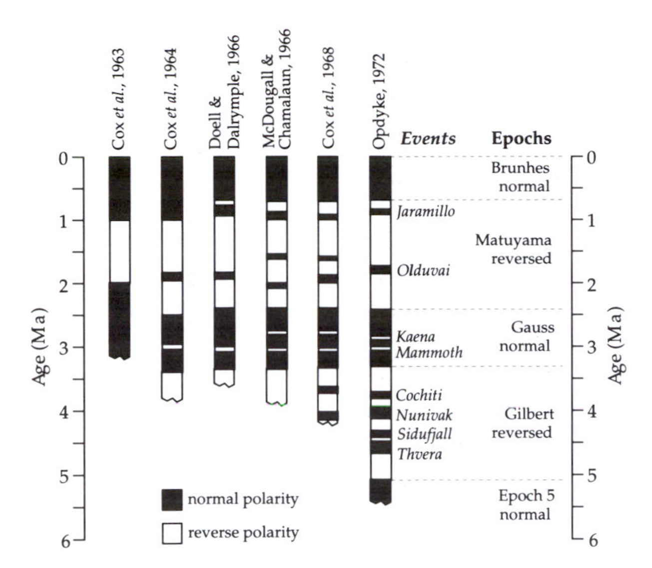

Fig. 5.73 Progressive evolution and refinement of the magnetic polarity timescale.

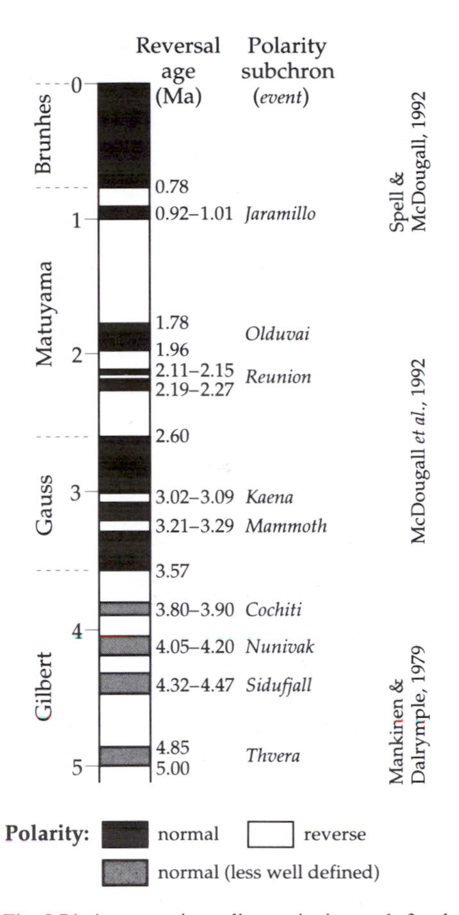

Fig. 5.74 A composite radiometric timescale for the past 5 Ma, compiled from several sources (Mankinen and Dalrymple, 1979; Spell and McDougall, 1992; McDougall *et al.*, 1992).

5.7.2 Magnetostratigraphy in lavas and sediments

In the 1950s the methods of dating rocks took a giant step forward with the development of improved techniques for dating rocks radiometrically. The potassium–argon method (§4.1.4.4) was applied to the determination of accurate ages for Pliocene and Pleistocene lava samples from flows that were also sampled for paleomagnetic purposes. The polarity of the thermoremanent magnetization of the lava was found to correlate with its age. There were distinct intervals of time in which the field polarity was the same as at present, and these were separated by intervals of exactly opposite polarity. At first the data were sparse, and in the earliest interpretations it was thought that the field changed polarity quite regularly, roughly once every million years (Fig. 5.73). Gradually, however, a more complex history evolved. Long *epochs* of a given polarity were found to contain much shorter *events* of opposite polarity. The polarity epochs were named after important investigators of paleomagnetic polarity (Brunhes, Matuyama) and geomagnetism (Gauss, Gilbert) while the polarity events were named after the geographical location where they were first discovered (Jaramillo creek in New Mexico, Olduvai gorge in Africa, etc.). Countless studies of magnetic polarity in radiometrically dated lavas have established the history of geomagnetic polarity in the last 5 Ma (Fig. 5.74). If the polarity record in a rock sequence can be identified, its age can be determined by comparison with the dated sequence. For this reason a dated polarity sequence is called a *geomagnetic polarity timescale*.

There is a practical limit to the application of this technique. As can be seen by quick inspection, some of the polarity events last less than 50 ka. If a reasonable precision of 1–2% is assumed for potassium–argon dating, the error in determining the age of a lava sample that is about 5 Ma old amounts to 50–100 ka. This is longer than the duration of many short events. The dating error makes it impossible to associate the lava sample unambiguously with the correct polarity interval. Extension of the magnetic polarity timescale beyond 5 Ma requires resorting to other methods.

In the middle 1960s the polarity record from lavas was augmented by a large amount of high-quality data acquired from young deep-sea sediments. The deep ocean basins provide a tranquil depositional environment, where sediments are deposited at rather uniform rates. Marine geologists routinely take cores of sediment with a special coring device (see Fig. 4.24) for sedimentological, geophysical and paleontological studies. The magnetostratigraphy of a core is studied by measuring the direction of magnetization in small oriented samples from different depths in the core.

Although deep-sea cores have vertical axes, they are not oriented in azimuth, so the declinations can only be determined relative to an arbitrary reference value. Polarity determinations are often based only on the inclination records. The boundaries between normal and reverse magnetozones are interpolated at the depths where the inclination is zero (Fig. 5.75). In equatorial cores, where inclinations are nearly zero, the relative changes in declination often give a good polarity record. The polarity records of numerous cores correlate well with the sequence found in contemporaneous young lavas. The sediment magnetic polarity records are independent of lithology; the same reversal occurs at different depths from core to core because of different sedimentation rates (Fig. 5.76). By correlating reversals with the radiometrically dated lava record the sediment ages at the reversal depths are obtained. From the depths and ages it is a simple matter to calculate the incremental sedimentation rates in the core. In addition to providing sedimentation rates, magnetic polarity stratigraphy also yields the absolute ages of the first and last appearances of key fossils, and so gives absolute dates for paleontological fossil zones. A recent innovation is the use of astrochronology, based on the identification of Milankovitch cycles (§2.3.4.5), to provide refined dating of sediments and their polarity record.

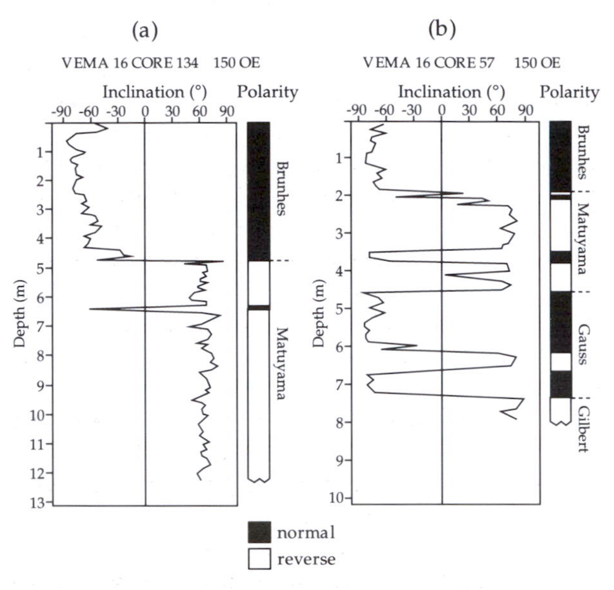

Fig. 5.75 Variations in magnetic inclination and inferred polarity with depth in two deep-sea sediment cores (based on data from Opdyke *et al.*, 1968).

The magnetostratigraphic data from deep-sea sediment cores eliminated the lingering doubt that reversals observed in lavas may be due to a self-reversal mechanism. Lavas and sediments acquire their magnetizations by quite different mechanisms; the thermoremanent magnetization of a lava is acquired rapidly during cooling from high temperature, whereas the depositional or post-depositional remanent magnetization in a sediment is acquired slowly at constant

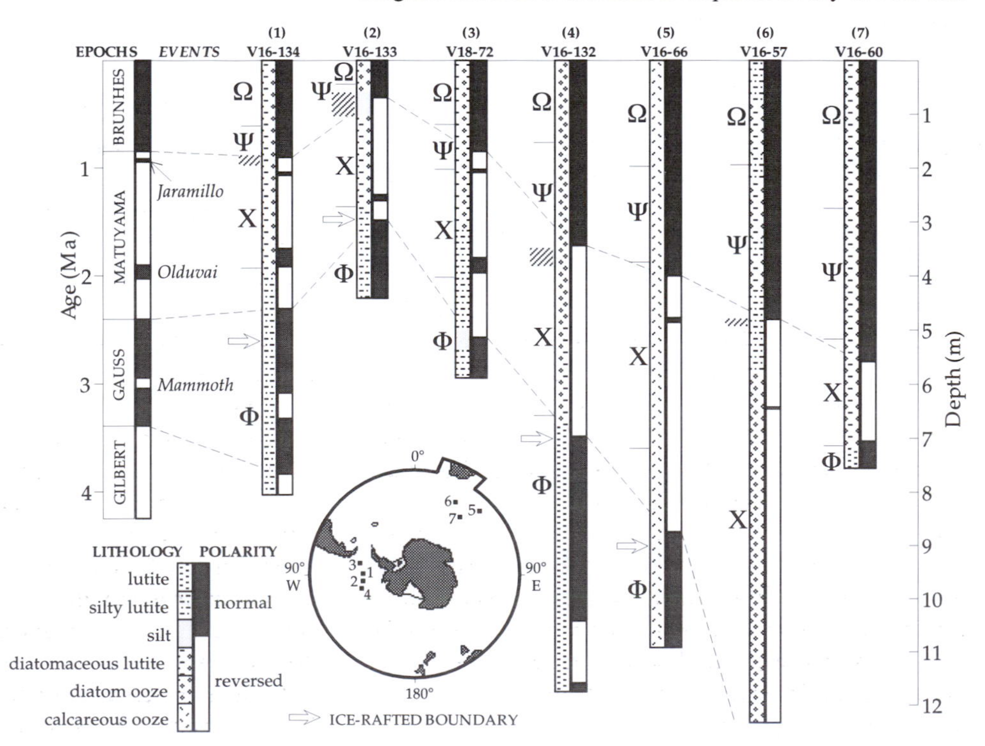

Fig. 5.76 Magnetic reversals in Antarctic deep-sea sediment cores correlate with the radiometric polarity timescale. This allows fossil zones (Greek letters) to be dated. Tie-lines between reversals illustrate the effects of different sedimentation rates (based on data from Opdyke *et al.*, 1968).

ambient temperature. Although a self-reversal mechanism might be invoked to cast doubt on polarity changes in lavas, the argument is invalid for the remanence of deep-sea sediments. The common polarity sequence in lavas and sediments can only be explained as a record of the alternations of polarity of the Earth's magnetic field. Moreover, the same pattern of reversals is found regardless of geographical location, emphasizing that reversals are a global phenomenon.

5.7.3 **Marine magnetic anomalies and geomagnetic polarity history**

The striped magnetic anomaly patterns formed at oceanic ridges contribute to the compelling geophysical evidence in favor of the theory of global plate tectonics (see §1.2.5). Marine magnetic surveys and independent investigations of the rock magnetic properties of marine rocks and sediments have identified the source of the magnetic anomalies to be the basaltic Layer 2A of the oceanic crust.

Seismic evidence indicates that the oceanic crust has a vertically stratified structure (see Fig. 3.79). The uppermost part, seismic Layer 1, consists of a layer of slowly accumulating marine sediments; the thickness of the sediments increases progressively away from the ridge crest. The sediments are so weakly magnetic that they are essentially transparent to the Earth's magnetic field. The seismic Layer 2A consists of a 500 m thick layer of oceanic basalts that are extruded as submarine lava flows or intruded as dikes. These basalts are strongly magnetic and are chiefly responsible for the strong magnetic anomalies observed at the ocean surface. The metamorphosed basalts of the underlying Layer 2B are too weakly magnetic to have much signature. The rocks of the deeper gabbroic Layer 3 may be sufficiently magnetic to add to the skewness of the magnetic anomalies.

5.7.3.1 *Marine magnetic anomalies*

The origin of marine magnetic anomalies was explained by the Vine–Matthews–Morley hypothesis in 1963 (§1.2.5.1). Oceanic basalts were found to have strong and stable remanent magnetizations. Their Königsberger ratios are much larger than unity, so the remanent magnetizations are more important than the magnetization induced by the present-day geomagnetic field. The conventional method of interpreting surveyed magnetic anomalies assumed that the anomaly was due to the *susceptibility contrast* between adjacent crustal blocks. According to the Vine–Matthews–Morley hypothesis the oceanic magnetic anomalies arise from the contrast in *remanent magnetizations* between adjacent, oppositely magnetized crustal blocks. The remanent magnetization is acquired thermally by the basalts in oceanic crustal Layer 2A.

The main magnetic mineral in oceanic basalts is titanomagnetite (§5.3.2.1). A basaltic lava is initially at a temperature well above 1000 °C. Its titanomagnetite grains frequently have skeletal structures, indicating that they cooled and solidified so rapidly that there was not enough time for the formation of normal crystals. Eventually the temperature of the lava sinks below the Curie point of the titanomagnetite (around 200–300 °C) and the lava acquires a thermoremanent magnetization (TRM) in the direction of the Earth's magnetic field at that time. Basalts formed contemporaneously along an active spreading ridge acquire the same polarity of magnetization. Long thin strips of similarly magnetized crust form on opposite sides of the spreading center. These elongated 'crustal blocks' may be hundreds of kilometers in length parallel to the ridge axis and several tens of kilometers wide normal to the ridge, while Layer 2A – the strongly magnetic upper part – is only 0.5 km thick.

Sea-floor spreading persists for millions of years at an oceanic ridge. During this time the magnetic field changes polarity many times. The alternating field polarity leaves some blocks of oceanic crust normally magnetized while their neighbors are reversely magnetized. When the total intensity of the field is measured from a survey ship or aircraft, an alternating sequence of positive and negative anomalies is observed (see Fig. 1.13), which can be interpreted in terms of the crustal magnetization. The anomalies can be correlated almost linearly between parallel profiles across a ridge system; consequently, the stripe-like anomalies are often referred to as *magnetic lineations*.

5.7.3.2 *Uniformity of sea-floor spreading*

Each anomaly in a set of magnetic lineations derives from a crustal block (or stripe) that formed at a ridge and was subsequently transported away from the spreading center. A magnetized crustal block forms during a period of sea-floor spreading when the geomagnetic polarity was constantly normal or reversed, and therefore represents a polarity chron or subchron. The width of a particular block depends on the duration of the chron and the spreading rate at the ocean ridge.

The spreading rate can be determined easily close to a ridge (see §1.2.5.2). The edges of the magnetized crustal stripes correspond to the occurrences of polarity reversals, which can be correlated directly with the radiometrically dated sequence for the last 3–4 Ma determined in lavas on the continents or islands. A plot of the distance of a given polarity reversal from the spreading axis against the age of the reversal is nearly linear near the ridge; the slope of a best-fitting straight line gives the average *half-rate of spreading* at the ridge (see Fig. 1.15). This is half the full rate of

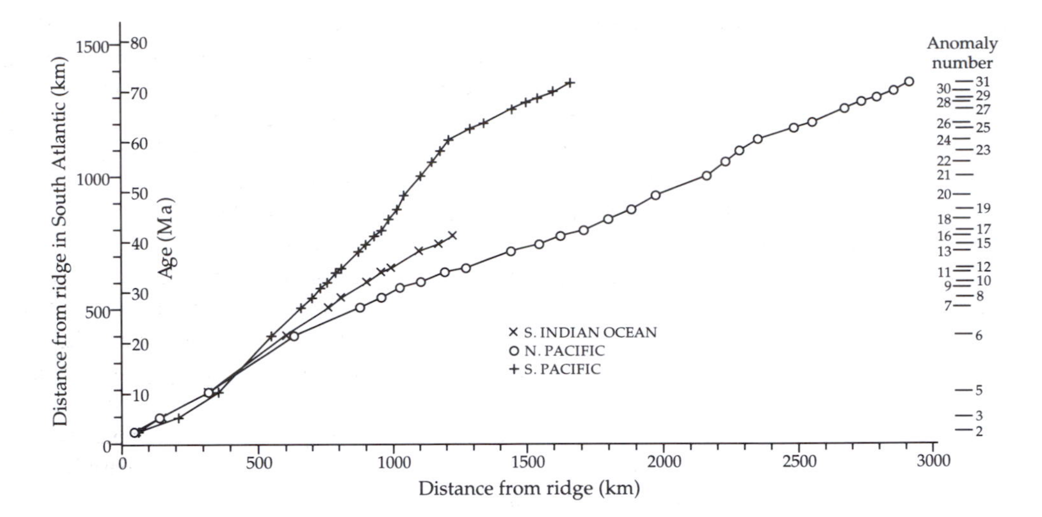

Fig. 5.77 Distances of anomalies from the ridge axis in the South Atlantic plotted against distances to the same anomalies from spreading centers in the Indian, North and South Pacific Oceans (after Heirtzler *et al.*, 1968).

plate separation, assuming that sea-floor spreading has been symmetric on each side of the ridge, which is often the case.

Accumulated evidence from marine magnetic profiles allows us to assess the constancy of sea-floor at different spreading ridge systems. It is thought to have been uniform for the longest time in the South Atlantic. A plot of the distance to a given anomaly in the South Atlantic against the distance to the same anomaly in the Indian, North Pacific and South Pacific oceans contains several long linear segments, representing constant rates of sea-floor spreading in both oceans defining the line (Fig. 5.77). A change in gradient indicates a change in spreading rate in one ocean relative to the other. The plot does not exclude a synchronous world-wide change of spreading rate in all oceans, but this would be a rather unlikely occurrence. Clearly the rate of sea-floor spreading changes from time to time, but for long intervals it is a remarkably constant process.

5.7.3.3 *The marine record of geomagnetic polarity history*

Investigations of magnetic anomalies in all major oceanic areas have given a clear, consistent record of the history of geomagnetic polarity during the past 155–160 Ma. It consists of two sequences of polarity reversals represented by magnetic lineations and a long interval of constant normal polarity (Fig. 5.78). The sequences of chrons derived from magnetic lineations have been confirmed by magnetostratigraphic research.

The most prominent positive magnetic anomalies are numbered in turn from the youngest (anomaly 1 at an active ridge axis) to the oldest (anomalies 33–34 in the Late Cretaceous). The associated polarity chrons are identified by the same number and a letter to indicate the polarity. The polarity chrons in the latest sequence fall largely in the

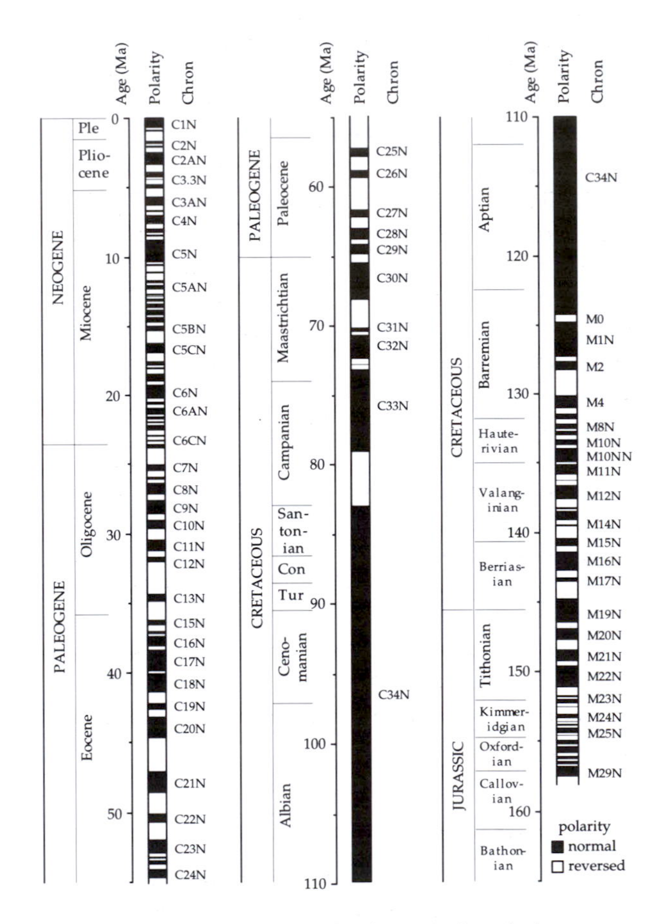

Fig. 5.78 The geomagnetic polarity timescale since the late Jurassic (based upon Harland *et al.*, 1990). Designations for normal polarity chrons are listed.

Cenozoic (see Fig. 4.2) and are identified by a leading letter C, which may be taken to stand for either 'chron' or 'Cenozoic'. The current Brunhes normal polarity interval corresponds to chron C1N; the reversed interval older than it is labelled chron C1R. Anomaly 2 corresponds to the normal Olduvai event, which interrupts the reversed Matuyama interval and is identified as polarity chron C2N; the reversed interval older than C2N is called polarity chron C2R, etc. The current reversal sequence began in the Late Cretaceous. The oldest normal polarity chron in the sequence is C33N; it is preceded by reversed polarity chron C33R, which ended a long interval (lasting about 35 Ma) in which no polarity reversals took place. This interval in which the geomagnetic field had a constant normal polarity is variously called the Cretaceous Quiet Interval, the Cretaceous Normal Polarity Superchron, or chron C34N.

A phase of alternating polarity giving lineated magnetic anomalies precedes the Cretaceous Quiet Interval. It began late in the Middle Jurassic and continued until the middle of the Early Cretaceous. These Late Mesozoic oceanic anomalies are referred to as the M-sequence. To distinguish them from the later sequence the numbered chrons are identified by a leading letter M. The youngest anomalies M0–M5 are numbered sequentially regardless of the magnetization polarity; older than M6 only *reverse* polarity chrons were numbered originally. The oldest securely identified chron in the sequence is M29. Some older anomalies with low amplitudes have been interpreted as polarity chrons, but it has not yet been established that they represent polarity reversals rather than geomagnetic intensity fluctuations.

The oldest regions of the modern oceans correspond to oceanic crust formed approximately 180 Ma ago, when Pangaea broke up and the current episode of sea-floor spreading was initiated. The marine magnetic anomalies over these areas have subdued amplitudes and they do not form lineations. Either no reversals happened during the period of initial spreading or the oceanic crust has not been able to retain the record. The character of the geomagnetic field during this part of the Early Jurassic has not yet been definitively established.

5.7.4 Geomagnetic polarity timescales

The interpretation of marine magnetic anomalies provides the most continuous and reliable record of geomagnetic polarity since the Middle Jurassic. The length of the record greatly exceeds the length of the securely established, radiometrically dated magnetic polarity timescale, which covers only about the past 5 Ma. Thus, it is not possible to date most of the marine magnetic anomalies by direct correlation with a radiometrically dated polarity sequence. Knowledge of the spreading rate at a ridge system provides an alternative way of determining the age of the oceanic crust. Assuming that the rate of sea-floor spreading at the spreading center is constant, the age of a given anomaly can be computed by dividing its distance from the spreading center by the spreading rate. However, this is an unsatisfactory method because the extrapolation is many times longer than the baseline. A further method of dating the polarity record is by establishing the same polarity sequence in sedimentary rocks that are dated paleontologically. This has been achieved in several investigations of the magnetic polarity stratigraphy in pelagic carbonate rocks. Using known absolute ages of major stage boundaries as tie-levels the ages of magnetic polarity chrons that are too old to be dated directly are calculated by interpolation.

5.7.4.1 *Magnetostratigraphic calibration of polarity sequences*

The independent confirmation and dating of marine magnetic anomalies 29–34, the oldest found in the Cenozoic to Late Cretaceous sequence, illustrates the power of combined magnetostratigraphical and paleontological studies in suitable rock formations. Along magnetic profiles in the Indian, North Pacific and South Atlantic oceans, the shapes of anomalies 29–34 are very different because of the different directions of the survey profiles and the orientations of the spreading axes. However, interpretation of the anomalies gives nearly identical crustal magnetization patterns (Fig. 5.79).

The Scaglia Rossa pelagic limestone in the Umbrian Apennines of Central Italy was deposited almost continuously from the Late Cretaceous to the Eocene. Rock magnetic analysis showed that the limestone contained an easily defined stable component of characteristic remanent magnetization. Samples were taken at approximately 0.5–1 m stratigraphic intervals in a long section through the limestone exposed in the Bottaccione gorge near Gubbio. The declinations and inclinations of the stable magnetization, after simple tectonic corrections, were used to calculate the latitude of the virtual geomagnetic pole (VGP) during deposition of the limestone. In times of normal polarity the VGP latitude is near 90°N, during reversed polarity it is 90°S. The fluctuations of VGP latitude clearly define magnetozones of normal and reversed polarity (Fig. 5.80). The Gubbio polarity record in a 200 m thick section of pelagic limestone correlates almost perfectly with the oceanic polarity sequence derived from anomalies 29–34, measured in different oceans on profiles that are hundreds of kilometers long. The limestone magnetostratigraphy confirms independently this part of the oceanic magnetic polarity record.

Paleontological studies in the Gubbio section gave the

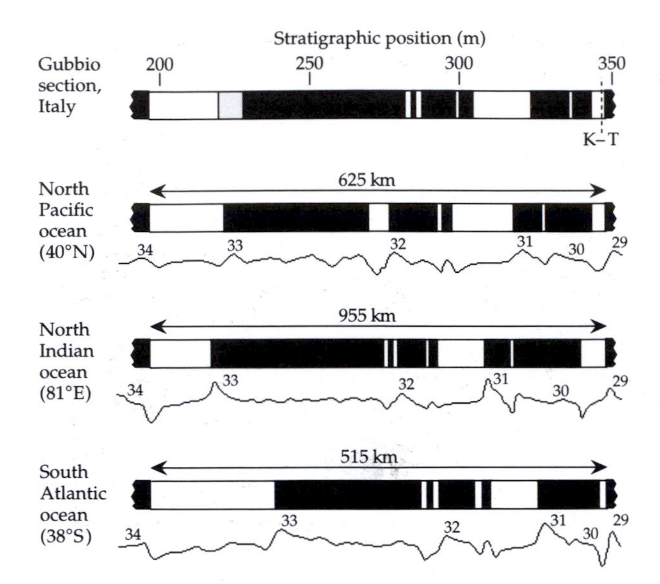

Fig. 5.79 Crustal magnetization patterns for anomalies 29–34 interpreted from magnetic profiles in the Indian, North Pacific and South Atlantic oceans, and comparison with the magnetic polarity stratigraphy in the Gubbio Bottaccione section (after Lowrie and Alvarez, 1977). K–T indicates the position of the Cretaceous–Tertiary boundary, which falls within reversed polarity chron C29R.

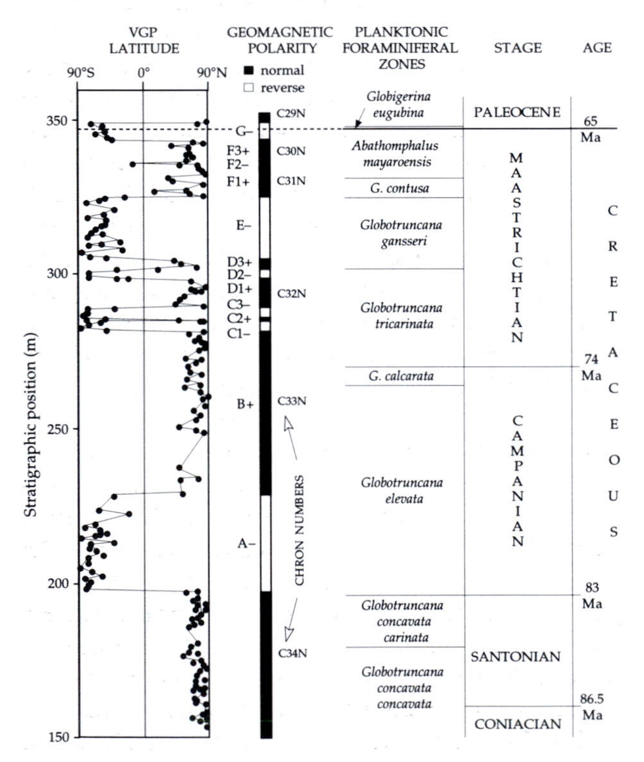

Fig. 5.80 Magnetostratigraphy and biostratigraphy in the Bottaccione section at Gubbio, Italy (after Lowrie and Alvarez, 1977).

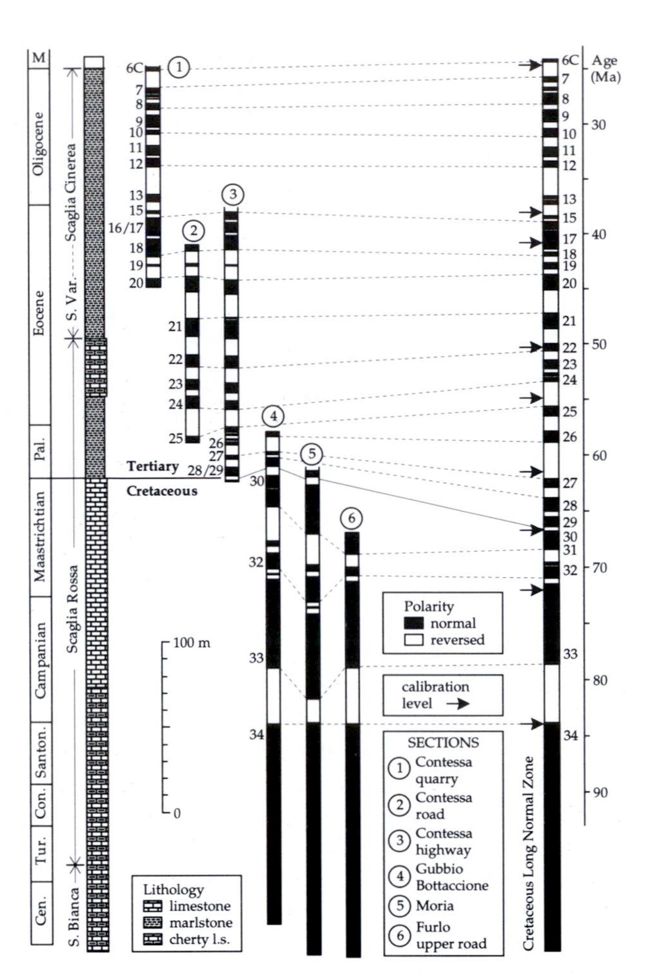

Fig. 5.81 Confirmation and calibration of the oceanic magnetic reversal record in Paleogene and Cretaceous magnetostratigraphic sections in Umbria, Italy (after Lowrie and Alvarez, 1981).

locations of important planktonic foraminifera zones and enabled the positions of the major stage boundaries to be located relative to the polarity sequence. The absolute ages of some important stage boundaries are known from independent radiometric and stratigraphic work. This enabled calculation of the absolute ages of the polarity reversals in the Gubbio section, and, by correlation, the ages of the corresponding parts of the ocean floor. For example, the Santonian–Campanian boundary (with an age of about 83 Ma) lies close to the old edge of the reversed polarity chron C33R; the geologically important Cretaceous–Tertiary boundary (age 65 Ma) falls within reversed polarity chron C29R.

In this way the geomagnetic polarity sequence in the Late Cretaceous and Paleogene has been confirmed in magnetostratigraphic sections, and the locations of many major and minor stage boundaries have been correlated to the polarity sequence paleontologically (Fig. 5.81). Reliable

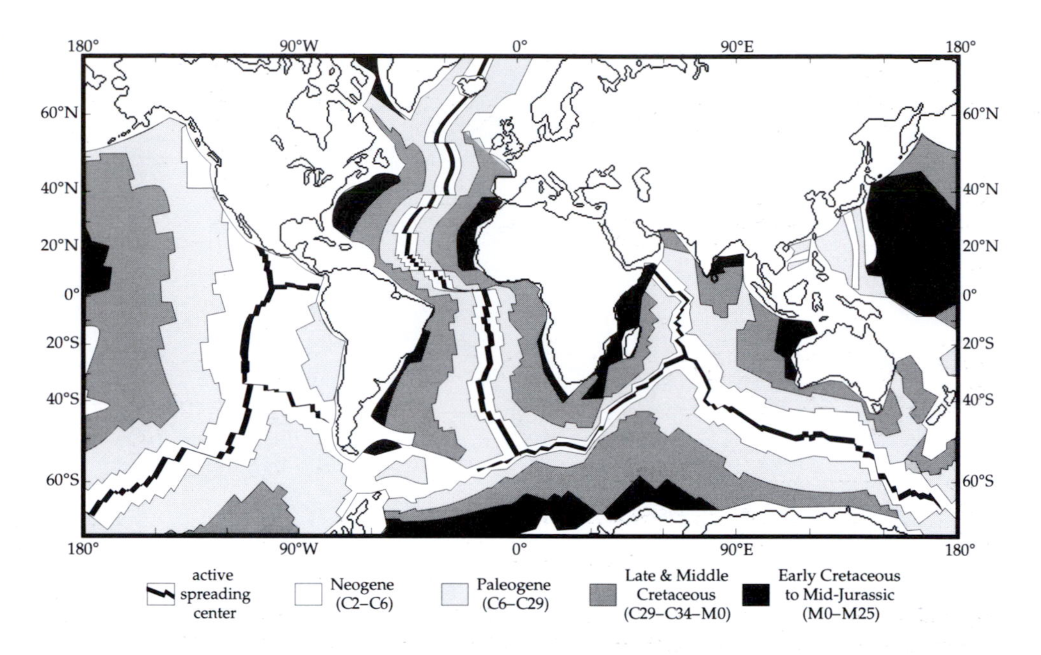

Fig. 5.82 Map of the age of oceanic crust, as interpreted from marine magnetic anomalies (simplified after Scotese *et al.*, 1988).

absolute ages are available for some of the stage boundaries, which can then be used as calibration levels. The ages of magnetic reversals between the dated tie-points are computed by interpolation, giving a numerical geomagnetic polarity timescale (see Fig. 5.78). In the same way, overlapping magnetostratigraphic sections of Early Cretaceous and Late Jurassic age have permitted independent confirmation and dating of the M-sequence polarity record.

5.7.4.2 *Reconstruction of plate tectonic motions*

Once the ages of magnetic anomalies are known, a map showing the positions of dated anomalies is equivalent to a chronological map of the ocean basins (Fig. 5.82). The different rates of sea-floor spreading are evident from the separations of the isochrons. The oldest domains of the oceans (about 180 Ma) are found in the Atlantic ocean adjacent to the coastlines of North America and Africa, and in the North Pacific ocean. They are very much younger than the oldest rocks from the continents, which are up to 3.6 Ga old. The oceanic crust has been entirely produced since the onset of sea-floor spreading and the anomaly patterns reflect plate motions.

The past motions of the major plates can be obtained in some detail from the anomaly ages given by a geomagnetic polarity timescale. The relative positions of the continents at any time since the late Mid-Jurassic can be reconstructed by matching coeval marine magnetic anomalies formed at the same spreading center. The procedure is similar to the reconstruction of supercontinents by matching coastlines, 500 fathom depth contours or apparent polar wander paths.

The method is illustrated by the sea-floor spreading between North America and Africa in the Central Atlantic (Fig. 5.83). Anomaly 33 forms a long stripe on each side of the Mid-Atlantic Ridge and subparallel to it. The anomaly is due to the magnetization contrast between chron C33N and the reverse polarity chron C33R which marks the end of the Cretaceous Quiet Interval. Its age is about 81 Ma. The anomaly on the west side of the ridge was formed at the same time as the anomaly on the east side, when the newly formed crust was magnetized at the ridge axis. If the African and North American plates are moved closer together until the east and west anomalies overlap, or until they match along their lengths with minimum misfit, the continents will be brought into the same positions relative to each other that they occupied around 81 Ma ago. By repeating this process of matching dated anomalies it is possible to reconstruct the successive relative positions of the African and North American plates as they separated.

Marine magnetic anomalies in the North Atlantic can be used likewise to describe the relative plate motions between Europe and North America. A picture evolves of the separate histories of separation of Africa and Europe from North America. The differences between the European–North American and African–North American plate motions permit the history of relative motion between Africa and Europe to be inferred. In the Late Cretaceous and Early Tertiary Africa moved eastwards in a giant shear motion relative to Europe, but since the Middle Tertiary the motion of Africa has been one of convergence and collision with the European plate. This is compatible with the formation of the Alpine fold belt and the present-day seismicity in the Alpine region.

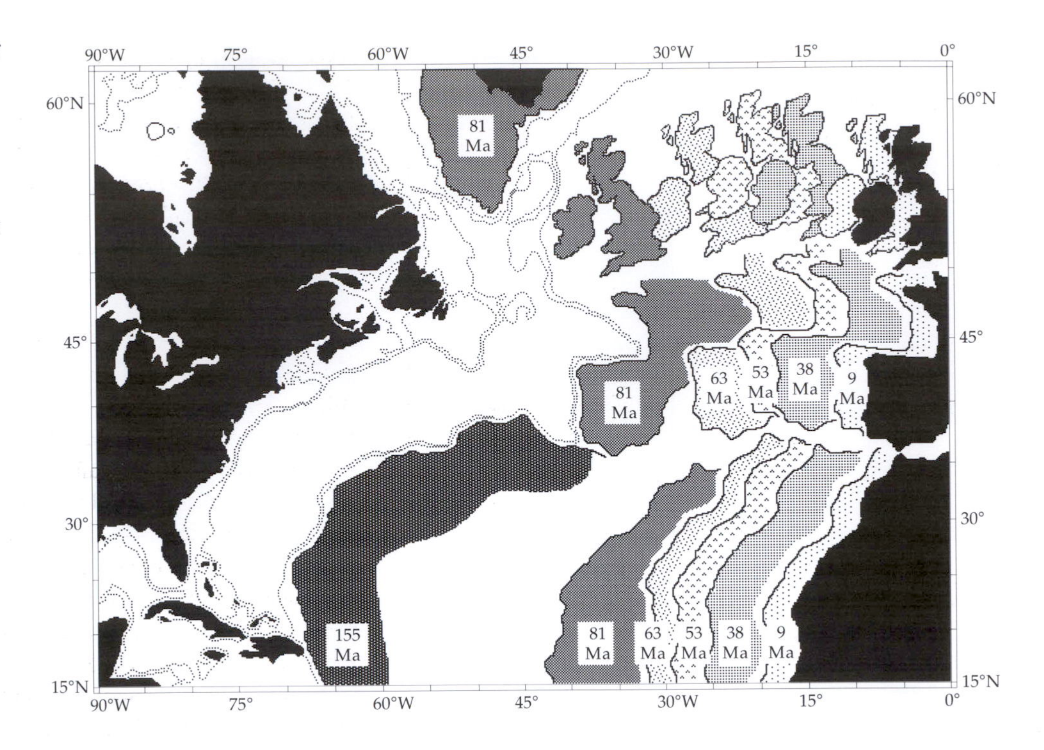

Fig. 5.83 Reconstruction of the history of opening of the North and Central Atlantic oceans. The figure shows the relative positions of Europe and Africa with respect to North America at specific times before the present (after Pitman and Talwani, 1972).

5.7.5 Frequency of polarity reversals

The availability of a well-dated polarity timescale makes it possible to analyse statistically the rate at which reversals happen, from which inferences can be drawn about the processes that govern their occurrence. The most extensive analyses have treated the Cenozoic–Late Cretaceous reversal sequence, which is dated more securely than the older M-sequence. Visual examination of the spacing of reversals (see Fig. 5.78) shows that the polarity intervals at the end of the Cretaceous lasted longer than in the Late Tertiary. In other words, the reversal rate has speeded up since the Late Cretaceous. A simple way to portray the long-term trend in the reversal record is to plot the ages of successive reversals against their numerical order, counting backwards from the present. The plot is not linear, confirming that the reversal rate has not remained constant. However, a low-order polynomial curve fits the data well. If the polynomial is differentiated, its gradient represents the average rate of reversals, which can then be plotted as a function of age (Fig. 5.84). The mean reversal rate was very low when the reversal process recommenced following the Cretaceous Quiet Interval. It increased to a maximum of about 5 reversals per Ma about 10–12 Ma ago, and has since been decreasing again.

In 1968 A. Cox theorized that reversals are random events, triggered by unknown mechanisms that affect the fluid motions in the Earth's liquid outer core. With a random reversal process there would be no continuity between successive reversals; as soon as a reversal was completed, the next one would be as likely to occur immediately as at any time later. Because there would be no waiting period, this type of mechanism would generate a large number of very short polarity chrons and a small number of long chrons. The frequency distribution of polarity chron durations should decrease exponentially with increasing chron duration. In fact this model does not fit the observed lengths of polarity intervals very well. The polarity sequence contains comparatively few short chrons, a lot of medium-duration chrons and few long chrons. It seems more likely that the process that causes reversals is not completely random. A distinct length of time must elapse after a reversal for the fluid motions to recover sufficiently to allow the next reversal to happen. However, the mechanism that causes a reversal is inadequately understood.

5.7.6 Early Mesozoic and Paleozoic reversal history

Because of the availability of the excellent marine magnetic anomaly record, it has been possible to construct a well-dated history of geomagnetic polarity in the last 155–160 Ma (see Fig. 5.78). The determination of a geomagnetic polarity timescale for eras older than the Middle Jurassic is more complicated, because no comparable oceanic record exists. Paleomagnetic results show numerous

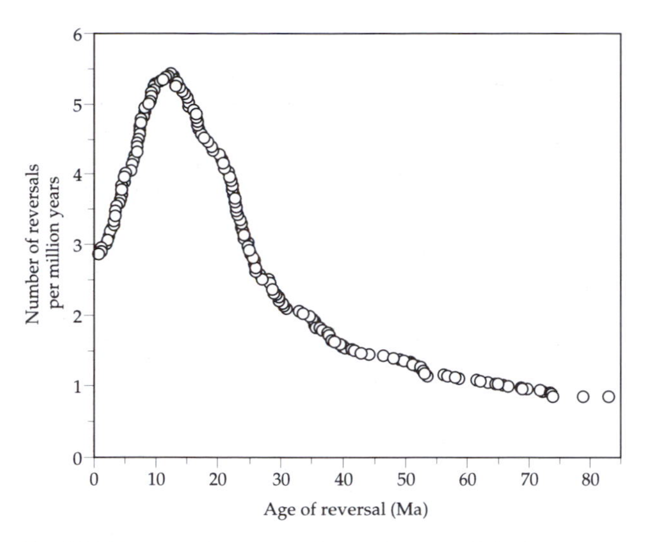

Fig. 5.84 The rate of geomagnetic polarity reversals in the Cenozoic and Late Cretaceous, based upon the polarity timescale of Cande and Kent (1992).

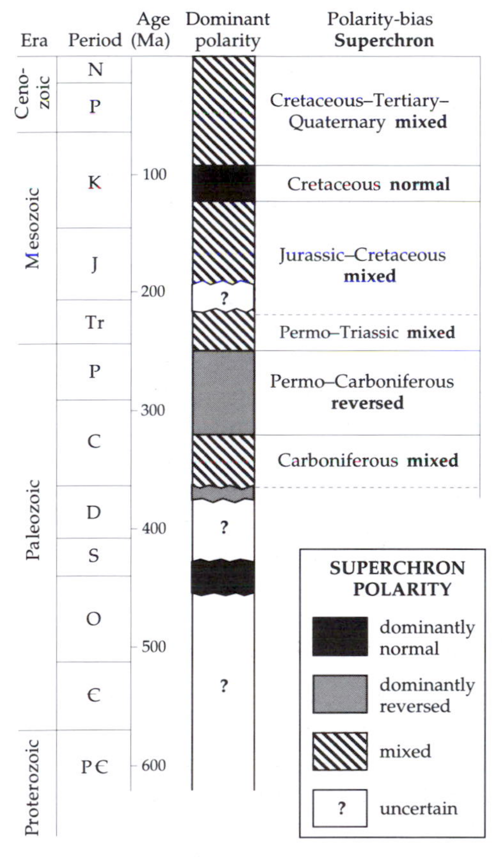

Fig. 5.85 Geomagnetic polarity bias superchrons in the Paleozoic, Mesozoic and Cenozoic (after Harland *et al.*, 1982).

reversals during the Triassic, but the Permian and Late Carboniferous were dominated by reversed polarity. A long interval of constant reversed polarity – the Kiaman interval – is a distinctive feature of the Paleozoic polarity record. Earlier in the Paleozoic reversals were common. Although magnetostratigraphic investigations of many formations are in progress, no unique record of the polarity succession is available for the Early Mesozoic or Paleozoic.

At present it is only possible to analyze Paleozoic polarity history in terms of the bias toward normal or reverse polarity. The polarity bias broadly defines superchrons (Fig. 5.85). When as many reversed as normal magnetizations are found it is assumed that the field polarity has been reversing; the interval is called a mixed polarity superchron. The Late Cretaceous and Cenozoic and the Triassic illustrate mixed polarity superchrons. Sometimes, for unknown reasons, the polarity was constant for long intervals. This was the case during the Cretaceous Quiet Interval, which in terms of polarity bias is called the Cretaceous Normal Polarity Superchron. The Kiaman reversed polarity interval is the same as the Permo–Carboniferous Reversed Polarity Superchron. The derivation of a more detailed history of geomagnetic polarity for Early Mesozoic and Paleozoic time is a massive task for paleomagnetists and biostratigraphers. Because of the absence of a marine magnetic record each polarity sequence will have to be confirmed by repetition in several magnetostratigraphic sections before it can be accepted as globally significant and representative of dipole field behavior.

5.8 SUGGESTIONS FOR FURTHER READING

INTRODUCTORY LEVEL

Hailwood, E. A. 1989. *Magnetostratigraphy, Geological Society Special Report No. 19*, Blackwell Scientific Publ., Oxford.

INTERMEDIATE LEVEL

Butler, R. F. 1992. *Paleomagnetism: Magnetic Domains to Geologic Terranes*, Blackwell Scientific Publ., Boston.

Irving, E. 1964. *Paleomagnetism and its Application to Geological and Geophysical Problems*, John Wiley & Sons, New York.

McElhinny, M. W. 1973. *Paleomagnetism and Plate Tectonics*, Cambridge Univ. Press, Cambridge.

Tarling, D. H. 1983. *Palaeomagnetism: Principles and Applications in Geology, Geophysics and Archeology*, Chapman and Hall, London.

ADVANCED LEVEL

Collinson, D. W. 1983. *Methods in Rock Magnetism and Palaeomagnetism*, Chapman and Hall, London.

Harland, W. B., Armstrong, R. L., Cox, A. V., Craig, L. E., Smith, A. G. and Smith, D. G. 1990. *A Geologic Time Scale 1989*, Cambridge Univ. Press, Cambridge.

Merrill, R. T. and McElhinny, M. W. 1983. *The Earth's Magnetic Field*, Academic Press, London.

Nagata, T. 1961. *Rock Magnetism*, Maruzen, Tokyo.

O'Reilly, W. 1984. *Rock and Mineral Magnetism*, Blackie, Glasgow.

Stacey, F. D. and Banerjee, S. K. 1974. *The Physical Principles of Rock Magnetism*, Elsevier, Amsterdam.

Van der Voo, R. 1993. *Paleomagnetism of the Atlantic, Tethys and Iapetus Oceans*, Cambridge Univ. Press, Cambridge.

6 Geodynamics

6.1 ISOSTASY

6.1.1 The discovery of isostasy

Newton formulated the law of universal gravitation in 1687 and confirmed it with Kepler's laws of planetary motion. However, in the 17th and 18th centuries the law could not be used to calculate the mass or mean density of the Earth, because the value of the gravitational constant was not yet known (it was first determined by Cavendish in 1798). Meanwhile, eighteenth century scientists attempted to estimate the mean density of the Earth by various means. They involved comparing the attraction of the Earth with that of a suitable mountain, which could be calculated. Inconsistent results were obtained.

During the French expedition to Peru in 1737–1740, Pierre Bouguer measured gravity with a pendulum at different altitudes, applying the elevation correction term which now bears his name. If the density of crustal rocks is ρ and the mean density of the Earth is ρ_0, the ratio of the Bouguer plate correction (see §2.5.4.3) for elevation h to mean gravity for a spherical Earth of radius R is

$$\frac{\Delta g_{BP}}{g}=\frac{2\pi G\rho h}{\frac{4}{3}\pi G\rho_0 R}=\frac{3}{2}\left(\frac{\rho}{\rho_0}\right)\left(\frac{h}{R}\right) \tag{6.1}$$

From the results he obtained near Quito, Bouguer estimated that the mean density of the Earth was about 4.5 times the density of crustal rocks.

The main method employed by Bouguer to determine the Earth's mean density consisted of measuring the deflection of the plumb-line (vertical direction) by the mass of a nearby mountain (Fig. 6.1). Suppose the elevation of a known star is measured relative to the local vertical direction at points N and S on the same meridian. The elevations should be α_N and α_S, respectively. Their sum is α, the angle subtended at the center of the Earth by the radii to N and S, which corresponds to the difference in latitude. If N and S lie on opposite sides of a large mountain, the plumb-line at each station is deflected by the attraction of the mountain. The measured elevations of the star are β_N and β_S, respectively, and their sum is β. The local vertical directions now intersect at the point D instead of at C, the center of the (assumed spherical) Earth. The difference $\delta=\beta-\alpha$ is the sum of the deviations of the vertical direction caused by the mass of the mountain.

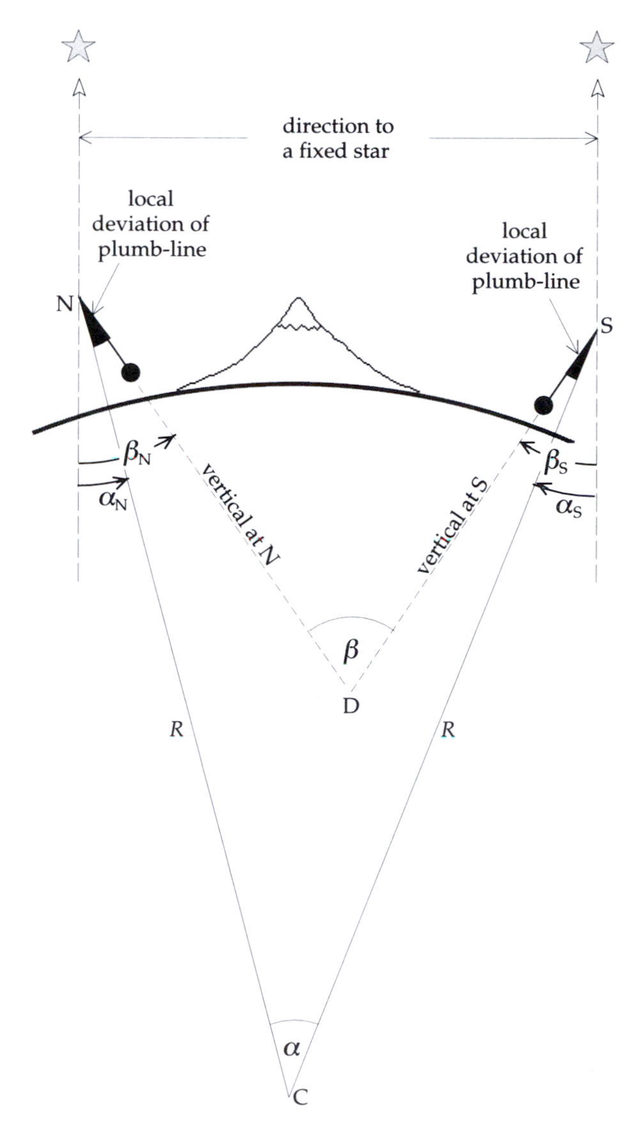

Fig. 6.1 Deviations of the local plumb-line at N and S on opposite sides of a large mountain cause the local vertical directions to intersect at the point D instead of at the center of the Earth C.

The horizontal attraction f of the mountain can be calculated from its shape and density, with a method that resembles the computation of the topographic correction in the reduction of gravity measurements. Dividing the mountain into vertical cylindrical elements, the horizontal attraction of each element is calculated and its component $(G\rho h_i)$ towards the center of mass of the mountain is found. Summing up the effects of all the cylindrical elements in the mountain gives the horizontal attraction f towards its center of mass. Comparing f with mean gravity g, we then write

$$\tan\delta = \frac{f}{g} = \frac{G\rho\sum_i h_i}{\frac{4}{3}\pi G\rho_0 R} = \frac{\sum_i h_i}{\frac{4}{3}\pi R}\left(\frac{\rho}{\rho_0}\right) \qquad (6.2)$$

For very small angles, $\tan\delta$ is equal to δ, and so the deflection of the vertical is proportional to the ratio ρ/ρ_0 of the mean densities of the mountain and the Earth. Bouguer measured the deflection of the vertical caused by Mt. Chimborazo (6272 m), Ecuador's highest mountain. His results gave a ratio of ρ/ρ_0 around 12, which is unrealistically large and quite different from the values he had obtained near Quito. The erroneous result indicated that the deflection of the vertical caused by the mountain was much too small for its estimated mass.

In 1774 Bouguer's Chimborazo experiment was repeated in Scotland by Neville Maskelyne on behalf of the Royal Society of London. Measurements of the elevations of stars were made on the north and south flanks of Mt. Schiehallion at sites that differed in latitude by 42.9″ of arc. The observed angle between the plumb-lines was 54.6″. The analysis gave a ratio of ρ/ρ_0 equal to 1.79, suggesting a mean density for the Earth of 4500 kg m^{-3}. This was more realistic than Bouguer's result, which still needed explanation.

Further results accumulated in the first half of the 19th century. From 1806 to 1843 the English geodesist George Everest carried out triangulation surveys in India. He measured by triangulation the separation of a site at Kalianpur on the Indo–Ganges plain from a site at Kaliana in the foothills of the Himalayas. The distance differed substantially from the separation of the sites computed from the elevations of stars, as in Fig. 6.1. The discrepancy of 5.23″ of arc (162 m) was attributed to deflection of the plumb-line by the mass of the Himalayas. This would affect the astronomic determination but not the triangulation measurement. In 1855 J. H. Pratt computed the minimum deflection of the plumb-line that might be caused by the mass of the Himalayas and found that it should be 15.89″ of arc, about three times larger than the observed deflection. Evidently the attraction of the mountain range on the plumb-line was not as large as it should be.

The anomalous deflections of the vertical were first understood in the middle of the 19th century, when it was realized that there are regions beneath mountains – 'root-zones' – in which rocks have a lower density than expected. The deflection of a plumb-line is not caused only by the horizontal attraction of the visible part of a mountain. The deficiency of mass at depth beneath the mountain means that the 'hidden part' exerts a reduced lateral attraction, which partly offsets the effect of the mountain and diminishes the deflection of the vertical. In 1889 C. E. Dutton referred to the compensation of a topographic load by a less-dense subsurface structure as *isostasy*.

6.1.2 **Models of isostasy**

Separate explanations of the anomalous plumb-line deflections were put forward by G. B. Airy in 1855 and J. H. Pratt in 1859. Airy was the Astronomer Royal and director of the Greenwich Observatory. Pratt was an archdeacon of the Anglican church at Calcutta, India, and a devoted scientist. Their hypotheses have in common the compensation of the extra mass of a mountain above sea-level by a less-dense region (or *root*) below sea-level, but they differ in the way the compensation is achieved. In the Airy model, when isostatic compensation is complete, the mass deficiency of the root equals the excess load on the surface. At and below a certain *compensation depth* the pressure exerted by all overlying vertical columns of crustal material is then equal. The pressure is then *hydrostatic*, as if the interior acted like a fluid. Hence, isostatic compensation is equivalent to applying Archimedes' principle to the uppermost layers of the Earth.

The Pratt and Airy models achieve compensation locally by equalization of the pressure below vertical columns under a topographic load. The models were very successful and became widely used by geodesists, who developed them further. In 1909–10, J. F. Hayford in the United States derived a mathematical model to describe the Pratt hypothesis. As a result, this theory of isostasy is often called the Pratt–Hayford scheme of compensation. Between 1924 and 1938 W. A. Heiskanen derived sets of tables for calculating isostatic corrections based on the Airy model. This concept of isostatic compensation has since been referred to as the Airy–Heiskanen scheme.

It became apparent that both models had serious deficiencies in situations that required compensation over a larger region. In 1931 F. A. Vening Meinesz, a Dutch geophysicist, proposed a third model, in which the crust acts as an elastic plate. As in the other models, the crust floats buoyantly on a substratum, but its inherent rigidity spreads topographic loads over a broader region.

6.1.2.1 *The Airy–Heiskanen model*

According to the Airy–Heiskanen model of isostatic compensation (Fig. 6.2a) an upper layer of the Earth 'floats' on a denser magma-like substratum, just as icebergs float in water. The upper layer is equated with the crust and the substratum with the mantle. The height of a mountain above sea-level is much less than the thickness of the crust under-

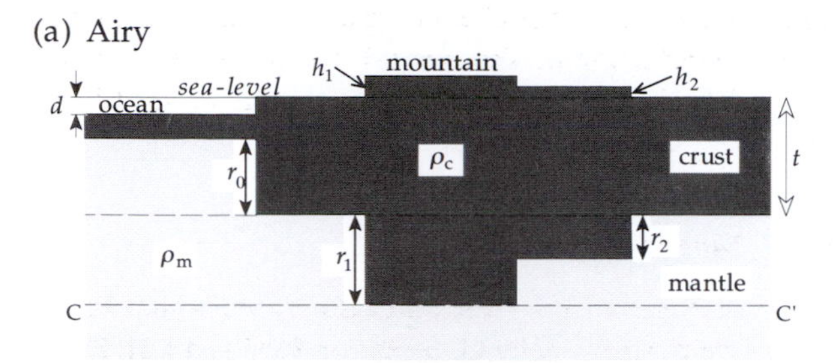

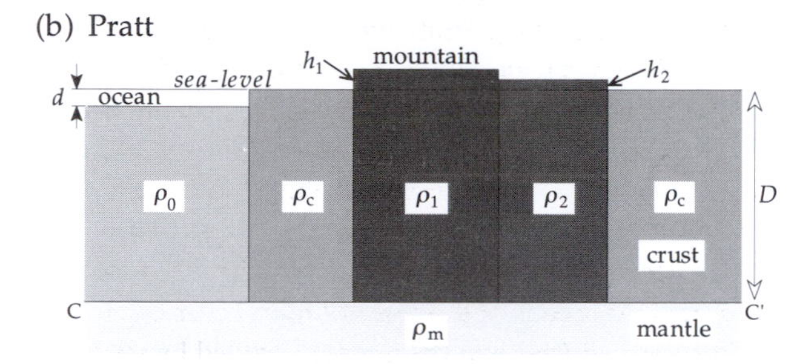

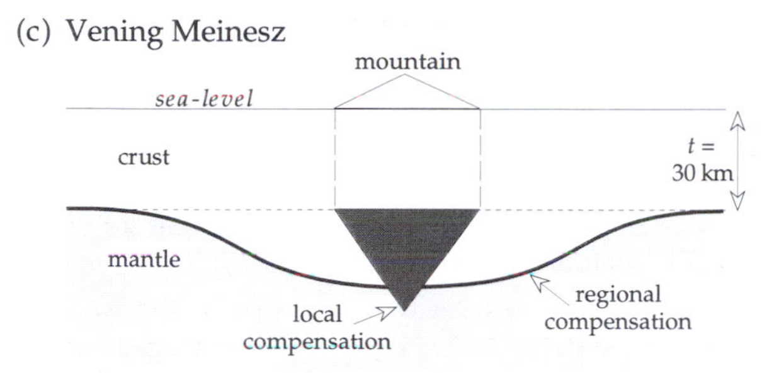

Fig. 6.2 Local isostatic compensation according to (a) the Airy–Heiskanen model and (b) the Pratt–Hayford model; (c) regional compensation according to the elastic plate model of Vening Meinesz.

neath it, just as the visible tip of an iceberg is much smaller than the subsurface part. The densities of the crust and mantle are assumed to be constant; the thickness of the root-zone varies in proportion to the elevation of the topography.

The analogy to an iceberg is not exact, because under land at sea-level the 'normal' crust is already about 30–35 km thick; the compensating root-zone of a mountain lies below this depth. Oceanic crust is only about 10 km thick, thinner than the 'normal' crust. The mantle between the base of the oceanic crust and the normal crustal depth is sometimes called the *anti-root* of the ocean basin.

The Airy–Heiskanen model assumes local isostatic compensation, i.e., the root-zone of a mountain lies directly under it. Isostasy is assumed to be complete, so that hydrostatic equilibrium exists at the compensation depth, which is equivalent to the base of the deepest mountain root. The pressure at this level is due to the weight of the rock material in the overlying vertical column (of basal area one square meter) extending to the Earth's surface. The vertical column for the mountain of height h_1 in Fig. 6.2a contains only crustal rocks of density ρ_c. The pressure at CC' due to the mountain, 'normal' crust of thickness t, and a root-zone of thickness r_1 amounts to $(h_1+t+r_1)\rho_c$. The vertical column below the 'normal' crust contains a thickness t of crustal rocks and thickness r_1 of mantle rocks; it exerts a pressure of $(t\rho_c+r_1\rho_m)$. For hydrostatic equilibrium the pressures are equal. Equating, and noting that each expression contains the term $t\rho_c$, we get

$$r_1=\frac{\rho_c}{\rho_m-\rho_c}h_1 \tag{6.3}$$

with a similar expression for the root of depth r_2 under the hill of height h_2. The thickness r_0 of the anti-root of the oceanic crust under an ocean basin of water depth d and density ρ_w is given by

$$r_0=\frac{\rho_c-\rho_w}{\rho_m-\rho_c}d \tag{6.4}$$

The Airy–Heiskanen model assumes an upper layer of constant density floating on a more-dense substratum. It has root-zones of variable thickness proportional to the overlying topography. This scenario agrees broadly with seismic evidence for the thickness of the Earth's crust (see §3.7). The continental crust is much thicker than the oceanic crust. Its thickness is very variable, being largest below mountain chains, although the greatest thickness is not always under the highest topography. Airy-type compensation suggests hydrostatic balance between the crust and the mantle.

6.1.2.2 *The Pratt–Hayford model*

The Pratt–Hayford isostatic model incorporates an outer layer of the Earth that rests on a weak magmatic substratum. Differential expansion of the material in vertical columns of the outer layer accounts for the surface topography, so that the higher the column above a common base the lower the mean density of rocks in it. The vertical columns have constant density from the surface to their base at depth D below sea-level (Fig. 6.2b). If the rock beneath a mountain of height h_i $(i=1, 2...)$ has density ρ_i, the pressure at CC' is $\rho_i(h_i+D)$. Beneath a continental region at sea-level the pressure of the rock column of density ρ_c is $\rho_c D$. Under an ocean basin the pressure at CC' is due to water of depth d and density ρ_w on top of a rock column of thickness $(D-d)$ and density ρ_0; it is equal to $\rho_w d+\rho_0(D-d)$. Equating these pressures, we get

$$\rho_i = \frac{D}{h_i + D}\rho_c \tag{6.5}$$

for the density below a topographic elevation h_i, and

$$\rho_0 = \frac{\rho_c D - \rho_w d}{D - d} \tag{6.6}$$

for the density under an oceanic basin of depth d. The compensation depth D is about 100 km.

The Pratt–Hayford and Airy–Heiskanen models represent *local isostatic compensation*, in which each column exerts an equal pressure at the compensation level. At the time these models were proposed very little was yet known about the internal structure of the Earth. This was only deciphered after the development of seismology in the late 19th and early 20th century. Each model is idealized, both with regard to the density distributions and the behavior of Earth materials. For example, the upper layer is assumed to offer no resistance to shear stresses arising from vertical adjustments between adjacent columns. Yet the layer has sufficient strength to resist stresses due to horizontal differences in density. It is implausible that small topographic features require compensation at large depths; more likely, they are entirely supported by the strength of the Earth's crust.

6.1.2.3 *Vening Meinesz elastic plate model*

In the 1920s F. A. Vening Meinesz made extensive gravity surveys at sea. His measurements were made in a submarine to avoid the disturbances of wave motions. He studied the relationship between topography and gravity anomalies over prominent topographic features, such as the deep sea trenches and island arcs in southeastern Asia, and concluded that isostatic compensation is often not entirely local. In 1931 he proposed a model of *regional isostatic compensation* which, like the Pratt–Hayford and Airy–Heiskanen models, envisages a light upper layer that floats on a denser fluid substratum. However, in the Vening Meinesz model the upper layer behaves like an elastic plate overlying a weak fluid. The strength of the plate distributes the load of a surface feature (e.g., an island or seamount) over a horizontal distance wider than the feature (Fig. 6.2c). The topographic load bends the plate downward into the fluid substratum, which is pushed aside. The buoyancy of the displaced fluid forces it upward, giving support to the bent plate at distances well away from the central depression. The bending of the plate which accounts for the regional compensation in the Vening Meinesz model depends on the elastic properties of the lithosphere.

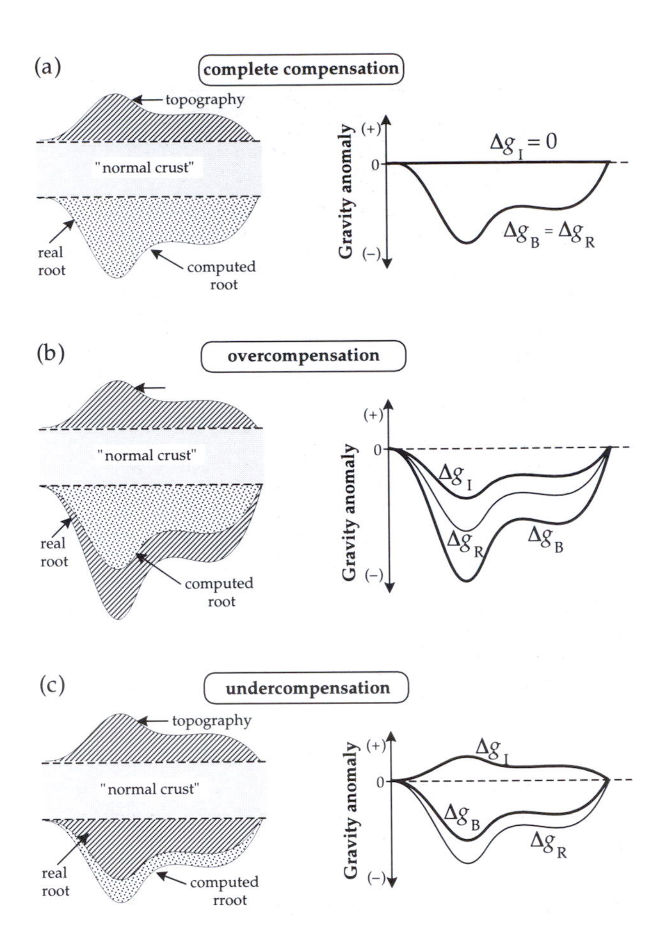

Fig. 6.3 Explanation of the isostatic gravity anomaly (Δg_I) as the difference between the Bouguer gravity anomaly (Δg_B) and the computed anomaly (Δg_R) of the root-zone estimated from the topography for (a) complete isostatic compensation, (b) isostatic overcompensation and (c) isostatic undercompensation.

6.1.3 **Isostatic compensation and vertical crustal movements**

In the Pratt–Hayford and Airy–Heiskanen models the lighter crust floats freely on the denser mantle. The system is in hydrostatic equilibrium, and local isostatic compensation is a simple application of Archimedes principle. A 'normal' crustal thickness for sea-level coastal regions is assumed (usually 30–35 km) and the additional depths of the root-zones below this level are exactly proportional to the elevations of the topography above sea-level. The topography is then *completely compensated* (Fig. 6.3a). However, isostatic compensation is often incomplete. The geodynamic imbalance leads to vertical crustal movements.

Mountains are subject to erosion, which can disturb isostatic compensation. If the eroded mountains are no longer high enough to justify their deep root-zones, the topography is isostatically *overcompensated* (Fig. 6.3b). Buoyancy forces

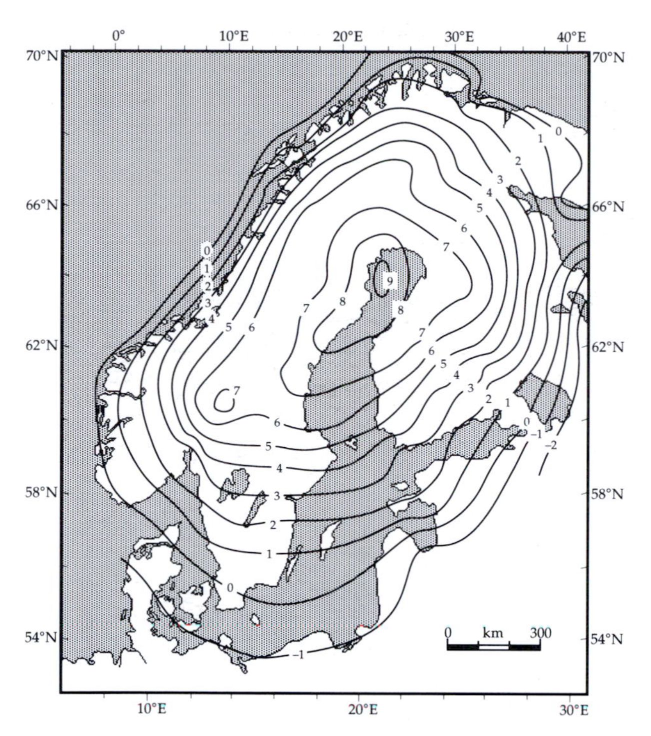

Fig. 6.4 Fennoscandian rates of vertical crustal movement (in mm yr^{-1}) relative to mean sea-level. Positive rates correspond to uplift, negative rates to subsidence (after Kakkuri, 1992).

are created, just as when a wooden block floating in water is pressed downward by a finger; the underwater part becomes too large in proportion to the amount above the surface. If the finger pressure is removed, the block rebounds in order to restore hydrostatic equilibrium. Similarly, the buoyancy forces that result from overcompensation of mountainous topography cause vertical *uplift*. The opposite scenario is also possible. When the visible topography has roots that are too small, the topography is isostatically *undercompensated* (Fig. 6.3c). This situation can result, for example, when tectonic forces thrust crustal blocks on top of each other. Hydrostatic equilibrium is now achieved by *subsidence* of the uplifted region.

The most striking and best-observed examples of vertical crustal movements due to isostatic imbalance are related to the phenomenon of glacial rebound observed in northern Canada and in Fennoscandia. During the latest ice-age these regions were covered by a thick ice-cap. The weight of ice depressed the underlying crust. Subsequently melting of the ice-cap removed the extra load on the crust, and it has since been rebounding. At stations on the Fennoscandian shield modern tide-gauge observations and precision leveling surveys made years apart allow the present uplift rates to be calculated (Fig. 6.4). The contour lines of equal uplift rate are inexact over large areas due to the incompleteness of data from inaccessible regions. Nevertheless, the general pattern of glacial rebound is clearly recognizable, with uplift rates of up to 9 mm yr^{-1}.

6.1.4 Isostatic gravity anomalies

The different degrees of isostatic compensation find expression in gravity anomalies. As explained in §2.5.6 the free-air gravity anomaly Δg_F is small near the center of a large region that is isostatically compensated; the Bouguer anomaly Δg_B is strongly negative. Assuming complete isostatic compensation, the size and shape of the root-zone can be determined from the elevations of the topography. With a suitable density contrast the gravity anomaly Δg_R of the modeled root-zone can be calculated; because the root-zone has lower density than adjacent mantle rocks Δg_R is also negative. The *isostatic gravity anomaly* Δg_I is defined as the difference between the Bouguer gravity anomaly and the computed anomaly of the root-zone, i.e.,

$$\Delta g_I = \Delta g_B - \Delta g_R \tag{6.7}$$

Examples of the isostatic gravity anomaly for the three types of isostatic compensation are shown schematically in Fig. 6.3. When isostatic compensation is complete, the topography is in hydrostatic equilibrium with its root-zone. Both Δg_B and Δg_R are negative but equal; consequently, the isostatic anomaly is everywhere zero ($\Delta g_I = 0$). In the case of overcompensation the eroded topography suggests a root-zone that is smaller than the real root-zone. The Bouguer anomaly is caused by the larger real root, so Δg_B is numerically larger than Δg_R. Subtracting the smaller negative anomaly of the computed root-zone leaves a negative isostatic anomaly ($\Delta g_I < 0$). On the other hand, with undercompensation the topography suggests a root-zone that is larger than the real root-zone. The Bouguer anomaly is caused by the smaller real root, so Δg_B is numerically smaller than Δg_R. Subtracting the larger negative anomaly of the root-zone leaves a positive isostatic anomaly ($\Delta g_I > 0$).

A national gravity survey of Switzerland carried out in the 1970s gave a high-quality map of Bouguer gravity anomalies (see Fig. 2.55). Seismic data gave representative parameters for the Central European crust and mantle: a crustal thickness of 32 km without topography, and mean densities of 2670 kg m^{-3} for the topography, 2810 kg m^{-3} for the crust and 3310 kg m^{-3} for the mantle. Using the Airy–Heiskanen model of compensation, a map of isostatic gravity anomalies in Switzerland was derived (Fig. 6.5) after correcting the gravity map for the effects of low-density sediments in the Molasse Basin north of the Alps and high-density material in the anomalous Ivrea body in the south.

The pattern of isostatic anomalies reflects the different

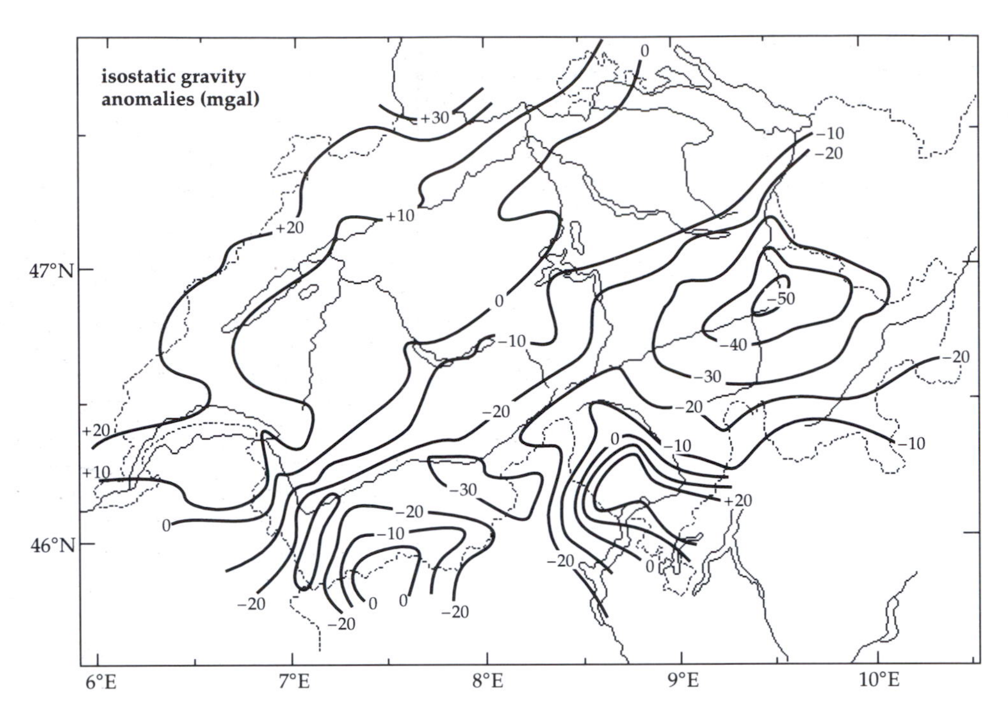

Fig. 6.5 Isostatic gravity anomalies in Switzerland (after Klingelé and Kissling, 1982), based on the national gravity map (Klingelé and Olivier, 1980), corrected for the effects of the Molasse Basin and the Ivrea body.

structures beneath the Jura mountains, which do not have a prominent root-zone, and the Alps, which have a low-density root that extends to more than 55 km depth in places. The dominant ENE–WSW trend of the isostatic gravity anomaly contour lines is roughly parallel to the trends of the mountain chains. In the northwest, near the Jura mountains, positive isostatic anomalies exceed 20 mgal. In the Alps isostatic anomalies are mainly negative, reaching more than − 50 mgal in the east.

A computation based on the Vening Meinesz model gave an almost identical isostatic anomaly map. The agreement of maps based on the different concepts of isostasy is somewhat surprising. It may imply that vertical crustal columns are not free to adjust relative to one another without friction as assumed in the Airy–Heiskanen model. The friction is thought to result from horizontal compressive stresses in the Alps, which are active in the on-going mountain-building process.

Comparison of the isostatic anomaly map with one of recent vertical crustal movements (Fig. 6.6) illustrates the relevance of isostatic gravity anomalies for tectonic interpretation. Precise leveling surveys have been carried out since the early 1900s along major valleys transecting and parallel to the mountainous topography of Switzerland. Relative rates of uplift or subsidence are computed from the differences between repeated surveys. The results have not been tied to absolute tide-gauge observations and so are relative to a base station at Aarburg in the canton of Aargau, in the northeast.

The rates of relative vertical movement in northeastern Switzerland are smaller than the confidence limits on the data and may not be significant, but the general tendency suggests subsidence. This region is characterized by mainly positive isostatic anomalies. The rates of vertical movement in the southern part of Switzerland exceed the noise level of the measurements and are significant. The most notable characteristic of the recent crustal motions is vertical uplift of the Alpine part of Switzerland relative to the central plateau and Jura mountains. The Alpine uplift rates are up to 1.5 mm yr^{-1}, considerably smaller than the rates observed in Fennoscandia. The most rapid uplift rates are observed in the region where isostatic anomalies are negative. The constant erosion of the mountain topography relieves the crustal load and the isostatic response is uplift. However, the interpretation is complicated by the fact that compressive stresses throughout the Alpine region acting on deep-reaching faults can produce non-isostatic uplift of the surface. The separation of isostatic and non-isostatic vertical crustal movements in the Alps will require detailed and exact information about the structure of the lithosphere and asthenosphere in this region.

6.2 RHEOLOGY

6.2.1 Brittle and ductile deformation

Rheology is the science of the deformation and flow of solid materials. This definition appears at first sight to contradict

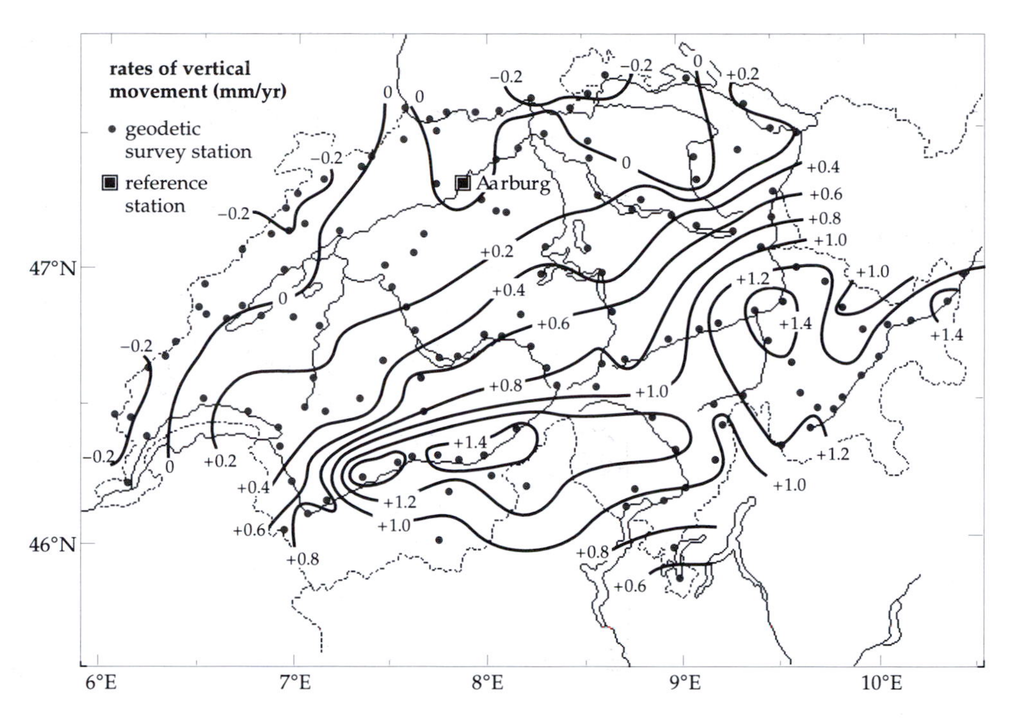

Fig. 6.6 Rates of vertical crustal motion in Switzerland deduced from repeated precise leveling. Positive rates correspond to uplift, negative rates to subsidence (data source: Gubler, 1991).

itself. A solid is made up of particles that cohere to each other; it is rigid and resists a change of shape. A fluid has no rigidity; its particles can move about comparatively freely. So how can a solid flow? In fact, the way in which a solid reacts to stress depends on how large the stress is and the length of time for which it is applied. Provided the applied stress does not exceed the yield stress (or elastic limit) the short-term behavior is elastic. This means that any deformation caused by the stress is completely recoverable when the stress is removed, leaving no permanent change in shape. However, if the applied stress exceeds the yield stress, the solid may experience either brittle or ductile deformation.

Brittle deformation consists of rupture without other distortion. This is an abrupt process that causes faulting in rocks and earthquakes, accompanied by the release of elastic energy in the form of seismic waves. Brittle fracture occurs at much lower stresses than the intrinsic strength of a crystal lattice. This is attributed to the presence of cracks, which modify the local internal stress field in the crystal. Fracture occurs under either extension or shear. Extensional fracture occurs on a plane at right angles to the direction of maximum tension. Shear fracture occurs under compression on one of two complementary planes which, reflecting the influence of internal friction, are inclined at an angle of less than 45° (typically about 30°) to the maximum principal compression. Brittle deformation is the main mechanism in tectonic processes that involve the uppermost 5–10 km of the lithosphere.

Ductile deformation is a slow process in which a solid acquires strain (i.e., it changes shape) over a long period of time. A material may react differently to a stress that is applied briefly than to a stress of long duration. If it experiences a large stress for a long period of time a solid can slowly and permanently change shape. The time-dependent deformation is called plastic flow and the capacity of the solid to flow is called its *ductility*. The ductility of a solid above its yield stress depends on temperature and confining pressure, and materials that are brittle under ordinary conditions may be ductile at high temperature and pressure. The behavior of rocks and minerals in the deep interior of the Earth is characterized by ductile deformation.

The transition from brittle to ductile types of deformation is thought to occur differently in oceanic and continental lithosphere (Fig. 6.7). The depth of the transition depends on several parameters, including the composition of the rocks, the local geothermal gradient, initial crustal thickness and the strain rate. Consequently it is sensitive to the vertically layered structure of the lithosphere. The oceanic lithosphere has a thin crust and shows a gradual increase in strength with depth, reaching a maximum in the upper mantle at about 30–40 km depth. At greater depths the lithosphere gradually becomes more ductile, eventually grading into the low-rigidity asthenosphere below about 100 km depth. The continental crust is much thicker than the oceanic crust and has a more complex layering. The upper crust is brittle, but the minerals of the lower crust are weakened by high temperature. As a result the lower crust becomes ductile near to the Moho at about 30–35 km depth.

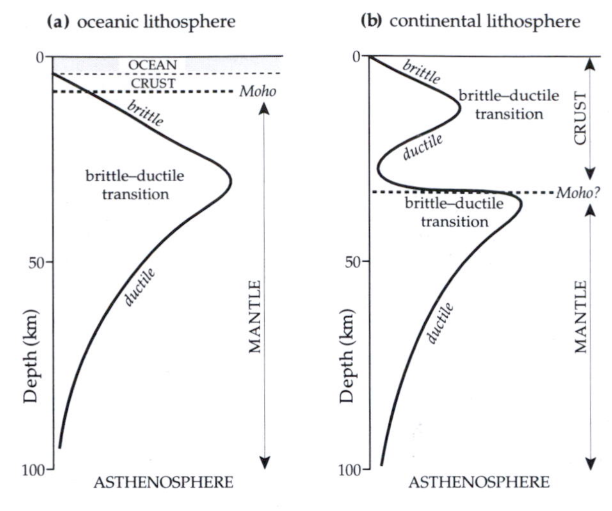

Fig. 6.7 Hypothetical vertical profiles of rigidity in (a) oceanic lithosphere and (b) continental lithosphere with the estimated depths of brittle–ductile transitions (after Molnar, 1988).

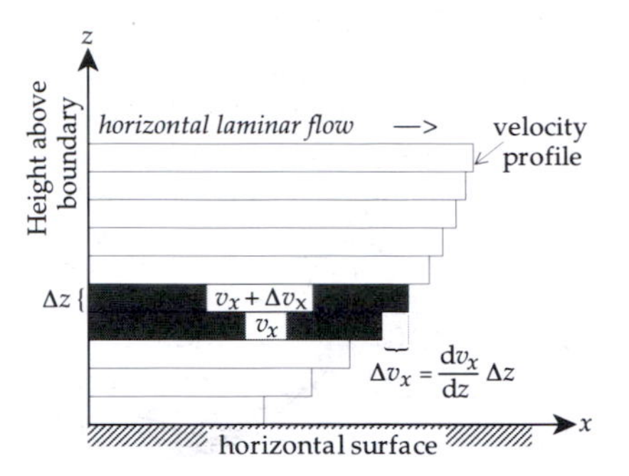

Fig. 6.8 Schematic representation of laminar flow of a fluid in infinitesimally thin layers parallel to a horizontal surface.

In the upper mantle the strength increases again, leading to a second brittle–ductile transition at about 40–50 km depth. The difference in rheological layering of continental and oceanic lithosphere is important in collisions between plates. The crustal part of the continental lithosphere may detach from the mantle. The folding, underthrusting and stacking of crustal layers produce folded mountain ranges in the suture zone and thickening of the continental crust. For example, great crustal thicknesses under the Himalayas are attributed to underthrusting of crust from the Indian plate beneath the crust of the Eurasian plate.

6.2.2 Viscous flow in liquids

Consider the case when a liquid or gas flows in thin layers parallel to a flat surface (Fig. 6.8). The *laminar flow* exists as long as the speed stays below a critical value, above which the flow becomes turbulent. Turbulent flow does not interest us here, because the rates of flow in solid earth materials are very slow.

Suppose that the velocity of laminar flow along the horizontal x-direction increases with vertical height z above the reference surface. The molecules of the fluid may be regarded as having two components of velocity. One component is the velocity of flow in the x-direction, but in addition there is a random component with a variable velocity whose root-mean-square value is determined by the temperature (§4.2.2). Because of the random component one-sixth of the molecules in a unit volume are moving upward and one-sixth downward on average at any time. This causes a transfer of molecules between adjacent layers in the laminar flow. The number of transfers per second depends on the size of the random velocity component, i.e. on temperature. The influx of molecules from the slower-moving layer reduces the momentum of the faster-moving layer. In turn, molecules transferred downward from the faster-velocity layer increase the momentum of the slower-velocity layer. This means that the two layers do not move freely past each other. They exert a shear force – or drag – on each other and the fluid is said to be *viscous*.

The magnitude of the shear force F_{xz} depends on how much momentum is transferred from one layer to the next. If all the molecules in a fluid have the same mass, the momentum transfer is determined by the change in the velocity v_x between the layers; this depends on the vertical gradient of the flow velocity (dv_x/dz). The momentum exchange depends also on the number of molecules that cross the boundary between adjacent layers and so is proportional to the surface area, A. We can bring these observations together, as did Newton in the 17th century, and derive the following proportionality relationship for F_{xz}:

$$F_{xz} \propto A \frac{dv_x}{dz} \tag{6.8}$$

If we divide both sides by the area A, the left side becomes the shear stress σ_{xz}. Introducing a proportionality constant η we get the equation

$$\sigma_{xz} = \eta \frac{dv_x}{dz} \tag{6.9}$$

This equation is Newton's law of viscous flow and η is the *coefficient of viscosity*. If η is constant, the fluid is called a *Newtonian fluid*. The value of h depends on the transfer rate of molecules between layers and so on temperature. Substituting the units of stress (pascal) and velocity gradient ((m s^{-1})/m=s^{-1}) we find that the unit of η is a pascal-second (Pa s). A shear stress applied to a fluid with low viscosity causes a large velocity gradient; the fluid flows easily. This is the case in a gas (η in air is of the order 2×10^{-5} Pa s) or in a liquid (η in water is 1.005×10^{-3} Pa s at 20 °C). The same shear stress applied to a very viscous fluid (with a large value of η) produces only a small velocity gradient transverse to the flow direction. The layers in the laminar flow are reluctant to move past each other. The viscous fluid is 'sticky' and it resists flow. For example, η in a viscous liquid like engine oil is around 0.1–10 Pa s, three or four orders of magnitude higher than in water.

6.2.3 Flow in solids

The response of a solid to an applied load depends upon whether the stress exceeds the elastic limit (§3.2.1) and for how long it is applied. When the yield stress (elastic limit) is reached, a solid may deform continuously without further increase in stress. This is called *plastic deformation*. In *perfectly plastic* behavior the stress–strain curve has zero slope, but the stress–strain curves of plastically deformed materials usually have a small positive slope (see Fig. 3.2a). This means that the stress must be increased above the yield stress for plastic deformation to advance. This effect is called *strain-hardening*. When the stress is removed after a material has been strain-hardened, a permanent residual strain is left.

Consider the effects that ensue if a stress is suddenly applied to a material at time t_0, held constant until time t_1 and then abruptly removed (Fig. 6.9a). As long as the applied stress is lower than the yield stress, the solid deforms elastically. The elastic strain is acquired immediately and remains constant as long as the stress is applied. Upon removal of the stress, the object at once recovers its original shape and there is no permanent strain (Fig. 6.9b).

If a constant load greater than the yield stress is applied, the resulting strain consists of a constant elastic strain and a changing *plastic* strain, which increases with time. After removal of the load at t_1 the plastic deformation does not disappear but leaves a permanent strain (Fig. 6.9c). In some plastic materials the deformation increases slowly at a decreasing rate, eventually reaching a limiting value for any specific value of the stress. This is called *viscoelastic* deformation (Fig. 6.9d). When the load is removed at t_1, the elastic part of the deformation is at once restored, followed by a slow decrease of the residual strain. This phase is called recovery or delayed elasticity. Viscoelastic behavior is an important rheological process deep in the Earth, for example in the asthenosphere and deeper mantle.

The analogy to the viscosity of liquids is apparent by inspection of Eq. (6.9). Putting $v_x = \mathrm{d}x/\mathrm{d}t$ and changing the order of differentiation, the equation becomes

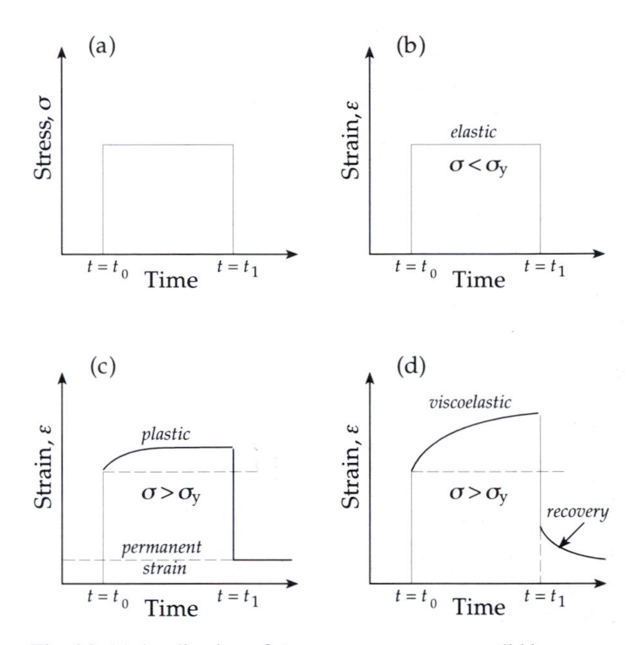

Fig. 6.9 (a) Application of a constant stress σ to a solid between times t_0 and t_1; (b) variation of elastic strain below the yield point; (c) plastic strain and (d) viscoelastic deformation at stresses above the yield point σ_y.

$$\sigma_{xz} = \eta \frac{\mathrm{d}}{\mathrm{d}z}\frac{\mathrm{d}x}{\mathrm{d}t} = \eta \frac{\mathrm{d}}{\mathrm{d}t}\frac{\mathrm{d}x}{\mathrm{d}z} = \eta \frac{\mathrm{d}}{\mathrm{d}t}\epsilon_{xz} \tag{6.10}$$

This equation resembles the elastic equation for shear deformation, which relates stress and strain through the shear modulus (see Eq. (3.15)). However, in the case of the 'viscous flow' of a solid the shear stress depends on the strain *rate*. The parameter η for a solid is called the *viscosity modulus*, or *dynamic viscosity*. It is analogous to the viscosity coefficient of a liquid but its value in a solid is many orders of magnitude larger. For example, the viscosity of the asthenosphere is estimated to be of the order of 10^{20}–10^{21} Pa s. Plastic flow in solids differs from true flow in that it only occurs when the stress exceeds the yield stress of the solid. Below this stress the solid does not flow. However, internal defects in a metal or crystal can be mobilized and reorganized by stresses well below the yield stress. As a result the solid may change shape over a long period of time.

6.2.3.1 *Viscoelastic model*

Scientists have tried to understand the behavior of rocks under stress by devising models based on mechanical analogs. In 1890 Lord Kelvin modelled viscoelastic deformation by combining the characteristics of a perfectly elastic solid and a viscous liquid. An applied stress causes both elastic and viscous effects. If the elastic strain is ϵ, the corresponding elastic part of the stress is $E\epsilon$, where E is

Young's modulus. Similarly, if the rate of change of strain with time is $d\epsilon/dt$, the viscous part of the applied stress is $\eta\, d\epsilon/dt$, where η is the viscosity modulus. The applied stress σ is the sum of the two parts and can be written

$$\sigma = E\epsilon + \eta\frac{d\epsilon}{dt} \tag{6.11}$$

To solve this equation we first divide throughout by E, then define the *retardation time* $\tau = \eta/E$, which is a measure of how long it takes for viscous strains to exceed elastic strains. Substituting and rearranging the equation we get

$$\epsilon + \tau\frac{d\epsilon}{dt} = \frac{\sigma}{E} \tag{6.12}$$

$$\frac{\epsilon}{\tau} + \frac{d\epsilon}{dt} = \frac{\epsilon_m}{\tau} \tag{6.13}$$

where $\epsilon_m = \sigma/E$. Multiplying throughout by the integrating factor $e^{t/\tau}$ gives

$$\frac{\epsilon}{\tau}e^{t/\tau} - \frac{d\epsilon}{dt}e^{t/\tau} = \frac{\epsilon_m}{\tau}e^{t/\tau} \tag{6.14}$$

$$\frac{d}{dt}\{\epsilon\, e^{t/\tau}\} = \frac{\epsilon_m}{\tau}e^{t/\tau} \tag{6.15}$$

Integrating both sides of this equation with respect to t gives

$$\epsilon\, e^{t/\tau} = \epsilon_m e^{t/\tau} + C \tag{6.16}$$

where C is a constant of integration determined by the boundary conditions. Initially, the strain is zero, i.e., at $t=0$, $\epsilon=0$; substituting in Eq. (6.16) gives $C = \epsilon_m$. The solution for the strain at time t is therefore

$$\epsilon = \epsilon_m[1 - e^{-t/\tau}] \tag{6.17}$$

The strain rises exponentially to a limiting value given by $\epsilon_m = \sigma/E$. This is characteristic of viscoelastic deformation.

6.2.4 Creep

Many solid materials deform slowly at room temperature when subjected to small stresses well below their brittle strength for long periods of time. The slow time-dependent deformation is known as *creep*. This is an important mechanism in the deformation of rocks because of the great intervals of time involved in geological processes. It is hard enough to approximate the conditions of pressure and temperature in the real Earth, but the time factor is an added difficulty in investigating the phenomenon of creep in laboratory experiments.

The results of observations on rock samples loaded by a constant stress typically show three main regimes of creep

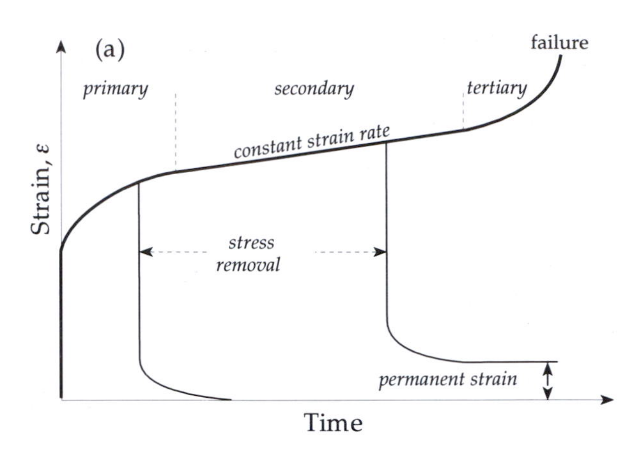

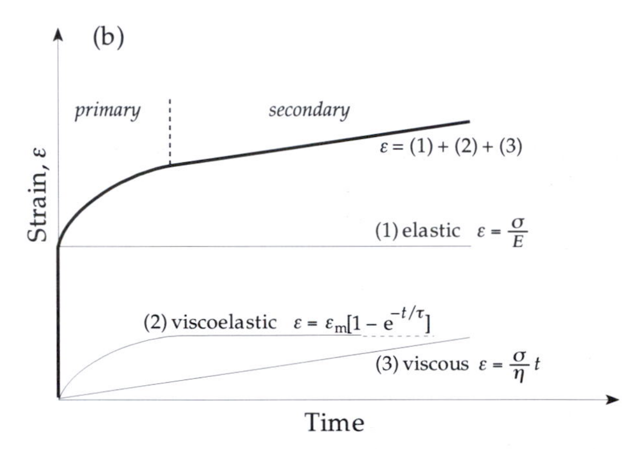

Fig. 6.10 Hypothetical strain–time curve for a material exhibiting creep under constant stress (after Ramsay, 1967).

(Fig. 6.10a). At first, the rock at once strains elastically. This is followed by a stage in which the strain initially increases rapidly (i.e. the strain rate is high) and then levels off. This stage is known as *primary creep* or *delayed elastic creep*. If the stress is removed within this regime, the deformation drops rapidly as the elastic strain recovers, leaving a deformation that sinks progressively to zero. Beyond the primary stage creep progresses at a slower and nearly constant rate. This is called *secondary creep* or *steady-state creep*. The rock deforms plastically, so that if the stress is removed a permanent strain is left after the elastic and delayed elastic recoveries. After the secondary stage the strain rate increases ever more rapidly in a stage called *tertiary creep*, which eventually leads to failure.

The primary and secondary stages of the creep curve can be modelled by combining elastic, viscoelastic and viscous elements (Fig. 6.10b). Below the yield stress only elastic and delayed elastic (viscoelastic) deformation occur and the solid does not flow. The strain flattens off at a limiting value $\epsilon_m = \sigma/E$. The viscous component of strain rises linearly

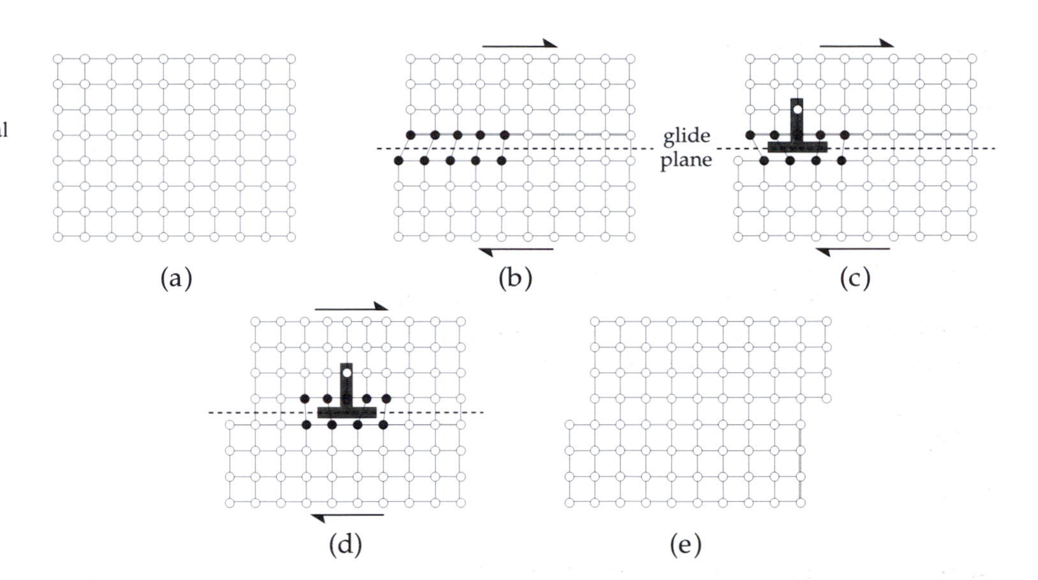

Fig. 6.11 Permanent (plastic) shear deformation produced by motion of a dislocation through a crystal in response to shear stress: (a) undeformed crystal lattice, (b) entry of dislocation at left edge, (c) accommodation of dislocation into lattice, (d) passage of dislocation across the crystal, and (e) sheared lattice after dislocation leaves the crystal.

with time, corresponding to a constant strain rate. In practice the stress must exceed the yield stress σ_y for flow to occur. In this case the viscous component of strain is proportional to the excess stress $(\sigma-\sigma_y)$ and to the time t. Combining terms gives the expression

$$\epsilon=\frac{\sigma}{E}+\epsilon_m[1-e^{-t/\tau}]+\frac{(\sigma-\sigma_y)}{\eta}t \tag{6.18}$$

This simple model explains the main features of experimentally observed creep curves. It is in fact very difficult to ensure that laboratory observations are representative of creep in the Earth. Conditions of pressure and temperature in the crust and upper mantle can be achieved or approximated. The major problems arise from the differences in time-scales and creep rates. Even creep experiments conducted over months or years are far shorter than the lengths of time in a geological process. The creep rates in nature (e.g., around 10^{-14} s^{-1}) are many orders of magnitude slower than the slowest strain rate used in laboratory experiments (around 10^{-8} s^{-1}). Nevertheless, the experiments have provided a better understanding of the rheology of the Earth's interior and the physical mechanisms active in different depths.

6.2.4.1 *Crystal defects*

Deformation in solids does not take place homogeneously. Laboratory observations on metals and minerals have shown that crystal defects play an important role. The atoms in a metal or crystal are arranged regularly to form a lattice with a simple symmetry. In some common arrangements the atoms are located at the corners of a cube or a hexagonal prism, defining a *unit cell* of the crystal. The lattice is formed by stacking unit cells together. Occasionally an imperfect cell may lack an atom. The space of the missing atom is called a vacancy. Vacancies may be distributed throughout the crystal lattice, but they can also form long chains called dislocations.

There are several types of dislocation, the simplest being an edge dislocation. It is formed when an extra plane of atoms is introduced in the lattice (Fig. 6.11). The edge dislocation terminates at a plane perpendicular to it called the *glide plane*. It clearly makes a difference if the extra plane of atoms is above or below the glide plane, so edge dislocations have a sign. This can be represented by a T-shape, where the cross-bar of the T is parallel to the glide plane and the stalk is the extra plane of atoms. If oppositely signed edge dislocations meet, they form a complete plane of atoms and the dislocations annihilate each other. The displacement of atoms in the vicinity of a dislocation increases the local internal stress in the crystal. As a result, the application of a small external stress may be enough to mobilize the dislocation, causing it to glide through the undisturbed part of the crystal (Fig. 6.11). If the dislocation glide is not blocked by an obstacle, the dislocation migrates out of the crystal, leaving a shear deformation.

Another common type of dislocation is the screw dislocation. It also is made up of atoms that are displaced from their regular positions, in this case forming a spiral about an axis.

The deformation of a crystal lattice by dislocation glide requires a shear stress; hydrostatic pressure does not cause plastic deformation. The shear stress needed to actuate dislocations is two or three orders of magnitude less than the shear stress needed to break the bonds between layers of atoms in a crystal. Hence, the mobilization of dislocations is an important mechanism in plastic deformation at low stress. As deformation progresses it is accompanied by an increase in the dislocation density (the number of dislocations per unit area normal to their lengths). The dislocations move along a glide plane until it intersects the glide plane of another set of dislocations. When several sets of dislocations are mobilized they may interfere and block each other, so that the stress must be increased to mobilize them further. This is manifest as strain-hardening, which is a thermodynamically unstable situation. At any stage of strain-hardening, given enough time, the dislocations redistribute themselves to a configuration with lower energy, thereby reducing the strain. The time-dependent strain relaxation is called *recovery*.

Recovery can take place by several processes, each of which requires thermal energy. These include the annihilation of oppositely signed edge dislocations moving on parallel glide planes (Fig. 6.12a) and the climb of edge dislocations past obstacles against which they have piled up (Fig. 6.12b). Edge dislocations with the same sign may align to form walls between domains of a crystal that have low dislocation density (Fig. 6.12c), a process called polygonization. These are some of the ways in which thermal energy promotes the migration of lattice defects, eventually driving them out of the crystal and leaving an annealed lattice.

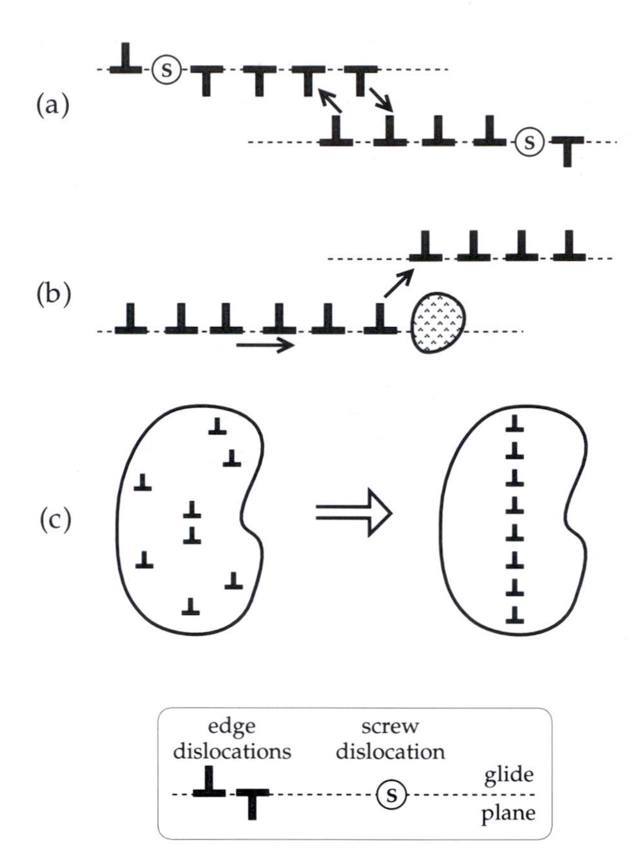

Fig. 6.12 Thermally activated processes that assist recovery: (a) annihilation of oppositely signed edge dislocations moving on parallel glide planes, (b) climb of edge dislocations past obstacles, and (c) polygonization by the alignment of edge dislocations with the same sign to form walls separating regions with low dislocation density (after Ranalli, 1987).

6.2.4.2 *Creep mechanisms in the Earth*

Ductile flow in the Earth's crust and mantle takes place by one of three mechanisms: low-temperature plastic flow; power-law creep; or diffusion creep. Each mechanism is a thermally activated process. This means that the strain rate depends on the temperature T according to an exponential function with the form $e^{-E_a/kT}$. Here k is Boltzmann's constant, while E_a is the energy needed to activate the type of flow; it is called the *activation energy*. At low temperatures, where $T \ll E_a/k$, the strain rate is very slow and creep is insignificant. Because of the exponential function, the strain rate increases rapidly with increasing temperature above $T=E_a/k$. The type of flow at a particular depth depends on the local temperature and its relationship to the melting temperature T_{mp}. Above T_{mp} the interatomic bonds in the solid break down and it flows as a true liquid.

Plastic flow at low temperature takes place by the motion of dislocations on glide planes. When the dislocations encounter an internal obstacle or a crystal boundary they pile up and some rearrangement is necessary. The stress must be increased to overcome the obstacle and reactuate dislocation glide. Plastic flow can produce large strains and may be an important mechanism in the bending of the oceanic lithosphere near some subduction zones. It is likely to be most effective at depths below the brittle–ductile transition.

Power-law creep, or hot creep, also takes place by the motion of dislocations on glide planes. It occurs at higher temperatures than low-temperature plastic flow, so internal obstacles to dislocation migration are thermally activated and diffuse out of the crystal as soon as they arise. The strain rate in power-law creep is proportional to the nth power of the stress σ and has the form

$$\frac{d\epsilon}{dt}=A\left(\frac{\sigma}{\mu}\right)^{n} e^{-E_a/kT} \tag{6.19}$$

where μ is the rigidity modulus and A is a constant with the dimensions of strain rate; typically $n \geq 3$. This relationship means that the strain rate increases much more rapidly than the stress. From experiments on metals, power-law creep is understood to be the most important mechanism of flow at temperatures between 0.55 T_{mp} and 0.85 T_{mp}. The temperature throughout most of the mantle probably exceeds half the melting point, so power-law creep is probably the flow mechanism that permits mantle convection. It is also likely to be the main form of deformation in the lower lithosphere, where the relationship of temperature to melting point is also suitable.

Diffusion creep consists of the thermally activated migration of crystal defects in the presence of a stress field. There are two main forms. *Nabarro–Herring creep* consists of diffusion of defects through the body of a grain; *Coble creep* takes place by migration of the defects along grain boundaries. In each case the strain rate is proportional to the stress, as in a Newtonian fluid. It is therefore possible to regard ductile deformation by diffusion creep as the slow flow of a very viscous fluid. Diffusion creep has been observed in metals at temperatures $T > 0.85\ T_{mp}$. In the Earth's mantle the temperature approaches the melting point in the asthenosphere. As indicated schematically in Fig. 6.7 the transition from the rigid lithosphere to the soft, viscous underlying asthenosphere is gradational. There is no abrupt boundary, but the concept of rigid lithospheric plates moving on the soft, viscous asthenosphere serves well as a geodynamic model.

6.2.5 Rigidity of the lithosphere

Lithospheric plates are thin compared to their horizontal extents. However, they evidently react rigidly to the forces that propel them. The lithosphere does not easily buckle under horizontal stress. A simple analogy may be made to a thin sheet of paper resting on a flat pillow. If pushed on one edge, the page simply slides across the pillow without crumpling. Only if the leading edge encounters an obstacle does the page bend, buckling upward some distance in front of the hindrance, while the leading edge tries to burrow under it. This is what happens when an oceanic lithospheric plate collides with another plate. A small forebulge develops on the oceanic plate and the leading edge bends downward into the mantle, forming a subduction zone.

The ability to bend is a measure of the rigidity of the plate. This is also manifest in its reaction to a local vertical load. If, in our analogy, a small weight is placed in the middle of the page, it is pressed down into the soft pillow. A large area around the weight takes part in this process, which may be compared with the Vening Meinesz type of regional isostatic compensation (§6.1.2.3). The weight of a seamount or chain of islands has a similar effect on the oceanic lithosphere. By studying the flexure due to local vertical loads, information is obtained about a *static property* of the lithosphere, namely its resistance to bending.

In our analogy the locally loaded paper sheet would not bend if it lay on a hard flat table. It is only able to flex if it rests on a soft, yielding surface. After the weight is removed, the page is restored to its original flat shape. The restoration after unloading is a measure of the properties of the pillow as well as the page. A natural example is the rebound of regions (such as the Canadian shield or Fennoscandia) that have been depressed by now-vanished ice-sheets. The analysis of the rates of glacial rebound provides information about a *dynamic property* of the mantle beneath the lithosphere. The depression of the surface forces mantle material to flow away laterally to make way for it; when the load is removed, the return mantle flow presses the concavity back upward. The ease with which the mantle material flows is described by its dynamic viscosity.

The resistance to bending of a thin elastic plate overlying a weak fluid is expressed by an elastic parameter called the *flexural rigidity* and denoted D. For a plate of thickness h,

$$D = \frac{E}{12(1-\nu^2)} h^3 \tag{6.20}$$

where E is Young's modulus and ν is Poisson's ratio (see §3.2.3 and §3.2.4 for the definition of these elastic constants). The dimensions of E are N m^{-2} and ν is dimensionless; hence, the dimensions of D are those of a bending moment (N m). D is a fundamental parameter of the elastic plate, which describes how easily it can be bent; a large value of D corresponds to a stiff plate.

Here we consider two situations of particular interest for the rigidity of the oceanic lithosphere. The first is the bending of the lithosphere by a topographic feature such as an oceanic island or seamount; only the vertical load on the elastic plate is important. The second is the bending of the lithosphere at a subduction zone. In this case vertical and horizontal forces are located along the edge of the plate and the plate experiences a bending moment which deflects it downward.

6.2.5.1 *Lithospheric flexure caused by oceanic islands*

The theory for elastic bending of the lithosphere is derived from the bending of thin elastic plates and beams. Only an outline of the theory can be given here. Consider the bending of a thin isotropic elastic plate of thickness h carrying a surface load $L(x, y)$ and supported by a substratum of density ρ_m (Fig. 6.13a). Let the deflection of the plate at a position (x, y) relative to the center of the load be $w(x, y)$. Applying Archimedes principle, we get a buoyant force equal to $(\rho_m - \rho_i)\, gw$, where ρ_i is the density of the material that fills in the depression caused by the deflection of the plate. The elasticity of the beam causes a force proportional to a fourth-order differential of the deflection w. Balancing the elastic and buoyancy forces against the deforming load we get

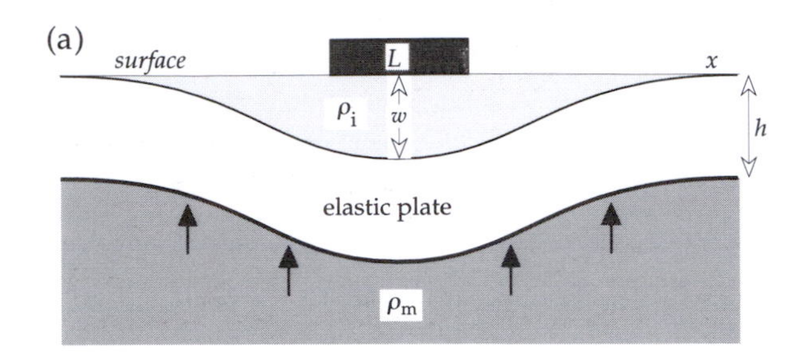

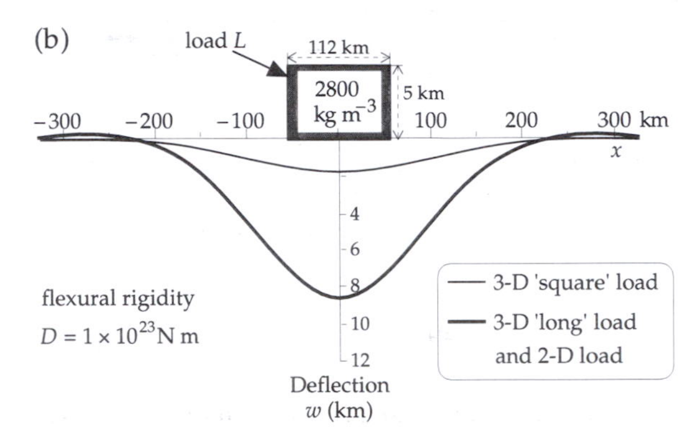

Fig. 6.13 (a) Geometry for the elastic bending of a thin plate of thickness h supported by a denser substratum: the surface load L causes a downward bending w. (b) Comparison of 2-D and 3-D elastic plate models. The load is taken to be a topographic feature of density 2800 kg/m^3, height 5 km and cross-sectional width 112 km (after Watts *et al.*, 1975).

$$D\left\{\frac{\partial^4 w}{\partial x^4}+2\frac{\partial^4 w}{\partial x^2 \partial y^2}+\frac{\partial^4 w}{\partial y^4}\right\}+(\rho_m-\rho_i)gw=L(x,y) \qquad (6.21)$$

For a linear topographic feature, such as a mountain range, oceanic island or chain of seamounts, the bending geometry is the same in any cross-section normal to its length and the problem reduces to the two-dimensional elastic bending of a thin beam. If the load is a linear feature in the y-direction, the variation of w with y disappears and the differential equation becomes:

$$D\frac{\partial^4 w}{\partial x^4}+(\rho_m-\rho_i)gw=L \qquad (6.22)$$

An important example is a linear load L concentrated along the y-axis at $x=0$. The solution is a damped sinusoidal function:

$$w=w_0 e^{-x/\alpha}\left(\cos\frac{x}{\alpha}+\sin\frac{x}{\alpha}\right) \qquad (6.23)$$

where w_0 is the amplitude of the maximum deflection underneath the load (at $x=0$). The parameter α is called the *flexural parameter*; it is related to the flexural rigidity D of the plate by

$$\alpha^4=\frac{4D}{(\rho_m-\rho_i)g} \qquad (6.24)$$

The elasticity of the plate (or beam) distributes the load of the surface feature over a large lateral distance. The fluid beneath the load is pushed aside by the penetration of the plate. The buoyancy of the displaced fluid forces it upward, causing uplift of the surface adjacent to the central depression. Eq. (6.23) shows that this effect is repeated with increasing distance from the load, the wavelength of the fluctuation is $\lambda=2\pi\alpha$. The amplitude of the disturbance diminishes rapidly because of the exponential attenuation factor. Usually it is only necessary to consider the central depression and the first uplifted region. The wavelength λ is equal to the distance across the central depression. Substituting in Eq. (6.24) gives D, which is then used with the parameters E and ν in Eq. (6.20) to obtain h, the thickness of the elastic plate. The computed values of h are greater than the thickness of the crust, i.e., the elastic plate includes part of the upper mantle. The value of h is equated with the thickness of the elastic lithosphere.

The difference between the deflection caused by a two-dimensional load (i.e., a linear feature) and that due to a three-dimensional load (i.e., a feature that has limited extent in the x- and y-directions) is illustrated in Fig. 6.13b. If the length of the three-dimensional load normal to the cross-section is more than about ten times its width, the deflection is the same as for a two-dimensional load. A load with a square base (i.e., extending the same distance along both x- and y-axes) causes a central depression that is less than a quarter the effect of the linear load.

The validity of the lithospheric flexural model of isostasy can be tested by comparing the computed gravity effect of the model with the observed free-air gravity anomaly Δg_F. The isostatic compensation of the Great Meteor seamount in the North Atlantic provides a suitable test for a three-dimensional model (Fig. 6.14). The shape of the free-air gravity anomaly obtained from detailed marine gravity surveys was found to be fitted best by a model in which the effective flexural rigidity of the deformed plate was assumed to be 6×10^{22} N m.

6.2.5.2 *Lithospheric flexure at a subduction zone*

The bathymetry of an oceanic plate at a subduction zone is typified by an oceanic trench, which can be many kilometers deep (Fig. 6.15a). Seaward of the trench axis the plate develops a small upward bulge (the outer rise) which can extend for 100–150 km away from the trench and reach heights of several hundred meters. The lithospheric plate bends sharply downward in the subduction zone. This bending can also be modelled with a thin elastic plate.

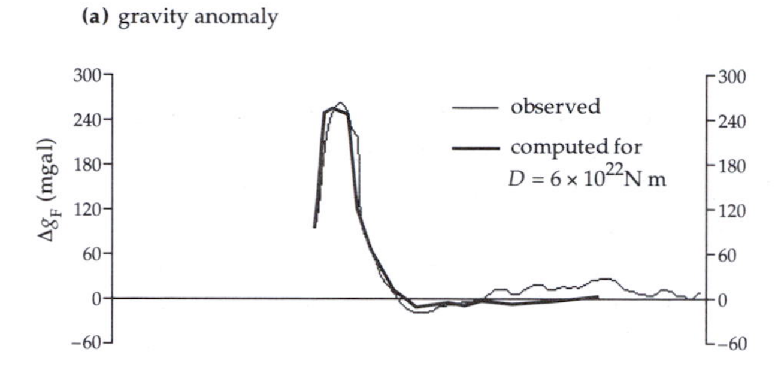

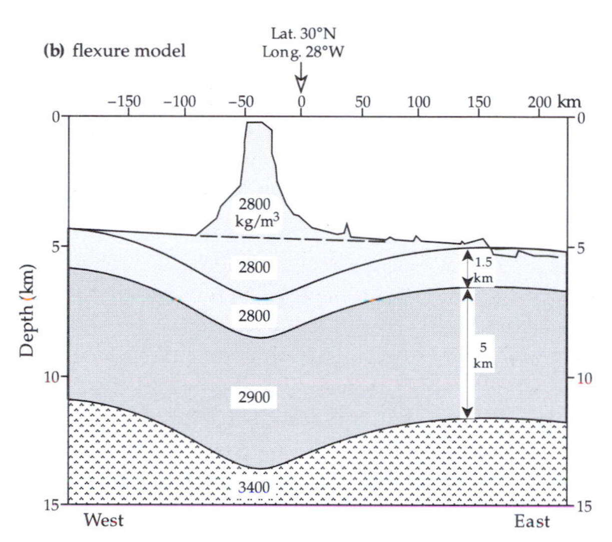

Fig. 6.14 (a) Comparison of observed free-air gravity anomaly profile across the Great Meteor seamount with the anomaly computed for (b) a lithospheric flexure model of isostatic compensation (after Watts *et al.*, 1975)

In the model, a horizontal force P pushes a plate of thickness h toward the subduction zone, the leading edge carries a vertical load L, and the plate is bent by a bending moment M (Fig. 6.15b). The horizontal force P is negligible in comparison to the effects of M and L. The vertical deflection of the plate must satisfy Eq. (6.22), with the same parameters as in the previous example. Choosing the origin to be the point nearest the trench where the deflection is zero simplifies the form of the solution, which is

$$w = Ae^{-x/\alpha}\sin\frac{x}{\alpha} \tag{6.25}$$

where A is a constant and α is the flexural parameter as in Eq. (6.24). The value of the constant A is found from the position x_b of the forebulge, where $dw/dx=0$:

$$\frac{dw}{dx} = A\left(-\frac{1}{\alpha}e^{-x/\alpha}\sin\frac{x}{\alpha} + \frac{1}{\alpha}e^{-x/\alpha}\cos\frac{x}{\alpha}\right) = 0 \tag{6.26}$$

from which

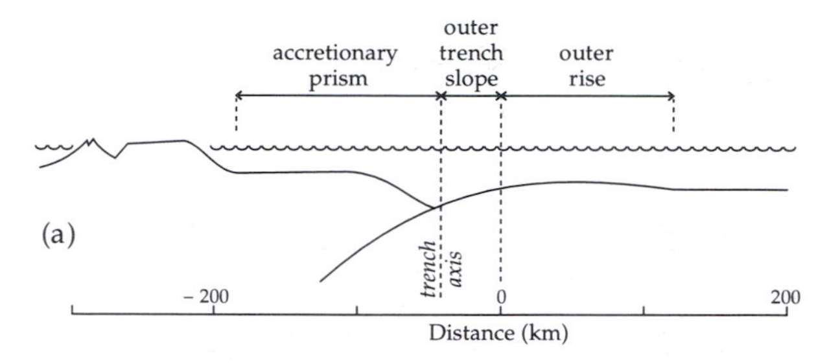

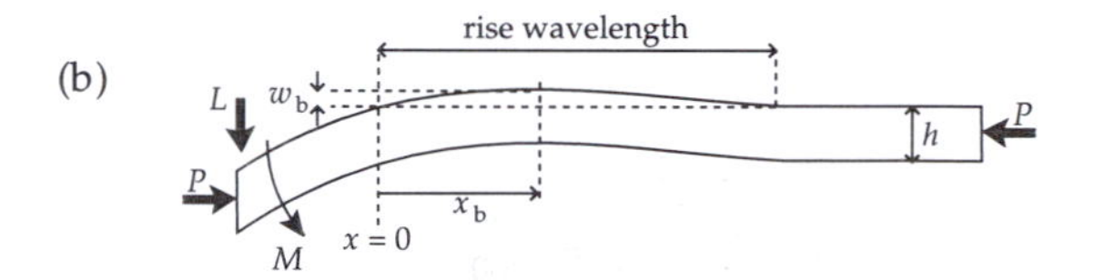

Fig. 6.15 (a) Schematic structural cross-section at a subduction zone (after Caldwell and Turcotte, 1979), and (b) the corresponding thin-plate model (after Turcotte *et al.*, 1978).

$$x_b = \frac{\pi}{4}\alpha \text{ and } A = w_b\sqrt{2}e^{\pi/4} \tag{6.27}$$

It is convenient to normalize the horizontal distance and vertical displacement: writing $x' = x/x_b$ and $w' = w/w_b$, the generalized equation for the elastic bending at an oceanic trench is obtained:

$$w' = \sqrt{2}\sin\left(\frac{\pi}{4}x'\right)\exp\left[\frac{\pi}{4}(1-x')\right] \tag{6.28}$$

The theoretical deflection of oceanic lithosphere at a subduction zone obtained from the elastic bending model agrees well with the observed bathymetry on profiles across oceanic trenches (Fig. 6.16a, b). The calculated thicknesses of the elastic lithosphere are of the order 20–30 km and the flexural rigidity is around 10^{23} N m. However, at some trenches the assumption of a completely elastic upper lithosphere is evidently inappropriate. At the Tonga trench the model curve deviates from the observed bathymetry inside the trench (Fig. 6.16c). It is likely that the elastic limit is exceeded at parts of the plate where the curvature is high. These regions may yield, leading to a reduction in the effective rigidity of the plate. This effect can be taken into account by assuming that the inelastic deformation is perfectly plastic. The deflection calculated within the trench for an elastic–perfectly plastic model agrees well with the observed bathymetry.

6.2.5.3 *Thickness of the lithosphere*

The rheological response of a material to stress depends on the duration of the stress. The reaction to a short-lasting stress, as experienced during the passage of a seismic wave,

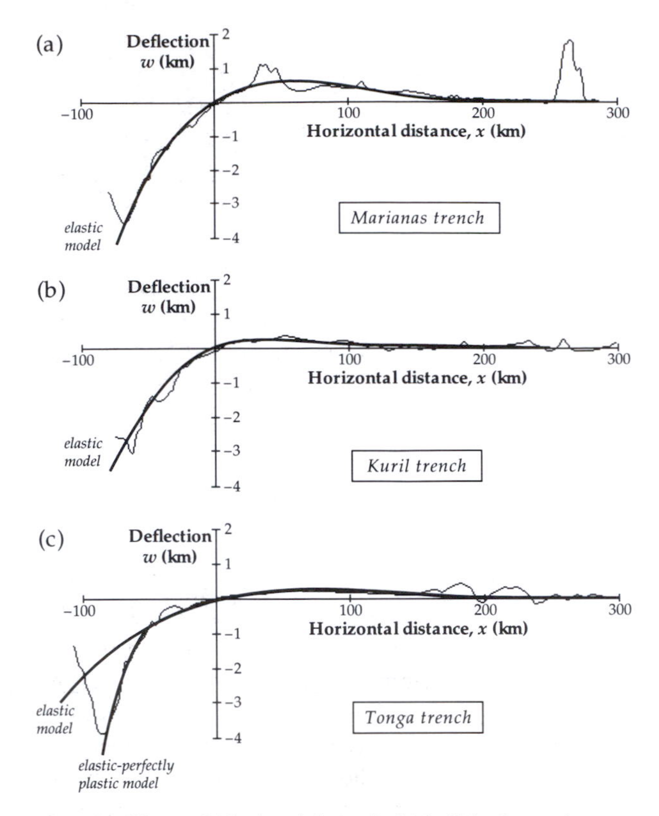

Fig. 6.16 Observed (*thin*) and theoretical (*thick*) bathymetric profiles for elastic flexure of the lithosphere at (a) the Marianas trench and (b) the Kuril trench. The flexure at (c) the Tonga trench is best explained by an elastic–perfectly plastic model (after Turcotte *et al.*, 1978).

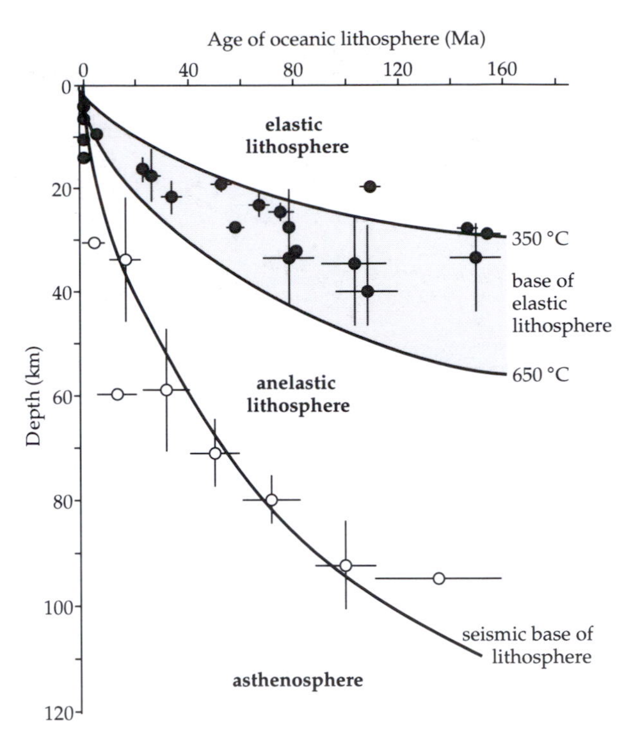

Fig. 6.17 Seismic and elastic thicknesses of oceanic lithosphere as a function of age (after Watts *et al.*, 1980).

may be quite different from the reaction of the same material to a steady load applied for a long period of time. This is evident in the different thicknesses obtained for the lithosphere in seismic experiments and in elastic plate modelling. Long-period surface waves penetrate well into the upper mantle. Long wavelengths are slowed down by the low rigidity of the asthenosphere, so the dispersion of surface waves allows estimates of the seismic thickness of the lithosphere. For oceanic lithosphere the seismic thickness increases with age of the lithosphere (i.e., with distance from the spreading center), increasing to more than 100 km at ages older than 100 Ma (Fig. 6.17). The lithospheric thicknesses obtained from elastic modelling of the bending caused by seamounts and island chains or at subduction zones also increase with distance from the ridge, but are only one-third to one-half of the corresponding seismic thickness. The discrepancy shows that only the upper part of the lithosphere is elastic. Indeed, if the entire lithosphere had the flexural rigidity found in elastic models ($D \approx 10^{21}$–10^{23} N m), it would bend by only small amounts under topographic loads or at subduction zones. The base of the elastic lithosphere agrees well with the modelled depths of the 300–600 °C oceanic isotherms. At greater depths the increase of temperature results in inelastic behavior of the lower lithosphere.

The elastic thickness of the continental lithosphere is much thicker than that of the oceanic lithosphere, except in rifts, passive continental margins and young orogenic belts. Precambrian shield areas generally have a flexural thickness greater than 100 km and a high flexural rigidity of around 10^{25} N m. Rifts, on the other hand, have a flexural thickness less than 25 km. Both continental and oceanic lithosphere grade gradually into the asthenosphere, which has a much lower rigidity and is able to flow in a manner determined by mantle viscosity.

6.2.6 Mantle viscosity

As illustrated by the transition from brittle to ductile behavior (§6.2.1), the Earth's rheology changes with depth. The upper part of the lithosphere behaves elastically. It has a constant and reversible response to both short-term and long-term loads. The behavior is characterized by the rigidity or shear modulus μ, which relates the strain to the applied shear stress and so has the dimensions of stress

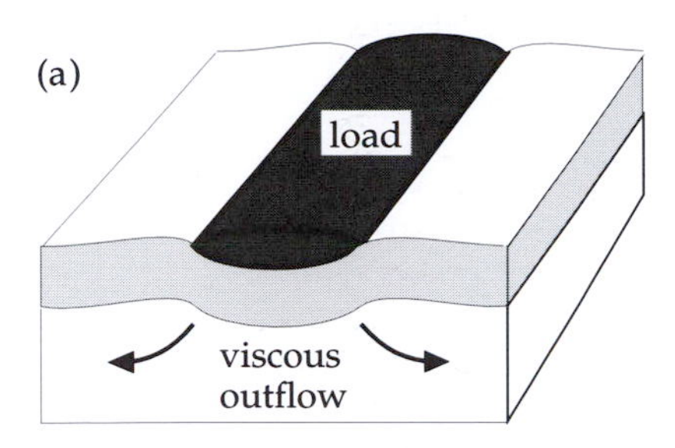

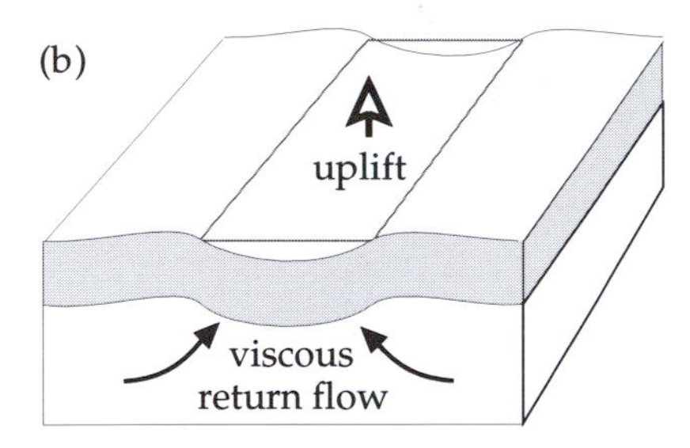

Fig. 6.18 (a) Depression of the lithosphere due to a surface load (ice-sheet) and accompanying viscous outflow in the underlying mantle; (b) return flow in the mantle and surface uplift after removal of the load.

(N m^{-2}, or Pa). The resistance of the lithosphere to flexure is described by the flexural rigidity D, which has the dimensions of a bending moment (N m). In a subduction zone the tight bending may locally exceed the elastic limit, causing parts of the plate to yield.

The deeper part of the lithosphere does not behave elastically. Although it has an elastic response to abrupt stress changes, it reacts to long-lasting stress by ductile flow. This kind of rheological behavior also characterizes the asthenosphere and the deeper mantle. Flow takes place with a strain rate that is proportional to the stress or a power thereof. In the simplest case, the deformation occurs by Newtonian flow governed by a viscosity coefficient h, whose dimensions (Pa s) express the time-dependent nature of the process.

Under a surface load, such as an ice-sheet, the elastic lithosphere is pushed down into the viscous mantle (Fig. 6.18a). This causes an outflow of mantle material away from the depressed region. When the ice-sheet melts, removing the load, hydrostatic equilibrium is restored and there is a return flow of the viscous mantle material (Fig. 6.18b). Thus, in modelling the time-dependent reaction of the Earth to a surface load at least two and usually three layers must be taken into account. The top layer is an elastic lithosphere up to 100 km thick; it has infinite viscosity (i.e., it does not flow) and a flexural rigidity around 5×10^{24} N m. Beneath it lies a low-viscosity 'channel' 75–250 km thick that includes the asthenosphere, with a low viscosity of typically 10^{19}–10^{20} Pa s. The deeper mantle which makes up the third layer has a higher viscosity around 10^{21} Pa s.

Restoration of the surface after removal of a load is accompanied by uplift, which can be expressed well by a simple exponential relaxation equation. If the initial depression of the surface is w_0, the deflection $w(t)$ after time t is given by

$$w(t)=w_0 e^{-t/\tau} \qquad (6.29)$$

Here τ is the relaxation time, which is related to the mantle viscosity by

$$\tau=\frac{4\pi}{\rho_m g \lambda}\eta \qquad (6.30)$$

where ρ_m is the mantle density, g is gravity at the depth of the flow and λ is the wavelength of the depression, a dimension appropriate to the scale of the load (as will be seen in examples below). A test of these relationships requires data that give surface elevations in the past. These data come from analyses of sea-level changes, present elevations of previous shorelines and directly observed rates of uplift. The ancient horizons have been dated by radiometric methods as well as by sedimentary methods such as varve chronology, which consists of counting the annually deposited pairs of silt and clay layers in laminated sediments. A good example of the exponential restoration of a depressed region is the history of uplift in central Fennoscandia since the end of the last ice age some 10,000 yr ago (Fig. 6.19). If it is assumed that about 30 m of uplift still remain, the observed uplift agrees well with Eq. (6.29) and gives a relaxation time of 4400 yr.

An important factor in modelling uplift is whether the viscous response is caused by the mantle as a whole, or whether it is confined to a low-viscosity layer (or 'channel') beneath the lithosphere. Seismic shear-wave velocities are reduced in a low-velocity channel about 100–150 km thick, whose thickness and seismic velocity are however variable from one locality to another. Interpreted as the result of softening or partial melting due to temperatures near the melting point, the seismic low-velocity channel is called the asthenosphere. It is not sharply bounded, yet it must be represented by a distinct layer in viscosity models, which indicate that its viscosity must be at least 25 times less than in the deeper mantle.

The larger the ice-sheet (or other type of surface load), the deeper the effects reach into the mantle. By studying the uplift following removal of different loads, information is obtained about the viscosity at different depths in the mantle.

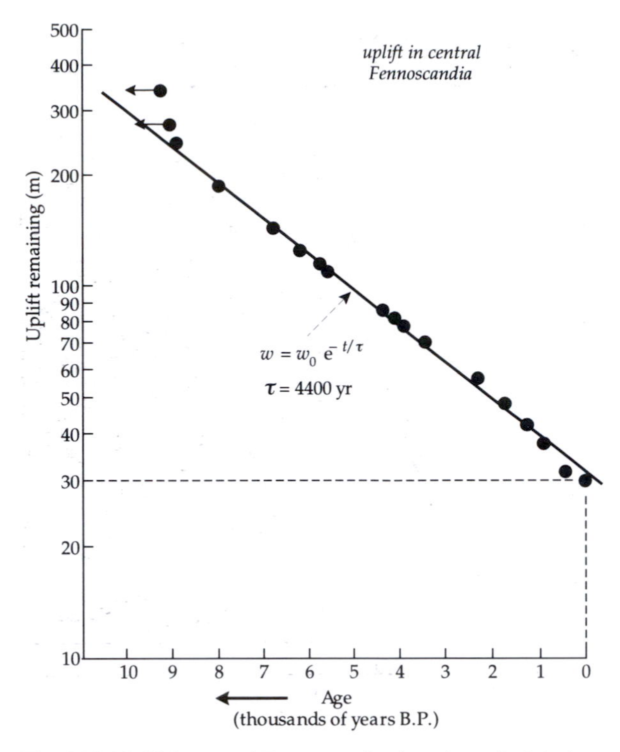

Fig. 6.19 Uplift in central Fennoscandia since the end of the last ice age illustrates exponential viscous relaxation with a time constant of 4400 yr (after Cathles, 1975).

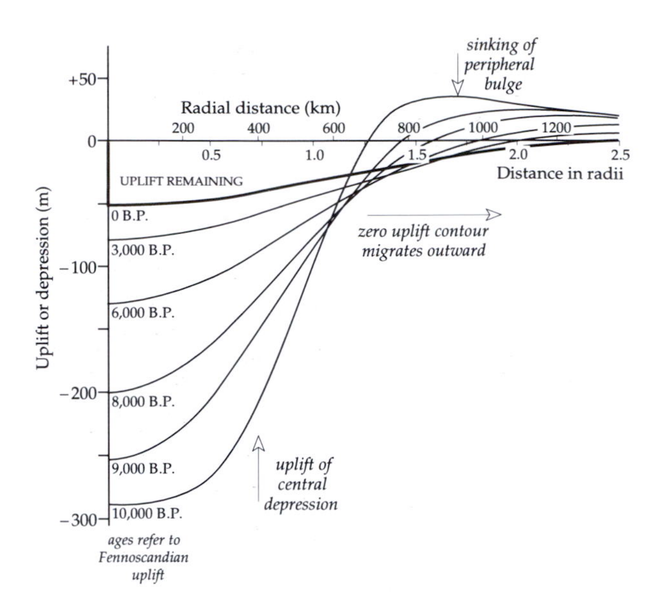

Fig. 6.20 Model calculations of the relaxation of the deformation caused by the Fennoscandian ice-sheet following its disappearance 10,000 yr ago (after Cathles, 1975).

6.2.6.1 *Viscosity of the upper mantle*

About 18,000–20,000 years ago Lake Bonneville, the predecessor of the present Great Salt Lake in Utah, U.S.A., had a radius of about 95 km and an estimated depth of about 305 m. The water mass depressed the lithosphere, which was later uplifted by isostatic restoration after the lake drained and dried up. Observations of the present heights of ancient shorelines show that the central part of the lake has been uplifted by about 65 m. Two parameters are involved in the process: the flexural rigidity of the lithosphere; and the viscosity of the mantle beneath. The elastic response of the lithosphere is estimated from the geometry of the depression that would be produced in isostatic equilibrium. The maximum flexural rigidity that would allow a 65 m deflection under a 305 m water load is found to be about 5×10^{23} N m.

The surface load can be modelled as a heavy vertical right cylinder with radius r, which pushes itself into the soft mantle. The underlying viscous material is forced aside so that the central depression is surrounded by a circular uplifted 'bulge'. After removal of the load, restorative uplift takes place in the central depression, the peripheral bulge subsides and the contour of zero-uplift migrates outward. The wavelength λ of the depression caused by a load with this geometry has been found to be about 2.6 times the diameter of the cylindrical load. In the case of Lake Bonneville $2r=192$ km, so λ is about 500 km. The mantle viscosity is obtained by assuming that the response time of the lithosphere was short enough to track the loading history quite closely. This implies that the viscous relaxation time τ must have been 4000 yr or less. Substitution of these values for λ and τ in Eq. (6.30) suggests a maximum mantle viscosity η of about 2×10^{20} Pa s in the top 250 km of the mantle. A lower value of η would require a thinner low-viscosity channel beneath the lithosphere.

The Fennoscandian uplift can be treated in the same way (Fig. 6.20), but the weight and lateral expanse of the load were much larger. The ice-cap is estimated to have been about 1100 m thick and to have covered most of Norway, Sweden and Finland. Although the load was therefore somewhat elongate, it is possible to model it satisfactorily by a vertical cylinder of radius $r=550$ km centered on the northern Gulf of Bothnia (Fig. 6.21). The load was applied for about 20,000 yr before being removed 10,000 yr ago. This caused an initial depression of about 300 m, which has subsequently been relaxing (Fig. 6.20). The uplift rates allow upper-mantle viscosities to be estimated. The data are compatible with different models of mantle structure, two of which are compared in Fig. 6.21. Each model has an elastic lithosphere with a flexural rigidity of 5×10^{24} N m underlain by a low-viscosity channel and the rest of the mantle. The first model has a 75 km thick channel ($\eta=4\times10^{19}$ Pa s) over a viscous mantle ($\eta=10^{21}$ Pa s). The

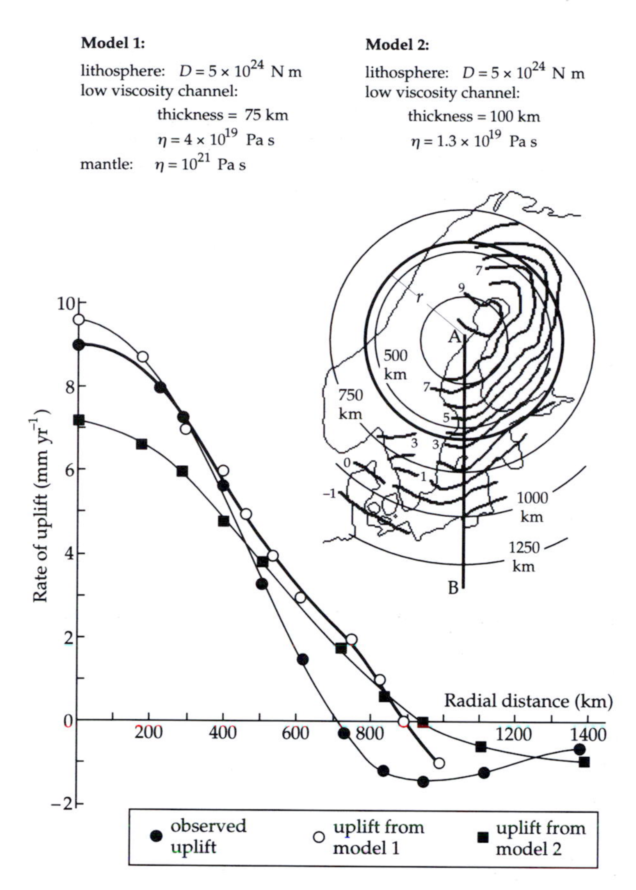

Fig. 6.21 Comparison of Fennoscandian uplift rates interpreted along profile AB (inset) with uplift rates calculated for two different models of mantle viscosity, assuming the ice-sheet can be represented by a vertical right cylindrical load centered on the northern Gulf of Bothnia (after Cathles, 1975).

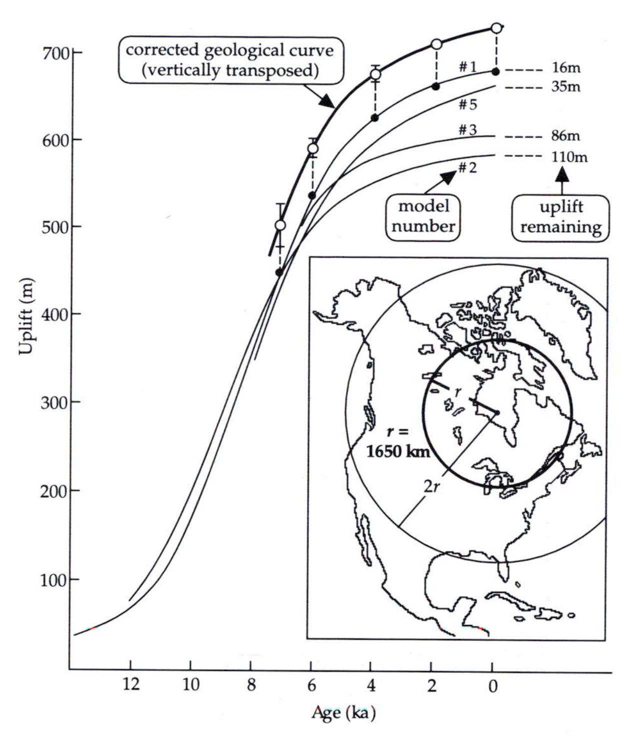

Fig. 6.22 Comparison of uplift history in the James Bay area with predicted uplifts for various Earth models after disappearance of the Wisconsin ice-sheet over North America, represented as a vertical right cylindrical load with radius 1650 km as in the inset (after Cathles, 1975). For model parameters see Table 6.1.

alternative model has a 100 km thick channel (viscosity coefficient $\eta=1.3\times10^{19}$ Pa s) over a rigid mantle. Both models yield uplift rates that agree quite well with the observed uplift rates. However, results from North America indicate that the lower mantle is not rigid and so model 1 fits the data better.

6.2.6.2 *Viscosity of the lower mantle*

Geologists have developed a coherent picture of the last glacial stage, the Wisconsin, during which much of North America was covered by an ice-sheet over 3500 m thick. The ice persisted for about 20,000 yr and melted about 10,000 yr ago. It caused a surface depression of about 600 m. The subsequent history of uplift in the James Bay area near the center of the feature has been reconstructed using geological indicators and dated by the radiocarbon method (see §4.1.4.1). For modelling purposes the ice-sheet can be represented as a right cylindrical load with radius $r=1650$ km (Fig. 6.22, *inset*). A load as large as this affects the deep mantle. The central uplift following removal of the load has been calculated for several Earth models. Each model has elastic parameters and density distribution obtained from seismic velocities, including a central dense core. The models differ from each other in the number of viscous layers in the mantle and the amount by which the density gradient departs from the adiabatic gradient (Table 6.1). The curvature of the observed uplift curve only fits models in which the viscosity of the lower mantle is around 10^{21} Pa s (Fig. 6.22). A highly viscous lower mantle ($\eta=10^{23}$ Pa s, model #4) is incompatible with the observed uplift history. The best fit is obtained with models #1 or #5. Each has an adiabatic density gradient, but model #1 has a uniform lower mantle with viscosity 10^{21} Pa s, and model #5 has η increasing from 10^{21} Pa s below the lithosphere to 3×10^{21} Pa s just above the core.

The viscoelastic properties of the Earth's interior influence the Earth's rotation, causing changes in the position of the instantaneous rotation axis. The motion of the axis is traced by repeated photo-zenith tube measurements, in which the zenith is located by photographing the stars verti-

Table 6.1 *Parameters of Earth models used in computing uplift rates. All models are for an elastic Earth with a dense core (after Cathles, 1975)*

Model	Density gradient	Viscosity (10^{21} Pa s)	Depth interval
#1	adiabatic	1	entire mantle
#2	adiabatic, except 335–635 km	1	entire mantle
#3	adiabatic, except 335–635 km	0.1	0–335 km
		1	335 km to core
#4	adiabatic	1	0–985 km
		100	985 km to core
#5	adiabatic	1	0–985 km
		2	985–2185 km
		3	2185 km to core

cally above an observatory. Photo-zenith tube measurements reveal systematic movements of the rotation axis relative to the axis of figure (Fig. 6.23). Decomposed into components along the Greenwich meridian (*X*-axis) and the 90°W meridian (*Y*-axis), the polar motion exhibits a fluctuation with cyclically varying amplitude superposed on a linear trend. The amplitude modulation has a period of approximately seven years and is due to the interference of the 12-month annual wobble and the 14-month Chandler wobble. The linear trend represents a slow drift of the pole toward northern Canada at a rate of 0.95 degrees per million years. It is due to the melting of the Fennoscandian and Laurentide icesheets. The subsequent uplift constitutes a redistribution of mass which causes modifications to the Earth's moments and products of inertia, thus affecting the rotation.

The observed polar drift can be modelled with different viscoelastic Earth structures. The models assume a 120 km thick elastic lithosphere and take into account different viscosities in the layers bounded by the seismic discontinuities at 400 km and 670 km depths (§3.7, Table 3.4) and the core–mantle boundary. An Earth that is homogeneous below the lithosphere (without a core) is found to give imperceptible drift. Inclusion of the core, with a density jump across the core–mantle boundary and assuming the mantle viscosity to be around 1×10^{21} Pa s, gives a drift that is perceptible but much slower than that observed. Introduction of a density change at the 670 km discontinuity increases the drift markedly; the 400 km discontinuity does not noticeably change the drift further. The optimum model has an upper-mantle viscosity of about 1×10^{21} Pa s and a lower mantle viscosity of about 3×10^{21} Pa s. The model satisfies both the rate of drift and its direction (Fig. 6.23). The viscosities are comparable to values found by modelling post-glacial uplift (Table 6.1, model #5).

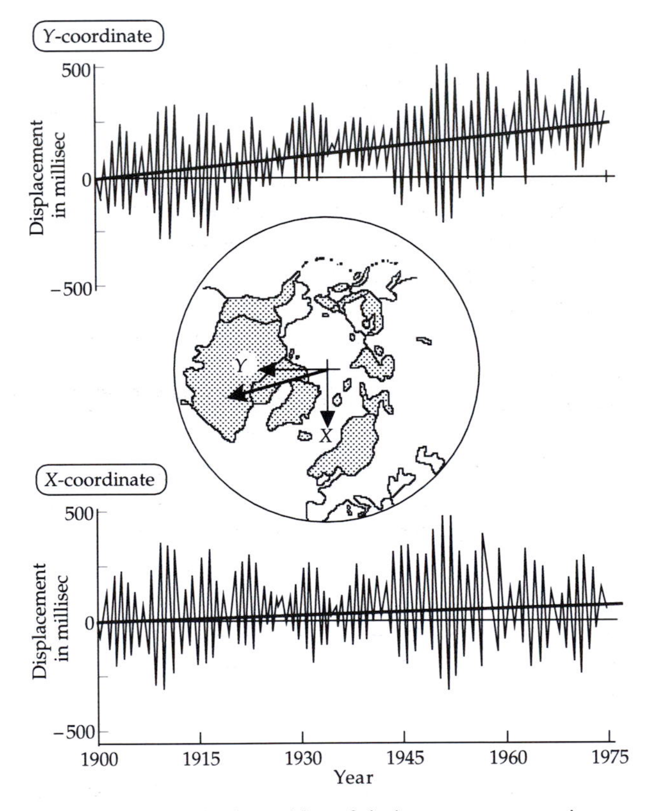

Fig. 6.23 Changes in the position of the instantaneous rotation axis from 1900 to 1975 relative to axes defined in the inset (after Peltier, 1989).

6.3 PLATE DYNAMICS

6.3.1 Mantle convection

It has gradually become accepted that thermally driven convection takes place in the mantle and that it is probably the most important mechanism in geodynamic processes. There are several reasons for these conclusions. The evidence summarized in §6.2 demonstrates that the mantle has a viscoelastic rheology. The passage of seismic compressional and shear waves through the mantle attest that it reacts as a solid to abrupt stress changes. Yet, observations of post-glacial isostatic uplift and long-term movements of the rotation axis indicate that the mantle is capable of viscous flow when stressed over long time intervals. The surmised temperature distribution in the mantle implies that, although conduction is mainly responsible for heat transfer in the lithosphere, convection is the predominant process deeper in the mantle, involving mass transfer by sub-solidus creep. Applying the theory of thermal convection to the mantle and using the best-available estimates of physical parameters indicates that robust convection must be taking place.

6.3.1.1 *Thermal convection*

The conditions for convection to occur (see §4.2.4.2) reflect a balance between causal forces due to thermal expansion and resistive effects due to viscosity and thermal diffusivity. When a fluid is heated, thermal expansion gives rise to an upward buoyancy force. This produces instability, which is partly counteracted by diffusion of heat into the surrounding fluid by thermal conduction. As soon as a volume of the fluid starts to rise in response to the buoyancy force its motion is resisted by viscous forces. The effects are familiar to anyone who has heated a pan of thick soup or porridge. If the pan is heated too rapidly, or the instructions to 'stir constantly' are ignored, the soup may stick to the bottom of the pan and become charred. This happens because the viscosity of the fluid is initially too large to allow convection. Despite the large temperature gradient between the hot bottom of the pan and the cool surface of the liquid, conduction is unable to transport heat away from the bottom of the pan fast enough to avoid charring. When heat flow by conduction reaches a critical limit, convection can begin.

The onset of convection in a fluid layer heated from beneath was first described in 1900 by H. Bénard on the basis of laboratory experiments. He noted that a hexagonal pattern of cells forms on the surface of the layer (Fig. 6.24). Hot fluid rises to the surface in the middle of each cell; at the surface it spreads out and cools. Adjoining cells come in contact at narrow margins, where the cooled fluid sinks back into the layer. Each cell has a rectangular cross-section in the vertical plane. A satisfactory theory of Bénard's observations was derived in 1916 by Lord Rayleigh. Although it applies to an ideal scenario (a horizontal layer with stress-free upper and lower boundaries, heated from below, and with a constant temperature on the upper surface) the theory permits approximate estimates for more complex convection in the spherical Earth.

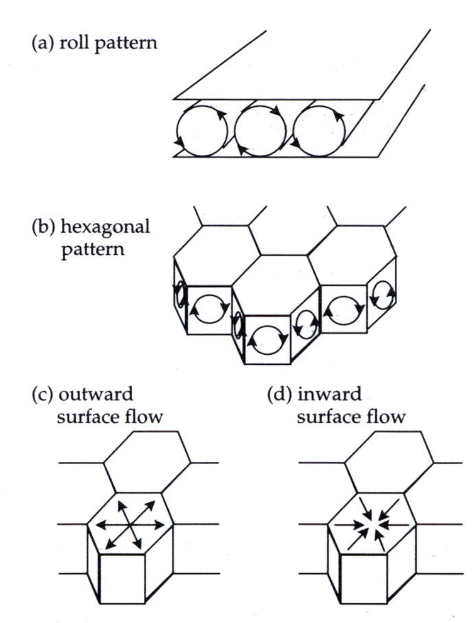

Fig. 6.24 Some patterns of steady convection in a plane layer heated from below. (a) Convection rolls, (b) vertical flow in hexagonal patterns, for which the surface flow may be (c) outward away from or (d) inward toward the center of the cell (after Busse, 1989).

The flow of a viscous fluid is governed by the Navier–Stokes equation, one of the most important equations in geophysics. It describes the conservation of momentum in the fluid, which in its simplest form means balancing several terms that express the driving forces exerted by pressure gradient and buoyancy against the viscous and inertial forces that resist motion. The ratio of the other forces to the inertial forces is expressed by the dimensionless Prandtl number:

$$\mathrm{Pr} = \frac{\nu}{\kappa} \tag{6.31}$$

where ν is the kinematic viscosity, and κ is the thermal diffusivity. In the mantle $\nu \approx 10^{18}$ m^2 s^{-1} and $\kappa \approx 10^{-6}$ m^2 s^{-1}, so that Pr $\approx 10^{24}$. The virtually infinite Prandtl number means that inertial forces are insignificant. Hence, mantle convection depends only on the conditions of pressure, temperature and viscosity.

Thermal diffusivity and viscosity act as stabilizing influences in a heated fluid. If heating is slow enough, the temperature gradient adjusts to transfer the heat by conduction, remaining close to the adiabatic gradient. Convection becomes possible when the real temperature gradient exceeds the adiabatic gradient; the difference θ is the *superadiabatic gradient*. The excess heat expands the fluid,

Table 6.2 *Some physical parameters for mantle convection models (mostly from Jarvis and Peltier, 1989)*
The Rayleigh numbers (Ra_T) for the onset of convection in each part of the mantle are calculated assuming a superadiabatic temperature gradient $\theta = 0.1$ K km^{-1} and a mean gravity $g = 10$ m s^{-2}. Lower-mantle parameters are interpolated from the upper- and whole-mantle values.

Physical parameter	Units	Upper mantle (70–670 km)	Lower mantle (670–2890 km)	Whole mantle (70–2890 km)
Layer thickness (H)	km	600	2,220	2,820
Expansion coefficient (α)	K^{-1}	2×10^{-5}	1.0×10^{-5}	1.4×10^{-5}
Density (ρ)	kg m^{-3}	3,700	5,500	4,700
Specific heat (c_p)	J kg^{-1} K^{-1}	1,260	1,260	1,260
Thermal conductivity (k)	W m^{-1} K^{-1}	6.7	20	15
Thermal diffusivity (κ)	m^2 s^{-1}	1.4×10^{-6}	3×10^{-6}	2.5×10^{-6}
Dynamic viscosity (η)	kg m^{-1} s^{-1}	1×10^{21}	2.5×10^{21}	2×10^{21}
Kinematic viscosity (ν)	m^2 s^{-1}	2.7×10^{17}	4.5×10^{17}	4.3×10^{17}
Rayleigh number (Ra_T)	—	7,000	180,000	820,000

causing the buoyancy force. When this becomes larger than the viscous resistance, convection ensues. The ratio of the competing forces is embodied in the *Rayleigh number* (Eq. (4.42)). The Rayleigh number (Ra_T) for convection due to the superadiabatic temperature gradient in a fluid layer of thickness D is

$$Ra_T = \frac{g\alpha\theta}{\kappa\nu}D^4 \qquad (6.32)$$

where g is gravity and α is the coefficient of thermal expansion.

The superadiabatic gradient is not the only source of power for convection. Although radioactive heat generation in mantle materials is small (see §4.2.5.1), it can still contribute to convection. If Q is the radiogenic heat production in a layer of thickness D, we can invoke Eq. (4.39) and Eq. (4.45) and replace q in the above equation by (QD/k), where k is the thermal conductivity. This allows us to define a second Rayleigh number (Ra_Q) for convection driven by radiogenic heat:

$$Ra_Q = \frac{g\alpha Q}{k\kappa\nu}D^5 \qquad (6.33)$$

Convection is initiated when the Rayleigh number exceeds a critical value, Ra_c, which is dependent on the geometry of the flow and the boundary conditions on the upper and lower surfaces. In Rayleigh–Bénard convection the top and bottom of the horizontal layer are stress-free; the critical Rayleigh number is $Ra_c=658$. If the top and bottom of the layer are rigid boundaries at which the horizontal velocity vanishes, $Ra_c=1708$. Table 6.2 shows computed Rayleigh numbers Ra_T for convection driven by the superadiabatic temperature gradient for viscous flow in the upper, lower and whole mantle, assuming representative values from the literature for the parameters in Eq. (6.32). Reasonable estimates of the radiogenic heat produced in the mantle give even larger values for Ra_Q.

6.3.1.2 *Convection at high Rayleigh numbers*

The computed Rayleigh numbers greatly exceed the critical values for convection throughout the entire mantle or in separate layers. Thus, each region of the sub-lithospheric mantle is capable of convection. The Rayleigh number for whole-mantle convection is so much larger than the critical value Ra_c that vigorous mantle convection must be expected. This does not imply rapid flow in normal terms. The speed of flow in the mantle is usually assumed to be of the same order as the rate of motion of tectonic plates, about 5–10 cm yr^{-1} on average. As long as the flow rate v is low, adjacent lamina of the fluid move past each other under the conditions for Newtonian viscosity (§6.2.2). At faster flow rates this condition breaks down, and the flow becomes turbulent. The conditions favoring turbulence are high momentum (ρv) and large scale D of the flow, whereas it is inhibited by high viscosity η. These factors are contained in the Reynolds number, Re, defined as

$$Re = \frac{\rho v D}{\eta} \qquad (6.34)$$

Reasonable values for the mantle are $\rho=5000$ kg m^{-3}, $D=2900$ km$=2.9\times10^6$ m, $v=5$ cm yr$^{-1}=1.5\times10^{-9}$ m s^{-1}, and $\eta=1.5\times10^{21}$ Pa s. The Reynolds number is found to be $Re=1.5\times10^{-20}$, which is so small that turbulence is negligible. Similar results are found by considering the upper or lower mantle alone. Clearly, although mantle convection involves high Rayleigh numbers (implying vigorous convection on a geological timescale), it takes place by laminar flow.

The effect of convection is to replace conduction as the principal mechanism of heat transfer. A measure of the relative effectiveness of the two processes of heat transfer is the Nusselt number, Nu. This is defined as the ratio of the heat transport in the presence of convection to the heat transport without convection. In the absence of radiogenic heat sources, the heat transport with convection is determined by the Rayleigh number Ra_T, while the non-convective heat transport is expressed by the critical Rayleigh number Ra_c. The Nusselt number depends on the ratio of these two numbers and can be written

$$\mathrm{Nu}=\beta\left(\frac{\mathrm{Ra}_T}{\mathrm{Ra}_c}\right)^S \tag{6.35}$$

where the coefficient β and the exponent S are functions of the aspect ratio of the convection cells. Mathematical evaluation of the problem of Rayleigh–Bénard convection with stress-free upper and lower boundaries gives $\beta\approx 1$ and $S=1/3$, and, since in this case $\mathrm{Ra}_c\approx 10^3$, the Nusselt number has the simpler form

$$\mathrm{Nu}\approx 0.1(\mathrm{Ra}_T)^{1/3} \tag{6.36}$$

Using the estimated values of Ra_T in Table 6.2 gives Nusselt numbers of 19 for layered convection in the upper mantle and 97 for whole-mantle convection. Hence, heat transfer by convection is dominant in the mantle.

Once convection has been initiated the boundary conditions determine the shapes of the convection cells. In Rayleigh–Bénard convection the aspect ratio of a cell – the ratio of its horizontal dimension to its vertical one – is $2^{1/2}=1.41$; when the layer has rigid boundaries the cell aspect ratio is 1.01. Hence, the horizontal extent of a convection cell is comparable with the layer thickness. This has implications for convection in the mantle. If we assume that the scale of mantle convection is represented by the pattern of the plate boundaries (Fig. 1.11), we can estimate the horizontal dimensions of the convection cells. It is evident that they must have very different sizes. The ridge-to-trench horizontal distances across major plates are in the range 2000–10,000 km, with an average of about 5000 km. This is larger than the maximum thickness of the convecting layer, whether we assume convection to be restricted to the upper mantle or to occupy the entire 2900 km thickness of the sub-lithospheric mantle. Thus, if convection is uniform through the whole mantle, the aspect ratios of at least some cells must be much larger than unity (Table 6.3). The reason for this is the rigidity of the cold upper boundary formed by the lithosphere, which inhibits the breakup of the fluid flow into cells with smaller horizontal extents.

Table 6.3 *Approximate aspect ratios of some mantle convection cells, estimated from the horizontal dimensions of the overlying lithospheric plates (after Turcotte and Schubert, 1982, Table 7.5)*

Plate	Upper mantle convection	Whole mantle convection
Pacific	14	3.3
North American	11	2.6
South American	11	2.6
Indian	8	2.1
Nazca	6	1.6

6.3.1.3 *Models of mantle convection*

The feasibility of mantle convection is accepted but there is still some doubt as to the form it takes. This is in part due to uncertainty as to the role played by the seismic discontinuities at 400 km and 670 km depth, which bound the upper-mantle transition zone. The discontinuities are not sharp, and are understood to represent mineral phase changes rather than compositional differences (as, for example, the crust–mantle and core–mantle boundaries). The upper discontinuity marks the olivine–spinel phase change, the lower one represents the phase change from spinel to perovskite structure (§3.7.5.2), with accompanying changes in density and elastic parameters. In principle, mass can be carried by convection currents across these discontinuities. The 670 km discontinuity is close to the maximum depth of seismicity in subduction zones, and may be where the subducting plate is absorbed into the mantle.

There are two main models of mantle convection, each with an interface at the 670 km seismic discontinuity. An important change in viscosity occurs at this level. In *whole-mantle convection* (Fig. 6.25a) the viscosity doubles from the upper mantle to the lower mantle (see Table 6.2) and there is a net flow of material across the boundary. In this model, convection ensures that the entire mantle is well mixed mechanically, and the phase changes at 400 and 670 km have only a small effect on the temperature gradient. This model agrees with much of the available evidence.

The alternative *layered convection* model has distinct convecting layers in the upper and lower mantle (Fig. 6.25b). There are two ways in which this can take place. The upper and lower convection patterns in a vertical section may represent circulations in the same sense (e.g., both clockwise or both anticlockwise) or in opposite senses

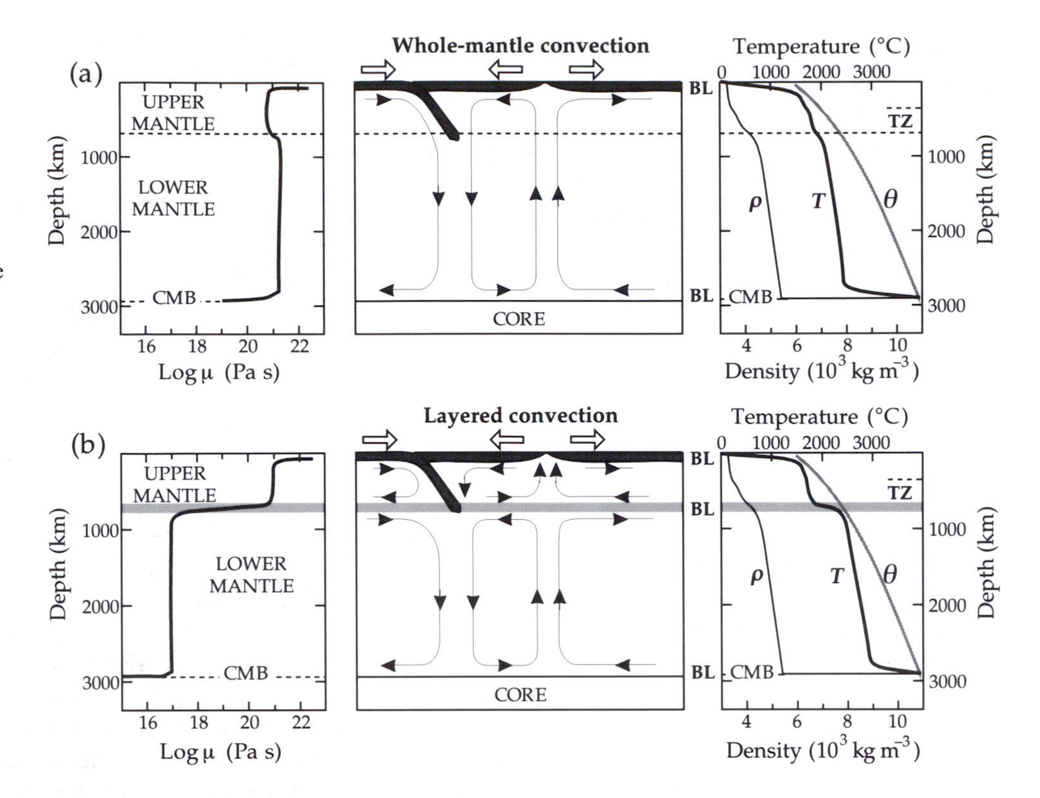

Fig. 6.25 Possible convection flow pattern (*center*) and profiles of viscosity μ (*left*), and density ρ, temperature T and solidus temperature ϑ (*right*) for (a) whole-mantle convection and (b) layered mantle convection. TZ is the upper-mantle transition zone, BL are boundary layers, CMB is the core-mantle boundary (based upon Peltier *et al.*, 1989).

(e.g., one clockwise and the other anticlockwise). In each case the radial velocity is zero at 670 km depth and there is no mass transfer across the discontinuity; the material in each flow pattern spreads out along the boundary. However, the models imply different types of coupling between the layers. Opposite senses of circulation in the layers would cause little or no shear between the tangential flows at the boundary, resulting in mechanical coupling between the layers. Cold material sinking in the upper mantle would overlie hot material rising in the lower mantle. However, if the layered flow patterns have the same sense of circulation (as in Fig. 6.25b), hot material rising in the upper mantle overlies hot material rising in the lower mantle, so that the flow regimes are coupled thermally. This model has a strong velocity shear across the 670 km discontinuity, which requires a large and abrupt change in viscosity at this depth; viscosity in the lower mantle would need to be at least two orders of magnitude smaller than in the upper mantle. Estimates of mantle viscosity (§6.2.6) indicate the opposite: viscosity is higher in the lower mantle than in the upper mantle.

A model of layered convection assumes that there is no mass transfer across the discontinuity. The upper and lower mantles are well mixed individually, but the separation of the flow patterns at the discontinuity means that they may have distinct chemical compositions. Because there is no convective flow across it, heat can only cross the boundary by conduction. The 670 km discontinuity therefore acts as a thermal boundary, with a large temperature change of perhaps 500–1000 deg K across it. Thus, the temperature profile in the lower mantle, although maintained adiabatic by the convection, would be 500–1000 deg K higher than in whole-mantle convection. This would result in a smaller temperature change across the CMB, a less-steep temperature gradient in the D"-layer, and so a lower heat flux from the core. The long-term rate of cooling of the Earth would thereby be reduced.

The problem of understanding mantle convection is complicated by the non-uniform structure and rheology of the mantle. As yet, there is no complete picture of how the various factors that influence convection act together. The convection pattern depends strongly on what happens physically and thermodynamically at the 670 km discontinuity. This can only be inferred indirectly. Our understanding of the discontinuity is incomplete, but it is essential to resolving the real pattern of mantle convection.

6.3.1.4 *Mantle plumes*

The viscosity in the upper mantle is inferred from post-glacial rebound studies to be around 10^{21} Pa s, but lower-mantle viscosity is less well known. The sub-solidus creep in the mantle implies a temperature-dependent viscosity,

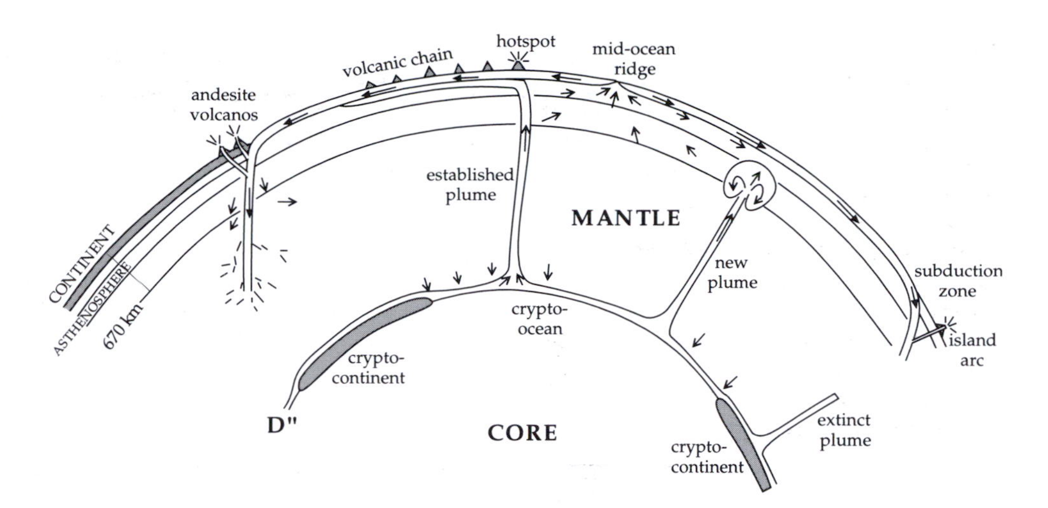

Fig. 6.26 An idealized cross-section through the mantle, showing convective flow and the relationship of mantle plumes to the D"-layer (after Stacey, 1992).

which allows thermal boundary layers at the top and bottom of the mantle to influence the patterns of convective flow.

The lithosphere constitutes an upper, cold boundary layer. It accretes at high temperature at spreading ridges, where upwelling magma from the mantle reaches the surface. The eruptive lavas issue from magma chambers beneath ridge crests, in which magma from the deeper mantle undergoes differentiation. As part of the plate tectonic cycle the lithosphere rapidly cools and hardens as it spreads away from the ridge. Its high viscosity (i.e., rigidity) inhibits internal convection, but at subduction zones the plate (by now old) flexes downward and carries cold material into the underlying mantle, altering its thermal balance. Seismic tomography (§3.7.6) has revealed broad regions of raised seismic velocity in the deep mantle below subduction zones, giving rise to the surmise that the material in the cold subducted plate eventually sinks to the bottom of the mantle. The material must eventually take part in a broad-scale return flow, completing the convective cycle, but how this takes place is not clear.

The core–mantle boundary (CMB) at 2890 km depth constitutes a lower, hot boundary layer. The D"-layer at the base of the mantle (§3.7.5.3) is characterized by reductions in seismic velocities between about 2740 km depth and the CMB. It evidently has different physical properties than the mantle above it, and appears to play an important thermodynamic role. The heat flux from the core to the mantle diminishes the rigidity of the layer, thus reducing the seismic velocities. The viscosity of the hot thin D"-layer is presumed to be much lower than that of the overlying mantle. The topography of the CMB has been explored by seismic waves reflected from the core or passing it at grazing incidence. The thickness of the D"-layer appears to be uneven, and has been interpreted by analogy to the crust. Thick segments have been designated as crypto-continents and thinner regions as crypto-oceans (Fig. 6.26).

The low-viscosity material in the D"-layer is thought to supply relatively fast-flowing narrow mantle plumes. This name is given to vertical features, thin in cross-section, that facilitate the upwelling of low-viscosity hot magma through the more viscous mantle. A new plume melts its way to the surface behind a larger head. Some plumes may not reach the surface but intrude their material into the asthenosphere or lower lithosphere. Other mature plumes may penetrate the entire mantle and reach the surface, where they are evident as places of persistent volcanism, high regional topography and local high heat flow, called *hotspots*. These areas of anomalous volcanism are found in the oceans and on the continents, within plates and on plate margins. The plumes that feed them are thought to remain fixed in position for long periods of time, and so the hotspots are anchored to the mantle below the lithosphere. As a result they have important consequences for studies of plate tectonic motions.

6.3.2 Hotspots

In 1958 S. W. Carey coined the term 'hot spot' – now often reduced to 'hotspot' – to refer to a long-lasting center of surface volcanism and locally high heat flow. At one time more than 120 of these thermal anomalies were proposed. Application of more stringent criteria has reduced their number to about 40 (Fig. 6.27). The hotspots may occur on the continents (e.g., Yellowstone), but are more common in the ocean basins. The oceanic hotspots are associated with depth anomalies. If the observed depth is compared with the depth predicted by cooling models of the oceanic lithosphere, the hotspots are found to lie predominantly in broad shallow regions, where the lithosphere apparently swells

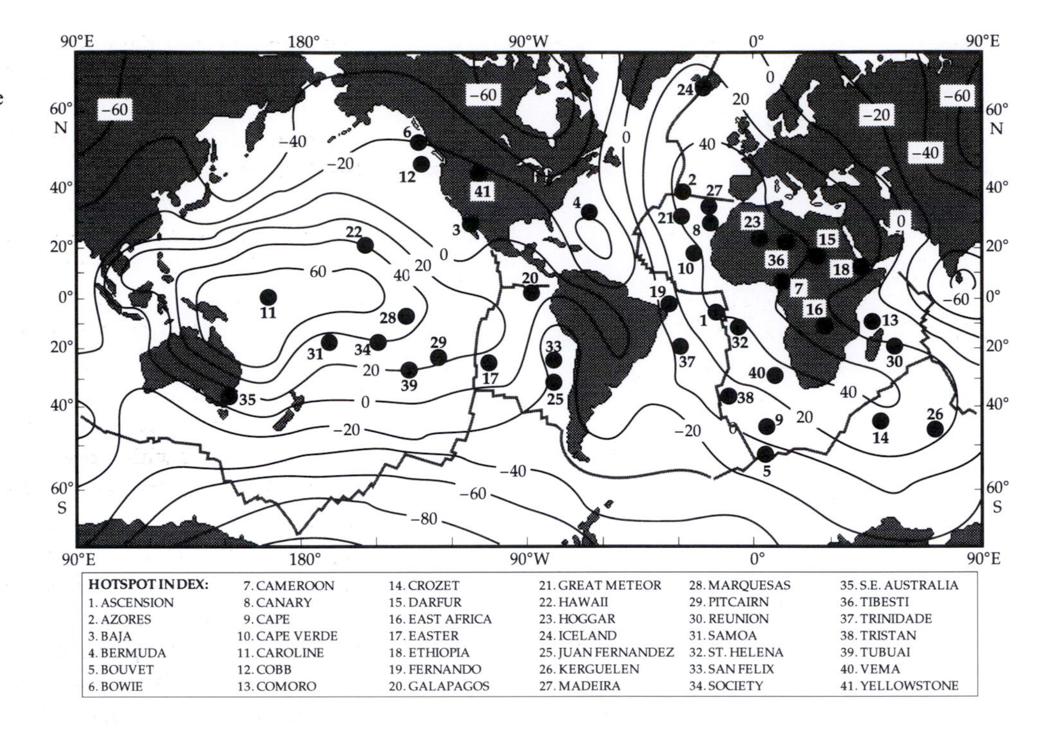

Fig. 6.27 The global distribution of 41 hotspots and their relationship to the residual geoid obtained by correcting geoid heights (shown in meters above the reference ellipsoid) for the effects of cold subducting slabs (after Crough and Jurdy, 1980).

upward. This elevates denser mantle material, which creates a mass anomaly and disturbs the geoid; the effect is partially mitigated by reduced density of material in the hot, rising plume. The geoid surface is also displaced by subduction zones. The residual geoid obtained by removing the effects associated with cold subducting slabs shows a remarkable correlation with the distribution of hotspots (Fig. 6.27). The oceanic hotspots are found in conjunction with intraplate island chains, which provide clues to the origin of hotspots and allow them to be used for measuring geodynamic processes.

Two types of volcanic island chains are important in plate tectonics. The arcuate chains of islands associated with deep oceanic trenches at consuming plate margins are related to the process of subduction and have an arcuate shape. Nearly linear chains of volcanic islands are observed within oceanic basins far from active plate margins. These intraplate features are particularly evident on a bathymetric map of the Pacific Ocean. The Hawaiian, Marquesas, Society and Austral Islands form subparallel chains that trend approximately perpendicular to the axis of ocean-floor spreading on the East Pacific rise. The most closely studied is the Hawaiian Ridge (Fig. 6.28a). The volcanism along this chain decreases from present-day activity at the southeast, on the island of Hawaii, to long extinct seamounts and guyots towards the northwest along the Emperor Seamount chain. The history of development of the chain is typical of other linear volcanic island chains in the Pacific basin (Fig. 6.28b). It was explained in 1963 by J. T. Wilson, before the modern theory of plate tectonics was formulated.

A hotspot is a long-lasting magmatic center rooted in the mantle below the lithosphere. A volcanic complex is built up above the magmatic source, forming a volcanic island or, where the structure does not reach sea-level, a seamount. The motion of the plate transports the island away from the hotspot and the volcanism becomes extinct. The upwelling material at the hotspot elevates the ocean floor by up to 1500 m above the normal depth of the ocean floor, creating a depth anomaly. As they move away from the hotspot the by now extinct volcanic islands sink beneath the surface; some are truncated by erosion to sea-level and become guyots. Coral atolls may accumulate on some guyots. The volcanic chain is aligned with the motion of the plate.

Confirmation of this theory is obtained from radiometric dating of basalt samples from islands and seamounts along the Hawaiian Ridge portion of the Hawaiian–Emperor chain. The basalts increase in age with distance from the active volcano Kilauea on the island of Hawaii (Fig. 6.29). The trend shows that the average rate of motion of the Pacific plate over the Hawaiian hotspot has been about 10 cm yr^{-1} during the last 20–40 Ma. The change in trend between the Hawaiian Ridge and the Emperor Seamount chain indicates a change in direction and speed of the Pacific plate about 43 Ma ago, at which time there was a global re-organization of plate motions. The earlier rate of

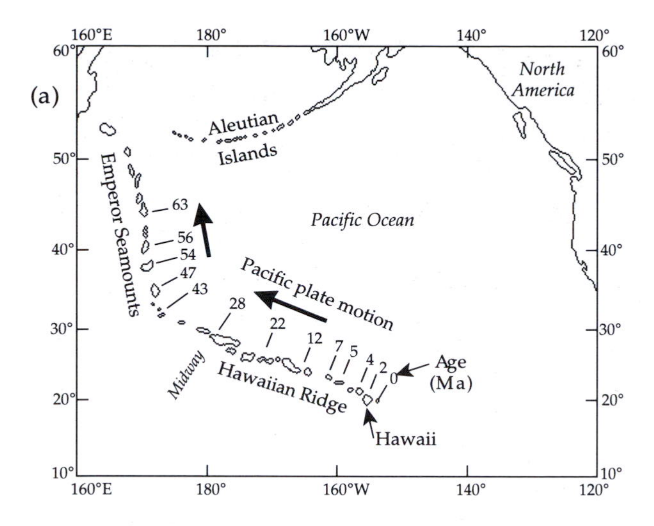

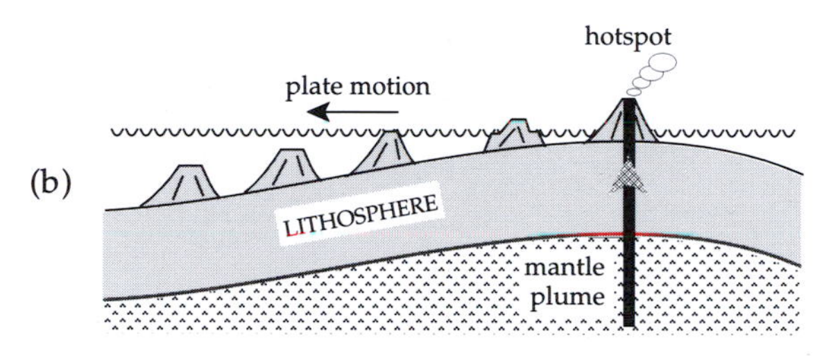

Fig. 6.28 (a) The Hawaiian Ridge and Emperor Seamount volcanic chains trace the motion of the Pacific plate over the Hawaiian hotspot; numbers give the approximate age of volcanism; note the change of direction about 43 Ma ago (after Van Andel, 1992). (b) Sketch illustrating the formation of volcanic islands and seamounts as a lithospheric plate moves over a hotspot (based on Wilson, 1963).

motion along the Emperor chain is less well determined but is estimated to be about 6 cm yr^{-1}.

Radiometric dating of linear volcanic chains in the Pacific basin gives almost identical rates of motion over their respective hotspots. This suggests that the hotspots form a stationary network, at least relative to the lithosphere. The velocities of plate motions over the hotspots are therefore regarded as absolute velocities, in contrast to the velocities derived at plate margins, which are the relative velocities between neighboring plates. The assumption that the hotspots are indeed stationary has been contested by studies that have yielded rates of interhotspot motion of the order of 1.5–2 cm yr^{-1} (comparable to present spreading rates in the Atlantic). Thus, the notion of a stationary hotspot reference frame may only be valid for a limited time interval. Nevertheless, any motions between hotspots are certainly much slower than the motions of plates, so the hotspot reference frame provides a useful guide to absolute plate motions over the typical time interval (~10 Ma) in which incremental sea-floor spreading is constant.

As well as geophysical evidence there are geochemical anomalies associated with hotspot volcanism. The type of basalt extruded at a hotspot is different from the andesitic basalts formed in subduction zone magmatism. It also has a different petrology from the mid-oceanic ridge basalts (MORB) formed during sea-floor spreading and characteristic of the ocean floor. The hotspot source is assumed to be a mantle plume that reaches the surface. Mantle plumes are fundamental features of mantle dynamics, but they remain poorly understood. Although they are interpreted as long-term features it is not known for how long they persist, or how they interact with convective processes in the mantle. Their role in heat transport and mantle convection, with consequent influence on plate motions, is believed to be important but is uncertain. Their sources are controversial. Some interpretations favor a comparatively shallow origin above the 670 km discontinuity, but the prevailing opinion appears to be that the plumes originate in the D″-layer at the core–mantle boundary. This requires the mantle plume to penetrate the entire thickness of the mantle (see Fig. 6.26). In either case the stationary nature of the hotspot network relative to the lithosphere provides a reference frame for determining absolute plate motions, and for testing the hypothesis of true polar wander.

6.3.3 Plate motion on the surface of a sphere

One of the great mathematicians of the 18th century was Leonhard Euler (1707–1783) of Switzerland. He made numerous fundamental contributions to pure mathematics, including to spherical trigonometry. A corollary of one of his theorems shows that the displacement of a rigid body on the surface of a sphere is equivalent to a rotation about an axis that passes through its center. This is applicable to the motion of a lithospheric plate.

Any motion restricted to the surface of a sphere takes place along a curved arc that is a segment of either a great circle (centered, like a ‘circle of longitude’, at the Earth's center) or a small circle. Small circles are defined relative to a pole of rotational symmetry (such as the geographical pole, when we define ‘circles of latitude’). A point on the surface of the sphere can be regarded as the end-point of a radius vector from the center of the Earth to the point; the position is given by two angles, akin to latitude and longitude. As a result of Euler's theorem any displacement of a point along a small circle is equivalent to rotating the radius vector about the pole of symmetry, which is called the Euler pole of the rotation. A displacement along a great circle – the shortest distance between two points on the surface of the sphere

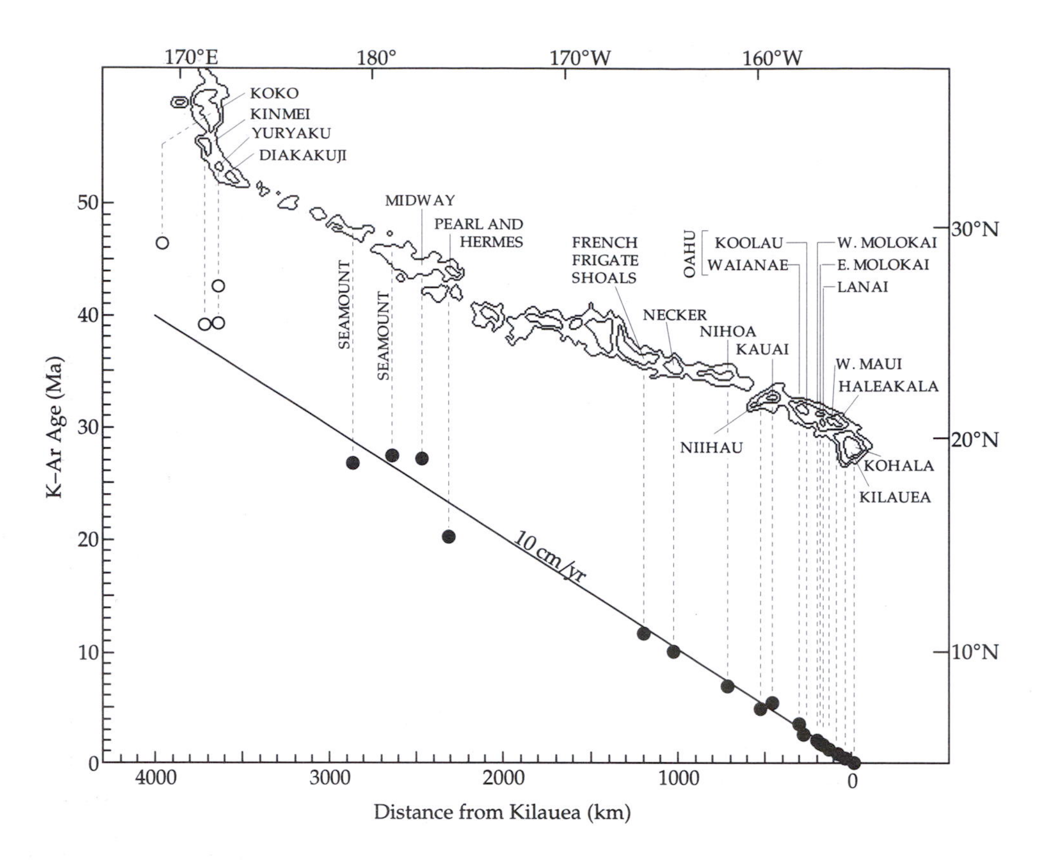

Fig. 6.29 Age of basaltic volcanism along the Hawaiian Islands as a function of distance from the active volcano Kilauea (based on Dalrymple *et al.*, 1977).

– is a rotation about an Euler pole 90° away from the arcuate path. Euler poles have already been encountered in discussions of conservative plate margins (§1.2.6.3) and paleogeographic reconstructions using apparent polar wander paths (§5.6.4.3).

6.3.3.1 *Euler poles of rotation*

Geophysical evidence does not in itself yield absolute plate motions. Present-day seismicity reflects relative motion between contiguous plates, oceanic magnetic anomaly patterns reveal long-term motion between neighboring plates, and paleomagnetism does not resolve displacements in longitude about a paleopole. The relative motion between plates is described by keeping one plate fixed and moving the other one relative to it; that is, we rotate it away from (or toward) the fixed plate (Fig. 6.30). The geometry of a rigid plate on the surface of a sphere is outlined by a set of bounding points, which maintain fixed positions relative to each other. Provided it remains rigid, each point of a moving plate describes an arc of a different small circle about the same Euler pole. Thus, the motion between plates is equivalent to a relative rotation about their mutual Euler rotation pole.

The traces of past and present-day plate motions are recorded in the geometries of transform faults and fracture zones, which mark, respectively, the present-day and earlier locations of conservative plate margins. A segment of a transform fault represents the local path of relative motion between two plates. As such, it defines a small circle about the Euler pole of relative rotation between the plates. Great circles drawn normal to the strike of the small circle (transform fault) should meet at the Euler pole (Fig. 6.31a), just as, at the present-day, circles of longitude are perpendicular to circles of latitude and converge at the geographic pole. In 1968, W. J. Morgan first used this method to locate the Euler rotation pole for the present-day plate motion between America and Africa (Fig. 6.31b). The Caribbean plate may be absorbing slow relative motion, but the absence of a well-defined seismic boundary between North and South America indicates that these plates are now moving essentially as one block. The great circles normal to transform faults in the Central Atlantic converge and intersect close to 58°N 36°W, which is an estimate of the Euler pole of recent motion between Africa and South America. The longitude of the Euler pole is determined more precisely than its latitude, the errors being ±2° and ±5°, respectively. When additional data from earthquake first motions and spreading rates are included, an Euler pole at 62°N 36°W is obtained, which is within the error of the first location.

The 'Bullard-type fit' of the African and South

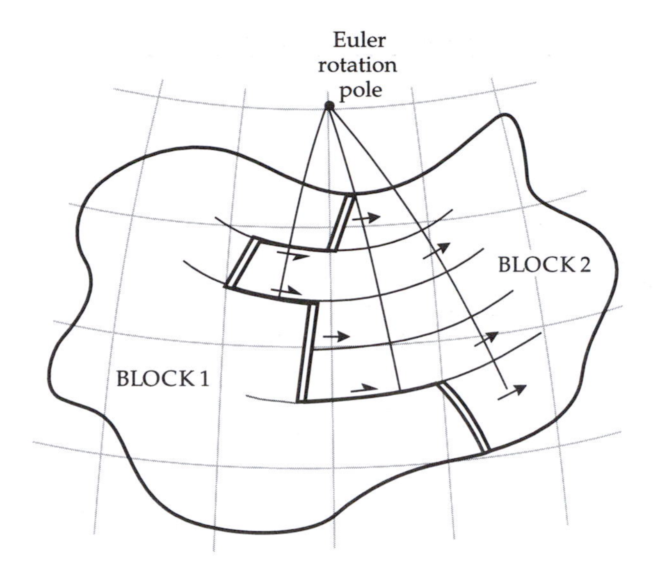

Fig. 6.30 Illustration that the displacement of a rigid plate on the surface of a sphere is equivalent to the rotation of the plate about an Euler pole (after Morgan, 1968)

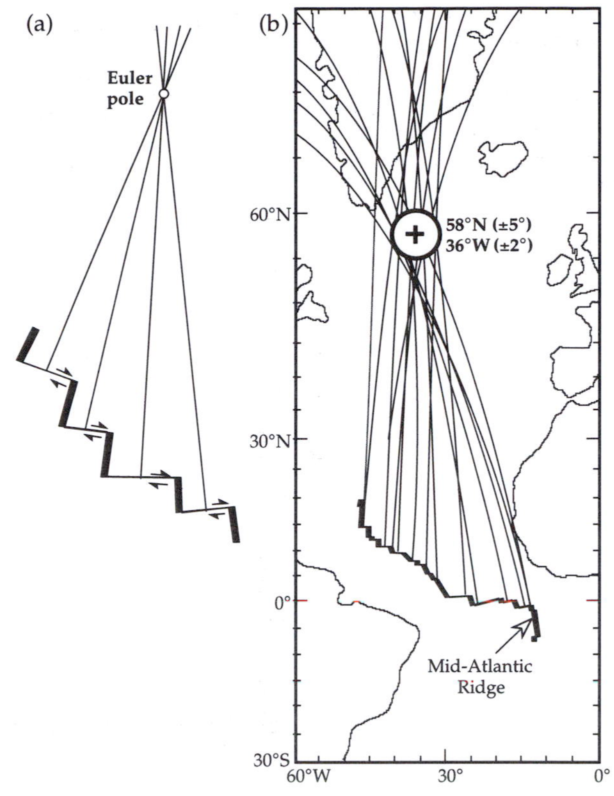

Fig. 6.31 (a) Principle of the method for locating the Euler pole of rotation between two plates where great circles normal to transform faults on the plate boundary intersect (after Kearey and Vine, 1990). (b) Location of the Euler pole of rotation for the *long-term* motion between Africa and South America, using transform faults on the Mid-Atlantic Ridge in the Central Atlantic (after Morgan, 1968,).

American coastlines (§1.2.2.2) is obtained by a rotation about a pole at 44°N 31°W. This pole reflects the average *long-term* motion between the continents. A rotation which matches a starting point with an end-point is a *finite rotation*. As the difference between the present-day and age-averaged Euler poles illustrates, a finite rotation is a mathematical formality not necessarily related to the actual motion between the plates, which may consist of a number of incremental rotations about different poles.

6.3.3.2 *Absolute plate motions*

The axial dipole hypothesis of paleomagnetism states that the mean geomagnetic pole – averaged over several tens of thousands of years – agrees with the contemporaneous geographic pole (i.e. rotation axis). Paleomagnetic directions allow the calculation of the apparent pole position at the time of formation of rocks of a given age from the same continent. By connecting the pole positions successively in order of their age an apparent polar wander (APW) path is derived for the continent. Viewed from the continent it appears that the pole (i.e., the rotation axis) has moved along the APW path. In fact, the path records the motion of the lithospheric plate bearing the continent, and differences between APW paths for different plates reflect motions of the plates relative to each other.

During the displacement of a plate (i.e., when it rotates about an Euler pole), the paleomagnetic pole positions obtained from rocks on the plate describe a trajectory which is the arc of a small circle about the Euler pole (Fig. 6.32). The motion of the plate over an underlying hotspot leaves a trace that is also a small circle arc about the same hotspot. The paleomagnetic record gives the motion of plates relative to the rotation axis, whereas the hotspot record shows the plate motion over a fixed point in the mantle. If the mantle moves relative to the rotation axis, the network of hotspots – each believed to be anchored to the mantle – shifts along with it. This motion of the mantle deeper than the mobile lithosphere is called true polar wander (TPW). The term is rather a misnomer, because it refers to motion of the mantle relative to the rotation axis.

Paleomagnetism provides a means of detecting whether long-term true polar wander has taken place. It involves comparing paleomagnetic poles from hotspots with contemporary poles from the stable continental cratons. Consider first the possibility that TPW does not take place: each hotspot maintains its position relative to the rotation axis. A lava that is magnetized at an active hotspot acquires a direction appropriate to the distance from the pole. If the

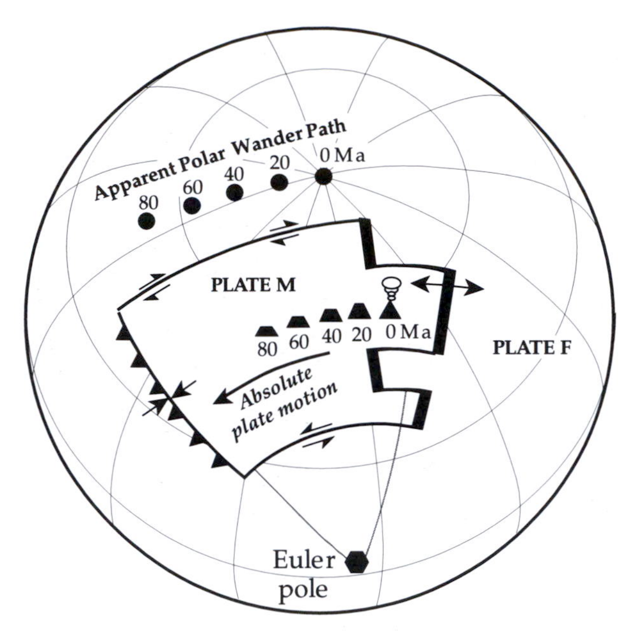

Fig. 6.32 Development of an arcuate apparent polar wander path and hotspot trace as small circles about the same Euler pole, when a mobile plate M moves relative to a fixed plate F (after Butler, 1992).

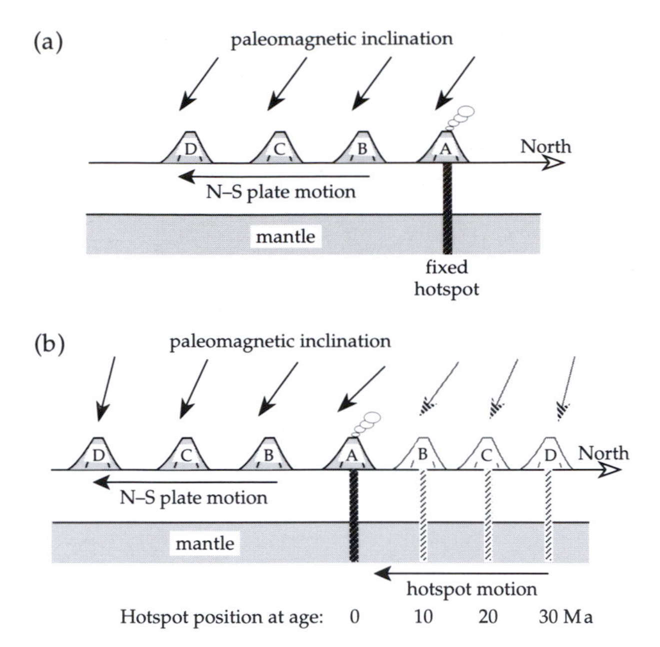

Fig. 6.33 Illustration of the effect of true polar wander on paleomagnetic inclination: (a) north–south plate motion over a stationary hotspot, (b) same plate motion over a north–south migrating hotspot. A, B, C and D are sequential positions.

plate moves from north to south over the stationary hotspot, a succession of islands and seamounts (A–D) is formed, which, independently of their age, have the same magnetization direction (Fig. 6.33a). Next, suppose that true polar wander does takes place: each hotspot moves with time relative to the rotation axis. For simplicity, let the hotspot migration also be from north to south (Fig. 6.33b). Seamount A is being formed at present and its magnetization direction corresponds to the present-day distance from the pole. However, older seamounts B, C and D were formed closer to the pole and have progressively steeper inclinations the further south they are. The change of paleomagnetic directions with age of the volcanism along the hotspot trace is evidence for true polar wander.

To test such a hypothesis adequately a large number of data are needed. The amount of data from a single plate, such as Africa, can be enlarged by using data from other plates. For example, in reconstructing Gondwanaland, South America is rotated into a matching position with Africa by a finite rotation about an Euler pole. The same rotation applied to the APW path of South America allows data from both continents to be combined. Likewise, rotations about appropriate Euler poles make the paleomagnetic records for North America and Eurasia accessible. Averaging the pooled data for age-windows 10 Ma apart gives a reconstructed paleomagnetic APW path for Africa (Fig. 6.34a). The next step is to determine the motions of plates over the network of hotspots, assuming the hotspots have not moved relative to each other. A 'hotspot' apparent polar wander path is obtained, which is the track of an axis in the hotspot reference frame presently at the north pole. The appearance of this track relative to Africa is shown in Fig. 6.34b.

We now have records of the motion of the lithosphere relative to the pole, and of the motion of the lithosphere relative to the hotspot reference frame. The records coincide for the present time, both giving pole positions at the present-day rotation axis, but they diverge with age as a result of true polar wander. A paleomagnetic pole of a given age is now moved along a great circle (i.e., rotated about an Euler pole in the equatorial plane) until it lies on the rotation axis. If the same rotation is applied to the hotspot pole of the same age, it should fall on the rotation axis also. The discrepancy is due to motion of the hotspot reference frame relative to the rotation axis. Joining locations in order of age gives a true polar wander path (Fig. 6.34c). This exercise can be carried out for only the last 200 Ma, in which plate reconstructions can be confidently made. The results show that TPW has indeed taken place but that its amplitude has remained less than 15° for the last 150 Ma.

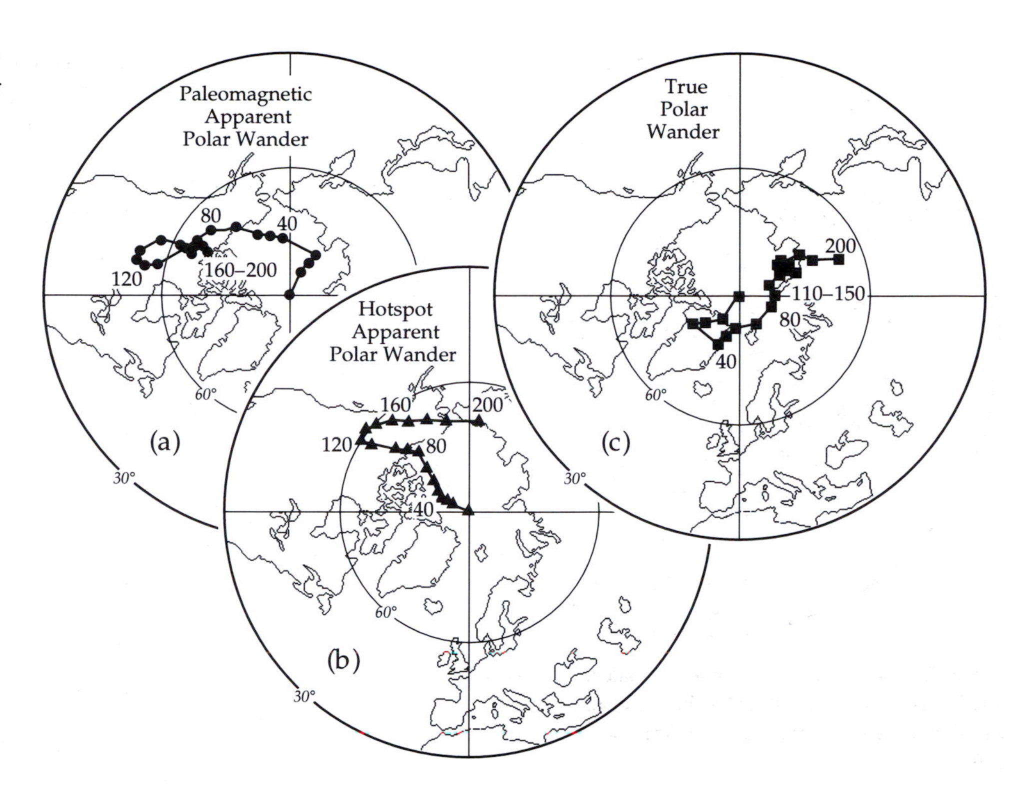

Fig. 6.34 (a) Paleomagnetic APW path reconstructed for Africa using data from several plates. (b) Hotspot APW path (motion of an axis at the geographic pole relative to the hotspot reference frame). (c) Computed true polar wander path (based on data from Courtillot and Besse, 1987, and Morgan, 1982). Values represent age in Ma.

6.3.4 Forces acting on lithospheric plates

An unresolved problem of plate tectonics is what mechanism drives plate motions. The forces acting on plates may be divided into forces that act on their bottom surfaces and forces that act on their margins. The bottom forces arise due to relative motion between the lithospheric plate and the viscous asthenosphere. In this context it is less important whether mantle flow takes place by whole-mantle convection or layered convection. For plate tectonics the important feature of mantle rheology is that viscous flow in the upper mantle is possible. The motion vectors of lithospheric plates do not reveal directly the mantle flow pattern, but some general inferences can be drawn. The flow pattern must include the mass transport involved in moving lithosphere from a ridge to a subduction zone, which has to be balanced by return flow deep in the mantle. Interactions between the plates and the viscous substratum necessarily influence the plate motions. In order to assess the importance of these effects we need to compare them to the other forces that act on plates, especially at their boundaries (Fig. 6.35).

Some forces acting on lithospheric plates promote motion while others resist it. Mantle flow could fall into either category. If the convective flow is faster than plate velocities, the plates are dragged along by the flow, but if the opposite is true there is exerted on the plate a *mantle drag force* (F_{DF}), opposing its motion. Plate velocities are observed to be inversely related to the area of continent on the plate, which suggests that the greater lithospheric thickness results in an additional *continental drag force* (F_{CD}) on the plate. The velocity of a plate also depends on the length of its subduction zone but not on the length of its spreading ridge. This suggests that subduction forces may be more important than spreading forces. However, we need to consider the forces at all three types of plate margin.

The most readily understood edge forces on a plate are those at a spreading ridge. The intrusion of magma may push the plates away from the ridge, and, since the ridges are elevated above the oceanic abyss, potential energy may encourage gravitational sliding toward the trenches. We can consider these two effects together as making up the *ridge push force* (F_{RP}).

The high seismicity on transform faults is evidence of interactive forces where the plates move past each other. A *transform force* (F_{TF}) can be envisioned as representing frictional resistance in the contact zone. Its magnitude may be different at a transform connecting ridge segments, where the plates are hot, than at a transform between subduction zones, where the plates are cold.

In subduction zones the descending slab is colder and denser than the surrounding mantle. This creates a positive mass anomaly, or negative buoyancy, which is accentuated by intraplate phase transitions. A *slab pull force* (F_{SP}) ensues that pulls the slab downwards into the mantle. Transferred

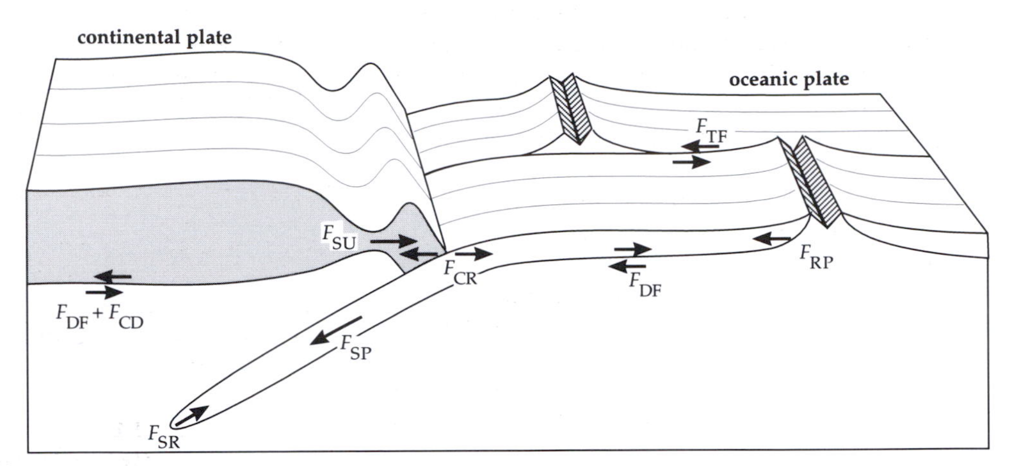

Fig. 6.35 Diagram illustrating some of the different forces acting on lithospheric plates (after Forsyth and Uyeda, 1975; Uyeda, 1978).

to the entire plate it acts as a force toward the subduction zone. However, the subducting plate eventually sinks to depths where it approaches thermal equilibrium with the surrounding mantle, loses its negative buoyancy and experiences a *slab resistance force* (F_{SR}) as it tries to penetrate further into the stiffer mantle.

The collision between plates results in both driving and resistive forces. The vertical pull on the descending plate may cause the bend in the lower plate to migrate away from the subduction zone, effectively drawing the upper plate toward the trench. The force on the *upper plate* has also been termed 'trench suction' (F_{SU}). The colliding plates also impede each other's motion and give rise to a *collision-resistance force* (F_{CR}). This force consists of separate forces due to the effects of mountains or trenches in the zone of convergence. Finally, the transfer of mantle material to the lithosphere at hotspots may result in a *hotspot force* (F_{HS}) on the plate.

In summary, the driving forces on plates are slab pull, ridge push and the trench pull force on the upper plate. The motion is opposed by slab resistance, collision resistance, and transform fault forces. Whether the forces between plate and mantle (mantle drag, continental drag) promote or oppose motion depends on the sense of the relative motion between the plate and the mantle. The motive force of plate tectonics is clearly a composite of these several forces. Some can be shown to be more important than others, and some are insignificant.

In order to evaluate the relative importance of the forces it is necessary to take into account their different directions. This is achieved by converting the forces to torques about the center of the Earth. Different mathematical analyses lead to similar general conclusions regarding the relative magnitudes of the torques. It turns out that the push exerted by hotspots, the resistance at transform faults and the drag between the mantle and oceanic lithosphere are negligible in

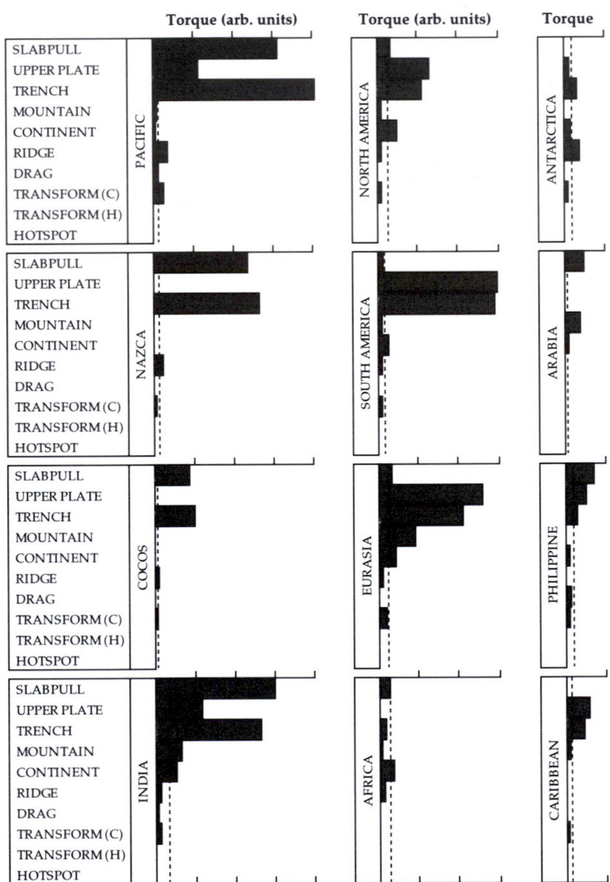

Fig. 6.36 Comparison of the magnitudes of torques acting on the 12 major lithospheric plates (after Chapple and Tullis, 1977).

comparison to the other forces (Fig. 6.36). The strongest driving forces are the pull of a descending slab on its plate and the force that pulls the upper plate toward a trench. A strong opposing force arises from the collision between the plates, but this resistance is consistently smaller than the upper plate force. The ridge push force is much smaller than the forces at a converging margin, which makes its significance questionable. However, the resistance experienced by the slab to deep mantle penetration serves to diminish the slab pull force, to the extent that the difference may be comparable to the ridge push force. Although the drag of convective mantle flow on oceanic lithosphere appears to be negligible, the drag on plates that have a continental root is significant.

Where do these observations leave us in trying to determine the driving mechanism of plate tectonics? Many factors that affect the balance of forces on plates remain uncertain. For example, although slab pull and ridge push forces have their respective adherents as the major motive forces, the North American plate has no converging margin along most of its length and the Philippine plate has no active spreading ridge. However, the drag forces exerted by relative motion between the asthenosphere and the base of the lithosphere are virtually unknown and may play a greater role than suspected, because they act over a large area of contact. Possibly no single force is dominant, and plate motions take place as a result of several effects acting jointly.

6.4 SUGGESTIONS FOR FURTHER READING

INTRODUCTORY LEVEL

Brown, G. C., Hawkesworth, C. J. and Wilson, R. C. L. (editors), 1992. *Understanding the Earth*, Cambridge Univ. Press, Cambridge.

Cox, A. and Hart, R. B. 1986. *Plate Tectonics*, Blackwell Scientific Publ., Oxford.

Kearey, P. and Vine, F. J. 1990. *Global Tectonics*, Blackwell Scientific Publ., Oxford.

INTERMEDIATE LEVEL

Bott, M. H. P. 1982. *The Interior of the Earth (2nd ed.)*, Edward Arnold, London.

Fowler, C. M. R. 1990. *The Solid Earth: an Introduction to Global Geophysics*, Cambridge Univ. Press, Cambridge.

Turcotte, D. L. and Schubert, G. 1982. *Geodynamics: Applications of Continuum Physics to Geological Problems*, John Wiley & Sons, New York.

ADVANCED LEVEL

Cathles, L. M. 1975. *The Viscosity of the Earth's Mantle*, Princeton Univ. Press, Princeton, N.J.

Garland, G. D. 1979. *Introduction to Geophysics*, W.B. Saunders., Philadelphia.

Mörner, N. A. (editor) 1980. *Earth Rheology, Isostasy and Eustasy*, John Wiley & Sons, New York.

Officer, C. B. 1974. *Introduction to Theoretical Geophysics*, Springer Verlag, New York.

Peltier, W. R. (editor) 1989. *Mantle Convection: Plate Tectonics and Global Dynamics*, Gordon and Breach, New York

Ramsay, J. G. 1967. *Folding and Fracturing of Rocks*, McGraw Hill, New York.

Ranalli, G. 1987. *Rheology of the Earth: Deformation and Flow Processes in Geophysics and Geodynamics*, Allen and Unwin, Winchester, Mass.

Stacey, F. D. 1992. *Physics of the Earth (3rd ed.)*, Brookfield Press, Brisbane, Australia

References

Achache, J., Le Mouël, J. L. and Courtillot, V. 1981. Long-period variations and mantle conductivity: an inversion using Bailey's method. *Geophys. J. R. Astr. Soc.*, **65**, 579–601.

Aki, K. and Richards, P. G. 1980. *Quantitative Seismology: Theory and Methods*, W.H. Freeman, San Francisco.

Anderson, R. N., Langseth, M. G. and Sclater, J. G. 1977. The mechanisms of heat transfer through the floor of the Indian Ocean. *J. Geophys. Res.*, **82**, 3391–409.

Anderson, R. N. and Skilbeck, J. N. 1981. Oceanic heat flow. In *The Sea, vol. 7, The Oceanic Lithosphere*, Emiliani, C., ed., J. Wiley-Interscience, New York, pp. 489–523.

Atwater, T. 1970. Implications of plate tectonics for the Cenozoic tectonic evolution of western North America. *Geol. Soc. Amer. Bull.*, **81**, 3513–36.

Banks, R. J. 1969. Geomagnetic variations and the electrical conductivity of the upper mantle. *Geophys. J. R. Astr. Soc.*, **17**, 457–87.

Barazangi, M. and Dorman, J. 1969. World seismicity maps compiled from ESSA, Coast and Geodetic Survey, epicenter data, 1961–1967. *Bull. Seism. Soc. Amer.*, **59**, 369–80.

Barton, C. E. 1989. Geomagnetic secular variation: direction and intensity. In *The Encyclopedia of Solid Earth Sciences*, James, D. E., ed., Van Nostrand Reinhold, New York, pp. 560–77.

Båth, M. 1966. Earthquake energy and magnitude. In *Physics and Chemistry of the Earth*, Ahrens, L. H., Press, F. and Runcorn, S. K., eds., Pergamon Press, Oxford, pp. 115–65.

Beatty, J. K. and Chaikin, A. (eds.) 1990. *The New Solar System*. Cambridge University Press and Sky Publishing Corp., Cambridge, 326 pp.

Benioff, H. 1935. A linear strain seismograph. *Bull. Seism. Soc. Amer.*, **25**, 283–309.

Benioff, H. 1958. Long waves observed in the Kamchatka earthquake of November 4, 1952. *J. Geophys. Res.*, **63**, 589–93.

Birch, F. 1964. Density and composition of mantle and core. *J. Geophys. Res.*, **69**, 4377–88.

Blackett, P. M. S. 1952. A negative experiment relating to magnetism and the earth's rotation. *Phil. Trans. Roy. Soc. London, Ser. A*, **245**, 309–70.

Bodell, J. M. and Chapman, D. S. 1982. Heat flow in the north-central Colorado Plateau. *J. Geophys. Res.*, **87**, 2869–84.

Bolt, B. A. 1988. *Earthquakes*, W. H. Freeman, New York.

Bott, M. H. P. 1982. *The Interior of the Earth (2nd ed.)*, Edward Arnold, London.

Bullard, E. C., Everett, J. E. and Smith, A. G. 1965. A Symposium on Continental Drift, IV. The fit of the continents around the Atlantic. *Phil. Trans. Roy. Soc. London, Ser. A*, **258**, 41–51.

Bullen, K. E. 1940. The problem of the Earth's density variation. *Bull. Seism. Soc. Amer.*, **30**, 235–50.

Bullen, K. E. 1942. The density variation of the Earth's central core. *Bull. Seism. Soc. Amer.*, **32**, 19–29.

Busse, F. H. 1989. Fundamentals of thermal convection. In *Mantle Convection: Plate Tectonics and Global Dynamics*, Peltier, W. R., ed., Gordon and Breach, New York, pp. 23–95.

Butler, R. F. 1992. *Paleomagnetism: Magnetic Domains to Geologic Terranes*, Blackwell Scientific Publ., Boston.

Cain, J. C. 1989. Geomagnetic field analysis. In *The Encyclopedia of Solid Earth Geophysics*, James, D. E., ed., Van Nostrand Reinhold, New York, pp. 517–22.

Caldwell, J. G. and Turcotte, D. L. 1979. Dependence of the thickness of the elastic oceanic lithosphere on age. *J. Geophys. Res.*, **84**, 7572–76.

Cande, S. C. and Kent, D. V. 1992. A new geomagnetic polarity time scale for the Late Cretaceous and Cenozoic. *J. Geophys. Res.*, **97**, 13,917–51.

Carey, S. W. 1958. A tectonic approach to continental drift. In *Continental drift, a symposium*, Carey, S. W., ed., Univ. Tasmania, Hobart, pp. 172–355.

Carter, W. E. 1989. Earth orientation. In *The Encyclopedia of Solid Earth Geophysics*, James, D. E., ed., Van Nostrand Reinhold, New York, pp. 231–39.

Cathles, L. M. 1975. *The Viscosity of the Earth's Mantle*, Princeton Univ. Press, Princeton, N.J.

Chapman, S. and Bartels, J. 1940. *Geomagnetism, Vol. I: Geomagnetic and Related Phenomena, Vol. II: Analysis of Data and Physical Theories*, Oxford Univ. Press, Oxford.

Chapple, W. M. and Tullis, T. E. 1977. Evaluation of the forces that drive plates. *J. Geophys. Res.*, **82**, 1967–84.

Chevallier, R. 1925. L'aimantation des lavas de l'Etna et l'orientation du champs terrestre en Sicile du 12e au 17e siècle. *Ann. Phys. Ser. 10*, **4**, 5–162.

Clague, D. A., Dalrymple, G. B. and Moberly, R. 1975. Petrography and K-Ar ages of dredged volcanic rocks from the western Hawaiian Ridge and the southern Emperor Seamount chain. *Geol. Soc. Amer. Bull.*, **86**, 991–8.

Cohen, E. R. and Taylor, B. N. 1986. *The 1986 Adjustment of the Fundamental Physical Constants, Report of the CODATA Task Group on Fundamental Constants*, Pergamon Press, Elmsford, N.Y.

Conrad, V. 1925. Laufzeitkurven des Tauernbebens vom 28. November, 1923. *Mitt. Erdb.-Komm. Wien*, **59**, 1–23.

Courtillot, V. and Besse, J. 1987. Magnetic field reversals, polar wander, and core-mantle coupling. *Science*, **237**, 1140–7.

Cox, A. 1968. Lengths of geomagnetic polarity intervals. *J. Geophys. Res.*, **73**, 3247–60.

Cox, A., Doell, R. R. and Dalrymple, G. B. 1963. Geomagnetic polarity epochs: Sierra Nevada II. *Science*, **142**, 382–5.

Cox, A., Doell, R. R. and Dalrymple, G. B. 1964. Reversals of the earth's magnetic field. *Science*, **144**, 1537–43.

Cox, A., Doell, R. R. and Dalrymple, G. B. 1968. Radiometric time scale for geomagnetic reversals. *Quart. J. Geol. Soc. London*, **124**, 53–66.

Cox, A. and Hart, R. B. 1986. *Plate Tectonics*, Blackwell Scientific Publications, Boston.

Cox, A. V. 1982. Magnetostratigraphic time scale. In *A Geologic Time Scale*, Harland, W. B., Cox, A. V., Llewellyn, P. G., Pickton, C. A. G., Smith, A. G. and Walters, R., eds., Cambridge Univ. Press, Cambridge, pp. 63–84.

Crough, S. T. 1983. Hotspot swells. *Annu. Rev. Earth Planet. Sci.*, **11**, 165–93.

Crough, S. T. and Jurdy, D. M. 1980. Subducted lithosphere, hotspots, and the geoid. *Earth Planet. Sci. Lett.*, **48**, 15–22.

Dalrymple, G. B. 1991. *The Age of the Earth*, Stanford Univ. Press, Stanford, California.

Dalrymple, G. B., Clague, D. A. and Lanphere, M. A. 1977. Revised age for Midway volcano, Hawaiian volcanic chain. *Earth Planet. Sci. Lett.*, **37**, 107–16.

Davis, J. C. 1973. *Statistics and Data Analysis in Geology*, John Wiley, New York.

DeMets, C., Gordon, R. G., Argus, D. F. and Stein, S. 1990. Current plate motions. *Geophys. J. Int.*, **101**, 425–78.

Dietz, R. S. 1961. Continent and ocean basin evolution by spreading of the sea floor. *Nature*, **190**, 854–7.

Dobrin, M. B. 1976. *Introduction to Geophysical Prospecting (3rd ed.)*, McGraw-Hill, New York.

Dobrin, M. B. and Savit, C. H. 1988. *Introduction to Geophysical Prospecting (4th ed.)*, McGraw-Hill, New York.

Doell, R. R. and Dalrymple, G. B. 1966. Geomagnetic polarity epochs: a new polarity event and the age of the Brunhes-Matuyama boundary. *Science*, **152**, 1060–1.

du Toit, A. L. 1937. *Our Wandering Continents*, Oliver and Boyd, Edinburgh.

Dziewonski, A. M. 1984. Mapping the lower mantle: determination of lateral heterogeneity in P velocity up to degree and order 6. *J. Geophys. Res.*, **89**, 5929–52.

Dziewonski, A. M. 1989. Earth structure, global. In *The Encyclopedia of Solid Earth Geophysics*, James, D. E., ed., Van Nostrand Reinhold, New York, pp. 331–59.

Dziewonski, A. M. and Anderson, D. L. 1981. Preliminary Reference Earth Model (PREM). *Phys. Earth Planet. Inter.*, **25**, 297–356.

Engeln, J. F., Wiens, D. A. and Stein, S. 1986. Mechanisms and depths of Atlantic transform earthquakes. *J. Geophys. Res.*, **91**, 548–77.

European Seismological Commission 1993. *European Macroseismic Scale 1992 (up-dated MSK-scale)*, Centre Européen de Géodynamique et de Séismologie, Luxembourg.

Evans, M. E. and McElhinny, M. W. 1969. An investigation of the origin of stable remanence in magnetite-bearing igneous rocks. *J. Geomag. Geoelect.*, **21**, 757–73.

Fisher, R. A. 1953. Dispersion on a sphere. *Proc. Roy. Soc. Lond., Ser. A*, **217**, 295–305.

Forsyth, D. and Uyeda, S. 1975. On the relative importance of the driving forces of plate motion. *Geophys. J. R. Astr. Soc.*, **43**, 163–200.

Grow, J. A. and Bowin, C. O. 1975. Evidence for high-density crust and mantle beneath the Chile Trench due to the descending lithosphere. *J. Geophys. Res.*, **80**, 1449–58.

Gubler, E. 1991. Recent crustal movements in Switzerland: vertical movements. In *Report on the Geodetic Activities in the Years 1987 to 1991: 5. Geodynamics*, Kahle, H. G. and Jeanrichard, F., eds., Swiss Geodetic Commission and Federal Office of Topography, Zürich, pp. 36 and Map 5.

Gutenberg, B. 1914. Über Erdbebenwellen. VIIA. Beobachtungen an Registrierungen von Fernbeben in Göttingen und Folgerungen über die Konstitution des Erdkörpers. *Nachr. Ges. Wiss. Göttingen, Math.-Phys. Klasse*, 1–52 and 125–76.

Gutenberg, B. 1945. Magnitude determination for deep focus earthquakes. *Bull. Seism. Soc. Amer.*, **35**, 117–30.

Gutenberg, B. 1956. The energy of earthquakes. *Quart. J. Geol. Soc. Lond.*, **112**, 1–14.

Gutenberg, B. 1959. *Physics of the Earth's Interior*, Academic Press, New York.

Gutenberg, B. and Richter, C. F. 1939. On seismic waves. *Beitr. Geophys.*, **54**, 94–136.

Gutenberg, B. and Richter, C. F. 1954. *Seismicity of the Earth and Associated Phenomena*, Princeton Univ. Press, Princeton.

Harland, W. B., Armstrong, R. L., Cox, A. V., Craig, L. E., Smith, A. G. and Smith, D. G. 1990. *A Geologic Time Scale 1989*, Cambridge Univ. Press, Cambridge.

Harland, W. B., Cox, A. V., Llewellyn, P. G., Pickton, C. A. G., Smith, A. G. and Walters, R. 1982. *A Geologic Time Scale*, Cambridge Univ. Press, Cambridge.

Hasegawa, A. 1989. Seismicity: subduction zone. In *The Encyclopedia of Solid Earth Geophysics*, James, D. E., ed., Van Nostrand Reinhold, New York, pp. 1054–61.

Heirtzler, J. R., Dickson, G. O., Herron, E. M., Pitman III, W. C. and Le Pichon, X. 1968. Marine magnetic anomalies, geomagnetic field reversals, and motions of the ocean floor and continents. *J. Geophys. Res.*, **73**, 2119–36.

Heirtzler, J. R., Le Pichon, X. and Baron, J. G. 1966. Magnetic anomalies over the Reykjanes Ridge. *Deep Sea Res.*, **13**, 427–43.

Hess, H. H. 1962. History of ocean basins. In *Petrologic studies: a volume in honor of A. F. Buddington*, Engel, A. E., James, H. L. and Leonard, B. F., eds., Geological Society of America, pp. 599–620.

Holliger, K. and Kissling, E. 1992. Gravity interpretation of a unified 2–D acoustic image of the central Alpine collision zone. *Geophys. J. Int.*, **111**, 213–25.

Huang, P. Y., Solomon, S. C., Bergman, E. A. and Nabelek, J. L. 1986. Focal depths and mechanisms of Mid-Atlantic Ridge earthquakes from body waveform inversion. *J. Geophys. Res.*, **91**, 579–98.

Hurst, R. W., Bridgwater, D., Collerson, K. D. and Weatherill, G. W. 1975. 3600 m.y. Rb-Sr ages from very early Archean gneisses from Saglek Bay, Labrador. *Earth Planet. Sci. Lett.*, **27**, 393–403.

Irving, E. 1957. Rock Magnetism: a new approach to some palaeogeographic problems. *Phil. Mag. Suppl. Adv. Phys.*, **6**, 194–218.

Irving, E. 1977. Drift of the major continental blocks since the Devonian. *Nature*, **270**, 304–9.

Irving, E. and Major, A. 1964. Post-depositional detrital remanent magnetization in artificial and natural sediments. *Sedimentology*, **3**, 135–43.

Irving, E. and Naldrett, A. J. 1977. Paleomagnetism in Abitibi greenstone belt, and Abitibi and Matachewan diabase belts: evidence of the Archean geomagnetic field. *J. Geology*, **85**, 157–76.

Isacks, B. and Molnar, P. 1969. Mantle earthquake mechanisms and the sinking of the lithosphere. *Nature*, **223**, 1121–4.

Isacks, B. L. 1989. Seismicity and plate tectonics. In *The Encyclopedia of Solid Earth Geophysics*, James, D. E., ed., Van Nostrand Reinhold, New York, pp. 1061–71.

Jarvis, G. T. and Peltier, W. R. 1989. Convection models and geophysical observations. In *Mantle Convection: Plate Tectonics and Global Dynamics*, Peltier, W. R., ed., Gordon and Breach, New York, pp. 479–593.

Jeffreys, H. and Bullen, K. E. 1940. *Seismological Tables*, British Association, Gray-Milne Trust, London.

Johnson, H. P. and Carlson, R. L. 1992. Variation of sea floor depth with age: A test of models based on drilling results. *Geophys. Res. Lett.*, **19**, 1971–4.

Joly, J. 1899. An estimation of the geological age of the Earth. *Ann. Rept. Smithson. Inst. 1899*, 247–88.

Kahle, H. G., Müller, M. V., Geiger, A., Danuser, G., Mueller, S., Veis, G., Billiris, H. and Paradissis, D. 1995. The strain field in northwestern Greece and the Ionian Islands: results inferred from GPS measurements. *Tectonophysics*, **249**, 41–52.

Kakkuri, J. 1992. Recent vertical crustal movement (Atlas map 6). In *A Continent Revealed: the European Geotraverse. Atlas of Compiled Data*, Freeman, R. and Mueller, S., eds., Cambridge Univ. Press, Cambridge.

Karnik, V. 1969. *Seismicity of the European Area, Part 1*, D. Reidel and Company, Dordrecht, Holland.

Kearey, P. and Vine, F. J. 1990. *Global Tectonics*, Blackwell Scientific Publ., London.

Keen, C. E. and Tramontini, C. 1970. A seismic refraction survey on the Mid-Atlantic Ridge. *Geophys. J. R. Astr. Soc.*, **20**, 473–91.

Kennett, B. L. N. and Engdahl, E. R. 1991. Traveltimes for global earthquake location and phase identification. *Geophys. J. Int.*, **105**, 429–65.

Kissling, E. 1993. Seismische Tomographie: Erdbebenwellen durchleuchten unseren Planeten. *Vierteljahresschrift Naturforsch. Ges. Zürich*, **138**, 1–20.

Klingelé, E. and Kissling, E. 1982. Zum Konzept der isostatischen Modelle in Gebirgen am Beispiel der Schweizer Alpen, Geodätisch-Geophysikalische Arbeiten in der Schweiz. *Schweiz. Geodät. Komm.*, **35**, 3–36.

Klingelé, E. and Olivier, R. 1980. Die neue Schwere-Karte der Schweiz (Bouguer-Anomalien). *Beitr. Geologie der Schweiz, Serie Geophys.*, **20**, 93p.

Köppen, W. and Wegener, A. 1924. *Die Klimate der geologischen Vorzeit*, Bornträger, Berlin.

Kurtz, R. D., DeLaurier, J. M. and Gupta, J. C. 1986. A magnetotelluric sounding across Vancouver Island detects the subducting Juan de Fuca plate. *Nature*, **321**, 596–9.

Lay, T. and Wallace, T. C. 1995. *Modern Global Seismology*, Academic Press, San Diego.

Le Pichon, X. 1968. Sea-floor spreading and continental drift. *J. Geophys. Res.*, **73**, 3661–97.

LeFevre, L. V. and Helmberger, D. V. 1989. Upper mantle P velocity structure of the Canadian shield. *J. Geophys. Res.*, **94**, 17,749–65.

Lehmann, I. 1936. P'. *Bur. Centr. Seism. Internat. A*, **14**, 3–31.

Lerch, F. J., Klosko, S. M., Laubscher, R. E. and Wagner, C. A. 1979. Gravity model improvement using Geos 3 (GEM 8 and 10). *J. Geophys. Res.*, **84**, 3897–3916.

Lindner, H., Militzer, H., Rösler, R. and Scheibe, R. 1984. Bearbeitung und Interpretation der gravimetrischen und magnetischen Messergebnisse. In *Angewandte Geophysik, I: Gravimetrie und Magnetik*, Militzer, H. and Weber, F., eds., Springer-Verlag, Vienna, pp. 226–83.

Love, A. E. H. 1911. *Some Problems of Geodynamics*, Cambridge University Press, Cambridge.

Lowrie, W. 1990. Identification of ferromagnetic minerals in a rock by coercivity and unblocking temperature properties. *Geophys. Res. Lett.*, **17**, 159–62.

Lowrie, W. and Alvarez, W. 1975. Paleomagnetic evidence for rotation of the Italian peninsula. *J. Geophys. Res.*, **80**, 1579–92.

Lowrie, W. and Alvarez, W. 1977. Upper Cretaceous-Paleocene magnetic stratigraphy at Gubbio, Italy. III. Upper Cretaceous magnetic stratigraphy. *Geol. Soc. Amer. Bull.*, **88**, 374–7.

Lowrie, W. and Alvarez, W. 1981. One hundred million years of geomagnetic polarity history. *Geology*, **9**, 392–7.

Lowrie, W. and Heller, F. 1982. Magnetic properties of marine limestones. *Rev. Geophys. Space Phys.*, **20**, 171–92.

Ludwig, W. J., Nafe, J. E. and Drake, C. L. 1970. Seismic refraction. In *The Sea, Vol. 4*, Maxwell, A. E., ed., Wiley-Interscience, New York, pp. 53–84.

MacDonald, K. L. 1957. Penetration of the geomagnetic secular field through a mantle with variable conductivity. *J. Geophys. Res.*, **62**, 117–41.

Mankinen, E. A. and Dalrymple, G. B. 1979. Revised geomagnetic polarity timescale for the interval 0–5 m.y. B.P. *J. Geophys. Res.*, **84**, 615–26.

Marsh, J. G., Koblinsky, C. J., Zwally, H. J., Brenner, A. C. and Beckley, B. D. 1992. A global mean sea surface based upon GEOS 3 and Seasat altimeter data. *J. Geophys. Res.*, **97**, 4915–21.

Mayer Rosa, D. and Müller, S. 1979. Studies of seismicity and selected focal mechanisms of Switzerland. *Schweiz. miner. petr. Mitt.*, **59**, 127–32.

McCann, W. R., Nishenko, S. P., Sykes, L. R. and Krause, J. 1979. Seismic gaps and plate tectonics. *Pageoph*, **117**, 1082–147.

McDougall, I., Brown, F. H., Cerling, T. E. and Hillhouse, J. W. 1992. A reappraisal of the geomagnetic polarity time scale to 4 Ma using data from the Turkana basin, East Africa. *Geophys. Res. Lett.*, **19**, 2349–52.

McDougall, I. and Chamalaun, F. H. 1966. Geomagnetic polarity scale of time. *Nature*, **212**, 1415–8.

McDougall, I. and Duncan, R. A. 1980. Linear volcanic chains – recording plate motions? *Tectonophysics*, **63**, 275–95.

McElhinny, M. W. 1973. *Paleomagnetism and Plate Tectonics*, Cambridge Univ. Press, Cambridge.

McKenzie, D. P. and Morgan, W. J. 1969. Evolution of triple junctions. *Nature*, **224**, 125–33.

McKenzie, D. P. and Parker, R. L. 1967. The North Pacific: an example of tectonics on a sphere. *Nature*, **216**, 1276–80.

Merrill, R. T. and McElhinny, M. W. 1983. *The Earth's Magnetic Field*, Academic Press, London.

Militzer, H., Scheibe, R. and Seiberl, W. 1984. Angewandte Magnetik. In *Angewandte Geophysik, I: Gravimetrie und Magnetik*, Militzer, H. and Weber, F., eds., Springer-Verlag, Vienna, pp. 127–89.

Mohorovičić, A. 1909. Das Beben vom 8.x.1909. *Jb. met. Obs. Zagreb (Agram)*, **9**, 1–63.

Molnar, P. 1988. Continental tectonics in the aftermath of plate tectonics. *Nature*, **335**, 131–7.

Morel, P. and Irving, E. 1981. Paleomagnetism and the evolution of Pangea. *J. Geophys. Res.*, **86**, 1858–72.

Morgan, P. 1989. Heat flow in the Earth. In *The Encyclopedia of Solid Earth Sciences*, James, D. E., ed., Van Nostrand Reinhold, New York, pp. 634–46.

Morgan, P. and Sass, J. H. 1984. Thermal regime of the continental lithosphere. *J. Geodynamics*, **1**, 143–66.

Morgan, W. J. 1968. Rises, trenches, great faults, and crustal blocks. *J. Geophys. Res.*, **73**, 1959–82.

Morgan, W. J. 1982. Hotspot tracks and the opening of the Atlantic and Indian Oceans. In *The Sea, Vol. 7: The Oceanic Lithosphere*, Emiliani, C., ed., Wiley-Interscience, New York, pp. 443–87.

Morley, L. W. and Larochelle, A. 1964. Paleomagnetism as a means of dating geological events. In *Geochronology in Canada. Roy. Soc. Canada Spec. Publ. No. 8*, Osborne, F. F., ed., Toronto Univ. Press, Toronto, pp. 39–51.

Mueller, S. 1977. A new model of the continental crust. In *Geophysical Monograph 20: The Earth's Crust*, Heacock, J. G., ed., Amer. Geophys. Union, Washington, D.C., pp. 289–317.

Nagata, T. 1961. *Rock Magnetism*, Maruzen, Tokyo.

Ni, J. and Barazangi, M. 1984. Seismotectonics of the Himalayan collision zone: Geometry of the underthrusting Indian plate beneath the Himalaya. *J. Geophys. Res.*, **89**, 1147–63.

Nuttli, O. W. 1973. The Mississippi Valley earthquakes of 1811 and 1812: Intensities, ground motion and magnitudes. *Bull. Seism. Soc. Amer.*, **63**, 227–48.

Ogilvie, K. W., Scudder, J. D., Vasyliunas, V. M., Hartle, R. E. and Siscoe, G. L. 1977. Observations of the planet Mercury by the plasma electron experiment Mariner 10. *J. Geophys. Res.*, **82**, 1807–24.

Oldham, R. D. 1906. The constitution of the interior of the Earth, as revealed by earthquakes. *Q. Jl. geol. Soc. London*, **62**, 456–75.

Ondoh, T. and Maeda, H. 1962. Geomagnetic storm correlation between the northern and southern hemisphere. *J. Geomag. Geoelectr.*, **14**, 22–32.

Opdyke, N. D. 1972. Paleomagnetism of deep-sea cores. *Rev. Geophys.*, **10**, 213–49.

Opdyke, N. D., Glass, B., Hays, J. D. and Foster, J. 1968. Paleomagnetic study of Antarctic deep-sea cores. *Science*, **154**, 349–57.

Opdyke, N. D. and Henry, K. W. 1969. A test of the dipole hypothesis. *Earth Planet. Sci. Lett.*, **6**, 139–51.

Parker, R. L. and Oldenburg, D. W. 1973. Thermal models of mid-ocean ridges. *Nature Phys. Sci.*, **242**, 137–9.

Parsons, B. and McKenzie, D. 1978. Mantle convection and the thermal structure of the plates. *J. Geophys. Res.*, **83**, 4485–96.

Parsons, B. and Sclater, J. G. 1977. An analysis of the variation of ocean floor bathymetry and heat flow with age. *J. Geophys. Res.*, **82**, 803–27.

Pavoni, N. 1977. Erdbeben im Gebiet der Schweiz. *Eclogae geol. Helv.*, **70/2**, 351–71.

Peltier, W. R. 1989. Mantle viscosity. In *Mantle Convection: Plate Tectonics and Global Dynamics*, Peltier, W. R., ed., Gordon and Breach, New York, pp. 389–478.

Peltier, W. R., Jarvis, G. T., Forte, A. M. and Solheim, L. P. 1989. The radial structure of the mantle general circulation. In *Mantle Convection: Plate Tectonics and Global Dynamics*, Peltier, W. R., ed., Gordon and Breach, New York, pp. 765–815.

Pidgeon, R. T. 1978. 3450–m.y.-old volcanics in the Archean layered greenstone succession of the Pilbara Block, Western Australia. *Earth Planet. Sci. Lett.*, **37**, 421–8.

Pitman III, W. C. and Heirtzler, J. R. 1966. Magnetic anomalies over the Pacific-Antarctic ridge. *Science*, **154**, 1164–71.

Pitman III, W. C. and Talwani, M. 1972. Sea-floor spreading in the North Atlantic. *Geol. Soc. Amer. Bull.*, **83**, 619–46.

Pollack, H. N., Hurter, S. J. and Johnson, J. R. 1993. Heat flow from the Earth's interior: analysis of the global data set. *Rev. Geophys.*, **31**, 267–80.

Powell, W. G., Chapman, D. S., Balling, N. and Beck, A. E. 1988. Continental heat-flow density. In *Handbook of Terrestrial Heat-Flow Density Determinations*, Haenel, R., Rybach, L. and Stegena, L., eds., Kluwer Academic Publishers, Dordrecht, pp. 167–222.

Press, F. and Siever, R. 1985. *Earth*, W.H. Freeman, San Francisco.

Prévot, M., Mankinen, E. A., Grommé, C. S. and Coe, R. S. 1985. How the geomagnetic field vector reverses polarity. *Nature*, **316**, 230–4.

Ramsay, J. G. 1967. *Folding and Fracturing of Rocks*, McGraw Hill, New York.

Ranalli, G. 1987. *Rheology of the Earth: Deformation and Flow Processes in Geophysics and Geodynamics*, Allen and Unwin, Winchester, Mass., USA.

Rayleigh, Lord. 1916. On convection currents in a horizontal layer of fluid, when the higher temperature is on the under side. *Phil. Mag.*, **32**, 529–46.

Reid, H. F. 1906. The elastic-rebound theory of earthquakes. *Bull. Dep. Geol. Univ. Calif.*, **6**, 413–44.

Richards, P. G. 1989. Seismic monitoring of nuclear explosions. In *The Encyclopedia of Solid Earth Geophysics*, James, D. E., ed., Van Nostrand Reinhold, New York, pp. 1071–89.

Richards, T. C. 1961. Motion of the ground on arrival of reflected longitudinal and transverse waves at wide-angle reflection distances. *Geophysics*, **26**, 277–97.

Robinson, E. S. and Çoruh, C. 1988. *Basic Exploration Geophysics*, John Wiley, New York.

Roy, R. F., Blackwell, D. D. and Birch, F. 1968. Heat generation of plutonic rocks and continental heat flow provinces. *Earth Planet. Sci. Lett.*, **5**, 1–12.

Runcorn, S. K. 1956. Paleomagnetic comparisons between Europe and North America. *Proc. Geol. Assoc. Canada*, **8**, 77–85.

Russell, C. T. 1980. Planetary magnetism. *Rev. Geophys. Space Phys.*, **18**, 77–106.

Rutherford, E. and Soddy, F. 1902. The cause and nature of radioactivity. *Phil. Mag.*, **4**, 370–96 and 569–85.

Ryan, J. W., Clark, T. A., Ma, C., Gordon, D., Caprette, D. S. and Himwich, W. E. 1993. Global scale tectonic plate motions with CDP VLBI data. In *Contributions to Space Geodesy: Crustal Dynamics*, Smith, D. E. and Turcotte, D. L., eds., American Geophysical Union, Washington, D.C., pp. 37–50.

Rybach, L. 1976. Radioactive heat production in rocks and its relation to other petrophysical parameters. *Pure Appl. Geophys.*, **114**, 309–18.

Rybach, L. 1988. Determination of heat production rate. In *Handbook of Terrestrial Heat Flow Density Determination*, Haenel, R., Rybach, L. and Stegena, L., eds., Kluwer Academic Publishers, Dordrecht, pp. 486.

Schubert, G., Yuen, D. A. and Turcotte, D. L. 1975. Role of phase transitions in a dynamic mantle. *Geophys. J. R. Astr. Soc.*, **42**, 705–35.

Sclater, J. C., Jaupart, C. and Galson, D. 1980. The heat flow through oceanic and continental crust and the heat loss of the earth. *Rev. Geophys. Space Phys.*, **18**, 269–311.

Sclater, J. C., Parsons, B. and Jaupart, C. 1981. Oceans and continents: similarities and differences in the mechanisms of heat loss. *J. Geophys. Res.*, **86**, 11535–52.

Scotese, C. R., Gahagan, L. M. and Larson, R. L. 1988. Plate tectonic reconstructions of the Cretaceous and Cenozoic ocean basins. *Tectonophysics*, **155**, 27–48.

Serson, P. H. 1973. Instrumentation for induction studies on land. *Phys. Earth Planet. Int.*, **7**, 313–22.

Singh, S. K., Dominguez, T., Castro, R. and Rodriguez, M. 1984. P waveform of large shallow earthquakes along the Mexico subduction zone. *Bull. Seism. Soc. Amer.*, **74**, 2135–56.

Slack, H. A., Lynch, V. M. and Langan, L. 1967. The geomagnetic elements. *Geophysics*, **32**, 877–92.

Smith, A. G. and Hallam, A. 1970. The fit of the southern continents. *Nature*, **225**, 139–44.

Spell, T. L. and McDougall, I. 1992. Revisions to the age of the Brunhes-Matuyama boundary and the Pleistocene geomagnetic polarity timescale. *Geophys. Res. Lett.*, **19**, 1181–84.

Stacey, F. D. 1992. *Physics of the Earth (3rd ed.)*, Brookfield Press, Brisbane, Australia

Steiger, R. H. and Jaeger, E. 1977. Subcomission on geochemistry: Convention on the use of decay constants in geo- and cosmochemistry. *Earth Planet. Sci. Lett.*, **36**, 359–62.

Stein, C. and Stein, S. 1992. A model for the global variation in oceanic depth and heat flow with lithospheric age. *Nature*, **359**, 123–9.

Stein, C. A. and Stein, S. 1994. Constraints on hydrothermal heat flux through the oceanic lithosphere from global heat flow. *J. Geophys. Res.*, **99**, 3081–95.

Stephenson, F. R. and Morrison, L. V. 1984. Long-term changes in the rotation of the earth: 700 B.C. to A.D. 1980. *Phil. Trans. Roy. Soc. London, Ser. A*, **313**, 47–70.

Strahler, A. N. 1963. *The Earth Sciences*, Harper and Row, New York.

Swisher, C. C., *et al.* 1992. Coeval 40Ar/39Ar ages of 65.0 million years ago from Chicxulub crater melt rock and Cretaceous-Tertiary boundary tektites. *Science*, **257**, 954–8.

Sykes, L. R. 1967. Mechanism of earthquakes and nature of faulting on the mid-ocean ridges. *J. Geophys. Res.*, **72**, 2131–53.

Sykes, L. R., Kisslinger, J. B., House, L., Davies, J. N. and Jacob, K. H. 1981. Rupture zones and repeat times of great earthquakes along the Alaska-Aleutian arc, 1784–1980. In *Earthquake Prediction: An International Review, Maurice Ewing Ser. 4*, Simpson, D. W. and Richards, P. G., eds., American Geophysical Union, Washington, D.C., pp. 73–92.

Talwani, M. and Ewing, M. 1960. Rapid computation of gravitational attraction of three-dimensional bodies of arbitrary shape. *Geophysics*, **25**, 203–25.

Talwani, M., Le Pichon, X. and Ewing, M. 1965. Crustal structure of the mid-ocean ridges. 2: Computed model from gravity and seismic refraction data. *J. Geophys. Res.*, **70**, 341–52.

Talwani, M., Worzel, J. L. and Landisman, M. 1959. Rapid gravity computations for two-dimensional bodies with application to the Mendocino submarine fracture zone. *J. Geophys. Res.*, **64**, 49–59.

Tapley, B. D., Schutz, B. E. and Eanes, R. J. 1985. Station coordinates, baselines and earth rotation from LAGEOS laser ranging: 1976–1984. *J. Geophys. Res.*, **90**, 9235–48.

Telford, W. M., Geldart, L. P. and Sheriff, R. E. 1990. *Applied Geophysics*, Cambridge Univ. Press, Cambridge.

Thellier, E. 1937. Aimantation des terres cuites: application à la recherche de l'intensité du champ magnétique terrestre dans le passé. *C. R. Acad. Sci. Paris*, **204**, 184–6.

Toksöz, M. N., Minear, J. W. and Julian, B. R. 1971. Temperature field and geophysical effects of a downgoing slab. *J. Geophys. Res.*, **76**, 1113–38.

Turcotte, D. L., McAdoo, D. C. and Caldwell, J. G. 1978. An elastic-perfectly plastic analysis of the bending of the lithosphere at a trench. *Tectonophysics*, **47**, 193–205.

Turcotte, D. L. and Schubert, G. 1982. *Geodynamics: Applications of Continuum Physics to Geological Problems*, J. Wiley, New York.

Turner, G., Enright, M. C. and Cadogan, P. H. 1978. The early history of chondrite parent bodies inferred from 40Ar-39Ar ages. *Proceedings of the Ninth Lunar and Planetary Science Conference*, 989–1025.

Uyeda, S. 1978. *The New View of the Earth*, W. H. Freeman, San Francisco.

Van Andel, T. H. 1992. Seafloor spreading and plate tectonics. In *Understanding the Earth*, Brown, C. J., Hawkesworth, C. J. and Wilson, R. C. L., eds., Cambridge Univ. Press, Cambridge., pp. 167–86.

Van der Voo, R. 1990. Phanerozoic paleomagnetic poles from Europe and North America and comparisons with continental reconstructions. *Rev. Geophys.*, **28**, 167–206.

Van der Voo, R. 1993. *Paleomagnetism of the Atlantic, Tethys and Iapetus Oceans*, Cambridge Univ. Press, Cambridge.

Van der Voo, R. and French, R. B. 1974. Apparent polar wandering for the Atlantic-bordering continents: Late Carboniferous to Eocene. *Earth Sci. Rev.*, **10**, 99–119.

Van Nostrand, R. G. and Cook, K. L. 1966. *Interpretation of resistivity data*, U.S.G.S. Prof. Paper No. 499.

Vestine, E. H. 1962. Space geomagnetism, radiation belts, and auroral zones. In *Proceedings of the Benedum Earth Magnetism Symposium*, pp. 11–29.

Vine, F. J. 1966. Spreading of the ocean floor: new evidence. *Science*, **154**, 1405–15.

Vine, F. J. and Matthews, D. H. 1963. Magnetic anomalies over oceanic ridges. *Nature*, **199**, 947–9.

Vitorello, I. and Pollack, H. N. 1980. On the variation of continental heat flow with age and the thermal evolution of continents. *J. Geophys. Res.*, **85**, 983–95.

Ward, S. H. 1990. Resistivity and induced polarization methods. In *Geotechnical and Environmental Geophysics, Vol.1*, Ward, S. H., ed., Society of Exploration Geophysicists, Tulsa, Oklahoma, pp. 147–90.

Watanabe, H. and Yukutake, T. 1975. Electromagnetic core-mantle coupling associated with changes in the geomagnetic dipole field. *J. Geomag. Geoelect.*, **27**, 153–73.

Watts, A. B., Bodine, J. H. and Steckler, M. S. 1980. Observations of flexure and the state of stress in the oceanic lithosphere. *J. Geophys. Res.*, **85**, 6369–76.

Watts, A. B., Cochran, J. R. and Selzer, G. 1975. Gravity anomalies and flexure of the lithosphere: a three-dimensional study of the Great Meteor seamount, Northeast Atlantic. *J. Geophys. Res.*, **80**, 1391–98.

Wegener, A. 1922. *Die Entstehung der Kontinente und Ozeane (3rd ed.)*, Hamburg.

Wiechert, E. 1897. Ueber die Massenverteilung im Innern der Erde. *Nachr. Ges. Wiss. Göttingen*, 221–43.

Williamson, E. D. and Adams, L. H. 1923. Density distribution in the Earth. *J. Wash. Acad. Sci.*, **13**, 413–28.

Wilson, J. T. 1963. A possible origin of the Hawaiian Islands. *Can. J. Phys.*, **41**, 863–70.

Wilson, J. T. 1965. A new class of faults and their bearing on continental drift. *Nature*, **207**, 907–10.

Woodhouse, J. H. and Dziewonski, A. M. 1984. Mapping the upper mantle: three dimensional modelling of earth structure by inversion of seismic waveforms. *J. Geophys. Res.*, **89**, 5953–86.

York, D. and Farquhar, R. M. 1972. *The Earth's Age and Geochronology*, Pergamon Press, Oxford.

Yukutake, T. 1967. The westward drift of the earth's magnetic field in historic times. *J. Geomag. Geoelect.*, **19**, 103–16.

Yukutake, T. and Tachinaka, H. 1968. The non-dipole part of the Earth's magnetic field. *Bull. Earthqu. Res. Inst. Tokyo*, **46**, 1027–62.

Zijderveld, J. D. A. 1967. A.C. demagnetization of rocks: Analysis of results. In *Methods in Palaeomagnetism*, Collinson, D. W., Creer, K. M. and Runcorn, S. K., eds., Elsevier, Amsterdam, pp. 254–86.

Index